Algebra and Trigonometry

Important Formulas

Quadratic Formula

If $ax^2 + bx + c = 0$, for $a \neq 0$, then

$$x = \frac{-b \pm \sqrt{b^2 - 4ac}}{2a}$$

Distance Formula

The distance d between $P_1 = (x_1, y_1)$ and $P_2 = (x_2, y_2)$ is

$$d = \sqrt{(x_2 - x_1)^2 + (y_2 - y_1)^2}$$

Slope Formula

The slope m of the line segment $\overline{AB}$, where $A = (x_1, y_1)$ and $B = (x_2, y_2)$ is

$$m = \frac{y_2 - y_1}{x_2 - x_1}$$

Geometry

Assume A = area, C = circumference, V = volume, S = surface area, r = radius, h = altitude, l = length, w = width, b (or a) = length of a base, and s = length of a side.

1	Square	$A = s^2$
2	Rectangle	$A = lw$
3	Parallelogram	$A = bh$
4	Triangle	$A = \frac{1}{2}bh$
5	Circle	$A = \pi r^2$; $C = 2\pi r$
6	Trapezoid	$A = \frac{1}{2}(a + b)h$
7	Cube	$S = 6s^2$; $V = s^3$
8	Rectangular Box	$S = 2(lw + wh + lh)$; $V = lwh$
9	Cylinder	$S = 2\pi rh$; $V = \pi r^2 h$
10	Sphere	$S = 4\pi r^2$; $V = \frac{4}{3}\pi r^3$
11	Cone	$S = \pi r\sqrt{r^2 + h^2}$; $V = \frac{1}{3}\pi r^2 h$

Algebra and Trigonometry

with Applications

Mustafa A. Munem

Macomb County Community College

David J. Foulis

University of Massachusetts

Worth Publishers, Inc.

ALGEBRA AND TRIGONOMETRY WITH APPLICATIONS

EDITOR: GORDON BECKHORN

PRODUCTION: GEORGE TOULOUMES

DESIGN: MALCOLM GREAR DESIGNERS

ILLUSTRATOR: FELIX COOPER

COPYEDITOR: TRUMBULL ROGERS

TYPOGRAPHER: SYNTAX INTERNATIONAL

PRINTING AND BINDING: RAND McNALLY AND COMPANY

COVER: COMPUTER GRAPHICS BY THOMAS BANCHOFF AND DAVID SALESIN

WORTH PUBLISHERS, INC.

444 PARK AVENUE SOUTH

NEW YORK, NEW YORK 10016

Preface

Purpose This book was written to provide students with a straightforward, readable text that presents algebra and trigonometry in an appealing manner. We have tried to "talk" with our readers, without patronizing or lecturing them; and we have avoided sophisticated "mathematical elegance" in favor of an intuitive approach to algebra and trigonometry.

Prerequisites The material in this book can be understood by the student who has had the equivalent of two years of college-preparatory mathematics in high school, including algebra and some plane geometry, or who has taken a college-level course in introductory algebra. Determined students with less preparation may be able to master the contents of this book, particularly if they use the accompanying *Study Guide* as an aid.

Objectives This book has three objectives:

First, to provide the reader with a competent working knowledge of college-level algebra and trigonometry. We have taken pains to ensure adequate coverage of the traditional topics and techniques, especially those that will be needed in other courses, such as calculus, linear algebra, physics, chemistry, and engineering. Furthermore, we have tried to attune the textbook to the two outstanding mathematical trends of our time—the computer revolution and the burgeoning use of mathematical models in the life sciences, the social sciences, economics, and business.

Second, to complement and enhance effective classroom teaching. Our own classroom experience has shown that successful teaching is fostered by a textbook that provides clear and complete explanations, cogent visual presentations, step-by-step procedures, numerous illustrative examples, and a multitude of topical, real-world applications. We have written the textbook with these criteria in mind.

Third, to allay "math anxiety." Many students enrolling in mathematics courses approach the subject with little or no self-confidence. In order to help students build such confidence, we begin with familiar material, explain new concepts in terms of ideas already well understood, offer worked-out examples that show in detail how problems are to be solved, and provide problem sets in which the level of difficulty increases gradually.

Special Features

1. *Use of calculators*   In keeping with the recommendations of the National Council of Teachers of Mathematics (NCTM) and the Mathematical Association of America (MAA), we have de-emphasized the use of logarithmic and trigonometric tables in favor of the use of scientific calculators. As classroom teachers, we feel obliged to prepare our students to function in the real world, where virtually everyone who uses mathematics in a practical way—from the actuary to the zoologist—routinely employs a calculator. Today, a scientific calculator can be purchased inexpensively and used throughout the student's tenure in college or university—and beyond.

2. *Logarithmic and Trigonometric Tables*   For those teachers who feel that some instruction in the use of tables is desirable, we have included in the appendices tables of logarithmic, exponential, and trigonometric functions, along with examples illustrating their use and the technique of linear interpolation.

3. *Mathematical Models*   Emphasis on the notion of mathematical models is a unique feature of this book. For instance, the idea of a mathematical model is effectively used in the section on biological growth in Chapter 5, as part of the discussion of oscillatory phenomena in Chapter 6, and in connection with examples of predator–prey relationships in Chapter 7.

4. *Trigonometric Functions*   Trigonometric functions are first defined in terms of right triangles. This simple, geometrically appealing approach provides a firm foundation for the ensuing treatment of trigonometric functions of general angles and of circular functions and their graphs.

5. *Examples, Figures, and Problems*   The textbook contains 568 examples, worked in detail, with all substitutions shown. The 477 figures in the book are placed next to the text that they illustrate. Problem sets, which appear at the end of each section, contain a total of 3692 problems, most of which correspond to worked-out examples in the text. Problems in each set progress from simple, drill-type exercises to more demanding conceptual questions. Odd-numbered problems require a level of understanding sufficient for most purposes, whereas even-numbered problems, especially those toward the end of each problem set, are more challenging and probe for deeper understanding. Answers to nearly all of the odd-numbered problems are provided in the back of the book.

6. *Homework*   You will notice that certain problems at the end of every section are identified with numerals printed in color. This is to indicate a group of problems that could serve as a homework assignment for the section.

7. *Review Problem Sets*   Review problem sets at the end of each chapter contain a total of 1425 additional problems. These review problems highlight the essential material in the chapter and serve a variety of purposes: Instructors may wish to use them for supplementary or extra-credit assignments or as a source of problems for quizzes and exams; students may wish to scan them to pinpoint areas where further study is needed.

8. *Sections*   The book is divided into 72 sections, with an average length of about 5 pages of text (exclusive of problem sets). Each section is designed to be covered in one 50-minute class period. Sections that are considered to be review material can, of course, be covered more rapidly.

9. *Index of Applications*   For convenience, an index of applications is provided at the back of the book. The usefulness of algebra and trigonometry is amply demonstrated by the abundant applications, not only to engineering, geometry, and the physical sciences, but also to biology, business, earth sciences, ecology, economics, medicine, navigation, and the social sciences.

Pace   The book has been designed for use in either one or two semesters or quarters. The pace of the course, as well as the choice of which topics to cover and which to emphasize, will vary from school to school. There is a more detailed discussion of these matters in the accompanying *Instructor's Test Manual*.

Study Guide   A *Study Guide* is available for students who need more drill or more assistance. It contains study objectives, carefully graded fill-in statements and problems that are broken down into simple units, and self-tests. Answers to all problems in the *Study Guide* are included, in order to provide immediate reinforcement for correct responses.

Acknowledgments   We wish to thank the following people, who reviewed the manuscript and offered many helpful suggestions: William P. Bair, *Pasadena City College;* Ann S. Bumpus, *Langley High School, McLean, Virginia;* Kenneth Chapman, *Michael J. Owens Technical College, Toledo;* Henry Cohen, *University of Pittsburgh;* Albert G. Fadell, *State University of New York at Buffalo;* Gerald E. Gannon, *California State University, Fullerton;* Thomas F. Gordon, *North Carolina State University at Raleigh;* B. C. Horne, Jr., *Virginia Polytechnic Institute and State University;* Gary Lippman, *California State University, Hayward;* Stanley M. Lukawecki, *Clemson University;* Eldon L. Miller, *University of Mississippi;* Philip R. Montgomery, *University of Kansas;* John Riner, *Ohio State University;* Robert G. Savage, *North Carolina State University at Raleigh;* Mary W. Scott, *University of Florida;* Howard E. Taylor, *West Georgia College;* Paul D. Trembeth, *Delaware Valley College of Science and Agriculture, Doylestown, Pennsylvania.*

   We would also like to express our appreciation to Professor Steve Fasbinder of Oakland University for solving all the problems in the book and for assisting in the proofreading; to Paula Ashley and to David B. Foulis for help in the proofreading; and to Hyla Gold Foulis for reviewing each successive stage of the manuscript, and for proofreading, solving problems, and assisting with the preparation of the *Study Guide*. Finally, we would like to thank the staff at Worth Publishers, especially Bob Andrews and Sally Immerman, for their constant help and encouragement.

Mustafa A. Munem
David J. Foulis

A Note to the Instructor on the Use of Calculators

Problems and examples for which the use of a calculator is recommended are marked with the symbol ⓒ. Answers to these problems and solutions for these examples were obtained using an HP-67 calculator—other calculators may give slightly different results because they use different internal routines.

The rule for rounding off numbers presented in Section 8 of Chapter 1 is consistent with the operation of most calculators with round-off capability. Some instructors may wish to mention the popular alternative round-off rule: If the first dropped digit is 5 and there are no nonzero digits to its right, round off so that the digit retained is even.

Because there are so many different calculators on the market, we have made no attempt to give detailed instructions for calculator operation in this textbook. Students should be urged to consult the instruction manuals furnished with their calculators.

Conscientious instructors will wish to encourage their students to learn to use calculators *efficiently;* for instance, to do chain calculations using the memory features of the calculator. In some of our examples we have shown the intermediate results of chain calculations so the students can check their calculator work; however, it should be emphasized that it is not necessary to write down these intermediate results when using the calculator.

As important as it is to encourage students to use their calculators for the examples and problems marked ⓒ, it is perhaps more important to *restrain* them from attempting to use their calculators when the symbol ⓒ is not present. For instance, a student who routinely uses a calculator to determine the algebraic sign of a trigonometric function of an angle may fail to learn the connection between the algebraic sign and the quadrant containing the angle.

Finally, we have made no attempt to provide a systematic discussion of the inaccuracies inherent in computations with a calculator. However, in order to make students aware that such inaccuracies exist, we have carried out some computations to the full ten places available on our calculator. (See, for instance, the example on the top of page 225.) An excellent account of calculator inaccuracy can be found in "Calculator Calculus and Roundoff Errors," by George Miel in *The American Mathematical Monthly,* 1980, Vol. 87, No. 4, pp. 243–52.

Contents

Algebra and Trigonometry

1

Concepts of Algebra

This chapter is designed as a review of the basic concepts and methods of algebra. Its purpose is to help you attain the algebraic skills that are required throughout the textbook. Topics covered include the language and symbols of algebra, polynomials, fractions, exponents, radicals, and the use of a calculator.

1 The Language and Notation of Algebra

Algebra begins with a systematic study of the operations and rules of arithmetic. The operations of addition, subtraction, multiplication, and division serve as a basis for all arithmetic calculations. In order to achieve generality, letters of the alphabet are used in algebra to represent numbers. A letter such as x, y, a, or b can stand for a particular number (known or unknown), or it can stand for any number at all. The sum, difference, product, and quotient of two numbers, x and y, can then be written as

$$x + y, \quad x - y, \quad x \times y, \quad \text{and} \quad x \div y.$$

In algebra, the notation $x \times y$ for the product of x and y is not often used because of the possible confusion of the letter x with the multiplication sign $\times$. The preferred notation is $x \cdot y$ or simply xy. Similarly, the notation $x \div y$ is usually avoided in favor of the fraction $\dfrac{x}{y}$ or x/y.

Algebraic notation—the "shorthand" of mathematics—is designed to clarify ideas and simplify calculations by permitting us to write expressions compactly and efficiently. For instance, $x + x + x + x + x$ can be written simply as $5x$. The use of exponents provides an economy of notation for products; for instance, $x \cdot x$ can be written simply as x^2 and $x \cdot x \cdot x$ as x^3. In general, if n is a positive integer,

$$x^n \quad \text{means} \quad \overbrace{x \cdot x \cdot x \cdots x}^{n \text{ times}}.$$

In using the **exponential notation x^n,** we refer to x as the **base** and n as the **exponent,** or the **power** to which the base is raised.

Example   Rewrite each expression using exponential notation.

(a) $3 \cdot 3 \cdot 3 \cdot 3 \cdot 3 \cdot 3 \cdot 3$

(b) $(-y)(-y)(-y)(-y)(-y)$

(c) $d \cdot d \cdot d \cdot e \cdot e \cdot e \cdot e$

(d) $(3a - 2b)(3a - 2b)(3a - 2b)$

Solution   (a) 3^7 (b) $(-y)^5$

(c) $d^3 e^4$ (d) $(3a - 2b)^3$

By writing an *equals sign* (=) between two algebraic expressions, we obtain an **equation** or **formula** stating that the two expressions represent the same number. Using equations and formulas, we can express mathematical facts in compact, easily remembered forms.

The important **principle of substitution** states that, *if $a = b$, we may substitute b for a in any formula or expression that involves a.* A second important rule, called the **reflexive principle,** states that *any expression may be set equal to itself.* Many standard algebraic procedures, such as *adding or subtracting the same quantity to both sides of an equation,* can be justified on the basis of the reflexive and substitution principles.

Example   Given that $a = b$, show that $a + c = b + c$.

Solution   By the reflexive principle,

$$a + c = a + c.$$

Since $a = b$, we can substitute b for a on the right side of this equation to obtain

$$a + c = b + c.$$

In such fields as geometry, physics, engineering, statistics, geology, business, medicine, economics, and the life sciences, formulas are used to express relationships among various quantities.

Example 1   Write a formula for the volume V of a cube that has sides of length x units.

Solution $V = x \cdot x \cdot x = x^3$ cubic units.

Example 2   A certain type of living cell divides every hour. Starting with one such cell in a culture, the number N of cells present at the end of t hours is given by the formula $N = 2^t$. Find the number of cells in the culture after 6 hours.

Solution   Substituting $t = 6$ in the formula $N = 2^t$, we find that

$$N = 2^6 = 2 \cdot 2 \cdot 2 \cdot 2 \cdot 2 \cdot 2 = 64 \text{ cells.}$$

1.1 Basic Algebraic Properties of Real Numbers

The numbers used to measure real-world quantities such as length, area, volume, speed, electrical charge, efficiency, probability of rain, intensity of earthquakes, profit, body temperature, gross national product, growth rate, and so forth are called **real numbers.** They include such numbers as

$$5, \quad -17, \quad \frac{27}{13}, \quad -\frac{2}{3}, \quad 0, \quad 2.71828, \quad \sqrt{2}, \quad -\frac{\sqrt{3}}{2}, \quad 3 \times 10^8, \quad \text{and} \quad \pi.$$

The basic algebraic properties of the real numbers can be expressed in terms of the two fundamental operations of addition and multiplication.

Basic Algebraic Properties of the Real Numbers

Let a, b, and c denote real numbers.

1. *The Commutative Properties*

(i) $a + b = b + a$	(ii) $a \cdot b = b \cdot a$

2. *The Associative Properties*

(i) $a + (b + c) = (a + b) + c$	(ii) $a \cdot (b \cdot c) = (a \cdot b) \cdot c$

3. *The Distributive Properties*

(i) $a \cdot (b + c) = a \cdot b + a \cdot c$	(ii) $(b + c) \cdot a = b \cdot a + c \cdot a$

4. *The Identity Properties*

(i) $a + 0 = 0 + a = a$	(ii) $a \cdot 1 = 1 \cdot a = a$

5. *The Inverse Properties*

(i) For each real number a, there is a real number $-a$, called the **additive inverse** of a, such that

$$a + (-a) = (-a) + a = 0.$$

(ii) For each real number $a \neq 0$, there is a real number $1/a$, called the **multiplicative inverse** of a, such that

$$a \cdot \frac{1}{a} = \frac{1}{a} \cdot a = 1.$$

The commutative properties say that the *order* in which we either add or multiply real numbers doesn't matter. Similarly, the associative properties tell us that the way real numbers are *grouped* when they are either added or multiplied doesn't matter. Because of the associative properties, expressions such as $a + b + c$

or $a \cdot b \cdot c$ make sense without parentheses. The distributive properties can be used to expand a product into a sum, such as

$$a(b + c + d) = ab + ac + ad,$$

or the other way around, to rewrite a sum as a product:

$$ax + bx + cx + dx + ex = (a + b + c + d + e)x.$$

Although the additive inverse of a, namely $-a$, is usually called the **negative** of a, you must be careful because $-a$ isn't necessarily a negative number. For instance, if $a = -2$, then $-a = -(-2) = 2$. Notice that the multiplicative inverse $1/a$ is assumed to exist only if $a \neq 0$. The real number $1/a$ is also called the **reciprocal** of a and is often written as a^{-1}.

Example   State one basic algebraic property of the real numbers to justify each statement.

(a) $7 + (-2) = (-2) + 7$ (b) $x + (3 + y) = (x + 3) + y$

(c) $a + (b + c)d = a + d(b + c)$ (d) $x[y + (z + w)] = xy + x(z + w)$

(e) $(x + y) + [-(x + y)] = 0$ (f) $(x + y) \cdot 1 = x + y$

(g) If $x + y \neq 0$, then $(x + y)[1/(x + y)] = 1$

Solution   (a) Commutative property for addition   (b) Associative property for addition

(c) Commutative property for multiplication

(d) Distributive property   (e) Additive inverse property

(f) Multiplicative identity property   (g) Multiplicative inverse property

Many of the important properties of the real numbers can be *derived* as results of the basic properties, although we shall not do so here. Among the more important derived properties are the following.

6. *The Cancellation Properties*

> (i) If $a + x = a + y$, then $x = y$.
>
> (ii) If $a \neq 0$ and $ax = ay$, then $x = y$.

7. *The Zero-Factor Properties*

> (i) $a \cdot 0 = 0 \cdot a = 0$
>
> (ii) If $a \cdot b = 0$, then $a = 0$ or $b = 0$ (or both).

8. *Properties of Negation*

> (i) $-(-a) = a$
>
> (ii) $(-a)b = a(-b) = -(ab)$
>
> (iii) $(-a)(-b) = ab$

Properties 1 through 8 are used throughout this textbook.
The operations of subtraction and division are defined as follows.

Definition 1   **Subtraction and Division**

Let a and b be real numbers.

   (i) The **difference** $a - b$ is defined by $a - b = a + (-b)$.

   (ii) The **quotient** $a \div b$ or $\dfrac{a}{b}$ is defined only if $b \neq 0$. If $b \neq 0$, then by definition
$$\frac{a}{b} = a \cdot \frac{1}{b}.$$

For instance, $7 - 4$ means $7 + (-4)$ and $\frac{7}{4}$ means $7 \cdot \frac{1}{4}$. Note that *division by zero is not allowed*. When $a \div b$ is written in the form a/b, it is called a **fraction** with **numerator a** and **denominator b**. Although the denominator can't be zero, there's nothing wrong with having a zero in the numerator. In fact, if $b \neq 0$,

$$\frac{0}{b} = 0 \cdot \frac{1}{b} = 0$$

by the zero-factor property 7(i).

PROBLEM SET 1

In problems 1–10, rewrite each expression using exponential notation.

1. $5 \cdot 5 \cdot 5 \cdot 5 \cdot 5 \cdot 5 \cdot 5 \cdot 5$
2. $(-7)(-7)(-7)(-7)$
3. $3 \cdot 3 \cdot 3 \cdot 3 \cdot 4 \cdot 4 \cdot 4$
4. $8 \cdot 8 + (-6)(-6)(-6)$
5. $x \cdot x \cdot x \cdot x \cdot y \cdot y \cdot y$
6. $4 \cdot 4 \cdot 4 \cdot 4 \cdot y \cdot y \cdot y$
7. $(-x)(-x)(-y)(-y)(-y)$
8. $y^2 y^2 y^2 z^3 z^3$
9. $(2a + 1)(2a + 1)(2a + 1)$
10. $(x + \frac{1}{2})(x + \frac{1}{2})(x - \frac{1}{2})(x - \frac{1}{2})(x - \frac{1}{2})$

In problems 11–22, write a formula for the given quantity.

11. The number z that is twice the sum of x and y.
12. The area A of a rectangle with length a units and width b units.
13. The area A of a circle of radius r units.
14. The perimeter P of a rectangle with length a units and width b units.
15. The number x that is 5% of a number n.
16. The volume V of a rectangular box with length L units, width W units, and height H units.
17. The surface area A of a cube with sides of length x units. [Hint: A cube has six faces, each of which is a square.]
18. The volume V of a sphere of radius r units. [Consult a geometry book if you don't know the formula.]
19. The area A of a triangle with base b units and height h units.
20. The number N of living cells in a culture after t hours if there are N_0 cells when $t = 0$, if each cell divides into two cells at the end of each hour, and if no cells die.
21. The amount A dollars you owe after t years if you borrow p dollars at a simple interest rate r per year.

22. The number L of board feet of lumber in a tree d feet in diameter and h feet high. Assume for simplicity that the tree is a right circular cylinder. [Note: One board foot is the volume of a board with dimensions 1 foot by 1 foot by 1 inch.]

ⓒ **23.** If p dollars is invested at a nominal annual interest rate r compounded n times per year, the investment will be worth $p\left(1 + \dfrac{r}{n}\right)^{nt}$ dollars at the end of t years. Suppose you invest \$5000 at 10% interest ($r = 0.10$) compounded semiannually ($n = 2$). What is your investment worth at the end of 2 years? [Use a calculator if you wish.]

24. The *half-life* of a radioactive substance is the period of time T during which exactly half of the substance will undergo radioactive disintegration. If q represents the quantity of such a substance at time t, then $q = \dfrac{q_0}{2^{t/T}}$, where q_0 is the original amount of the substance when $t = 0$. Potassium-42, which is used as a biological tracer, has a half-life of $T = 12.5$ hours. If a certain quantity of Potassium-42 is injected into an organism, and if none is lost by excretion, what percentage will remain after $t = 50$ hours?

In problems 25–40, state one basic algebraic property of the real numbers to justify each statement.

25. $5 + (-3) = (-3) + 5$

26. $5 \cdot (3 + 7) = (3 + 7) \cdot 5$

27. $(-13)(-12) = (-12)(-13)$

28. $5 \cdot (3 + 7) = 5 \cdot 3 + 5 \cdot 7$

29. $1 \cdot (3 + 7) = 3 + 7$

30. $[(3)(4)](5) = (3)[(4)(5)]$

31. $4 + (x + z) = (4 + x) + z$

32. $5 \cdot \frac{1}{5} = 1$

33. $4 \cdot (x + 0) = 4x$

34. $(x - y) + [-(x - y)] = 0$

35. $x[(-y) + y] = x(-y) + xy$

36. $(-4) \cdot \dfrac{1}{(-4)} = 1$

37. $(4 + x) + [-(4 + x)] = 0$

38. $x + [y + (z + w)] = (x + y) + (z + w)$

39. $(ab)(cd) = [(ab)c]d$

40. $(a + b)(c + d) = a(c + d) + b(c + d)$

In problems 41–48, state one of the derived algebraic properties (6–8, page 4) to justify each statement.

41. $-(-5) = 5$

42. $(-3)(-x) = 3x$

43. $5(-6) = -(5)(6)$

44. $(x + y) \cdot 0 = 0$

45. If $7 + x = 7 + x^2$, then $x = x^2$.

46. If $2y = 2y^{-1}$, then $y = y^{-1}$.

47. If $(2x + 3)(x + 1) = 0$, then $2x + 3 = 0$ or $x + 1 = 0$.

48. If $x^2 = 0$, then $x = 0$.

In problems 49 and 50, a *mistake* has been made. Find the mistake and make the correct calculation.

49. $(3 + 5)^2 = 3^2 + 5^2 = 9 + 25 = 34$?

50. $\dfrac{1}{3 + 5} = \dfrac{1}{3} + \dfrac{1}{5} = \dfrac{5}{15} + \dfrac{3}{15} = \dfrac{8}{15}$?

51. Does the operation of subtraction have the associative property; that is, is it always true that $a - (b - c) = (a - b) - c$?

52. Does the operation of division have the associative property?

53. Prove that both sides of an equation can be multiplied by the same quantity; that is, prove that if $a = b$, then $c \cdot a = c \cdot b$. [Hint: Use the reflexive and substitution principles.]

54. Prove that both sides of an equation can be interchanged; that is, prove that if $a = b$, then $b = a$. This is called the **symmetric property of equality**. [Hint: Use the reflexive and substitution principles.]

2 Sets of Real Numbers

Grouping or **classifying** is a familiar technique in the natural sciences for dealing with the immense diversity of things in the real world. For instance, in biology, plants and animals are divided into various divisions or phyla, and then into classes, orders, families, genera, and species. In much the same way, real numbers can be grouped or classified by singling out important features possessed by some numbers but not by others. By using the idea of a *set*, classification of real numbers can be accomplished with clarity and precision.

A **set** may be thought of as a collection of objects. Most sets considered in this textbook are sets of real numbers. Any one of the objects in a set is called an **element,** or **member,** of the set. Sets are denoted either by capital letters such as A, B, C or else by braces $\{\cdots\}$ enclosing symbols for the elements in the set. Thus, if we write $\{1, 2, 3, 4, 5\}$, we mean the set whose elements are the numbers 1, 2, 3, 4, and 5. Two sets are said to be *equal* if they contain precisely the same elements.

Example Write the set A consisting of all the odd numbers between 2 and 8.

Solution $A = \{3, 5, 7\}$

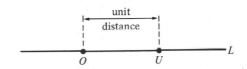

Figure 1

Sets of numbers and relations among such sets can often be visualized by the use of a **number line** or **coordinate axis.** A number line is constructed by fixing a point O called the **origin** and another point U called the **unit point** on a straight line L (Figure 1). The distance between O and U is called the **unit distance,** and may be 1 inch, 1 centimeter, or 1 unit of whatever measure you choose. If the straight line L is horizontal, it is customary to place U to the right of O.

Each point P on the line L is now assigned a "numerical address" or **coordinate** x representing its signed distance from the origin, measured in terms of the given unit. Thus, $x = \pm d$, where d is the distance between O and P; the plus sign or minus sign is used to indicate whether P is to the right or left of O (Figure 2). Of course, the origin O is assigned the coordinate 0 (zero), and the unit point U is assigned the coordinate 1. On the resulting number scale (Figure 3), each point P has a corresponding numerical coordinate x and each real number x is the coordinate of a uniquely determined point P. It is convenient to use an arrowhead on the number line to indicate the direction in which the numerical coordinates are increasing (to the right in Figure 3).

Figure 2

Figure 3

A set of numbers can be illustrated on a number line by shading or coloring the points whose coordinates are members of the set. For instance:

1. The **natural numbers,** also called the **counting numbers,** or **positive integers,** are the numbers 1, 2, 3, 4, 5, and so on, obtained by adding 1 to itself over and over again. The set $\{1, 2, 3, 4, 5, \cdots\}$ of all natural numbers, denoted by the symbol $\mathbb{N}$, is illustrated in Figure 4.

Figure 4

2. The **integers** consist of all the natural numbers, the negatives of the natural numbers, and zero. The set $\{\cdots -4, -3, -2, -1, 0, 1, 2, 3, 4, \cdots\}$ of all integers, denoted by the symbol **I,** is illustrated in Figure 5.

Figure 5

3. The **positive real numbers** correspond to points to the right of the origin (Figure 6a), and the **negative real numbers** correspond to points to the left of the origin (Figure 6b). The set of all real numbers is denoted by the symbol $\mathbb{R}$ (Figure 6c).

Figure 6

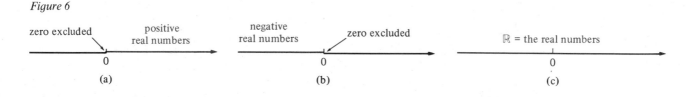

4. The **rational numbers** are those real numbers that can be written in the form a/b, where a and b are integers and $b \neq 0$. Since b may equal 1, every integer is a rational number. Other examples of rational numbers are $\frac{13}{2}$, $\frac{3}{4}$, and $-\frac{22}{7}$. The set of all rational numbers is denoted by the symbol $\mathbb{Q}$ (which reminds us that rational numbers are *quotients* of integers).

5. The **irrational numbers** are the real numbers that are not rational. A real number is irrational if and only if its decimal representation is **nonterminating** and **nonrepeating.** Examples are $\sqrt{2} = 1.4142135 \cdots$, $\sqrt{3} = 1.7320508 \cdots$, and $\pi = 3.1415926 \cdots$.

2.1 Inequalities and Intervals

If the point with coordinate x lies to the left of the point with coordinate y (Figure 7), we say that y is **greater than** x (or equivalently, that x is **less than** y) and we write $y > x$ (or $x < y$). In other words, $y > x$ (or $x < y$) means that $y - x$ is positive. A statement of the form $y > x$ (or $x < y$) is called an **inequality.**

Figure 7

$$x < y$$

$$\begin{array}{c} \bullet \qquad\qquad \bullet \\ x \qquad\qquad y \end{array}$$

Sometimes we know only that a certain inequality does *not* hold. If $x < y$ does not hold, then either $x > y$ or $x = y$. In this case, we say that x is *greater than or equal to y* and we write $x \geq y$ (or $y \leq x$). If $y \leq x$, we say that y is *less than or equal to x*. Statements of the form $y < x$ (or $x > y$) are called **strict** inequalities, whereas those of the form $y \leq x$ (or $x \geq y$) are called **nonstrict** inequalities.

If we write $x < y < z$, we mean that $x < y$ *and* $y < z$. Likewise, $x \geq y > z$ means that $x \geq y$ *and* $y > z$. Notice that this notation for combined inequalities is only used when the inequalities run in the *same direction*. If you are ever tempted to write something like $x < y > z$, resist the urge—such notation is improper and confusing.

In Section 5 of Chapter 2 we shall study inequalities in more detail. Here we use inequalities to define sets called **intervals.**

Definition 1   **Bounded Intervals**

> Let a and b be fixed real numbers with $a < b$.
>
> (i)   The **open interval** (a, b) with *endpoints* a and b is the set of all real numbers x such that $a < x < b$.
>
> (ii)   The **closed interval** $[a, b]$ with *endpoints* a and b is the set of all real numbers x such that $a \leq x \leq b$.
>
>
>
> (iii) The **half-open interval** $[a, b)$ with *endpoints* a and b is the set of all real numbers x such that $a \leq x < b$.
>
> (iv) The **half-open interval** $(a, b]$ with *endpoints* a and b is the set of all real numbers x such that $a < x \leq b$.
>
>

Notice that a closed interval contains its endpoints, but an open interval does not. A half-open interval (also called a **half-closed** interval) contains one of its endpoints, but not the other.

Unbounded intervals, which extend indefinitely to the right or left, are written with the aid of the special symbols $+\infty$ and $-\infty$, called **positive infinity** and **negative infinity,** respectively.

Definition 2   **Unbounded Intervals**

Let a be a fixed real number.

(i) $(a, +\infty)$ is the set of all real numbers x such that $a < x$

(ii) $(-\infty, a)$ is the set of all real numbers x such that $x < a$

(iii) $[a, +\infty)$ is the set of all real numbers x such that $a \leq x$

(iv) $(-\infty, a]$ is the set of all real numbers x such that $x \leq a$

It must be emphasized that $+\infty$ and $-\infty$ are just convenient symbols—*they are not real numbers* and should not be treated as if they were. In the notation for unbounded intervals, we usually write ∞ rather than $+\infty$. For instance, $(5, \infty)$ denotes the set of all real numbers that are greater than 5.

Example   Illustrate each set on a number line.

(a) $(2, 5]$ (b) $(3, \infty)$

(c) The set A of all real numbers that belong to both intervals $(2, 5]$ and $(3, \infty)$.

(d) The set B of all real numbers that belong to at least one of the intervals $(2, 5]$ and $(3, \infty)$.

Solution   (a) The interval $(2,5]$ consists of all real numbers between 2 and 5, including 5, but excluding 2.

(b) The interval $(3, \infty)$ consists of all real numbers that are greater than 3.

(c) The set A of all real numbers that belong to both intervals $(2,5]$ and $(3, \infty)$ is the interval $(3, 5]$.

(d) The set B consists of all numbers in the interval $(2,5]$ together with all numbers in the interval $(3, \infty)$, so $B = (2, \infty)$.

2.2 Rational Numbers and Decimals

By using long division, you can express a rational number as a decimal. For instance, if you divide 2 by 5, you will obtain $\frac{2}{5} = 0.4$, a terminating decimal. Similarly, if you divide 2 by 3, you will obtain $\frac{2}{3} = 0.66666 \cdots$, a nonterminating, repeating decimal. A repeating decimal, such as $0.66666 \cdots$, is often written as $0.\overline{6}$, where the overbar indicates the number or numbers that repeat; hence $\frac{2}{3} = 0.\overline{6}$.

Example   Express each rational number as a decimal.

(a) $-\frac{3}{5}$ (b) $\frac{3}{8}$ (c) $\frac{17}{6}$ (d) $\frac{3}{7}$

Solution   (a) $-\frac{3}{5} = -0.6$ (b) $\frac{3}{8} = 0.375$

(c) $\frac{17}{6} = 2.83333 \cdots = 2.8\overline{3}$

(d) $\frac{3}{7} = 0.428571428571428571 \cdots = 0.\overline{428571}$

Every rational number can be expressed as a decimal that is either terminating or repeating. Conversely, every terminating or repeating decimal represents a rational number. The following example illustrates how you can rewrite a terminating decimal as a quotient of integers.

Example   Express each terminating decimal as a quotient of integers.

(a) 0.7   (b) -0.63 (c) 1.075

Solution   (a) $0.7 = \frac{7}{10}$ (b) $-0.63 = -\frac{63}{100}$ (c) $1.075 = \frac{1075}{1000} = \frac{43}{40}$

In Section 1.1 of Chapter 2 you will see how to rewrite a repeating decimal as a quotient of integers.

Fractions or decimals are often expressed as percentages; for instance, 3% means $\frac{3}{100}$ or 0.03.

Example 1   Rewrite each percent as a decimal.
(a) 5.7%   (b) 0.003%

Solution   (a) $5.7\% = \frac{5.7}{100} = 0.057$ (b) $0.003\% = \frac{0.003}{100} = 0.00003$

Example 2   Rewrite each rational number as a percent.
(a) $\frac{4}{5}$ (b) $\frac{1}{3}$

Solution   (a) $\frac{4}{5} = 0.8 = 0.8 \times 100\% = 80\%$ (b) $\frac{1}{3} = 0.\overline{3} = 0.\overline{3} \times 100\% = 33.\overline{3}\%$

Example 3   What percent is 40 of 2000?

Solution $\frac{40}{2000} = 0.02 = 0.02 \times 100\% = 2\%$

If a number increases, then the **percent of increase** is given by

$$\frac{\text{amount of increase}}{\text{original value}} \times 100\%.$$

If a number decreases, then the **percent of decrease** is given by

$$\frac{\text{amount of decrease}}{\text{original value}} \times 100\%.$$

Example 1   Juanita's weekly salary increases from $205 to $213.20. What is the percent of increase?

Solution   The amount of increase is $213.20 - $205 = $8.20. Since the original salary before the increase was $205, the percent of increase is given by

$$\frac{8.20}{205} \times 100\% = 0.04 \times 100\% = 4\%.$$

Example 2   A small town decreases its annual budget from $800,000 to $600,000. What is the percent of decrease?

Solution $\frac{800{,}000 - 600{,}000}{800{,}000} \times 100\% = \frac{200{,}000}{800{,}000} \times 100\% = 0.25 \times 100\% = 25\%$

PROBLEM SET 2

In problems 1–4, list all of the elements that belong to each set.

1. A is the set of all even integers between 3 and 11.

2. B is the set of all natural numbers x such that $-\frac{1}{2} \le x \le \frac{7}{2}$.

3. C is the set of all real numbers that are in the closed interval $[2, 4]$ but not in the open interval $(2, 4)$.

4. D is the set of all real numbers x such that $x(x - 1)(x + 1) = 0$.

In problems 5–16, indicate whether the statement is true or false.

5. $\frac{1}{2}$ is an element of $\mathbb{N}$. **6.** $\frac{1}{2}$ is an element of $\mathbb{Q}$. **7.** $\frac{1}{2}$ is an element of $(\frac{1}{2}, \frac{2}{3})$.

8. $\frac{1}{2}$ is an element of $(-\frac{2}{3}, \frac{2}{3})$. **9.** $\sqrt{2}$ is an element of $\mathbb{R}$. **10.** $\sqrt{2}$ is an element of $\mathbb{Q}$.

11. $\frac{1}{2} + \frac{2}{3}$ is an element of $\mathbb{Q}$. **12.** $\frac{1}{2} + \sqrt{2}$ is an element of $\mathbb{Q}$. **13.** Every natural number is an integer.

14. Every natural number is a rational number.

15. Every positive real number is a rational number.

16. $\sqrt{\pi}$ is a rational number.

In problems 17–24, illustrate each set on a number line.

17. The set of odd positive integers. **18.** The set of even integers that are not positive.

19. (a) $(2, 5)$ (b) $[-1, 3)$ (c) $[-4, 0]$ (d) $(-\infty, -3)$
(e) $[1, \infty)$ (f) $(-\frac{1}{2}, \infty)$ (g) $[-\frac{3}{2}, \frac{5}{2}]$ (h) $(-4, 0]$

20. The set of real numbers x such that $-3 < x < 3$ and $2x$ is in **I**.

21. The set A of all real numbers that belong to both of the intervals $(0, 3]$ and $(1, \infty)$.

22. The set B of all real numbers that belong to both of the intervals $(-1, \frac{1}{2}]$ and $[\frac{1}{2}, \frac{3}{4}]$.

23. The set C of all real numbers that belong to at least one of the intervals $[-2, -1]$ and $[1, 2]$.

24. The set D of all real numbers that belong to at least one of the intervals $(-1, 0]$ and $(0, 1]$.

In problems 25–38, express each rational number as a decimal.

25. $\frac{4}{5}$ **26.** $-\frac{7}{25}$ **27.** $-\frac{3}{8}$ **28.** $\frac{3}{50}$ **29.** $-\frac{13}{50}$
30. $\frac{7}{100}$ **31.** $\frac{-17}{200}$ **32.** $\frac{200}{500}$ **33.** $-\frac{5}{8}$ **34.** $\frac{75}{30}$
35. $-\frac{5}{3}$ **36.** $\frac{11}{12}$ **37.** $\frac{15}{7}$ **38.** $\frac{-23}{7}$

In problems 39–48, express each terminating decimal as a quotient of integers.

39. 0.41 **40.** −0.54 **41.** −0.032 **42.** 22.61 **43.** −0.581
44. 2.691 **45.** −0.913 **46.** 0.00012 **47.** 1.0451 **48.** −2.00002

In problems 49–56, rewrite each percent as a decimal.

49. 11% **50.** 99.44% **51.** 1.03% **52.** 315% **53.** 432%
54. 0.001% **55.** 0.0006% **56.** 1.00001%

In problems 57–64, rewrite each rational number as a percent.

57. $\frac{1}{2}$ **58.** $\frac{7}{50}$ **59.** $\frac{3}{125}$ **60.** 0.042 **61.** $\frac{2}{3}$
62. $\frac{5}{8}$ **63.** $\frac{2}{7}$ **64.** 1.245

65. What percent is 3 of 20?

66. In the United States, 5 people out of every 1000 are in the army and 8 people out of every 10,000 are army officers. What percentage of army personnel are officers?

67. The cost of a book increases from $20 to $24. What is the percent of increase?

68. In April, Pedro's salary is $180 per week. In May, his April salary is increased by 5%. In June, his May salary is decreased by 5%. What is Pedro's salary in June?

69. The cost of a calculator decreases from $24 to $20. What is the percent of decrease?

3 Polynomials

An **algebraic expression** is an expression formed from any combination of numbers and variables by using the operations of addition, subtraction, multiplication, division, exponentiation (raising to powers), or extraction of roots. For instance,

$$7, \quad x, \quad 2x - 3y + 1, \quad \frac{5x^3 - 1}{4xy + 1}, \quad \pi r^2, \quad \text{and} \quad \pi r\sqrt{r^2 + h^2}$$

are algebraic expressions. By an algebraic expression *in* certain variables, we mean an expression that contains only those variables, and by a **constant,** we mean an algebraic expression that contains no variables at all. If numbers are substituted for the variables in an algebraic expression, the resulting number is called the **value** of the expression for these values of the variables.

Example Find the value of $\dfrac{2x - 3y + 1}{xy^2}$ when $x = 2$ and $y = -1$.

Solution Substituting $x = 2$ and $y = -1$, we obtain

$$\frac{2(2) - 3(-1) + 1}{2(-1)^2} = \frac{4 + 3 + 1}{2(1)} = \frac{8}{2} = 4.$$

If an algebraic expression consists of parts connected by plus or minus signs, it is called an **algebraic sum,** and each of the parts, together with the sign preceding it, is called a **term.** For instance, in the algebraic sum

$$3x^2y - \frac{4xz^2}{y} + \pi x^{-1}y,$$

the terms are $3x^2y$, $-4xz^2/y$, and $\pi x^{-1}y$.

Any part of a term that is multiplied by the remaining part is called a **coefficient** of the remaining part. For instance, in the term $-4xz^2/y$, the coefficient of z^2/y is $-4x$, whereas the coefficient of xz^2/y is -4. A coefficient such as -4, which involves no variables, is called a *numerical* coefficient. Terms such as $5x^2y$ and $-12x^2y$, which differ only in their numerical coefficients, are called **like terms** or **similar terms.**

An algebraic expression such as $4\pi r^2$ can be considered an algebraic sum consisting of just one term. Such a one-termed expression is called a **monomial.** An algebraic sum with two terms is called a **binomial,** and an algebraic sum with three terms is called a **trinomial.** For instance, the expression $3x^2 + 2xy$ is a binomial, whereas $-2xy^{-1} + 3\sqrt{x} - 4$ is a trinomial. An algebraic sum with two or more terms is called a **multinomial.**

A **polynomial** is an algebraic sum in which no variables appear in denominators or under radical signs, and all variables that do appear are raised only to positive-integer powers. For instance, the trinomial

$$-2xy^{-1} + 3\sqrt{x} - 4$$

is *not* a polynomial; however, the trinomial

$$3x^2y^4 + \sqrt{2}xy - \tfrac{1}{2}$$

is a polynomial in the variables x and y. A term such as $-\tfrac{1}{2}$, which contains no variables, is called a **constant term** of the polynomial. The numerical coefficients of the terms in a polynomial are called the **coefficients** of the polynomial. The coefficients of the polynomial above are

$$3, \quad \sqrt{2}, \quad \text{and} \quad -\tfrac{1}{2}.$$

The **degree** of a term in a polynomial is the sum of all the exponents of the variables in the term. In adding exponents, you should regard a variable with no exponent as being a first power. For instance, in the polynomial

$$9xy^7 - 12x^3yz^2 + 3x - 2,$$

the term $9xy^7$ has degree $1 + 7 = 8$, the term $-12x^3yz^2$ has degree $3 + 1 + 2 = 6$, and the term $3x$ has degree 1. The constant term, if it is nonzero, is always regarded as having degree 0.

The highest degree of all terms that appear with nonzero coefficients in a polynomial is called the **degree** of the polynomial. For instance, the polynomial considered above has degree 8. Although the constant monomial 0 is regarded as a polynomial, this particular polynomial is not assigned a degree.

Example In each case, identify the algebraic expression as a monomial, binomial, trinomial, multinomial and/or polynomial, and specify the variables involved. For any polynomials, give the degree and the coefficients.

(a) $\dfrac{4x}{y}$

(b) $4x + 3y^{-1}$

(c) $-\tfrac{5}{3}x^2y + 8xy - 11$

Solution (a) Monomial in x and y (*not* a polynomial because of the variable y in the denominator)

(b) Binomial in x and y, multinomial (*not* a polynomial because of the negative exponent on y)

(c) Trinomial, multinomial, polynomial in x and y of degree 3 with coefficients $-\tfrac{5}{3}, 8,$ and -11

A polynomial of degree n in a single variable x can be written in the **general form**

$$a_nx^n + a_{n-1}x^{n-1} + \cdots + a_2x^2 + a_1x + a_0,$$

in which $a_n, a_{n-1}, \ldots, a_2, a_1, a_0$ are the numerical coefficients, $a_n \neq 0$ (although any of the other coefficients can be zero), and a_0 is the constant term.

3.1 Addition and Subtraction of Polynomials

To find the sum of two or more polynomials, we use the associative and commutative properties of addition to group like terms together, and then we combine the like terms by using the distributive property (see Problem 74).

Example   Find the following sum:
$$(2x^2 + 7x - 5) + (3x^2 - 11x + 8).$$

Solution

$$(2x^2 + 7x - 5) + (3x^2 - 11x + 8)$$
$$= (2x^2 + 3x^2) + (7x - 11x) + (-5 + 8)$$
$$= 5x^2 - 4x + 3$$

To find the difference of two polynomials, we change the signs of all the terms in the polynomial being subtracted, and then add.

Example   Find the following difference:
$$(3x^3 - 5x^2 + 8x - 3) - (5x^3 - 7x + 11).$$

Solution

$$(3x^3 - 5x^2 + 8x - 3) - (5x^3 - 7x + 11)$$
$$= (3x^3 - 5x^2 + 8x - 3) + (-5x^3 + 7x - 11)$$
$$= (3x^3 - 5x^3) - 5x^2 + (8x + 7x) + (-3 - 11)$$
$$= -2x^3 - 5x^2 + 15x - 14$$

In adding or subtracting polynomials, you may prefer to use a vertical arrangement with like terms in the same columns.

Example   Perform the indicated operations:
$$(4x^3 + 7x^2y + 2xy^2 - 2y^3) + (2x^3 + xy^2 + 4y^3) + (4x^2y - 8xy^2 - 9y^3) - (y^3 - 7x^2y).$$

Solution   First we use a vertical arrangement to perform the addition.

$$
\begin{array}{r}
4x^3 + \ 7x^2y + 2xy^2 - 2y^3 \\
(+) \quad 2x^3 \qquad\quad + \ xy^2 + 4y^3 \\
(+) \qquad\qquad 4x^2y - 8xy^2 - 9y^3 \\
\hline
6x^3 + 11x^2y - 5xy^2 - 7y^3
\end{array}
$$

Then we perform the subtraction vertically.

$$
\begin{array}{r}
6x^3 + 11x^2y - 5xy^2 - 7y^3 \\
(-) \qquad\ - \ 7x^2y \qquad\quad + \ y^3 \\
\hline
6x^3 + 18x^2y - 5xy^2 - 8y^3
\end{array}
$$

3.2 Multiplication of Polynomials

To multiply two or more monomials, we use the commutative and associative properties of multiplication along with the following properties of exponents.

Properties of Exponents

Let a and b denote real numbers. Suppose that m and n are positive integers. Then:

$$\text{(i) } a^m a^n = a^{m+n} \qquad \text{(ii) } (a^m)^n = a^{mn} \qquad \text{(iii) } (ab)^n = a^n b^n$$

We verify (i) as follows and leave (ii) and (iii) as exercises (Problems 75 and 76):

$$a^m a^n = \overbrace{(a \cdot a \cdots a)}^{m \text{ factors}}\overbrace{(a \cdot a \cdots a)}^{n \text{ factors}} = \overbrace{a \cdot a \cdots a \cdot a \cdot a \cdots a}^{m + n \text{ factors}} = a^{m+n}.$$

Properties (i), (ii), and (iii) are useful for simplifying algebraic expressions containing exponents. In general, when the properties of real numbers are used to rewrite an algebraic expression as compactly as possible, or in a form so that further calculations are made easier, we say that the expression has been **simplified.** Thus, although the word "simplify" has no precise mathematical definition, its meaning is usually clear from the context in which it is used.

Example Use the properties of exponents to simplify each expression.

(a) $x^5 x^4$ (b) $2y^4 y^6 y^2 z^2$

(c) $(5x^3 y)(-3x^2 y^4)$ (d) $(3x^2 y^4)^2$

(e) $(-2x^4 y^2)(3x^2 y^3)(5xy^4)$ (f) $(x + 3)^2 (x + 3)^4$

Solution

(a) $x^5 x^4 = x^{5+4} = x^9$ [by Exponent Property (i)]

(b) $2y^4 y^6 y^2 z^2 = 2y^{4+6+2} z^2 = 2y^{12} z^2$ [by Exponent Property (i)]

(c) $(5x^3 y)(-3x^2 y^4) = (5)(-3)(x^3 x^2)(y^1 y^4)$

                   $= -15x^5 y^5$ [by Exponent Property (i)]

(d) $(3x^2 y^4)^2 = 3^2 (x^2)^2 (y^4)^2$ [by Exponent Property (iii)]

            $= 9x^{(2)(2)} y^{(4)(2)}$ [by Exponent Property (ii)]

            $= 9x^4 y^8$

(e) $(-2x^4 y^2)(3x^2 y^3)(5xy^4) = (-2)(3)(5)(x^4 x^2 x)(y^2 y^3 y^4) = -30x^7 y^9$

(f) $(x + 3)^2 (x + 3)^4 = (x + 3)^{2+4} = (x + 3)^6$

To multiply a polynomial by a monomial we use the distributive property. Thus, we multiply each term of the polynomial by the monomial, and then simplify the resulting products by using the properties of exponents.

Example   Find the product $(-3x^2y^3 + 5xy + 7)(4x^3y)$.

Solution

$$(-3x^2y^3 + 5xy + 7)(4x^3y) = (-3x^2y^3)(4x^3y) + (5xy)(4x^3y) + 7(4x^3y)$$
$$= -12x^5y^4 + 20x^4y^2 + 28x^3y$$

To multiply two polynomials, we again employ the distributive property. Thus, we multiply each term of the first polynomial by each term of the second and combine like terms.

Example   Find the product $(x^4 - 5x^2 + 7)(3x^2 + 2)$.

Solution

$$(x^4 - 5x^2 + 7)(3x^2 + 2) = (x^4 - 5x^2 + 7)(3x^2) + (x^4 - 5x^2 + 7)(2)$$
$$= (3x^6 - 15x^4 + 21x^2) + (2x^4 - 10x^2 + 14)$$
$$= 3x^6 - 13x^4 + 11x^2 + 14$$

The same work is arranged vertically as follows:

$$
\begin{array}{l}
\qquad x^4 - 5x^2 + 7 \\
(\times) \qquad 3x^2 + 2 \\
\hline
3x^6 - 15x^4 + 21x^2 \longleftarrow (x^4 - 5x^2 + 7)(3x^2) \\
(+) \qquad 2x^4 - 10x^2 + 14 \longleftarrow (x^4 - 5x^2 + 7)(2) \\
\hline
3x^6 - 13x^4 + 11x^2 + 14 \longleftarrow \text{the sum of the above}
\end{array}
$$

Certain products of polynomials occur so often that it is useful to know the expanded forms by heart. The following list contains some of these **special products.**

Special Products

If a, b, c, d, x, and y are real numbers, then:

1. $(a - b)(a + b) = a^2 - b^2$
2. $(a + b)^2 = a^2 + 2ab + b^2$
3. $(a - b)^2 = a^2 - 2ab + b^2$
4. $(a + b)(a^2 - ab + b^2) = a^3 + b^3$
5. $(a - b)(a^2 + ab + b^2) = a^3 - b^3$
6. $(a + b)^3 = a^3 + 3a^2b + 3ab^2 + b^3$
7. $(a - b)^3 = a^3 - 3a^2b + 3ab^2 - b^3$
8. $(ax + by)(cx + dy) = acx^2 + (ad + bc)xy + bdy^2$.

You should verify the expansions in the list above by actually doing the multiplication (Problem 78). They should become so familiar that you use them automatically in your calculations.

Example   Use Special Products 1–8 to perform each multiplication.

(a) $(3x - 2y)(3x + 2y)$ (b) $(7x^2 + 3y^3)^2$

(c) $(3x + 4)(9x^2 - 12x + 16)$ (d) $[4(p + q) - 5]^2$

(e) $(5r + 6s)^3$ (f) $(2p - 3q + 4r)^2$

Solution

(a) $(3x - 2y)(3x + 2y) = (3x)^2 - (2y)^2$ (by Special Product 1)
$$= 9x^2 - 4y^2$$

(b) $(7x^2 + 3y^3)^2 = (7x^2)^2 + 2(7x^2)(3y^3) + (3y^3)^2$ (by Special Product 2)
$$= 49x^4 + 42x^2y^3 + 9y^6$$

(c) $(3x + 4)(9x^2 - 12x + 16)$
$$= (3x + 4)[(3x)^2 - (3x)(4) + 4^2]$$
$$= (3x)^3 + 4^3 \qquad \text{(by Special Product 4)}$$
$$= 27x^3 + 64$$

(d) $[4(p + q) - 5]^2$
$$= 16(p + q)^2 - 2[4(p + q)(5)] + 5^2 \qquad \text{(by Special Product 3)}$$
$$= 16(p^2 + 2pq + q^2) - 40(p + q) + 25 \qquad \text{(by Special Product 2)}$$
$$= 16p^2 + 32pq + 16q^2 - 40p - 40q + 25$$

(e) $(5r + 6s)^3$
$$= (5r)^3 + 3(5r)^2(6s) + 3(5r)(6s)^2 + (6s)^3 \qquad \text{(by Special Product 6)}$$
$$= 125r^3 + 450r^2s + 540rs^2 + 216s^3$$

(f) $(2p - 3q + 4r)^2$
$$= [(2p - 3q) + 4r]^2$$
$$= (2p - 3q)^2 + 2(2p - 3q)(4r) + (4r)^2 \qquad \text{(by Special Product 2)}$$
$$= 4p^2 - 12pq + 9q^2 + 16pr - 24qr + 16r^2 \qquad \text{(Why?)}$$

PROBLEM SET 3

In problems 1–4, find the value of the algebraic expression for the given values of the variables.

1. $-3xy + 5x + 2$ when $x = 2$ and $y = -1$

2. prt when $p = 1000$, $r = 0.08$, and $t = 3$

3. $\frac{1}{2}gt^2$ when $g = 32$ and $t = 5$

4. $\frac{1}{2}(-b + \sqrt{b^2 - 4c})$ when $b = 1$ and $c = -2$

In problems 5–14, identify the algebraic expression as a monomial, binomial, trinomial, multi-nomial and/or polynomial, and specify the variables involved. For the polynomials, give the degree and the coefficients.

5. $-4x^2$

6. $2\dfrac{x^4}{y} - 3\dfrac{xy^2}{z} + 2xyz - 17$

7. $5x^3y^{-1} + 8x^2 + 1$

8. $\sqrt{x} + \sqrt{y} + \sqrt{x^2 + w^2}$

9. $3xz^2 - \frac{6}{11}x^2z - x - z + 2$

10. $-\dfrac{b}{2a} + \dfrac{\sqrt{b^2 - 4ac}}{2a}$

11. $\sqrt{2}x^5 + \pi x^4 - \sqrt{\pi}x^3 + \frac{12}{13}x^2 - 5$

12. $pq - \pi$

13. $x^3 + y^3 + z^3 + w^3$

14. $x^{-1}y^{-1}z^{-1}$

In problems 15–28, perform the indicated operations.

15. $(-5x + 1) + (7x + 11)$

16. $(4z^2 + 2) + (-3z^2 + z)$

17. $(7x^2 - x + 9) + (11x^2 + 2x - 4)$

18. $5s^2 + (\pi r^2 + 2s^2) - 2s^2$

19. $(3z^2 + 7z + 5) - (-z^2 - 3z + 2)$

20. $3\pi r^2 + (2\pi^2 h - \pi r^2)$

21. $(n^3 - 5n^2 + 3n + 4) - (-2n^3 + 4n - 3)$

22. $(4t^3 - 3t^2 + 2t + 1) + (-2t^3 + 5t^2 + 3t + 4)$

23. $(5x^2y - 3xy^2 + 7xy - 11) + (-3x^2y + 7xy^2 + 4xy + 8)$

24. $(-4pq^2 + 5p^2q + 11pq + 7) - (-7pq^2 + 4pq - 5p + 4)$

25. $(5t^2 + 4t - 3) + (-9t^2 + 2t + 1) - (3t^2 - 8t + 7)$

26. $(5uv - 3u + 5v) - (2uv^2 + 7u + 2v) + (6uv - 3uv^2 - 8v + 7u)$

27. $(3x^2 - xy + y^2 - 2x + y - 5) + (7x^2 + 2y^2 - x + 2) - (8x^2 + 4xy - 3y^2 - x)$

28. $x - \{[xy + (1 + x)] - [-xy + x^2]\}$

In problems 29–44, use the properties of exponents to simplify each expression.

29. $(3x^4)(2x^3)$

30. $(-y)^4(-y)^5$

31. $r^{7n}r^{2n}r^{4n}$

32. $(4t^{2n})(3t)^n$

33. $(3x + y)^2(3x + y)^4$

34. $(5r - 3t)^5(3t - 5r)^2(3t - 5r)$

35. $(t^4)^{12}$

36. $(-x^4)^{11}$

37. $(u^n)^{5n}$

38. $[-(3x + y)^4]^3$

39. $(3v)^4(2v)^2$

40. $(-5r^4)^3$

41. $(5a^2b)^2(3ab^2)^3$

42. $(rs)^n(r^2s)^{2n}(rs^2t)^{3n}$

43. $[2(x + 3y)^2]^5[3(x + 3y)^4]^3$

44. $(n^n)^n$

In problems 45–56, expand each product.

45. $5x^3y(3x - 4xy + 4z)$

46. $-a^2b(-3a^2 + 4b + 2b^2)$

47. $(3x + 2y)(-4x + 5y)$

48. $(2p^3 + 5q)(3p^2 - q)$

49. $(x + 2y)(x^2 - 2xy + 4y^2)$

50. $(r^2 + 3r + s^2)(r^2 + 3r - s^2)$

51. $(5c + d)(5c - d)(25c^2 + d^2)$

52. $(6x^3 + 2x^2 - 3x + 1)(2x^2 - 4x + 3)$

53. $(x^2 + 7xy - y^2)(3x^2 - xy + y^2 - 2x - y + 2)$

54. $(2x^{2n} - x^n - 1)(3x^{2n} + 2x^n + 2)$

55. $(4p^2q - 3pq^2 + 5pq)(p^2q - 2pq^2 - 3pq)$

56. $(3x^3 - 2x^2 + x - 1)(x^4 - x^3 + 5x^2 - 2x + 1)$

In problems 57–72, use the appropriate special products (Section 3.2) to perform each multiplication.

57. $(4 + 3x)^2$

58. $(5p^2 - 2q^2)^2$

59. $[(4t^2 + 1) - s]^2$

60. $(5u + 2v + w)^2$

61. $(3r - 2s)(3r + 2s)$

62. $(4a^n - b^n)(4a^n + b^n)$

63. $[(2a - 3b) - 4c][(2a - 3b) + 4c]$

64. $[(x + 3y) - 2z][(x + 3y) + 2z]$

65. $(2 + t)(4 - 2t + t^2)$

66. $(3 - u)(9 + 3u + u^2)$

67. $(2x^2 + 3y)^3$

68. $(p^n - 2q^n)^3$

69. $(2x - y + z)^2$

70. $[1 + y + z][(1 + y)^2 - (1 + y)z + z^2]$

71. $(t^3 - 2t^2 + 5t + 2)^2$

72. $[(x + y)^2]^3$

73. Explain why the degree of the product of two polynomials is the sum of their degrees.

74. Explain how the procedure of combining like terms (for instance, $3x^2y + 4x^2y = 7x^2y$) is justified by the distributive property.

75. If a is a real number and n and m are positive integers, show that $(a^m)^n = a^{mn}$.

76. If a and b are real numbers and n is a positive integer, show that $(ab)^n = a^nb^n$.

77. Obtain a formula for the expansion of $(a + b)^4$.

78. Verify Special Products 1–8 in Section 3.2 by actually carrying out the multiplication.

79. If a and b are real numbers, $x = a^2 - b^2$, $y = 2ab$, and $z = a^2 + b^2$, show that $x^2 + y^2 = z^2$.

80. Using the result of problem 79, find several triples of positive integers x, y, z such that $x^2 + y^2 = z^2$. [Hint: Substitute positive integers for a and b.]

4 Factoring Polynomials

When two or more algebraic expressions are multiplied, each expression is called a **factor** of the product. For instance, in the product

$$(x - y)(x + y)(2x^2 - y)x,$$

the factors are $x - y$, $x + y$, $2x^2 - y$, and x. Often we are given a product in its expanded form and we need to find the original factors. The process of finding these factors is called **factoring.**

In this section, we confine our study of factoring to polynomials with integer coefficients. Thus, we shall not yet consider such possibilities as $5x^2 - y^2 = (\sqrt{5}x - y)(\sqrt{5}x + y)$, because $\sqrt{5}$ isn't an integer.

Of course, we can factor any polynomial "trivially" by writing it as 1 times itself or as -1 times its negative. A polynomial with integer coefficients that cannot be factored (except trivially) into two or more polynomials with integer coefficients is said to be **prime.** When a polynomial is written as a product of prime factors, we say that it is **factored completely.**

The distributive property can be used to factor a polynomial in which all the terms contain a common factor. The following example illustrates how to "remove the common factor."

Example   Factor each polynomial by removing the common factor.

(a) $20x^2y + 8xy$ (b) $u(v + w) + 7v(v + w)$

Solution   (a) Here $4xy$ is a common factor of the two terms, since

$$20x^2y = (4xy)(5x) \quad \text{and} \quad 8xy = (4xy)(2).$$

Therefore,

$$20x^2y + 8xy = (4xy)(5x) + (4xy)(2) = 4xy(5x + 2).$$

(b) Here the common factor is $v + w$, and we have

$$u(v + w) + 7v(v + w) = (u + 7v)(v + w).$$

4.1 Factoring by Recognizing Special Products

Success in factoring depends on your ability to recognize patterns in the polynomials to be factored—an ability that grows with practice. Special Products 1–8 in Section 3.2, read from right to left, suggest useful patterns; for instance, Special Product 5, read as $a^3 - b^3 = (a - b)(a^2 + ab + b^2)$, shows that a **difference of two cubes** can always be factored.

Example   Use the special products on page 18 to factor each expression.

(a) $25t^2 - 16s^2$ (b) $x^3 + 64y^3$ (c) $27r^3 - 8c^3$

(d) $16x^4 - (3y + 2z)^2$ (e) $81b^4 - 1$ (f) $(x + y)^3 - (z - w)^3$

Solution

(a) Notice that $25t^2 = (5t)^2$ and $16s^2 = (4s)^2$, so the given expression is the **difference of two squares;** that is,

$$25t^2 - 16s^2 = (5t)^2 - (4s)^2.$$

Using Special Product 1 in the form $a^2 - b^2 = (a - b)(a + b)$, with $a = 5t$ and $b = 4s$, we have

$$25t^2 - 16s^2 = (5t)^2 - (4s)^2 = (5t - 4s)(5t + 4s).$$

(b) Since $64 = 4^3$, the expression $x^3 + 64y^3$ is a **sum of two cubes.** Using Special Product 4 in the form $a^3 + b^3 = (a + b)(a^2 - ab + b^2)$ with $a = x$ and $b = 4y$, we have

$$x^3 + 64y^3 = x^3 + (4y)^3 = (x + 4y)(x^2 - 4xy + 16y^2).$$

(c) Here we have a difference of two cubes, so we use Special Product 5:

$$27r^3 - 8c^3 = (3r)^3 - (2c)^3 = (3r - 2c)[(3r)^2 + (3r)(2c) + (2c)^2]$$
$$= (3r - 2c)(9r^2 + 6rc + 4c^2).$$

(d) The expression is a difference of two squares, so

$$16x^4 - (3y + 2z)^2 = (4x^2)^2 - (3y + 2z)^2$$
$$= [4x^2 + (3y + 2z)][4x^2 - (3y + 2z)]$$
$$= (4x^2 + 3y + 2z)(4x^2 - 3y - 2z).$$

(e) Using Special Product 1 *twice*, we have

$$81b^4 - 1 = (9b^2)^2 - 1$$
$$= (9b^2 - 1)(9b^2 + 1)$$
$$= (3b - 1)(3b + 1)(9b^2 + 1).$$

(f) $(x + y)^3 - (z - w)^3$
$$= [(x + y) - (z - w)][(x + y)^2 + (x + y)(z - w) + (z - w)^2]$$
$$= (x + y - z + w)(x^2 + 2xy + y^2 + xz - xw + yz - yw + z^2 - 2zw + w^2).$$

4.2 Factoring Trinomials of the Type $ax^2 + bx + c$

If a, b, and c are integers and $a \neq 0$, it may be possible to factor a trinomial of the type $ax^2 + bx + c$ into a product of two binomials. For instance,

$$8x^2 + 22x + 15 = (2x + 3)(4x + 5).$$

Notice the relationship between the coefficients of the trinomial and the coefficients of the factors. The coefficients 8 and 15 are obtained as follows:

$$8x^2 + 22x + 15 = (2x + 3)(4x + 5).$$

The coefficient of the middle term in the trinomial, namely 22, is obtained as follows:

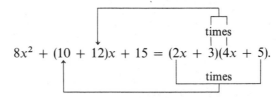

$$8x^2 + (10 + 12)x + 15 = (2x + 3)(4x + 5).$$

Thus, to factor a trinomial $ax^2 + bx + c$, begin by writing

$$ax^2 + bx + c = (\;x +\;\;)(\;x +\;\;),$$
$$\uparrow\qquad\uparrow\;\;\uparrow\qquad\uparrow$$
$$?\qquad\;?\;\;?\qquad\;?$$

where the blanks are to be filled in with integers. The product of the unknown coefficients of x must be the integer a, and the product of the unknown constant terms must be the integer c. Just try all possible choices of such integers until you find a combination that gives the desired middle coefficient b. If all of the coefficients of the trinomial are positive, you need only try combinations of positive integers.

Example   Factor the trinomial $2x^2 + 9x + 4$.

Solution   Since the only two positive integers whose product is 2 are 2 and 1, we can begin by writing

$$2x^2 + 9x + 4 = (2x +\;\;)(x +\;\;).$$
$$\uparrow\qquad\quad\uparrow$$
$$?\qquad\quad?$$

Because 4 can be factored as $4 = 4 \cdot 1$ or as $4 = 2 \cdot 2$ or as $4 = 1 \cdot 4$, there are just three possible ways to fill the remaining blanks:

$$(2x + 4)(x + 1) \quad \text{or} \quad (2x + 2)(x + 2) \quad \text{or} \quad (2x + 1)(x + 4).$$

Now, $(2x + 4)(x + 1) = 2x^2 + 6x + 4$, and that isn't what we want. Also, $(2x + 2)(x + 2) = 2x^2 + 6x + 4$, and that isn't what we want either. But, $(2x + 1)(x + 4) = 2x^2 + 9x + 4$, and that *is* what we want! Therefore,

$$2x^2 + 9x + 4 = (2x + 1)(x + 4).$$

If some of the coefficients of the trinomial $ax^2 + bx + c$ are negative, you will have to try combinations with negative integers.

Example   Factor the trinomial $6x^2 + 13x - 5$.

Solution   Because -5 can be factored as $-5 = 5 \cdot (-1)$ or as $-5 = (-5) \cdot 1$, we can begin by writing

$$6x^2 + 13x - 5 = (\;x \pm 5)(\;x \mp 1),$$
$$\uparrow\qquad\quad\uparrow$$
$$?\qquad\quad?$$

where the two blanks must still be filled in, and we must choose the correct algebraic signs. The possibilities for the blanks are given by $6 = 6 \cdot 1 = 3 \cdot 2 = 2 \cdot 3 = 1 \cdot 6$.

Hence, the possible combinations are

$$(6x + 5)(x - 1) \quad \text{or} \quad (6x - 5)(x + 1)$$

or

$$(3x + 5)(2x - 1) \quad \text{or} \quad (3x - 5)(2x + 1)$$

or

$$(2x + 5)(3x - 1) \quad \text{or} \quad (2x - 5)(3x + 1)$$

or

$$(x + 5)(6x - 1) \quad \text{or} \quad (x - 5)(6x + 1).$$

We try each of these, one at a time, until we find that $(2x + 5)(3x - 1)$ works. Therefore,

$$6x^2 + 13x - 5 = (2x + 5)(3x - 1).$$

The method illustrated above can also be used to factor trinomials of the form $ax^2 + bxy + cy^2$.

Example   Factor the trinomial $6x^2 - 19xy + 3y^2$.

Solution   We begin by writing

$$6x^2 - 19xy + 3y^2 = (\quad x - 3y)(\quad x - y).$$

$$\uparrow \qquad \qquad \uparrow$$
$$? \qquad \qquad ?$$

The two minus signs provide for the positive coefficient $+3$ of y^2 and for the negative coefficient -19 of the middle term. Since $6 = 6 \cdot 1 = 3 \cdot 2 = 2 \cdot 3 = 1 \cdot 6$, there are four possible ways to fill in the two blanks. We try them one at a time until we find that $(x - 3y)(6x - y)$ works. Therefore,

$$6x^2 - 19xy + 3y^2 = (x - 3y)(6x - y).$$

If, in attempting to factor a trinomial $ax^2 + bx + c$ or $ax^2 + bxy + cy^2$, you find that none of the possible combinations works, you can conclude that the trinomial is prime. However, you can test to see if it is prime without bothering to try all these combinations just by evaluating the expression

$$b^2 - 4ac.$$

If $b^2 - 4ac$ is the square of an integer, then the trinomial can be factored; otherwise, it is prime. (You will see why this test works in Section 3.4 of Chapter 2.)

Example   Test each trinomial to see whether it is factorable or prime. If it can be factored, do so.

(a) $4x^2 - 8x + 3$ (b) $3x^2 - 5xy + y^2$

Solution (a) Here $a = 4$, $b = -8$, $c = 3$, and

$$b^2 - 4ac = (-8)^2 - 4(4)(3) = 64 - 48 = 16 = 4^2.$$

Therefore, the trinomial can be factored. Indeed,

$$4x^2 - 8x + 3 = (2x - 1)(2x - 3).$$

(b) Here $a = 3$, $b = -5$, $c = 1$, and

$$b^2 - 4ac = (-5)^2 - 4(3)(1) = 25 - 12 = 13.$$

Because 13 is not the square of an integer, $3x^2 - 5xy + y^2$ is prime.

4.3 Factoring by Grouping

Sometimes we have to group the terms of a polynomial in a certain way in order to see how it can be factored. The following example illustrates the technique.

Example Factor each expression by grouping the terms in a suitable way.

(a) $ac + d - c - ad$ (b) $x^2 + 4xy - 9c^2 + 4y^2$
(c) $4x^3 - 8x^2 - x + 2$ (d) $x^4 + 6x^2y^2 + 25y^4$

Solution (a) $ac + d - c - ad = (ac - c) + (d - ad) = c(a - 1) + d(1 - a)$
$\qquad = c(a - 1) - d(a - 1) = (c - d)(a - 1)$

(b) $x^2 + 4xy - 9c^2 + 4y^2 = (x^2 + 4xy + 4y^2) - 9c^2 = (x + 2y)^2 - (3c)^2$
$\qquad = [(x + 2y) - 3c][(x + 2y) + 3c]$
$\qquad = (x + 2y - 3c)(x + 2y + 3c)$

(c) $4x^3 - 8x^2 - x + 2 = 4x^2(x - 2) - x + 2 = 4x^2(x - 2) - (x - 2)$
$\qquad = (4x^2 - 1)(x - 2) = (2x - 1)(2x + 1)(x - 2)$

(d) If the middle term were $10x^2y^2$ rather than $6x^2y^2$, we could factor the expression as $x^4 + 10x^2y^2 + 25y^4 = (x^2 + 5y^2)^2$. But the middle term can be changed to $10x^2y^2$ by adding $4x^2y^2$ to the $6x^2y^2$ already there and then subtracting $4x^2y^2$ at the end of the expression. Thus,

$$x^4 + 6x^2y^2 + 25y^4 = (x^4 + 10x^2y^2 + 25y^4) - 4x^2y^2$$

$$= (x^2 + 5y^2)^2 - (2xy)^2$$

$$= [(x^2 + 5y^2) - 2xy][(x^2 + 5y^2) + 2xy]$$

$$= (x^2 - 2xy + 5y^2)(x^2 + 2xy + 5y^2).$$

PROBLEM SET 4

In problems 1–8, factor the polynomial by removing the common factor.

1. $10x^3y - 5x^2y^2$ 2. $x^3y^2z - x^2y^3z + xy^4z^3$ 3. $a^3b + 2a^2b + a^2b^2$
4. $14a^2b - 35ab - 63ab^2$ 5. $2(x - y)r^2 + 2(x - y)rh$ 6. $a^2(s + 2t)^2 + a(-s - 2t)$
7. $t(a - b) - r(b - a)$ 8. $9a^{n+1}b^2 - 3a^nb$

In problems 9–24, use the special products (page 18) to factor each expression.

9. $x^2 - 36$

10. $4u^2 - 25v^2$

11. $p^2q^2 - 64$

12. $49s^2 - 25t^2$

13. $n^2 - 36m^2$

14. $121s^{2n} - 81t^2$

15. $(x + y)^2 - 49z^2$

16. $100a^4 - (3a + 2b)^2$

17. $81y^4 - z^4$

18. $(x - 2y)^4 - (2x - y)^4$

19. $8x^3 + 1000$

20. $8a^3 + (b + c)^3$

21. $a^3 - (3b + c)^3$

22. $m^9 + n^9$

23. $64x^6 + (p - q)^3$

24. $8r^{12} - 27(s + 5t)^3$

In problems 25–42, factor the trinomial.

25. $x^2 - 8x + 15$

26. $r^2 - 12r + 35$

27. $y^2 - 3y - 10$

28. $u^2 + 2uv - 35v^2$

29. $t^2 - 7t - 18$

30. $8 - 2t - t^2$

31. $x^2 - xy - 20y^2$

32. $x^{2n} - 9x^n - 22$

33. $8x^2 + 10x - 7$

34. $30 - 49w + 6w^2$

35. $2v^2 + 3v - 20$

36. $8t^4 - 14t^2 - 15$

37. $4r^2 - 12rs + 9s^2$

38. $6(u + v)^2 - 5(u + v) - 6$

39. $6x^4 + 13x^2 + 6$

40. $8x^2 + 22x(y + 2z) + 5(y + 2z)^2$

41. $3(3a - 2b)^2 + (3a - 2b) - 14$

42. $3s^2t^{2n} - st^n - 10$

In problems 43–48, test the trinomial to see whether it is factorable or prime. If it can be factored, do so.

43. $x^2 + x + 1$

44. $16x^2 - 12xy + y^2$

45. $12r^2 - 11r + 2$

C **46.** $52a^2 - 37ab - 35b^2$

47. $2a^2 - 6a + 5$

C **48.** $33(x + y)^2 - 22(x + y) + 3$

In problems 49–58, factor each expression by grouping the terms in a suitable way.

49. $3x - 2y - 6 + xy$

50. $3rs - 2t - 3rt + 2s$

51. $a^2x + a^2d - x - d$

52. $x^2 + 3x - y^2 + 3y$

53. $4x^2 + 12x - 9y^2 + 9$

54. $9a^4 + 8a^2b^2 + 4b^4$

55. $12u^2v + 9 - 4u^2 - 27v$

56. $p^4 + q^4 - pq^3 - p^3q$

57. $x^2 + x - 9y^2 - 3y$

58. $a^3b^3 + a^3 - b^3 - 1$

In problems 59–68, factor each expression completely.

59. $x^4 + x^2y^2 + y^4$

60. $9r^4 + 2r^2s^2 + s^4$

61. $t^4 - 8t^2 + 16$

62. $u^4 + 5u^2v^2 + 9v^4$

63. $(2s + 1)a^3 - (2s + 1)b^3$

64. $s^8 - 82s^4 + 81$

65. $x^8 - 256y^8$

66. $4(x + y)^2z - 25w - 25z + 4(x + y)^2w$

67. $10x^2y^3z - 13xy^4z - 3y^5z$

68. $x^6 - x^4 + 4x^2 - 4$

69. The following expressions were obtained as solutions to problems in a popular calculus textbook. Factor each expression and then simplify the factors.

(a) $2(3x^2 + 7)(6x)(5 - 3x)^3 + 3(3x^2 + 7)^2(-3)(5 - 3x)^2$

(b) $2(5t^2 + 1)(10t)(3t^4 + 2)^4 + 4(5t^2 + 1)(12t^3)(3t^4 + 2)^3$

5 Fractions

If p and q are algebraic expressions, the quotient p/q is called a **fractional expression** (or simply a *fraction*) with **numerator p** and **denominator q**. Always remember that *the denominator of a fraction cannot be zero.* If $q = 0$, the expression p/q simply has no meaning. Therefore, whenever we use a fractional expression, we shall automatically assume that the variables involved are restricted to numerical values that will give a nonzero denominator.

A fractional expression in which both the numerator and the denominator are polynomials is called a **rational expression.** Examples of rational expressions are

$$\frac{2x}{1}, \quad \frac{3}{y}, \quad \frac{5x+3}{17}, \quad \frac{2st-t^3}{3s^4-t^4}, \quad \text{and} \quad \frac{x^2-1}{y^2-1}.$$

Notice that a fraction such as $3x/\sqrt{y}$ isn't a rational expression because its denominator isn't a polynomial. We say that a rational expression is **reduced to lowest terms** or **simplified** if its numerator and denominator have no common factors (other than 1 and -1). Thus, to simplify a rational expression, we factor both the numerator and denominator into prime factors and then **cancel** common factors by using the following property.

The Cancellation Property for Fractions

If $q \neq 0$ and $k \neq 0$, then $\dfrac{pk}{qk} = \dfrac{p}{q}$.

Cancellation is usually indicated by slanted lines drawn through the canceled factors; for instance,

$$\frac{x^2-1}{x^2-3x+2} = \frac{(x-1)(x+1)}{(x-1)(x-2)} = \frac{x+1}{x-2}.$$

If one fraction can be obtained from another by canceling common factors or by multiplying numerator and denominator by the same nonzero expression, then the two fractions are said to be **equivalent.** Thus, the calculation above shows that

$$\frac{x^2-1}{x^2-3x+2} \quad \text{and} \quad \frac{x+1}{x-2}$$

are equivalent fractions.

Example   Reduce each fraction to lowest terms.

(a) $\dfrac{14x^7y}{56x^3y^2}$ (b) $\dfrac{cd-c^2}{c^2-d^2}$

(c) $\dfrac{5x^2-14x-3}{2x^2+x-21}$ (d) $\dfrac{16x-32}{8y(x-2)-4(x-2)}$

Solution   (a) $\dfrac{14x^7y}{56x^3y^2} = \dfrac{14x^3y \cdot x^4}{14x^3y \cdot 4y} = \dfrac{x^4}{4y}$

(b) First, we factor the numerator and denominator, and then we use the fact that $d-c = -(c-d)$:

$$\frac{cd-c^2}{c^2-d^2} = \frac{c(d-c)}{(c-d)(c+d)} = \frac{-c(c-d)}{(c-d)(c+d)} = \frac{-c}{c+d}.$$

(c) $\dfrac{5x^2 - 14x - 3}{2x^2 + x - 21} = \dfrac{(5x + 1)(x - 3)}{(2x + 7)(x - 3)} = \dfrac{5x + 1}{2x + 7}$

(d) $\dfrac{16x - 32}{8y(x - 2) - 4(x - 2)} = \dfrac{\overset{4}{16(x - 2)}}{4(x - 2)(2y - 1)} = \dfrac{4}{2y - 1}$

5.1 Multiplication and Division of Fractions

The following rules for multiplication and division of fractions can be derived from the basic algebraic properties of the real numbers and the definition of a quotient.

1. *Rule for Multiplication of Fractions*

> If p, q, r, and s are real numbers, $q \neq 0$, and $s \neq 0$, then
> $$\frac{p}{q} \cdot \frac{r}{s} = \frac{pr}{qs}.$$

For instance,

$$\frac{2}{3} \cdot \frac{5}{7} = \frac{2 \cdot 5}{3 \cdot 7} = \frac{10}{21}.$$

2. *Rule for Division of Fractions*

> If p, q, r, and s are real numbers, $q \neq 0$, $r \neq 0$, and $s \neq 0$, then
> $$\frac{p}{q} \div \frac{r}{s} = \frac{p/q}{r/s} = \frac{p}{q} \cdot \frac{s}{r} = \frac{ps}{qr}.$$

For instance,

$$\frac{2}{3} \div \frac{5}{7} = \frac{2}{3} \cdot \frac{7}{5} = \frac{14}{15} \quad \text{and} \quad \frac{3/4}{4/5} = \frac{3}{4} \cdot \frac{5}{4} = \frac{15}{16}.$$

Of course, the same rules apply to multiplying or dividing fractional expressions in general. Before multiplying fractions, it's a good idea to factor the numerators and denominators completely to reveal any factors that can be canceled.

Example   Perform the indicated operation and simplify the result.

(a) $\dfrac{x + 4}{y} \cdot \dfrac{y^3}{5x + 20}$ (b) $\dfrac{x^2 - 1}{5x^2 - 26x + 5} \cdot \dfrac{5x^2 + 9x - 2}{x^2 + x - 2}$

(c) $\dfrac{t^2 - 49}{t^2 - 5t - 14} \div \dfrac{2t^2 + 15t + 7}{2t^2 - 13t - 7}$

Solution   (a) $\dfrac{x + 4}{y} \cdot \dfrac{y^3}{5x + 20} = \dfrac{x + 4}{y} \cdot \dfrac{y \cdot y^2}{5(x + 4)} = \dfrac{(x + 4)y \cdot y^2}{y(5)(x + 4)} = \dfrac{y^2}{5}$

(b) $\dfrac{x^2 - 1}{5x^2 - 26x + 5} \cdot \dfrac{5x^2 + 9x - 2}{x^2 + x - 2} = \dfrac{(x - 1)(x + 1)}{(5x - 1)(x - 5)} \cdot \dfrac{(5x - 1)(x + 2)}{(x + 2)(x - 1)}$

$$= \frac{\cancel{(x - 1)}(x + 1)\cancel{(5x - 1)}\cancel{(x + 2)}}{\cancel{(5x - 1)}(x - 5)\cancel{(x + 2)}\cancel{(x - 1)}} = \frac{x + 1}{x - 5}$$

(c) $\dfrac{t^2 - 49}{t^2 - 5t - 14} \div \dfrac{2t^2 + 15t + 7}{2t^2 - 13t - 7} = \dfrac{t^2 - 49}{t^2 - 5t - 14} \cdot \dfrac{2t^2 - 13t - 7}{2t^2 + 15t + 7}$

$$= \frac{(t - 7)(t + 7)}{(t + 2)(t - 7)} \cdot \frac{(2t + 1)(t - 7)}{(2t + 1)(t + 7)}$$

$$= \frac{\cancel{(t - 7)}\cancel{(t + 7)}\cancel{(2t + 1)}(t - 7)}{(t + 2)\cancel{(t - 7)}\cancel{(2t + 1)}\cancel{(t + 7)}} = \frac{t - 7}{t + 2}$$

5.2 Addition and Subtraction of Fractions

Two or more fractions with the same denominator are said to have a **common denominator.** The following rules for adding and subtracting fractions with a common denominator can be derived from the basic algebraic properties of the real numbers and the definition of a quotient.

  1. *Rule for Addition of Fractions with a Common Denominator*

If p, q, and r are real numbers and $q \neq 0$, then

$$\frac{p}{q} + \frac{r}{q} = \frac{p + r}{q}.$$

  2. *Rule for Subtraction of Fractions with a Common Denominator*

If p, q, and r are real numbers and $q \neq 0$, then

$$\frac{p}{q} - \frac{r}{q} = \frac{p - r}{q}.$$

For instance,

$$\frac{3}{7} + \frac{2}{7} = \frac{3 + 2}{7} = \frac{5}{7} \quad \text{and} \quad \frac{3}{7} - \frac{2}{7} = \frac{3 - 2}{7} = \frac{1}{7}.$$

Again, the same rules apply to adding or subtracting fractional expressions.

Example   Perform each operation.

(a) $\dfrac{5x}{2x - 1} + \dfrac{3x}{2x - 1}$

(b) $\dfrac{5x}{(3x - 2)^2} - \dfrac{3x}{(3x - 2)^2}$

Solution

(a) $\dfrac{5x}{2x-1} + \dfrac{3x}{2x-1} = \dfrac{5x+3x}{2x-1} = \dfrac{8x}{2x-1}$

(b) $\dfrac{5x}{(3x-2)^2} - \dfrac{3x}{(3x-2)^2} = \dfrac{5x-3x}{(3x-2)^2} = \dfrac{2x}{(3x-2)^2}$

To add or subtract fractions that do not have a common denominator, you must rewrite the fractions so they do have the same denominator. To do this, multiply the numerator and denominator of each fraction by an appropriate quantity. For instance,

$$\frac{2}{3} + \frac{4}{5} = \frac{2\cdot 5}{3\cdot 5} + \frac{3\cdot 4}{3\cdot 5} = \frac{10}{15} + \frac{12}{15} = \frac{10+12}{15} = \frac{22}{15}.$$

More generally, if $q \neq s$, you can always add p/q and r/s as follows:

$$\frac{p}{q} + \frac{r}{s} = \frac{p\cdot s}{q\cdot s} + \frac{q\cdot r}{q\cdot s} = \frac{ps+qr}{qs}.$$

Example   Add the expressions $\dfrac{3x}{4x-1}$ and $\dfrac{2x}{3x-5}$.

Solution

$\dfrac{3x}{4x-1} + \dfrac{2x}{3x-5} = \dfrac{3x(3x-5)}{(4x-1)(3x-5)} + \dfrac{(4x-1)(2x)}{(4x-1)(3x-5)}$

$\qquad = \dfrac{3x(3x-5)+(4x-1)(2x)}{(4x-1)(3x-5)} = \dfrac{9x^2-15x+8x^2-2x}{(4x-1)(3x-5)}$

$\qquad = \dfrac{17x^2-17x}{(4x-1)(3x-5)} = \dfrac{17x(x-1)}{(4x-1)(3x-5)}$

When adding or subtracting fractions, it's usually best to use the **least common denominator,** abbreviated L.C.D. The L.C.D. of two or more fractions is found as follows:

Step 1. Factor each denominator completely.

Step 2. Form the product of all the different prime factors in the denominators of the fractions, each taken the greatest number of times it occurs in any of the denominators.

Example   Find the L.C.D. of the fractions, perform the indicated operations, and simplify the result.

(a) $\dfrac{8x}{x^2-9} + \dfrac{4}{5x-15}$ (b) $\dfrac{x}{x^3+x^2+x+1} - \dfrac{1}{x^3+2x^2+x} - \dfrac{1}{x^2+2x+1}$

Solution   (a) We factor the denominators to obtain

$$\frac{8x}{x^2-9} + \frac{4}{5x-15} = \frac{8x}{(x-3)(x+3)} + \frac{4}{5(x-3)}.$$

Here, the L.C.D. is $5(x - 3)(x + 3)$. Now we rewrite each fraction as an equivalent fraction with denominator $5(x - 3)(x + 3)$. We do this by multiplying the numerator and denominator of the first fraction by 5 and the numerator and denominator of the second fraction by $x + 3$. Thus,

$$\frac{8x}{x^2 - 9} + \frac{4}{5x - 15} = \frac{8x}{(x - 3)(x + 3)} + \frac{4}{5(x - 3)}$$

$$= \frac{5(8x)}{5(x - 3)(x + 3)} + \frac{4(x + 3)}{5(x - 3)(x + 3)}$$

$$= \frac{5(8x) + 4(x + 3)}{5(x - 3)(x + 3)} = \frac{44x + 12}{5(x - 3)(x + 3)} = \frac{4(11x + 3)}{5(x - 3)(x + 3)}.$$

(b) Factoring the denominators, we have

$$\frac{x}{x^3 + x^2 + x + 1} - \frac{1}{x^3 + 2x^2 + x} - \frac{1}{x^2 + 2x + 1}$$

$$= \frac{x}{x^2(x + 1) + (x + 1)} - \frac{1}{x(x^2 + 2x + 1)} - \frac{1}{x^2 + 2x + 1}$$

$$= \frac{x}{(x^2 + 1)(x + 1)} - \frac{1}{x(x + 1)^2} - \frac{1}{(x + 1)^2},$$

so the L.C.D. is $x(x^2 + 1)(x + 1)^2$. Therefore,

$$\frac{x}{(x^2 + 1)(x + 1)} - \frac{1}{x(x + 1)^2} - \frac{1}{(x + 1)^2}$$

$$= \frac{x \cdot x(x + 1)}{x(x^2 + 1)(x + 1)^2} - \frac{x^2 + 1}{x(x^2 + 1)(x + 1)^2} - \frac{x(x^2 + 1)}{x(x^2 + 1)(x + 1)^2}$$

$$= \frac{x^2(x + 1) - (x^2 + 1) - x(x^2 + 1)}{x(x^2 + 1)(x + 1)^2} = \frac{x^3 + x^2 - x^2 - 1 - x^3 - x}{x(x^2 + 1)(x + 1)^2}$$

$$= \frac{-1 - x}{x(x^2 + 1)(x + 1)^2} = -\frac{\cancel{(x + 1)}}{x(x^2 + 1)(x + 1)^{\cancel{2}}} = -\frac{1}{x(x^2 + 1)(x + 1)}.$$

5.3 Complex Fractions

A fraction that contains one or more fractions in its numerator or denominator is called a **complex fraction.** Examples are

$$\frac{\dfrac{3}{xy^2}}{\dfrac{2}{x^2 y}}, \quad \frac{\dfrac{x}{y} - \dfrac{y}{x}}{\dfrac{1}{x} + \dfrac{1}{y}}, \quad \text{and} \quad \frac{x^2 - \dfrac{1}{x}}{x + \dfrac{1}{x} + 1}.$$

A complex fraction may be simplified by reducing its numerator and denominator (separately) to simple fractions and then dividing.

Example   Simplify the complex fraction $\dfrac{1 + \dfrac{1}{x}}{x - \dfrac{1}{x}}$.

Solution $\dfrac{1 + \dfrac{1}{x}}{x - \dfrac{1}{x}} = \dfrac{\dfrac{x + 1}{x}}{\dfrac{x^2 - 1}{x}} = \dfrac{x + 1}{x} \div \dfrac{x^2 - 1}{x} = \dfrac{x + 1}{x} \cdot \dfrac{x}{x^2 - 1}$

$$= \dfrac{(x + 1)\cancel{x}}{(x^2 - 1)\cancel{x}} = \dfrac{\cancel{x + 1}}{(x - 1)\cancel{(x + 1)}} = \dfrac{1}{x - 1}$$

An alternative method for simplifying a complex fraction is to multiply its numerator and denominator by the L.C.D. of all fractions occurring in its numerator *and* denominator. The resulting fraction may then be simplified by cancellation.

Example   Simplify the complex fraction $\dfrac{\dfrac{1}{x^3} + \dfrac{2}{x^2 y} + \dfrac{1}{xy^2}}{\dfrac{y}{x^2} - \dfrac{1}{y}}$.

Solution   The L.C.D. of x^3, $x^2 y$, xy^2, x^2, and y is $x^3 y^2$. Therefore,

$$\dfrac{\dfrac{1}{x^3} + \dfrac{2}{x^2 y} + \dfrac{1}{xy^2}}{\dfrac{y}{x^2} - \dfrac{1}{y}} = \dfrac{x^3 y^2 \left(\dfrac{1}{x^3} + \dfrac{2}{x^2 y} + \dfrac{1}{xy^2}\right)}{x^3 y^2 \left(\dfrac{y}{x^2} - \dfrac{1}{y}\right)} = \dfrac{y^2 + 2xy + x^2}{xy^3 - x^3 y} = \dfrac{(y + x)^2}{xy(y^2 - x^2)}$$

$$= \dfrac{(y + x)^{\cancel{2}}}{xy(y - x)\cancel{(y + x)}} = \dfrac{y + x}{xy(y - x)}.$$

PROBLEM SET 5

In problems 1–18, reduce each fraction to lowest terms.

1. $\dfrac{2a^2 bxy^2}{6a^2 xy}$

2. $\dfrac{18x^4(-y)(-z)^5}{30x(-y)^2(-z)}$

3. $\dfrac{cy + cz}{2y + 2z}$

4. $\dfrac{(2t + 6)^2}{4t^2 - 36}$

5. $\dfrac{t - 5}{25 - t^2}$

6. $\dfrac{c^2 - cd}{bc - bd}$

7. $\dfrac{r - 3}{r^2 + 3r - 18}$

8. $\dfrac{a^2 - 9a + 18}{3a^2 - 5a - 12}$

9. $\dfrac{6y^2 - y - 1}{2y^2 + 9y - 5}$

10. $\dfrac{r^2 - s^2}{s^4 - r^4}$

11. $\dfrac{(a + b)^2 - 4ab}{(a - b)^2}$

12. $\dfrac{(x - y)^2 - z^2}{(x + z)^2 - y^2}$

13. $\dfrac{6c^2 - 7c - 3}{4c^2 - 8c + 3}$

14. $\dfrac{6m(m - 1) - 12}{m^3 - 8 - (m - 2)^2}$

15. $\dfrac{3(r + t)^2 + 17(r + t) + 10}{2(r + t)^2 + 7(r + t) - 15}$

16. $\dfrac{c^2 - d^2 + c - d}{c^2 + 2cd + d^2 - 1}$

17. $\dfrac{(c + d)^2 - 4(c + d)x + 3x^2}{c^2 + 2cd + d^2 - x^2}$

18. $\dfrac{x^{2n+2} + x^{2n+1}y + x^{2n}y^2}{x^{n+3} - x^n y^3}$

In problems 19–36, perform the indicated operations and simplify the result.

19. $\dfrac{c^2 - 8c}{c - 8} \cdot \dfrac{c + 2}{c}$

20. $\dfrac{81 - p^2}{p + q} \cdot \dfrac{p}{9 - p}$

21. $\dfrac{2u - v}{4u} \cdot \dfrac{2u - v}{4u^2 - 4uv + v^2}$

22. $\dfrac{r^4 - 16}{(r - 2)^2} \cdot \dfrac{r - 2}{4 - r^2}$

23. $\dfrac{3x^2 + 15}{x^2 + 6x + 15} \cdot \dfrac{x^2 + 2x + 1}{x^2 - 1}$

24. $\dfrac{x(2x - 9) - 5}{20 + x(1 - x)} \cdot \dfrac{x(3x + 14) + 8}{x(2x - 11) - 6}$

25. $\dfrac{c^2}{a^2 - 1} \div \dfrac{c^2}{a - 1}$

26. $\dfrac{z^2 - 49}{z^2 - 5z - 14} \div \dfrac{z + 7}{2z^2 - 13z - 7}$

27. $\dfrac{2t^2 - t}{4t^2 - 4t + 1} \div \dfrac{t^2}{8t - 4}$

28. $\dfrac{y - 1}{y^2 + 1} \div \dfrac{(y - 1)^2}{y^4 - 1}$

29. $\dfrac{x^3 - y^3}{x - y} \div \dfrac{x^2 + xy + y^2}{x^2 - 2xy + y^2}$

30. $\dfrac{xy^3 - 4x^2y^2}{y - x} \div \dfrac{16x^2y^2 - y^4}{4x^2 - 3xy - y^2}$

31. $\dfrac{x(3x + 22) - 16}{x(x + 11) + 24} \div \dfrac{x(3x + 13) - 10}{x(2x + 13) + 15}$

32. $\dfrac{(t + 1)^2 - 9}{(t + \frac{1}{2})^2 - \frac{49}{4}} \div \dfrac{(t - \frac{1}{2})^2 - \frac{9}{4}}{t^2 - 3t}$

33. $\dfrac{14u^2 + 23u + 3}{2u^2 + u - 3} \div \dfrac{7u^2 + 15u + 2}{2u^2 - 3u + 1}$

34. $\dfrac{w(3w - 17) + 10}{3w(3w - 4) + 4} \div \dfrac{20 + w(1 - w)}{w(6w - 7) + 2}$

35. $\left(\dfrac{2t^2 + 9t - 5}{t^2 + 10t + 21} \div \dfrac{2t^2 - 13t + 6}{3t^2 + 11t + 6}\right) \cdot \dfrac{t^2 + 3t - 28}{3t^2 - 10t - 8}$

36. $\left(\dfrac{a(3a + 17) + 10}{a(a + 10) + 25} \cdot \dfrac{a^2 - 25}{a(3a + 11) + 6}\right) \div \dfrac{a(5a - 24) - 5}{a(5a + 16) + 3}$

In problems 37–44, perform the indicated operations and simplify the result.

37. $\dfrac{a}{b} + \dfrac{2a - 1}{b}$

38. $\dfrac{m}{m + n} - \dfrac{m + n}{m + n}$

39. $\dfrac{x^2}{x + 1} - \dfrac{1}{x + 1}$

40. $\dfrac{x^2}{x - 2} + \dfrac{3x}{x - 2} - \dfrac{10}{x - 2}$

41. $\dfrac{2}{x + 2} + \dfrac{1}{x - 5}$

42. $1 - \dfrac{1}{x + 1}$

43. $\dfrac{4u}{2u + 1} - \dfrac{3}{2u - 1} + \dfrac{1}{4}$

44. $\dfrac{3 + p}{2 - p} + \dfrac{1 + 2p}{1 + 3p + 2p^2}$

In problems 45–54, find the L.C.D. of the fractions, perform the indicated operations, and simplify the result.

45. $\dfrac{y^2 - 2}{y^2 - y - 2} + \dfrac{y + 1}{y - 2}$

46. $\dfrac{4z}{xy} - \dfrac{9x}{yz}$

47. $\dfrac{3}{x^2 - 9} - \dfrac{2}{x^2 + 6x + 9}$

48. $\dfrac{t + 3}{t^2 - t - 2} + \dfrac{2t - 1}{t^2 + 2t - 8}$

49. $\dfrac{2}{t - 2} + \dfrac{3}{t + 1} - \dfrac{t - 8}{t^2 - t - 2}$

50. $\dfrac{y}{y - z} - \dfrac{2y}{y + z} + \dfrac{3yz}{z^2 - y^2}$

51. $\dfrac{u}{(u^2 + 3)(u - 1)} - \dfrac{3u^2}{(u - 1)^2(u + 2)}$

52. $\dfrac{1}{(a - b)(a - c)} + \dfrac{1}{(b - a)(c - b)} - \dfrac{1}{(a - c)(b - c)}$

53. $\dfrac{2x}{4x^3 - 4x^2 + x} - \dfrac{x}{2x^3 - x^2} + \dfrac{1}{x^3}$

54. $\dfrac{3}{t - 3} - \dfrac{2}{t^2 + 3t} + \dfrac{10}{t^3 - 9t}$

In problems 55–64, simplify each complex fraction.

55. $\dfrac{\dfrac{1}{d}-\dfrac{1}{c}}{c^2-d^2}$

56. $\dfrac{\dfrac{x}{y}+1}{(x+y)^2}$

57. $\dfrac{(a+b)^2}{\dfrac{1}{a}+\dfrac{1}{b}}$

58. $\dfrac{\dfrac{1}{a^2}+\dfrac{2}{ab}+\dfrac{1}{b^2}}{\dfrac{1}{a^2}-\dfrac{1}{b^2}}$

59. $\dfrac{\dfrac{1}{p^2}-\dfrac{1}{q^2}}{\dfrac{1}{p^3}-\dfrac{1}{q^3}}$

60. $\dfrac{\dfrac{a^3}{b^2}+b}{1-\dfrac{a}{b}+\dfrac{a^2}{b^2}}$

61. $\dfrac{\dfrac{1}{x+y}-\dfrac{1}{x-y}}{\dfrac{2y}{x^2-y^2}}$

62. $\dfrac{\dfrac{6}{p^2+3p-10}-\dfrac{1}{p-2}}{\dfrac{1}{p-2}+1}$

63. $\dfrac{\dfrac{x-y}{x+y}-\dfrac{y}{y-x}}{1-y\left(\dfrac{3}{x-y}-\dfrac{2}{x+y}\right)}$

64. $\dfrac{\dfrac{x^2-y^2}{x^2+y^2}+\dfrac{x^2+y^2}{x^2-y^2}}{\dfrac{x-y}{x+y}-\dfrac{x+y}{x-y}}$

The expressions in problems 65 and 66 were obtained as solutions to problems in a popular calculus textbook. Simplify each expression.

65. $\dfrac{\dfrac{1}{(x+h)^2}-\dfrac{1}{x^2}}{h}$

66. $3\left(\dfrac{x+1}{x^2+1}\right)^2\dfrac{x^2+1-(x+1)(2x)}{(x^2+1)^2}$

67. If $q\neq0$ and n is a positive integer, explain why $(p/q)^n=p^n/q^n$.

68. If $q\neq0$ and m and n are positive integers, explain why

$$\frac{q^n}{q^m}=\begin{cases}q^{n-m}&\text{if }n>m\\1&\text{if }n=m\\\dfrac{1}{q^{m-n}}&\text{if }n<m\end{cases}$$

69. In optics, the distance x of an object from the focus of a thin converging lens of focal length f and the distance y of the corresponding image from the focus of the lens satisfy the relation $xy=f^2$ (Figure 1). If $p=x+f$ and $q=y+f$ are the distances from the object to the lens and from the image to the lens, show that

$$\frac{1}{p}+\frac{1}{q}=\frac{1}{f}.$$

Figure 1

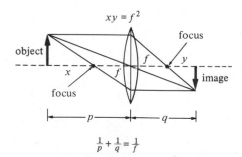

$xy=f^2$

$\frac{1}{p}+\frac{1}{q}=\frac{1}{f}$

70. In electronics, if two resistors of resistance R_1 and R_2 ohms are connected in parallel (Figure 2), then the resistance R of the combination is given by

$$R = \frac{1}{(1/R_1) + (1/R_2)}$$

Simplify this formula for R.

Figure 2

In problems 71–75, an error in calculation has been made. Find the error.

71. $\dfrac{4 - x}{4} = 1 - x$?

72. $\dfrac{3 + 4x}{5 - 3x} = \dfrac{7}{2}$?

73. $\dfrac{(a + b)c}{d + c} = \dfrac{a + b}{d}$?

74. $\dfrac{a}{x + y} = \dfrac{a}{x} + \dfrac{a}{y}$?

75. $(a + b)\left(\dfrac{1}{a} + \dfrac{1}{b}\right) = a\left(\dfrac{1}{a}\right) + b\left(\dfrac{1}{b}\right) = 2$?

6   Radical Expressions

If n is a positive integer and a is a real number, then any real number x such that $x^n = a$ is called an ***n*th root of *a*.** For $n = 2$ an nth root is called a **square root** and for $n = 3$ it is called a **cube root.** For instance, 4 is a cube root of 64, since $4^3 = 64$, and -3 is a cube root of -27, since $(-3)^3 = -27$. Because $2^2 = 4$ and $(-2)^2 = 4$, *both* 2 and -2 are square roots of 4.

If n is an *odd* positive integer, each real number a has exactly one (real) nth root x, and x has the same algebraic sign as a. If n is an *even* positive integer, each positive number has *two* (real) nth roots—one positive and one negative. However, if n is even, negative numbers have no real nth roots (why?). Thus, we make the following definition.

Definition 1   **The Principal *n*th Root of a Number**

> Let n be a positive integer and suppose that a is a real number.
>
> (i)   If a is a positive number, the **principal *n*th root of *a*,** $\sqrt[n]{a}$, is the positive nth root of a.
>
> (ii)   The **principal *n*th root of zero** is zero, that is, $\sqrt[n]{0} = 0$.
>
> (iii)   If a is a negative number, the **principal *n*th root of *a*,** $\sqrt[n]{a}$, is defined only when n is odd, in which case it is the negative number whose nth power is a.

Notice that $\sqrt[n]{a}$ is defined except when a is negative and n is even. Furthermore, if $\sqrt[n]{a}$ is defined, it represents just *one* real number. For instance, $\sqrt{4} = 2$ (we usually write $\sqrt{}$ rather than $\sqrt[2]{}$), $\sqrt[3]{27} = 3$, $\sqrt[5]{32} = 2$, and $\sqrt[5]{-32} = -2$, whereas $\sqrt{-4}$ is undefined as a real number. Although it is correct to say that 2 and -2 are square roots of 4, it is *incorrect* to write $\sqrt{4} = \pm 2$ because the symbol $\sqrt{}$ denotes the **principal square root.**

In the expression $\sqrt[n]{a}$, the symbol $\sqrt[n]{}$ is called the **radical sign** (or simply the **radical**); the positive integer n is called the **index**; and the real number a under the radical is called the **radicand**:

$$\underset{\text{radical}}{\overset{\text{index}}{\sqrt[4]{625}}} = 5 \quad \begin{array}{l} \text{principal fourth root of 625} \\ \end{array}$$
radicand

Algebraic expressions containing radicals are called **radical expressions.** Examples are

$$\sqrt[4]{625}, \quad 5 + \sqrt{x}, \quad \text{and} \quad \sqrt{t} - \sqrt[7]{t^4 - 1}.$$

A positive integer is called a **perfect nth power** if it is the nth power of an integer. Obviously, the nth root of a perfect nth power is an integer. For instance,

$$25 = 5^2 \text{ is a perfect square and } \sqrt{25} = 5$$
$$27 = 3^3 \text{ is a perfect cube and } \sqrt[3]{27} = 3$$
$$625 = 5^4 \text{ is a perfect fourth power and } \sqrt[4]{625} = 5.$$

Example   Find each principal root (if it is defined).

(a) $\sqrt{36}$ (b) $\sqrt[3]{64}$

(c) $\sqrt[3]{-64}$ (d) $\sqrt[6]{-64}$

Solution   (a) $\sqrt{36} = 6$ (b) $\sqrt[3]{64} = 4$
(c) $\sqrt[3]{-64} = -4$ (d) $\sqrt[6]{-64}$ is undefined.

If both of the positive integers h and k are perfect nth powers, then $\sqrt[n]{h/k} = \sqrt[n]{h}/\sqrt[n]{k}$ (why?), and it follows that $\sqrt[n]{h/k}$ is a rational number. However, if the fraction h/k is reduced to lowest terms and if either h or k (or both) fails to be a perfect nth power, it can be shown that $\sqrt[n]{h/k}$ is an irrational number.

Example   Determine which of the indicated principal roots are rational numbers and evaluate those that are:

(a) $\sqrt[3]{\frac{8}{125}}$ (b) $\sqrt{\frac{2}{50}}$ (c) $\sqrt[4]{75}$

Solution   (a) Both the numerator $8 = 2^3$ and the denominator $125 = 5^3$ are perfect cubes, so $\sqrt[3]{\frac{8}{125}} = \frac{2}{5}$.

(b) The fraction $\frac{2}{50}$ isn't reduced to lowest terms. But $\frac{2}{50} = \frac{1}{25}$ and both $1 = 1^2$ and $25 = 5^2$ are perfect squares, so $\sqrt{\frac{2}{50}} = \sqrt{\frac{1}{25}} = \frac{1}{5}$.

(c) The number 75 isn't a perfect 4th power, so $\sqrt[4]{75}$ isn't a rational number.

Radical expressions can often be simplified by using the following properties.

Properties of Radicals

Let a and b be real numbers and suppose that m and n are positive integers. Then, provided that all expressions are defined:

(i) $\sqrt[n]{ab} = \sqrt[n]{a}\sqrt[n]{b}$ (ii) $\sqrt[n]{\dfrac{a}{b}} = \dfrac{\sqrt[n]{a}}{\sqrt[n]{b}}$ (iii) $\sqrt[n]{\sqrt[m]{a}} = \sqrt[nm]{a}$

(iv) $\sqrt[n]{a^m} = (\sqrt[n]{a})^m$ (v) $\sqrt[n]{a^n} = (\sqrt[n]{a})^n = a$

Furthermore, if n is odd,

$$\sqrt[n]{-a} = -\sqrt[n]{a}.$$

In using these properties to simplify radical expressions, it's a good idea to factor the radicand in order to reveal any exponents that are multiples of the index.

Example   Use the properties of radicals to simplify each expression. Assume that variables are restricted to values for which all expressions are defined.

(a) $\sqrt{9x^3}$ (b) $\sqrt[3]{-16x^7y^4}$ (c) $\sqrt{\dfrac{8x^7}{9y^4}}$

(d) $\sqrt[3]{(-x^5)^2}$ (e) $\sqrt[7]{\sqrt[5]{x^{35}}}$ (f) $\sqrt[5]{a^4}\sqrt[5]{a^3}$

Solution   (a) $\sqrt{9x^3} = \sqrt{9x^2 \cdot x} = \sqrt{9x^2}\sqrt{x} = 3x\sqrt{x}$

(b) $\sqrt[3]{-16x^7y^4} = -\sqrt[3]{16x^7y^4} = -\sqrt[3]{8x^6y^3 \cdot 2xy}$
$= -\sqrt[3]{8x^6y^3}\sqrt[3]{2xy} = -2x^2y\sqrt[3]{2xy}$

[Here we used $\sqrt[3]{x^6} = \sqrt[3]{(x^2)^3} = x^2$.]

(c) $\sqrt{\dfrac{8x^7}{9y^4}} = \sqrt{\dfrac{4x^6}{9y^4}(2x)} = \sqrt{\dfrac{4x^6}{9y^4}}\sqrt{2x} = \dfrac{\sqrt{4x^6}}{\sqrt{9y^4}}\sqrt{2x} = \dfrac{2x^3}{3y^2}\sqrt{2x}$

(d) $\sqrt[3]{(-x^5)^2} = \sqrt[3]{x^{10}} = \sqrt[3]{x^9x} = \sqrt[3]{x^9}\sqrt[3]{x} = \sqrt[3]{(x^3)^3}\sqrt[3]{x} = x^3\sqrt[3]{x}$

(e) $\sqrt[7]{\sqrt[5]{x^{35}}} = \sqrt[35]{x^{35}} = x$

(f) $\sqrt[5]{a^4}\sqrt[5]{a^3} = \sqrt[5]{a^4a^3} = \sqrt[5]{a^7} = \sqrt[5]{a^5a^2} = \sqrt[5]{a^5}\sqrt[5]{a^2} = a\sqrt[5]{a^2}$

Two or more radical expressions are said to be *like* or *similar* if, after being simplified, they contain the same index and radicand. For instance, $3\sqrt{5}$, $x\sqrt{5}$, and $\sqrt{5}/2$ are similar, but $3\sqrt{5}$ and $5\sqrt[3]{5}$ are not similar. To simplify a sum whose terms are radical expressions, we simplify each term and then combine similar terms.

Example   Simplify each sum. Assume that variables are restricted to values for which all expressions are defined.

(a) $7\sqrt{12} + \sqrt{75} - 5\sqrt{27}$

(b) $\sqrt{2x^2y} + x\sqrt{18y} - 3\sqrt{x^3y}$

(c) $\sqrt[3]{4a^5} + a\sqrt[3]{32a^2}$

Solution   (a) $7\sqrt{12} + \sqrt{75} - 5\sqrt{27} = 7\sqrt{4\cdot 3} + \sqrt{25\cdot 3} - 5\sqrt{9\cdot 3}$
$$= 7\sqrt{4}\sqrt{3} + \sqrt{25}\sqrt{3} - 5\sqrt{9}\sqrt{3}$$
$$= 7\cdot 2\sqrt{3} + 5\sqrt{3} - 5\cdot 3\sqrt{3}$$
$$= 14\sqrt{3} + 5\sqrt{3} - 15\sqrt{3}$$
$$= (14 + 5 - 15)\sqrt{3} = 4\sqrt{3}$$

(b) $\sqrt{2x^2 y} + x\sqrt{18y} - 3\sqrt{x^3 y} = \sqrt{x^2}\sqrt{2}\sqrt{y} + x\sqrt{9}\sqrt{2}\sqrt{y} - 3\sqrt{x^2}\sqrt{x}\sqrt{y}$
$$= x\sqrt{2}\sqrt{y} + 3x\sqrt{2}\sqrt{y} - 3x\sqrt{x}\sqrt{y}$$
$$= 4x\sqrt{2}\sqrt{y} - 3x\sqrt{x}\sqrt{y}$$
$$= x\sqrt{y}(4\sqrt{2} - 3\sqrt{x})$$

(c) $\sqrt[3]{4a^5} + a\sqrt[3]{32a^2} = \sqrt[3]{a^3 \cdot 4a^2} + a\sqrt[3]{8\cdot 4a^2} = \sqrt[3]{a^3}\sqrt[3]{4a^2} + a\sqrt[3]{8}\sqrt[3]{4a^2}$
$$= a\sqrt[3]{4a^2} + 2a\sqrt[3]{4a^2} = 3a\sqrt[3]{4a^2}$$

Sums involving radicals are multiplied the same way polynomials are.

Example   Expand each product and simplify the result. Assume that $x \geq 0$ and $y \geq 0$.
(a) $\sqrt{3}(\sqrt{27} + \sqrt{15})$
(b) $(3\sqrt{x} - 7\sqrt{y})(5\sqrt{x} + 2\sqrt{y})$
(c) $(\sqrt[3]{s} - \sqrt[3]{2})(\sqrt[3]{s^2} + \sqrt[3]{2s} + \sqrt[3]{4})$

Solution   (a) $\sqrt{3}(\sqrt{27} + \sqrt{15}) = \sqrt{3}\sqrt{27} + \sqrt{3}\sqrt{15}$
$$= \sqrt{3}\sqrt{3}\sqrt{9} + \sqrt{3}\sqrt{3}\sqrt{5} = 3\cdot 3 + 3\sqrt{5} = 9 + 3\sqrt{5}$$

(b) $(3\sqrt{x} - 7\sqrt{y})(5\sqrt{x} + 2\sqrt{y}) = 15(\sqrt{x})^2 + (3\cdot 2 - 7\cdot 5)\sqrt{x}\sqrt{y} - 14(\sqrt{y})^2$
$$= 15x - 29\sqrt{xy} - 14y$$

(c) $(\sqrt[3]{s} - \sqrt[3]{2})(\sqrt[3]{s^2} + \sqrt[3]{2s} + \sqrt[3]{4}) = (\sqrt[3]{s} - \sqrt[3]{2})[(\sqrt[3]{s})^2 + \sqrt[3]{s}\sqrt[3]{2} + (\sqrt[3]{2})^2]$
$$= (\sqrt[3]{s})^3 - (\sqrt[3]{2})^3 = s - 2$$

[Here we used Special Product 5 on page 18, with $a = \sqrt[3]{s}$ and $b = \sqrt[3]{2}$.]

In part (c) of the last example, the product of two radical expressions contains no radicals. Whenever the product of two radical expressions is free of radicals, we say that the two expressions are **rationalizing factors** for each other. For instance,
$$(2\sqrt{3} - 3\sqrt{x})(2\sqrt{3} + 3\sqrt{x}) = (2\sqrt{3})^2 - (3\sqrt{x})^2 = 12 - 9x,$$
so $2\sqrt{3} - 3\sqrt{x}$ is a rationalizing factor for $2\sqrt{3} + 3\sqrt{x}$.

Fractions containing radicals are sometimes easier to deal with if their denominators are free of radicals. To rewrite a fraction so that there are no radicals in the denominator, we multiply the numerator and denominator by a rationalizing factor for the denominator. This is called **rationalizing the denominator.** For instance,
$$\frac{3}{\sqrt{5}} = \frac{3\sqrt{5}}{\sqrt{5}\sqrt{5}} = \frac{3\sqrt{5}}{5}.$$

Rationalizing factors can often be found by using the special products that yield the difference of two squares or the sum or difference of two cubes.

Example Rationalize the denominator of each fraction and simplify the result. Assume that x is restricted to values for which all expressions are defined.

(a) $\dfrac{\sqrt{3}-1}{\sqrt{2}+1}$ (b) $\dfrac{3}{7\sqrt[3]{x}}$ (c) $\dfrac{\sqrt{x}-3}{\sqrt{x}+3}$ (d) $\dfrac{2}{\sqrt[3]{x}+2}$

Solution (a) Since $(\sqrt{2}+1)(\sqrt{2}-1) = (\sqrt{2})^2 - 1^2 = 2 - 1 = 1$, it follows that $\sqrt{2}-1$ is a rationalizing factor for the denominator. Thus,

$$\frac{\sqrt{3}-1}{\sqrt{2}+1} = \frac{(\sqrt{3}-1)(\sqrt{2}-1)}{(\sqrt{2}+1)(\sqrt{2}-1)} = \frac{\sqrt{6}-\sqrt{3}-\sqrt{2}+1}{1}$$
$$= \sqrt{6}-\sqrt{3}-\sqrt{2}+1.$$

(b) Since $\sqrt[3]{x}(\sqrt[3]{x^2}) = \sqrt[3]{x^3} = x$, we have

$$\frac{3}{7\sqrt[3]{x}} = \frac{3\sqrt[3]{x^2}}{7\sqrt[3]{x}\sqrt[3]{x^2}} = \frac{3\sqrt[3]{x^2}}{7x}.$$

(c) $\dfrac{\sqrt{x}-3}{\sqrt{x}+3} = \dfrac{(\sqrt{x}-3)(\sqrt{x}-3)}{(\sqrt{x}+3)(\sqrt{x}-3)}$

$$= \frac{(\sqrt{x})^2 - 6\sqrt{x} + 9}{(\sqrt{x})^2 - 9}$$
$$= \frac{x - 6\sqrt{x} + 9}{x - 9}$$

(d) Using Special Product 4 on page 18, with $a = \sqrt[3]{x}$ and $b = 2$, we see that a rationalizing factor for the denominator is provided by

$$a^2 - ab + b^2 = (\sqrt[3]{x})^2 - (\sqrt[3]{x})(2) + 2^2 = \sqrt[3]{x^2} - 2\sqrt[3]{x} + 4.$$

Thus,

$$(\sqrt[3]{x}+2)(\sqrt[3]{x^2}-2\sqrt[3]{x}+4) = (a+b)(a^2-ab+b^2) = a^3 + b^3$$
$$= (\sqrt[3]{x})^3 + 2^3 = x + 8,$$

and we have

$$\frac{2}{\sqrt[3]{x}+2} = \frac{2(\sqrt[3]{x^2}-2\sqrt[3]{x}+4)}{(\sqrt[3]{x}+2)(\sqrt[3]{x^2}-2\sqrt[3]{x}+4)} = \frac{2\sqrt[3]{x^2}-4\sqrt[3]{x}+8}{x+8}.$$

In calculus, it is sometimes necessary to rationalize the *numerator* of a fraction.

Example Rationalize the numerator of $\dfrac{\sqrt{x+h+1}-\sqrt{x+1}}{h}$.

Solution

$$\frac{\sqrt{x+h+1}-\sqrt{x+1}}{h} = \frac{(\sqrt{x+h+1}-\sqrt{x+1})(\sqrt{x+h+1}+\sqrt{x+1})}{h(\sqrt{x+h+1}+\sqrt{x+1})}$$

$$= \frac{(\sqrt{x+h+1})^2-(\sqrt{x+1})^2}{h(\sqrt{x+h+1}+\sqrt{x+1})}$$

$$= \frac{(x+h+1)-(x+1)}{h(\sqrt{x+h+1}+\sqrt{x+1})}$$

$$= \frac{\not{h}}{\not{h}(\sqrt{x+h+1}+\sqrt{x+1})}$$

$$= \frac{1}{\sqrt{x+h+1}+\sqrt{x+1}}.$$

PROBLEM SET 6

Throughout this problem set, assume that variables are restricted to values for which all radical expressions are defined.

In problems 1–12, determine which of the indicated principal roots are rational numbers and evaluate those that are.

1. $\sqrt{121}$

2. $\sqrt[4]{256}$

3. $\sqrt[3]{\dfrac{27}{64}}$

4. $\sqrt[5]{-32}$

5. $\sqrt{24}$

6. $\sqrt{0.04}$

7. $\sqrt[3]{\dfrac{16}{250}}$

8. $\sqrt[3]{-\dfrac{18}{54}}$

9. $\sqrt[7]{\dfrac{1}{128}}$

10. $\sqrt[4]{\dfrac{-1}{81}}$

11. $\sqrt[4]{\dfrac{81}{256}}$

12. $\sqrt[4]{(81)(256)}$

In problems 13–41, simplify each radical expression.

13. $\sqrt{27x^3}$

14. $\sqrt[3]{4a^5}$

15. $\sqrt{75x^4y^9}$

16. $\sqrt[4]{81x^5y^{16}}$

17. $\sqrt[3]{24a^4b^5}$

18. $\sqrt[5]{-u^6v^7}$

19. $(\sqrt{x-1})^4$

20. $(\sqrt[3]{2a+b})^6$

21. $\dfrac{\sqrt{16x^4y^3}}{\sqrt{64x^{12}y}}$

22. $\sqrt[5]{\dfrac{1}{243b^5}}$

23. $\sqrt{\dfrac{6y}{7}}\sqrt{\dfrac{35}{72y^2}}$

24. $\sqrt[3]{\dfrac{u^3v^6}{(3uv^2)^3}}$

25. $\sqrt[3]{\dfrac{-27x^4}{y^{21}}}$

26. $\dfrac{\sqrt[6]{u^{15}v^{21}}}{\sqrt[6]{u^3v^6}}$

27. $(2\sqrt[3]{11})^3(-\tfrac{1}{2}\sqrt[5]{24})^5$

28. $\dfrac{(\tfrac{1}{2}\sqrt[3]{9}\sqrt{3})^4}{(3\sqrt[5]{3}\sqrt{3})^5}$

29. $\sqrt[4]{125t^2\sqrt{25t^4}}$

30. $\sqrt[5]{\sqrt[3]{x^{30}}}$

31. $\sqrt[4]{x^2y^{10}}\sqrt[4]{x^6y^9}$

32. $\sqrt[5]{s^3t^5}\sqrt[5]{st^5}\sqrt[5]{st}$

33. $\sqrt[3]{ab^2}\sqrt[3]{a^2b^5}$

34. $\sqrt{\dfrac{5a+b}{3a^2b^2}}\sqrt{\dfrac{54a^5b^5}{25a^2-b^2}}$

35. $\sqrt{\dfrac{u-v}{u+v}}\sqrt{\dfrac{u^2+2uv+v^2}{u^2-v^2}}$

36. $\dfrac{\sqrt{3xy^3z}\sqrt{2x^2yz^4}}{\sqrt{6x^3y^4z^3}}$

37. $\dfrac{\sqrt[3]{54x^2yz^4}}{\sqrt[3]{16xy^4z^3}}$

38. $\sqrt[5]{\dfrac{(a+b)^4(c+d)^3}{8}}\sqrt[5]{\dfrac{(a+b)^6}{4(c+d)^8}}$

39. $\dfrac{\sqrt[3]{x^2y^3}\sqrt[3]{125x^3y^2}}{\sqrt[3]{8x^3y^4}}$

40. $\sqrt[4]{u\sqrt{u}\sqrt[3]{u}}$

41. $\dfrac{\sqrt[3]{a^2b^4}\sqrt[3]{a^4b}\sqrt[3]{a^3b^4}}{\sqrt[3]{ab^2}\sqrt[3]{a^2b^7}}$

42. If n is an odd positive integer, show that $\sqrt[n]{-a} = -\sqrt[n]{a}$.

In problems 43–52, simplify each algebraic sum.

43. $6\sqrt{2} - 3\sqrt{8} + 98$

44. $2\sqrt{54} - \sqrt{216} - 7\sqrt{24}$

45. $5\sqrt[3]{81} - 3\sqrt[3]{24} + \sqrt[3]{192}$

46. $8xy\sqrt{x^2y} - 3\sqrt{x^4y^3} + 5x^2\sqrt{y^3}$

47. $4t^3\sqrt{180t} - 2t^2\sqrt{20t^3} - \sqrt{5t^7}$

48. $\sqrt[3]{375} - \sqrt[6]{576} - 4\sqrt[3]{27}$

49. $\sqrt[4]{\dfrac{625}{216}} - \sqrt[4]{\dfrac{32}{27}}$

50. $\sqrt{\dfrac{a^3}{3b^3}} + ab\sqrt{\dfrac{a}{3b^5}} - \dfrac{b^2}{3}\sqrt{\dfrac{3a^3}{b^7}}$

51. $\sqrt[3]{27x^4y^5} + xy\sqrt[3]{-8xy^2}$

52. $4w^3\sqrt{180w} - 2w^2\sqrt{20w^3} - \sqrt{5w^7}$

In problems 53–66, expand each product and simplify the result.

53. $\sqrt{6}(2\sqrt{3} - 3\sqrt{2})$

54. $\sqrt{2x}(x\sqrt{2} - 2\sqrt{x})$

55. $(2 + \sqrt{3})(1 - \sqrt{3})$

56. $(5\sqrt{x} - 1)(3\sqrt{x} + 2)$

57. $(\sqrt{x} + \sqrt{3})^2$

58. $(\sqrt{5} - \sqrt{3})(\sqrt{7} - 2)\sqrt{3}$

59. $(\sqrt{5} - 3\sqrt{2})(\sqrt{5} + 3\sqrt{2})$

60. $(3\sqrt{x} - 2\sqrt{y})(3\sqrt{x} + 2\sqrt{y})$

61. $\sqrt{\sqrt{21} + \sqrt{5}}\sqrt{\sqrt{21} - \sqrt{5}}$

62. $\sqrt[3]{\sqrt{33} - \sqrt{6}}\sqrt[3]{\sqrt{33} + \sqrt{6}}$

63. $(\sqrt{3} + \sqrt{2} + 1)(\sqrt{3} + \sqrt{2} - 1)$

64. $(\sqrt{a} + \sqrt{b} - \sqrt{c})(\sqrt{a} + \sqrt{b} + \sqrt{c})$

65. $(\sqrt[3]{a + b} + \sqrt[3]{a - b})[(\sqrt[3]{a + b})^2 - \sqrt[3]{a^2 - b^2} + (\sqrt[3]{a - b})^2]$

66. $(2\sqrt{p^2 - 1} - \sqrt{p^2 + 1})(2\sqrt{p^2 - 1} + \sqrt{p^2 + 1})\left(\dfrac{1}{\sqrt{3p} - \sqrt{5}}\right)$

In problems 67–84, rationalize the denominator and simplify the result.

67. $\dfrac{5\sqrt{2}}{3\sqrt{5}}$

68. $\dfrac{3x}{\sqrt{21x}}$

69. $\dfrac{\sqrt{2}}{3\sqrt{x + 2y}}$

70. $\dfrac{10t}{3\sqrt{5(3t + 7)}}$

71. $\dfrac{20}{\sqrt[3]{5}}$

72. $\dfrac{a^2b^3}{3\sqrt[4]{ab}}$

73. $\dfrac{\sqrt{3}}{\sqrt{3} + 5}$

74. $\dfrac{5}{1 - \sqrt{p}}$

75. $\dfrac{\sqrt{5} + \sqrt{2}}{\sqrt{5} - \sqrt{2}}$

76. $\dfrac{2\sqrt{x} - 3\sqrt{y}}{3\sqrt{x} - 4\sqrt{y}}$

77. $\dfrac{3\sqrt{p} + \sqrt{q}}{4\sqrt{p} - 3\sqrt{q}}$

78. $\dfrac{5\sqrt{3} - 4\sqrt{5}}{2\sqrt{5} + 3\sqrt{5}}$

79. $\dfrac{\sqrt{a + 5} + \sqrt{a}}{\sqrt{a + 5} - \sqrt{a}}$

80. $\dfrac{2\sqrt{x^2 - 1} + \sqrt{x^2 + 1}}{3\sqrt{x^2 - 1} + \sqrt{x^2 + 1}}$

81. $\dfrac{5}{2 - \sqrt[3]{x}}$

82. $\dfrac{2}{1 + \sqrt{x} - \sqrt{y}}$

83. $\dfrac{\sqrt[3]{a} - \sqrt[3]{b}}{\sqrt[3]{a} + \sqrt[3]{b}}$

84. $\dfrac{1}{\sqrt{x} + \sqrt{y} + \sqrt{z}}$

In problems 85–88, rationalize the *numerator* of each fraction and simplify the result.

85. $\dfrac{\sqrt{x + h} - \sqrt{x}}{h}$

86. $\dfrac{\sqrt[3]{x} - \sqrt[3]{a}}{x - a}$

87. $\dfrac{\sqrt{(x + h)^2 + 1} - \sqrt{x^2 + 1}}{h}$

88. $\dfrac{\dfrac{1}{\sqrt{x + h}} - \dfrac{1}{\sqrt{x}}}{h}$

89. Show that the expression

$$\dfrac{\sqrt{x}}{\sqrt{x + a}} - \dfrac{\sqrt{x + a}}{\sqrt{x}}$$

can be rewritten in the form

$$\dfrac{-a}{\sqrt{x(x + a)}}.$$

90. In astronomy, the relativistic Doppler shift formula

$$v = v_0 \frac{\sqrt{1 - (v^2/c^2)}}{1 + (v/c)}$$

is used to determine the frequency v of light from a distant galaxy as measured at an observatory on earth. (v is the Greek letter nu.) In this formula, c is the speed of light, v is the speed with which the galaxy is receding from the earth, and v_0 is the frequency of the light emitted by the galaxy. The fact that the observed frequency v is smaller than the original frequency is popularly known as the "red shift." If $b = v/c$, show that

$$\frac{v}{v_0} = \sqrt{\frac{1 - b}{1 + b}}.$$

HALE OBSERVATORIES

7 Rational Exponents

In Section 3.2 we established the following properties of exponents:

(i) $a^m a^n = a^{m+n}$ (ii) $(a^m)^n = a^{mn}$ (iii) $(ab)^n = a^n b^n$

These properties follow from the definition

$$a^n = \overbrace{aaa \cdots a,}^{n \text{ factors}}$$

which makes sense only if n is a positive integer. In the present section we extend this definition so that zero, negative integers, and rational numbers can be used as exponents. The key idea is to extend the definition in such a way that Properties (i), (ii), and (iii) continue to hold.

We begin with the question of how to define a^0. If zero as an exponent is to obey Property (i), we must have

$$a^0 a^n = a^{0+n} = a^n.$$

If $a \neq 0$, this equation can hold only if $a^0 = 1$, which leads us to the following definition.

Definition 1   **Zero as an Exponent**

> If a is any nonzero real number, we define $a^0 = 1$.

Notice that 0^0 is not defined.

Now we consider how to define a^{-n} when n is a positive integer. Again, if Property (i) is to hold, we must have

$$a^{-n} a^n = a^{-n+n} = a^0 = 1 \qquad (a \neq 0).$$

If $a \neq 0$, this equation can hold only if $a^{-n} = 1/a^n$, which leads us to the following definition.

Definition 2   **Negative Integer Exponents**

> If a is any nonzero real number and n is a positive integer,
> $$a^{-n} = \frac{1}{a^n}.$$

In other words, a^{-n} is the reciprocal of a^n. In particular,

$$a^{-1} = \frac{1}{a^1} = \frac{1}{a}.$$

Definitions 1 and 2 enable us to use *any integers*—positive, negative, or zero—as exponents, with the exception that 0^n is defined only when n is positive. You can verify that Properties (i), (ii), and (iii) hold for all integer exponents if

$a \neq 0$ and $b \neq 0$. The following properties also hold for all integers m and n and all nonzero real numbers a and b (Problems 73–79):

(iv) $\left(\dfrac{a}{b}\right)^n = \dfrac{a^n}{b^n}$	(v) $\dfrac{a^m}{a^n} = a^{m-n}$	(vi) $\dfrac{a^{-m}}{b^{-n}} = \dfrac{b^n}{a^m}$	(vii) $\left(\dfrac{a}{b}\right)^{-1} = \dfrac{b}{a}$

Property (vi) permits us to move a factor in a numerator to the denominator of a fraction, or vice versa, simply by changing the sign of the exponent of the factor. By Property (vii), the reciprocal of a nonzero fraction is obtained by inverting the fraction.

Example   Rewrite each expression so it contains only positive exponents and simplify the result.

(a) 2^{-3}

(b) $(7x^0)^{-2}$

(c) $(x^4)^{-2}$

(d) $\left(\dfrac{3}{x}\right)^{-4}$

(e) $\left(\dfrac{3}{2^{-1}}\right)^{-1}$

(f) $(x-y)^{-4}(x-y)^{13}$

(g) $(x^{-2}y^{-3})^{-4}$

(h) $\dfrac{5x^{-3}(a+b)^2}{15x^4(a+b)^{-5}}$

(i) $\dfrac{x^{-1}-y^{-1}}{x-y}$

Solution   (a) $2^{-3} = \dfrac{1}{2^3} = \dfrac{1}{8}$ $\qquad\qquad\qquad\qquad\qquad\qquad$ (by Definition 2)

(b) $(7x^0)^{-2} = (7 \cdot 1)^{-2}$ $\qquad\qquad\qquad\qquad\qquad$ (by Definition 1)

$\qquad\qquad = 7^{-2} = \dfrac{1}{7^2} = \dfrac{1}{49}$ $\qquad\qquad\qquad$ (by Definition 2)

(c) $(x^4)^{-2} = x^{-8} = \dfrac{1}{x^8}$ $\qquad\qquad\qquad\qquad$ [by Property (ii) and Definition 2]

(d) $\left(\dfrac{3}{x}\right)^{-4} = \dfrac{3^{-4}}{x^{-4}} = \dfrac{x^4}{3^4} = \dfrac{x^4}{81}$ $\qquad\qquad$ [by Properties (iv) and (vi)]

(e) $\left(\dfrac{3}{2^{-1}}\right)^{-1} = \dfrac{2^{-1}}{3} = \dfrac{1}{2 \cdot 3} = \dfrac{1}{6}$ $\qquad\quad$ [by Properties (vii) and (vi)]

(f) $(x-y)^{-4}(x-y)^{13} = (x-y)^{-4+13} = (x-y)^9$ $\quad$ [by Property (i)]

(g) $(x^{-2}y^{-3})^{-4} = (x^{-2})^{-4}(y^{-3})^{-4}$ $\qquad\qquad\qquad$ [by Property (iii)]

$\qquad\qquad\qquad = x^{(-2)(-4)}y^{(-3)(-4)} = x^8y^{12}$ $\qquad$ [by Property (ii)]

(h) $\dfrac{5x^{-3}(a+b)^2}{15x^4(a+b)^{-5}} = \dfrac{(a+b)^2(a+b)^5}{3x^4x^3} = \dfrac{(a+b)^7}{3x^7}$ $\quad$ [by Properties (vi) and (i)]

(i) $\dfrac{x^{-1}-y^{-1}}{x-y} = \dfrac{\dfrac{1}{x}-\dfrac{1}{y}}{x-y} = \dfrac{\dfrac{y-x}{xy}}{x-y} = \dfrac{y-x}{xy} \cdot \dfrac{1}{x-y} = \dfrac{y-x}{xy(x-y)}$

$\qquad\qquad = \dfrac{-(x-y)}{xy(x-y)} = \dfrac{-1}{xy}$

If m and n are integers with $n \neq 0$, how shall we define $a^{m/n}$? If Property (ii) is to hold, we must have

$$a^{m/n} = a^{(1/n)m} = (a^{1/n})^m,$$

so the basic question is how to define $a^{1/n}$. But, again, if Property (ii) is to hold, we must have

$$(a^{1/n})^n = a^{(1/n)n} = a^1 = a;$$

in other words, $a^{1/n}$ must be an nth root of a. Thus, we have the following definition.

Definition 3   **Rational Exponents**

> Let a be a nonzero real number. Suppose that m and n are integers, that n is positive, and that the fraction m/n is reduced to lowest terms. Then, if $\sqrt[n]{a}$ exists,
>
> $$a^{1/n} = \sqrt[n]{a} \quad \text{and} \quad a^{m/n} = (\sqrt[n]{a})^m = (a^{1/n})^m.$$
>
> Also, if m/n is a positive rational number,
>
> $$0^{m/n} = 0.$$

Notice that $a^{1/n}$ is just an alternative notation for $\sqrt[n]{a}$, the principal nth root of a. For instance,

$$25^{1/2} = \sqrt{25} = 5, \quad 0^{1/7} = \sqrt[7]{0} = 0, \quad (-27)^{1/3} = \sqrt[3]{-27} = -3,$$

and so forth. It is important for you to keep in mind that Definition 3 is to be used *only when $n > 0$ and m/n is reduced to lowest terms.* For instance, it is *incorrect* to write $(-8)^{2/6} = (\sqrt[6]{-8})^2$ because $2/6$ isn't reduced to lowest terms. Instead, we must first reduce $2/6$ to lowest terms and write

$$(-8)^{2/6} = (-8)^{1/3} = \sqrt[3]{-8} = -2.$$

Example   Find the value of each expression (if it is defined).

(a) $8^{4/3}$ (b) $81^{-3/4}$

(c) $(-7)^{3/2}$ (d) $(-64)^{8/12}$

Solution   Using Definition 3, we have

(a) $8^{4/3} = (\sqrt[3]{8})^4 = 2^4 = 16$

(b) $81^{-3/4} = 81^{(-3)/4} = (\sqrt[4]{81})^{-3} = 3^{-3} = \dfrac{1}{3^3} = \dfrac{1}{27}$

(c) $(-7)^{3/2}$ is undefined since $\sqrt{-7}$ does not exist (as a real number).

(d) $(-64)^{8/12} = (-64)^{2/3} = (\sqrt[3]{-64})^2 = (-4)^2 = 16$

For convenience, we now summarize the Properties of Rational Exponents.

Properties of Rational Exponents

Let a and b be real numbers, suppose that p and q are rational numbers, and let n be a positive integer. Then, provided that all expressions are defined (as real numbers):

(i) $a^p a^q = a^{p+q}$ (ii) $(a^p)^q = a^{pq}$ (iii) $(ab)^p = a^p b^p$

(iv) $\left(\dfrac{a}{b}\right)^p = \dfrac{a^p}{b^p}$ (v) $\dfrac{a^p}{a^q} = a^{p-q}$ (vi) $\dfrac{a^{-p}}{b^{-q}} = \dfrac{b^q}{a^p}$

(vii) $\left(\dfrac{a}{b}\right)^{-1} = \dfrac{b}{a}$ (viii) $\sqrt[n]{a^p} = (\sqrt[n]{a})^p = a^{p/n}$ (ix) $a^{-p} = \dfrac{1}{a^p}$

As illustrated by the following example, these properties are especially useful for simplifying algebraic expressions containing rational exponents.

Example Simplify each expression and write the answer so that it contains only positive exponents. You may assume that variables are restricted to values for which all expressions are defined.

(a) $5^{-1/3} 5^{7/3}$

(b) $(x^{-3/4})^{-8}$

(c) $(64x^{-3})^{-2/3}$

(d) $\left[\dfrac{(27x^2 y^3)^{1/3}}{9x^{-2} y^4}\right]^{-1}$

(e) $\dfrac{(3x + 2)^{1/2}(3x + 2)^{-1/4}}{(3x + 2)^{-3/4}}$

(f) $(a^{-1/2} - b^{-1/2})(a^{-1/2} + b^{-1/2})$

Solution

(a) $5^{-1/3} 5^{7/3} = 5^{-1/3 + 7/3} = 5^{6/3} = 5^2 = 25$

(b) $(x^{-3/4})^{-8} = x^{(-3/4)(-8)} = x^6$

(c) $(64x^{-3})^{-2/3} = 64^{-2/3}(x^{-3})^{-2/3}$

$= (\sqrt[3]{64})^{-2} x^{(-3)(-2/3)}$

$= 4^{-2} x^2 = x^2/4^2 = x^2/16$

(d) $\left[\dfrac{(27x^2 y^3)^{1/3}}{9x^{-2} y^4}\right]^{-1} = \dfrac{9x^{-2} y^4}{(27x^2 y^3)^{1/3}} = \dfrac{9x^{-2} y^4}{27^{1/3} x^{2/3} y^1}$

$= \dfrac{9y^4 y^{-1}}{3x^{2/3} x^2} = \dfrac{3y^{4-1}}{x^{2/3 + 2}} = \dfrac{3y^3}{x^{8/3}}$

(e) $\dfrac{(3x + 2)^{1/2}(3x + 2)^{-1/4}}{(3x + 2)^{-3/4}} = (3x + 2)^{1/2}(3x + 2)^{-1/4}(3x + 2)^{3/4}$

$= (3x + 2)^{1/2 - 1/4 + 3/4} = (3x + 2)^1 = 3x + 2$

(f) $(a^{-1/2} - b^{-1/2})(a^{-1/2} + b^{-1/2}) = (a^{-1/2})^2 - (b^{-1/2})^2$

$= a^{-1} - b^{-1}$

$= \dfrac{1}{a} - \dfrac{1}{b} = \dfrac{b - a}{ab}$

To factor an algebraic sum in which each term contains a rational power of the same expression, begin by taking as a common factor the *smallest* rational power to which the expression is raised.

Example   Factor $(1 - 3x)(3x + 4)^{-1/2} + (3x + 4)^{1/2}$ and simplify the result.

Solution   We factor out $(3x + 4)^{-1/2}$ because $-1/2$ is the smaller exponent of $3x + 4$. Thus,

$$(1 - 3x)(3x + 4)^{-1/2} + (3x + 4)^{1/2} = (3x + 4)^{-1/2}[(1 - 3x) + (3x + 4)^{1/2 - (-1/2)}]$$
$$= (3x + 4)^{-1/2}[1 - 3x + (3x + 4)^1]$$
$$= (3x + 4)^{-1/2}(1 - 3x + 3x + 4)$$
$$= 5(3x + 4)^{-1/2} = \frac{5}{\sqrt{3x + 4}}.$$

PROBLEM SET 7

In problems 1–34, rewrite each expression so it contains only positive exponents, and simplify the result. Assume that n is a positive integer.

1. $\left(\frac{1}{3}\right)^{-3}$

2. $\frac{1}{7^{-2}}$

3. $(8x^0)^{-2}$

4. $(2^0 y^{-2})^{-5}$

5. $x^{-2} y^4 z^{-1}$

6. $[(3^0)/(4^{-2})]^{-1}$

7. $(-1)^{-1}$

8. $(x^{-3})^6 (x^0)^{-2}$

9. $(4c^{-4})(-7c^6)$

10. $(a + b)^{-4}(a + b)^9$

11. $(m^4)^{-2} m^{11}$

12. $[(c + 3d)^{-1}]^{-5}$

13. $\frac{3x^{-5}}{6y^{-2}}$

14. $\frac{c^{-1} + d^{-1}}{cd}$

15. $(t^{-1} + 3^{-2})^{-1}$

16. $\frac{a^{-1} - b^{-1}}{(a + b)^{-2}}$

17. $\frac{a^{-1}}{b^{-1}} + \left(\frac{b}{a}\right)^{-1}$

18. $(1 + x^{-1})^{-1} + (1 + x)^{-1}$

19. $x^{n+3} x^{n-4}$

20. $t^n t^{n-1} t^{n+1}$

21. $[(x^{-1})^{-1}]^{-1}$

22. $[(-y^{-2})^{-1}]^{-1}$

23. $1 - (p - 1)^{-1} + (p + 1)^{-1}$

24. $\frac{x^{-2} - y^{-2}}{x^{-1} + y^{-1}}$

25. $(t + 2)^{-1}(t + 2)^{-2}(t^2 - 4)^2$ 26. $(x^{-2} y^{-2} z^{-3})^{-2n}$

27. $\left(\frac{a^{-3}}{b^{-3}}\right)^{-n}$

28. $\left(\frac{5x^{-1}}{y}\right)^{-1} \frac{y}{5x^{-1}}$

29. $\frac{(a + 8b)^3}{(a + 8b)^{-n}}$

30. $\left[\frac{(cd)^{-2n}}{c^{-2n} d^{-2n}}\right]^{5n}$

31. $\frac{(c + d)^{-2}}{(r + s)^{-2}}\left(\frac{r + s}{c + d}\right)^2$

32. $\left[\frac{(5x)^{-1}(3x)^2 y^{-3}}{15x^{-2}(25y^{-4})}\right]^{-4}$

33. $\frac{(p + q)^{-1}(p - q)^{-1}}{(p + q)^{-1} - (p - q)^{-1}}$

34. $\frac{2(x + 5y)^{-1} + 3(x + 5y)^{-1} z^{-2}}{4(x + 5y)^{-2} - 9(x + 5y)^{-2} z^{-4}}$

In problems 35–46, find the value of each expression (if it is defined). Do not use a calculator.

35. $9^{3/2}$

36. $16^{-5/4}$

37. $(-8)^{5/3}$

38. $(-4)^{7/8}$

39. $32^{0.6}$

40. $(-1)^{-10/6}$

41. $(-8)^{0.3}$

42. $\left(\frac{-8}{27}\right)^{4/6}$

43. $\left(-\frac{1}{8}\right)^{-6/9}$

44. $\left(\frac{-1}{32}\right)^{1.8}$

45. $6^{1/2} 15^{1/2} 10^{1/2}$

46. $(-0.125)^{-2/6}$

In problems 47–64, simplify each expression and write the answer so it contains only positive exponents. (You may assume that variables are restricted to values for which all expressions are defined.)

47. $a^{1/2}a^{3/2}$

48. $m^{-1.4}m^{2.4}m^{-2}$

49. $y^{1/3}y^{2/3}y^{4/3}$

50. $(x+3)^{-1/2}(x+3)^{5/2}$

51. $(x^{-7/9})^{27/14}$

52. $[(3t+5)^{-7/5}]^{-10/7}$

53. $(16x^{-4})^{-3/4}$

54. $[(a^4b^{-3})^{1/5}]^{-10}$

55. $\left(\dfrac{x^{-1/3}}{x^{3/2}}\right)^6$

56. $\left(\dfrac{x^{1/2}}{y^2}\right)^4\left(\dfrac{y^{-1/3}}{x^{2/3}}\right)^3$

57. $\left(\dfrac{x^{m/3}y^{-3m/2}}{x^{-2m/3}y^{m/2}}\right)^{-2/m}$, $m>0$

58. $(x^{3/2}-y^{3/2})^2$

59. $(2p+q)^{-1/4}(2p+q)^{1/2}(2p+q)^{3/4}$

60. $[(s+2t)^{1/n}(s+2t)^{1/m}]^{nm/(n+m)}$, $n>0$, $m>0$

61. $\left[\dfrac{(3x+2y)^{1/2}(4r+3t)^{1/3}}{(4r+3t)^{1/2}(3x+2y)^{1/3}}\right]^6$

62. $[(2x+7)^{-3/4}]^{4/3}-[(2x+7)^{4/3}]^{-3/4}$

63. $(m+n)^{1/3}(m-n)^{1/3}(m^2-n^2)^{-2/3}$

64. $[(x^2+1)^{1/3}-1][(x^2+1)^{2/3}+(x^2+1)^{1/3}+1]$

In problems 65–72, factor each expression and simplify the result.

65. $x^{3/2}+2x^{1/2}y+x^{-1/2}y^2$

66. $(2x-1)^{-1/2}(6x-3)+(2x-1)^{1/2}$

67. $(p-1)^{-2}-2(p-1)^{-3}(p+1)$ [Hint: -3 is smaller than -2.]

68. $-2(1-5a)^{-3}+3(3a-4)^{-1}(1-5a)^{-4}$

69. $2(4t-1)^{-1}(2t+1)^{-2}+4(4t-1)^{-2}(2t+1)^{-1}$ [Hint: $-2<-1$.]

70. $2(2t+3)(4t-3)^{-1/4}+4(4t-3)^{3/4}$

71. $x^{-2/3}(x-1)^{2/3}+2x^{1/3}(x-1)^{-1/3}$

72. $3(1-x)^{-1/5}(1+x^2)^{-2/3}+5x(1-x)^{4/5}(1+x^2)^{-5/3}$

In problems 73–78, assume that a and b are nonzero real numbers and that m and n are integers. By considering all possible cases in which m or n are positive, negative, or zero, verify each property.

73. $a^na^m=a^{n+m}$

74. $(a^n)^m=a^{nm}$

75. $(ab)^n=a^nb^n$

76. $\left(\dfrac{a}{b}\right)^n=\dfrac{a^n}{b^n}$

77. $\dfrac{a^m}{a^n}=a^{m-n}$

78. $\dfrac{a^{-m}}{b^{-n}}=\dfrac{b^n}{a^m}$

79. If $a\neq0$ and $b\neq0$, show that $\left(\dfrac{a}{b}\right)^{-1}=\dfrac{b}{a}$.

C 80. Using a calculator, find approximate numerical values for
(a) $2^{1/2}$ (b) $3^{5/2}$ (c) $5^{-1.5}$ (d) π^{-2}.

The expressions in problems 81–84 were obtained as answers to problems in a popular calculus textbook. Factor each expression and simplify the result.

81. $2(3x+x^{-1})(3-x^{-2})(6x-1)^5+30(3x+x^{-1})^2(6x-1)^4$

82. $-3(3t-1)^{-2}(2t+5)^{-3}-6(3t-1)^{-1}(2t+5)^{-4}$

83. $-14(7y+3)^{-3}(2y-1)^4+8(7y+3)^{-2}(2y-1)^3$

84. $-5(6u+u^{-1})^{-6}(6-u^{-2})(2u-2)^7+14(6u+u^{-1})(2u-2)^6$

In problems 85–89, an *error in calculation* has been made. Find the error.

85. $(-1)^{2/4}=[(-1)^2]^{1/4}=1^{1/4}=1$?

86. $\sqrt[4]{(-4)^2}=(\sqrt[4]{-4})^2$ is undefined?

87. $[(-2)(-8)]^{3/2}=(-2)^{3/2}(-8)^{3/2}$ is undefined?

88. $(-32)^{0.2}=(-32)^{2/10}=(\sqrt[10]{-32})^2$ is undefined?

89. $[(-1)^2]^{1/2}=(-1)^{2(1/2)}=(-1)^1=-1$?

8 Calculators, Scientific Notation, and Approximations

Today many students own or have access to an electronic calculator. A scientific calculator with keys for exponential, logarithmic, and trigonometric functions costs less than many college textbooks and will expedite some of the calculations required in this book. Problems or groups of problems for which the use of a calculator is recommended are marked with the symbol ⓒ. If you don't have access to a calculator, you can still work most of these problems by using the tables provided in the appendixes at the back of the book.

There are two types of calculators available, those using **algebraic notation (AN)** and those using **reverse Polish notation (RPN).** Advocates of AN claim that it is more "natural," while supporters of RPN say that RPN is just as "natural" and avoids the parentheses required when sequential calculations are made in AN. Before purchasing a scientific calculator, you should familiarize yourself with both AN and RPN so that you can make an intelligent decision based on your own preferences.

After acquiring any calculator, learn to use it properly by studying the instruction booklet furnished with it. In particular, practice performing chain calculations so you can do them as efficiently as possible, using whatever "memory" features your calculator may possess to store intermediate results. After you learn *how* to use a calculator, it is imperative that you learn *when* to use it and especially when *not* to use it. Attempts to use a calculator for problems that are *not* marked with the symbol ⓒ can lead to bad habits, which not only waste time, but actually hinder understanding.

8.1 Scientific Notation

In applied mathematics, very large and very small numbers are written in compact form by using integer powers of 10. For instance, the speed of light in vacuum,

$$c = 300,000,000 \text{ meters per second (approximately)}$$

can be written more compactly as

$$c = 3 \times 10^8 \text{ meters per second.}$$

More generally, a positive real number x is said to be expressed in **scientific notation** if it is written in the form

$$x = p \times 10^n,$$

where n is an integer and $1 \leq p < 10$.

Many calculators automatically switch to scientific notation whenever the number is too large or too small to be displayed in ordinary decimal form. When a number such as 2.579×10^{-13} is displayed, the multiplication sign and the base 10 usually do not appear and the display shows simply

$$2.579 \qquad -13.$$

To change a number from ordinary decimal form to scientific notation, move the decimal point to obtain a number between 1 and 10 and multiply by 10^n or by 10^{-n}, where n is the number of places the decimal point was moved to the left or to the right, respectively. Final zeros after the decimal point can be dropped unless it is necessary to retain them to indicate the accuracy of an approximation.

Example   Rewrite each statement so that all numbers are expressed in scientific notation.

(a) The volume of the earth is approximately

$$1{,}087{,}000{,}000{,}000{,}000{,}000{,}000 \text{ cubic meters.}$$

(b) The earth rotates about its axis with an angular speed of approximately 0.00417 degree per second.

Solution   (a) We move the decimal point 21 places to the left

$$1.087000000000000000000$$

to obtain a number between 1 and 10 and multiply by 10^{21}, so that

$$1{,}087{,}000{,}000{,}000{,}000{,}000{,}000 = 1.087 \times 10^{21}.$$

Thus, the volume of the earth is approximately 1.087×10^{21} cubic meters.

(b) We move the decimal point three places to the right

$$0.00417$$

to obtain a number between 1 and 10 and multiply by 10^{-3}, so that

$$0.00417 = 4.17 \times 10^{-3}.$$

Thus, the earth rotates about its axis with an angular speed of approximately 4.17×10^{-3} degree per second.

The procedure above can be reversed whenever a number is given in scientific notation and we wish to rewrite it in ordinary decimal form.

Example   Rewrite the following numbers in ordinary decimal form:

(a) 7.71×10^5 (b) 6.32×10^{-8}

Solution   (a) $7.71 \times 10^5 = 771000. = 771{,}000$
(b) $6.32 \times 10^{-8} = 0.0000000632 = 0.000{,}000{,}0632$

8.2 Approximations

Numbers produced by a calculator are often inexact, because the calculator can work only with a finite number of decimal places. For instance, a 10-digit calculator gives

$$2 \div 3 = 6.666666667 \times 10^{-1} \quad \text{and} \quad \sqrt{2} = 1.414213562,$$

both of which are **approximations** of the true values. Therefore, unless we explicitly ask for numerical approximations or indicate that a calculator is recommended, it's usually best to leave answers in fractional form or as radical expressions. Don't be too quick to pick up your calculator—answers such as $2/3$, $\sqrt{2}$, $(\sqrt{2} + \sqrt{3})/7$, and $\pi/4$ are often *preferred* to much more lengthy decimal expressions that are only approximations.

Most numbers obtained from measurements of real-world quantities are subject to error and also have to be regarded as approximations. If the result of a measurement (or any calculation involving approximations) is expressed in scientific notation, $p \times 10^n$, it is usually understood that p should contain only **significant digits,** that is, digits that, except possibly for the last, are known to be correct or reliable. (The last digit may be off by one unit because the number was rounded off.) For instance, if we read in a physics textbook that

$$\text{one electron volt} = 1.60 \times 10^{-19} \text{ joule,}$$

we understand that the digits 1, 6, and 0 are significant and we say that, *to an accuracy of three significant digits*, one electron volt is 1.60×10^{-19} joule.

To emphasize that a numerical value is only an approximation, we often use a wave-shaped equal sign, $\approx$. For instance,

$$\sqrt{2} \approx 1.414.$$

However, we sometimes use ordinary equals signs when dealing with inexact quantities, simply because it becomes tiresome to indicate repeatedly that approximations are involved.

8.3 Rounding Off

Some scientific calculators can be set to round off all displayed numbers to a particular number of decimal places or significant digits. However, it's easy enough to round off numbers without a calculator: Simply drop all unwanted digits to the right of the digits that are to be retained, and increase the last retained digit by 1 if the first dropped digit is 5 or greater. It may be necessary to replace dropped digits by zeros in order to hold the decimal point; for instance, we round off 5157.3 to the nearest hundred as 5200.

Rounding off should be done in one step, rather than digit by digit. Digit-by-digit rounding off may produce an incorrect result. For instance, if 8.2347 is rounded off to four significant digits as 8.235, which in turn is rounded off to three significant digits, the result would be 8.24. However, 8.2347 is correctly rounded off in one step to three significant digits as 8.23.

Example   Round off the given number as indicated.

(a) 1.327 to the nearest tenth

(b) -19.8735 to the nearest thousandth

(c) 4671 to the nearest hundred

(d) 9.22345×10^7 to four significant digits

Solution   (a) To the nearest tenth,

$$1.327 \approx 1.3$$

(b) To the nearest thousandth,

$$-19.8735 \approx -19.874$$

thousandths place

(c) To the nearest hundred,

$$4671 \approx 4700$$

hundreds place

(d) To four significant digits, $9.22345 \times 10^7 \approx 9.223 \times 10^7$.

If approximate numbers expressed in ordinary decimal form are added or subtracted, the result is accurate only to as many decimal places as the least accurate of the numbers, and it should be rounded off accordingly.

Example   © Suppose that each of the quantities $a = 1.7 \times 10^{-2}$, $b = 2.711 \times 10^{-2}$, and $c = 6.213455 \times 10^2$ is accurate only to the number of displayed digits. Find $a - b + c$ and express the result in scientific notation rounded off to an appropriate number of significant digits.

Solution   Since we are adding and subtracting, we begin by rewriting

$$a = 0.017, \quad b = 0.02711, \quad \text{and} \quad c = 621.3455$$

in ordinary decimal form. The least accurate of these numbers is a, which is accurate only to the nearest thousandth. Using a calculator, we find that

$$a - b + c = 621.33539,$$

but we must round off this answer to the nearest thousandth and write

$$a - b + c = 621.335 = 6.21335 \times 10^2.$$

If approximate numbers are multiplied or divided, the result is accurate only to the number of significant digits in the least accurate of the numbers, and it should be rounded off accordingly.

Example   © Suppose that each of the quantities $a = 2.15 \times 10^{-3}$ and $b = 2.874 \times 10^2$ is accurate only to the number of displayed digits. Calculate the indicated quantity and express it in scientific notation rounded off to an appropriate number of significant digits.

(a) ab (b) b/a (c) b^2

Solution

(a) Using a calculator, we find that $ab = 6.1791 \times 10^{-1}$. Since a, the least accurate of the two factors, is accurate only to three significant digits, we must round off our answer to three significant digits and write $ab = 6.18 \times 10^{-1}$.

(b) Here we have $b/a = 1.336744186 \times 10^5$, but again we must round off our answer to three significant digits and write $b/a = 1.34 \times 10^5$.

(c) Here we have $b^2 = 8.259876 \times 10^4$, but, since b is accurate only to four significant digits, we must round off our answer to four significant digits and write $b^2 = 8.260 \times 10^4$.

PROBLEM SET 8

In problems 1–8, rewrite each number in scientific notation.

1. 15,500
2. 0.0043
3. 58,761,000
4. 77 million
5. 186,000,000,000
6. 420 trillion
7. 0.000,000,901
8. $(0.025)^{-5}$

In problems 9–14, rewrite each number in ordinary decimal form.

9. 3.33×10^4
10. 1.732×10^{10}
11. 4.102×10^{-5}
12. -8.255×10^{-11}
13. 1.001×10^7
14. -2.00×10^9

In problems 15–20, rewrite each statement so that all numbers are expressed in scientific notation.

15. The image of one frame in a motion-picture film stays on the screen approximately 0.062 second.
16. One liter is defined to be 0.001,000,028 cubic meter.
17. An *astronomical unit* is defined to be the average distance between the earth and the sun, 92,900,000 miles, and a *parsec* is the distance at which one astronomical unit would subtend one second of arc, about 19,200,000,000,000 miles.
18. A *light year* is the distance that light, traveling at approximately 186,200 miles per second, traverses in one year. Thus, a light year is approximately 5,872,000,000,000 miles.
19. In physics, the average lifetime of a lambda particle is estimated to be 0.000,000,000,251 second.
20. In thermodynamics, the Boltzmann constant is 0.000,000,000,000,000,000,000,0138 joule per degree Kelvin.

In problems 21–24, convert the given numbers to scientific notation and calculate the indicated quantity. (Use the Properties of Exponents.) Do not round off your answers.

21. $(8,000)(2,000,000,000)(0.000,03)$
22. $(0.000,006)^3(500,000,000)^{-4}$
23. $\dfrac{(7,000,000,000)^3}{0.0049}$
24. $\dfrac{(0.000,000,039)^2(591,000)^3}{(197,000)^2}$

C **25.** In electronics, $P = I^2R$ is the formula for the power P in watts dissipated by a resistance of R ohms through which a current of I amperes flows. Calculate P if $I = 1.43 \times 10^{-4}$ ampere and $R = 3.21 \times 10^4$ ohms.

C **26.** The mass of the sun is approximately 1.97×10^{29} kilograms and our galaxy (the Milky Way) is estimated to have a total mass of 1.5×10^{11} suns. The mass of the known universe is at least 10^{11} times the mass of our galaxy. Calculate the approximate mass of the known universe.

In problems 27–30, specify the accuracy of the indicated value in terms of significant digits.

27. A drop of water contains 1.7×10^{21} molecules.

28. The binding energy of the earth to the sun is 2.5×10^{33} joules.

29. One mile $= 6.3360 \times 10^4$ inches.

30. One atmosphere $= 1.01 \times 10^5$ newtons per square meter.

In problems 31–38, round off the given number as indicated.

31. 5280 to the nearest hundred

32. 9.29×10^7 to the nearest million

33. 0.0145 to the nearest thousandth

34. 999 to the nearest ten

35. 111111.11 to the nearest ten thousand

36. 5.872×10^{12} to three significant digits

37. 2.1448×10^{-13} to three significant digits

38. π to four significant digits

C In problems 39–48, find the numerical value of the indicated quantity rounded off to an appropriate number of significant digits. Assume that the given values are accurate only to the number of displayed digits.

39. $a + b$ if $a = 2.0371 \times 10^2$ and $b = 2.7312 \times 10^1$

40. $a + b - c$ if $a = 1.450 \times 10^5$, $b = 7.63 \times 10^2$, and $c = 2.251 \times 10^3$

41. $a - b + c$ if $a = 4.900 \times 10^{-4}$, $b = 3.512 \times 10^{-6}$, and $c = 2.27 \times 10^{-7}$

42. $a + b - c + d$ if $a = 8.1370$, $b = 2.2 \times 10^1$, $c = 1 \times 10^{-3}$, and $d = 5.23 \times 10^{-4}$

43. ab if $a = 3.19 \times 10^2$ and $b = 4.732 \times 10^{-3}$

44. ab^2 if $a = 2.11 \times 10^4$ and $b = 1.009 \times 10^{-2}$

45. a^3 if $a = 1.02 \times 10^9$

46. $\dfrac{ab}{c}$ if $a = 7.71 \times 10^3$, $b = 3.250 \times 10^{-4}$, and $c = 1.09 \times 10^5$

47. $\dfrac{a^2}{b}$ if $a = 3.32 \times 10^2$ and $b = 3.18 \times 10^{-1}$

48. $\dfrac{a + b}{c}$ if $a = 4.163 \times 10^2$, $b = 2.142 \times 10^1$, and $c = 1.555 \times 10^3$

C **49.** According to the U.S. Bureau of Economic Analysis, the gross national product (GNP) of the United States in 1978 was $\$2.1076 \times 10^{12}$. According to the U.S. Office of Management and Budget, the national debt at the end of fiscal 1978 was $\$7.804 \times 10^{11}$. Round off your answers to the following questions in an appropriate manner: (a) How much more was the GNP than the national debt in 1978? (b) If we estimate the population of the United States in 1978 as 2.2×10^8, find the per capita GNP (that is, GNP ÷ population) in 1978.

50. Let x be a positive number and let r be the result of rounding off x to the nearest 10^{-n}, where n is an integer. Prove that $|x - r|$, the error made in estimating x by r, cannot exceed $5 \times 10^{-n-1}$.

REVIEW PROBLEM SET

In problems 1–4, rewrite each expression using exponential notation.

1. $5 \cdot 5 \cdot 5 \cdot 5 \cdot 5 \cdot 5 \cdot 5 \cdot 5 \cdot x \cdot x \cdot x$

2. $w^2 \cdot w^2 \cdot w^2 \cdot w^2 \cdot z^3 \cdot z^3 \cdot z^3 \cdot z^3$

3. $(-4)(-4)(-4)(-4)(-4)yyyyyy$

4. $(-a^4)(-a^4)(-a^4)(-a^4)(-a^4)$

In problems 5–10, write a formula for the given quantity.

5. w is three times the product of x and y, divided by z.

6. x is 7% less than the number n.

7. s is one-half of the perimeter of a triangle with sides a, b, and c.

8. The surface area A of a rectangular box, open at the top and closed at the bottom, with length l, width w, and height h.

9. The population P of a town n years from now, if the current population is 1000 and the population triples every year. [Hint: After $n = 1$ year, the population is 3000; after $n = 2$ years, it is 9000, and so forth.]

10. The number N of board feet of lumber in n "two-by-fours," each of which is l feet long. [Note: The dimensions of a cross section of a "two-by-four" are actually 1.5 inches by 3.5 inches; one board foot is the volume of a board with dimensions 1 foot by 1 foot by 1 inch.]

11. The formula $K = \frac{1}{2}mv^2$ gives the kinetic energy K, in joules, of an object of mass m kilograms moving with a speed of v meters per second. A jogger with a mass of 70 kilograms is running at a speed of 3 meters per second. Find the kinetic energy of the jogger.

C 12. If P dollars is borrowed and paid back in n equal periodic installments of R dollars each, including interest at the rate r per period on the unpaid balance, then

$$R = \frac{Pr}{1 - (1 + r)^{-n}}.$$

Find R if $20,000 is borrowed and paid back in 5 equal yearly installments including an interest of 10% per year ($r = 0.1$) on the unpaid balance.

In problems 13–22, state one basic algebraic property of the real numbers to justify each statement.

13. $3 \cdot (-7) = (-7) \cdot 3$

14. $3(x + 2) = 3x + 3 \cdot 2$

15. $(-3) + (5 + \pi) = [(-3) + 5] + \pi$

16. $0 + y^2 = y^2$

17. $15 \cdot (x + y) = (x + y) \cdot 15$

18. $3(\pi + 0) = 3\pi$

19. $1 \cdot (a - b) = a - b$

20. $6 \cdot (4 \cdot 3) = (6 \cdot 4) \cdot 3$

21. $(-3) \cdot \dfrac{1}{(-3)} = 1$

22. $\pi + (-\pi) = 0$

In problems 23–26, state one of the derived algebraic properties (6–8, page 4) to justify each statement.

23. $-[-(x + y)] = x + y$

24. $(-4)(-\pi) = 4\pi$

25. $(x^2 - y^2) \cdot 0 = 0$

26. If $(3x^2 - 5)(2x - 1) = 0$, then $3x^2 - 5 = 0$ or $2x - 1 = 0$.

In problems 27 and 28, find the mistake and correct the calculation.

27. $\frac{1}{2} + \frac{1}{3} = \frac{2}{5}$?

28. $\sqrt{9 + 16} = 3 + 4 = 7$?

In problems 29 and 30, list all the elements that belong to the set.

29. A is the set of all natural numbers x such that $-\frac{3}{2} \le x \le \frac{5}{2}$.

30. B is the set of all real numbers in $[0, 1]$ but not in $(0, 1)$.

In problems 31–34, illustrate each set on a number line.

31. (a) $(-1, 3)$ (b) $[-2, 5]$ (c) $[-7, \infty)$ (d) $(-\infty, 4]$
     (e) $[-\frac{1}{3}, 5]$ (f) $(-\frac{5}{2}, \frac{3}{2})$ (g) $[-\frac{2}{3}, \frac{1}{3})$ (h) $(-\infty, \sqrt{2})$

32. The set of all real numbers x such that $-3 \le x \le 0$ and $3x$ is an integer.

33. The set A of all real numbers that belong to both of the intervals $(-\infty, 5]$ and $(-\frac{2}{3}, 10]$.

34. The set B of all real numbers that belong to at least one of the intervals $(-3, -1]$ and $[1, 3)$.

In problems 35–40, express each rational number (a) as a decimal and (b) as a percent.

35. $\frac{11}{50}$ **36.** $\frac{-3}{1000}$ **37.** $-\frac{17}{200}$ **38.** $-\frac{7}{8}$
39. $\frac{130}{40}$ **40.** $-\frac{7}{3}$

In problems 41–46, express each percent (a) as a decimal and (b) as a quotient of integers.

41. 49.5%   **42.** 0.007%   **43.** 0.43%   **44.** 13.4 %
45. 140%   **46.** 215%

47. The price of an automobile increases from \$8000 to \$8500. What is the percent of increase?

48. Employment at a factory decreases from 400 workers to 375 workers. What is the percent of decrease?

In problems 49–60, specify the type of each algebraic expression (monomial, binomial, trinomial, multinomial, polynomial, constant, fraction, rational expression, or radical expression). For the polynomials, give the degree and the coefficients.

49. $-2x^2$ **50.** $\sqrt{7}$ **51.** $\dfrac{x + y}{x - y}$ **52.** $3x^3 - 2x^2 + 6x - 4$

53. $3x^2 - 5x - 1$ **54.** xy **55.** $xy + \sqrt{x}$ **56.** $xy + x^{-1} - 1$

57. $\sqrt{2}x^3 - \sqrt[5]{7}$ **58.** $\dfrac{x^2 - y^2}{x^2 + 1}$ **59.** $\sqrt{\pi} + \dfrac{x}{y}$ **60.** $x^2 + x + \sqrt{x^2 + 1}$

In problems 61–64, perform the indicated operations.

61. $(x^3 - 2x^2 + 7x - 5) + (2x^3 - x^2 - 5x + 11)$ **62.** $(3x^3 - 3x^2 - 8x - 17) - (4x^3 + 5x - 7)$
63. $(a^3 + 3a^2b + 2ab^2 + b^3) + (2a^3 + ab^2 - 3b^3) - (4a^2b - 3ab^2 - 5b^3)$
64. $(u^3 + 3u^2v + v^2) - (2u^3 - u^2v - 3uv - v^3) + (u^3 - 2uv - 4v^2)$

In problems 65–74, use the Properties of Exponents to simplify each expression.

65. $5y^2 \cdot 4y^3$ **66.** $(-6t^4)(-5t^6)t^{10}$ **67.** $(-x^2y)(x^4y^3)$ **68.** $t^{3n}t^{2n}t^n$
69. $(-p^2)^4$ **70.** $(-q^3)^5$ **71.** $(-x^2y)^7$ **72.** $[-(x + y)^2]^3$
73. $(ab^n)(ab)^n$ **74.** $(2x^n)^4$

In problems 75–84, expand each product.

75. $3x^2(x^3 + 2x - 3)$ **76.** $x^2y(2x + y + 7)$ **77.** $(2t + 3)(t - 4)$
78. $(2u^2 - v)(u^2 + 3v)$ **79.** $(xy^2 + 3)(2xy^2 + 1)$ **80.** $(s^3 + t^2)(s^3 - 2t^2)$
81. $(2p + 3)(p^2 - 4p + 1)$ **82.** $(x^2 + x - 2)(x^2 + 3x + 1)$ **83.** $(2x + 3)(x - 1)(x - 2)$
84. $(2x - y)(x + 3y)(3x + y)(3x - y)$

In problems 85–94, use the Special Products 1–8, page 18, to perform each multiplication.

85. $(3x + 5y)^2$ **86.** $(2q + 7r)^2$ **87.** $(2x^2 - 5yz)^2$
88. $(3a^n - b^n)^2$ **89.** $(2x - y + 3z)^2$ **90.** $(p - q)^2(p + q)^2$
91. $(3t^n - 11)(3t^n + 11)$ **92.** $(2x^2 + y^2)^3$ **93.** $(3x^3 - 2xy)^3$
94. $(p^n - 3)(p^{2n} + 3p^n + 9)$

In problems 95–124, factor each polynomial completely.

95. $9x^2y^2 - 12xy^4$ **96.** $18r^2s + 12r^3s^4$ **97.** $(a + b)^2c^2 - (a + b)c^4$

98. $(3p + q)^3 - (3p + q)^2 u$ **99.** $36c^2 - d^4$ **100.** $x^6 - 25y^4$

101. $(x - y)^2 - z^2$ **102.** $(t + 2s)^2 - 9u^2$ **103.** $x^{2n} - y^2$

104. $x^8 y^8 - 1$ **105.** $25x^2 - 49x^2 y^2$ **106.** $(c - 2d)^4 - (3c - d)^4$

107. $8p^3 + 27q^3$ **108.** $125x^3 - 64y^3$ **109.** $(a + 2b)^3 - (a - 2b)^3$

110. $t^3 + 125(u + v)^3$ **111.** $x^2 + 2x - 24$ **112.** $x^2 - 13x + 40$

113. $a^6 + 8a^3 + 16$ **114.** $t^4 - 5t^2 + 4$ **115.** $6x^2 + 5xy - 6y^2$

116. $x^{2n} + x^n - 6$ **117.** $4u^2 v^2 - 7uv - 2$ **118.** $4t^2 + 19tu - 30u^2$

119. $20 + 7x - 6x^2$ **120.** $2a^2 + 4ab + 2b^2 - a - b - 10$

121. $p^2 + 9q^2 - 4 + 6pq$ **122.** $x^2 + 2xy - 4x - 4y + y^2 + 4$

123. $(a + b)^4 - 7(a + b)^2 + 1$ **124.** $(x - y)^{2n} - 2(x - y)^n(x^2 + y^2) + (x^2 + y^2)^2$

In problems 125–130, reduce each fraction to lowest terms.

125. $\dfrac{(x + 3)^2(x - 1)}{5(x + 3)(x - 1)}$ **126.** $\dfrac{3(x - 2)^3(x + 1)^2}{12(x - 2)^2(x + 1)}$ **127.** $\dfrac{t^2 + 5t + 6}{t^2 + 4t + 4}$

128. $\dfrac{2a^2 - 3a - 2}{2a^2 + 3a + 1}$ **129.** $\dfrac{c(c - 2) - 3}{(c - 2)(c + 1)}$ **130.** $\dfrac{(b - 2)(b + 1) - 4}{(b + 2)(b - 2)}$

In problems 131–148, perform the indicated operations and simplify the result.

131. $\dfrac{x^2 - 9}{6x - 3} \cdot \dfrac{10x - 5}{x^2 + 3x}$ **132.** $\dfrac{2a + b}{a^2 - 2ab} \cdot \dfrac{a^3 - 2a^2 b}{4a^2 - b^2}$ **133.** $\dfrac{x^2 + 6x + 5}{2x^2 - 2x - 12} \cdot \dfrac{4x^2 - 36}{x^2 + 8x + 15}$

134. $\dfrac{2y^2 - 7y - 15}{5y^2 - 24y - 5} \cdot \dfrac{20y^2 + 14y + 2}{2y^2 + 11y + 12} \cdot \dfrac{3y^2 + y - 2}{10y^2 + 35y + 15}$

135. $\dfrac{x^2 y - xy^2}{3x^2 - 9xy + 6y^2} \div \dfrac{x^3 + x^2 y}{6x^3 - 6x^2 y - 12xy^2}$ **136.** $\dfrac{-p^3 + p}{p^2 - p - 2} \div \dfrac{p^3 - p^2}{p^2 - 5p + 6}$

137. $\dfrac{3t^2 + 9t - 54}{2t^2 - 2t - 12} \div \dfrac{3t^2 + 21t + 18}{4t^2 - 12t - 40}$ **138.** $\dfrac{14a^2 + 23a + 3}{2a^2 + a - 3} \div \dfrac{7a^2 + 15a + 2}{2a^2 - 3a + 1}$

139. $\dfrac{2x - y}{3x^2} + \dfrac{4x + y}{3x^2}$ **140.** $\dfrac{x}{x - 2} - \dfrac{x + 2}{x + 1}$

141. $\dfrac{t^2 - 2t + 1}{t^2 + t} - \dfrac{t - 3}{t + 1}$ **142.** $\dfrac{a - 2b}{2ab - 6b^2} - \dfrac{b}{a^2 - 4ab + 3b^2}$

143. $\dfrac{2}{c - 2} - \dfrac{1}{c + 3} - \dfrac{10}{c^2 + c - 6}$ **144.** $\dfrac{3}{p + 1} - \dfrac{3}{p^2 + p} + \dfrac{6}{p^2 - 1}$

145. $\dfrac{1 + \dfrac{6}{a - 3}}{a + 3}$ **146.** $\dfrac{2 - \dfrac{3}{y + 2}}{\dfrac{x}{y - 1} + \dfrac{x}{y + 2}}$

147. $\dfrac{\dfrac{6}{a^2 + 3a - 10} - \dfrac{1}{a - 2}}{\dfrac{1}{a - 2} + 1}$ **148.** $\dfrac{\dfrac{1}{x + y} - \dfrac{1}{x - y}}{\dfrac{2y}{x^2 - y^2}}$

In problems 149–154, determine which of the indicated principal roots are rational numbers and evaluate those that are. Do not use a calculator.

149. $\sqrt{169}$ **150.** $\sqrt[3]{\dfrac{375}{24}}$ **151.** $\sqrt[5]{0.6}$

152. $\sqrt[3]{-\dfrac{108}{32}}$ **153.** $\sqrt[4]{-81}$ **154.** $\sqrt[3]{0.216}$

In problems 155–162, use the properties of radicals to simplify each expression. Assume that variables are restricted to values for which all expressions are defined.

155. $\sqrt[3]{4x^2}\sqrt[3]{2x^4}$

156. $\sqrt[5]{(c+2d)^4}\sqrt[5]{c+2d}$

157. $\sqrt[3]{\dfrac{(a+b)^9}{27a^3}}$

158. $\dfrac{\sqrt[4]{64a^3b^2c}}{\sqrt[8]{16a^2b^{12}c^{10}}}$

159. $\sqrt{\sqrt[4]{p^{24}}}\sqrt[n]{p^{4n}}$

160. $\sqrt[n]{(a+b)^{4n}c^{2n}}\sqrt[m]{(a+b)^{2m}c^m}$

161. $\sqrt[3]{\dfrac{a^{14}\sqrt{a^6}}{a^7}}\sqrt[3]{\dfrac{5a^{12}}{a^{15}}}$

162. $\dfrac{\sqrt[4]{x^6y^3z^2}\sqrt[4]{x^3yz^6}}{\sqrt[4]{xy^2}}$

In problems 163–180, perform the indicated operations and simplify the result. Rationalize all denominators (whenever possible). Assume that variables are restricted to values for which all expressions are defined.

163. $\sqrt{50a} + 2\sqrt{32a} - \sqrt{2a}$

164. $\sqrt{8a^3} - 2\sqrt{18a^3} + 3\sqrt{50a^3}$

165. $5\sqrt[3]{2p} + 4\sqrt[3]{16p}$

166. $\sqrt[3]{250x^2} - 6\sqrt[3]{16x^2}$

167. $\sqrt{6}(5 - \sqrt{6}) + \sqrt[3]{216}$

168. $(\sqrt{y}+1)(\sqrt{y}-2)$

169. $(\sqrt{2a} - \sqrt{3})(\sqrt{2a} + \sqrt{3})$

170. $(\sqrt{y+z} - 3\sqrt{x})(\sqrt{y+z} + 3\sqrt{x})$

171. $(\sqrt{a+b} - \sqrt{a})^2$

172. $(\sqrt[3]{x+1} - \sqrt[3]{x-1})(\sqrt[3]{(x+1)^2} + \sqrt[3]{x^2-1} + \sqrt[3]{(x-1)^2})$

173. $\dfrac{6}{\sqrt{2x}}$

174. $\dfrac{5}{\sqrt[3]{3p}}$

175. $\dfrac{\sqrt{a}}{\sqrt{a}-\sqrt{b}}$

176. $\dfrac{\sqrt{c}}{\sqrt{c}+\sqrt{d}}$

177. $\dfrac{\sqrt{a-1}}{1+\sqrt{a-1}}$

178. $\dfrac{(x+1)^2}{x\sqrt{x+1}} - \dfrac{\sqrt{x-1}}{2x\sqrt{x}}$

179. $\dfrac{5}{\sqrt[3]{2}-1}$

180. $\dfrac{6}{\sqrt[3]{x+y} - \sqrt[3]{x}}$

In problems 181 and 182, rationalize the *numerator*.

181. $\dfrac{3\sqrt{x}+\sqrt{y}}{5}$

182. $\dfrac{\sqrt{x+h+2} - \sqrt{x+2}}{h}$

In problems 183–194, simplify each expression and write it in a form containing only positive exponents.

183. $[(ab^{-1})^{-2} + (c^{-1}d)^{-3}]^0$

184. $[(-5)^{-2} + 3^{-1}]^{-1}$

185. $\left(\dfrac{x}{y^{-2}}\right)^{-1} + \left(\dfrac{y}{x^{-2}}\right)^{-1}$

186. $\dfrac{(a+2)^{-1} - (a-2)^{-1}}{(a+2)^{-1} + (a-2)^{-1}}$

187. $x^{-3}(x - x^{-1})$

188. $\dfrac{c^{-1} + c^{-2}}{c^{-3}}$

189. $(-3a^{-3})(-a^{-1})^3$

190. $(a^2b^{-4})^{-1}(a^{-3}b^2)^{-2}$

191. $\left(\dfrac{x^{-2}}{y^3}\right)^{-2}\left(\dfrac{x^{-3}}{y^{-4}}\right)^{-3}$

192. $\dfrac{(xy^{-1})^{-2}}{x}\cdot\left(\dfrac{x}{y^{-1}}\right)^{-3}$

193. $\dfrac{(5p^2)^{-2}(5p^5)^{-2}}{(5^{-1}p^{-2})^2}$

194. $\dfrac{(a^3b^2c^4)^{-2}(a^4b^2c)^{-1}}{(abc)^{-1}(a^2bc^3)^2}$

In problems 195–198, find the value of each expression without using a calculator.

195. $\left(\dfrac{8}{27}\right)^{2/3}$

196. $243^{0.6}$

197. $32^{-1.8}$

198. $(64^{1/6} + 4069^{1/12})^{-2}$

In problems 199–206, use the properties of rational exponents to simplify each expression and write it in a form containing only positive exponents. Assume that variables are restricted to values for which all quantities are defined.

199. $y^{-3/4}y^{2/3}y^{4/3}y^{-1/4}$

200. $x^{-1/2}(x^{3/2} + x^{1/2})$

201. $(a^{-1/4})^8(a^{-1/15})^{-45}a^2$

202. $a^{1/3}b^{1/3}\left[\left(\dfrac{a + b}{2}\right)^2 - \left(\dfrac{a - b}{2}\right)^2\right]^{-2/3}$

203. $(x^2y^{-1})^{-1/2}(x^{-3})^{-1/3}(y^{-2})^{-1/2}$

204. $(a^{1/m}b^{-m})^{-m}(a^{-m}b^{1/m})^m, \qquad m > 0$

205. $\left(\dfrac{-64a^3}{b^6c^4}\right)^{-2/3}\left(\dfrac{8a^{1/3}b^{3/2}}{c^{1/3}}\right)^6$

206. $\left(\dfrac{a^{-3/5}b^{-1/3}c^{2/5}}{a^{-1/5}b^{-2/3}c^{1/5}}\right)^{15}$

In problems 207–210, factor each expression and simplify the result.

207. $y^{-12}(x - y)(x + y)^{-3} + y^{-10}(x + y)^{-4}$

208. $a^{7/5}b^{-2/3} - a^{2/5}b^{1/3}$

209. $(y + 2)^{-2/3}(y + 1)^{2/3} + 2(y + 2)^{1/3}(y + 1)^{-1/3}$

210. $2(x + 1)^{5/3}(x - 2)^{-1/3} + (x^2 - x - 2)^{2/3}$

In problems 211–214, rewrite each number in scientific notation.

211. 57,120,000,000

212. 731 billion

213. 0.000,000,714

214. 33 millionths

In problems 215–218, rewrite each number in ordinary decimal form.

215. 1.732×10^7

216. -1.066×10^4

217. 3.12×10^{-8}

218. -3.05×10^{-11}

In problems 219–222, rewrite each statement so that all numbers are expressed in scientific notation.

219. An amoeba weighs about 5 millionths of a gram.

220. A tobacco mosaic virus weighs about 0.000,000,000,000,000,066 gram.

221. The diameter of the star Betelgeuse is approximately 358,400,000 kilometers.

222. In a game of bridge there is one chance in approximately 158,800,000,000 that a player will be dealt a hand containing all cards of the same suit, and there is one chance in approximately 2,235,000,000,000,000,000,000,000 that all four players will be dealt such a hand.

C In problems 223 and 224, convert the given numbers to scientific notation and calculate the indicated quantity. Do not round off your answer.

223. $\dfrac{(40,320,000,000)(0.000,007,703)}{21,000}$

224. $\dfrac{(97,400,000)(705,000)(1,410,000)^2}{0.000,000,220,9}$

In problems 225–228, specify the accuracy of the indicated value in terms of significant digits.

225. The chances of being dealt a "full house" in five-card poker are about one in 6.94×10^2.

226. Five thousand miles is about 3×10^8 inches.

227. One British thermal unit is about 6.6×10^{21} electron volts.

228. The standard value of the acceleration of gravity is $g = 9.80665$ meters/second^2.

In problems 229–232, round off the given number as indicated.

229. 17,450 to the nearest thousand

230. 0.00251 to the nearest thousandth

231. 7.2283×10^5 to three significant digits

232. 2.71828 to four significant digits

C In problems 233–238, find the numerical value of the indicated quantity rounded off to an appropriate number of significant digits. Assume that the given values are accurate only to the number of displayed digits.

233. $R_1 + R_2 + R_3$ if $R_1 = 2.7 \times 10^4$, $R_2 = 1.5 \times 10^3$, and $R_3 = 7 \times 10^3$

234. $(R_1^{-1} + R_2^{-1} + R_3^{-1})^{-1}$ if $R_1 = 1.7 \times 10^3$, $R_2 = 3.1 \times 10^4$, and $R_3 = 5 \times 10^3$

235. $\frac{1}{2}mv^2$ if $m = 5.98 \times 10^{24}$ and $v = 2.9770 \times 10^4$

236. mc^2 if $c = 3.00 \times 10^8$ and $m = 9.11 \times 10^{-31}$

237. $\dfrac{IB}{nex}$ if $I = 2.05 \times 10^2$, $B = 1.5$, $n = 8.4 \times 10^{28}$, $e = 1.6 \times 10^{-19}$, and $x = 1.3 \times 10^{-3}$

238. $\frac{4}{3}\pi r^3$ if $r = 6.4 \times 10^6$

239. In adding approximate numbers, does it ever matter whether you round the numbers off (to the number of decimal places in the least accurate of them) before adding them, instead of adding them and then rounding off the sum?

2 Equations and Inequalities

The basic algebraic skills developed in Chapter 1 are especially useful in solving the equations and inequalities that arise in practical applications of mathematics. In this chapter we discuss methods for solving equations and inequalities that contain just one variable. Equations and inequalities containing more than one variable are considered later in the book.

1 Equations

An equation containing a variable is neither true nor false until a particular number is substituted for the variable. If a true statement results from such a substitution, we say that the substitution **satisfies** the equation. For instance, the substitution $x = 3$ satisfies the equation $x^2 = 9$, but the substitution $x = 4$ does not.

An equation that is satisfied by every substitution for which both sides are defined is called an **identity.** For instance, $(x + 1)^2 = x^2 + 2x + 1$ is an identity, as is $(\sqrt{x})^2 = x$. An equation that is not an identity is called a **conditional equation.** For instance, $2x = 6$ is a conditional equation because there is at least one substitution (say, $x = 4$) that produces a false statement.

If the substitution $x = a$ satisfies an equation, we say that the number a is a **solution** or a **root** of the equation. Thus, 3 is a root of the equation $2x = 6$, but 4 is not. Two equations are said to be **equivalent** if they have exactly the same roots. Thus, the equation $2x - 6 = 0$ is equivalent to the equation $2x = 6$ because both equations have one and the same root, namely, $x = 3$.

You can change an equation into an equivalent one by performing any of the following operations:

1. Add or subtract the same quantity on both sides of the equation.

2. Multiply or divide both sides of the equation by the same nonzero quantity.

3. Simplify one or both sides of the equation by using the methods described in Chapter 1.

4. Interchange the two sides of the equation.

To **solve** an equation means to find all of its roots. The usual method for solving an equation is to write a sequence of equations, starting with the given one, in which each equation is equivalent to the previous one, but "simpler" in some sense. The last equation should either express the solution directly, or be so simple that its solution is obvious. For example, to solve the equation

$$2x - 6 = 0,$$

we begin by adding 6 to both sides to get the equivalent equation

$$2x = 6,$$

then we divide both sides by 2 to produce the equivalent equation

$$x = 3.$$

The last equation shows that the root is 3.

Variables representing quantities whose value or values we wish to find by solving equations are called **unknowns.** A common practice is to use letters toward the end of the alphabet for unknowns, and letters toward the front of the alphabet for **constants** whose value we can assign at will. In particular, the letter x is often used for an "unknown quantity," and the letters a, b, and c are used for constants. A **literal** or **general equation** is an equation containing, in addition to one or more unknowns, at least one letter that stands for a constant. For instance,

$$ax + b = 0$$

is a literal equation in which x is the unknown and the constant coefficients a and b can be assigned whatever values we please. If we let $a = 2$ and $b = -6$, we obtain

$$2x - 6 = 0,$$

whose solution is $x = 3$.

In applied mathematics, we cannot always follow the convention that unknowns are represented by letters toward the end of the alphabet, because certain symbols are reserved for special quantities. For instance, in physics, c is used for the speed of light, m is used for mass, v is used for velocity, and so on. We shall specify which letters represent unknowns to be solved for, whenever it isn't clear from the context.

An equation such as $7x^3 + 3x^2 + x + 1 = 2x - 5$, in which both sides are polynomials in the unknown, is called a **polynomial equation.** By subtracting the polynomial on the right from both sides of the polynomial equation, we obtain an equivalent polynomial equation in **standard form:**

$$7x^3 + 3x^2 + x + 1 = 2x - 5$$

$$7x^3 + 3x^2 + x + 1 - (2x - 5) = 0 \qquad \text{(We subtracted } 2x - 5 \text{ from both sides.)}$$

$$7x^3 + 3x^2 - x + 6 = 0 \qquad \text{(We combined like terms.)}$$

The last equation is in standard form. The **degree** of a polynomial equation is defined as the degree of the polynomial on the left side when the equation is in standard form. For instance, $7x^3 + 3x^2 + x + 1 = 2x - 5$ is a third-degree polynomial equation because, after it is rewritten in standard form, $7x^3 + 3x^2 - x + 6 = 0$, the polynomial on the left side has degree 3.

1.1 First-Degree or Linear Equations

A **first-degree** or **linear equation** in x is written in standard form as

$$ax + b = 0 \qquad \text{with } a \neq 0.$$

This equation is solved as follows:

$$ax + b = 0$$

$$ax = -b \qquad \text{(We subtracted b from both sides.)}$$

$$x = \frac{-b}{a} \qquad \text{(We divided both sides by a.)}$$

In many cases, simple first-degree equations can be solved mentally. For example,

the solution of $5x = 10$ is $x = 2$,

and

the solution of $2x + 3 = 0$ is $x = -\frac{3}{2}$.

Sometimes it is convenient to use a calculator.

Example $\boxed{\text{c}}$ Solve the equation $2.35x - 3.337 = 0$.

Solution   The solution is $x = 3.337/2.35$. Using a calculator, we find that $x = 1.42$.

As the following examples illustrate, an equation that is not in the standard first-degree form can often be changed into this form and then solved.

Examples   Solve each equation.

1 $29 - 2x = 15x - 5$

Solution

$$29 - 2x = 15x - 5$$

$$29 - 2x - 15x = -5 \qquad \text{(We subtracted $15x$ from both sides.)}$$

$$29 - 17x = -5 \qquad \text{(We combined like terms.)}$$

$$-17x = -34 \qquad \text{(We subtracted 29 from both sides.)}$$

$$17x = 34 \qquad \text{(We multiplied both sides by -1.)}$$

$$x = \tfrac{34}{17} \qquad \text{(We divided both sides by 17.)}$$

$$x = 2$$

2 $(2n + 3)(6n - 1) - 9 = 15n^2 - (3n - 2)(n - 2)$

Solution   We begin by expanding the products on both sides of the equation:

$$(12n^2 + 16n - 3) - 9 = 15n^2 - (3n^2 - 8n + 4)$$

$$12n^2 + 16n - 12 = 12n^2 + 8n - 4 \qquad \text{(We collected like terms.)}$$

$$16n - 12 = 8n - 4 \qquad \text{(We subtracted } 12n^2 \text{ from both sides.)}$$

$$8n - 12 = -4 \qquad \text{(We subtracted } 8n \text{ from both sides.)}$$

$$8n = 8 \qquad \text{(We added 12 to both sides.)}$$

$$n = 1 \qquad \text{(We divided both sides by 8.)}$$

3 $\frac{1}{7}(3x - 1) - \frac{1}{5}(2x - 4) = 1$

Solution   In order to clear the equation of fractions, we begin by multiplying both sides by 35, the L.C.D. of the two fractions:

$$5(3x - 1) - 7(2x - 4) = 35$$

$$15x - 5 - 14x + 28 = 35 \qquad \text{(We expanded the products.)}$$

$$x + 23 = 35 \qquad \text{(We collected like terms.)}$$

$$x = 12 \qquad \text{(We subtracted 23 from both sides.)}$$

If, in solving an equation, you multiply both sides by an expression containing the unknown, you must always check the solution. The following example shows why.

Example   Solve the equation $\dfrac{1}{y(y - 1)} - \dfrac{1}{y} = \dfrac{1}{y - 1}$.

Solution   Multiplying both sides of the equation by the L.C.D. $y(y - 1)$ and simplifying, we have

$$\cancel{y(y-1)}\,\frac{1}{\cancel{y(y-1)}} - y(y-1)\frac{1}{\cancel{y}} = y\cancel{(y-1)}\,\frac{1}{\cancel{y-1}},$$

that is,

$$1 - (y - 1) = y \quad \text{or} \quad 2 - y = y.$$

Adding y to both sides of the last equation, we obtain

$$2 = 2y, \quad \text{that is,} \quad 2y = 2,$$

from which it follows that $y = 1$. We now check by substituting $y = 1$ in the original equation to obtain

$$\frac{1}{1(1 - 1)} - \frac{1}{1} = \frac{1}{1 - 1},$$

an equation in which neither side is defined because of the zeros in the denominators.

In other words, the substitution $y = 1$ doesn't make the equation true—it makes the equation meaningless! We conclude that the equation *has no root*.

In the example above, a solution $y = 1$ was found for the *final* equation, but this was not a solution of the *original* equation. What happened? Well, we multiplied both sides of the original equation by $y(y - 1)$, a quantity that equals zero when $y = 1$. But, multiplication of both sides of an equation by zero does not produce an equivalent equation! A fake "root" that is obtained in this way and that doesn't satisfy the original equation is called an **extraneous root.**

The following example illustrates an interesting application of linear equations.

Example Express the repeating decimal $0.32\overline{57}$ as a quotient of integers.

Solution Let $x = 0.32\overline{57}$. Then $100x = 32.5\overline{757}$. If we subtract $0.32\overline{57}$ from $32.5\overline{757}$, the repeating portion of the decimals cancels out:

$$
\begin{array}{r}
32.5\overline{757} \\
(-) \quad 0.32\overline{57} \\
\hline
32.25
\end{array}
$$

Therefore,

$$100x - x = 32.5\overline{757} - 0.32\overline{57} = 32.25,$$

that is,

$$99x = 32.25 \quad \text{or} \quad 9900x = 3225.$$

It follows that

$$x = \tfrac{3225}{9900} = \tfrac{43}{132}.$$

1.2 Literal Equations That Can Be Reduced to First-Degree Form

Literal equations containing one unknown can often be solved by using the methods illustrated in Section 1.1. It's usually a good idea to begin by trying to bring all terms containing the unknown to one side of the equation, and all terms not containing the unknown to the opposite side. As always, you must be careful not to divide by zero.

Example 1 Solve the equation $ax + 4c = b - 2x$ for x.

Solution

$$ax + 4c = b - 2x$$

$ax + 2x + 4c = b$ (We added $2x$ to both sides so that all terms containing x are on the left side.)

$ax + 2x = b - 4c$ (We subtracted $4c$ from both sides so that all terms not containing x are on the right side.)

$(a + 2)x = b - 4c$ (We used the distributive property.)

Now, provided that $a + 2 \neq 0$, we can divide both sides of the last equation by $a + 2$ to obtain the solution

$$x = \frac{b - 4c}{a + 2} \qquad \text{for } a \neq -2.$$

You can check this solution by substituting it into the original equation (Problem 71).

Example 2   The formula $S = 2\pi r^2 + 2\pi rh$ gives the total surface area S of a closed right-circular cylinder of radius r and height h (Figure 1). Solve for h in terms of S and r.

Solution

$$S = 2\pi r^2 + 2\pi rh$$

$2\pi r^2 + 2\pi rh = S$ (We interchanged the two sides.)

$2\pi rh = S - 2\pi r^2$ (We subtracted $2\pi r^2$ from both sides to isolate the term containing h on the left side.)

$$h = \frac{S - 2\pi r^2}{2\pi r} \qquad \text{(We divided both sides by } 2\pi r.\text{)}$$

Because the radius of a cylinder must be positive, the denominator $2\pi r$ is nonzero.

Figure 1

r—
area of top πr^2

lateral surface area $2\pi rh$

h

area of bottom πr^2

PROBLEM SET 1

In problems 1–6, solve each equation mentally for x.

1. $3x + 6 = 0$
2. $5x = -4$
3. $6x - 8 = 0$
4. $\frac{1}{2}x + 2 = 0$
5. $\frac{2}{3}x - 3 = 0$
6. $cx = d, \quad c \neq 0$

[c] In problems 7–10, solve each equation with the aid of a calculator. Round off all answers to the correct number of significant digits.

7. $31.02x + 47.71 = 0$
8. $2713x + (7.412 \times 10^4) = 0$
9. $0.1559x - 6.637 = 0$
10. $(3.442 \times 10^{-14})x + (2.193 \times 10^9) = 0$

In problems 11–30, solve each equation.

11. $3x - 2 = 7 + 2x$
12. $3x + 8 = 9 - 2x$
13. $2t + 3 = t + 6$
14. $-2c + 18 = 3c + 3$
15. $9 - 2y = 12 - 3y$
16. $5p + 6 = 3p + 5$
17. $3(y + 6) = y - 1$
18. $10x - 1 - 7x + 3 = 7x - 10$
19. $14 - (3x - 30) = 15x - 10$
20. $7(2n + 5) - 6(n + 8) = 7$
21. $\frac{3}{4}x - \frac{5}{2}x = -7$
22. $\frac{a}{3} + \frac{13}{6} = 3 - \frac{a}{2}$
23. $\frac{1}{2}y - \frac{2}{3}y = 7 - \frac{3}{4}y$
24. $\frac{n - 1}{3} + 3 = \frac{n + 14}{9}$
25. $\frac{5 + x}{6} - \frac{10 - x}{3} = 1$
26. $\frac{2(4x - 5)}{3} + 9 = \frac{3(x + 2)}{4} - \frac{13}{6}$
27. $(2x + 3)^2 = (2x - 1)(2x + 1)$
28. $(3x - 1)^2 - 2x(x + 1) = 7x^2 - 5x + 2$
29. $(u - 1)^2 - (u + 1)^2 = 1 - 5u$
30. $(t - 1)(2t + 3) + (t + 1)(t - 4) = 3t^2$

In problems 31 and 32, rewrite each repeating decimal as a quotient of integers.

31. (a) $0.\overline{21}$ (b) $3.41\overline{21}$ (c) $0.0\overline{39}$ (d) $-1.00\overline{17}$ (e) $0.00\overline{7}$

32. (a) $0.\overline{121}$ (b) $-3.\overline{321}$ (c) $0.1\overline{523}$ (d) $0.\overline{285714}$

In problems 33–40, solve each equation. Be sure to check for extraneous roots.

33. $\dfrac{10}{x} - 2 = \dfrac{5-x}{4x}$

34. $\dfrac{3-y}{3y} + \dfrac{1}{4} = \dfrac{1}{2y}$

35. $\dfrac{t}{t+4} = \dfrac{1}{2}$

36. $\dfrac{u-5}{u+5} + \dfrac{u+15}{u-5} = \dfrac{25}{25-u^2} + 2$

37. $\dfrac{2}{x-2} + \dfrac{1}{x+1} = \dfrac{1}{(x-2)(x+1)}$

38. $\dfrac{2n}{n+7} - 1 = \dfrac{n}{n+3} + \dfrac{1}{(n+7)(n+3)}$

39. $\dfrac{1}{y-3} - \dfrac{1}{3-y} = \dfrac{1}{y^2-9}$

40. $\dfrac{5}{y-1} + \dfrac{1}{4-3y} = \dfrac{3}{6y-8}$

In problems 41–50, solve each literal equation for the indicated unknown. Be careful not to divide by zero.

41. $5(2x + a) = bx - c$ for x

42. $7(2t + 5a) - 6(t + b) = 3a$ for t

43. $\dfrac{ax+b}{c} = d + \dfrac{x}{4c}$ for x, if $c \neq 0$

44. $\dfrac{y-3a}{b} = \dfrac{2a}{b} + y$ for y, if $b \neq 0$

45. $\dfrac{x}{m} - \dfrac{a-x}{m} = d$ for x, if $m \neq 0$

46. $\dfrac{3ap-2b}{3b} - \dfrac{ap-a}{2b} = \dfrac{ap}{b} - \dfrac{2}{3}$ for p, if $b \neq 0$

47. $\dfrac{mn}{x} - bc = d + \dfrac{1}{x}$ for x

48. $\dfrac{1}{a} + \dfrac{a}{a+x} = \dfrac{a+x}{ax}$ for x, if $a \neq 0$

49. $\dfrac{2x}{x-b} = 3 - \dfrac{x-b}{x}$ for x

50. $\dfrac{x-2r}{25+x} + \dfrac{x+2r}{25-x} = \dfrac{4rs}{625-x^2}$ for x

In problems 51–58, a formula used in the specified field of applied mathematics is given. In each case, solve for the indicated unknown.

51. $V = \pi r^2 h$ for h (geometry)

52. $F = \dfrac{mv^2}{r}$ for m (mechanics)

53. $A = P(1 + rt)$ for t (finance)

54. $PV = nRT$ for T (physics)

55. $\dfrac{1}{p} + \dfrac{1}{q} = \dfrac{1}{f}$ for f (optics)

56. $S = \dfrac{rl-a}{r-1}$ for r (economics)

57. $\dfrac{P_1 V_1}{T_1} = \dfrac{P_2 V_2}{T_2}$ for T_2 (thermodynamics)

58. $I = \dfrac{nE}{nr+R}$ for n (electrical engineering)

In problems 59–64, determine whether each equation is a conditional equation or an identity.

59. $(4x + 3)^2 = 16x^2 + 24x + 9$

60. $\dfrac{1}{(x+1)^2} = \dfrac{x}{x^3+2x^2+x}$

61. $\sqrt{x^2} = x$

62. $\sqrt{1+x^2} = 1 + x$

63. $\dfrac{1}{x} + \dfrac{1}{2} = \dfrac{2}{x+2}$

64. $\dfrac{1-x}{x^2-1} + \dfrac{x}{x+1} = \dfrac{x-1}{x+1}$

In problems 65–69, determine whether the given equations are equivalent. Give reasons for your answers.

65. $x = 6$ and $x^2 = 36$

66. $(x-1)(x+2) = x^2$ and $(x-1)(x+2)x = x^3$

67. $x = 3$ and $x^3 = 27$

68. $\dfrac{t^2-1}{t+1} = t$ and $t^2 - 1 = t(t+1)$

69. $x^2 = 1$ and $x^3 = 1$

70. For what value of the constant a is $x = -1$ a solution of $2(ax + 2) - ax = 1$?

71. Check the solution of Example 1 in Section 1.2.

72. The formula $H = (A + B\sqrt{V} - CV)(S - T)$ gives the heat loss (wind chill) H in Btu's per square foot of skin per hour if the air temperature is T degrees Fahrenheit and the wind speed is V miles per hour. Here $S = 91.4°F$ represents neutral skin temperature, and A, B, and C are constants determined experimentally to be $A = 2.14$, $B = 1.37$, and $C = 0.0916$. The *equivalent temperature* (**wind chill index**) is defined to be the air temperature T_E degrees Fahrenheit at which the same heat loss would occur if the wind speed were 4 miles per hour (the speed of a brisk walk). Thus,

$$(A + B\sqrt{V} - CV)(S - T) = [A + B\sqrt{4} - C(4)](S - T_E).$$

(a) By solving the last equation, find a formula for T_E in terms of T, V, S, A, B, and C.

$\boxed{c}$ (b) Find the equivalent temperature T_E if $T = 25°F$ and $V = 20$ miles per hour.

$\boxed{c}$ In problems 73 and 74, solve each equation with the aid of a calculator. Round off your answers to the correct number of significant digits.

73. $2.72x + 2.24 = 2.45x - 2.65$

74. $(6.86 \times 10^{-5})w - (7.14 \times 10^9) = (7.28 \times 10^{-5})w + (1.05 \times 10^9)$

---

2 Applications Involving First-Degree Equations

Questions that arise in the real world are usually expressed in words, rather than in mathematical symbols. For example: "What will be the monthly payment on my mortgage?" "How much insulation must I use in my house?" "What course should I fly to Boston?" "How safe is this new product?" In order to answer such questions, it is necessary to have certain pertinent information. For instance, to determine the monthly payment on a mortgage, you need to know the amount of the mortgage, the interest rate, and the time period involved.

Problems in which a question is asked and pertinent information is supplied in the form of words are called "word problems" or "story problems" by students and teachers alike. In this section, we study word problems that can be worked by setting up an equation containing the unknown and solving it by the methods illustrated in Section 1. For working these problems, we recommend the following systematic procedure:

Step 1. Begin by reading the problem carefully, several times if necessary, until you understand it well. Draw a diagram whenever possible. Look for the question or questions you are to answer.

Step 2. List all of the unknown numerical quantities involved in the problem. It may be useful to arrange these quantities in a table or chart along with related known quantities. Select one of the unknown quantities in your list, one that seems to play a prominent role in the problem, and call it x. (Of course, any other letter will do as well.)

Step 3. Using information given or implied in the wording of the problem, write algebraic relationships among the numerical quantities listed in step 2.

Relationships that express some of these quantities in terms of x are especially useful. Reread the problem, sentence by sentence, to make sure you have rewritten all the given information in algebraic form.

Step 4. Combine the algebraic relationships written in step 3 into a single equation containing only x and known numerical constants.

Step 5. Solve the equation for x. Use this value of x to answer the question or questions in step 1.

Step 6. Check your answer to make certain that it agrees with the facts in the problem.

Of course, a calculator is often useful to expedite arithmetic.

Example 1   One number is 15 less than a second number. Three times the first number added to twice the second number is 80. Find the two numbers.

Solution   We follow the procedure just outlined.

(1) Question: What are the two numbers?

(2) Unknown quantities: *The first number* and *the second number*. Let $x =$ *the first number*. (See the alternative solution below, where we choose x to represent the second number.)

(3) Information given:
(i) *The first number* = *the second number* − 15, that is,

$$x = \textit{the second number} - 15.$$

(ii) 3(*the first number*) + 2(*the second number*) = 80, that is,

$$3x + 2(\textit{the second number}) = 80.$$

(4) From relationship (i) in step (3), we have

$$\textit{the second number} = x + 15.$$

Therefore, relationship (ii) can be written as

$$3x + 2(x + 15) = 80.$$

(5) Solving the equation $3x + 2(x + 15) = 80$, we obtain

$$3x + 2x + 30 = 80$$
$$5x = 50$$
$$x = 10.$$

Therefore

$$\textit{the first number} = x = 10$$

and

$$\textit{the second number} = x + 15 = 10 + 15 = 25.$$

(6) *Check:* Indeed, 10 is 15 less than 25 and $3(10) + 2(25) = 80$.

Alternative Solution

In step (2) above, we could have let $x = $ *the second number.* With this assignment of the variable, relationship (i) in step (3) becomes *the first number* $= x - 15$, and relationship (ii) becomes $3(x - 15) + 2x = 80$, or $5x = 125$. Hence, *the second number* $= x = 25$ and *the first number* $= x - 15 = 10$.

Example 2

A suit is on sale for $195. What was the original price of the suit if it has been discounted 25%?

Solution

(1) Question: What was the original price of the suit?

(2) Unknown quantities: The *original price of the suit* and the *amount of the discount in dollars.* Let $x = $ *the original price.*

(3) (*original price*) $-$ *discount* $=$ sale price $= 195$ dollars, that is,

$$x - discount = 195.$$

Discount $= 25\%$ of *original price* $= 0.25x$.

(4) $x - 0.25x = 195$

(5) $0.75x = 195$

$$x = \frac{195}{0.75}$$

$$= 260.$$

The original price of the suit was $260.

(6) *Check:* If a $260 suit is discounted by 25%, the discount is $(0.25)(\$260) = \$65$ and the sale price is $260 - \$65 = \$195$.

Example 3

A businesswoman has invested a total of $30,000 in two certificates. The first certificate pays 10.5% annual simple interest, and the second pays 9% annual simple interest. At the end of one year, her combined interest on the two certificates is $2970. How much did she originally invest in each certificate?

Solution

In solving this problem, we must use the simple interest formula $I = Prt$, where I is the **simple interest,** P is the **principal** (the amount invested), r is the **rate** of interest per period, and t is the number of periods. Here, the period is $t = 1$ year, so that $I = Pr$.

(1) Question: What was the principal for each of the two certificates?

(2) Let $x = $ *the principal for the first certificate* in dollars. Quantities involved in the problem appear in the following table:

Certificate	Principal	Rate	Time	Simple Interest
First	x dollars	0.105	1	$0.105x$ dollars
Second	$30{,}000 - x$ dollars	0.09	1	$0.09(30{,}000 - x)$ dollars

(3) Most of the information given in the problem appears in the table. The only remaining fact is that

the combined simple interest = 2970 dollars.

(4) Because the sum of the simple interest on the two certificates is the combined simple interest,

$$0.105x + 0.09(30,000 - x) = 2970.$$

(5) $\quad 0.105x + 0.09(30,000 - x) = 2970$
$$0.105x + 2700 - 0.09x = 2970$$
$$0.105x - 0.09x = 2970 - 2700$$
$$0.015x = 270$$
$$15x = 270,000$$
$$x = \frac{270,000}{15} = 18,000.$$

Therefore, $18,000 was invested in the first certificate and

$$\$30,000 - \$18,000 = \$12,000$$

was invested in the second.

(6) *Check:* The simple interest on $18,000 for one year at 10.5% is 0.105($18,000) = $1890. The simple interest on $12,000 for one year at 9% is 0.09($12,000) = $1080. The total amount invested is

$$\$18,000 + \$12,000 = \$30,000$$

and the combined interest is $1890 + $1080 = $2970.

Many word problems involving mixtures of substances or items can be worked by solving first-degree equations. The following two examples illustrate how to solve typical mixture problems.

Example 1   A chemist has one solution containing a 10% concentration of acid and a second solution containing a 15% concentration of acid. How many milliliters of each should be mixed in order to obtain 10 milliliters of a solution containing a 12% concentration of acid?

Solution   Let x = *the number of milliliters of the first solution,* so that $10 - x$ = *the number of milliliters of the second solution.* The following table summarizes the given information:

	Milliliters of Solution	Acid Concentration	Milliliters of Acid in Solution
First Solution	x	0.10	$0.10x$
Second Solution	$10 - x$	0.15	$0.15(10 - x)$
Mixture	10	0.12	$0.12(10) = 1.2$

Since the amount of acid in the mixture is the sum of the amounts in the two solutions,

$$0.10x + 0.15(10 - x) = 1.2$$
$$0.10x + 1.5 - 0.15x = 1.2$$
$$0.10x - 0.15x = 1.2 - 1.5$$
$$-0.05x = -0.3$$
$$5x = 30$$
$$x = 6.$$

Hence, 6 milliliters of the first solution and $10 - 6 = 4$ milliliters of the second solution should be mixed.

Check: In 6 milliliters of the first solution there is $(0.10)6 = 0.6$ milliliter of acid. In 4 milliliters of the second solution there is $(0.15)4 = 0.6$ milliliter of acid. Thus, there are $0.6 + 0.6 = 1.2$ milliliters of acid in the $6 + 4 = 10$ milliliters of the mixture. Therefore, the acid concentration of the mixture is $\frac{1.2}{10} = 0.12 = 12\%$.

Example 2   A vending machine for chewing gum accepts nickels, dimes, and quarters. When the coin box is emptied, the total value of the coins is found to be $24.15. Find the number of coins of each kind in the box if there are twice as many nickels as quarters and five more dimes than nickels.

Solution   Let $n =$ *the number of nickels.* The following table summarizes the given information:

	Number of Coins	Individual Value	Total Value in Dollars
Nickels	n	$0.05	$0.05n$
Dimes	$n + 5$	$0.10	$0.10(n + 5)$
Quarters	$\frac{1}{2}n$	$0.25	$0.25(\frac{1}{2}n)$

Since the coins have a total value of $24.15,

$$0.05n + 0.10(n + 5) + 0.25(\tfrac{1}{2}n) = 24.15.$$

To solve this equation, we begin by multiplying both sides by 100 to remove the decimals:

$$5n + 10(n + 5) + 25(\tfrac{1}{2}n) = 2415$$
$$15n + \frac{25}{2}n = 2415 - 50 = 2365$$
$$30n + 25n = 4730$$
$$55n = 4730$$
$$n = \frac{4730}{55} = 86.$$

Therefore, there are 86 nickels, $86 + 5 = 91$ dimes, and $\frac{1}{2}(86) = 43$ quarters.

Check: $0.05(86) + \$0.10(91) + \$0.25(43) = \$24.15$.

Another type of applied problem involves objects that move a distance d at a constant rate r (also called speed) in t units of time. To solve these problems, use the formula

$$d = rt \qquad (distance = rate \times time).$$

This formula can be rewritten as $r = d/t$ or as $t = d/r$.

Example

A jogger and a bicycle rider leave a field house at the same time and set out for a nearby town. The jogger runs at a constant speed of 16 kilometers per hour. At the end of 2 hours, the bicycle rider is 19.2 kilometers ahead of the jogger. How fast is the bicycle rider traveling, assuming that her speed is constant?

Solution

Let $x =$ *the speed of the bicycle rider.* The following table summarizes the given information:

	Rate	Time	Rate × Time = Distance
Jogger	16 km/hr	2 hr	$2(16) = 32$ km
Bicyclist	x km/hr	2 hr	$2x$ km

Since the distance the bicyclist has traveled is 19.2 kilometers more than the jogger has run during the 2 hours,

$$2x = 32 + 19.2$$
$$2x = 51.2$$
$$x = 25.6.$$

Therefore, the speed of the bicyclist is 25.6 kilometers per hour.

Check: $2(25.6) = 51.2 = 32 + 19.2$.

Problems concerning a job that is done at a constant rate can often be solved by using the following principle: *If a job can be done in time t, then $1/t$ of the job can be done in one unit of time.*

Example

At a factory, smokestack A pollutes the air 1.25 times as fast as smokestack B. How long would it take smokestack B, operating alone, to pollute the air by as much as both smokestacks do in 20 hours?

Solution

Here the job in question is polluting the air by as much as both smokestacks do in 20 hours. Let t be the time in hours required for smokestack B operating alone to do this job. Then $1/t$ of the job is done by smokestack B in one hour, and $1.25(1/t)$

of the job is done by smokestack A in one hour. The two smokestacks together accomplish

$$\frac{1}{t} + 1.25\left(\frac{1}{t}\right) = \frac{2.25}{t}$$

of the job in one hour. We also know that both smokestacks accomplish $\frac{1}{20}$ of the job in one hour. Therefore,

$$\frac{2.25}{t} = \frac{1}{20}.$$

Solving this equation, we find that

$$20t\frac{2.25}{t} = 20t\frac{1}{20}$$

$$20(2.25) = t$$

$$45 = t.$$

Therefore, it requires 45 hours for smokestack B to do the job alone.

Check: Is it true that $\frac{1}{45} + 1.25(\frac{1}{45}) = \frac{1}{20}$? Yes.

PROBLEM SET 2

C In many of the following problems, a calculator may be useful to expedite the arithmetic.

1. The difference between two numbers is 12. If 2 is added to seven times the smaller number, the result is the same as if 2 is subtracted from three times the larger. Find the numbers.

2. Psychologists define the intelligence quotient (IQ) of a person to be 100 times the person's mental age divided by the chronological age. What is the chronological age of a person with an IQ of 150 and a mental age of 18?

3. At the end of the model year, a car dealer advertises that the list prices on all of last year's models have been discounted by 20%. What was the original list price of a car that has a discounted price of $6800?

4. A retail outlet sells wood-burning stoves for $675. At this price, the profit on the stove is one-third of its cost to the retailer. What is the amount of the retailer's profit on the stove? [Hint: cost + profit = selling price.]

5. A person invests part of $75,000 in a certificate that yields 8.5% simple annual interest, and the rest in a certificate that yields 9.2% simple annual interest. At the end of the year, the combined interest on the two certificates is $6606. How much was invested in each certificate?

6. To reduce their income tax, a couple invests a total of $140,000 in two municipal bonds, one that pays 6% tax-free simple annual interest and one that pays 6.5%. The total non-taxable income from both investments at the end of one year is $8775. How much was invested in each bond?

7. A family takes advantage of a state income tax credit of 15% of the cost of installing solar-heating equipment and 8% of the cost of upgrading insulation in their home. After spending a total of $6510 on insulation and solar heating, the family receives a state income tax credit of $854. How much was spent on solar heating and how much on insulation?

8. The manager of a trust fund invests $210,000 in three enterprises. She invests three times as much at 8% as she does at 9%, and commits the rest at 10%. Her total annual income from the three investments will be $17,850. How much does she invest at each rate?

9. A factory pays time and a half for all hours worked above 40 hours per week. An employee who makes $7.30 per hour grossed $478.15 in one week. How many hours did the employee work that week?

10. Because of inflation, the price of units in a condominium increases by 3% in April. The price increases again, by 2%, in July. What was the price of a unit before the April increase if its price after the July increase is $78,795?

11. A petroleum distributor has two gasohol storage tanks, the first containing 9% alcohol and the second containing 12% alcohol. An order is received for 300,000 gallons of gasohol containing 10% alcohol. How can this order be filled by mixing gasohol from the two storage tanks?

12. The cooling system of an automobile engine holds 16 liters of fluid. The system is filled with a mixture of 80% water and 20% antifreeze. It is necessary to increase the amount of antifreeze to 40%. This is to be done by draining some of the mixture and adding pure antifreeze to bring the total amount of fluid back up to 16 liters. How much of the mixture should be drained?

13. To generate hydrogen in a chemistry laboratory, a 40% solution of sulfuric acid is needed. How many milliliters of water must be mixed with 25 milliliters of an 88% solution of sulfuric acid to dilute it to the required 40% of acid?

14. At a certain factory, twice as many men as women apply for work. If 5% of the people who apply are hired and 3% of the men who apply are hired, what percentage of the women who apply are hired?

15. A bill of $7.45 was paid with 32 coins: half dollars, quarters, dimes, and nickels. If there were seven more dimes than nickels and twice as many dimes as quarters, how many coins of each type were used?

16. Three brothers, Joe, Jamal, and Gus, decided to contribute toward a present for their mother. Their father agreed to match their combined contribution and purchase the gift. Joe's contribution was entirely in nickels, Jamal's was in dimes, and Gus's was in quarters. Jamal contributed three times as many coins as Gus, and Joe contributed 10 more coins than Jamal. The gift cost $10.80. How much did each person contribute?

17. A cross-country skier starts from a certain point and travels at a constant speed of 4 kilometers per hour. A snowmobile starts from the same point 45 minutes later, follows the skier's tracks at a constant speed, and catches up to the skier in 10 minutes. What is the speed of the snowmobile in kilometers per hour?

18. Two airplanes leave an airport at the same time and travel in opposite directions. One plane is traveling 64 kilometers per hour faster than the other. After 2 hours, they are 3200 kilometers apart. How fast is each plane traveling?

19. A jogger takes 3.5 minutes to run the same distance that a second jogger can run in 3 minutes. What is this distance, if the second jogger runs 2 feet per second faster than the first jogger?

20. A driver plans to average 50 miles per hour on a trip from A to B. Her average speed for the first half of the distance from A to B is 45 miles per hour. How fast must she drive for the rest of the way?

21. Factory A pollutes a lake twice as fast as factory B. The two factories operating together emit a certain amount of pollutant in 18 hours. How long would it take for factory A, operating alone, to produce the same amount of pollutant?

22. Student activists at Curmudgeon College have planned to distribute leaflets on the campus. The leaflets were to be run off on two machines, one electrically driven and the other operated by a hand crank. The electric machine produces copies four times as fast as the hand-cranked machine. The students figured that with both machines operating together they could run off the number of leaflets they needed in a total of 1.5 hours. However, because of a power blackout, they can use only the hand-cranked machine. How long will it take to run off the leaflets?

23. A computer can do a biweekly payroll in 10 hours. If a second computer is added, the two computers can do the job in only 3 hours. How long would it take the second computer alone to do the payroll?

24. A solar collector can generate 50 Btu's in 8 minutes. (One Btu is the amount of energy necessary to raise the temperature of one pound of water by one degree Fahrenheit.) A second solar collector can generate 50 Btu's in 5 minutes. The first collector is operated by itself for 1 minute, then the second collector is activated and both operate together. How long, after the second collector is activated, will it take to generate a total of 50 Btu's?

25. The length of a rectangular playground is twice its width and the perimeter is 900 meters. Find the length and width of the playground.

26. The circumference of the earth at the equator is 1.315×10^8 feet. Suppose that a steel belt is fitted tightly around the equator. If an additional 10 feet is added onto this belt and the slack is uniformly distributed around the earth, would you be able to crawl under the belt?

27. The Fahrenheit temperature corresponding to a particular Celsius temperature can be found by adding 32 to $\frac{9}{5}$ of the Celsius temperature. Find the temperature at which the reading is the same on both the Fahrenheit and the Celsius scales.

28. A commuter is picked up by her husband at the train station every afternoon. The husband leaves the house at the same time every day, always drives at the same speed, and regularly arrives at the station just as his wife's train pulls in. One day she takes a different train and arrives at the station one hour earlier than usual. She starts immediately to walk home at a constant speed. Her husband sees her along the road, picks her up, and drives her the rest of the way home. They arrive there 10 minutes earlier than usual. How many minutes did she spend walking?

29. In problem 27, when is the Celsius reading three times the Fahrenheit reading?

30. In problem 28, if the wife walks 4 miles an hour, how fast does the husband drive?

31. The primary (P) wave of an earthquake travels 1.7 times faster than the secondary (S) wave. Assuming that the S wave travels at 275 kilometers per minute and that the P wave is recorded at a seismic station 5.07 minutes before the S wave, how far from the station was the earthquake?

3   Second-Degree or Quadratic Equations

A **second-degree** or **quadratic equation** in x is written in standard form as

$$ax^2 + bx + c = 0, \quad \text{with } a \neq 0.$$

We discuss three methods for solving quadratic equations: **factoring, completing the square,** and using the **quadratic formula.**

3.1 Solution by Factoring

When a quadratic equation is in standard form, it may be possible to factor its left side as a product of two first-degree polynomials. The equation can then be solved by setting each factor equal to zero and solving the resulting first-degree equations. This procedure is justified by the fact that a product of real numbers is zero if and only if at least one of the factors is zero. (See the zero-factor properties on page 4.)

Examples   Solve each equation by factoring.

1 $15x^2 + 14x = 8.$

Solution   We begin by subtracting 8 from both sides of the equation to change it into standard form

$$15x^2 + 14x - 8 = 0.$$

Factoring the polynomial on the left, we obtain

$$(3x + 4)(5x - 2) = 0.$$

Now we set each factor equal to zero and solve the resulting first-degree equations:

$$3x + 4 = 0 \qquad\qquad 5x - 2 = 0$$
$$x = -\tfrac{4}{3} \qquad\qquad x = \tfrac{2}{5}$$

Therefore, the roots are $-\tfrac{4}{3}$ and $\tfrac{2}{5}$.

2 $\dfrac{x - 6}{3x + 4} - \dfrac{2x - 3}{x + 2} = 0.$

Solution   We begin by multiplying both sides of the equation by $(3x + 4)(x + 2)$, the L.C.D. of the two fractions:

$$(3\cancel{x + 4})(x + 2)\frac{x - 6}{3\cancel{x + 4}} - (3x + 4)(\cancel{x + 2})\frac{2x - 3}{\cancel{x + 2}} = 0$$

$$(x + 2)(x - 6) - (3x + 4)(2x - 3) = 0$$

$$x^2 - 4x - 12 - (6x^2 - x - 12) = 0$$

$$x^2 - 4x - 12 - 6x^2 + x + 12 = 0$$

$$-5x^2 - 3x = 0$$

$$5x^2 + 3x = 0.$$

Factoring the left side of the last equation and setting the factors equal to zero, we obtain:

$$x(5x + 3) = 0$$

$$x = 0 \qquad\qquad 5x + 3 = 0$$
$$x = -\tfrac{3}{5}$$

Because we multiplied both sides of the original equation by the expression $(3x + 4)(x + 2)$, which contains the unknown, we must check to be sure that $x = 0$ and $x = -\tfrac{3}{5}$ aren't extraneous roots. Substituting $x = 0$ in the original equation, we obtain

$$\frac{0 - 6}{3(0) + 4} - \frac{2(0) - 3}{0 + 2} = \frac{-6}{4} - \frac{-3}{2} = -\frac{3}{2} + \frac{3}{2} = 0,$$

so $x = 0$ is a solution. Substituting $x = -\frac{3}{5}$, we have

$$\frac{(-\frac{3}{5}) - 6}{3(-\frac{3}{5}) + 4} - \frac{2(-\frac{3}{5}) - 3}{(-\frac{3}{5}) + 2} = \frac{-33}{11} - \frac{-21}{7} = -3 + 3 = 0,$$

so $x = -\frac{3}{5}$ is also a solution. Hence, the roots are $-\frac{3}{5}$ and 0.

3 $x^2 - 4x + 4 = 0.$

Solution   Factoring, and setting the factors equal to zero, we have

$$x^2 - 4x + 4 = 0$$

$$(x - 2)(x - 2) = 0$$

$x - 2 = 0$	$x - 2 = 0$
$x = 2$	$x = 2$

Therefore, there is only one solution, namely, $x = 2$.

When the two first-degree polynomials obtained by factoring have the same root, as in Example 3 above, we call the result a **double root.**

3.2 Solution by Completing the Square

A second method for solving quadratic equations is based on the idea of a **perfect square**—a polynomial that is the square of another polynomial. For example, $x^2 + 6x + 9$ is a perfect square because it is the square of $x + 3$. The polynomial $x^2 + 6x$ isn't a perfect square, but if we add 9 to it, we will get the perfect square $x^2 + 6x + 9$.

More generally, by adding $(k/2)^2$ to $x^2 + kx$ we obtain a perfect square:

$$x^2 + kx + \left(\frac{k}{2}\right)^2 = \left(x + \frac{k}{2}\right)^2.$$

The process of creating a perfect square from an expression of the form $x^2 + kx$ by adding the square of half the coefficient of x is called **completing the square.** Be careful—this procedure can be used *only when the coefficient of x^2 is 1.*

Example   Complete the square by adding a constant to each expression.

(a) $x^2 + 8x$ (b) $x^2 - 4x$ (c) $x^2 - 3x$

Solution   (a) $x^2 + 8x + \left(\frac{8}{2}\right)^2 = x^2 + 8x + 16 = (x + 4)^2$

(b) $x^2 - 4x + \left(\frac{-4}{2}\right)^2 = x^2 - 4x + 4 = (x - 2)^2$

(c) $x^2 - 3x + \left(\dfrac{-3}{2}\right)^2 = x^2 - 3x + \dfrac{9}{4} = \left(x - \dfrac{3}{2}\right)^2$

The following example illustrates how to solve a quadratic equation by completing the square.

Example   Solve the equation $2x^2 - 6x - 5 = 0$.

Solution   We begin by adding 5 to both sides:

$$2x^2 - 6x = 5.$$

Next, we divide both sides of the equation by 2 so that the coefficient of x^2 will be 1:

$$x^2 - 3x = \tfrac{5}{2}.$$

Now we can complete the square for the expression on the left by adding $\left(\dfrac{-3}{2}\right)^2 = \dfrac{9}{4}$ to both sides of the equation:

$$x^2 - 3x + \dfrac{9}{4} = \dfrac{5}{2} + \dfrac{9}{4}$$

$$\left(x - \dfrac{3}{2}\right)^2 = \dfrac{19}{4}.$$

It follows that $x - \dfrac{3}{2}$ is a square root of $\dfrac{19}{4}$. But there are two square roots of $\dfrac{19}{4}$, the principal square root, $\sqrt{\dfrac{19}{4}} = \dfrac{\sqrt{19}}{2}$, and its negative, $-\sqrt{\dfrac{19}{4}} = -\dfrac{\sqrt{19}}{2}$. Therefore,

$$x - \dfrac{3}{2} = \dfrac{\sqrt{19}}{2} \quad \text{or else} \quad x - \dfrac{3}{2} = -\dfrac{\sqrt{19}}{2},$$

that is,

$$x = \dfrac{3}{2} + \dfrac{\sqrt{19}}{2} = \dfrac{3 + \sqrt{19}}{2} \quad \text{or else} \quad x = \dfrac{3}{2} - \dfrac{\sqrt{19}}{2} = \dfrac{3 - \sqrt{19}}{2}.$$

These two solutions can be written in the compact form

$$x = \dfrac{3 \pm \sqrt{19}}{2}.$$

3.3 The Quadratic Formula

Let's apply the method of completing the square to solve for the unknown x in the literal equation

$$ax^2 + bx + c = 0, \qquad \text{with } a \neq 0.$$

First, we subtract c from both sides:

$$ax^2 + bx = -c.$$

Then, to prepare for completing the square, we divide both sides by a:

$$x^2 + \frac{b}{a}x = -\frac{c}{a}.$$

To complete the square, we add $\left[\frac{1}{2}\left(\frac{b}{a}\right)\right]^2 = \frac{b^2}{4a^2}$ to both sides:

$$x^2 + \frac{b}{a}x + \frac{b^2}{4a^2} = \frac{b^2}{4a^2} - \frac{c}{a}$$

$$\left(x + \frac{b}{2a}\right)^2 = \frac{b^2 - 4ac}{4a^2}.$$

Now, provided that $b^2 - 4ac \geq 0$,

$$x + \frac{b}{2a} = \pm\sqrt{\frac{b^2 - 4ac}{4a^2}}$$

$$x = -\frac{b}{2a} \pm \frac{\sqrt{b^2 - 4ac}}{2a}.$$

Therefore, the roots of the quadratic equation $ax^2 + bx + c = 0$ can be found by using the **quadratic formula**:

$$x = \frac{-b \pm \sqrt{b^2 - 4ac}}{2a}.$$

Although the method of factoring is the quickest way to solve a quadratic equation when the factors are easily recognized, *we recommend using the quadratic formula in all other cases.* Quadratic equations are rarely solved by completing the square—this method was introduced primarily to show how the quadratic formula is derived.

Example 1   Use the quadratic formula to solve the equation $2x^2 - 5x + 1 = 0$.

Solution   The equation is in the standard form $ax^2 + bx + c = 0$, with $a = 2$, $b = -5$, and $c = 1$. Substituting these values into the quadratic formula, we obtain

$$x = \frac{-b \pm \sqrt{b^2 - 4ac}}{2a} = \frac{-(-5) \pm \sqrt{(-5)^2 - 4(2)(1)}}{2(2)}$$

$$= \frac{5 \pm \sqrt{25 - 8}}{4} = \frac{5 \pm \sqrt{17}}{4}.$$

In other words, the two roots are $\dfrac{5 - \sqrt{17}}{4}$ and $\dfrac{5 + \sqrt{17}}{4}$.

Example 2   [c] Using the quadratic formula and a calculator, find the roots of the quadratic equation $-1.32x^2 + 2.78x + 9.37 = 0$. Round off the answers to two decimal places.

Solution   According to the quadratic formula, the roots are

$$x = \frac{-2.78 - \sqrt{(2.78)^2 - 4(-1.32)(9.37)}}{2(-1.32)} \approx 3.92$$

and

$$x = \frac{-2.78 + \sqrt{(2.78)^2 - 4(-1.32)(9.37)}}{2(-1.32)} \approx -1.81.$$

     Quadratic equations have many applications in the sciences, business, economics, medicine, and engineering.

Example   A telephone company is placing telephone poles along a road. If the distance between successive poles were increased by 10 meters, 5 fewer poles per kilometer would be required. How many telephone poles is the company now placing along each kilometer of the road?

Solution   Let x be the number of poles now being placed along each kilometer. Then the distance between successive poles is

$$\frac{1}{x} \text{ kilometer} = \frac{1000}{x} \text{ meters.}$$

If the distance between poles were increased by 10 meters, then the (new) number of poles per kilometer would be $x - 5$. Therefore, the (new) distance between poles would be

$$\frac{1}{x - 5} \text{ kilometer} = \frac{1000}{x - 5} \text{ meters.}$$

We know that this new distance between poles would be 10 meters more than the current distance, so the new distance would also be given by $(1000/x) + 10$ meters. Therefore,

$$\frac{1000}{x - 5} = \frac{1000}{x} + 10.$$

We multiply both sides of this equation by the L.C.D., $x(x - 5)$, and simplify:

$$x(x-5)\frac{1000}{x-5} = x(x - 5)\frac{1000}{x} + 10x(x - 5)$$

$$1000x = 1000x - 5000 + 10x^2 - 50x$$

$$10x^2 - 50x - 5000 = 0$$

$$x^2 - 5x - 500 = 0.$$

Using the quadratic formula, with $a = 1$, $b = -5$, and $c = -500$, we find that

$$x = \frac{-b \pm \sqrt{b^2 - 4ac}}{2a} = \frac{-(-5) \pm \sqrt{(-5)^2 - 4(1)(-500)}}{2(1)}$$

$$= \frac{5 \pm \sqrt{25 + 2000}}{2} = \frac{5 \pm \sqrt{2025}}{2} = \frac{5 \pm 45}{2}.$$

The two solutions of the quadratic equation are

$$x = \frac{5 - 45}{2} = -20 \quad \text{and} \quad x = \frac{5 + 45}{2} = 25.$$

Since we cannot have a negative number of poles per kilometer, only the second solution has meaning, so we conclude that 25 poles are currently being placed per kilometer.

3.4 The Discriminant and Complex Roots

The expression $b^2 - 4ac$, which appears under the radical sign in the quadratic formula

$$x = \frac{-b \pm \sqrt{b^2 - 4ac}}{2a},$$

is called the **discriminant** of the quadratic equation

$$ax^2 + bx + c = 0.$$

We can use the algebraic sign of the discriminant to determine the number and the nature of the roots of the quadratic equation.

Case 1. $b^2 - 4ac > 0$. In this case, the quadratic equation has *two real and unequal roots*,

$$x = \frac{-b - \sqrt{b^2 - 4ac}}{2a} \quad \text{and} \quad x = \frac{-b + \sqrt{b^2 - 4ac}}{2a}.$$

(See Examples 1 and 2 in Section 3.3.)

Case 2. $b^2 - 4ac = 0$. In this case, the quadratic equation has *only one root, a double root*, $x = \dfrac{-b}{2a}$. (See Example 3 in Section 3.1.)

Case 3. $b^2 - 4ac < 0$. In this case, the quadratic equation has *no real root*. [However, as we shall soon see, it does have two nonreal "complex roots."]

Example   Use the discriminant to determine the nature of the roots of each quadratic equation without actually solving it.
(a) $5x^2 - x - 3 = 0$
(b) $9x^2 + 42x + 49 = 0$
(c) $x^2 - x + 1 = 0$

Solution   (a) Here $a = 5$, $b = -1$, $c = -3$, and $b^2 - 4ac = (-1)^2 - 4(5)(-3) = 61 > 0$; hence, by case 1, there are two unequal real roots.

(b) Here $a = 9$, $b = 42$, $c = 49$, and $b^2 - 4ac = 42^2 - 4(9)(49) = 0$; hence, by case 2, the equation has just one root—a double root—and this root is a real number.

(c) Here $a = 1$, $b = -1$, $c = 1$, and $b^2 - 4ac = (-1)^2 - 4(1)(1) = -3 < 0$; hence, by case 3, the equation has no real roots. (As we shall see, it has two "complex conjugate" roots.)

In order to deal with case 3, in which the discriminant is negative, we use the so-called **imaginary number**

$$i = \sqrt{-1},$$

which has the property that $i^2 = -1$. Of course, there is no real number that has a negative square. The symbol i or $\sqrt{-1}$ represents an "invented number" with a wide variety of practical uses in fields ranging from electrical engineering to statistics. By writing an expression of the form

$$a + bi,$$

where a and b are real numbers, we obtain a more general type of "number" called a **complex number.** Two complex numbers of the form

$$a + bi \quad \text{and} \quad a - bi$$

are called **conjugates** of each other. Complex numbers are discussed in more detail in Section 1 of the final chapter.

Using i, we can define the **principal square root** of a negative number as follows:

$$\text{If } c > 0, \quad \sqrt{-c} = \sqrt{c(-1)} = \sqrt{c}\,i.$$

For instance, $\sqrt{-3} = \sqrt{3}\,i$ and $\sqrt{-4} = \sqrt{4}\,i = 2i$. This allows us to write the roots of a quadratic equation with a negative discriminant as a pair of complex numbers that are conjugates of each other.

Example Find the roots of the quadratic equation $x^2 - x + 1 = 0$.

Solution Using the quadratic formula, with $a = 1$, $b = -1$, and $c = 1$, we have

$$x = \frac{-(-1) \pm \sqrt{(-1)^2 - 4(1)(1)}}{2(1)}$$

$$= \frac{1 \pm \sqrt{-3}}{2}$$

$$= \frac{1 \pm \sqrt{3}\,i}{2} = \frac{1}{2} \pm \frac{\sqrt{3}}{2}\,i.$$

Thus, the roots are the complex conjugates

$$\frac{1}{2} - \frac{\sqrt{3}}{2}\,i \quad \text{and} \quad \frac{1}{2} + \frac{\sqrt{3}}{2}\,i.$$

A more detailed and more general discussion of complex roots of polynomial equations can also be found in the final chapter.

PROBLEM SET 3

In problems 1–24, solve each equation by factoring.

1. $x^2 - 7x = 0$

2. $2x^2 - 5x = 0$

3. $3t^2 - 48 = 0$

4. $9 - y^2 = 2y^2$

5. $x^2 + 2x = 3$

6. $z^2 - 2z = 35$

7. $x^2 - 4x = 21$

8. $2t^2 + 5t = -3$

9. $2z^2 - 7z - 15 = 0$

10. $10r^2 + 19r + 6 = 0$

11. $6y^2 - 13y = 5$

12. $54z^2 - 9z = 30$

13. $15x^2 + 4 = 23x$

14. $24u^2 + 94u = 25$

15. $(8x + 19)x = 27$

16. $30(y^2 + 1) = 61y$

17. $25x - 40 + \dfrac{16}{x} = 0$

18. $36 + \dfrac{60}{s} = \dfrac{-25}{s^2}$

19. $x - 16 = \dfrac{105}{x}$

20. $\dfrac{10y + 19}{y} = \dfrac{15}{y^2}$

21. $\dfrac{5}{x + 4} - \dfrac{3}{x - 2} = 4$

22. $\dfrac{2y + 11}{2y + 8} = \dfrac{3y - 1}{y - 1}$

23. $\dfrac{2y - 5}{2y + 1} + \dfrac{6}{2y - 3} = \dfrac{7}{4}$

24. $\dfrac{t - 4}{t + 1} - \dfrac{15}{4} = \dfrac{t + 1}{t - 4}$

In problems 25–30, complete the square by adding a constant to each expression.

25. $x^2 + 6x$

26. $x^2 - 6x$

27. $x^2 - 5x$

28. $x^2 + \dfrac{b}{a}x$

29. $x^2 + \dfrac{3}{4}x$

30. $x^2 + \sqrt{2}x$

In problems 31–36, complete the square to solve each equation.

31. $x^2 + 4x - 15 = 0$

32. $x^2 - 5x - 5 = 0$

33. $x^2 + 4 = 6x$

34. $2y^2 + y = 3$

35. $3r^2 + 6r = 4$

36. $5u^2 + 9u = 3$

In problems 37–48, use the quadratic formula to solve each equation.

37. $5x^2 - 7x - 6 = 0$

38. $6x^2 - x - 2 = 0$

39. $3x^2 + 5x + 1 = 0$

40. $12y^2 + y - 1 = 0$

41. $2x^2 - x - 2 = 0$

42. $4u^2 - 11u + 3 = 0$

43. $4y^2 - 3y - 3 = 0$

44. $5x^2 + 17x - 3 = 0$

45. $9x^2 - 6x + 1 = 0$

46. $\dfrac{2}{3}x^2 - \dfrac{8}{9}x = 1$

47. $\dfrac{x + 2}{x} + \dfrac{x}{x - 2} = 5$

48. $\dfrac{7u + 4}{u^2 - 6u + 8} - \dfrac{5}{2 - u} = \dfrac{u + 5}{u - 4}$

C In problems 49–52, use the quadratic formula and a calculator to find the roots of each quadratic equation. Round off all answers to two decimal places.

49. $44.04x^2 + 64.72x - 31.23 = 0$

50. $8.85x^2 - 71.23x + 94.73 = 0$

51. $4.59x^2 - 90.29x + 118.85 = 0$

52. $-1.47x^2 + 9.06x + 6.57\pi = 0$

In problems 53–64, solve each quadratic equation by any method you wish.

53. $x^2 = 5$

54. $25y^2 - 20y + 4 = 0$

55. $4x(2x - 1) = 3$

56. $2x^2 + 7x + 6 = 0$

57. $x^2 + 6x = -9$

58. $3z^2 - 2z = 5$

59. $x^2 + 4x + 1 = 0$

60. $15 - 7u - 4u^2 = 0$

61. $9t^2 - 6t + 1 = 0$

62. $0.2x^2 - 1.2x + 1.7 = 0$

63. $z^2 + 2z = 8$

64. $8t^2 - 10t + 3 = 0$

65. The staff members of an athletic department at a college agreed to contribute equal amounts to make up a scholarship fund of $200. Since then, two new members have been added to the staff; as a result, each member's share has been reduced by $5. How many members are now on the staff?

66. A cable television company plans to begin operations in a small town. The company foresees that about 600 people will subscribe to the service if the price per subscriber is $5 per month, but that for each 5-cent increase in the monthly subscription price, 4 of the original

600 people will decide not to subscribe. The company begins operations, and its total revenue for the first month is $1500. How many people have subscribed to the cable television service?

67. A gardener sets 180 plants in rows. Each row contains the same number of plants. If there were 40 more plants in each row, the gardener would need 6 fewer rows. How many rows are there?

68. In a medical laboratory, the quantity x milligrams of antigen present during an antigen–antibody reaction is related to the time t in minutes required for the precipitation of a fixed amount of antigen–antibody by the equation

$$3t = 140 - 50x + 5x^2.$$

Find x if $t = 20$ minutes.

69. In order to support a solar collector at the correct angle, the roof trusses for a house are designed as right triangles. Rafters form the right angle, and the base of the truss is the hypotenuse (Figure 1). If the rafter on the same side as the solar collector is 4 feet shorter than the other rafter and if the base of each truss is 36 feet long, how long are each of the rafters?

Figure 1

70. Carlos, who is training to run in the Boston Marathon, runs 18 miles every Saturday afternoon. His goal is to cut his running time by one-half hour; he figures that to accomplish this he will have to increase his average speed by 1.2 miles per hour. What is his current average speed for the 18 mile run?

C 71. Neglecting air resistance, a projectile shot straight upward with an initial velocity of v meters per second will be at a height of $vt - 4.9t^2$ meters t seconds later. If $v = 30$ meters per second, how long will it take the projectile to reach a height of 28 meters? (Use a calculator and round off your answer to the nearest tenth of a second.)

In problems 72–74, a formula from applied mathematics is given. Solve each equation for the indicated unknown variable. State any necessary restrictions on the values of the remaining variables.

72. $s = vt + \frac{1}{2}gt^2$ for t (mechanics)

73. $LI^2 + RI + \dfrac{1}{C} = 0$ for I (electronics)

74. $P = EI - RI^2$ for I (electrical engineering)

In problems 75–80, use the discriminant to determine the nature of the roots of each quadratic equation without solving it.

75. $x^2 - 5x - 7 = 0$ 76. $6t^2 - 2 = 7t$ 77. $9x^2 - 6x + 1 = 0$
78. $4x^2 + 20x + 25 = 0$ 79. $6y^2 + y + 3 = 0$ 80. $10r^2 + 2r + 8 = 0$

In problems 81–88, use the quadratic formula to find the complex roots of each quadratic equation.

81. $6x^2 + x + 3 = 0$ 82. $10x^2 + 2x + 8 = 0$ 83. $5x^2 - 4x + 2 = 0$
84. $4x^2 + 4x + 5 = 0$ 85. $x^2 + x + 1 = 0$ 86. $x^2 - 2x + \frac{3}{2} = 0$
87. $4x^2 - 2x + 1 = 0$ 88. $x^2 - 6x + 13 = 0$

4 Miscellaneous Equations

In this section, we discuss special types of equations that can be solved using slight variations of the methods presented in previous sections. Here we consider only real roots of equations.

4.1 Equations of the Form $x^p = a$

If the exponent p is an integer, you can solve the equation $x^p = a$ as shown in the following example.

Example Solve each equation.

(a) $x^3 = 5$ (b) $x^{-3} = 5$ (c) $x^4 = 5$ (d) $x^4 = -5$

Solution (a) Since the exponent is odd, there is just one solution, $x = \sqrt[3]{5}$.

(b) We begin by rewriting the equation in the equivalent form

$$\frac{1}{x^3} = 5 \quad \text{or} \quad x^3 = \frac{1}{5}.$$

The solution is

$$x = \sqrt[3]{\frac{1}{5}} = \frac{1}{\sqrt[3]{5}} = \frac{\sqrt[3]{25}}{5}.$$

(c) Since the exponent is even, there are two solutions, $x = \sqrt[4]{5}$ and $x = -\sqrt[4]{5}$.

(d) A real number raised to an even power cannot be negative, so the equation $x^4 = -5$ has no real root.

The following example shows how to solve the equation $x^p = a$ when the exponent p is the reciprocal of an integer.

Example Solve each equation.

(a) $x^{1/6} = 5$ (b) $x^{1/3} = -5$ (c) $x^{-1/3} = 5$ (d) $x^{1/6} = -5$

Solution (a) The equation can be rewritten as $\sqrt[6]{x} = 5$, so that $x = 5^6 = 15{,}625$.

(b) Here we have $\sqrt[3]{x} = -5$, so that $x = (-5)^3 = -125$.

(c) We begin by rewriting the equation in the equivalent form

$$\frac{1}{x^{1/3}} = 5 \quad \text{or} \quad x^{1/3} = \frac{1}{5},$$

that is, $\sqrt[3]{x} = \frac{1}{5}$. Thus, $x = (\frac{1}{5})^3 = \frac{1}{125}$.

(d) The equation can be rewritten as $\sqrt[6]{x} = -5$. Since a principal sixth root cannot be negative, the equation has no real solution.

If n and m are positive integers and the fraction n/m is reduced to lowest terms, an equation of the form

$$x^{n/m} = a$$

can be rewritten as

$$\sqrt[m]{x^n} = a$$

and solved using the methods illustrated above.

Example Solve each equation.

(a) $x^{3/2} = -8$ (b) $(t - 3)^{2/5} = 4$ (c) $x^{-5/2} = -\frac{1}{4}$

Solution (a) The equation is equivalent to $\sqrt{x^3} = -8$. Because a principal square root of a real number cannot be negative, the equation has no real solution.

(b) The equation can be rewritten as $\sqrt[5]{(t - 3)^2} = 4$ or $(t - 3)^2 = 4^5$. Hence,

$$t - 3 = \pm\sqrt{4^5} = \pm(\sqrt{4})^5 = \pm 2^5 = \pm 32.$$

Therefore, $t = 3 \pm 32$, and the two roots are $t = -29$ and $t = 35$.

(c) We begin by rewriting the equation as

$$\frac{1}{x^{5/2}} = -\frac{1}{4} \quad \text{or} \quad x^{5/2} = -4.$$

The last equation is equivalent to $\sqrt{x^5} = -4$, so it has no real solution.

4.2 Radical Equations

An equation in which the unknown appears in a radicand is called a **radical equation.** For instance,

$$\sqrt[4]{x^2 - 5} = x + 1 \quad \text{and} \quad \sqrt{3t + 7} + \sqrt{t + 2} = 1$$

are radical equations.

To solve a radical equation, begin by isolating the most complicated radical expression on one side of the equation, and then eliminate the radical by raising both sides of the equation to a power equal to the index of the radical. You may have to repeat this technique in order to eliminate all radicals. When the equation is radical-free, simplify and solve it. Since extraneous roots may be introduced when both sides of an equation are raised to an even power, all roots must be checked in the original equation whenever a radical with an *even index* is involved.

Examples Solve each equation.

1 $\sqrt[3]{x - 1} - 2 = 0$

Solution To isolate the radical, we add 2 to both sides of the equation:

$$\sqrt[3]{x - 1} = 2.$$

Now we raise both sides to the power 3 and obtain

$$(\sqrt[3]{x-1})^3 = 2^3 \quad \text{or} \quad x - 1 = 8.$$

It follows that $x = 9$. Since we did not raise both sides of the equation to an even power, it isn't necessary to check our solution, but it's good practice to do so anyway. Substituting $x = 9$ in the original equation, we obtain

$$\sqrt[3]{9-1} - 2 = \sqrt[3]{8} - 2 = 2 - 2 = 0;$$

hence, the solution $x = 9$ is correct.

2 $\quad \sqrt{3t+7} + \sqrt{t+2} = 1$

Solution   We add $-\sqrt{t+2}$ to both sides of the equation to isolate $\sqrt{3t+7}$ on the left side. Thus,

$$\sqrt{3t+7} = 1 - \sqrt{t+2}.$$

Now we square both sides of the equation to obtain

$$(\sqrt{3t+7})^2 = (1 - \sqrt{t+2})^2$$
$$3t + 7 = 1 - 2\sqrt{t+2} + (t+2).$$

The equation still contains a radical, so we simplify and isolate this radical:

$$2t + 4 = -2\sqrt{t+2}$$
$$-(t+2) = \sqrt{t+2}.$$

Again, we square both sides, so

$$[-(t+2)]^2 = (\sqrt{t+2})^2$$
$$t^2 + 4t + 4 = t + 2$$
$$t^2 + 3t + 2 = 0$$
$$(t+2)(t+1) = 0$$

$$t + 2 = 0 \qquad \qquad t + 1 = 0$$
$$t = -2 \qquad \qquad t = -1$$

Check: For $t = -2$,

$$\sqrt{3t+7} + \sqrt{t+2} = \sqrt{3(-2)+7} + \sqrt{-2+2} = \sqrt{1} + \sqrt{0} = 1;$$

hence, $t = -2$ is a solution.

For $t = -1$,

$$\sqrt{3t+7} + \sqrt{t+2} = \sqrt{3(-1)+7} + \sqrt{-1+2} = \sqrt{4} + \sqrt{1} \neq 1.$$

Therefore, $t = -1$ is an extraneous root that was introduced by squaring both sides of the equation. The only solution is $t = -2$.

4.3 Equations of Quadratic Type

An equation with x as the unknown is said to be of **quadratic type** if it can be rewritten in the form

$$au^2 + bu + c = 0,$$

where u is an expression containing x. For instance,

$$x^4 - 5x^2 + 4 = 0$$

can be rewritten as

$$(x^2)^2 - 5(x^2) + 4 = 0,$$

so it is of quadratic type with $u = x^2$. Other examples are

$$x^{2/3} - x^{1/3} - 12 = 0 \qquad (u = x^{1/3}),$$

$$(x^2 - x)^2 - 8(x^2 - x) + 12 = 0 \qquad (u = x^2 - x),$$

$$x^2 + 3x + \sqrt{x^2 + 3x} = 6 \qquad (u = \sqrt{x^2 + 3x}).$$

If such an equation is first solved for u, any resulting values can be used to solve for x; thus the solution of the original equation can be found.

Examples Solve each equation.

1 $x^{2/3} - x^{1/3} - 12 = 0$

Solution   Let $u = x^{1/3}$. Then the equation becomes

$$u^2 - u - 12 = 0$$

$$(u - 4)(u + 3) = 0.$$

Therefore, $u = 4$ or $u = -3$. Since $u = x^{1/3}$, we have

$$x^{1/3} = 4 \qquad\qquad x^{1/3} = -3$$

$$x = 4^3 = 64 \qquad\qquad x = (-3)^3 = -27.$$

Hence, the two solutions are $x = 64$ and $x = -27$.

2 $t^2 + 3t - 2 + \sqrt{t^2 + 3t - 2} - 20 = 0$

Solution   Let $u = \sqrt{t^2 + 3t - 2}$. Then the equation becomes

$$u^2 + u - 20 = 0$$

$$(u - 4)(u + 5) = 0.$$

Therefore, $u = 4$ or $u = -5$. Since $u = \sqrt{t^2 + 3t - 2}$, we have

$$\sqrt{t^2 + 3t - 2} = 4 \quad \text{or} \quad \sqrt{t^2 + 3t - 2} = -5.$$

The second equation has no roots because a principal square root cannot be negative. Squaring both sides of the first equation, we have

$$t^2 + 3t - 2 = 16$$
$$t^2 + 3t - 18 = 0$$
$$(t - 3)(t + 6) = 0.$$

Hence, $t = 3$ or $t = -6$. Because we squared both sides of an equation, we must check for extraneous roots. Substituting $t = 3$ in the original equation, we obtain

$$t^2 + 3t - 2 + \sqrt{t^2 + 3t - 2} - 20 = 3^2 + 3(3) - 2 + \sqrt{3^2 + 3(3) - 2} - 20$$
$$= 9 + 9 - 2 + \sqrt{9 + 9 - 2} - 20$$
$$= 16 + \sqrt{16} - 20 = 0,$$

so $t = 3$ is a solution. Substituting $t = -6$ in the original equation, we get

$$t^2 + 3t - 2 + \sqrt{t^2 + 3t - 2} - 20 = (-6)^2 + 3(-6) - 2$$
$$+ \sqrt{(-6)^2 + 3(-6) - 2} - 20$$
$$= 36 - 18 - 2 + \sqrt{36 - 18 - 2} - 20$$
$$= 16 + \sqrt{16} - 20 = 0,$$

so $t = -6$ is also a solution. Hence, the two solutions are $t = 3$ and $t = -6$.

4.4 Nonquadratic Equations That Can Be Solved by Factoring

The method of solving an equation by rewriting it as an equivalent equation with zero on the right side, then factoring the left side and equating each factor to zero, is not limited to quadratic equations. Indeed, this method works whenever the required factorization can be accomplished.

Examples   Solve each equation by factoring.

1 $4x^3 - 8x^2 - x + 2 = 0$

Solution   We begin by grouping the terms so the common factor $x - 2$ emerges:

$$4x^2(x - 2) - (x - 2) = 0$$
$$(x - 2)(4x^2 - 1) = 0$$
$$(x - 2)(2x - 1)(2x + 1) = 0.$$

Setting each factor equal to zero, we have

$x - 2 = 0$	$2x - 1 = 0$	$2x + 1 = 0$
$x = 2$	$x = \frac{1}{2}$	$x = -\frac{1}{2}$

Therefore, the three solutions are $x = 2$, $x = \frac{1}{2}$, and $x = -\frac{1}{2}$.

2          $x^{5/6} - 3x^{1/3} - 2x^{1/2} + 6 = 0$

Solution   Again, we group and factor:

$$x^{1/3}[x^{(5/6)-(1/3)} - 3] - 2[x^{1/2} - 3] = 0$$

$$x^{1/3}(x^{1/2} - 3) - 2(x^{1/2} - 3) = 0$$

$$(x^{1/2} - 3)(x^{1/3} - 2) = 0.$$

Setting each factor equal to zero, we have

$x^{1/2} - 3 = 0$	$x^{1/3} - 2 = 0$
$x^{1/2} = 3$	$x^{1/3} = 2$
$x = 3^2 = 9$	$x = 2^3 = 8$

Therefore, the two solutions are $x = 9$ and $x = 8$.

PROBLEM SET 4

In this problem set, we consider only real roots of equations.

In problems 1–24, solve each equation.

1. $x^2 = 9$
2. $x^4 = -16$
3. $x^6 = 2$
4. $(x - 1)^4 = 81$
5. $x^{-2} = 4$
6. $u^{-4} = \frac{1}{9}$
7. $x^3 = 27$
8. $x^5 = -3$
9. $x^7 = -4$
10. $t^{-3} = -\frac{1}{8}$
11. $x^{-5} = \frac{-1}{32}$
12. $x^{-7} = 3$
13. $y^{1/3} = 2$
14. $x^{-1/2} = -4$
15. $x^{-1/3} = -\frac{1}{3}$
16. $t^{-2/3} = 9$
17. $z^{5/2} = 3$
18. $(x - 3)^{3/2} = 8$
19. $(2t - 1)^{-2/5} = 4$
20. $(3z - 7)^{4/3} = -1$
21. $(3x)^{-5/3} = \frac{1}{32}$
22. $(x^{1/2})^{3/5} = \frac{1}{27}$
23. $(x^2 - 7x)^{4/3} = 16$
24. $(x^2 + 12x)^{2/3} = 9$

In problems 25–46, solve each equation. Check for extraneous roots.

25. $\sqrt{2t + 9} = t - 13$
26. $\sqrt[3]{2y - 3} = \sqrt[3]{y + 1}$
27. $\sqrt{10 - x} - 8 = 2x$
28. $t = -\sqrt{19 - 2t} - 8$
29. $\sqrt{16 - m} + 4 = m$
30. $\sqrt{u + \sqrt{u - 4}} = 2$
31. $\sqrt[3]{2x + 3} = -1$
32. $\sqrt[5]{3x - 1} = 2$
33. $\sqrt[4]{y^4 + 2y - 6} = y$
34. $\sqrt[3]{x^3 + 2x^2 - 9x - 26} = x + 1$
35. $\sqrt{3t + 12} - 2 = \sqrt{2t}$

36. $\sqrt{5y + 1} = 1 + \sqrt{3y}$
37. $6\sqrt{x} - 3 = \sqrt{10x - 1}$
38. $2\sqrt{x - 1} - (x/2) = \frac{3}{2}$
39. $\sqrt{2y - 5} - \sqrt{y - 2} = 2$
40. $\sqrt{3t + 8} = 3\sqrt{t} - 2\sqrt{2}$
41. $\sqrt{5x + 6} = \sqrt{x + 1} + 1$
42. $(2\sqrt{n})^{-1} + (\sqrt{4n + 9})^{-1} = 9(\sqrt{16n^2 + 36n})^{-1}$
43. $\sqrt{2x + 5} - \sqrt{5x - 1} = \sqrt{2x - 4}$
44. $x(\sqrt{x^2 + 1})^{-1} = (4 - x)(\sqrt{x^2 - 8x + 17})^{-1}$
45. $\sqrt{5t - 10} = \sqrt{8t + 2} - \sqrt{3t + 12}$
46. $\dfrac{\sqrt{x + 1} + \sqrt{x - 1}}{\sqrt{x + 1} - \sqrt{x - 1}} = \dfrac{4x - 1}{2}$

In problems 47–64, solve each equation.

47. $x^4 - 7x^2 + 12 = 0$
48. $6x^4 + 3 = 19x^2$
49. $(x^2 + x)^2 + (x^2 + x) - 6 = 0$
50. $(y^2 + y - 2)^2 + (y^2 + y - 2) = 20$
51. $t^{1/3} - 4t^{1/6} + 3 = 0$
52. $y^{-2/3} + 2y^{-1/3} = 8$
53. $x^{1/4} + 2x^{1/2} = 3$
54. $10t^{-2} - 9t^{-1} - 1 = 0$

55. $9x^{-4} - 145x^{-2} + 16 = 0$

56. $z^2 + z + \dfrac{12}{z^2 + z} = 8$

57. $2x^2 - 5x + 7 - 8\sqrt{2x^2 - 5x + 7} + 7 = 0$

58. $26 - 8x = 6x^2 - \sqrt{3x^2 + 4x + 1}$

59. $y^2 + 2y - 3\sqrt{y^2 + 2y + 4} = 0$

60. $t^2 + 6t - 6\sqrt{t^2 + 6t - 2} + 3 = 0$

61. $\sqrt{t + 20} - 4\sqrt[4]{t + 20} + 3 = 0$

62. $3(1 - x)^{1/6} + 2(1 - x)^{1/3} = 2$

63. $2r^2 - 3r - 2\sqrt{2r^2 - 3r} - 3 = 0$

64. $2s^2 + s - 4\sqrt{2s^2 + s + 4} = 1$

In problems 65–72, solve each equation by factoring.

65. $x^3 + x^2 - x - 1 = 0$

66. $x^5 - 4x^3 - x^2 + 4 = 0$

67. $x^6 - x^4 + 4x^2 - 4 = 0$

68. $z^7 - 27z^4 - z^3 + 27 = 0$

69. $x^{5/6} + 2x^{1/2} - x^{1/3} - 2 = 0$

70. $x^{7/12} - 2x^{1/3} + 5x^{1/4} - 10 = 0$

71. $x + x^{2/3} - x^{1/3} - 1 = 0$

72. $4x^{3/2} - 8x - x^{1/2} + 2 = 0$

73. The formula for the lateral surface area A of a right circular cone with height h and base radius r is $A = \pi r \sqrt{r^2 + h^2}$. Solve for r in terms of A and h.

74. Prove that the equation $x^3 = a$ has exactly one real solution, $x = \sqrt[3]{a}$, by rewriting the equation in the form $x^3 - (\sqrt[3]{a})^3 = 0$ and factoring.

75. Find the dimensions of a rectangle that has an area of 60 square centimeters and a diagonal that is 13 centimeters long.

76. The frequency ω of a certain electronic circuit is given by

$$\omega = \frac{1}{2\pi \sqrt{\dfrac{LC_1 C_2}{C_1 + C_2}}}.$$

Solve for C_2 in terms of ω, L, and C_1.

77. The deflection d of a certain beam is given by $d = l\sqrt{a/(2l + a)}$. Solve for l in terms of a and d.

78. In meteorology the heat loss H (wind chill) is given by

$$H = (A + B\sqrt{V} - CV)(S - T).$$

Solve for the wind speed V in terms of H, the neutral skin temperature S, the air temperature T, and the constants A, B, and C.

5   Inequalities

We introduced the inequality symbols $<$, $\leq$, $>$, and $\geq$, in Section 2.1 of Chapter 1. By writing one of these symbols between two expressions, we obtain an *inequality*, and the two expressions are called the **members** or **sides** of the inequality. By connecting more than two expressions with these symbols, we can form **compound inequalities** such as $x < y \leq z$ and $1 \geq 2x - 3 \geq 5$. A compound inequality can always be broken down into simple inequalities: For instance, $x < y \leq z$ means $x < y$ and $y \leq z$.

In Section 1.1 of Chapter 1, we presented the *Basic Algebraic Properties* of the real numbers. All of our work up to now has been founded upon these properties. To deal with inequalities, we must introduce some further properties.

Basic Order Properties of the Real Numbers

Let a, b, and c denote real numbers.

1. *The Trichotomy Property*

> One and only one of the following is true:
>
> $$a < b \quad \text{or} \quad b < a \quad \text{or} \quad a = b.$$

2. *The Transitivity Property*

> If $a < b$ and $b < c$, then $a < c$.

3. *The Addition Property*

> If $a < b$, then $a + c < b + c$.

4. *The Multiplication Property*

> If $a < b$ and $c > 0$, then $ac < bc$.

Figure 1

The trichotomy property has a simple geometric interpretation on the number line; namely, if a and b are distinct, then one must lie to the left of the other. Similarly, the transitivity property means that if a lies to the left of b and b lies to the left of c, then a lies to the left of c (Figure 1). The addition and multiplication properties can also be illustrated geometrically (Problem 44).

Example   State one Basic Order Property of the real numbers to justify each statement.

(a) The discriminant of a quadratic equation is either positive, negative, or zero.

(b) $-1 < 0$ and $0 < 1$, so it follows that $-1 < 1$.

(c) $5 < 6$, so it follows that $5 + (-11) < 6 + (-11)$.

(d) $4 < 7$ and $\frac{1}{2} > 0$, so it follows that $2 < \frac{7}{2}$.

Solution   (a) The trichotomy property.   (b) The transitivity property.

(c) The addition property.   (d) The multiplication property.

Many of the important properties of inequalities can be *derived* from the Basic Order Properties, although we shall not derive them here. Among the more important derived properties are the following.

5. *The Subtraction Property*

$$\text{If } a < b, \text{ then } a - c < b - c.$$

6. *The Division Property*

$$\text{If } a < b \text{ and } c > 0, \text{ then } \frac{a}{c} < \frac{b}{c}.$$

According to the multiplication and division properties, you can multiply or divide both sides of an inequality by a **positive** number. According to the following property, if you multiply or divide both sides of an inequality by a **negative** number, you must **reverse the inequality.**

7. *The Order-Reversing Properties*

(i)   If $a < b$ and $c < 0$, then $ac > bc$.

(ii) If $a < b$ and $c < 0$, then $\dfrac{a}{c} > \dfrac{b}{c}$.

For instance, if you multiply both sides of the inequality $2 < 5$ by -3, you must reverse the inequality and write $-6 > -15$, or $-15 < -6$. Indeed, on a number line, -15 lies to the left of -6.

Statements analogous to Properties 2–7 can be made for nonstrict inequalities (those involving $\leq$ or $\geq$) and for compound inequalities. For instance, if you know that

$$-1 \leq 3 - \frac{x}{2} < 5,$$

you can multiply all members by -2, provided that you reverse the inequalities:

$$2 \geq x - 6 > -10$$

or

$$-10 < x - 6 \leq 2.$$

An inequality containing a variable is neither true nor false until a particular number is substituted for the variable. If a true statement results from such a substitution, we say that the substitution *satisfies* the inequality. If the substitution $x = a$ satisfies an inequality, we say that the real number a is a **solution** of the inequality. The set of all solutions of an inequality is called its **solution set.** Two inequalities are said to be **equivalent** if they have exactly the same solution set.

To **solve** an inequality—that is, to find its solution set—you proceed in much the same way as in solving equations, except of course that when you multiply or divide all members by a negative quantity, you must reverse all signs of inequality. The usual approach is to try to isolate the unknown in one member of the inequality.

Examples   Solve each inequality and sketch its solution set on a number line.

1 $5x + 3 < 18$

Solution

$$5x + 3 < 18$$

$$5x < 15 \qquad \text{(We subtracted 3 from both sides.)}$$

$$x < 3 \qquad \text{(We divided both sides by 5.)}$$

Figure 2

Since the last inequality is equivalent to the original one, we conclude that the solution of $5x + 3 < 18$ consists of all numbers x such that $x < 3$. In other words, the solution set is the interval $(-\infty, 3)$ (Figure 2).

2 $3 - 2x \geq 4x - 9$

Solution

$$3 - 2x \geq 4x - 9$$

$$3 - 6x \geq -9 \qquad \text{(We subtracted } 4x \text{ from both sides.)}$$

$$-6x \geq -12 \qquad \text{(We subtracted 3 from both sides.)}$$

$$x \leq 2 \qquad \text{(We divided both sides by } -6 \text{ and reversed the inequality.)}$$

Figure 3

Therefore, the solution set is $(-\infty, 2]$ (Figure 3).

3 $\frac{5}{6}x - \frac{3}{4} \geq \frac{1}{2}x - \frac{2}{3}$

Solution   We begin by multiplying both sides by 12, the L.C.D. of the fractional coefficients:

$$10x - 9 \geq 6x - 8$$

$$4x - 9 \geq -8 \qquad \text{(We subtracted } 6x \text{ from both sides.)}$$

$$4x \geq 1 \qquad \text{(We added 9 to both sides.)}$$

$$x \geq \tfrac{1}{4} \qquad \text{(We divided both sides by 4.)}$$

Figure 4

Therefore, the solution set is $\left[\frac{1}{4}, \infty\right)$ (Figure 4).

4 $-11 \leq 2x - 3 < 7$

Solution   We begin by adding 3 to all members to help isolate x in the middle:

$$-8 \leq 2x < 10$$

$$-4 \leq x < 5 \qquad \text{(We divided all members by 2.)}$$

Therefore, the solution set is $[-4, 5)$ (Figure 5).

Figure 5

To solve a compound inequality in which the unknown appears only in the middle member, you can always proceed as in Example 4 above. Otherwise, you can break the compound inequality up into simple inequalities and proceed as in the following example.

Example   Solve the inequality $3x - 10 \leq 5 < x + 3$ and sketch its solution set on a number line.

Solution   The given inequality holds if and only if *both* the inequalities $3x - 10 \leq 5$ and $5 < x + 3$ hold.

Solution of $3x - 10 \leq 5$	Solution of $5 < x + 3$
$3x - 10 \leq 5$	$5 < x + 3$
$3x \leq 5 + 10$	$5 - 3 < x$
$3x \leq 15$	$2 < x$
$x \leq 5$	
Here, the solution set is $(-\infty, 5]$ (Figure 6a).	Here, the solution set is $(2, \infty)$ (Figure 6b).

Therefore, the solution set of $3x - 10 \leq 5 < x + 3$ consists of all real numbers x that belong to *both* the intervals $(-\infty, 5]$ and $(2, \infty)$; that is, the solution set is the interval $(2, 5]$ (Figure 6c).

Figure 6

The following example illustrates one of the many practical applications of inequalities.

Example   A technician determines that an electronic circuit fails to operate because the resistance between points A and B, 1200 ohms, exceeds the specifications, which call for a resistance of no less than 400 ohms and no greater than 900 ohms. The circuit can be made to satisfy the specifications by adding a shunt resistor R ohms, $R > 0$ (Figure 7). After adding the shunt resistor, the resistance between points A and B will be $\dfrac{1200R}{1200 + R}$ ohms. What are the possible values of R?

Figure 7

Solution   To satisfy the specifications, we must have

$$400 \leq \frac{1200R}{1200 + R} \leq 900.$$

Because R must be positive, it follows that $1200 + R$ is positive, so we can multiply all members of the inequality by $1200 + R$:

$$400(1200 + R) \leq 1200R \leq 900(1200 + R)$$

$$480{,}000 + 400R \leq 1200R \leq 1{,}080{,}000 + 900R.$$

Now we break the compound inequality into two simple inequalities.

$480{,}000 + 400R \leq 1200R$	$1200R \leq 1{,}080{,}000 + 900R$
$480{,}000 \leq 1200R - 400R$	$1200R - 900R \leq 1{,}080{,}000$
$480{,}000 \leq 800R$	$300R \leq 1{,}080{,}000$
$\dfrac{480{,}000}{800} \leq R$	$R \leq \dfrac{1{,}080{,}000}{300}$
$600 \leq R$	$R \leq 3600$

Therefore, we must have both $600 \leq R$ and $R \leq 3600$; that is,

$$600 \leq R \leq 3600.$$

The shunt resistance R must be no less than 600 ohms and no greater than 3600 ohms.

PROBLEM SET 5

1. State one Basic Order Property of the real numbers to justify each statement.
   (a) $3 < 5$, so it follows that $3 + 2 < 5 + 2$.
   (b) $-3 < -2$ and $-2 < 2$, so it follows that $-3 < 2$.
   (c) $-3 < -2$ and $\frac{1}{3} > 0$, so it follows that $-1 < -\frac{2}{3}$.
   (d) We know that $\sqrt{2} \neq 1.414$, so it follows that either $\sqrt{2} < 1.414$ or else $\sqrt{2} > 1.414$.
2. State one derived order property of the real numbers to justify each statement.
   (a) $-1 < 0$, so it follows that $-2 < -1$.
   (b) $-5 < -3$ and $5 > 0$, so it follows that $-1 < -\frac{3}{5}$.
   (c) $3 < 5$ and $-2 < 0$, so it follows that $-6 > -10$.
   (d) $3 < 5$ and $-2 < 0$, so it follows that $-\frac{3}{2} > -\frac{5}{2}$.
3. Given that $x > -3$, what equivalent inequality is obtained if (a) 4 is added to both sides?
   (b) 4 is subtracted from both sides? (c) Both sides are multiplied by 5? (d) Both sides are multiplied by -5?
4. Given that $-2 \leq 3 - 4x < 1$, what equivalent inequality is obtained if (a) 3 is subtracted from all members? (b) 2 is added to all members? (c) All members are divided by 2? (d) All members are divided by -4?

In problems 5–36, solve each inequality and sketch its solution set on a number line.

5. $2x - 5 < 5$
6. $7 - 3x < 22$
7. $2x - 10 \leq 15 - 3x$
8. $5x + 3 \leq 8x - 6$
9. $15 + 6x > 9x + 9$
10. $x - 11 > 2x - 13$

11. $-5(1 - x) \geq 15$

12. $4(x - 4) - 3 \geq 17$

13. $7(x + 2) \geq 2(2x - 4)$

14. $21 - 3(x - 7) \leq x + 20$

15. $\dfrac{x}{3} < 2 - \dfrac{x}{4}$

16. $\dfrac{2x}{3} > \dfrac{23}{24} - \dfrac{5x}{4}$

17. $\frac{1}{2}(x + 1) \geq \frac{1}{3}(x - 5)$

18. $\frac{1}{5}(x + 1) \geq \frac{1}{3}(x - 5)$

19. $\frac{1}{3}(x + 1) \geq -\frac{1}{3}(2 - x)$

20. $\frac{2}{3}x - \frac{3}{5}(5 - x) \geq 0$

21. $\frac{1}{6}(x + 3) - \frac{1}{2}(x - 1) \leq \frac{1}{3}$

22. $\dfrac{x + 5}{6} - \dfrac{10 - x}{-3} < \pi$

23. $5 \leq 2x + 3 \leq 13$

24. $14 \leq 5 - 3x < 20$

25. $4 \leq 5x - 1 < 9$

26. $0 < 3x - 5 \leq 16$

27. $-3 < 5 - 4x \leq 17$

28. $3 + x < 3(x - 1) \leq x - 7$

29. $4x - 1 \leq 3 \leq 7 + 2x$

30. $3x + 1 \leq 2 - x < 18 + 7x$

31. $1 - 2x < 5 - x \leq 25 - 6x$

32. $2x - 1 \leq 5 \leq x + 2$

33. $1 \leq \dfrac{10x}{10 + x} \leq 5, \quad x > 0$

34. $1 \leq \dfrac{4x}{4 + x} \leq 3, \quad x > -4$

35. $\dfrac{4}{x + 1} \leq 3 \leq \dfrac{6}{x + 1}, \quad x > 0$

36. $\dfrac{8}{x + 3} \leq 2 \leq \dfrac{13}{x + 3}, \quad x > -3$

37. A student's scores on the first three tests in a sociology course are 65, 78, and 84. What range of scores on a fourth test will give the student an average that is less than 80 but no less than 70 for the four tests?

38. An object thrown straight upward with an initial velocity of 100 feet per second will have a velocity v, given by $v = 100 - 32t$, exactly t seconds later. Over what interval of time will its velocity be greater than 50 feet per second but no greater than 60 feet per second?

39. A shunt resistor R ohms is to be added to a resistance of 600 ohms in order to reduce the resistance to a value strictly between 540 ohms and 550 ohms. After the shunt is added, the resistance is given by $600R/(600 + R)$. Find the possible values of R. (Assume that $R > 0$.)

40. A realtor charges the seller of a house a 7% commission on the price for which the house is sold. Suppose that the seller wants to clear at least $60,000 after paying the commission, and that similar houses are selling for $70,000 or less.
(a) What is the range of selling prices for the house?
(b) If the house is sold, what is the range of amounts for the realtor's commission?

41. A rectangular solar collector is to have a height of 1.5 meters, but its length is still to be determined. What is the range of values for this length if the collector provides 400 watts per square meter, and if it must provide a total of between 2000 and 3500 watts?

42. Suppose that families with an income, after deductions, of over $19,200 but not over $23,200 must pay an income tax of $3260 plus 28% of the amount over $19,200. What is the range of possible values for the income, after deductions, of families whose income tax liability is between $3484 and $4044?

43. An operator-assisted station-to-station phone call from Boston, MA, to Tucson, AZ, costs $2.25 plus $0.38 for each additional minute after the first 3 minutes. A group of such calls were made, each costing between $6.05 and $8.71, inclusive. Give the range of minutes for the lengths of these calls.

44. Sketch diagrams to illustrate the addition and multiplication order properties of the real numbers.

45. If $a \leq b$, prove that $a + c \leq b + c$. [Hint: Consider separately the two possibilities $a < b$ and $a = b$.]

46. If $c \neq 0$, prove that $c^2 > 0$. [Hint: Consider separately the two cases $c > 0$ and $c < 0$.]

47. Give an example to show that, in general, you *cannot* square both sides of an inequality and get an equivalent inequality.

48. If $c > 0$, show that $1/c > 0$. [Hint: If $1/c$ is not positive it must be negative, since it cannot be zero.]

6 Equations and Inequalities Involving Absolute Values

Many calculators have an *absolute-value* key marked |x| or ABS. If you enter a *nonnegative* number and press this key, the same number appears in the display. However, if you enter a *negative* number and press the absolute-value key, the display shows the number without the negative sign. For instance,

Enter	Press ABS Key	Display Shows
3.2	$\longmapsto\longrightarrow$	3.2
-3.2	$\longmapsto\longrightarrow$	3.2

This feature is useful whenever you are interested only in the "magnitude" of a number, without regard to its algebraic sign.

On a calculator without an absolute-value key, you can obtain the absolute value |x| of a number x by entering x, pressing the x^2 key, and then pressing the square-root key. In other words, if you square a number and take the principal square root of the result, you will remove any negative sign that may have been present, so that

$$\sqrt{x^2} = |x|.$$

For instance, $\sqrt{(-5)^2} = \sqrt{25} = 5 = |-5|$.

The absolute value is defined formally as follows.

Definition 1   **Absolute Value**

If x is a real number, then |x|, the **absolute value** of x, is defined by

$$|x| = \begin{cases} x, & \text{if } x \geq 0 \\ -x, & \text{if } x < 0. \end{cases}$$

For instance,

$$|5| = 5 \quad \text{because } 5 \geq 0,$$
$$|0| = 0 \quad \text{because } 0 \geq 0,$$
$$|-5| = -(-5) = 5 \quad \text{because } -5 < 0.$$

Notice that $|x| \geq 0$ is always true. (Why?)

On a number line, |x| *is the number of units of distance between the point with coordinate x and the origin*, regardless of whether the point is to the right or left of the origin (Figure 1a). For instance, both the point with coordinate -3 and the point with coordinate 3 are 3 units from the origin (Figure 1b), so $|-3| = 3 = |3|$.

Figure 1 (a)   (b)

More generally, $|x - y|$ *is the number of units of distance between the point with coordinate x and the point with coordinate y.* This statement holds no matter which point is to the left of the other (Figure 2a and b), and of course it also holds when $x = y$. For simplicity, we refer to the distance $|x - y|$ as the **distance between the numbers x and y.**

Figure 2   (a)

(b)

Example   Find the distance between the numbers.

(a) 4 and 7   (b) 7 and 4   (c) -4 and 7

(d) -4 and -7 (e) -4 and 0   (f) -4 and 4

Solution   (a) $|4 - 7| = |-3| = 3$ (b) $|7 - 4| = |3| = 3$

(c) $|-4 - 7| = |-11| = 11$ (d) $|-4 - (-7)| = |-4 + 7| = |3| = 3$

(e) $|-4 - 0| = |-4| = 4$ (f) $|-4 - 4| = |-8| = 8$

The absolute value of a product of two numbers is equal to the product of their absolute values; that is,

$$|xy| = |x||y|.$$

This rule can be derived as follows:

$$|xy| = \sqrt{(xy)^2} = \sqrt{x^2 y^2} = \sqrt{x^2}\sqrt{y^2} = |x||y|.$$

A similar rule applies to quotients; if $y \neq 0$,

$$\left|\frac{x}{y}\right| = \frac{|x|}{|y|}$$

(Problem 75). However, there is no such rule for sums or differences:

$$|x + y| \quad \text{*need not be the same as*} \quad |x| + |y|;$$

for instance,

$$|5 + (-2)| = |3| = 3, \quad \text{but} \quad |5| + |-2| = 5 + 2 = 7.$$

Similarly,

$$|x - y| \quad \text{*need not be the same as*} \quad |x| - |y|.$$

(See Problems 74 and 78.)

Example   Let $x = -3$ and $y = 5$. Find the value of each expression.

(a) $|6x + y|$ (b) $|6x| + |y|$ (c) $|xy|$

(d) $|x||y|$ (e) $\dfrac{|x|}{x}$ (f) $\left|\dfrac{x}{y}\right|$

Solution   (a) $|6x + y| = |6(-3) + 5| = |-18 + 5| = |-13| = 13$

(b) $|6x| + |y| = |6(-3)| + |5| = |-18| + |5| = 18 + 5 = 23$

(c) $|xy| = |(-3)5| = |-15| = 15$

(d) $|x||y| = |-3||5| = 3 \cdot 5 = 15$

(e) $\dfrac{|x|}{x} = \dfrac{|-3|}{-3} = \dfrac{3}{-3} = -1$

(f) $\left|\dfrac{x}{y}\right| = \left|\dfrac{-3}{5}\right| = \dfrac{3}{5}$

6.1 Equations Involving Absolute Values

Geometrically, the equation $|x| = 3$ means that the point with coordinate x is 3 units from 0 on the number line. Obviously, the number line contains *two* points that are 3 units from the origin—one to the right of the origin and the other to the left (Figure 3). Thus, the equation $|x| = 3$ has two solutions, $x = 3$ and $x = -3$. More generally, we have the following theorem, whose simple proof is left as an exercise (Problem 77).

Figure 3

Theorem 1   **Solution of the Equation $|u| = a$**

Let a be a real number.

    (i)   If $a > 0$, then $|u| = a$ if and only if $u = -a$ or $u = a$.

    (ii) If $a < 0$, then the equation $|u| = a$ has no solution.

    (iii) $|u| = 0$ if and only if $u = 0$.

Examples   Solve each equation.

1 $|4x| = 12$

Solution   By part (i) of Theorem 1, with $u = 4x$, we have $4x = -12$ or $4x = 12$; that is, $x = -3$ or $x = 3$.

2 $|3t| - 1 = 17$

Solution   The equation is equivalent to $|3t| = 18$, so that $3t = -18$ or $3t = 18$; that is, $t = -6$ or $t = 6$.

3 $|7x + 1| = -3$

Solution   By part (ii) of Theorem 1, the equation has no solution.

4 $|1 - 2x| = 7$

Solution   The equation holds if and only if

$$1 - 2x = 7 \quad \text{or} \quad 1 - 2x = -7$$
$$-2x = 6 \quad \text{or} \quad -2x = -8$$
$$x = -3 \quad \text{or} \quad x = 4.$$

5 $|x^2 + x - 6| = 0$

Solution   By part (iii) of Theorem 1, the equation is equivalent to

$$x^2 + x - 6 = 0, \quad \text{that is,} \quad (x + 3)(x - 2) = 0.$$

The solutions are $x = -3$ and $x = 2$.

An equation of the form $|u| = |v|$ can be solved by using the following theorem, whose proof is left as an exercise (Problem 79).

Theorem 2   **Solution of the Equation $|u| = |v|$**

$$|u| = |v| \quad \text{if and only if} \quad u = -v \quad \text{or} \quad u = v.$$

Example   Solve the equation $|p - 5| = |3p + 7|$.

Solution   By Theorem 2 with $u = p - 5$ and $v = 3p + 7$, the given equation holds if and only if

$$p - 5 = 3p + 7 \quad \text{or} \quad p - 5 = -(3p + 7)$$
$$-2p = 12 \quad \text{or} \quad 4p = -2$$
$$p = -6 \quad \text{or} \quad p = -\tfrac{1}{2}.$$

6.2 Inequalities Involving Absolute Values

Referring once again to the number line (Figure 4), we see that the inequality $|x| < 3$ means that the point with coordinate x is less than 3 units from 0, that is, $-3 < x < 3$. More generally, we have the following theorem (Problem 80).

Figure 4

Theorem 3   **Solution of the Inequalities $|u| < a$ and $|u| \leq a$**

> If $a > 0$, then:
>
> (i) $|u| < a$ if and only if $-a < u < a$.
>
> (ii) $|u| \leq a$ if and only if $-a \leq u \leq a$.

Examples   Solve each inequality and sketch its solution set on a number line.

1 $|2x + 3| < 5$

Solution

Figure 5

$|2x+3|<5$
$-6\ -5\ -4\ -3\ -2\ -1\quad 0\quad 1\quad 2\quad 3\quad 4$

$$|2x + 3| < 5$$
$$-5 < 2x + 3 < 5 \quad \text{[By part (i) of Theorem 3.]}$$
$$-8 < 2x < 2 \quad \text{(We subtracted 3 from all members.)}$$
$$-4 < x < 1 \quad \text{(We divided all members by 2.)}$$

Hence, the solution set is the interval $(-4, 1)$ (Figure 5).

2 $|2x| + 1 \leq 5$

Solution

Figure 6

$|2x|+1\leq 5$
$-5\ -4\ -3\ -2\ -1\quad 0\quad 1\quad 2\quad 3\quad 4\quad 5$

$$|2x| + 1 \leq 5$$
$$|2x| \leq 4 \quad \text{(We subtracted 1 from both sides.)}$$
$$-4 \leq 2x \leq 4 \quad \text{[By part (ii) of Theorem 3.]}$$
$$-2 \leq x \leq 2 \quad \text{(We divided all members by 2.)}$$

Hence, the solution set is the interval $[-2, 2]$ (Figure 6).

Figure 7 $|x|>3$ if and only if x
belongs to one of these intervals

$\cdots\ -5\ -4\ -3\ -2\ -1\quad 0\quad 1\quad 2\quad 3\quad 4\quad 5\quad 6$

On a number line, the inequality $|x| > 3$ means that the point with coordinate x is more than 3 units from 0 (Figure 7); that is, either $x > 3$ or $x < -3$. More generally, we have the following theorem (Problem 82).

Theorem 4   **Solution of the Inequalities $|u| > a$ and $|u| \geq a$**

> If $a > 0$, then:
>
> (i) $|u| > a$ if and only if $u < -a$ or $u > a$.
>
> (ii) $|u| \geq a$ if and only if $u \leq -a$ or $u \geq a$.

Example   Solve the inequality $|\frac{1}{2}x - 1| \geq 2$ and sketch its solution set on a number line.

Solution   By part (ii) of Theorem 4 with $u = \frac{1}{2}x - 1$, we have

$$\frac{1}{2}x - 1 \leq -2 \quad \text{or} \quad \frac{1}{2}x - 1 \geq 2$$
$$\frac{1}{2}x \leq -1 \quad \text{or} \quad \frac{1}{2}x \geq 3$$
$$x \leq -2 \quad \text{or} \quad x \geq 6.$$

Thus, the solution set consists of all numbers that belong to either of the intervals $(-\infty, -2]$ or $[6, \infty)$ (Figure 8).

Figure 8

PROBLEM SET 6

In problems 1–10, find the distance between the numbers and illustrate on a number line.

1. 5 and 3
2. −3 and 0
3. −3 and −5
4. −5 and 3
5. $\frac{5}{2}$ and $\frac{3}{2}$
© 6. 412.7 and 152.3
© 7. −3.311 and 2.732
© 8. $-\sqrt{3}$ and $\sqrt{2}$
© 9. π and 1
© 10. π and $\frac{22}{7}$

In problems 11–20, find the value of each expression if $x = -4$ and $y = 6$.

11. $|5x + 2y|$
12. $|5x| + |2y|$
13. $|5x - 2y|$
14. $|5x| - |2y|$
15. $\big||5x| - |2y|\big|$
16. $\big||2y| - |5x|\big|$
17. $|3xy| - 3|x|\,|y|$
18. $\left|\dfrac{5x}{2y}\right|$
19. $\left|\dfrac{5x}{2y}\right| - \dfrac{5|x|}{2|y|}$
20. $\dfrac{|x|}{x} - \dfrac{x}{|x|}$

In problems 21–44, solve each equation.

21. $|4x| = 16$
22. $|-5y| = 15$
23. $|5y| + 2 = 17$
24. $|2t| - 3 = 5$
25. $|2x| + |-4| = 4$
26. $|7x| = |-14|$
27. $|x^2 - x - 20| = 0$
28. $|2p - 1| = 3$
29. $|4 - y| = 2$
30. $|11 - 5x| = 16$
31. $|3p| = 3p$
32. $|7y - 3| = 3 - 7y$
33. $\left|\dfrac{3|x|}{5} + 1\right| = 0$
34. $\left|\dfrac{5c}{2} - 3\right| = -1$
35. $\left|\dfrac{x}{2} - \dfrac{3}{4}\right| = \dfrac{1}{12}$
36. $\left|\dfrac{13}{12} + \dfrac{2x}{3}\right| = \dfrac{11}{12}$
37. $|t + 1| = |4t|$
38. $|3x - 1| = x + 5$
39. $|1 - x| = |2x + 3|$
40. $|2t - 7| = t^2$
41. $|4 - 3y| = |10 - 5y|$
42. $|2x - 1| = x^2$
43. $|4x - 7| = |2 + x|$
44. $|x + 1| + |x - 1| = 4$

In problems 45–68, solve each inequality and sketch its solution set on a number line.

45. $|2x| < 4$
46. $|x| - 3 \leq 2$
47. $|5x| - 1 \leq 9$
48. $|3t| > 15$
49. $|x - 3| < 2$
50. $|x| + 2 \geq 3$
51. $|x| - 3 \leq 2$
52. $|-3y| - 2 \geq 4$
53. $|4x - 6| \geq 3$
54. $|2x - 5| < 1$
55. $|3 - 4x| > 9$
56. $|3y + 5| \leq |-2|$
57. $|5x - 8| \leq 2$
58. $|5 - 3p| \geq |-11|$
59. $|2x + 3| \geq \frac{1}{4}$
60. $|x + 5| \geq x + 1$
61. $|1 - \frac{7}{6}x| \leq \frac{1}{3}$
62. $|3x + 6| > 0$
63. $|7 - 4x| \geq |-3|$
64. $|6y - 3| > -3$
65. $|\frac{11}{2} - 2x| > \frac{3}{2}$
66. $|3x + 5| < |2x + 1|$
67. $|2x - 5| \geq 0$
68. $|5x + 2| < |3x - 4|$

In problems 69–72, describe each interval as the set of all real numbers x that satisfy an inequality containing an absolute value.

69. $(-1, 1)$
70. $(-3, 7)$
71. $[-3, 3]$
72. $[-2, 3]$

73. Show that $-|x| \leq x \leq |x|$ holds for all real numbers x.
74. Although the absolute value of a sum need not be the same as the sum of the absolute values, the inequality $|x + y| \leq |x| + |y|$, called the **triangle inequality,** always holds. Prove the triangle inequality by considering the case in which $x + y \geq 0$ separately from the case in which $x + y < 0$. Use the facts that $-|x| \leq x \leq |x|$ and $-|y| \leq y \leq |y|$. (See problem 73.)
75. If $y \neq 0$, show that $\left|\dfrac{x}{y}\right| = \dfrac{|x|}{|y|}$.
76. Use the triangle inequality (problem 74) to establish the following: (a) If $|x - 2| < \frac{1}{2}$ and $|y + 2| < \frac{1}{3}$, then $|x + y| < \frac{5}{6}$. (b) If $|x - 2| < \frac{1}{2}$ and $|z - 2| < \frac{1}{3}$, then $|x - z| < \frac{5}{6}$.
77. Prove Theorem 1.
78. Using the triangle inequality (problem 74), prove that $\big||x| - |y|\big| \leq |x - y|$.
79. Prove Theorem 2.
80. Prove Theorem 3.
81. Show that if $a < b$, then $a < \dfrac{a + b}{2} < b$ and the distance between a and $\dfrac{a + b}{2}$ is the same as the distance between $\dfrac{a + b}{2}$ and b.
82. Prove Theorem 4.
83. Show that the expression $\frac{1}{2}(x + y + |x - y|)$ always equals the larger of the two numbers x and y (or their common value if they are equal).
84. Find an expression that is similar to the one given in problem 83, but that gives the smaller of the two numbers x and y (or their common value if they are equal).
85. Show that if $a < b$, then the solution set of $|a + b - 2x| < b - a$ is the open interval (a, b).
86. Statisticians often report their findings in terms of so-called **confidence intervals**—intervals that they believe (with a certain level of confidence) contain the unknown quantity with which they are concerned. Suppose that a statistician determines that (with a certain level of confidence) the average weight x pounds of newborn babies in a large city satisfies the inequality $|x - 6.8| < 0.85$. Find the confidence interval for x by solving this inequality.

7 Polynomial and Rational Inequalities

Up to this point, we have dealt mainly with inequalities containing only first-degree polynomials in the unknown. In this section, we study inequalities containing higher degree polynomials or rational expressions in the unknown.

7.1 Polynomial Inequalities

An inequality such as $4x^3 - 3x^2 + x - 1 > x^2 + 3x - x$, in which all members are polynomials in the unknown, is called a **polynomial inequality.** Before solving such inequalities, we shall rewrite them in **standard form** with zero on the right side and a polynomial on the left side.

Suppose we have to solve the polynomial inequality

$$4x^2 - x - 8 < 3x^2 - 4x + 2.$$

We begin by subtracting $3x^2 - 4x + 2$ from both sides to bring it into the standard form

$$x^2 + 3x - 10 < 0.$$

Now, let's consider how the algebraic sign of the polynomial on the left changes as we vary the values of the unknown. As we move x along the number line, the quantity $x^2 + 3x - 10$ is sometimes positive, sometimes negative, and sometimes zero. To solve the inequality, we must find the values of x for which $x^2 + 3x - 10$ is negative.

It seems plausible (and is, in fact, true—see Problem 52) that *intervals where $x^2 + 3x - 10$ is positive are separated from intervals where it is negative by values of x for which it is zero.* To locate these intervals, we begin by solving the equation

$$x^2 + 3x - 10 = 0.$$

Factoring, we obtain $(x + 5)(x - 2) = 0$, so that $x = -5$ or $x = 2$. The two roots -5 and 2 divide the number line into three open intervals, $(-\infty, -5)$, $(-5, 2)$, and $(2, \infty)$ (Figure 1). Therefore, *the quantity $x^2 + 3x - 10$ will have a constant algebraic sign over each of these intervals.*

Figure 1

To find out whether $x^2 + 3x - 10$ is positive or negative over the first interval, $(-\infty, -5)$, we select any convenient test number in this interval, say, $x = -6$, and substitute it into $x^2 + 3x - 10$:

$$(-6)^2 + 3(-6) - 10 = 36 - 18 - 10 = 8 > 0.$$

Because $x^2 + 3x - 10$ is positive at one number in $(-\infty, -5)$ and cannot change its algebraic sign over this interval, we can conclude that $x^2 + 3x - 10 > 0$ for all values of x in $(-\infty, -5)$.

Similarly, to find the algebraic sign of $x^2 + 3x - 10$ over the interval $(-5, 2)$, we substitute a test number, say, $x = 0$, to obtain

$$0^2 + 3(0) - 10 = -10 < 0,$$

and conclude that $x^2 + 3x - 10 < 0$ for all values of x in the interval $(-5, 2)$. Finally, using another test number, say, $x = 3$, from the interval $(2, \infty)$, we find that

$$3^2 + 3(3) - 10 = 9 + 9 - 10 = 8 > 0,$$

so that $x^2 + 3x - 10 > 0$ for all values of x in the interval $(2, \infty)$.

The information we now have about the algebraic sign of $x^2 + 3x - 10$ in each interval is summarized in Figure 2. Thus, $x^2 + 3x - 10 < 0$ if and only if $-5 < x < 2$; in other words, the solution set is the open interval $(-5, 2)$.

Figure 2   algebraic sign of $x^2 + 3x - 10$

value is zero   value is zero

$$\cdots + + + + + + + + \,|\, - - - - - - - - \,|\, + + + + + + + + \cdots$$

$$\quad\quad\quad\quad -5 \quad\quad\quad\quad\quad 2$$

The method illustrated above can be used to solve any polynomial inequality in one unknown. This method is summarized in the following step-by-step procedure.

Procedure for Solving a Polynomial Inequality

Step 1. Bring the inequality into standard form, with zero on the right side and a polynomial on the left side.

Step 2. Set the polynomial equal to zero and find all real roots of the resulting equation.

Step 3. Arrange the roots obtained in step 2 in increasing order on a number line. These roots will divide the number line into open intervals.

Step 4. The algebraic sign of the polynomial cannot change over any of the intervals obtained in step 3. Determine this sign for each interval by selecting a test number on the interval and substituting it for the unknown in the polynomial. The algebraic sign of the resulting value is the sign of the polynomial over the entire interval.

Step 5. Draw a figure showing the information obtained in step 4 about the algebraic sign of the polynomial over the various open intervals. The solution set of the inequality can be read from this figure.

If it turns out in step 2 that the polynomial has no real roots, there is just one open interval involved, namely the entire number line $\mathbb{R}$.

Examples   Solve each polynomial inequality.

1 $4x^2 + 8x \geq 5$

Solution   We carry out the steps in the procedure above.

(1) The given inequality is equivalent to $4x^2 + 8x - 5 \geq 0$.

(2) We write $4x^2 + 8x - 5 = 0$ and factor the left side to obtain

$$(2x + 5)(2x - 1) = 0.$$

The roots are $x = -\frac{5}{2}$ and $x = \frac{1}{2}$.

(3) The open intervals over which the algebraic sign of $4x^2 + 8x - 5$ is constant are $(-\infty, -\frac{5}{2})$, $(-\frac{5}{2}, \frac{1}{2})$, and $(\frac{1}{2}, \infty)$ (Figure 3).

Figure 3

(4) For the interval $(-\infty, -\frac{5}{2})$, we substitute a test number, say, $x = -3$, in the polynomial $4x^2 + 8x - 5$ to obtain

$$4(-3)^2 + 8(-3) - 5 = 7 > 0.$$

We conclude that $4x^2 + 8x - 5 > 0$ for all values of x in the interval $(-\infty, -\frac{5}{2})$. For the interval $(-\frac{5}{2}, \frac{1}{2})$, we substitute a test number, say, $x = 0$, in the polynomial to get

$$4(0)^2 + 8(0) - 5 = -5 < 0.$$

Therefore, $4x^2 + 8x - 5 < 0$ for all values of x in the interval $(-\frac{5}{2}, \frac{1}{2})$. For the interval $(\frac{1}{2}, \infty)$, we substitute a test number, say, $x = 1$, and find that

$$4(1)^2 + 8(1) - 5 = 7 > 0.$$

Hence, $4x^2 + 8x - 5 > 0$ for all values of x in the interval $(\frac{1}{2}, \infty)$.

(5) The information obtained in step 4 is summarized in Figure 4. From this figure, we see that $4x^2 + 8x - 5 \geq 0$ for $x \leq -\frac{5}{2}$ and also for $x \geq \frac{1}{2}$. In other words, the solution set consists of all real numbers that belong to either one of the intervals $(-\infty, -\frac{5}{2}]$ or $[\frac{1}{2}, \infty)$. Note that the *nonstrict* inequality $4x^2 + 8x - 5 \geq 0$ requires that we include the values $-\frac{5}{2}$ and $\frac{1}{2}$ in the solution set.

Figure 4

2 $5x^2 + x + 16 < -4x^2 - 23x$

Solution   The given inequality is equivalent to

$$9x^2 + 24x + 16 < 0; \quad \text{that is,} \quad (3x + 4)^2 < 0,$$

which is impossible. Therefore, the inequality has no solution.

7.2 Rational Inequalities

An inequality such as $\dfrac{2x + 5}{x + 1} < \dfrac{x + 1}{x - 1}$, in which all members are rational expressions in the unknown, is called a **rational inequality.** To rewrite such an inequality in *standard form*, subtract the expression on the right side from both sides and then simplify the left side. Be careful—don't *multiply* both sides of an inequality by an expression containing the unknown unless you are certain that the multiplier can have only positive values. (Why not?)

Example   Rewrite the rational inequality $\dfrac{2x+5}{x+1} < \dfrac{x+1}{x-1}$ in standard form.

Solution   We begin by subtracting $\dfrac{x+1}{x-1}$ from both sides:

$$\frac{2x+5}{x+1} - \frac{x+1}{x-1} < 0.$$

Using the L.C.D. $(x+1)(x-1)$, we combine the fractions on the left side to obtain

$$\frac{(2x+5)(x-1) - (x+1)(x+1)}{(x+1)(x-1)} < 0 \quad \text{or} \quad \frac{x^2 + x - 6}{(x+1)(x-1)} < 0.$$

Finally, factoring the numerator, we can write the inequality as

$$\frac{(x+3)(x-2)}{(x+1)(x-1)} < 0.$$

When you have rewritten a rational inequality in standard form, you can solve it by considering how the algebraic sign of the expression on the left side changes as the unknown is varied. Notice that a fraction can change its algebraic sign only if its numerator or denominator changes its algebraic sign. Hence, *intervals where a rational expression is positive are separated from intervals where it is negative by values of the variable for which the numerator or denominator is zero.* (Don't forget that the fraction itself is undefined when its denominator is zero.) It is thus possible to solve rational inequalities by a simple modification of the step-by-step procedure for solving polynomial inequalities. The method is illustrated in the following example.

Example   Solve the rational inequality $\dfrac{(x+3)(x-2)}{(x+1)(x-1)} < 0$.

Solution   The numerator of $\dfrac{(x+3)(x-2)}{(x+1)(x-1)}$ is zero when $x = -3$ and when $x = 2$; its denominator is zero when $x = -1$ and when $x = 1$. We arrange these four numbers in order along a number line to determine five open intervals over which the algebraic sign of $\dfrac{(x+3)(x-2)}{(x+1)(x-1)}$ does not change, namely, $(-\infty, -3)$, $(-3, -1)$, $(-1, 1)$, $(1, 2)$, and $(2, \infty)$ (Figure 5). We select test numbers, one from each of these intervals,

Figure 5

intervals over which the algebraic sign of $\dfrac{(x+3)(x-2)}{(x+1)(x-1)}$ doesn't change say, $-4, -2, 0, \frac{3}{2}$, and 3. Substituting these test numbers into $\dfrac{(x + 3)(x - 2)}{(x + 1)(x - 1)}$, we obtain

$$\frac{(-4 + 3)(-4 - 2)}{(-4 + 1)(-4 - 1)} = \frac{2}{5} > 0, \qquad \frac{(-2 + 3)(-2 - 2)}{(-2 + 1)(-2 - 1)} = -\frac{4}{3} < 0,$$

$$\frac{(0 + 3)(0 - 2)}{(0 + 1)(0 - 1)} = 6 > 0, \qquad \frac{(\frac{3}{2} + 3)(\frac{3}{2} - 2)}{(\frac{3}{2} + 1)(\frac{3}{2} - 1)} = -\frac{9}{5} < 0,$$

$$\frac{(3 + 3)(3 - 2)}{(3 + 1)(3 - 1)} = \frac{3}{4} > 0.$$

The algebraic signs obtained for the test numbers determine the algebraic signs of $\dfrac{(x + 3)(x - 2)}{(x + 1)(x - 1)}$ over each of the five open intervals, as shown in Figure 6. From this figure, we see that the given inequality $\dfrac{(x + 3)(x - 2)}{(x + 1)(x - 1)} < 0$ holds for values of x in the intervals $(-3, -1)$ and $(1, 2)$, and that it fails to hold for all other values of x. Therefore, the solution set consists of all real numbers that belong to either of the intervals $(-3, -1)$ or $(1, 2)$.

Figure 6

PROBLEM SET 7

In problems 1–4, rewrite the polynomial inequality in standard form.

1. $6x^2 - x + 3 < 5x^2 + 5x - 5$

2. $17x^2 - 12x + 33 > 14x^2 - 15x + 28$

3. $\dfrac{x^2}{2} + \dfrac{25}{2} \geq 5x$

4. $4x^3 + x \leq 3x^3 + 3x^2 - x$

In problems 5–30, use the step-by-step procedure of Section 7.1 to solve the polynomial inequality.

5. $(x - 2)(x - 4) < 0$

6. $x^2 - 2x < 15$

7. $(x + 4)(x + 6) > 0$

8. $x^2 > x + 6$

9. $x^2 + 49 \leq 14x$

10. $x^2 + 32 \geq 12x + 6$

11. $3x^2 + 22x + 35 \geq 0$

12. $x^2 + 25 \geq 10x$

13. $6x^2 + 1 \leq 5x$

14. $2t^2 + 3t \leq 5$

15. $25y^2 + 4 \geq 20y$

16. $x(6x - 13) > -6$

17. $x(10 - 3x) < 8$

18. $y(10y + 19) \leq 15$

19. $1 - y \geq 2y^2$

20. $2x^2 < x + 2$

21. $(x + 2)^2 > (3x + 1)^2$

22. $(x - 1)(x + 1)(x + 2) \geq 0$

23. $x^2 - x + 1 > 0$

24. $4x^3 - 8x^2 + 2 < x$

25. $5x^3 \geq x^2 + 3x$

26. $x^3 + x^2 \leq x + 1$

27. $x^3 > x$

28. $x^4 + 12 > 7x^2$

29. $(x + 4)(x - 2) \leq (2x - 3)(x + 1)$

30. $(2x + 1)(3x - 1)(x - 2) \leq 0$

In problems 31–34, rewrite the rational inequality in standard form.

31. $\dfrac{2x}{x+3} > \dfrac{1}{x+3}$

32. $\dfrac{1-5x}{1+2x} \le \dfrac{9-3x}{x-3}$

33. $\dfrac{x^2-3}{x^2+5x+6} \le \dfrac{1}{x+2}$

34. $\dfrac{3x+1}{x+2} > \dfrac{2x-1}{x+3}$

In problems 35–48, solve the rational inequality.

35. $\dfrac{3x+6}{x+2} > 0$

36. $\dfrac{2x-1}{3x-2} < 0$

37. $\dfrac{3y-5}{y+3} \le 0$

38. $\dfrac{(t-4)^2}{3t+1} \ge 0$

39. $\dfrac{8x}{2x-3} > 0$

40. $\dfrac{x^2-2x-15}{2x^2+3x-5} \le 0$

41. $\dfrac{x+1}{x-2} \le 1$

42. $\dfrac{x+2}{x-3} \ge 1$

43. $\dfrac{6x+8}{x+2} > \dfrac{3x-1}{x-1}$

44. $\dfrac{x+1}{x+2} < \dfrac{x+3}{x+4}$

45. $\dfrac{(x-3)(x+3)}{(x+1)(x-2)} > 0$

46. $\dfrac{(x+1)(x-2)(x+2)}{(x-3)(x-4)(x+5)} \le 0$

47. $\dfrac{(x-1)(x+2)(3-x)}{(x+3)(x^2-x-2)} \le 0$

48. $\dfrac{(x-2)^2(x+1)(2x-1)}{(x+3)(x-1)^2(x+4)} \ge 0$

49. A projectile is fired straight upward from ground level with an initial velocity of 480 feet per second. Its distance above ground level t seconds later is given by the expression $480t - 16t^2$. For what time interval is the projectile more than 3200 feet above ground level?

50. In economics, the **supply equation** for a commodity relates the selling price, p dollars per unit, to the total number of units q that the manufacturers would be willing to put on the market at that selling price. If the manufacturers' cost of producing one unit is c dollars, then, assuming that all units are sold, the total profit P dollars to the manufacturers is given by $P = pq - cq$. The supply equation for a small manufacturer of souvenirs in a resort area is $q = 8p + 90$, and each souvenir costs the manufacturer $c = \$4$ to produce. Assuming that all souvenirs are sold, what price per souvenir should be charged to bring in a total profit of at least \$315?

51. David, Dean, and Scott have grown 100 pumpkins. They believe they can sell all their pumpkins at a price of 75 cents each, but that each 5-cent increase in the price per pumpkin will result in the sale of 4 fewer pumpkins. What is the maximum price they can charge per pumpkin and still each make \$25 on the sale?

52. By examining the algebraic signs of the factors $x + 5$ and $x - 2$, prove that the intervals where $x^2 + 3x - 10$ is positive are separated from intervals where it is negative by values of x for which it is zero.

53. In an experimental solar power plant, at least 60 mirrors that will focus sunlight on a boiler are to be arranged in rows, each row containing the same number of mirrors. For optimal operation, it is determined that the number of mirrors per row must be 2 more than twice the number of rows. At least how many rows must there be?

54. By examining the algebraic signs of $x + 3$, $x - 2$, $x + 1$, and $x - 1$, prove that the intervals where $\dfrac{(x+3)(x-2)}{(x+1)(x-1)}$ is positive are separated from intervals where it is negative by values of x for which it is zero or undefined.

Ⓒ In problems 55–58, solve each inequality with the aid of a calculator.

55. $23.8x^2 - 21.1x + 3.3 \ge 0$

56. $2.12x^2 - 9.87x + 6.79 < 0$

57. $\dfrac{3.4x - 2.7}{2.1x^2 + 8.9x - 1.3} < 0$

58. $\dfrac{22.71x^2 + 11.11x - 8.23}{7.23x^2 - 41.02x + 6.66} \le 0$

REVIEW PROBLEM SET

In problems 1–16, solve each equation.

1. $6x - 30 = 50 - 10x$

2. $10(t + 1) = 6t - 3(2t + 5)$

3. $3(2y - 3) = 4(y + 1) - 3$

4. $7(w + 2) - 4(w + 1) = -21$

5. $5\left(\dfrac{t}{3} + \dfrac{2}{5}\right) = 2\left(\dfrac{4t}{5} + \dfrac{3}{2}\right)$

6. $\frac{1}{5}(6a - 7) = \frac{1}{2}(3a - 1)$

7. $\dfrac{5x - 2}{6} - \dfrac{2x - 5}{3} = 1$

8. $\dfrac{5 + u}{6} - \dfrac{10 - u}{3} = 1$

9. $\dfrac{3t}{4} - 12 = \dfrac{3(t - 12)}{5}$

10. $\dfrac{10 - r}{2} - \dfrac{5 - r}{3} = \dfrac{5}{2}$

11. $\dfrac{3 - y}{3y} + \dfrac{1}{4} = \dfrac{1}{2y}$

12. $\dfrac{5}{2x} = \dfrac{9 - 2x}{8x} + 3$

13. $\dfrac{n - 5}{n + 5} + \dfrac{n + 15}{n - 5} = \dfrac{25}{25 - n^2} + 2$

14. $\dfrac{2}{x - 2} + \dfrac{1}{x + 1} = \dfrac{1}{x^2 - x - 2}$

15. $\dfrac{2x}{x + 7} - 1 = \dfrac{x}{x + 3} + \dfrac{1}{x^2 + 10x + 21}$

16. $\dfrac{9}{y - 3} - \dfrac{4}{y - 6} = \dfrac{18}{y^2 - 9y + 18}$

In problems 17–20, solve each equation for the indicated unknown.

17. $\dfrac{x}{a + b} - b = a$ for x

18. $\dfrac{ay}{b} + \dfrac{by}{a} = 2$ for y

19. $\dfrac{m}{x} + \dfrac{x - m}{x} - m = 1$ for x

20. $\dfrac{a - a^2}{x} = 1 - \dfrac{x - 2a}{2x}$ for x

© In problems 21–24, solve each equation with the aid of a calculator. Round off all answers to the correct number of significant digits.

21. $71.42x - 33.21 = 0$

22. $\dfrac{35x + 17.05}{-3x - 21.14} = 37.01$

23. $(3.004 \times 10^{17})x + (2.173 \times 10^{15}) = 0$

24. $13.031 = 2\pi(11.076 + x)$

25. A pharmacist has 8 liters of a 15% solution of acid. How much distilled water must she add to it to reduce the concentration of acid to 10%?

26. A family has spent a total of $3500 on a solar water heater and insulation for their home. Federal income tax credits of 30% of the cost of the solar water heater and 15% of the cost of the insulation will amount to $825. How much was spent on the water heater and how much on insulation?

27. A total of $10,000 is invested, part at 7% annual simple interest, and part at 8%. If a return of $735 is expected at the end of one year, how much was invested at each rate?

28. In a certain district, 25% of the voters eligible for grand-jury service are members of minorities. Records show that during a 5-year period, 3% of all eligible people in that district were selected for grand-jury service, whereas 4% of the eligible minority members were selected. What percent of the eligible nonminority members in that district were selected for grand-jury duty over the 5-year period?

29. By installing a $120 thermostat that automatically reduces the temperature setting at night, a family hopes to cut its annual bill for heating oil by 8%, and thereby recover the cost of the thermostat in fuel savings after 2 years. What was the family's annual fuel bill before installing the thermostat?

30. An archaeologist discovers a crown weighing 800 grams in the tomb of an ancient king. Evidence indicates that the crown is made of a mixture of gold and silver. Gold weighs 19.3 grams per cubic centimeter; silver weighs 10.5 grams per cubic centimeter; and it is found that the crown weighs 16.2 grams per cubic centimeter. How many grams of gold does the crown contain?

31. Juanita and Pedro run a race from point A to point B and Juanita wins by 5 meters. She calculates that Pedro will have a fair chance to win a second race to point B if he starts at point A and she starts 6 meters behind him. What is the distance between point A and point B?

32. Plant A produces methane from biomass three times as fast as plant B. The two plants operating together yield a certain amount of methane in 12 hours. How long would it take plant A operating alone to yield the same amount of methane?

In problems 33–36, complete the square by adding a constant to each quantity.

33. $x^2 + x$

34. $x^2 - 24x$

35. $x^2 - 9x$

36. $x^2 + \sqrt{3}x$

In problems 37–50, find all roots (real or complex) of each equation. (In problems 43 and 44, round off your answers to three decimal places.)

37. $2x^2 - 5x - 7 = 0$

38. $2t^2 - 7t + 5 = 0$

39. $10y^2 = 3 + y$

40. $r(3r + 1) = 4$

41. $5x^2 - 7x - 2 = 0$

42. $x^2 + 2x = 5$

[C] 43. $2.35t^2 + 6.42t - 0.91 = 0$

[C] 44. $1.7y^2 + 0.33y = \dfrac{\pi}{3}$

45. $4x^2 + 3 = 3x$

46. $4s^2 + \frac{11}{2}s + 4 = 0$

47. $2r = 1 - \dfrac{2}{r}$

48. $5x = \dfrac{3}{x - 2}$

49. $\dfrac{1 - 6x}{2x} = 2x$

50. $x^2 + \sqrt{2}x - 1 = 0$

In problems 51–54, use the discriminant to determine the nature of the roots of each quadratic equation without solving it.

51. $2x^2 - 3x + 1 = 0$

[C] 52. $169x^2 + 286x + 121 = 0$

53. $4x^2 + 9x + 6 = 0$

54. $\sqrt{3}x^2 + 2x - \sqrt{3} = 0$

55. A commercial jet could decrease the time needed to cover a distance of 2475 nautical miles by one hour if it were to increase its present speed by 100 knots (1 knot = 1 nautical mile per hour). What is its present speed?

[C] 56. A biologist finds that the number N of water mites in a sample of lake water is related to the water temperature T in degrees Celsius by the equation

$$N = 5.5T^2 - 19T.$$

At what temperature is $N = 2860$?

57. The diagonal of a rectangular television screen measures 20 inches and the height of the screen is 4 inches less than its width. Find the dimensions of the screen.

58. All of the electricity in an alternative-energy home is supplied by batteries that are charged by a combination of solar panels and a wind-powered generator. Working together on a sunny, windy day, the solar panels and the generator can charge the batteries in 2.4 hours. Working alone on a windless, sunny day, it takes the solar panels 2 hours longer to charge the batteries than it takes the generator to do the job alone on a windy night. How long does it take the solar panels to charge the batteries on a windless, sunny day?

In problems 59–92, find all real roots of each equation. Check for extraneous roots when necessary.

59. $\sqrt{x - 3} - 6 = 0$

60. $8 - \sqrt[3]{y - 1} = 6$

61. $t + \sqrt{t + 1} = 7$

62. $\sqrt{2y - 5} = 10 - y$

63. $\sqrt[4]{x^2 + 5x + 6} = \sqrt{x + 4}$

64. $\sqrt[3]{t + 1} = \sqrt{t + 1}$

65. $\sqrt{t + 1} + \sqrt{2t + 3} = 5$

66. $\sqrt{5x - 1} - \sqrt{2x} = \sqrt{x - 1}$

67. $\sqrt{t + 4} + \sqrt{2t + 10} = 3$

68. $\sqrt{2r + 5} - \sqrt{r + 2} = \sqrt{3r - 5}$

69. $t^4 - 5t^2 + 4 = 0$

70. $24v^4 - v^2 - 10 = 0$

71. $9c^4 - 18c^2 + 8 = 0$

72. $6a^4 + a^2 - 15 = 0$

73. $y - 2y^{1/2} - 15 = 0$

74. $t^{2/3} + 2t^{1/3} - 48 = 0$

75. $\left(x - \dfrac{8}{x}\right)^2 + \left(x - \dfrac{8}{x}\right) = 42$

76. $(t^2 + 2t)^2 - 2(t^2 + 2t) = 3$

77. $z^{-2} + z^{-1} = 6$

78. $x^{-2} = 2x^{-1} + 8$

79. $y + 2 + \sqrt{y + 2} = 2$

80. $x + 7 = 2 + \sqrt{x + 7}$

81. $\sqrt{x + 20} + 3 = 4\sqrt{x + 20}$

82. $2x^2 + x - 4\sqrt{2x^2 + x + 4} = 1$

83. $x^5 - 2x^3 - x^2 + 2 = 0$

84. $(u - 1)^3 - (u - 1)^2 = u - 2$

85. $t^4 - t^3 - t + 1 = 0$

86. $x^{7/12} + x^{1/4} - x^{1/3} = 1$

87. $x^{-3/5} + x^{-2/5} - 2x^{-1/5} - 2 = 0$

88. $u - 4u^{5/7} + 32u^{2/7} = 128$

89. $\sqrt{5t + 2\sqrt{t^2 - t + 7}} - 4 = 0$

90. $\dfrac{\sqrt{v + 16}}{\sqrt{4 - v}} + \dfrac{\sqrt{4 - v}}{\sqrt{v + 16}} = \dfrac{5}{2}$

91. $\dfrac{\sqrt{1 + x} + \sqrt{1 - x}}{\sqrt{1 + x} - \sqrt{1 - x}} = 2$

92. $\sqrt{13 + 3\sqrt{y + 5 + 4\sqrt{y + 1}}} = 5$

93. In engineering, the equation

$$L\sqrt[3]{F} = d\sqrt[3]{4bwY}$$

can be used to determine the bending of a beam with a rectangular cross section that is freely supported at the ends and loaded at the middle with a force F. Here L is the length of the beam, d is its depth, w is its width, b is the amount of bending in the middle, and Y is Young's modulus of elasticity for the material of the beam. Solve the equation for b in terms of L, F, d, w, and Y.

[c] **94.** The *equally-tempered* musical scale, used by Bach in his "*Well-Tempered Clavier*" of 1722, consists of thirteen notes, C, C$^\#$, D, D$^\#$, E, F, F$^\#$, G, G$^\#$, A, A$^\#$, B, and high C. In this scale, the frequency of high C is twice the frequency of C, and the ratios of the frequencies of successive notes are equal. If the frequency of A is taken to be 440 Hz ("concert A"), find the frequencies of the remaining notes.

95. Square roots can be found on a four-function calculator as follows: To find $\sqrt{x}$, begin by making a reasonable estimate. Add half of the estimated number to x divided by twice the estimated number. Call the result the second estimate. Repeat the procedure with the second estimate to obtain a third estimate. Continue in this way until the estimate no longer changes when you carry out the procedure. The resulting number is $\sqrt{x}$, to within the calculator's limits of accuracy. Prove this algebraically.

96. Government economists monitoring the rate of inflation calculate the percentage of increase in the cost of living at the end of each month. If m is the percentage increase in a given month, then the equivalent annual percentage increase a is given by

$$a = 100\left(1 + \frac{m}{100}\right)^{12} - 100.$$

(a) Solve this equation for m in terms of a.

[c] (b) Find the equivalent annual percent of increase corresponding to $m = 1\%$ increase per month.

97. Given that $-3 \le 2 - 5x < 4$, what equivalent inequality is obtained if

(a) 3 is added to all members?

(b) All members are divided by -5?

98. If $0 \le a < b$, prove that $\sqrt{a} < \sqrt{b}$.

In problems 99–104, solve each inequality and sketch its solution set on a number line.

99. $3x + 1 \ge 16$

100. $-4x + 3 < 15$

101. $-9t - 3 < -30$

102. $-8y - 4 \ge -32$

103. $\dfrac{4x}{3} - 2 > \dfrac{x}{4}$

104. $\dfrac{3x}{2} + 13 \le \dfrac{x}{8}$

105. A student on academic probation must earn a C in his algebra course in order to avoid being expelled from college. Since he is lazy and has no interest in earning a grade any higher than necessary, he wants his final numerical grade for algebra to lie on the interval $[70, 80)$. This grade will be determined by taking three-fifths of his classroom average, which is 68, plus two-fifths of his score on the final exam. Since he has always had trouble with inequalities, he can't figure out the range in which his score on the final exam should fall to allow him to stay in school. Can you help him?

106. A runner leaves a starting point and runs along a straight road at a steady speed of 8.8 miles per hour. After a while, she turns around and begins to run back to the starting point. How fast must she run on the way back so that her average speed for the entire run will be greater than 8 miles per hour, but no greater than 8.5 miles per hour? (Caution: The average speed for her entire run *cannot* be found by adding 8.8 miles per hour to her return speed and dividing by 2.)

In problems 107–112, find the distance between the numbers and illustrate on a number line.

107. -3 and 4 **108.** 0 and -5 C **109.** -2.735 and $-\pi$

110. -3.2 and 4.1 **111.** $-\frac{2}{3}$ and $\frac{5}{2}$ C **112.** 1.42 and $\sqrt{2}$

In problems 113–118, determine which equations or inequalities are true for all values of the variables.

113. $|x^2 - 4| = |x - 2||x + 2|$ **114.** $|x^2 - 4| = x^2 + 4$ **115.** $|-x|^2 = x^2$

116. $|x - y| = |y - x|$ **117.** $|x^2 + 3x| \le x^2 + |3x|$ **118.** $|x - y| \le |x| + |y|$

In problems 119–146, solve each equation or inequality. Sketch the solution set of each inequality on a number line.

119. $|2x - 3| = 6$ **120.** $|2y + 1| = 5$ **121.** $|t - 1| = 3t$

122. $|2t + 3| = |t - 2|$ **123.** $|3x - 6| > 3$ **124.** $|2x - 3| \le 5$

125. $|4x - 3| \ge 5$ **126.** $|2x + 1| < 3$ **127.** $|2 - 3y| \le 1$

128. $|x - 1| < |x + 1|$ **129.** $|5 - 3z| < \frac{1}{2}$ **130.** $|2 - s| > |5 - 3s|$

131. $6x^2 - 11x - 10 < 0$ **132.** $6x^2 - x \ge 2$ **133.** $6 - 5x - 6x^2 \ge 0$

134. $6x^2 \le 15x$ **135.** $5 > x(3x + 14)$ **136.** $x^5 + 2x^4 < 2x^3 + x$

137. $x(x + 3)(x + 5) < 0$ **138.** $x^4 + 1 \ge x^3 + x$ **139.** $\dfrac{3x - 1}{2x - 5} > 0$

140. $\dfrac{t + 1}{t - 2} \le 0$ **141.** $\dfrac{3x - 1}{5x - 7} \ge 2$ **142.** $\dfrac{5x - 3}{7x + 1} \le 1$

143. $\dfrac{x(x + 1)(x - 2)}{(x + 2)(x + 3)} < 0$ **144.** $\dfrac{(x - 1)(x - 2)(x + 3)}{(x - 4)(x + 5)(x + 1)} > 0$ **145.** $\dfrac{x^2(x^2 - 1)}{x^2 - x - 2} \ge 0$

146. $\dfrac{1 - 5x}{9 - 3x} \le \dfrac{2x + 1}{x - 3}$

C **147.** A company specializes in buying complimentary copies of textbooks from professors and selling the books to students. The company has purchased 500 copies of a popular calculus book from professors at $5 per copy. Suppose that all of these copies will be sold to students if the selling price is $10 per copy, but that each $1 increase in the selling price will result in 100 fewer copies being sold. If the profit to the company from the sale of these calculus books is to be between $1000 and $1100, inclusive, what will the selling price be? [Hint: Profit = Revenue − Cost.]

148. It is predicted that the number N of cars to be produced by an automobile factory for the coming year will satisfy the inequality $|N - 4{,}000{,}000| < 500{,}000$. Describe the anticipated production as an interval of real numbers.

149. Engineers estimate the percentage p of the heat stored in wood that will be delivered by a free-standing stove with good air control satisfies the inequality $|p - 65| \leq 10$. Find an interval containing p according to this estimate.

150. The diameter d, in millimeters, of a human capillary satisfies $|d - 10^{-2}| \leq 5 \times 10^{-3}$. According to this inequality, what are the maximum and minimum diameters of a human capillary?

151. It has been predicted that over the next decade, the number Q of quadrillions of Btu's of energy used per year in the United States will satisfy the inequality $|Q - 95| \leq 10$. Find an interval containing Q according to this estimate.

152. Automakers in Detroit calculate that for each car they produce, they will have to replace a weight W of not less than 400 pounds and not more than 500 pounds of steel by lightweight materials, in order to meet mileage goals set by the government. Find numbers A and B such that W satisfies the inequality $|W - A| \leq B$.

3 Functions and Their Graphs

Ever since the Italian scientist Galileo Galilei (1564–1642) first used quantitative methods in the study of dynamics, advances in our understanding of the natural world have become more and more dependent on the use of mathematics. Important quantities have been identified, methods of measurement have been developed, and relationships among quantities have been formulated as scientific laws and principles. The idea of a *function* enables us to express relationships among observable quantities with efficiency and precision. In this chapter we discuss the cartesian coordinate system, equations of straight lines and circles, graphs of functions, the algebra of functions, and inverse functions.

1   The Cartesian Coordinate System

In Section 2 of Chapter 1, we saw that a point P on a number line can be specified by a real number x called its **coordinate.** Similarly, by using a *cartesian coordinate system,* named in honor of the French philosopher and mathematician René Descartes (1596–1650), we can specify a point P in the plane with *two* real numbers, also called **coordinates.**

A **cartesian coordinate system** consists of two perpendicular number lines, called **coordinate axes,** which meet at a common origin O (Figure 1). Ordinarily, one of the number lines, called the ***x* axis,** is horizontal, and the other, called the ***y* axis,** is vertical. Numerical coordinates increase to the right along the x axis and upward along the y axis. We usually use the same scale (that is, the same unit distance) on the two axes, although in some of our figures, space considerations make it convenient to use different scales.

Figure 1

117

Figure 2

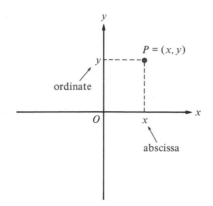

If P is a point in the plane, the **coordinates** of P are the coordinates x and y of the points where perpendiculars from P meet the two axes (Figure 2). The x coordinate is called the **abscissa** of P and the y coordinate is called the **ordinate** of P. The coordinates of P are traditionally written as an ordered pair (x, y) enclosed in parentheses, with the abscissa first and the ordinate second. (Unfortunately, this is the same symbolism used for an open interval; however, it is always clear from the context what is intended.)

To **plot** the point P with coordinates (x, y) means to draw cartesian coordinate axes and to place a dot representing P at the point with abscissa x and ordinate y. You can think of the ordered pair (x, y) as the numerical "address" of P. The correspondence between P and (x, y) seems so natural that in practice we identify the point P with its "address" (x, y) by writing $P = (x, y)$. With this identification in mind, we call an ordered pair of real numbers (x, y) a **point,** and we refer to the set of all such ordered pairs as the **cartesian plane** or the **xy plane.**

Figure 3

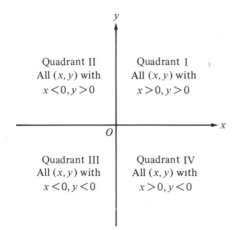

The x and y axes divide the plane into four regions called **quadrants I, II, III,** and **IV** (Figure 3). Quadrant I consists of all points (x, y) for which both x and y are positive, quadrant II consists of all points (x, y) for which x is negative and y is positive, and so forth, as shown in Figure 3. Notice that a point on a coordinate axis belongs to no quadrant.

Example 1 Plot each point and indicate which quadrant or coordinate axis contains the point.

(a) $(4, 3)$ (b) $(-3, 2)$ (c) $(-5, -1)$ (d) $(2, -4)$
(e) $(-3, 0)$ (f) $(0, 4)$ (g) $(0, -\frac{3}{2})$ (h) $(0, 0)$

Figure 4

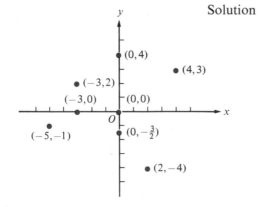

Solution   The points are plotted in Figure 4.
(a) $(4, 3)$ lies in quadrant I
(b) $(-3, 2)$ lies in quadrant II
(c) $(-5, -1)$ lies in quadrant III
(d) $(2, -4)$ lies in quadrant IV
(e) $(-3, 0)$ lies on the x axis
(f) $(0, 4)$ lies on the y axis
(g) $(0, -\frac{3}{2})$ lies on the y axis
(h) $(0, 0)$, the origin, lies on both axes

Example 2   Plot the point $P = (-4, 2)$ and determine the coordinates of the point Q if the line segment $\overline{PQ}$ is perpendicular to the x axis and is bisected by it.

Solution   We begin by plotting the point $P = (-4, 2)$ (Figure 5). We see that P is 2 units directly above the point with coordinate -4 on the x axis; hence, Q is 2 units directly below the same point. Therefore, $Q = (-4, -2)$.

Figure 5

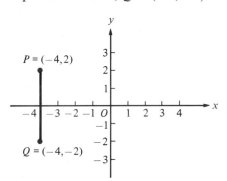

1.1 The Distance Formula

One of the attractive features of the cartesian coordinate system is that there is a simple formula that gives the distance between two points in terms of their coordinates. If P_1 and P_2 are two points in the cartesian plane, we denote the distance between P_1 and P_2 by $|\overline{P_1P_2}|$.

Theorem 1   **The Distance Formula**

> If $P_1 = (x_1, y_1)$ and $P_2 = (x_2, y_2)$ are two points in the cartesian plane, then the distance between P_1 and P_2 is given by
> $$|\overline{P_1P_2}| = \sqrt{(x_2 - x_1)^2 + (y_2 - y_1)^2}.$$

Proof   The *horizontal* distance between P_1 and P_2 is the same as the distance between the points with coordinates x_1 and x_2 on the x axis (Figure 6a). From the discussion of absolute value in Section 6 of Chapter 2, it follows that the horizontal distance between P_1 and P_2 is $|x_2 - x_1|$ units. Similarly, the *vertical* distance between P_1 and P_2 is $|y_2 - y_1|$ units (Figure 6b). If the line segment $\overline{P_1P_2}$ is neither horizontal

Figure 6

Figure 7

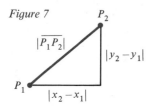

nor vertical, it forms the hypotenuse of a right triangle with legs of lengths $|x_2 - x_1|$ and $|y_2 - y_1|$ (Figure 7). Therefore, by the Pythagorean theorem,

$$|\overline{P_1P_2}|^2 = |x_2 - x_1|^2 + |y_2 - y_1|^2 = (x_2 - x_1)^2 + (y_2 - y_1)^2,$$

and it follows that

$$|\overline{P_1P_2}| = \sqrt{(x_2 - x_1)^2 + (y_2 - y_1)^2}.$$

We leave it to you to check that the formula works even if the line segment $\overline{P_1P_2}$ is horizontal or vertical (Problem 28).

Because $(x_2 - x_1)^2 = (x_1 - x_2)^2$ and $(y_2 - y_1)^2 = (y_1 - y_2)^2$, the distance formula can also be written as

$$|\overline{P_1P_2}| = \sqrt{(x_1 - x_2)^2 + (y_1 - y_2)^2}.$$

In other words, the order in which you subtract the abscissas or the ordinates does not affect the result.

Example 1   Let $A = (-2, -1)$, $B = (1, 3)$, $C = (-1, 2)$, and $D = (3, -2)$. Find
(a) the distance $|\overline{AB}|$ (b) the distance $|\overline{CD}|$.

Solution   By the distance formula (Theorem 1), we have
(a) $|\overline{AB}| = \sqrt{[1 - (-2)]^2 + [3 - (-1)]^2} = \sqrt{3^2 + 4^2} = \sqrt{25} = 5$ units
(b) $|\overline{CD}| = \sqrt{[3 - (-1)]^2 + (-2 - 2)^2} = \sqrt{32} = 4\sqrt{2}$ units.

Example 2   Ⓒ Let $P = (31.42, -17.04)$ and $Q = (13.75, 11.36)$. Using a calculator, find $|\overline{PQ}|$. Round off your answer to four significant digits.

Solution $$|\overline{PQ}| = \sqrt{(13.75 - 31.42)^2 + [11.36 - (-17.04)]^2} = 33.45 \text{ units.}$$

Example 3   Let $A = (-5, 3)$, $B = (6, 0)$, and $C = (5, 5)$.
(a) Plot the points A, B, and C, and draw the triangle ABC.
(b) Find the distances $|\overline{AB}|$, $|\overline{AC}|$, and $|\overline{BC}|$.
(c) Show that ACB is a right triangle.
(d) Find the area of triangle ACB.

Solution   (a) The points A, B, and C are plotted and the triangle ABC is drawn in Figure 8.

Figure 8

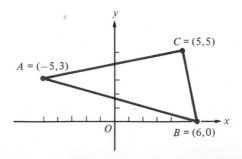

(b) $|\overline{AB}| = \sqrt{[6 - (-5)]^2 + (0 - 3)^2} = \sqrt{11^2 + (-3)^2} = \sqrt{130}$

$|\overline{AC}| = \sqrt{[5 - (-5)]^2 + (5 - 3)^2} = \sqrt{10^2 + 2^2} = \sqrt{104} = 2\sqrt{26}$

$|\overline{BC}| = \sqrt{(5 - 6)^2 + (5 - 0)^2} = \sqrt{1^2 + 5^2} = \sqrt{26}$

(c) Figure 8 leads us to suspect that the angle at vertex C is a right angle. To confirm this, we use the *converse* of the Pythagorean theorem; that is, we check to see if $|\overline{AC}|^2 + |\overline{BC}|^2 = |\overline{AB}|^2$. From (b), we have

$$|\overline{AC}|^2 + |\overline{BC}|^2 = 104 + 26 = 130 = |\overline{AB}|^2.$$

Therefore, ACB is, indeed, a right triangle.

(d) Taking $|\overline{AC}| = 2\sqrt{26}$ as the base of the triangle and $|\overline{BC}| = \sqrt{26}$ as its altitude, we find that area $= \frac{1}{2}$ base $\times$ altitude $= \frac{1}{2}(2\sqrt{26})\sqrt{26} = 26$ square units.

1.2 Graphs in the Cartesian Plane

The **graph** of an equation or inequality in two unknowns x and y is defined to be the set of all points $P = (x, y)$ in the cartesian plane whose coordinates x and y satisfy the equation or inequality. Many (but not all) equations in x and y have graphs that are smooth curves in the plane. For instance, consider the equation $x^2 + y^2 = 9$. We can rewrite this equation as $\sqrt{x^2 + y^2} = \sqrt{9}$ or as

$$\sqrt{(x - 0)^2 + (y - 0)^2} = 3.$$

Figure 9

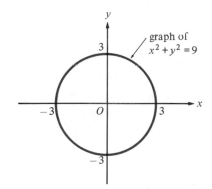

graph of
$x^2 + y^2 = 9$

By the distance formula (Theorem 1), this last equation holds if and only if the point $P = (x, y)$ is 3 units from the origin $O = (0,0)$. Therefore, the graph of $x^2 + y^2 = 9$ is a circle of radius 3 units with its center at the origin O (Figure 9).

If we are given a curve in the cartesian plane, we can ask whether there is an equation for which it is the graph. Such an equation is called an **equation for the curve** or an **equation of the curve**. For instance, $x^2 + y^2 = 9$ is an equation for the circle in Figure 9. Two equations or inequalities in x and y are said to be **equivalent** if they have the same graph. For example, the equation $x^2 + y^2 = 9$ is equivalent to the equation $\sqrt{x^2 + y^2} = 3$. We often use an equation for a curve to designate the curve; for instance, if we speak of "the circle $x^2 + y^2 = 9$," we mean "the circle for which $x^2 + y^2 = 9$ is an equation."

If $r > 0$, then the circle of radius r with center (h, k) consists of all points (x, y) such that the distance between (x, y) and (h, k) is r units (Figure 10). Using the distance formula (Theorem 1), we can write an equation for this circle as $\sqrt{(x - h)^2 + (y - k)^2} = r$, or, equivalently,

Figure 10

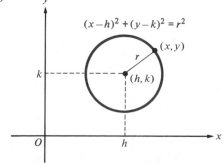

$(x - h)^2 + (y - k)^2 = r^2$

(x, y)

r

(h, k)

$$(x - h)^2 + (y - k)^2 = r^2.$$

This last equation is called the **standard form** for the equation of a circle in the xy plane.

Example 1   Find an equation for the circle of radius 5 with center at the point $(3, -2)$.

Solution   Here $r = 5$ and $(h, k) = (3, -2)$, so, in standard form, the equation of the circle is

$$(x - 3)^2 + [y - (-2)]^2 = 5^2 \quad \text{or} \quad (x - 3)^2 + (y + 2)^2 = 25.$$

If desired, we can expand the squares, combine like terms, and rewrite the equation in the equivalent form

$$x^2 + y^2 - 6x + 4y - 12 = 0.$$

Example 2   Sketch the graph of the equation $(x - 1)^2 + (y + 3)^2 = 16$.

Solution   The equation has the form $\qquad (x - h)^2 + (y - k)^2 = r^2,$

with $h = 1$, $k = -3$, and $r = 4$. Therefore, its graph is a circle of radius $r = 4$ with center

$$(h, k) = (1, -3)$$

(Figure 11).

Figure 11

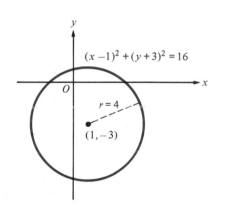

We study circles and closely related curves called *ellipses* in more detail in the final chapter.

PROBLEM SET 1

1. Plot each point and indicate which quadrant or coordinate axis contains it.
   (a) $(1, 6)$ (b) $(-2, 3)$ (c) $(4, -1)$ (d) $(4, -2)$
   (e) $(-1, -4)$ (f) $(0, 2)$ (g) $(-3, 0)$ (h) $(0, -4)$
2. Plot the points $A = (\pi, \sqrt{2})$, $B = (-\sqrt{3}, \sqrt{2})$, $C = (\sqrt{5}, -\sqrt{2})$, and $D = (\frac{3}{4}, -\frac{27}{5})$.

In problems 3–6, plot the point P and determine the coordinates of points Q, R, and S such that: (a) The line segment $\overline{PQ}$ is perpendicular to the x axis and is bisected by it. (b) The line segment $\overline{PR}$ is perpendicular to the y axis and is bisected by it. (c) The line segment $\overline{PS}$ passes through the origin and is bisected by it.

3. $P = (3, 2)$ 4. $P = (-4, -3)$ 5. $P = (-1, 3)$ 6. $P = \left(\dfrac{\sqrt{3}}{2}, -\dfrac{1}{2} \right)$

In problems 7–18, find the distance between the two points.

7. $(7, 10)$ and $(1, 2)$ **8.** $(-1, 7)$ and $(2, 11)$ **9.** $(7, -1)$ and $(7, 3)$
10. $(-4, 7)$ and $(0, -8)$ **11.** $(-6, 3)$ and $(3, -5)$ **12.** $(0, 4)$ and $(-4, 0)$
13. $(0, 0)$ and $(-8, -6)$ **14.** $(t, 4)$ and $(t, 8)$ **15.** $(-3, -5)$ and $(-7, -8)$
16. $(-\frac{1}{2}, -\frac{3}{2})$ and $(-3, -\frac{5}{2})$ **17.** $(2, -t)$ and $(5, t)$ **18.** $(a, b+1)$ and $(a+1, b)$

C In problems 19 and 20, use a calculator to find the distance between the two points. Round off your answer to four significant digits.

19. $(-2.714, 7.111)$ and $(3.135, 4.982)$ **20.** $(\pi, \frac{53}{4})$ and $(-\sqrt{17}, \frac{211}{5})$

In problems 21–24, (a) use the distance formula and the converse of the Pythagorean theorem to show that triangle ABC is a right triangle, and (b) find the area of triangle ABC.

21. $A = (1, 1)$, $B = (5, 1)$, $C = (5, 7)$ **22.** $A = (-1, -2)$, $B = (3, -2)$, $C = (-1, -7)$
23. $A = (0, 0)$, $B = (-3, 3)$, $C = (2, 2)$ **24.** $A = (-2, -5)$, $B = (9, \frac{1}{2})$, $C = (4, \frac{21}{2})$

25. Show that the points $A = (-2, -3)$, $B = (3, -1)$, $C = (1, 4)$, and $D = (-4, 2)$ are the vertices of a square.
26. Show that the distance between the points (x_1, y_1) and (x_2, y_2) is the same as the distance between the point $(x_1 - x_2, y_1 - y_2)$ and the origin.
27. If $A = (-5, 1)$, $B = (-6, 5)$, and $C = (-2, 4)$, determine whether or not triangle ABC is isosceles.
28. Verify that the distance formula (Theorem 1) holds even if the line segment is horizontal or vertical.
29. Find all values of t so that the distance between the points $(-2, 3)$ and (t, t) is 5 units.
30. If P_1, P_2, and P_3 are points in the plane, then P_2 lies on the line segment $\overline{P_1P_3}$ if and only if $|\overline{P_1P_3}| = |\overline{P_1P_2}| + |\overline{P_2P_3}|$. Illustrate this geometric fact with diagrams.

In problems 31–33, determine whether or not P_2 lies on the line segment $\overline{P_1P_3}$ by checking to see if $|\overline{P_1P_3}| = |\overline{P_1P_2}| + |\overline{P_2P_3}|$ (see problem 30).

31. $P_1 = (1, 2)$, $P_2 = (0, \frac{5}{2})$, $P_3 = (-1, 3)$
32. $P_1 = (-\frac{7}{2}, 0)$, $P_2 = (-1, 5)$, $P_3 = (2, 11)$
33. $P_1 = (2, 3)$, $P_2 = (3, -3)$, $P_3 = (-1, -1)$

34. Show that the point $P_2 = \left(\frac{a+c}{2}, \frac{b+d}{2}\right)$ is the midpoint of the line segment joining $P_1 = (a, b)$ and $P_3 = (c, d)$. [Hint: Use the condition in problem 30 to show that P_2 actually belongs to the line segment $\overline{P_1P_3}$. Then show that $|\overline{P_1P_2}| = |\overline{P_2P_3}|$.]
35. Use the result of problem 34 to find the coordinates of the midpoint P_2 of the line segment joining $P_1 = (-2, 3)$ and $P_3 = (8, 5)$. Plot the points P_1, P_2, and P_3.
36. On a cartesian coordinate grid, an aircraft carrier is detected by radar at point $A = (52, 71)$ and a submarine is detected by sonar at point $S = (47, 83)$. If distances are measured in nautical miles, how far is the carrier from a point on the surface of the water directly over the submarine?
37. Find the equation of (a) the circle of radius 4 with center at the origin, and (b) the circle of radius 2 with center at the point $(-1, 3)$.
38. Sketch the graph of the equation $x^2 + y^2 = 36$.
39. Sketch the graph of the equation $(x - 3)^2 + (y - 5)^2 = 49$.
40. Sketch the graph of the equation $x^2 - 4x + y^2 - 10y + 4 = 0$. [Hint: Complete the squares for both $x^2 - 4x$ and $y^2 - 10y$.]
41. Sketch the graph of the equation $(x + 2)^2 + (y - 1)^2 = 64$.
42. Sketch the graph of the inequality $x^2 + y^2 \le 9$.

2   The Slope of a Line

Figure 1

Figure 1

In ordinary language, the word "slope" refers to a steepness, an incline, or a deviation from the horizontal. For instance, we speak of a ski slope or the slope of a roof. In mathematics, the word "slope" has a precise meaning. Consider the line segment $\overline{AB}$ in Figure 1. The horizontal distance between A and B is called the **run** and the vertical distance between A and B is called the **rise.** The ratio of rise to run is called the **slope** of the line segment $\overline{AB}$ and is traditionally denoted by the symbol m:

$$m = \text{the slope of } \overline{AB} = \frac{\text{rise}}{\text{run}}.$$

Figure 2

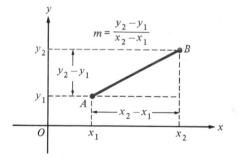

If the line segment $\overline{AB}$ is turned so that it becomes more nearly vertical, then the rise increases, the run decreases, and the slope $m = \text{rise/run}$ becomes larger. Therefore, the slope m really does give a numerical measure of the inclination or steepness of the line segment $\overline{AB}$—the greater the inclination, the greater the slope.

If the line segment $\overline{AB}$ is horizontal, its rise is zero, so its slope $m = \text{rise/run}$ is zero. Thus, *horizontal line segments have slope zero.* If $\overline{AB}$ slants downward to the right, as in Figure 2, its rise is considered to be negative; hence, its slope $m = \text{rise/run}$ is negative. (The run is always considered to be nonnegative.) Thus, *a line segment that slants downward from left to right has a negative slope. Notice that the slope $m = \text{rise/run}$ of a vertical line segment is undefined because the denominator is zero.*

Figure 3

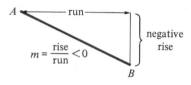

Now let $A = (x_1, y_1)$ and $B = (x_2, y_2)$, and consider the line segment $\overline{AB}$ (Figure 3). If B is above and to the right of A, the line segment $\overline{AB}$ has rise $= y_2 - y_1$ and run $= x_2 - x_1$, so its slope is

$$m = \frac{y_2 - y_1}{x_2 - x_1}.$$

Even if B is not above and to the right of A, the slope m of $\overline{AB}$ is given by the same formula (Problem 25), and we have the following theorem.

Theorem 1   **The Slope Formula**

Let $A = (x_1, y_1)$ and $B = (x_2, y_2)$ be two points in the cartesian plane. Then, if $x_1 \neq x_2$, the slope m of the line segment $\overline{AB}$ is

$$m = \frac{y_2 - y_1}{x_2 - x_1}.$$

Notice that $\dfrac{y_2 - y_1}{x_2 - x_1} = \dfrac{y_1 - y_2}{x_1 - x_2}$ (why?), so the slope of a line segment is the same regardless of which endpoint is called (x_1, y_1) and which is called (x_2, y_2).

Example In each case, sketch the line segment $\overline{AB}$ and find its slope m by using the slope formula (Theorem 1).

(a) $A = (-3, -2)$, $B = (4, 1)$ (b) $A = (-2, 3)$, $B = (5, 1)$

(c) $A = (-2, 4)$, $B = (5, 4)$ (d) $A = (3, -1)$, $B = (3, 6)$

Solution The line segments are sketched in Figure 4.

(a) $m = \dfrac{y_2 - y_1}{x_2 - x_1} = \dfrac{1 - (-2)}{4 - (-3)} = \dfrac{1 + 2}{4 + 3} = \dfrac{3}{7}$

(b) $m = \dfrac{y_2 - y_1}{x_2 - x_1} = \dfrac{1 - 3}{5 - (-2)} = \dfrac{-2}{5 + 2} = -\dfrac{2}{7}$

(c) $m = \dfrac{y_2 - y_1}{x_2 - x_1} = \dfrac{4 - 4}{5 - (-2)} = \dfrac{0}{5 + 2} = 0$

(d) m is undefined, since $x_2 - x_1 = 0$

Figure 4

(a)

(b)

(c)

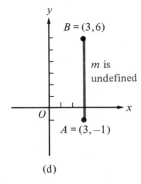

(d)

From the similar triangles APB and CQD in Figure 5, you can see that *two parallel line segments $\overline{AB}$ and $\overline{CD}$ have the same slope*. Likewise, *if two line segments $\overline{AB}$ and $\overline{CD}$ lie on the same straight line L, then they have the same slope* (Figure 6). The common slope of all the segments lying on a line L is called the **slope** of L.

Figure 5

Figure 6

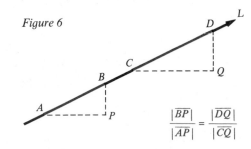

Example   Sketch the line L that contains the point $P = (1, 2)$ and has slope

(a) $m = \frac{2}{3}$ (b) $m = -\frac{2}{3}$.

Solution   (a) The condition $m = 2/3$ means that, for every 3 units we move to the right from a point on L, we must move up 2 units to get back to L. If we start at the point $P = (1, 2)$ on L, move 3 units to the right and 2 units up, we arrive at the point $Q = (1 + 3, 2 + 2) = (4, 4)$ on L. Because any two points on a line determine the line, we simply plot $P = (1, 2)$ and $Q = (4, 4)$, and use a straightedge to draw L (Figure 7a).

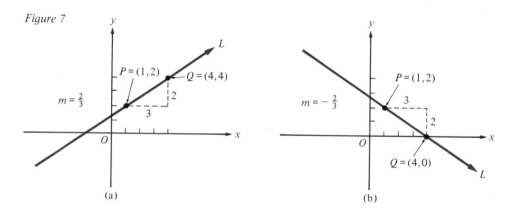

Figure 7

(a)   (b)

(b) The condition $m = -2/3$ means that, for every 3 units we move to the right from a point on L, we must move down 2 units to get back to L. If we start at the point $P = (1, 2)$ on L, move 3 units to the right and 2 units down, we arrive at the point $Q = (1 + 3, 2 - 2) = (4, 0)$ on L. Thus, we plot $P = (1, 2)$ and $Q = (4, 0)$, and use a straightedge to draw L (Figure 7b).

From the fact that two parallel line segments have the same slope, it follows that *two parallel lines have the same slope*. Conversely, you can prove by elementary geometry that *two distinct lines with the same slope are parallel* (Problem 26). Thus, we have the following theorem.

Theorem 2   **Parallelism Condition**

> Two distinct nonvertical straight lines in the cartesian plane are parallel if and only if they have the same slope.

Example   Sketch the line L that contains the point $P = (3, 4)$ and is parallel to the line segment $\overline{AB}$, where $A = (-1, 2)$ and $B = (4, -5)$.

Solution   By the slope formula, the line segment $\overline{AB}$ has slope

$$m = \frac{y_2 - y_1}{x_2 - x_1} = \frac{-5 - 2}{4 - (-1)} = -\frac{7}{5};$$

hence, by the parallelism condition (Theorem 2), L also has slope $m = -\frac{7}{5}$. Starting at the point $P = (3, 4)$ on L, we move 5 units to the right and 7 units down to the point $Q = (3 + 5, 4 - 7) = (8, -3)$. Using a straightedge, we draw the line L through P and Q (Figure 8).

Figure 8

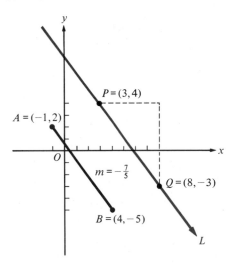

The following theorem gives a condition for two lines to be perpendicular.

Theorem 3   **Perpendicularity Condition**

> Two nonvertical straight lines in the cartesian plane are perpendicular if and only if the slope of one of the lines is the negative of the reciprocal of the slope of the other.

Proof   Let the two lines be L_1 and L_2, with slopes m_1 and m_2, respectively. The condition that the slope of either one of the lines is the negative of the reciprocal of the slope of the other can be written as $m_1 m_2 = -1$. Neither the angle between the lines nor their slopes are affected if we place the origin O at the point where L_1 and L_2

Figure 9

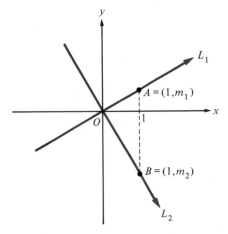

intersect (Figure 9). Starting at O on L_1, we move 1 unit to the right and $|m_1|$ units vertically to arrive at the point $A = (1, m_1)$ on L_1. Likewise, the point $B = (1, m_2)$ is on line L_2. By the Pythagorean theorem and its converse, triangle AOB is a right triangle if and only if

$$|\overline{AB}|^2 = |\overline{OA}|^2 + |\overline{OB}|^2.$$

Using the distance formula (Theorem 1, Section 1), we find that

$$|\overline{AB}|^2 = (1 - 1)^2 + (m_1 - m_2)^2 = m_1^2 - 2m_1m_2 + m_2^2,$$

$$|\overline{OA}|^2 = (1 - 0)^2 + (m_1 - 0)^2 = 1 + m_1^2, \quad \text{and}$$

$$|\overline{OB}|^2 = (1 - 0)^2 + (m_2 - 0)^2 = 1 + m_2^2.$$

Therefore, the condition $|\overline{AB}|^2 = |\overline{OA}|^2 + |\overline{OB}|^2$ is equivalent to

$$m_1^2 - 2m_1m_2 + m_2^2 = 1 + m_1^2 + 1 + m_2^2.$$

The last equation simplifies to $m_1m_2 = -1$, and the proof is complete.

Example 1   Find the slope m_1 of a line that is perpendicular to the line segment $\overline{AB}$, where $A = (-1, 2)$ and $B = (4, -5)$.

Solution   The slope m_2 of the line containing the points A and B is given by

$$m_2 = \frac{y_2 - y_1}{x_2 - x_1} = \frac{-5 - 2}{4 - (-1)} = -\frac{7}{5}.$$

Therefore, by the perpendicularity condition (Theorem 3),

$$m_1 = -\frac{1}{m_2} = -\frac{1}{(-7/5)} = \frac{5}{7}.$$

Example 2   Sketch the line L that contains the point $P = (2, 1)$ and is perpendicular to the line segment $\overline{AB}$, where $A = (-\frac{5}{3}, 0)$ and $B = (0, 5)$.

Solution   Slope of $\overline{AB} = \dfrac{y_2 - y_1}{x_2 - x_1} = \dfrac{5 - 0}{0 - (-5/3)} = \dfrac{5}{5/3} = 3.$

Therefore, the slope m of L is $m = -\frac{1}{3}$. Starting at the point $P = (2, 1)$ on L, we move 3 units to the right and 1 unit down, to the point $Q = (2 + 3, 1 - 1) = (5, 0)$ on L. Using a straightedge, we draw the line L through P and Q (Figure 10).

Figure 10

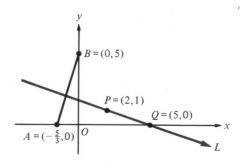

PROBLEM SET 2

In problems 1–10, sketch the line segment $\overline{AB}$ and find its slope using the slope formula (Theorem 1).

1. $A = (-1, 8), B = (4, 3)$
2. $A = (6, -1), B = (-3, 4)$
3. $A = (-2, 2), B = (2, -2)$
4. $A = (-3, 7), B = (-1, 0)$
5. $A = (4, 0), B = (0, -1)$
6. $A = (-1, 3), B = (0, 0)$
7. $A = (-3, -\frac{1}{2}), B = (1, \frac{3}{2})$
8. $A = (-\frac{1}{3}, \frac{1}{4}), B = (\frac{1}{4}, -\frac{1}{3})$
Ⓒ 9. $A = (2.6, -5.3), B = (1.7, -1.1)$
Ⓒ 10. $A = (73.24, 31.53), B = (1.71, 2.4)$

In problems 11–16, sketch the line L that contains the point P and has slope m.

11. $P = (1, 1), m = 2$
12. $P = (-4, 1), m = 0$
13. $P = (-3, 2), m = -\frac{2}{5}$
14. $P = (\frac{1}{2}, 0), m = \frac{1}{2}$
15. $P = (-\frac{2}{3}, -3), m = -2$
16. $P = (-\frac{4}{3}, \frac{1}{2}), m = -\frac{3}{4}$

In problems 17 and 18, sketch the line L that contains the point P and is parallel to the line segment $\overline{AB}$.

17. $P = (4, -3), A = (-2, 3), B = (3, -7)$
18. $P = (\frac{2}{3}, \frac{5}{3}), A = (\frac{1}{5}, \frac{3}{5}), B = (-\frac{2}{5}, \frac{4}{5})$

In problems 19 and 20, find the slope m_1 of a line L that is perpendicular to the line segment $\overline{AB}$.

19. $A = (-1, 4), B = (-2, -1)$
20. $A = (2, \frac{7}{5}), B = (1, \frac{12}{5})$

In problems 21–24, sketch the line L that contains the point P and is perpendicular to the line segment $\overline{AB}$.

21. $P = (1, 2), A = (-7, -3), B = (-5, 0)$
22. $P = (5, 12), A = (3, -2), B = (\frac{4}{3}, -\frac{4}{3})$
23. $P = (\frac{3}{2}, \frac{5}{2}), A = (4, -\frac{1}{3}), B = (\frac{1}{2}, 6)$
24. $P = (0, 4), A = (\frac{22}{7}, \sqrt{2}), B = (\frac{22}{7}, \sqrt{3})$

25. Show that the slope formula in Theorem 1 holds in all cases—even if the point B is not above and to the right of the point A.
26. Show that two distinct straight lines with the same slope are necessarily parallel. [Hint: If they weren't parallel, they would meet at some point P.]
27. Use slopes to show that the triangle with vertices $(-4, -2)$, $(2, -8)$, and $(4, 6)$ is a right triangle.
28. Use slopes to show that the quadrilateral with vertices $(-5, -2)$, $(1, -1)$, $(4, 4)$, and $(-2, 3)$ is a parallelogram. [Hint: Show that opposite sides have the same slope.]
29. (a) Determine d so that the line containing the points $(d, 3)$ and $(-2, 1)$ is perpendicular to the line containing the points $(5, -2)$ and $(1, 4)$.
    (b) Determine k so that the line containing the points $(k, 3)$ and $(-2, 1)$ is parallel to the line containing the points $(5, -2)$ and $(1, 4)$.
30. Consider the quadrilateral $ABCD$ with $A = (3, 1)$, $B = (2, 4)$, $C = (7, 6)$, and $D = (8, 3)$.
    (a) Use the concept of slope to determine whether or not the diagonals $\overline{AC}$ and $\overline{BD}$ are perpendicular.
    (b) Is the quadrilateral $ABCD$ a parallelogram? A rectangle? A square? A rhombus? Justify your answer.

3   Equations of Straight Lines in the Cartesian Plane

Consider a nonvertical line L having slope m and containing the point $P_0 = (x_0, y_0)$ (Figure 1). If $P = (x, y)$ is any other point on L, then, by the slope formula (Theorem 1, Section 2),

$$m = \frac{y - y_0}{x - x_0},$$

Figure 1

that is,

$$y - y_0 = m(x - x_0).$$

Notice that this last equation holds even if $P = P_0$, when it simply reduces to $0 = 0$. Using the parallelism condition (Theorem 2, Section 2) and elementary geometry, you can see that, conversely, any point $P = (x, y)$ whose coordinates satisfy the equation $y - y_0 = m(x - x_0)$ lies on the line L (Problem 32). Therefore, we have the following theorem.

Theorem 1   **Point-Slope Equation of a Straight Line**

> In the cartesian plane, the straight line L that contains the point $P_0 = (x_0, y_0)$ and has slope m is the graph of the equation
>
> $$y - y_0 = m(x - x_0).$$

You can write the equation $y - y_0 = m(x - x_0)$ as soon as you know the coordinates (x_0, y_0) of one point P_0 on the line L and the slope m of L. For this reason, the equation is called a **point-slope form.**

Examples   Find an equation in point-slope form of the given line L.

1 L contains the point $(3, 4)$ and has slope $m = 5$.

Solution   Substituting $x_0 = 3$, $y_0 = 4$, and $m = 5$ in the equation $y - y_0 = m(x - x_0)$, we have $y - 4 = 5(x - 3)$.

2 L contains the two points $(-1, 3)$ and $(4, -2)$.

Solution   By the slope formula, L has slope

$$m = \frac{-2 - 3}{4 - (-1)} = \frac{-5}{5} = -1.$$

Since the point $(-1, 3)$ belongs to L, we can use $(x_0, y_0) = (-1, 3)$ in Theorem 1. The resulting point-slope equation of L is

$$y - 3 = -1[x - (-1)] \quad \text{or} \quad y - 3 = -(x + 1).$$

Figure 2

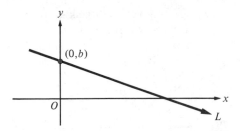

Now suppose that L is any nonvertical line with slope m. Since L is not parallel to the y axis, it must intersect this axis at some point $(0, b)$ (Figure 2). The ordinate b of the intersection point is called the **y intercept** of L. Since $(0, b)$ belongs to L, a point-slope equation of L is $y - b = m(x - 0)$. This equation simplifies to

$$y = mx + b,$$

which is called the **slope-intercept form** of the equation for L. In this form, the coefficient of x is the slope and the constant term on the right is the y intercept.

Example   Show that the graph of the equation $3x - 5y - 15 = 0$ is a straight line by rewriting the equation in slope-intercept form. Find the slope m and the y intercept b, and sketch the graph.

Solution   We begin by solving the equation for y in terms of x:

$$3x - 5y - 15 = 0$$
$$-5y = -3x + 15$$
$$y = \tfrac{3}{5}x - 3.$$

Figure 3

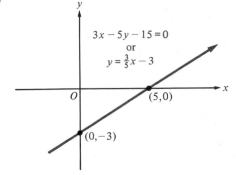

$$3x - 5y - 15 = 0$$
or
$$y = \tfrac{3}{5}x - 3$$

Thus, we have the equation of a straight line in slope-intercept form with slope $m = \tfrac{3}{5}$ and y intercept $b = -3$. We can obtain the graph by drawing the line with slope $m = \tfrac{3}{5}$ through the point $(0, b) = (0, -3)$. An alternative method is to find the **x intercept** of the line; that is, the abscissa x of the point $(x, 0)$ where the line intersects the x axis. This is accomplished by putting $y = 0$ in the original equation $3x - 5y - 15 = 0$ and then solving for x:

$$3x - 15 = 0 \quad \text{or} \quad x = 5.$$

Thus, we obtain the graph by drawing the line through the point $(0, -3)$ on the y axis and the point $(5, 0)$ on the x axis (Figure 3).

Figure 4

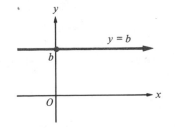

$$y = b$$

A horizontal line has slope $m = 0$; hence, in slope-intercept form, its equation is $y = 0(x) + b$, or simply $y = b$ (Figure 4). The equation $y = b$ places no restriction at all on the abscissa x of a point (x, y) on the horizontal line, but it requires that all of the ordinates y have the same value b.

Figure 5

A vertical line has an undefined slope, so you can't write its equation in slope-intercept form. However, since all points on a vertical line have the same abscissa, say, a, an equation of such a line is $x = a$ (Figure 5).

If A, B, and C are constants and if A and B are not both zero, an equation of the form

$$Ax + By + C = 0$$

represents a straight line. If $B \neq 0$, the equation can be rewritten in the slope-intercept form as

$$y = \left(\frac{-A}{B}\right)x + \left(\frac{-C}{B}\right),$$

with slope $m = \dfrac{-A}{B}$ and y intercept $b = \dfrac{-C}{B}$. If $B = 0$, then $A \neq 0$ and the equation can be rewritten as

$$x = \frac{-C}{A},$$

the equation of a vertical line. The equation $Ax + By + C = 0$ is called the **general form** of the equation of a line.

Example   Let L be the line that contains the point $(-1, 2)$ and is perpendicular to the line L_1 whose equation is $3x - 2y + 5 = 0$. Find an equation of L in (a) point-slope form, (b) slope-intercept form, and (c) general form.

Solution   We obtain the slope m_1 of L_1 by solving the equation $3x - 2y + 5 = 0$ for y in terms of x. The result is $y = \frac{3}{2}x + \frac{5}{2}$, so the slope of L_1 is $m_1 = \frac{3}{2}$. By the perpendicularity condition (Theorem 3, Section 2), the slope m of L is

$$m = -\frac{1}{m_1} = -\frac{1}{3/2} = -\frac{2}{3}.$$

(a) Since L has slope $m = -\frac{2}{3}$ and contains the point $(-1, 2)$, its equation in point-slope form is

$$y - 2 = -\tfrac{2}{3}[x - (-1)]$$

or

$$y - 2 = -\tfrac{2}{3}(x + 1).$$

(b) Solving the equation in (a) for y in terms of x, we obtain the equation of L in slope-intercept form,

$$y = -\tfrac{2}{3}x + \tfrac{4}{3}.$$

(c) Multiplying both sides of the equation in (b) by 3 and rearranging terms, we obtain an equation of L in general form:

$$2x + 3y - 4 = 0.$$

PROBLEM SET 3

In problems 1–6, find an equation in point-slope form of the given line L.

1. L contains the point $(3, 2)$ and has slope $m = \frac{3}{4}$.
2. L contains the point $(0, 2)$ and has slope $m = -\frac{2}{3}$.
3. L contains the points $(-3, 2)$ and $(4, 1)$.
4. L contains the points $(1, 3)$ and $(-1, 1)$.
5. L contains the point $(-3, 5)$ and is parallel to the line segment $\overline{AB}$, where $A = (3, 7)$ and $B = (-2, 2)$.
6. L contains the point $(7, 2)$ and is parallel to the line segment $\overline{AB}$, where $A = (\frac{1}{3}, 1)$ and $B = (-\frac{2}{3}, \frac{3}{5})$.

In problems 7–14, rewrite each equation in slope-intercept form, find the slope m and the y intercept b, and sketch the graph.

7. $3x - 2y = 6$
8. $5x - 2y - 10 = 0$
9. $y - 3x - 1 = 0$
10. $y + 1 = 0$
11. $x = -3y + 9$
12. $2x + y + 3 = 0$
13. $y - 2x - 3 = 0$
14. $x = -\frac{3}{5}y + \frac{7}{5}$

In problems 15–26, find an equation of the line L in (a) point-slope form, (b) slope-intercept form, and (c) general form.

15. L contains the point $(-5, 2)$ and has slope $m = 4$.
16. L contains the point $(3, -1)$ and has slope $m = 0$.
17. L has slope $m = -3$ and y intercept $b = 5$.
18. L has slope $m = \frac{4}{5}$ and intersects the x axis at $(-3, 0)$.
19. L intersects the x and y axes at $(3, 0)$ and $(0, 5)$.
20. L contains the points $(\frac{7}{2}, \frac{5}{3})$ and $(\frac{2}{5}, -6)$.
21. L contains the point $(4, -4)$ and is parallel to the line that has the equation $2x - 5y + 3 = 0$.
22. L contains the point $(-3, \frac{2}{3})$ and is perpendicular to the y axis.
23. L contains the point $(-3, \frac{2}{3})$ and is perpendicular to the line that has the equation $5x + 3y - 1 = 0$.
24. L contains the point $(-6, -8)$ and has y intercept $b = 0$.
25. L contains the point $(\frac{2}{3}, \frac{5}{7})$ and is parallel to the line that has the equation $7x + 3y - 12 = 0$.
26. L is the perpendicular bisector of the line segment $\overline{AB}$, where $A = (3, -2)$ and $B = (7, 6)$.

27. Find a real number B so that the graph of $3x + By - 5 = 0$ has y intercept $b = -4$.
28. Suppose that the line L intersects the axes at $(a, 0)$ and $(0, b)$. Show that the equation of L can be written in **intercept form**
$$\frac{x}{a} + \frac{y}{b} = 1.$$

29. A car-rental company leases automobiles for a charge of $22 per day plus $0.20 per mile. Write an equation for the cost y dollars in terms of the distance x miles driven if the car is leased for N days. If $N = 3$, sketch a graph of the equation.
30. If a piece of property is *depreciated linearly* over a period of n years, then its value y dollars at the end of x years is given by $y = c[1 - (x/n)]$, where c dollars is the original value of the property. An apartment building built in 1975 and originally worth $400,000 is being depreciated linearly over a period of 40 years. Sketch a graph showing the value y dollars of the apartment building x years after it was built and determine its value in the year 1995.
31. In 1980, the Solar Electric Company showed a profit of $3.45 per share, and it expects this figure to increase annually by $0.25 per share. If the year 1980 corresponds to $x = 0$, and successive years correspond to $x = 1, 2, 3$, and so on, find the equation $y = mx + b$ of the line that allows the company to predict its profit y dollars per share in future years. Sketch the graph of this equation and find the predicted profit per share in 1988.

32. Let L be the line that contains the point (x_0, y_0) and has slope m. Suppose that (x, y) is a point such that $y - y_0 = m(x - x_0)$. Show that (x, y) lies on the line L.

33. In 1980, tests showed that water in a lake was polluted with 7 milligrams of mercury compounds per 1000 liters of water. Cleaning up the lake became an immediate priority, and environmentalists determined that the pollution level would drop at the rate of 0.75 milligram of mercury compounds per 1000 liters of water per year if all of their recommendations were followed. If 1980 corresponds to $x = 0$ and successive years correspond to $x = 1, 2, 3$, and so on, find the equation $y = mx + b$ of the line that allows the environmentalists to predict the pollution level y in future years if their recommendations are followed. Sketch the graph of the equation and determine when the lake will be free of mercury pollution according to this graph.

---

4   Functions

Advances in our scientific understanding of the world often result from the discovery that things depend upon one another in definite ways. For instance, the gravitational attraction between two material bodies depends on the distance between them, and the pitch of a guitar string depends on its tension. If the numerical value of a variable quantity y depends on the numerical value of another variable quantity x, so that each value of x determines one and only one corresponding value of y, we say that y is a **function** of x. For example, if y denotes the area of a circle of radius x, then y depends on x according to the formula $y = \pi x^2$. Therefore, the area of a circle is a function of its radius.

More generally, we have the following definition.

Definition 1   **Function**

> A **function** is a rule or correspondence that assigns one and only one numerical value of a variable y to each numerical value of a variable x. The variable x, which is called the **independent** variable, can take on any value in a certain set of real numbers called the **domain** of the function. The variable y is called the **dependent** variable, since its value depends on the value of x. The set of values assumed by y as x runs through the domain is called the **range** of the function.

The notation

$$x \longmapsto y$$

is often used to indicate that there is a function (a definite rule) whereby to each value of x there corresponds a uniquely determined value of y, or, as mathematicians say, each value of x is "mapped onto" a corresponding value of y. For this reason, the symbolism $x \longmapsto y$ is called the **mapping notation** for a function. If $x \longmapsto y$, we say that y is the **image** of x under the function.

Scientific calculators have special keys for some of the more important functions. By entering a number x and touching, for instance, the $\sqrt{x}$ key, you obtain a vivid demonstration of the mapping $x \longmapsto \sqrt{x}$ as the display changes from x to its image $\sqrt{x}$ under the square-root function. For instance,

$$4 \longmapsto 2,$$
$$25 \longmapsto 5,$$
$$2 \longmapsto \sqrt{2} \approx 1.414.$$

Programmable calculators and home computers have "user-definable" keys that can be programmed for whatever function $x \longmapsto y$ may be required. The program for the required function is the actual rule whereby y is to be calculated from x. Each user-definable key is marked with a letter of the alphabet or other symbol, so that, after the key has been programmed for a particular function, the letter or symbol can be used as the "name" of the function.

The use of letters of the alphabet to designate functions is not restricted exclusively to calculating machines. Although any letters of the alphabet can be used to designate functions, the letters f, g, and h as well as F, G, and H are most common. (Letters of the Greek alphabet are also used.) For instance, if we wish to designate the square-root function $x \longmapsto \sqrt{x}$ by the letter f, we write

$$x \overset{f}{\longmapsto} \sqrt{x} \quad \text{or} \quad f : x \longmapsto \sqrt{x}.$$

If $f : x \longmapsto y$ is a function, it is customary to write the value of y that corresponds to x as $f(x)$, read "f of x." In other words, $f(x)$ is the image of x under the function f. For instance, if $f : x \longmapsto \sqrt{x}$ is the square-root function, then

$$f(4) = \sqrt{4} = 2,$$
$$f(25) = \sqrt{25} = 5,$$
$$f(2) = \sqrt{2} \approx 1.414,$$

and, in general, for any nonnegative value of x,

$$f(x) = \sqrt{x}.$$

If $f : x \mapsto y$, then, for every value of x in the domain of f, we have

$$y = f(x),$$

an equation relating the dependent variable y to the independent variable x. Conversely, an equation of the form

$$y = \text{an expression involving } x$$

determines a function $f : x \longmapsto y$, and we say that the function f is *defined by* or *given by* the equation. For instance, the equation

$$y = 3x^2 - 1$$

defines a function $f : x \longmapsto y$, so that

$$y = f(x) = 3x^2 - 1$$

or simply

$$f(x) = 3x^2 - 1.$$

When a function f is defined by an equation, you can determine, by substitution, the image $f(a)$ corresponding to a particular value $x = a$. For instance, if f is defined by

$$f(x) = 3x^2 - 1,$$

then

$$f(2) = 3(2)^2 - 1 = 11,$$
$$f(0) = 3(0)^2 - 1 = -1,$$
$$f(-1) = 3(-1)^2 - 1 = 2,$$

and

$$f(t + 1) = 3(t + 1)^2 - 1 = 3(t^2 + 2t + 1) - 1$$
$$= 3t^2 + 6t + 2.$$

Example 1 Let g be the function defined by $g(x) = 5x^2 + 3x$. Find the indicated values.

(a) $g(1)$ (b) $g(-2)$ (c) $g(t^3)$
(d) $[g(-1)]^2$ (e) $g(t + h)$ (f) $g(-x)$

Solution

(a) $g(1) = 5(1)^2 + 3(1) = 5 + 3 = 8$
(b) $g(-2) = 5(-2)^2 + 3(-2) = 20 - 6 = 14$
(c) $g(t^3) = 5(t^3)^2 + 3(t^3) = 5t^6 + 3t^3$
(d) $[g(-1)]^2 = [5(-1)^2 + 3(-1)]^2 = (5 - 3)^2 = 2^2 = 4$
(e) $g(t + h) = 5(t + h)^2 + 3(t + h) = 5t^2 + 10th + 5h^2 + 3t + 3h$
(f) $g(-x) = 5(-x)^2 + 3(-x) = 5x^2 - 3x$

Example 2 A small company finds that its profit depends on the amount of money it spends on advertising. If x dollars are spent on advertising, the corresponding profit $p(x)$ dollars is given by

$$p(x) = \frac{x^2}{100} - \frac{x}{50} + 100.$$

Determine the profit if the amount spent on advertising is (a) $50, (b)$100, (c) $300.

Solution

(a) $p(50) = \dfrac{50^2}{100} - \dfrac{50}{50} + 100 = 25 - 1 + 100 = 124$ dollars

(b) $p(100) = \dfrac{100^2}{100} - \dfrac{100}{50} + 100 = 100 - 2 + 100 = 198$ dollars

(c) $p(300) = \dfrac{300^2}{100} - \dfrac{300}{50} + 100 = 900 - 6 + 100 = 994$ dollars

Whenever a function $f : x \longmapsto y$ is defined by an equation, you may assume (unless we say otherwise) that its domain consists of all values of x for which the equation makes sense and determines a corresponding real number y. The range of the function is then automatically determined, since it consists of the set of all values of y that correspond, by the equation that defines the function, to values of x in the domain.

Examples   Find the domain of the function defined by each equation.

1 $h(x) = \dfrac{1}{x - 1}$

Solution   The domain of h is the set of all real numbers except 1. In other words, it consists of the interval $(-\infty, 1)$ together with the interval $(1, \infty)$.

2 $G(x) = \sqrt{4 - x}$

Solution   The expression $\sqrt{4 - x}$ represents a real number if and only if $4 - x \geq 0$; that is, if and only if $x \leq 4$. Therefore, the domain of G is the interval $(-\infty, 4]$.

3 $F(x) = 3x - 5$

Solution   Since the expression $3x - 5$ is defined for all real values of x, the domain of F is the set $\mathbb{R}$ of all real numbers.

Functions that arise in applied mathematics may have restrictions imposed on their domains by physical or geometrical circumstances. For instance, the function $x \longmapsto \pi x^2$ that expresses the correspondence between the radius x and the area πx^2 of a circle would have its domain restricted to the interval $(0, \infty)$, since a circle must have a positive radius.

For a function f, the expression

$$\frac{f(x + h) - f(x)}{h}, \qquad h \neq 0,$$

called the **difference quotient,** plays an important role in calculus.

Example   Find the difference quotient for the function f defined by $f(x) = \sqrt{x}$ and simplify the result.

Solution   Assuming that $h \neq 0$, $x \geq 0$, and $x + h \geq 0$, we have

$$\frac{f(x + h) - f(x)}{h} = \frac{\sqrt{x + h} - \sqrt{x}}{h}.$$

In calculus, this is "simplified" by rationalizing the numerator, so that

$$\frac{f(x + h) - f(x)}{h} = \frac{(\sqrt{x + h} - \sqrt{x})(\sqrt{x + h} + \sqrt{x})}{h(\sqrt{x + h} + \sqrt{x})}$$

$$= \frac{(x + h) - x}{h(\sqrt{x + h} + \sqrt{x})} = \frac{h}{h(\sqrt{x + h} + \sqrt{x})}$$

$$= \frac{1}{\sqrt{x + h} + \sqrt{x}}.$$

In dealing with a function f, it is important to distinguish among the *function itself*

$$f : x \longmapsto y,$$

which is a rule or correspondence; the *image*

$$y \quad \text{or} \quad f(x),$$

which is a number depending on x; and the *equation*

$$y = f(x),$$

which relates the dependent variable y to the independent variable x. Nevertheless, people tend to take shortcuts and speak, incorrectly, of "the function $f(x)$" or "the function $y = f(x)$." Similarly, in applied mathematics, people often say that "y is a function of x," for instance, "current is a function of voltage." Although we avoid these practices when absolute precision is required, we indulge in them whenever it seems convenient and harmless.

The particular letters used to denote the dependent and independent variables are of no importance in themselves—the important thing is the rule by which a definite value of the dependent variable is assigned to each value of the independent variable. In applied work, variables other than x and y are often used because physical and geometrical quantities are designated by conventional symbols. For instance, the radius of a circle is often denoted by r and its area by A. Thus, the function $f : r \longmapsto A$ that assigns to each positive value of r the corresponding value of A is given by

$$A = f(r) = \pi r^2.$$

4.1 The Graph of a Function

The **graph** of a function f is defined to be the graph of the corresponding equation $y = f(x)$. In other words, the graph of f is the set of all points (x, y) in the cartesian plane such that x is in the domain of f and $y = f(x)$.

For instance, if m and b are constants, then the graph of the function

$$f(x) = mx + b$$

is the same as the graph of the equation

$$y = mx + b,$$

a straight line with slope m and y intercept b. For this reason, a function of the form $f(x) = mx + b$ is called a **linear function.**

Example   Sketch the graph of the function $f(x) = \frac{3}{4}x + 2$.

Solution   The graph of f is the same as the graph of the equation

$$y = \frac{3}{4}x + 2,$$

a line with slope $m = \frac{3}{4}$ and y intercept $b = 2$ (Figure 1).

Figure 1

Figure 2

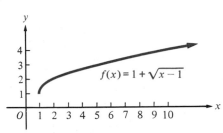

Graphs of functions that are not linear are often (but not always) smooth curves in the cartesian plane. For instance, the graph of $f(x) = 1 + \sqrt{x - 1}$ is shown in Figure 2. The problem of sketching such a graph can be considerably more challenging than sketching a straight line, although the use of a calculator to determine points on the graph will often give a good indication of its general shape. Incidentally, many of the home computers now on the market have graph-plotting capabilities. In Sections 5 and 6, we consider graph sketching in more detail.

In scientific work, a graph showing the relationship between two variable quantities is often obtained by means of actual measurement. For instance, Figure 3 shows the blood pressure p (in millimeters of mercury) in an artery of a healthy person plotted against time t in seconds. Such a curve may be regarded as the graph of a function $p = f(t)$, even though it may not be clear how to write a "mathematical formula" giving p in terms of t.

Figure 3

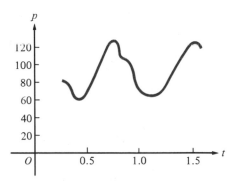

It is important to realize that *not every curve in the cartesian plane is the graph of a function.* Indeed, the definition of a function (Definition 1) requires that there be one and *only one* value of y corresponding to each value of x in the domain. Thus, on the graph of a function, *we cannot have two points (x, y_1) and (x, y_2) with the same abscissa x and different ordinates y_1 and y_2.* Hence, we have the following test.

Vertical Line Test

> A set of points in the cartesian plane is the graph of a function if and only if no vertical straight line intersects the set more than once.

Example   Which of the curves in Figure 4 is the graph of a function?

Solution   By the vertical line test, the curve in Figure 4a is the graph of a function, but the curve in Figure 4b is not.

Figure 4

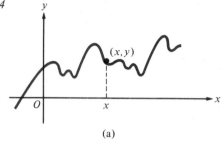

(a)   (b)

PROBLEM SET 4

In problems 1–42, let

$$f(x) = 2x + 1 \qquad g(x) = x^2 - 3x - 4 \qquad h(x) = \sqrt{3x + 5}$$

$$F(x) = \frac{x - 2}{3x + 7} \qquad G(x) = \sqrt[3]{x^3 - 4} \qquad H(x) = |2 - 5x|$$

Find the indicated values.

1. $f(-3)$

2. $g(-2)$

3. $h\left(-\dfrac{1}{3}\right)$

4. $F\left(\dfrac{7}{3}\right)$

5. $G(\sqrt[3]{31})$

6. $H(-4)$

7. $g(0)$

8. $G(-5)$

9. $F\left(-\dfrac{1}{2}\right)$

10. $f(-7)$

C 11. $g(4.718)$

C 12. $h(2.003)$

13. $F\left(-\dfrac{1}{3}\right)$

14. $\left[H\left(-\dfrac{1}{2}\right)\right]^2$

15. $[h(-1)]^2$

16. $[G(2)]^3$

17. $\sqrt{f(4)}$

18. $\sqrt{h(1)}$

19. $f(a + 1)$

20. $f\left(\dfrac{1}{a}\right)$

21. $g(-b)$

22. $F\left(\dfrac{a}{3}\right)$

23. $H(c + 2)$

24. $g(-\tfrac{1}{2}ab)$

25. $H\left(\dfrac{1}{a}\right)$

26. $h(2x - 1)$

27. $f(-x)$

28. $G(\sqrt{b})$

29. $h(x^4)$

30. $g(-x)$

31. $f\left(\dfrac{x}{2}\right)$

32. $g(x + t)$

33. $f\left(\dfrac{x - 1}{2}\right)$

34. $f(1) - f(0)$

35. $g(a) - g(b)$

36. $g(a) - g(-a)$

37. $h(x) - h(0)$

38. $F(x + t) - F(x)$

39. $\dfrac{f(x) - f(0)}{x}$

C 40. $f(g(2.341))$

C 41. $g(h(1.357))$

42. $f(g(x))$

In problems 43–54, find the domain of each function.

43. $f(x) = 1 - 4x^2$

44. $g(x) = (x + 2)^{-1}$

45. $h(x) = \sqrt{x}$

46. $F(x) = \sqrt{5 - 3x}$

47. $f(x) = -3x^{-3}$

48. $G(x) = 7/(5 - 6x)$

49. $K(x) = (4 - 5x)^{-1/2}$

50. $p(x) = \sqrt{9 - x^2}$

51. $g(x) = (9 + x^4)^{3/4}$

52. $h(x) = \dfrac{1}{x + |x|}$

53. $F(x) = \dfrac{x^3 - 8}{x^2 - 4}$

54. $h(x) = \sqrt{\dfrac{x - 2}{x - 4}}$

In problems 55–60, find the difference quotient $\dfrac{f(x + h) - f(x)}{h}$ and simplify the result.

55. $f(x) = 4x - 1$

56. $f(x) = 5$

57. $f(x) = x^2 + 3$

58. $f(x) = mx + b, \qquad m \neq 0$

59. $f(x) = 1/\sqrt{x}$

60. $f(x) = 1/x$

In problems 61–66, sketch the graph of the function.

61. $f(x) = \frac{2}{3}x - 5$

62. $f(x) = 2x \quad$ for $\quad x \geq 0$

63. $f(x) = x$

64. $f(x) = 4$

65. $f(x) = -2x + 1$

66. $f(x) = \sqrt{9 - x^2}$

67. Use the vertical line test to determine which of the curves in Figure 5 are graphs of functions.

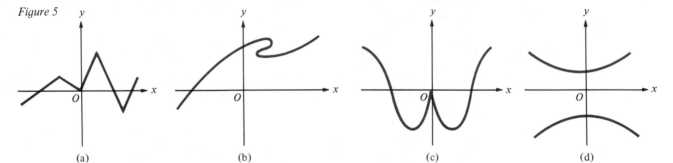

Figure 5

(a)   (b)   (c)   (d)

68. Is the graph of the equation $x^2 + y^2 = 9$ the graph of a function? Why or why not?

69. The function $f : C \longmapsto F$ given by the equation $F = \frac{9}{5}C + 32$ converts the temperature C in degrees Celsius to the corresponding temperature F in degrees Fahrenheit. Find $f(0)$, $f(15)$, $f(-10)$, and $f(55)$, and write the results using both function and mapping notation.

70. An ecologist investigating the effect of air pollution on plant life finds that the percentage, $p(x)$ percent, of diseased trees and shrubs at a distance of x kilometers from an industrial city is given by $p(x) = 32 - (3x/50)$ for $50 \leq x \leq 500$. Sketch a graph of the function p and find $p(50)$, $p(100)$, $p(200)$, $p(400)$, and $p(500)$.

71. An airline chart shows that the temperature T in degrees Fahrenheit at an altitude of $h = 15{,}000$ feet is $T = 5°$. At an altitude of $h = 20{,}000$ feet, $T = -15°$. Supposing that T is a linear function of h, obtain an equation that defines this function, sketch its graph, and find the temperature at an altitude $h = 30{,}000$ feet.

[c] 72. In physics, the (absolute) pressure P in newtons per square meter at a point h meters below the surface of a body of water is shown to be a linear function of h. When $h = 0$, $P = 1.013 \times 10^5$ newtons per square meter. When $h = 1$ meter, $P = 2.003 \times 10^5$ newtons per square meter. Obtain the equation that defines P as a function of h, and use it to find the pressure at a depth of $h = 100$ meters.

73. A rectangle of length x units and width y units has a perimeter of 24 units. Express the area A of the rectangle as a function of x.

[c] 74. The period $T(\ell)$ seconds of a simple pendulum of length ℓ meters swinging along a small arc is given by $T(\ell) = 2\pi\sqrt{\ell/9.807}$. Using a calculator and rounding off your answer to four significant digits, find $T(0.1)$, $T(1)$, $T(1.5)$, and $T(0.2484)$.

5   Graph Sketching and Properties of Graphs

In this section, we consider methods for sketching graphs of functions that may not be linear, properties of graphs, and graphs of some important functions. The basic graph-sketching procedure is as follows.

The Point-Plotting Method

> To sketch the graph of $y = f(x)$, select several values of x in the domain of f, calculate the corresponding values of $f(x)$, plot the resulting points, and connect the points with a smooth curve. The more points you plot, the more accurate your sketch will be.

Example   Use the point-plotting method to sketch the graph of $f(x) = x^2$, where the domain of f is restricted by the condition that $x > 0$.

Solution   The domain of f is the interval, $(0, \infty)$, so we begin by selecting several values of x in this interval and calculating the corresponding values of $f(x) = x^2$, as in the table in Figure 1. We then plot the points $(x, f(x))$ from the table and connect them with a smooth curve (Figure 1). Because the domain of f consists only of positive numbers, the point $(0, 0)$ is excluded from the graph. This excluded point is indicated by a small open circle. Notice that we have used different scales on the x and y axes to obtain a figure with reasonable dimensions.

Figure 1

x	$f(x) = x^2$
1	1
2	4
3	9
4	16
5	25
6	36

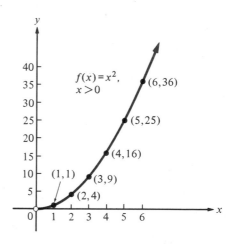

The point-plotting method requires us to *guess* about the shape of the graph between or beyond known points, and must therefore be used with caution (see Problems 43 and 44). If the function is fairly simple, the point-plotting method usually works pretty well; however, more complicated functions may require more advanced methods that are studied in calculus.

5.1 Geometric Properties of Graphs

The graph in Figure 2a is always *rising* as we move to the right, a geometric indication that the function f is **increasing**; that is, as x increases, so does the value of $f(x)$. On the other hand, the graph in Figure 2b is always *falling* as we move to the right, indicating that the function g is **decreasing**; that is, as x increases, the value of $g(x)$ decreases. In Figure 2c, the graph doesn't rise or fall, indicating that h is a **constant function** whose values $h(x)$ do not change as we increase x.

Figure 2

(a)

(b)

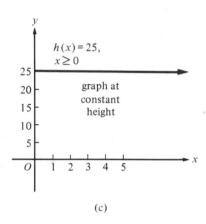

(c)

The domain and range of a function are easily found from its graph. Indeed, as Figure 3 illustrates, *the domain of a function is the set of all abscissas of points on its graph* (Figure 3a) and *the range of a function is the set of all ordinates of points on its graph* (Figure 3b).

Figure 3

(a)

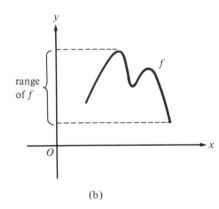

(b)

Example   For the function f whose graph is shown in Figure 4, indicate the intervals over which f is increasing, over which it is decreasing, and over which it is constant. Also, find the domain and range of f.

Figure 4

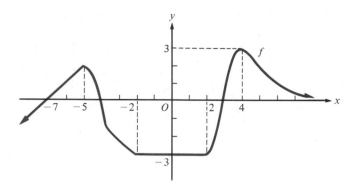

Solution   As we move from left to right, the function f is increasing over intervals where the graph is rising and decreasing over intervals where it is falling. Thus, assuming that the graph continues indefinitely to the left and right in the directions indicated by the arrowheads, we conclude that f is increasing on the intervals $(-\infty, -5]$ and $[2, 4]$ and that f is decreasing on the intervals $[-5, -2]$ and $[4, \infty)$. On the interval $[-2, 2]$, f is constant. The graph "covers" the entire x axis, so the domain of f is the set $\mathbb{R}$ of all real numbers. We assume that the graph keeps dropping as we move to the left of -5 on the x axis, but that it never climbs any higher than $y = 3$; hence, the range of f is the interval $(-\infty, 3]$.

Consider the graphs in Figure 5. The graph of f (Figure 5a) is **symmetric about the y axis**; that is, the portion of the graph to the right of the y axis is the mirror image of the portion to the left of it. Specifically, if the point (x, y) belongs to the graph of f, then so does the point $(-x, y)$, so that $f(-x) = f(x)$. Similarly, the graph of g (Figure 5b) is **symmetric about the origin** because, if the point (x, y) belongs to the graph, then so does the point $(-x, -y)$; that is, $g(-x) = -g(x)$.

Figure 5

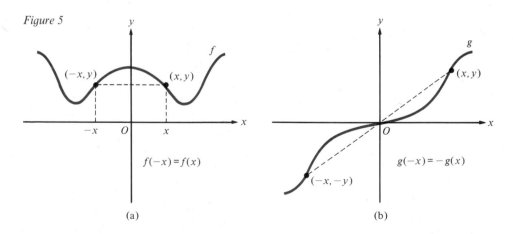

(a)   (b)

A function whose graph is symmetric about the y axis is called an **even function**; a function whose graph is symmetric about the origin is called an **odd function**. This is stated more formally in the following definition.

Definition 1   **Even and Odd Functions**

> (i) A function f is said to be **even** if, for every number x in the domain of f, $-x$ is also in the domain of f and
>
> $$f(-x) = f(x).$$
>
> (ii) A function f is said to be **odd** if, for every number x in the domain of f, $-x$ is also in the domain of f and
>
> $$f(-x) = -f(x).$$

Of course, there are many functions that are neither even nor odd.

Example 1   Determine which of the functions whose graphs are shown in Figure 6 are even, odd, or neither.

Figure 6

(a)

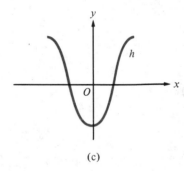

(b)   (c)

Solution   In Figure 6a, the graph of f is symmetric about the origin; thus, $f(-x) = -f(x)$ and f is an odd function. In Figure 6b, the graph of g is symmetric neither about the y axis nor about the origin, so g is neither even nor odd. In Figure 6c, the graph of h is symmetric about the y axis; thus, $h(-x) = h(x)$ and h is an even function.

Example 2   Determine whether each function is even, odd, or neither.

(a) $f(x) = x^4$ (b) $g(x) = x - 1$

(c) $h(x) = 2x^2 - 3|x|$ (d) $F(x) = x^5$

Solution   (a) $f(-x) = (-x)^4 = x^4 = f(x)$, so f is an even function.

(b) $g(-x) = -x - 1$, while $g(x) = x - 1$ and $-g(x) = -x + 1$. Since we have neither $g(-x) = g(x)$ nor $g(-x) = -g(x)$, g is neither even nor odd.

(c) $h(-x) = 2(-x)^2 - 3|-x| = 2x^2 - 3|x| = h(x)$, so h is an even function.

(d) $F(-x) = (-x)^5 = -x^5 = -F(x)$, so F is an odd function.

5.2 Graphs of Some Particular Functions

The following examples illustrate the ideas discussed above and exhibit the graphs of some important functions.

Examples   In each case find the domain of f, sketch its graph, discuss the symmetry of the graph, indicate the intervals where f is increasing or decreasing, and find the range of f. (Do these things in whatever order seems most convenient.)

1 $f(x) = x$ (the **identity** function)

Solution   This is a special case of a linear function $f(x) = mx + b$ with slope $m = 1$ and y intercept $b = 0$ (Figure 7). Note that the graph consists of all points for which the abscissa equals the ordinate. The domain and range of f are both the set $\mathbb{R}$ of all real numbers, and f is increasing over $\mathbb{R}$. Since $f(-x) = -x = -f(x)$, it follows that f is an odd function and that its graph is symmetric about the origin.

Figure 7

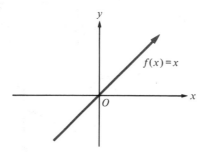

2 $f(x) = x^2$ (the **squaring** function)

Solution   The domain of f is the set $\mathbb{R}$. Because $f(-x) = (-x)^2 = x^2 = f(x)$, the function f is even and its graph is symmetric about the y axis. We have already sketched the portion of this graph for $x > 0$ in Figure 1. The full graph includes the mirror image of Figure 1 on the other side of the y axis and the point $(0,0)$ (Figure 8). From the graph we see that the range of f is the interval $[0, \infty)$, and that f is decreasing on the interval $(-\infty, 0]$ and increasing on the interval $[0, \infty)$.

Figure 8

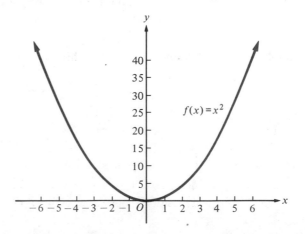

3 $f(x) = \sqrt{x}$ (the **square-root** function)

Solution   Here the domain of f is the interval $[0, \infty)$. We tabulate some points $(x, f(x))$, plot them, and connect them by a smooth curve to obtain a sketch of the graph (Figure 9). The function f is increasing on the interval $[0, \infty)$, and its graph rises higher and higher without bound, so the range of f is the interval $[0, \infty)$. The graph is neither symmetric about the y axis nor the origin, so f is neither even nor odd.

Figure 9

x	$f(x) = \sqrt{x}$
0	0
1	1
4	2
9	3
16	4
25	5

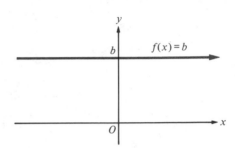

Figure 10

4 $f(x) = b$ (a **constant** function)

Solution   This is a special case of the linear function $f(x) = mx + b$, with slope $m = 0$ and y intercept b. Its graph is a line parallel to the x axis and containing the point $(0, b)$ on the y axis. All points on this line have the same ordinate b (Figure 10). The domain of f is the set $\mathbb{R}$ of all real numbers and the range is the set $\{b\}$. The constant function f is neither increasing nor decreasing. Since $f(-x) = b = f(x)$, the function f is even and its graph is symmetric about the y axis.

5 $f(x) = |x|$ (the **absolute-value** function)

Solution   The domain of f consists of all real numbers $\mathbb{R}$. For $x \geq 0$, $f(x) = x$, so the portion of the graph to the right of the y axis is the same as the graph of the identity function (Figure 7). Since $f(-x) = |-x| = |x| = f(x)$, the function f is even and its graph is symmetric about the y axis. Reflecting the portion of the graph of the identity function for $x \geq 0$ across the y axis, we obtain the V-shaped graph shown in Figure 11. Evidently, the range of f is the interval $[0, \infty)$; f is decreasing on the interval $(-\infty, 0]$ and increasing on the interval $[0, \infty)$.

Figure 11

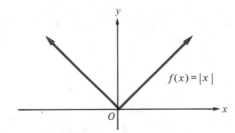

Sometimes a function is defined by using different equations in different intervals. Such a *piecewise-defined* function is illustrated in the following example.

Example   Sketch the graph of the function f defined by

$$f(x) = \begin{cases} -2 & \text{if } x < 0 \\ x^2 & \text{if } 0 \leq x < 2 \\ x & \text{if } x \geq 2. \end{cases}$$

Solution   For $x < 0$, we have $f(x) = -2$, so this portion of the graph of f will look like the graph of a constant function (Figure 10 with $b = -2$). In sketching this portion of the graph, we must be careful to leave out the point $(0, -2)$ on the y axis because of the strict inequality $x < 0$ (Figure 12). For $0 \leq x < 2$, we have $f(x) = x^2$; this part of our graph coincides with the graph of the squaring function (Figure 8), starting at the origin and ending at the point $(2,4)$, which does not belong to the graph of f (Figure 12). Finally, for $x \geq 2$, we have $f(x) = x$; this part of our graph will coincide with the graph of the identity function (Figure 7), starting at the point $(2,2)$, which does belong to the graph, and continuing indefinitely to the right (Figure 12).

Figure 12

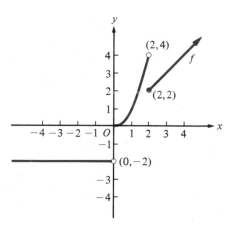

PROBLEM SET 5

In problems 1–8, use the point-plotting method to sketch the graph of each function. ⓒ You may wish to improve the accuracy of your sketch by using a calculator to determine the coordinates of some points on the graph.

1. $f(x) = x^3$ for $x \geq 0$

2. $g(x) = \sqrt[3]{x}$ for $-8 \leq x \leq 8$

3. $h(x) = \dfrac{1}{x}$ for $x \geq 1$

4. $F(x) = \dfrac{x-1}{x+1}$ for $x \geq 0$

5. $G(x) = x^2 + x + 1$

6. $H(x) = \dfrac{1}{x^2 + 1}$

7. $p(x) = \sqrt{25 - x^2}$

8. $q(x) = x + \sqrt{x}$

9. For each function in Figure 13, find the domain and range; indicate the intervals over which the function is increasing, decreasing, or constant; and determine whether the function is even, odd, or neither.

Figure 13

(a)

(b)

(c)

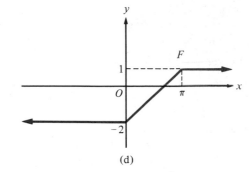

(d)

10. The graph of the function $f(x) = x^{1/3}(x - 1)^{2/3}$ is shown in Figure 14. Indicate the intervals over which f is increasing or decreasing, and determine whether f is even, odd, or neither.

Figure 14

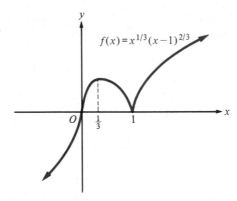

In problems 11–19, determine without drawing a graph whether the function is even, odd, or neither, and discuss any symmetry of the graph.

11. $f(x) = x^4 + 3x^2$

12. $g(x) = -5x^2 + 4$

13. $h(x) = 4x^3 - 5x$

14. $F(x) = x^2 + |x|$

15. $G(x) = \sqrt{8x^4 + 1}$

16. $H(x) = 5x^2 - x^3$

17. $q(x) = \sqrt[3]{x^3 - 9}$

18. $p(x) = \dfrac{\sqrt{x^2 + 1}}{|x|}$

19. $r(x) = \dfrac{1}{x}$

20. Let $f(x) = 3x^4 - 2x^3 + x^2 - 1$. Define functions g and h by $g(x) = \frac{1}{2}[f(x) + f(-x)]$ and $h(x) = \frac{1}{2}[f(x) - f(-x)]$. (a) Show that f is neither even nor odd. (b) Show that g is even. (c) Show that h is odd. (d) Show that $f(x) = g(x) + h(x)$ holds for all values of x.

In problems 21–35, find the domain of the function, sketch its graph, discuss the symmetry of the graph, indicate the intervals where the function is increasing or decreasing, and find the range of the function. Do these things in any convenient order. [C] Use a calculator if you wish.

21. $f(x) = 3x + 1$
22. $g(x) = -4x + 3$
23. $h(x) = 5$
24. $F(x) = -2\sqrt{x}$
25. $H(x) = 2 - x^2$
26. $G(x) = -7$
27. $f(x) = x|x|$
28. $g(x) = \sqrt{x - 4} + x$
29. $F(x) = 1 + \sqrt{x - 1}$
30. $h(x) = x^2 - 4x$
31. $H(x) = \dfrac{x}{|x|}$
32. $g(x) = x - |\frac{1}{2}x|$
33. $f(x) = x^3 + 2x$
34. $f(x) = |x| - x$
35. $g(x) = \left|\dfrac{1}{x}\right|$

36. The symbol $[\![x]\!]$ is often used to denote the **greatest integer** less than or equal to x; that is, $[\![x]\!]$ is the one and only integer such that $[\![x]\!] \le x < [\![x]\!] + 1$. For instance, $[\![1.7]\!] = 1$, $[\![3.14]\!] = 3$, $[\![2]\!] = 2$, $[\![-1.3]\!] = -2$, $[\![-3]\!] = -3$, and so forth. The function defined by $f(x) = [\![x]\!]$ is called the *greatest integer* function. Sketch a graph of the greatest integer function for $-4 \le x \le 4$.

In problems 37–42, sketch the graph of each piecewise-defined function.

37. $f(x) = \begin{cases} 2 & \text{if } x < 0 \\ -1 & \text{if } x \ge 0 \end{cases}$
38. $g(x) = \begin{cases} x & \text{if } x \ge 0 \\ 2 & \text{if } x < 0 \end{cases}$
39. $F(x) = \begin{cases} 5 + x & \text{if } x \le 3 \\ 9 - \dfrac{x}{3} & \text{if } x > 3 \end{cases}$
40. $h(x) = \begin{cases} x^2 & \text{if } x < 0 \\ -x^2 & \text{if } x \ge 0 \end{cases}$
41. $g(x) = \begin{cases} -x & \text{if } x < 0 \\ x^2 & \text{if } 0 \le x < 2 \\ 8 - 2x & \text{if } x \ge 2 \end{cases}$
42. $G(x) = \begin{cases} 0 & \text{if } x < -2 \\ 3 + x^2 & \text{if } -2 \le x < 2 \\ 0 & \text{if } x \ge 2 \end{cases}$

[C] **43.** A student sketches the graph of $f(x) = 4x^4 - 14x^3 + 20x^2 - 5x$ for $x \ge 0$ by plotting five points and connecting them with a smooth curve, as shown in Figure 15. However, the graph is not correct as shown. Find the error.

Figure 15

x	$f(x)$
0	0
0.5	1
1	5
1.5	10.5
2	22

C 44. A student sketches the graph of $f(x) = x^3 + 3x^2$ by plotting seven points and connecting them with a smooth curve, as shown in Figure 16. Criticize the student's work.

Figure 16

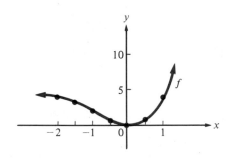

x	$f(x)$
-2	4
-1.5	3.38
-1	2
-0.5	0.63
0	0
0.5	0.88
1	4

45. The management of a city parking lot charges its customers $0.50 per hour or part of an hour for up to 12 hours, with a minimum charge of $2.00 and a maximum charge of $5.00. If $C(x)$ denotes the charge in dollars for parking a car for x hours, sketch a graph of the function C for $0 \le x \le 12$ hours.

46. Explain how to use the graph of the squaring function (Figure 8) to see that, if $a < b < 0$, then $a^2 > b^2$.

6 Shifting, Stretching, and Reflecting Graphs

If f is a function and k is a constant, then *the graph of the function F defined by*

$$F(x) = f(x) + k$$

is obtained by **shifting** *the graph of f* **vertically by** $|k|$ **units,** **upward** *if k > 0, and* **downward** *if k < 0* (Figure 1). Do you see why?

Figure 1

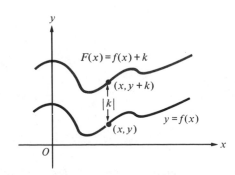

Example   Sketch the graphs of $F(x) = |x| + 2$ and $G(x) = |x| - 3$.

Solution   The graph of

$$f(x) = |x|$$

was shown in Figure 11 of Section 5. We shift this graph upward by 2 units to obtain the graph of $F(x) = |x| + 2$; we shift it downward by 3 units to obtain the graph of $G(x) = |x| - 3$ (Figure 2).

Figure 2

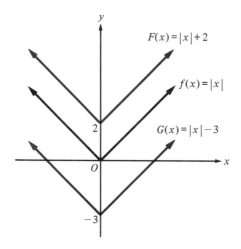

Now, let f be a function, suppose that h is a constant, and define the function F by

$$F(x) = f(x - h).$$

Thus, a point (x, y) belongs to the graph of F if and only if $(x - h, y)$ belongs to the graph of f (Figure 3). Therefore, *the graph of $F(x) = f(x - h)$ is obtained by* **shifting** *the graph of f* **horizontally** *by $|h|$ units, to the* **right** *if $h > 0$, and to the* **left** *if $h < 0$.*

Figure 3

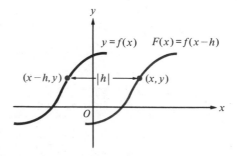

Example   Sketch the graph of $F(x) = (x - 3)^2$.

Solution   The graph of

$$f(x) = x^2$$

was shown in Figure 8 of Section 5. We shift this graph 3 units to the right to obtain the graph of $F(x) = (x - 3)^2$ (Figure 4).

Figure 4

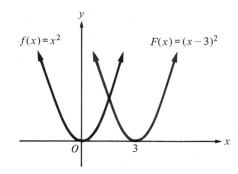

Of course, we can combine vertical and horizontal shifts.

Theorem 1   **Graph-Shifting**

If h and k are constants and f is a function, then the graph of

$$F(x) = f(x - h) + k$$

is obtained by shifting the graph of f horizontally by $|h|$ units and vertically by $|k|$ units. The horizontal shift is to the right if $h > 0$, and to the left if $h < 0$. The vertical shift is upward if $k > 0$, and downward if $k < 0$.

Example   Sketch the graph of $F(x) = |x + 2| - 3$.

Solution   The graph of

$$f(x) = |x|$$

was shown in Figure 11 of Section 5. Here $F(x) = f(x - h) + k$, with $h = -2$ and $k = -3$. Therefore, by the graph-shifting theorem (Theorem 1), we obtain the graph of $F(x) = |x + 2| - 3$ by shifting the graph of $f(x) = |x|$ to the left by 2 units and downward by 3 units (Figure 5).

Figure 5

If f is a function and c is a positive constant, then *the graph of the function F defined by*

$$F(x) = cf(x)$$

is obtained from the graph of f by multiplying each ordinate by c. If $c > 1$, the result is to "stretch" the graph vertically by a factor of c. If $0 < c < 1$, the result is to "flatten" the graph.

Example   Sketch the graphs of $F(x) = 2x^2$ and $G(x) = \frac{1}{2}x^2$.

Figure 6

(a)

(b)

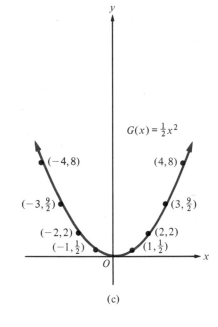

(c)

Solution   We begin with the graph of $f(x) = x^2$ (Figure 6a). In Figure 6b, we have doubled the ordinates of the points on the graph of $f(x) = x^2$ to obtain the graph of $F(x) = 2x^2$. In Figure 6c we have multiplied the ordinates of the points on the graph of $f(x) = x^2$ by $\frac{1}{2}$ to obtain the graph of $G(x) = \frac{1}{2}x^2$.

Figure 7

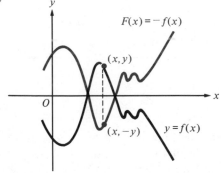

If f is a function, then *the graph of the function*

$$F(x) = -f(x)$$

is obtained by **reflecting** *the graph of f across the x axis* (Figure 7). Do you see why?

Example   Sketch the graphs of $F(x) = -|x|$ and $G(x) = 2 - |x|$.

Solution   We obtain the graph of $F(x) = -|x|$ by reflecting the graph of $f(x) = |x|$ across the x axis (Figure 8a). By shifting the graph of $F(x) = -|x|$ upward by 2 units, we obtain the graph of $G(x) = 2 - |x|$ (Figure 8b).

The techniques of shifting, stretching, and reflecting are often useful for determining the "general shape" of a graph before you begin to sketch it. Using this information and plotting a few points, you can rapidly sketch a reasonably accurate graph.

Figure 8

(a)

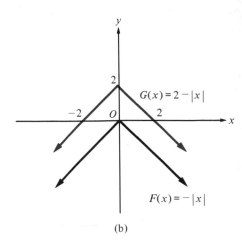

(b)

PROBLEM SET 6

In problems 1–18, use the techniques of shifting, stretching, and reflecting to sketch the graph of each function.

1. $F(x) = x^2 + 2$ 　　　**2.** $G(x) = x^2 - 1$ 　　　**3.** $H(x) = |x - 2|$ 　　　**4.** $f(x) = \sqrt{x - 1}$

5. $g(x) = (x - 2)^2$ 　　**6.** $p(x) = \sqrt{x + 4} - 2$ 　**7.** $q(x) = |x - 1| + 1$ 　**8.** $F(x) = 3\sqrt{x - 1}$

9. $G(x) = 3|x|$ 　　　　**10.** $H(x) = 3|x - 1| + 1$ 　**11.** $Q(x) = \frac{3}{4}x^2 + 1$ 　**12.** $R(x) = \frac{2}{3}(x - 1)^2 + 2$

13. $F(x) = -x^2$ 　　　**14.** $G(x) = 3 - \sqrt{x}$ 　　**15.** $H(x) = 1 - 3|x|$ 　**16.** $f(x) = 1 - 2|x + 4|$

17. $g(x) = 1 - 2x^2$ 　　**18.** $q(x) = 2 - \frac{3}{2}\sqrt{x + 4}$

In problems 19–26, compare the graph of the first function with the graph of the second by sketching them both on the same coordinate system and explaining in words how they are related.

19. $f(x) = x$ 　and 　$F(x) = x + 2$ 　　　　　　**20.** $g(x) = |x|$ 　and 　$G(x) = |x| - 5$

21. $p(x) = x^2$ 　and 　$P(x) = 1 - x^2$ 　　　　　**22.** $q(x) = |x|$ 　and 　$Q(x) = |x - 3| + 3$

23. $r(x) = |x|$ 　and 　$R(x) = 2|x| + 1$ 　　　　**24.** $s(x) = x^2$ 　and 　$S(x) = -\frac{1}{2}(x - 1)^2$

25. $t(x) = \sqrt{x}$ 　and 　$T(x) = 2\sqrt{x - 2} + 4$ 　**26.** $f(x) = x^3$ 　and 　$F(x) = 1 - \frac{2}{3}(x - 1)^3$

In problems 27–30, use the graph of the function f in the indicated figure to obtain the graph of

　(a) $F(x) = f(x) + 2$ 　　(b) $G(x) = f(x - 2)$ 　　(c) $H(x) = 2f(x)$ 　　(d) $Q(x) = -f(x)$

27.

28.

29.

30.

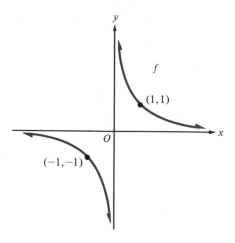

31. How is the graph of $F(x) = f(-x)$ related to the graph of f?

32. If c is a positive constant, how is the graph of $F(x) = f(cx)$ related to the graph of f?

33. If c is a constant and $c > 1$, the graph of $F(x) = cf(x/c)$ is said to be obtained from the graph of f by *magnification*, and c is called the *magnification factor*. Sketch the graph obtained if the graph of $f(x) = x^2$ is magnified by a factor of 2.

34. What happens if you magnify the graph of a linear function $f(x) = mx + b$ by a factor of c? [See problem 33.]

---

7   Algebra of Functions and Composition of Functions

In the business world, the *profit P* from the sale of goods is related to the *revenue R* from the sale and to the *cost C* of producing the goods by the equation

$$P = R - C.$$

In other words, $R = P + C$. Often, the quantities P, R, and C are variables that depend on another variable, for instance, on the number x of units of goods produced; that is, P, R, and C are functions of x. Thus, business applications of mathematics may involve the *addition and subtraction of functions;* other applications require the *multiplication and division of functions*. For example, the function that describes an amplitude-modulated (AM) radio signal is the product of the function that describes the audio signal and the function that describes the carrier wave. Therefore, we make the following definition.

Definition 1   **Sum, Difference, Product, and Quotient of Functions**

Let f and g be functions whose domains overlap. We define functions $f + g$, $f - g$, $f \cdot g$ and f/g as follows:

$$(f + g)(x) = f(x) + g(x)$$
$$(f - g)(x) = f(x) - g(x)$$
$$(f \cdot g)(x) = f(x) \cdot g(x)$$
$$\left(\frac{f}{g}\right)(x) = \frac{f(x)}{g(x)}, \qquad g(x) \neq 0.$$

Example 1   Let $f(x) = 2x^3 - 1$ and $g(x) = x^2 + 5$. Find

(a) $(f + g)(x)$ (b) $(f - g)(x)$ (c) $(f \cdot g)(x)$ (d) $\left(\frac{f}{g}\right)(x)$

Solution

(a) $(f + g)(x) = f(x) + g(x) = (2x^3 - 1) + (x^2 + 5) = 2x^3 + x^2 + 4$
(b) $(f - g)(x) = f(x) - g(x) = (2x^3 - 1) - (x^2 + 5) = 2x^3 - x^2 - 6$
(c) $(f \cdot g)(x) = f(x) \cdot g(x) = (2x^3 - 1)(x^2 + 5) = 2x^5 + 10x^3 - x^2 - 5$
(d) $\left(\frac{f}{g}\right)(x) = \frac{f(x)}{g(x)} = \frac{2x^3 - 1}{x^2 + 5}$

Example 2   The Molar Brush Company finds that the total production cost for manufacturing x toothbrushes is given by the function

$$C(x) = 5000 + 30\sqrt{x} \text{ dollars.}$$

These toothbrushes sell for \$1.50 each. (a) Write a formula for the revenue $R(x)$ dollars to the company if x toothbrushes are sold. (b) Write a formula for the profit $P(x)$ dollars if x toothbrushes are manufactured and sold. (c) Find $P(40,000)$.

Solution

(a) If x toothbrushes are sold at \$1.50 each, $R(x) = 1.50x = 1.5x$ dollars.
(b) $P(x) = R(x) - C(x) = 1.5x - (5000 + 30\sqrt{x})$ dollars.
(c) $P(40,000) = 1.5(40,000) - (5000 + 30\sqrt{40,000}) = 49,000$ dollars.

Geometrically, the graph of the sum, difference, product, or quotient of f and g has at each point an ordinate that is the sum, difference, product, or quotient, respectively, of the ordinates of f and g at the corresponding points. Thus, we can sketch graphs of $f + g$, $f - g$, $f \cdot g$, or f/g by the method of *adding, subtracting, multiplying,* or *dividing ordinates*, respectively. The following example illustrates the method of adding ordinates.

Example   Use the method of adding ordinates to sketch the graph of $h = f + g$ if $f(x) = x^2$ and $g(x) = x - 1$.

Solution   We begin by sketching the graph of f (Figure 1a) and the graph of g (Figure 1b). Then we add the ordinates $f(x)$ and $g(x)$ corresponding to selected values of x in order to obtain the corresponding ordinates $h(x) = f(x) + g(x)$. For instance, $f(2) = 2^2 = 4$ and $g(2) = 2 - 1 = 1$, so $h(2) = f(2) + g(2) = 4 + 1 = 5$. By calculating and plotting several such points and connecting them with a smooth curve, we obtain the graph of h (Figure 1c).

Figure 1

(a)

(b)

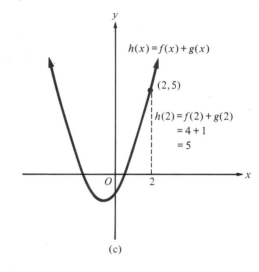

(c)

7.1 Composition of Functions

Suppose that we have three variable quantities y, t, and x. If y depends on t and if t, in turn, depends on x, then it is clear that y depends on x. In other words, if y is a function of t and t is a function of x, then y is a function of x. For instance, suppose that

$$y = t^2 \quad \text{and} \quad t = 3x - 1.$$

Then, substituting the value of t from the second equation into the first equation, we find that

$$y = (3x - 1)^2.$$

More generally, if

$$y = f(t) \quad \text{and} \quad t = g(x),$$

then, substituting t from the second equation into the first equation, we find that

$$y = f(g(x)).$$

To avoid a pileup of parentheses, we often replace the outside parentheses in the last equation by square brackets and write

$$y = f[g(x)].$$

If f and g are functions, the equation $y = f[g(x)]$ defines a new function h:

$$h(x) = f[g(x)].$$

The function h, obtained by "chaining" f and g together this way, is called the **composition** of f and g and is written as

$$h = f \circ g,$$

read, "h equals f composed with g." Thus, by definition,

$$(f \circ g)(x) = f[g(x)].$$

Example 1   Let $f(x) = 3x - 1$ and $g(x) = x^3$. Find

(a) $(f \circ g)(2)$ (b) $(g \circ f)(2)$ (c) $(f \circ g)(x)$
(d) $(g \circ f)(x)$ (e) $(f \circ f)(x)$ © (f) $(f \circ g)(3.007)$

Solution   (a) $(f \circ g)(2) = f[g(2)] = f(2^3) = f(8) = 3(8) - 1 = 23$
(b) $(g \circ f)(2) = g[f(2)] = g[3(2) - 1] = g(5) = 5^3 = 125$
(c) $(f \circ g)(x) = f[g(x)] = f(x^3) = 3x^3 - 1$
(d) $(g \circ f)(x) = g[f(x)] = g(3x - 1) = (3x - 1)^3$
(e) $(f \circ f)(x) = f[f(x)] = f(3x - 1) = 3(3x - 1) - 1 = 9x - 4$
(f) $(f \circ g)(3.007) = f[g(3.007)] = 3(3.007)^3 - 1 \approx 80.568324$

Example 2   Let $f(x) = 3x - 1$, $g(x) = x^3$, and $h(x) = \sqrt{x}$. Find $[f \circ (g \circ h)](x)$.

Solution $[f \circ (g \circ h)](x) = f[(g \circ h)(x)] = f[g(h(x))] = f[g(\sqrt{x})] = f[(\sqrt{x})^3]$
$= f(x^{3/2}) = 3x^{3/2} - 1$

Example 3   If $f(x) = 3x - 1$ and $g(x) = \frac{1}{3}(x + 1)$, show that f and g "undo" each other in the sense that both $f \circ g$ and $g \circ f$ are the same as the identity function (page 146).

Solution $(f \circ g)(x) = f[g(x)] = f[\frac{1}{3}(x + 1)] = 3[\frac{1}{3}(x + 1)] - 1 = x,$

so $f \circ g$ is the same as the identity function. Also,

$(g \circ f)(x) = g[f(x)] = g[3x - 1] = \frac{1}{3}[(3x - 1) + 1] = x.$

Therefore, $g \circ f$ is also the identity function.

Example 4   The labor force $F(x)$ persons required by a certain industry to manufacture x units of a product is given by $F(x) = \frac{1}{2}\sqrt{x}$. At present, there is a demand for 40,000 units of the product, and this demand is increasing at a constant rate of 10,000 units per year. (a) Write an equation for the function $u(t)$ that gives the number of units that will be demanded t years from now. (b) Write a formula for $(F \circ u)(t)$ and interpret the resulting expression.

Solution   (a) $u(t) = 40,000 + 10,000t$
(b) $(F \circ u)(t) = F[u(t)] = F(40,000 + 10,000t)$
$= \frac{1}{2}\sqrt{40,000 + 10,000t} = 50\sqrt{4 + t}$

gives the labor force that will be required t years from now.

Using a programmable calculator, you can see a vivid demonstration of function composition. Suppose that the f and g keys are programmed with functions of your choice. If you enter a number x and touch the g key, you see the mapping

$$x \longmapsto g(x)$$

take place, and the number $g(x)$ appears in the display. Now, if you touch the f key, the mapping

$$g(x) \longmapsto f[g(x)]$$

will take place. Therefore, after you enter the number x, you can perform the composite mapping

$$x \longmapsto (f \circ g)(x)$$

by *touching first the g key, then the f key*. This fact is represented by the diagram in Figure 2.

Figure 2

In using a programmable calculator or a computer, it is important to be able to tell when a complicated function can be obtained as a composition of simpler functions. The same skill is essential in calculus. If

$$h(x) = (f \circ g)(x) = f[g(x)],$$

let's agree to call g the "inside function" and f the "outside function" because of the positions they occupy in the expression $f[g(x)]$. In order to see that h can be obtained as a composition $h = f \circ g$, you must be able to recognize the inside function g and the outside function f in the equation that defines h.

Example Express the function $h(x) = (3x + 2)^2$ as a composition $h = f \circ g$ of two functions f and g.

Solution Here we can take the inside function to be $g(x) = 3x + 2$ and the outside function to be $f(t) = t^2$, so that

$$(f \circ g)(x) = f[g(x)] = f(3x + 2) = (3x + 2)^2 = h(x).$$

In the solution above, we have chosen to write the squaring function as $f(t) = t^2$ rather than as $f(x) = x^2$, since x was used for the independent variable of the function $g(x) = 3x + 2$. Of course, we could just as well have written $f(x) = x^2$, since it is the *rule f* that is important, and no special significance is attached to the symbols used to denote dependent and independent variables. We would still obtain $(f \circ g)(x) = h(x)$.

PROBLEM SET 7

In problems 1–10, find

    (a) $(f + g)(x)$ (b) $(f - g)(x)$ (c) $(f \cdot g)(x)$ (d) $\left(\dfrac{f}{g}\right)(x)$.

1. $f(x) = 5x + 2, g(x) = 2x - 5$

2. $f(x) = \dfrac{1}{2}x + 3, g(x) = 2x - 5$

3. $f(x) = x^2, g(x) = 4$
5. $f(x) = 2, g(x) = -3$

4. $f(x) = ax + b, g(x) = cx + d$
6. $f(x) = |x|, g(x) = x - |x|$

7. $f(x) = 2x - 5, g(x) = x^2 + 1$

8. $f(x) = 1 + \dfrac{1}{x}, g(x) = 1 - \dfrac{1}{x}$

9. $f(x) = \dfrac{5x}{2x - 1}, g(x) = \dfrac{3x + 1}{2x - 1}$

10. $f(x) = \sqrt{x - 3}, g(x) = \dfrac{1}{\sqrt{x - 3}}$

In problems 11–14, use the method of adding or subtracting ordinates to sketch the graph of h.

11. $f(x) = x^2 - 2, g(x) = 1 - \dfrac{x}{2}, h = f + g$

12. $f(x) = 2x^2 + 1, g(x) = x, h = f + g$

13. $f(x) = 1 - x^2, g(x) = x - 1, h = f - g$

14. $f(x) = (x - 2)^2 + 2, g(x) = x^2 + 1, h = f - g$

In problems 15–22, let $f(x) = x - 3$ and $g(x) = x^2 + 4$. Find the indicated value.

15. $(f \circ g)(4)$ **16.** $(f \circ g)(\sqrt{2})$ ⓒ **17.** $(g \circ f)(4.73)$ ⓒ **18.** $(g \circ f)(-2.08)$
19. $(f \circ f)(3)$ **20.** $(g \circ g)(-3)$ **21.** $[f \circ (g \circ f)](2)$ **22.** $[(f \circ g) \circ f](2)$

In problems 23–32, find

    (a) $(f \circ g)(x)$ (b) $(g \circ f)(x)$ (c) $(f \circ f)(x)$.

23. $f(x) = 3x, g(x) = x + 1$
25. $f(x) = x^2, g(x) = \sqrt{x}$
27. $f(x) = 1 - x, g(x) = x^2 + 2$
29. $f(x) = 1 - 5x, g(x) = |2x + 3|$

24. $f(x) = ax + b, g(x) = cx + d$
26. $f(x) = x^3 + 1, g(x) = \sqrt[3]{x - 1}$
28. $f(x) = x + x^{-1}, g(x) = \sqrt{x - 1}$
30. $f(x) = 2, g(x) = 7$

31. $f(x) = \dfrac{1}{2x - 3}, g(x) = 2x - 3$

32. $f(x) = \dfrac{Ax + B}{Cx + D}, g(x) = \dfrac{ax + b}{cx + d}$

In problems 33–40, let $f(x) = 4x, g(x) = x^2 - 3$, and $h(x) = \sqrt{x}$. Express each function as a composition of functions chosen from $f, g,$ and h.

33. $F(x) = \sqrt{x^2 - 3}$ **34.** $G(x) = (\sqrt{x})^2 - 3$ **35.** $H(x) = 2\sqrt{x}$ **36.** $K(x) = 4x^2 - 12$
37. $Q(x) = 4\sqrt{x}$ **38.** $q(x) = 16x^2 - 3$ **39.** $r(x) = \sqrt[4]{x}$ **40.** $s(x) = x^4 - 6x^2 + 6$

In problems 41–48, express each function h as a composition $h = f \circ g$ of two simpler functions f and g.

41. $h(x) = (2x^2 - 5x + 1)^{-7}$ **42.** $h(x) = |2x^2 - 3|$ **43.** $h(x) = \left(\dfrac{1 + x^2}{1 - x^2}\right)^5$ **44.** $h(x) = \dfrac{3(x^2 - 1)^2 + 4}{1 - (x^2 - 1)^2}$

45. $h(x) = \sqrt{\dfrac{x + 1}{x - 1}}$ **46.** $h(x) = \dfrac{1}{(4x + 5)^5}$ **47.** $h(x) = \dfrac{|x + 1|}{x + 1}$ **48.** $h(x) = \sqrt{1 - \sqrt{x - 1}}$

49. A company manufacturing integrated circuits finds that its production cost in dollars for manufacturing x of these circuits is given by the function $C(x) = 50{,}000 + 10{,}000 \sqrt[3]{x + 1}$.

It sells the integrated circuits to distributors for $10 each, and the demand is so high that all the manufactured circuits are sold.

(a) Write a formula for the revenue $R(x)$ in dollars to the company if x integrated circuits are sold.

(b) Write a formula for the profit function $P(x)$ in dollars.

[C] (c) Find the profit if $x = 46,655$.

50. An offshore oil well begins to leak and the oil slick begins to spread on the surface of the water in a circular pattern with a radius $r(t)$ kilometers. The radius depends on the time t in hours since the beginning of the leak, according to the equation $r(t) = 0.5 + 0.25t$.

(a) If $A(x) = \pi x^2$, obtain a formula for $(A \circ r)(t)$ and interpret the resulting expression.

[C] (b) Find the area of the oil slick 4 hours after the beginning of the leak. Round off your answer to two decimal places.

51. The area A of an equilateral triangle depends on its side length x, and the side length x depends on the perimeter p of the triangle. In fact, $A = f(x)$ and $x = g(p)$, where $f(x) = (\sqrt{3}/4)x^2$ and $g(p) = (1/3)p$. Obtain a formula for $(f \circ g)(p)$ and interpret the resulting expression.

52. Suppose that user-definable keys f and g on a programmable calculator are programmed so that $f : x \longmapsto 3x^2 - 2$ and $g : x \longmapsto 5x - 3$. Draw a mapping diagram (see Figure 2, page 160) to show the effect of entering a number x and (a) touching first the g key, then the f key; (b) touching first the f key, then the g key.

53. Let $f(x) = 2x + 3$, $g(x) = x^2$, and $F(x) = \frac{1}{2}(x - 3)$. Find

(a) $(f \circ F)(x)$ (b) $(F \circ f)(x)$ (c) $[f \circ (g \circ F)](x)$

(d) $[(f \circ g) \circ F](x)$ (e) $[(f \circ g) + (f \circ F)](x)$

54. Give examples of (a) a function f for which $f \circ f$ is the same as $f \cdot f$, and (b) a function g for which $g \circ g$ is not the same as $g \cdot g$.

55. Give an example to show that $f \circ g$ need not be the same as $g \circ f$.

56. (a) Is $f \cdot (g + h)$ always the same as $(f \cdot g) + (f \cdot h)$?

(b) Is $f \circ (g + h)$ always the same as $(f \circ g) + (f \circ h)$?

57. If f, g, and h are three functions, show that $(f \circ g) \circ h$ is the same function as $f \circ (g \circ h)$.

58. Prove that the composition of two linear functions is again a linear function.

59. Let $f(x) = 5x + 3$ and $g(x) = 3x + k$, where k is a constant. Find a value of k for which $f \circ g$ and $g \circ f$ are the same function.

60. If $f(x) = mx + b$ with $m \neq 0$, find a function F such that $F \circ f$ is the same as the identity function.

61. Let $f(x) = cx + 1$, where c is a constant. Find a value of c so that $f \circ f$ is the same as the identity function.

8   Inverse Functions

Many calculators have both a squaring key (marked x^2) and a square-root key (marked $\sqrt{x}$). For nonnegative numbers x, the functions represented by these keys "undo each other" in the sense that

$$x \xrightarrow{\text{squaring function}} x^2 \xrightarrow{\text{square-root function}} \sqrt{x^2} = x$$

and

$$x \xrightarrow{\text{square-root function}} \sqrt{x} \xrightarrow{\text{squaring function}} (\sqrt{x})^2 = x.$$

Two functions related in such a way that each "undoes" what the other "does" are said to be **inverse** to one another. Most scientific calculators not only have keys corresponding to a number of important functions, but they also can calculate the values of the inverses of these functions.

In order that two functions f and g be inverses of one another, we must have the following: For every value of x in the domain of f, $f(x)$ is in the domain of g and

$$x \xmapsto{\;f\;} f(x) \xmapsto{\;g\;} g[f(x)] = x;$$

likewise, for every value of x in the domain of g, $g(x)$ is in the domain of f and

$$x \xmapsto{\;g\;} g(x) \xmapsto{\;f\;} f[g(x)] = x.$$

In other words, f and g are inverses of each other if and only if

$$g[f(x)] = x$$

for every value of x in the domain of f, and

$$f[g(x)] = x$$

for every value of x in the domain of g. A function f for which such a function g exists is said to be **invertible.**

Examples   In each case, show that the functions f and g are inverses of each other.

1 $f(x) = 5x - 1$ and $g(x) = \dfrac{x + 1}{5}.$

Solution   The domain of f is the set $\mathbb{R}$ of all real numbers. For every real number x, we have

$$g[f(x)] = g(5x - 1) = \frac{(5x - 1) + 1}{5} = \frac{5x}{5} = x.$$

The domain of g is also the set $\mathbb{R}$ of all real numbers, and for every real number x, we have

$$f[g(x)] = f\left(\frac{x + 1}{5}\right) = 5\left(\frac{x + 1}{5}\right) - 1 = x + 1 - 1 = x.$$

2 $f(x) = \dfrac{2x - 1}{x}$ and $g(x) = \dfrac{1}{2 - x}.$

Solution   The domain of f is the set of all nonzero real numbers. If $x \neq 0$, we have

$$g[f(x)] = g\left(\frac{2x - 1}{x}\right) = \frac{1}{2 - \left(\dfrac{2x - 1}{x}\right)} = \frac{x}{\left[2 - \left(\dfrac{2x - 1}{x}\right)\right]x}$$

$$= \frac{x}{2x - (2x - 1)} = \frac{x}{1} = x.$$

The domain of g is the set of all real numbers except 2. If $x \neq 2$, we have

$$f[g(x)] = f\left(\frac{1}{2-x}\right) = \frac{2\left(\dfrac{1}{2-x}\right) - 1}{\left(\dfrac{1}{2-x}\right)} = \left[2\left(\frac{1}{2-x}\right) - 1\right](2-x)$$

$$= 2 - (2-x) = x.$$

Geometrically, f and g are inverses of each other if and only if the graph of g is the mirror image of the graph of f across the straight line $y = x$ (Figure 1). This will be clear if you note first that the mirror image of a point (a, b) across the line $y = x$ is the point (b, a). But, if f and g are inverses of each other and if (a, b) belongs to the graph of f, then $b = f(a)$, so $g(b) = g[f(a)] = a$; that is, (b, a) belongs to the graph of g. Similarly, you can show that, if (b, a) belongs to the graph of g, then (a, b) belongs to the graph of f (Problem 35).

Figure 1

Figure 2

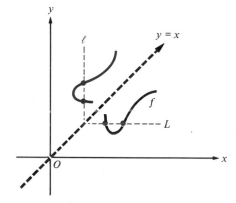

Not every function is invertible. Indeed, consider the function f whose graph appears in Figure 2. The mirror image of the graph of f across the line $y = x$ isn't the graph of a function, because there is a vertical line ℓ that intersects it more than once. (Recall the vertical line test, page 139.) Notice that the horizontal line L obtained by reflecting ℓ across the line $y = x$ intersects the graph of f more than once. These considerations provide the basis of the following test (see Problem 36).

Horizontal Line Test

A function f is invertible if and only if no horizontal straight line intersects its graph more than once.

Example   Use the horizontal line test to determine whether or not the functions graphed in Figure 3 are invertible.

Figure 3

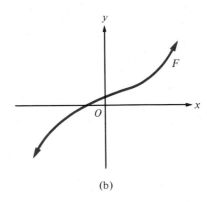

(a)   (b)

Solution   (a) Any horizontal line drawn above the origin will intersect the graph of f *twice*. Therefore, f is not invertible.

(b) No horizontal line intersects the graph of F more than once; hence, F is invertible.

Figure 4

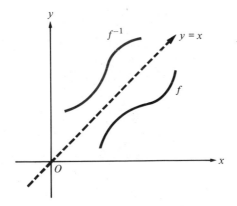

If a function f is invertible, there is *exactly one* function g such that f and g are inverses of each other; indeed, g is the one and only function whose graph is the mirror image of the graph of f across the line $y = x$. We call g the *inverse* of the function f and we write $g = f^{-1}$. The notation f^{-1} is read "f inverse." If you imagine that the graph of f is drawn with wet ink, then the graph of f^{-1} would be the imprint obtained by folding the paper along the line $y = x$ (Figure 4).

The fact that the graph of f^{-1} is the set of all points (x, y) such that (y, x) belongs to the graph of f provides the basis for the following procedure.

Algebraic Method for Finding f^{-1}

Step 1. Write the equation $y = f(x)$ that defines f.

Step 2. Interchange x and y in the equation obtained in step 1; that is, change y to x and change all x's to y's so that the equation becomes $x = f(y)$.

Step 3. Solve the equation in step 2 for y in terms of x to get $y = f^{-1}(x)$. This equation defines f^{-1}.

Example 1   Use the algebraic method to find the inverse of the function $f(x) = x^2$, for $x \geq 0$, and sketch the graphs of f and f^{-1} on the same coordinate system.

Solution   We carry out the algebraic procedure:

Step 1. $y = x^2$ for $x \geq 0$
Step 2. $x = y^2$ for $y \geq 0$
Step 3. Solving the equation $x = y^2$ for y and using the condition that $y \geq 0$, we obtain $y = \sqrt{x}$; therefore, $f^{-1}(x) = \sqrt{x}$ (Figure 5).

Figure 5

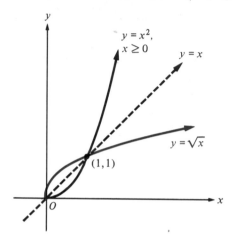

Example 2   Let $f(x) = 2x + 1$. (a) Use the algebraic method to find f^{-1}. (b) Check that $f^{-1}[f(x)] = x$ for all values of x in the domain of f. (c) Check that $f[f^{-1}(x)] = x$ for all values of x in the domain of f^{-1}.

Solution   (a) We carry out the algebraic procedure:

Step 1. $y = 2x + 1$
Step 2. $x = 2y + 1$
Step 3. $2y + 1 = x$
$2y = x - 1$
$y = \frac{1}{2}(x - 1)$; therefore, $f^{-1}(x) = \frac{1}{2}(x - 1)$.

(b) $f^{-1}[f(x)] = f^{-1}(2x + 1) = \frac{1}{2}[(2x + 1) - 1] = x$
(c) $f[f^{-1}(x)] = f[\frac{1}{2}(x - 1)] = 2[\frac{1}{2}(x - 1)] + 1 = x$

In working with inverses of functions, you must be careful not to confuse $f^{-1}(x)$ and $[f(x)]^{-1}$. Notice that $f^{-1}(x)$ is the value of the function f^{-1} at x, while $[f(x)]^{-1} = \dfrac{1}{f(x)}$ is the reciprocal of the value of the function f at x. For instance, in Example 2 above, $f^{-1}(x) = \frac{1}{2}(x - 1)$, whereas $[f(x)]^{-1} = \dfrac{1}{2x + 1}$.

PROBLEM SET 8

In problems 1–6, show that the functions f and g are inverses of each other.

1. $f(x) = 2x - 3$ and $g(x) = \dfrac{x + 3}{2}$

2. $f(x) = x^3$ and $g(x) = \sqrt[3]{x}$

3. $f(x) = \dfrac{1}{x}$ and $g(x) = \dfrac{1}{x}$

4. $f(x) = \dfrac{2x - 3}{3x - 2}$ and $g(x) = \dfrac{2x - 3}{3x - 2}$

5. $f(x) = \sqrt[3]{x + 8}$ and $g(x) = x^3 - 8$

6. $f(x) = x^2 + 1$ for $x \geq 1$ and $g(x) = \sqrt{x - 1}$

7. Use the horizontal line test to determine whether or not the functions graphed in Figure 6 are invertible.

Figure 6

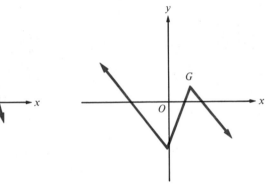

(a)   (b)   (c)

8. Under what conditions is a linear function $f(x) = mx + b$ invertible?

9. The three functions graphed in Figure 7 are invertible. In each case, obtain the graph of the inverse function by reflecting the given graph across the line $y = x$.

Figure 7

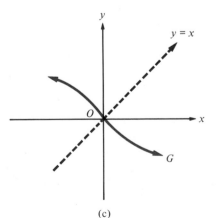

(a)   (b)   (c)

10. If a linear function f is invertible, show that f^{-1} is also a linear function.

In problems 11–20, use the algebraic method to find $f^{-1}(x)$, and sketch the graphs of f and f^{-1} on the same coordinate system.

11. $f(x) = 7x - 13$

12. $f(x) = 5 + 7x$

13. $f(x) = -\frac{1}{2}x + 3$

14. $f(x) = x^3$

15. $f(x) = -x^2$ for $x \le 0$

16. $f(x) = 4 - x^2$ for $x \ge 0$

17. $f(x) = 1 + \sqrt{x}$

18. $f(x) = (x - 2)^2$ for $x \ge 2$

19. $f(x) = (x - 1)^2 + 1$ for $x \ge 1$

20. $f(x) = (x + 3)^2 + 2$ for $x \le -3$

In problems 21–30, (a) use the algebraic method to find f^{-1}, (b) check that $f^{-1}[f(x)] = x$ for all values of x in the domain of f, and (c) check that $f[f^{-1}(x)] = x$ for all values of x in the domain of f^{-1}.

21. $f(x) = 2x - 5$

22. $f(x) = \dfrac{3x + 2}{5}$

23. $f(x) = (x + 2)^3$

24. $f(x) = (x + 1)^3 - 2$

25. $f(x) = 1 - 2x^3$

26. $f(x) = x^5 - 1$

27. $f(x) = -\dfrac{1}{x} - 1$

28. $f(x) = \dfrac{3}{x + 2}$

29. $f(x) = \dfrac{3x - 7}{x + 1}$

30. $f(x) = (x + 3)^5 - 2$

31. Sketch the graph of $f(x) = |x - 1|$ and determine whether or not f is invertible.

32. The linear function $f(x) = x$ is its own inverse. Find all linear functions that are their own inverses.

33. If $a, b, c,$ and d are constants such that $ad \ne bc$, use the algebraic method to find the inverse of

$$f(x) = \frac{ax + b}{cx + d}.$$

34. If $f(x) = \sqrt{4 - x^2}$ for $0 \le x \le 2$, show that f is its own inverse.

35. If f and g are inverses of each other and (b, a) belongs to the graph of g, show that (a, b) belongs to the graph of f.

36. Complete the argument to justify the horizontal line test by showing that, if no horizontal line intersects the graph of f more than once, then f is invertible.

37. A function f is said to be *one-to-one* if, whenever a and b are in the domain of f and $f(a) = f(b)$, it follows that $a = b$. By using the horizontal line test, show that f is invertible if and only if it is one-to-one.

38. If f is invertible, show that the domain of f coincides with the range of f^{-1}.

39. Suppose that the concentration C of an anesthetic in body tissues satisfies an equation of the form $t = f(C)$, where t is the elapsed time since the anesthetic was administered. If f is invertible, write an equation for C in terms of t.

40. Show that the functions f and g are inverses of each other if and only if $(g \circ f)(x) = x$ for every x in the domain of f and $(f \circ g)(x) = x$ for every x in the domain of g.

41. Sketch the graph of any invertible function f; then turn the paper over, rotate it $90°$ clockwise, and hold it up to the light. Through the paper you will see the graph of f^{-1}. Why does this procedure work?

REVIEW PROBLEM SET

In problems 1–6, plot each pair of points and find the distance between them by using the distance formula.

1. $(1, 1)$ and $(4, 5)$ **2.** $(-1, 2)$ and $(5, -7)$ **3.** $(-3, 2)$ and $(2, 14)$

C **4.** $(4.71, -3.22)$ and $(0, \pi)$ **5.** $(-2, -5)$ and $(-2, 3)$ C **6.** $(\sqrt{\pi}, \pi)$ and $\left(\dfrac{1 + \sqrt{2}}{2}, \dfrac{31}{7}\right)$

C **7.** What is the perimeter of the triangle with vertices $(-1, 5)$, $(8, -7)$, and $(4, 1)$? Round off your answer to two decimal places.

8. Are the points $(-5, 4)$, $(7, -11)$, $(12, -11)$, and $(0, 4)$ the vertices of a parallelogram?

In problems 9 and 10, sketch the graph of the equation.

9. $(x - 2)^2 + (y + 3)^2 = 25$ **10.** $x^2 + 2x + y^2 + 2y + 1 = 0$

In problems 11–14, find the slope of the line segment $\overline{AB}$ and find an equation in point-slope form of the line containing A and B.

11. $A = (3, -5)$ and $B = (2, 2)$ **12.** $A = (0, 7)$ and $B = (5, 0)$
13. $A = (1, 2)$ and $B = (-3, -4)$ **14.** $A = \left(\frac{3}{2}, \frac{2}{3}\right)$ and $B = \left(\frac{1}{6}, -\frac{5}{6}\right)$

In problems 15 and 16, sketch the line L that contains the point P and has slope m, and find an equation in point-slope form for L.

15. $P = (5, 2)$ and $m = -\frac{3}{5}$ **16.** $P = \left(-\frac{2}{3}, \frac{1}{2}\right)$ and $m = \frac{3}{2}$

In problems 17 and 18, (a) find an equation of the line that contains the point P and is parallel to the line segment $\overline{AB}$, and (b) find an equation of the line that contains the point P and is perpendicular to line segment $\overline{AB}$.

17. $P = (7, -5)$, $A = (1, 8)$, $B = (-3, 2)$ **18.** $P = \left(\frac{2}{5}, \frac{1}{3}\right)$, $A = \left(\frac{7}{3}, -\frac{3}{5}\right)$, $B = \left(1, \frac{2}{5}\right)$

In problems 19 and 20, rewrite each equation in slope-intercept form, find the slope m and the y intercept b of the graph, and sketch it.

19. $4x - 3y + 2 = 0$ **20.** $\frac{2}{3}x - \frac{1}{5}y + 3 = 0$

In problems 21–24, find an equation of the line L in (a) point-slope form, (b) slope-intercept form, and (c) general form.

21. L contains the point $(-7, 1)$ and has slope $m = 3$.
22. L contains the points $(2, 5)$ and $(1, -3)$.
23. L contains the point $(1, -2)$ and is parallel to the line whose equation is $7x - 3y + 2 = 0$.
24. L contains the point $(3, -4)$ and is perpendicular to the line $2x - 5y + 4 = 0$.

25. If (a, b) is a point on the circle $x^2 + y^2 = 9$, find an equation of the line that is tangent to the circle at (a, b). [Hint: The line tangent to a circle at a point is perpendicular to the radius drawn from the center to the point.]

26. Show that the line containing the two points $P = (a, b)$ and $Q = (c, d)$ has the equation $(b - d)x - (a - c)y + ad - bc = 0$.

27. In calculus, it is shown that the slope m of the line tangent to the graph of $y = x^2 + 4$ at the point (a, b) is given by $m = 2a$. (a) Write an equation of this tangent line. (b) Find an equation of the tangent line at $(2, 8)$. (c) Sketch the graph of $y = x^2 + 4$ and show the tangent line at $(2, 8)$.

28. An advertising agency claims that a furniture store's revenue will increase by $20 per month for each additional dollar spent on advertising. The current average monthly sales revenue is $140,000, with an expenditure of $100 per month for advertising. Find the equation that relates the store's expected average monthly sales revenue y to the total expenditure x for advertising. Find the value of y if $400 per month is spent for advertising.

In problems 29–40, let $f(x) = 3x^2 - 4$, $g(x) = 6 - 5x$, and $h(x) = 1/x$. Find the indicated values.

29. $f(-3)$

30. $h\left(\dfrac{1}{2}\right)$

31. $g\left(\dfrac{6}{5}\right)$

32. $h[h(x)]$

33. $f(x) - f(2)$

34. $f(x + k) - f(x)$

35. $f[g(x)]$

36. $g\left(\dfrac{1}{4 + k}\right)$

37. $g(x) + g(-x)$

38. $\sqrt{f(-|x|)}$

39. $\dfrac{h(x + k) - h(x)}{k}$

40. $\dfrac{1}{h(4 + k)}$

In problems 41–44, find the domain of each function.

41. $f(x) = \dfrac{1}{x - 1}$

42. $g(x) = \dfrac{1}{\sqrt{4 - x^2}}$

43. $h(x) = \sqrt{1 + x}$

44. $F(x) = \dfrac{3}{|x| - x}$

In problems 45 and 46, find the difference quotient $\dfrac{f(x + h) - f(x)}{h}$ and simplify the resulting expression.

45. $f(x) = 3x^2 - 2x + 1$

46. $f(x) = x^{-1/3}$

47. Which of the graphs in Figure 1 are graphs of functions?

Figure 1

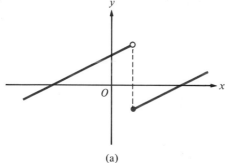

(a)

(b)

48. Which of the curves in Figure 2 are graphs of functions?

Figure 2

(a)

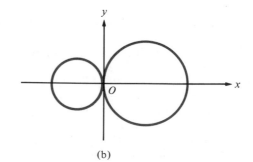

(b)

C **49.** The function $f : x \mapsto y$ given by $f(x) = 71.88 + 0.37x$ relates the atmospheric pressure x in centimeters of mercury to the boiling temperature $f(x)$ of water in degrees Celsius at that pressure. Find $f(74)$, $f(75)$, and $f(76)$. Round off your answers to two decimal places.

50. If $g(x) = \dfrac{2 + x}{2 - x}$, find and simplify $\dfrac{g(t) - g(-t)}{1 + g(t)g(-t)}$.

In problems 51 and 52, use the point-plotting method to sketch the graph of each function. **C** You may wish to improve the accuracy of your sketch by using a calculator to determine the coordinates of some points on the graph.

51. $f(x) = \dfrac{x^4}{16}$ for $-2 \leq x \leq 2$

52. $f(x) = \dfrac{1 + x}{1 + x^2}$ for $-2 \leq x \leq 2$

53. For each function in Figure 3, find the domain and range; indicate the intervals over which the function is increasing, decreasing, or constant; and determine whether the function is even, odd, or neither.

Figure 3

(a)

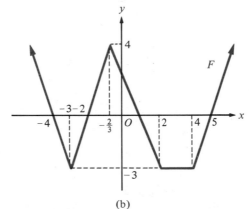

(b)

54. Let $g(x) = |x - 5| - |x + 5|$. (a) Sketch the graph of g. (b) Determine the intervals over which g is increasing or decreasing. [Hint: Express $g(x)$ without absolute value symbols for the three cases $x \leq -5$, $-5 < x < 5$, and $5 \leq x$.]

In problems 55–60, determine without drawing a graph whether the function is even, odd, or neither, and discuss any symmetry of the graph.

55. $f(x) = 5x^5 + 3x^3 + x$

56. $g(x) = (x^4 + x^2 + 1)^{-1}$

57. $h(x) = (x + 1)x^{-1}$

58. $F(x) = -x^3|x|$

59. $G(x) = x^{80} - 5x^6 + 9$

60. $H(x) = \dfrac{\sqrt{x}}{1 + x}$

In problems 61–66, find the domain of the function, sketch its graph, discuss any symmetry of the graph, indicate the intervals where the function is increasing or decreasing, and find the range of the function. Do these things in any convenient order. **C** Use a calculator if you wish.

61. $f(x) = 5x - 3$

62. $g(x) = 3 - (x/5)$

63. $F(x) = 2\sqrt{x - 2}$

64. $G(x) = x^{1/3}$

65. $h(x) = \begin{cases} x^2 & \text{if } x > 0 \\ -x^2 & \text{if } x \leq 0 \end{cases}$

66. $H(x) = \begin{cases} x & \text{if } x < 0 \\ 2x & \text{if } 0 \leq x \leq 1 \\ 3x^3 - 1 & \text{if } x > 1 \end{cases}$

67. The graph of a function f is shown in Figure 4. Sketch the graph of the function F defined by.
(a) $F(x) = f(x) + 1$ (b) $F(x) = f(x) - 2$ (c) $F(x) = f(x - 1)$
(d) $F(x) = f(x + 2)$ (e) $F(x) = -f(x)$.

Figure 4

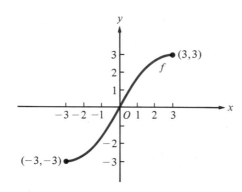

68. Compare the graph of the first function with the graph of the second by sketching them both on the same coordinate system and explaining in words how they are related.
(a) $F(x) = 3x + 2, f(x) = 3x - 1$ (b) $G(x) = x^2 - 5, g(x) = x^2$
(c) $H(x) = 1 - |x|, h(x) = |x|$ (d) $K(x) = \sqrt{x - 1} + 2, k(x) = \sqrt{x}$
(e) $Q(x) = \frac{1}{2}(x + 2)^2, q(x) = x^2$ (f) $R(x) = 1 - \frac{1}{2}|x + 1|, r(x) = |x|$

C **69.** The power delivered by a wind-powered generator is given by

$$P(x) = kx^3 \text{ horsepower,}$$

where x is the speed of the wind in miles per hour and k is a constant depending on the size of the generator and its efficiency. For a certain wind-powered generator, the constant $k = 3.38 \times 10^{-4}$. Sketch the graph of the function P for this generator and determine how many horsepower are generated when the wind speed is 35 miles per hour.

70. In economics, it is often assumed that the demand for a commodity is a function of selling price; that is, $q = f(p)$, where p dollars is the selling price per unit of the commodity and q is the number of units that will sell at that price. (a) Would you expect f to be an increasing or a decreasing function? (b) Write an equation for the total amount of money in dollars $F(p)$ spent by consumers for the commodity if the selling price per unit is p dollars. (c) What would it mean to say that there is a value p_0 dollars for which $f(p_0) = 0$?

71. In economics, it is often assumed that the number of units s of a commodity that producers will supply to the market place is a function of the selling price p dollars per unit; that is, $s = g(p)$. (a) Would you expect g to be an increasing or a decreasing function? (b) Write an equation for the total amount of money in dollars $G(p)$ spent by consumers for the commodity if the selling price per unit is p dollars and all supplied units are purchased. (c) What would it mean to say that there is a value p_1 dollars for which $g(p_1) = 0$?

72. A manufacturer finds that 100,000 electronic calculators are sold per month at a price of $50 each, but that only 60,000 are sold per month if the price is $75 each. Suppose that the demand function f for these calculators is linear (see problem 70). (a) Find a formula for $f(p)$, where p dollars is the selling price per calculator and $f(p)$ is the number that will sell at that price. (b) Find the selling price per calculator if the monthly demand for calculators is 80,000. (c) What price would be so high that no calculators would be sold?

In problems 73–82, find

(a) $(f + g)(x)$ (b) $(f - g)(x)$ (c) $(f \cdot g)(x)$ (d) $\left(\dfrac{f}{g}\right)(x)$ (e) $(f \circ g)(x)$.

73. $f(x) = x + 2, g(x) = 3x - 4$

74. $f(x) = x^2 + 2x, g(x) = x^2 - 2x$

75. $f(x) = \dfrac{1}{x - 1}, g(x) = \dfrac{1}{x + 1}$

76. $f(x) = \dfrac{x + 3}{x - 2}, g(x) = \dfrac{x}{x - 2}$

77. $f(x) = x^4, g(x) = \sqrt{x + 1}$

78. $f(x) = x, g(x) = |x - 2| - x$

79. $f(x) = |x|, g(x) = -x$

80. $f(x) = \sqrt{1 + x^2}, g(x) = \pi|x|$

81. $f(x) = x^{2/3} + 1, g(x) = \sqrt{x}$

82. $f(x) = \dfrac{|x|}{x}, g(x) = \dfrac{-x}{|x|}$

C In problems 83–86, let $f(x) = 1 + x^5$ and $g(x) = 1 - \sqrt{x}$. Use a calculator to find each value. Round off to four decimal places.

83. $(f \circ g)(2.7746)$

84. $[(f \circ g) \circ f](\pi)$

85. $(g \circ f)(2.7746)$

86. $(g \circ g)(0.0007)$

In problems 87–92, let $f(x) = x^2$, $g(x) = \sqrt{x}$, and $h(x) = x + 1$. Express each function as a composition of functions chosen from f, g, and h.

87. $F(x) = \sqrt{x + 1}$

88. $G(x) = |x|$

89. $H(x) = \sqrt{x} + 1$

90. $K(x) = x^4$

91. $P(x) = x^2 + 2x + 1$

92. $p(x) = x + 2$

In problems 93–96, express each function h as a composition $h = f \circ g$ of two simpler functions f and g.

93. $h(x) = (4x^3 - 2x + 5)^{-3}$

94. $h(x) = \sqrt[3]{\dfrac{4 + x^3}{4 - x^3}}$

95. $h(x) = \dfrac{2(x^2 + 1)^2 + x^2 + 1}{\sqrt{x^2 + 1}}$

96. $h(x) = \dfrac{(2x^2 - 5x + 1)^{-2/3}}{(2x^2 - 5x + 1)^2 + 2}$

97. If $f(x) = x^2 + 1, g(x) = x^2 - 1$, and $h(x) = \sqrt{x}$, find
   (a) $[(f + g) \circ h](x)$
   (b) $[(f - g) \circ h](x)$
   (c) $[(f \circ h) + (g \circ h)](x)$
   (d) $[(f \circ h) - (g \circ h)](x)$.

98. Find $(f \circ g)(x)$ if

$$f(x) = \begin{cases} x & \text{if } x \geq 0 \\ 2x & \text{if } x < 0 \end{cases} \quad \text{and} \quad g(x) = \begin{cases} -2x & \text{if } x \geq 0 \\ 4x & \text{if } x < 0. \end{cases}$$

99. A baseball diamond is a square that is 90 feet long on each side. A ball is hit from home plate directly toward third base at the rate of 50 feet per second. Let y denote the distance of the ball from first base in feet; let x denote its distance in feet from home plate; and let t denote the elapsed time in seconds since the ball was hit. Here y is a function of x, say, $y = f(x)$, and x is a function of t, say, $x = g(t)$.
   (a) Find formulas for $f(x)$ and $g(t)$.
   (b) Explain why $y = (f \circ g)(t)$.
   (c) Find a formula for $(f \circ g)(t)$.

100. In an alternative-energy home, solar cells and a wind-powered generator charge batteries that supply electric power to the home. The batteries have a constant internal resistance of r ohms and provide a fixed voltage, E volts. The current I amperes drawn from the batteries depends on the net resistance R ohms of the appliances being used in the home according to the equation $I = \dfrac{E}{r + R}$. The electric power P watts being consumed by these appliances is given by the equation $P = I^2 R$. Find a function f such that $P = f(R)$.

101. Two college students earn extra money on weekends by delivering firewood in their pickup truck. They have found that they can sell x cords per weekend at a price of p dollars per cord, where $x = 75 - \frac{3}{5}p$. The students buy the firewood from a supplier who charges them C dollars for x cords according to the equation $C = 500 + 15x + \frac{1}{5}x^2$. (a) Find a function f such that $P = f(p)$, where P dollars is the profit per weekend for the students if they charge p dollars per cord. (b) Find the profit P dollars if $p = \$95$.

102. Which of the following functions is invertible?

(a) $f(x) = x$ (b) $f(x) = 1/x$ (c) $f(x) = 3x + 5$ (d) $f(x) = 3x^2 + 5$

In problems 103–108, find $f^{-1}(x)$ for each function f, verify that $f^{-1}[f(x)] = x$ and $f[f^{-1}(x)] = x$, and sketch the graphs of f and f^{-1} on the same coordinate system.

103. $f(x) = 3x - 1$

104. $f(x) = (x - 1)^{-1}$

105. $f(x) = \frac{1}{5}x + 5$

106. $f(x) = \frac{1}{4}x^3$

107. $f(x) = 2\sqrt{x} - 1$

108. $f(x) = \dfrac{(x - 2)^2}{4} + 3, \quad x \geq 2.$

109. Sketch the graph of the inverse of f, g, and h in Figure 5.

Figure 5

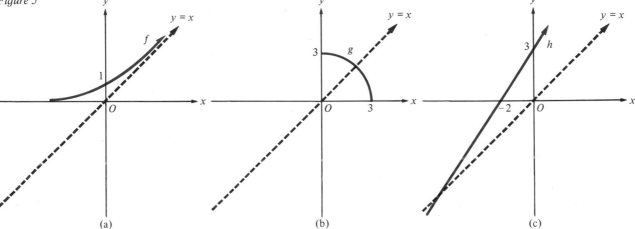

(a)   (b)   (c)

110. If $f(x) = x^2 - 3x + 2$ for $x \leq \frac{3}{2}$, find $f^{-1}(x)$.

111. If $f(x) = \dfrac{x - 1}{x}$, find $f^{-1}(x)$.

112. If f is invertible, show that f^{-1} is also invertible and that $(f^{-1})^{-1}$ is the same as f.

113. Suppose that f is invertible, that the domain of f is $\mathbb{R}$, and that the range of f is $\mathbb{R}$. What function is $f^{-1} \circ f$?

114. Are there values of the constants a, b, c, and d for which the function $f(x) = \dfrac{ax + b}{cx + d}$ is its own inverse?

115. Let A, B, and C be constants with $A > 0$, and suppose that $f(x) = Ax^2 + Bx + C$ for $x \geq \dfrac{-B}{2A}$. Find $f^{-1}(x)$.

116. Is there any invertible function f such that $f^{-1}(x) = [f(x)]^{-1}$ for all values of x in the domain of f?

117. If bacteria are allowed to reproduce in a culture, the number N of bacteria at time t satisfies an equation of the form $f(N) = kt$, where k is a constant and f is an invertible function. Write an equation involving f^{-1} that gives N in terms of t.

4 Polynomial and Rational Functions

In this chapter, we continue the study of functions and their graphs. Here we explore polynomial and rational functions, their "zeros," and some of the properties of their graphs. The chapter also includes the remainder theorem, synthetic division, and the rational zeros theorem; and it ends with a section on ratio, proportion, and variation.

1 Quadratic Functions

A function f of the form

$$f(x) = ax^2 + bx + c,$$

where a, b, and c are constants and $a \neq 0$, is called a **quadratic function.** Such functions often arise in applied mathematics. For instance, the height of a projectile is a quadratic function of time; the velocity of blood flow is a quadratic function of the distance from the center of the blood vessel; and the force exerted by the wind on the blades of a wind-powered generator is a quadratic function of the wind speed.

The simplest quadratic function is the squaring function $f(x) = x^2$, whose graph was sketched in Section 5.2 of Chapter 3 (page 146). As we saw in Section 6 of Chapter 3, the graph of $f(x) = ax^2$ is obtained from the graph of $y = x^2$ by vertical "stretching" if $a > 1$ or "flattening" if $0 < a < 1$. Furthermore, the graph of $f(x) = ax^2$ for negative values of a is obtained by reflecting the graph $y = |a|x^2$ across the x axis. Figure 1 shows the graph of $f(x) = ax^2$ for various values of a.

The graphs of equations of the form $y = ax^2$ (Figure 1) are examples of curves called **parabolas.**

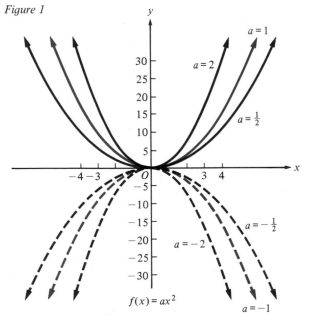

Figure 1

$f(x) = ax^2$

175

Figure 2

(a)

(b)

Figure 3

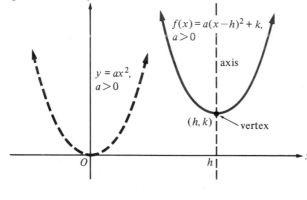

The particular parabolas with equations of the form $y = ax^2$ are symmetric about the y axis; they **open upward** and have a lowest point at $(0,0)$ if $a > 0$ (Figure 2a), and they **open downward** and have a highest point at $(0,0)$ if $a < 0$ (Figure 2b). The highest or lowest point of the graph of $y = ax^2$ is called the **vertex** of the parabola, and its line of symmetry (in this case, the y axis) is simply called the **axis** of the parabola. (We continue the study of parabolas in the final chapter.)

By the graph-shifting theorem (Chapter 3, Section 6, Theorem 1, page 153), the graph of

$$f(x) = a(x - h)^2 + k$$

is obtained by shifting the parabola $y = ax^2$ horizontally by $|h|$ units and vertically by $|k|$ units. Hence, its graph is a parabola with vertex at (h, k) (Figure 3). The parabola opens upward if $a > 0$ and downward if $a < 0$, and its axis is the vertical line $x = h$.

Figure 4

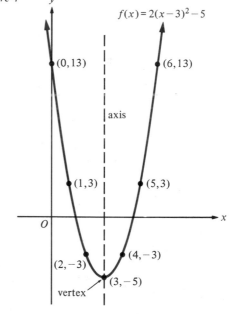

Example   Sketch the graph of $f(x) = 2(x - 3)^2 - 5$.

Solution   The function has the form $f(x) = a(x - h)^2 + k$, with $a = 2$, $h = 3$, and $k = -5$; hence, its graph is a parabola, opening upward, with vertical axis $x = 3$ and vertex $(h, k) = (3, -5)$. This information indicates the general appearance of the graph. By plotting a few points, we can obtain a reasonably accurate sketch of the parabola (Figure 4).

As the following theorem shows, you can use the technique illustrated in the example above to sketch the graph of any quadratic function.

Theorem 1   **The Graph of a Quadratic Function**

> The quadratic function
> $$f(x) = ax^2 + bx + c$$
> can be rewritten in the form
> $$f(x) = a(x - h)^2 + k,$$
> where
> $$h = -\frac{b}{2a} \quad \text{and} \quad k = f(h).$$
> Hence, the graph of f is a parabola with vertex at (h, k). The parabola opens upward if $a > 0$ and downward if $a < 0$, and its axis is the vertical line $x = h$.

Proof   The proof is based on the idea of completing the square (Section 3.2 of Chapter 2). Thus, we begin by writing

$$f(x) = ax^2 + bx + c = a\left(x^2 + \frac{b}{a}x\right) + c.$$

By adding $\left[\frac{1}{2}\left(\frac{b}{a}\right)\right]^2 = \frac{b^2}{4a^2}$ to $x^2 + \frac{b}{a}x$, we obtain the perfect square

$$x^2 + \frac{b}{a}x + \frac{b^2}{4a^2} = \left(x + \frac{b}{2a}\right)^2.$$

Therefore,

$$f(x) = a\left(x^2 + \frac{b}{a}x + \frac{b^2}{4a^2}\right) + c - a\left(\frac{b^2}{4a^2}\right)$$
$$= a\left(x + \frac{b}{2a}\right)^2 + c - \frac{b^2}{4a}$$
$$= a(x - h)^2 + k,$$

where $h = -\frac{b}{2a}$ and $k = c - \frac{b^2}{4a}$. But, from the equation

$$f(x) = a(x - h)^2 + k,$$

we have

$$f(h) = a(h - h)^2 + k = a(0)^2 + k = k,$$

that is, $k = f(h)$.

Example   Use Theorem 1 to rewrite $f(x) = -3x^2 - 12x - 1$ in the form $f(x) = a(x - h)^2 + k$ and sketch the graph of f.

Solution   By Theorem 1, with $a = -3$, $b = -12$, and $c = -1$, we have

$$h = -\frac{b}{2a} = -\frac{(-12)}{2(-3)} = -2$$

and

$$k = f(h) = f(-2)$$
$$= -3(-2)^2 - 12(-2) - 1 = 11.$$

Therefore,

$$f(x) = a(x - h)^2 + k$$
$$= -3(x + 2)^2 + 11.$$

The graph is a parabola, opening downward, with vertex $(-2, 11)$ and vertical axis $x = -2$ (Figure 5).

Figure 5

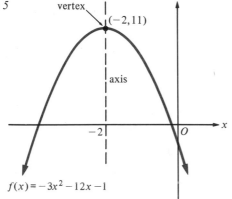

$$f(x) = -3x^2 - 12x - 1$$

Figure 6

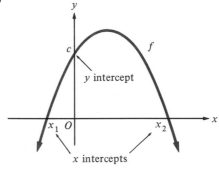

Notice that the graph of $f(x) = ax^2 + bx + c$ intersects the y axis at the point $(0, c)$ (Figure 6). We call c the **y intercept** of the graph. If the graph intersects the x axis at the points $(x_1, 0)$ and $(x_2, 0)$, we call x_1 and x_2 its **x intercepts.** Note that x_1 and x_2, if they exist, are the real roots of the quadratic equation $ax^2 + bx + c = 0$. (Why?)

Example   Find the vertex and the y and x intercepts of the graph of $f(x) = -2x^2 - 5x + 3$, determine whether the graph opens upward or downward, sketch the graph, and find the domain and range of f.

Solution   Here we have $a = -2$, $b = -5$, and $c = 3$; hence, by Theorem 1,

$$h = -\frac{b}{2a} = -\frac{(-5)}{2(-2)} = -\frac{5}{4}$$

and

$$k = f(h) = f(-\tfrac{5}{4})$$
$$= -2(-\tfrac{5}{4})^2 - 5(-\tfrac{5}{4}) + 3 = \tfrac{49}{8}.$$

Figure 7

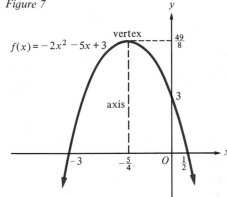

$f(x) = -2x^2 - 5x + 3$

vertex

axis

Therefore, the vertex is

$$(h, k) = (-\tfrac{5}{4}, \tfrac{49}{8}).$$

The y intercept is $c = 3$, and the x intercepts are the solutions of the quadratic equation

$$-2x^2 - 5x + 3 = 0 \quad \text{or} \quad (x + 3)(-2x + 1) = 0.$$

It follows that the x intercepts are -3 and $\tfrac{1}{2}$. Because $a = -2 < 0$, the graph opens downward. Using this information, we can sketch the parabola (Figure 7). (If a more accurate sketch is desired, a calculator can be used to determine additional points on the graph.) The domain of f is $\mathbb{R}$ and, as Figure 7 shows, its range is the interval $(-\infty, \tfrac{49}{8}]$.

1.1 Maximum and Minimum Values of Quadratic Functions

We know from Theorem 1 that the graph of $f(x) = ax^2 + bx + c$ is a parabola with a vertical axis. The vertex

$$\left(-\frac{b}{2a}, f\left(-\frac{b}{2a} \right) \right)$$

is the lowest point on the graph if $a > 0$ (Figure 8a) and the highest point on the graph if $a < 0$ (Figure 8b). Therefore, if $a > 0$, the number $f(-b/2a)$ is the **minimum** (smallest) value of the function f, and if $a < 0$, the number is the **maximum** (largest) value of f. This fact has many useful applications.

Figure 8

(a)

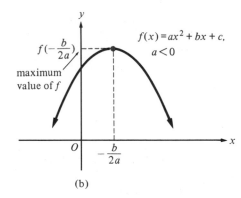

(b)

Example 1   A manufacturer of synfuel (synthetic fuel) from coal estimates that the cost $f(x)$ in dollars per barrel for a production run of x thousand barrels is given by $f(x) = 9x^2 - 180x + 940$. How many thousands of barrels should be produced during each run to minimize the cost per barrel, and what is the minimum cost per barrel of the synfuel?

Solution   The cost per barrel is given by

$$f(x) = 9x^2 - 180x + 940$$
$$= ax^2 + bx + c,$$

where $a = 9$, $b = -180$, and $c = 940$. Therefore, $f(x)$ attains its minimum value when

$$x = -\frac{b}{2a} = -\frac{(-180)}{2(9)} = 10 \text{ thousand barrels,}$$

and the minimum cost per barrel is given by

$$f(10) = 9(10)^2 - 180(10) + 940 = 40 \text{ dollars.}$$

Example 2   An orchard contains 30 apple trees, each of which yields approximately 400 apples over the growing season. The owner plans to add more trees to the orchard, but the State Agricultural Service advises that because of crowding, each new tree will reduce the average yield per tree by about 10 apples over the growing season. How many trees should be added to maximize the total yield of apples, and what is the maximum yield?

Solution   Let x denote the number of trees added. After x trees are added, the orchard will contain $30 + x$ trees, but the average yield of each tree per season will be reduced from 400 apples to $400 - 10x$ apples. Therefore, the total yield of apples per season is given by

$$f(x) = (30 + x)(400 - 10x) = 12{,}000 + 100x - 10x^2$$
$$= ax^2 + bx + c,$$

where $a = -10$, $b = 100$, and $c = 12{,}000$. Thus, $f(x)$ attains its maximum value when

$$x = -\frac{b}{2a} = -\frac{100}{2(-10)} = 5 \text{ trees.}$$

If 5 trees are added to the orchard, the total yield per growing season will be

$$f(5) = 12{,}000 + 100(5) - 10(5)^2$$
$$= 12{,}250 \text{ apples.}$$

PROBLEM SET 1

1. Sketch the graph of $f(x) = ax^2$ for (a) $a = 3$, (b) $a = -\frac{1}{3}$, and (c) $a = -3$.
2. Sketch the graph of $f(x) = 2x^2 + k$ for (a) $k = 0$, (b) $k = 1$, and (c) $k = -1$.

In problems 3–6, sketch the graph of each function.

3. $f(x) = 2(x - 1)^2 + 3$ 4. $g(x) = -2(x + 2)^2 - 3$ 5. $h(x) = -\frac{1}{2}(x + 1)^2 - 4$ 6. $p(x) = \frac{1}{2}(x + 1)^2 - \frac{1}{2}$

In problems 7–12, use Theorem 1 to rewrite each function f in the form $f(x) = a(x - h)^2 + k$, and sketch the graph of f.

7. $f(x) = x^2 + 2x - 4$ 8. $f(x) = x^2 - 12x + 5$ 9. $f(x) = 3x^2 - 10x - 2$
10. $f(x) = 2x^2 + 3x - 1$ 11. $f(x) = -2x^2 + 6x + 3$ 12. $f(x) = \frac{3}{2}x^2 - 6x - 7$

In problems 13–26, find the vertex and the intercepts of the graph of each function, determine whether the graph opens upward or downward, sketch the graph, and find the domain and range of the function. $\boxed{\text{C}}$(You may use a calculator to determine additional points on the graph if you want to obtain a more accurate sketch.)

13. $f(x) = x^2 - 2x - 3$

14. $g(x) = \frac{1}{2}x^2 - \frac{3}{4}$

15. $h(x) = x^2 + 4x$

16. $K(x) = -x^2 - 8x - 15$

17. $Q(x) = -2x^2 + x - 15$

18. $f(t) = -t^2 - 2t + 8$

19. $p(x) = 6x^2 + x - 2$

20. $g(x) = 4x - 12x^2 + 21$

21. $f(x) = -3x^2 + 12x + 15$

22. $H(x) = 30 - x(1 + 14x)$

23. $F(x) = \frac{1}{2}x^2 + x + 2$

24. $G(t) = t - \frac{2}{3}t^2 - \frac{1}{3}$

25. $f(x) = 2x^2 - 20x + 57, \quad 0 \le x \le 10$

26. $g(x) = -3x^2 + 24x - 50, \quad 2 < x < 6$

27. Find two positive numbers whose sum is 100 and whose product is as large as possible.

28. If an object is projected straight upward from ground level with an initial speed of 96 feet per second, then (neglecting air resistance) its height $h(t)$ in feet after t seconds is given by $h(t) = 96t - 16t^2$. Find the maximum height reached by the object and determine the time t at which it returns to ground level.

29. Find the length and width of a rectangular plot of land whose perimeter is 600 meters and whose area is the maximum for that perimeter.

30. Find the minimum value of $g(x) = (ax^2 + bx + c)^2$ in terms of a, b, and c.

31. If a manufacturer produces x thousand tons of a new lightweight alloy for engine blocks, each block will cost $3x^2 - 600x + 30{,}090$ dollars. How many thousand tons of the alloy should be produced to minimize the cost of the engine blocks, and what is the resulting minimum cost of one block?

32. In medicine, it is often assumed that a patient's *reaction* $R(x)$ to a drug dose of size x is given by an equation of the form $R(x) = Ax^2(B - x)$, where A and B are positive constants. It can then be shown that the body's *sensitivity* $S(x)$ to a dose of size x is given by $S(x) = 2ABx - 3Ax^2$. Find the reaction to the dose for which the sensitivity is maximum.

33. The management of a racquetball club foresees that 60 members will join the club if each membership is $5.00 per month, but that for each 50-cent increase in the membership price per month, 4 of the 60 potential members will decide not to join. The cost to the club per member is estimated to be $3.50 per month. What membership price per month will bring in the maximum profit to the club?

34. A real estate company manages an apartment building containing 80 units. When the rent for each unit is $250 per month, all apartments are occupied. However, for each $10 increase in monthly rent per unit, one of the units becomes vacant. Each vacant unit costs the management $15 per month for taxes and upkeep, and each occupied unit costs the management $65 per month for taxes, service, upkeep, and water. What rent should be charged for a maximum profit?

35. Suppose that the distance d in kilometers that a certain car can travel on one tank of gasoline at a speed of v kilometers per hour is given by $d = 12v - (v/4)^2$. What speed maximizes the distance d, and hence minimizes fuel consumption?

Figure 9

36. Ship A is 65 nautical miles due east of ship B and is sailing south at 15 knots (nautical miles per hour), while ship B is sailing east at 10 knots. Figure 9 shows the original positions P and Q of ships A and B, their positions t hours later, and the distance d between the ships at time t. Find the minimum distance between the ships and the time when it occurs. [Hint: Use the Pythagorean theorem and the fact that d is minimum when d^2 is minimum.]

2   Polynomial Functions

We have now discussed constant functions

$$f(x) = b,$$

linear functions

$$f(x) = ax + b,$$

and quadratic functions

$$f(x) = ax^2 + bx + c.$$

These are all special cases of **polynomial functions**; that is, functions whose values are given by polynomials in the independent variable.

A polynomial function f of the form

$$f(x) = ax^n,$$

Figure 1

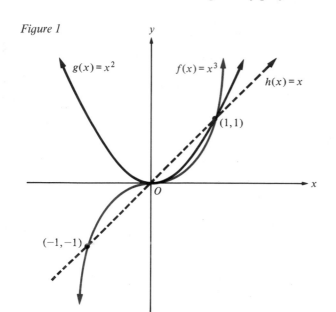

where $a \neq 0$ and n is a positive integer, is called a **power function** of *degree n*. For $n = 1$, the graph of f is a straight line that has slope a and contains the origin; for $n = 2$, the graph of f is a parabola that has its vertex at the origin and opens upward if $a > 0$ and downward if $a < 0$.

Figure 1 shows the graph of $f(x) = x^3$ and, for comparison, the graphs of $g(x) = x^2$ and $h(x) = x$. Notice that if x is less than 1 and positive, x^2 is smaller than x, and x^3 is even smaller than x^2. (For instance, if $x = \frac{1}{4}$, then $x^2 = \frac{1}{16} < \frac{1}{4}$ and $x^3 = \frac{1}{64} < \frac{1}{16}$.) It follows that on the open interval $(0, 1)$, the graph of $g(x) = x^2$ lies below the graph of $h(x) = x$, while the graph of $f(x) = x^3$ lies below the graph of $g(x) = x^2$. However, on the interval $(1, \infty)$, $x^3 > x^2 > x$, so the graph of $f(x) = x^3$ is above the graph of $g(x) = x^2$, which in turn is above the graph of $h(x) = x$. Notice that all three graphs contain the origin and the point $(1, 1)$. Since $f(x) = x^3$ is an odd function (why?), its graph is symmetric about the origin.

If n is an *even* positive integer, then the power function $f(x) = x^n$ is even (why?) and its graph is symmetric about the y axis (Figure 2). The graph contains the origin and the points $(-1, 1)$ and $(1, 1)$. It never falls below the x axis. If $n = 2$, the graph is a parabola; but, for $n > 2$, the graph of $f(x) = x^n$ falls below the parabola $y = x^2$ over the open intervals $(-1, 0)$ and $(0, 1)$ and rises above the parabola over the intervals $(-\infty, -1)$ and $(1, \infty)$.

As the even integer n becomes larger and larger, the graph of $f(x) = x^n$ becomes flatter and flatter on both sides of the origin and rises more and more

Figure 2

Figure 3

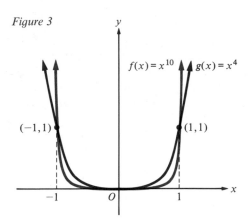

sharply through the two points $(-1,1)$ and $(1,1)$. Figure 3 shows the graph of $f(x) = x^{10}$ and contrasts it with the graph of $g(x) = x^4$. For large values of n, the graph of $f(x) = x^n$ may come so close to the x axis on both sides of the origin that it appears to coincide with a segment of the x axis. In reality, of course, the graph of $f(x) = x^n$ touches the x axis only at the origin.

If n is an *odd* positive integer, then the power function $f(x) = x^n$ is odd (why?) and its graph is symmetric about the origin (Figure 4). The graph contains the origin and the points $(-1, -1)$ and $(1, 1)$. For $n > 3$, the graph is similar to the graph of $y = x^3$, but is flatter near the origin, rises more sharply to the right of $x = 1$, and drops more rapidly to the left of $x = -1$ (Figure 4).

Example   Sketch each of the graphs of $f(x) = x^5$, $g(x) = 2x^5$, and $h(x) = -2x^5$ on the same coordinate system.

Solution   We begin by sketching the graph of $f(x) = x^5$. Then, by doubling the ordinates, we obtain the graph of $g(x) = 2x^5$. Finally, we reflect the graph of $g(x) = 2x^5$ across the x axis to obtain the graph of $h(x) = -2x^5$ (Figure 5).

Figure 4

Figure 5

Figure 6

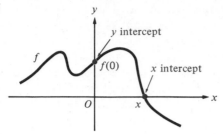

The **y intercept** of the graph of a function f is the ordinate $f(0)$ of the point $(0, f(0))$ where the graph intersects the y axis. Similarly, the abscissa of a point where the graph of f intersects the x axis is called an **x intercept** of the graph (Figure 6). Thus, the x intercepts of the graph of f are the real roots (if any) of the equation $f(x) = 0$.

Examples   Find the y and x intercepts and sketch the graph of each function.

1 $f(x) = -x^4 + 1$

Solution   The y intercept is $f(0) = 1$, and the x intercepts are the real roots of the equation $f(x) = 0$; that is, $-x^4 + 1 = 0$ or $x^4 = 1$. Hence, the x intercepts are $x = \sqrt[4]{1} = 1$ and $x = -\sqrt[4]{1} = -1$. By reflecting the graph of $y = x^4$ (Figure 3) across the x axis and then shifting it 1 unit upward, we obtain the graph of $f(x) = -x^4 + 1$ (Figure 7).

2 $G(x) = -\frac{1}{4}(x - 1)^5 + 8$

Solution   The y intercept is $G(0) = -\frac{1}{4}(0 - 1)^5 + 8 = \frac{1}{4} + 8 = \frac{33}{4}$, and the x intercept is the real root of the equation $G(x) = 0$; that is,

$$-\tfrac{1}{4}(x - 1)^5 + 8 = 0 \quad \text{or} \quad (x - 1)^5 = 32.$$

Thus, $x - 1 = \sqrt[5]{32} = 2$, so the x intercept is given by $x = 3$. By the graph-shifting theorem (page 153), the graph of $G(x) = -\frac{1}{4}(x - 1)^5 + 8$ is obtained by shifting the graph of $y = -\frac{1}{4}x^5$ upward by 8 units and to the right by 1 unit (Figure 8).

Figure 8

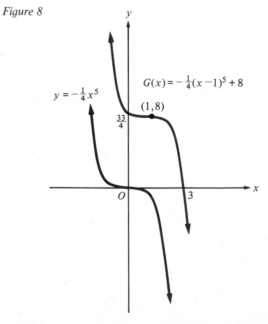

$G(x) = -\frac{1}{4}(x-1)^5 + 8$

$y = -\frac{1}{4}x^5$

$(1, 8)$

$\frac{33}{4}$

Figure 7

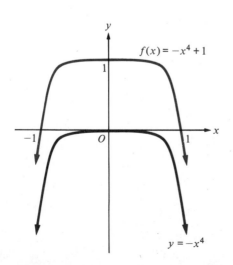

$f(x) = -x^4 + 1$

$y = -x^4$

In calculus, it is shown that the graph of every polynomial function is a smooth curve with no jumps or breaks. Although the graph of a polynomial function f can wiggle up and down as in Figure 9a, it cannot have the sharp corners or jumps illustrated in Figure 9b. The x intercepts of the graph of f divide the x axis into open intervals. You can determine the algebraic signs of $f(x)$ over these open intervals by using convenient test numbers, just as you did in Section 7.1 of Chapter 2. The following example illustrates how you can use this information to help sketch the graph of f.

Figure 9

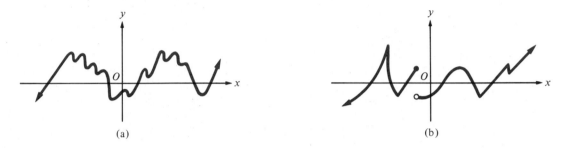

(a)   (b)

Example   Let $f(x) = (2x - 1)(x^2 - x - 2)$. (a) Find the x intercepts of the graph of f. (b) Sketch the graph of f. (c) Use the graph to solve the inequality $f(x) < 0$.

Solution   (a) The x intercepts are the real roots of $f(x) = 0$; that is, $(2x - 1)(x^2 - x - 2) = 0$ or $(2x - 1)(x + 1)(x - 2) = 0$. Setting each factor equal to zero, we obtain

$$
\begin{array}{c|c|c}
2x - 1 = 0 & x + 1 = 0 & x - 2 = 0 \\
x = \tfrac{1}{2} & x = -1 & x = 2.
\end{array}
$$

Thus, in increasing order, the x intercepts are $-1, \tfrac{1}{2}$, and 2.

(b) The x intercepts divide the x axis into four open intervals, as shown in Figure 10.

Figure 10

The graph of f touches the x axis only at the endpoints of these intervals; over each interval, the graph is either entirely above the x axis $[f(x) > 0]$ or entirely below it $[f(x) < 0]$. To see which is the case, we select convenient test numbers on each open interval (Figure 10) and evaluate $f(x) = (2x - 1)(x^2 - x - 2)$ at each test number. We obtain

$$f(-2) = -20, \quad f(0) = 2, \quad f(1) = -2, \quad \text{and} \quad f(3) = 20.$$

Since $f(-2) = -20$, the point $(-2, -20)$ belongs to the graph of f. This point is below the x axis, so the graph of f stays below the x axis over the first interval

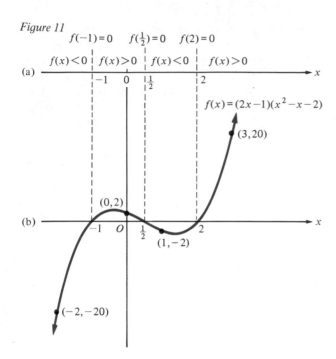

Figure 11

$f(-1)=0$ $f(\frac{1}{2})=0$ $f(2)=0$

(a)

$f(x)<0$ | $f(x)>0$ | $f(x)<0$ | $f(x)>0$

$f(x)=(2x-1)(x^2-x-2)$

$(3,20)$

$(0,2)$

(b)

$(1,-2)$

$(-2,-20)$

$(-\infty, -1)$. We apply similar reasoning to the remaining intervals and obtain the results in Figure 11a. Plotting the points $(-1,0)$, $(\frac{1}{2},0)$, and $(2,0)$ corresponding to the x intercepts and the points $(-2, -20)$, $(0, 2)$, $(1, -2)$, and $(3, 20)$ corresponding to the test numbers, and using the information in Figure 11a, we can sketch the graph of f (Figure 11b).

(c) From the graph of f (Figure 11b), it is clear that $f(x) < 0$ only for values of x in the two intervals $(-\infty, -1)$ and $(\frac{1}{2}, 2)$.

When you use the method illustrated in the example above, you should always ask whether the graph contains hidden peaks and valleys that are not indicated by the plotted points. Although methods studied in calculus are required to answer this question, the polynomial functions that we consider will be fairly simple, so that all peaks and valleys on their graphs can be detected by plotting a reasonable number of points.

PROBLEM SET 2

In problems 1–6, decide which functions are polynomial functions and which are not.

1. $f(x) = 2x^2 - 16x + 29$

2. $g(x) = \pi x - \sqrt{3}x^5 + \frac{1}{2}x^3$

3. $h(x) = \sqrt{x}$

4. $F(x) = (2x^2 - 1)(3x^3 + 2)$

5. $G(x) = 2x^{-3} + 7$

6. $H(x) = \dfrac{1}{x}$

In each of problems 7–10, sketch graphs of the functions f, g, and h on the same coordinate system. $\boxed{\text{C}}$(You may use a calculator to determine additional points on the graph if you want to obtain a more accurate sketch.)

7. $f(x) = x^4$, $g(x) = 2x^4$, $h(x) = -2x^4$

8. $f(x) = x^6$, $g(x) = \dfrac{x^6}{3}$, $h(x) = -\dfrac{x^6}{3}$

9. $f(x) = x^5$, $g(x) = \dfrac{1}{2}x^5$, $h(x) = -\dfrac{1}{2}x^5$

10. $f(x) = x^7$, $g(x) = \dfrac{x^7}{9}$, $h(x) = -\dfrac{x^7}{9}$

In problems 11–22, find the y and x intercepts and sketch the graph of each function. $\boxed{\text{C}}$(Use a calculator if you wish.)

11. $f(x) = x^4 + 1$

12. $g(x) = -\frac{2}{3}x^6 + 2$

13. $h(x) = 3x^5 - 1$

14. $F(x) = \frac{1}{16}(x + 2)^9$

15. $G(x) = 2(x + 1)^4$

16. $H(x) = -\frac{1}{2}(x - 1)^6$

17. $f(x) = -x^8 - 1$

18. $g(x) = -\frac{2}{3}x^7 + 1$

19. $h(x) = (x + \frac{1}{2})^3 - 1$

20. $F(x) = 7(x + \frac{1}{2})^3 + 2$

21. $G(x) = 2(x - 6)^4 + 1$

22. $H(x) = -20(x - 8)^8 + 8$

In problems 23–32, find all x intercepts of the graph of each function.

23. $f(x) = 5x + 10(x - 2) - 40$

24. $g(x) = 5x^2 + 17x - 12$

25. $h(x) = (x - 1)(4x^2 - 12x + 9)$

26. $k(x) = (3x + 1)(3x^2 - 4x)$

27. $p(x) = (4x^2 - 1)(6x^2 - 5x + 1)$

28. $P(x) = (x - 1)[x(x + 1) + 2x]$

29. $H(x) = (9x^2 - 25)(x^2 - 5x - 14)$

30. $q(x) = 3x^4 - 3x^2 - 6$

31. $F(x) = x^6 - x^4$

32. $f(x) = x^5 + 6x^3 + 9x$

In problems 33–40, (a) find the x intercepts of the graph of the function, (b) sketch the graph, and (c) use the graph to solve the given inequality. [C] (Use a calculator if you wish.)

33. $f(x) = x(x - 1)(x + 1); f(x) > 0$

34. $g(x) = (x^2 - x - 2)(x + 3); g(x) \geq 0$

35. $p(x) = -x(x + 2)(x - 1); p(x) \geq 0$

36. $F(x) = -(2x + 1)(x - 1)(3x + 4); F(x) < 0$

37. $h(x) = \frac{1}{6}(x^2 - 1)(x - 3); h(x) < 0$

38. $G(x) = x^2(x - 1)(x + 3); G(x) \leq 0$

39. $H(x) = (x + 4)(x^2 + x - 2); H(x) \leq 0$

40. $g(x) = (x + 1)^2(x^2 - 9); g(x) \geq 0$

[C] **41.** Equal squares are cut off at each corner of a rectangular piece of cardboard 8 inches wide by 15 inches long, and an open-topped box is formed by turning up the sides (Figure 12). If x is the length of the sides of the cutoff squares, then the volume $V(x)$ of the resulting box is given by $V(x) = x(8 - 2x)(15 - 2x)$ cubic inches. Sketch the graph of the polynomial function V for $x > 0$.

Figure 12

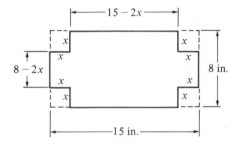

42. Use the graph obtained in problem 41 to determine the approximate value of x for which $V(x)$ is maximum.

3 Division of One Polynomial by Another

In elementary arithmetic, you learned to divide one integer by another to obtain a quotient and a remainder; for instance,

$$
\begin{array}{r}
71 \leftarrow\!\!\!-\text{quotient} \\
\text{divisor}\longrightarrow 32\,\overline{\smash{)}2277} \leftarrow\!\!\!-\text{dividend} \\
224 \\
\hline
37 \\
32 \\
\hline
5 \leftarrow\!\!\!-\text{remainder}
\end{array}
$$

The result of this calculation can be expressed as

$$2277 = (32)(71) + 5,$$

that is,

$$\text{dividend} = (\text{divisor})(\text{quotient}) + \text{remainder}.$$

In this section, we study a similar procedure, called **long division,** for dividing one polynomial by another, and we introduce a useful shortcut called **synthetic division.** These procedures will be used extensively in Section 4, where we continue our study of polynomial functions and their graphs.

When you divide one positive integer by another, you continue the procedure until you obtain a remainder that is less than the divisor. Likewise, when you divide one polynomial by another, you should continue the long-division procedure until the remainder is either the zero polynomial or a polynomial of lower degree than the divisor.

Examples   Perform the indicated long division to find a quotient polynomial and a remainder polynomial. Be sure that the remainder is either the zero polynomial or a polynomial of lower degree than the divisor. Check your work by verifying that

$$\text{dividend} = (\text{divisor})(\text{quotient}) + \text{remainder}.$$

1 $2x^2 + x - 1 \overline{\smash{\big)}\, 6x^4 + x^3 + 4x + 4}$

Solution   Note that both the divisor $2x^2 + x - 1$ and the dividend $6x^4 + x^3 + 4x + 4$ are arranged in descending powers of x. (If they weren't, we would begin by rewriting them so they were.) We divide $6x^4$, the leading term of the dividend, by $2x^2$, the leading term of the divisor, to obtain

$$\frac{6x^4}{2x^2} = 3x^2,$$

the first term of the quotient. Thus we write

$$
\begin{array}{r}
3x^2 \quad\quad\quad\quad \text{(first term of the quotient)} \\
2x^2 + x - 1 \overline{\smash{\big)}\, 6x^4 + x^3 + 4x + 4}
\end{array}
$$

Now, we multiply $3x^2$ by the divisor $2x^2 + x - 1$, write this product under the dividend, and subtract to obtain a first trial remainder:

$$
\begin{array}{r}
3x^2 \quad\quad\quad\quad\quad\quad\quad \\
2x^2 + x - 1 \overline{\smash{\big)}\, 6x^4 + x^3 \quad\quad + 4x + 4} \\
6x^4 + 3x^3 - 3x^2 \quad\quad\quad \\
\hline
-2x^3 + 3x^2 + 4x + 4 \quad \text{(first trial remainder)}
\end{array}
$$

(subtract)

(To make the subtraction easier, we have spaced the terms so that like terms are aligned vertically.) Because the degree of our first trial remainder isn't less than the degree of the divisor, we must repeat the procedure. Thus, we divide $-2x^3$, the leading term of the first trial remainder, by $2x^2$, the leading term of the divisor, to obtain $-x$, the second term of the quotient. We multiply $-x$ by the divisor $2x^2 + x - 1$, write this product under the first trial remainder, and subtract to obtain a second trail remainder:

$$
\begin{array}{r}
3x^2 - x \quad\quad\quad\quad \text{(second term of the quotient)} \\
2x^2 + x - 1 \overline{\smash{\big)}\, 6x^4 + x^3 \quad\quad\quad +4x + 4} \\
6x^4 + 3x^3 - 3x^2 \quad\quad\quad \\
\hline
-2x^3 + 3x^2 + 4x + 4 \\
-2x^3 - x^2 + x \quad\quad \\
\hline
4x^2 + 3x + 4 \quad \text{(second trial remainder)}
\end{array}
$$

(subtract)

The degree of the second trial remainder still isn't less than the degree of the divisor, so we divide its leading term $4x^2$ by $2x^2$ to obtain 2, the third term of the quotient. We multiply 2 by the divisor $2x^2 + x - 1$, write this product under the second trial remainder, and subtract to obtain a third trial remainder:

$$
\begin{array}{r}
\text{third term of} \\
\text{the quotient}
\end{array}
$$

$$
\begin{array}{l}
3x^2 - x + 2 \quad\leftarrow \\
2x^2 + x - 1\,\overline{\big)\,6x^4 + x^3 + 4x + 4} \\
6x^4 + 3x^3 - 3x^2 \\
\overline{-2x^3 + 3x^2 + 4x + 4} \\
-2x^3 - x^2 + x \\
\overline{4x^2 + 3x + 4} \\
\text{subtract} \longrightarrow 4x^2 + 2x - 2 \\
\overline{x + 6}
\end{array}
$$

third trial remainder (the remainder)

The third trial remainder is actually the remainder, since its degree is less than the degree of the divisor. We conclude that

$$3x^2 - x + 2 \quad \text{is the quotient polynomial}$$

and

$$x + 6 \quad \text{is the remainder polynomial.}$$

To check, we calculate

$$
\begin{aligned}
(\text{divisor})(\text{quotient}) + \text{remainder} &= (2x^2 + x - 1)(3x^2 - x + 2) + (x + 6) \\
&= 6x^4 + x^3 + 3x - 2 + x + 6 \\
&= 6x^4 + x^3 + 4x + 4 \\
&= \text{the dividend,}
\end{aligned}
$$

so our work is correct.

2 $x^3 - 2x^2 + 3x - 4\,\overline{\big)\,2x^5 - 7x^4 + 13x^3 - 19x^2 + 15x - 4}$

Solution

$$
\begin{array}{l}
2x^2 - 3x + 1 \quad\leftarrow \quad \text{quotient polynomial} \\
x^3 - 2x^2 + 3x - 4\,\overline{\big)\,2x^5 - 7x^4 + 13x^3 - 19x^2 + 15x - 4} \\
2x^5 - 4x^4 + 6x^3 - 8x^2 \\
\text{subtract}\overline{- 3x^4 + 7x^3 - 11x^2 + 15x - 4} \quad\leftarrow \text{first trial remainder} \\
- 3x^4 + 6x^3 - 9x^2 + 12x \\
\text{subtract}\overline{x^3 - 2x^2 + 3x - 4} \quad\leftarrow \text{second trial remainder} \\
x^3 - 2x^2 + 3x - 4 \\
\text{subtract}\overline{0} \quad\leftarrow \text{third trial remainder (the remainder)}
\end{array}
$$

The quotient polynomial is $2x^2 - 3x + 1$ and the remainder is the zero polynomial. To check, we calculate

$$
\begin{aligned}
(\text{divisor})(\text{quotient}) + \text{remainder} &= (x^3 - 2x^2 + 3x - 4)(2x^2 - 3x + 1) + 0 \\
&= 2x^5 - 7x^4 + 13x^3 - 19x^2 + 15x - 4 \\
&= \text{the dividend,}
\end{aligned}
$$

and our work is correct.

A systematic procedure that is guaranteed to work in a finite number of steps is called an **algorithm.** Long division is an algorithm because, as you carry it out, you accumulate the quotient polynomial term by term and you reduce the degree of the trial remainder at each stage. The process comes to an end when the trial remainder is either the zero polynomial or a polynomial of lower degree than the divisor. This is summarized by the following noteworthy theorem.

Theorem 1   **The Division Algorithm**

Let f and g be polynomial functions and suppose that g is not the constant zero polynomial. Then there exist unique polynomial functions q and r such that

$$f(x) = g(x)q(x) + r(x)$$

holds for all values of x, and r is either the constant zero polynomial or a polynomial of degree lower than g.

In the division algorithm, $f(x)$ is the dividend polynomial, $g(x)$ the divisor polynomial, $q(x)$ the quotient polynomial, and $r(x)$ the remainder polynomial. The identity

$$f(x) = g(x)q(x) + r(x)$$

expresses the fact that

$$\text{dividend} = (\text{divisor})(\text{quotient}) + \text{remainder}.$$

We often refer to the function f/g or to a value $f(x)/g(x)$ of such a function as a *quotient*, but the word used in this way must not be confused with the *quotient polynomial* $q(x)$ in the division algorithm. The relationship between the two types of quotients is expressed in the following theorem.

Theorem 2   **The Quotient Theorem**

Let $q(x)$ be the quotient polynomial and $r(x)$ be the remainder polynomial obtained by long division of the polynomial $f(x)$ by the nonzero polynomial $g(x)$. Then, for all values of x such that $g(x) \neq 0$,

$$\frac{f(x)}{g(x)} = q(x) + \frac{r(x)}{g(x)}$$

Proof   Divide both sides of the identity $f(x) = g(x)q(x) + r(x)$ by $g(x)$.

The quotient theorem is often used to write a rational expression $\dfrac{f(x)}{g(x)}$ as the sum of a polynomial $q(x)$ and a rational expression $\dfrac{r(x)}{g(x)}$ that is *proper* in the sense that its numerator is of lower degree than its denominator.

Example   Rewrite the rational expression $\dfrac{2x^3 - x^2 - 7}{x - 2}$ as the sum of a polynomial and a proper rational expression.

Solution   Let $f(x) = 2x^3 - x^2 - 7$ and $g(x) = x - 2$. By long division, we have

$$
\begin{array}{r}
2x^2 + 3x + 6 \qquad = q(x) \\
x - 2 \overline{\smash{\big)}\, 2x^3 - x^2 \qquad\ - 7} \\
\underline{2x^3 - 4x^2 \qquad\qquad} \\
3x^2 \qquad\ - 7 \\
\underline{3x^2 - 6x \qquad} \\
6x - 7 \\
\underline{6x - 12} \\
5 = r(x).
\end{array}
$$

By Theorem 2, $\dfrac{f(x)}{g(x)} = q(x) + \dfrac{r(x)}{g(x)}$; that is,

$$\frac{2x^3 - x^2 - 7}{x - 2} = 2x^2 + 3x + 6 + \frac{5}{x - 2}.$$

In the example above, the divisor is a *first-degree* polynomial of the form $g(x) = x - c$. In all such cases, the remainder will either be the zero polynomial or it will have degree zero—in other words, the remainder will be a *constant*. This particular type of division can be carried out by a shortcut called *synthetic division*.

Procedure for Synthetic Division

To find the quotient polynomial $q(x)$ and the remainder $r(x) = R$ when a dividend polynomial $f(x)$ of degree $n \geq 1$ is divided by a first-degree polynomial $g(x) = x - c$, do the following:

Step 1. Arrange the polynomial $f(x)$ in descending powers of x:

$$f(x) = a_n x^n + a_{n-1} x^{n-1} + a_{n-2} x^{n-2} + \cdots + a_1 x + a_0.$$

Represent all missing powers by using zero coefficients.

Step 2. Write down the value of c, then draw a vertical line and after it list the coefficients of $f(x)$:

$$
\begin{array}{c|ccccccc}
c & a_n & a_{n-1} & a_{n-2} & \cdots & a_1 & a_0
\end{array}
$$

Step 3. Leave some space below the row of coefficients, draw a horizontal line, and copy the leading coefficient a_n below the line:

$$
\begin{array}{c|ccccccc}
c & a_n & a_{n-1} & a_{n-2} & \cdots & a_1 & a_0 \\
& \downarrow & & & & & \\
\hline
& a_n & & & & &
\end{array}
$$

Step 4. Multiply a_n by c and write the product above the horizontal line under the second coefficient a_{n-1}; then add a_{n-1} to this product and write the result s_1 below the line:

$$
\begin{array}{c|ccccccc}
c & a_n & a_{n-1} & a_{n-2} & \cdots & a_1 & a_0 \\
& & ca_n & & & & \\
\hline
& a_n & s_1 & & & &
\end{array}
$$

Now multiply s_1 by c and write the product above the line under the third coefficient a_{n-2}; then add a_{n-2} to this product and write the result s_2 below the line:

$$
\begin{array}{c|ccccccc}
c & a_n & a_{n-1} & a_{n-2} & \cdots & a_1 & a_0 \\
& & ca_n & cs_1 & & & \\
\hline
& a_n & s_1 & s_2 & & &
\end{array}
$$

Continue in this way, multiplying each newly obtained number below the line by c, writing the product above the line under the next coefficient, and adding to produce the next number below the line. Do this until you have numbers below the line for every coefficient. Isolate the very last sum by drawing a short vertical line:

$$
\begin{array}{c|ccccccc}
c & a_n & a_{n-1} & a_{n-2} & \cdots & a_1 & a_0 \\
& & ca_n & cs_1 & \cdots & cs_{n-2} & cs_{n-1} \\
\hline
& a_n & s_1 & s_2 & \cdots & s_{n-1} & s_n
\end{array}
$$

Step 5. Conclude that the numbers $a_n, s_1, s_2, \ldots, s_{n-1}$ are the coefficients of the quotient polynomial

$$q(x) = a_n x^{n-1} + s_1 x^{n-2} + s_2 x^{n-3} + \cdots + s_{n-2}x + s_{n-1},$$

and that $s_n = R$, the remainder. [Note that $q(x)$ has degree 1 less than $f(x)$.]

Examples Use synthetic division to obtain the quotient polynomial $q(x)$ and the remainder R upon division of the polynomial $f(x)$ by the first-degree polynomial $g(x)$.

1 $f(x) = 2x^3 - x^2 - 7; g(x) = x - 2$

Solution The dividend $f(x) = 2x^3 - x^2 + 0x - 7$ has coefficients 2, -1, 0, and -7, and the divisor has the form $g(x) = x - c$ with $c = 2$. By synthetic division:

$$
\begin{array}{c|cccc}
2 & 2 & -1 & 0 & -7 \\
& & 4 & 6 & 12 \\
\hline
& 2 & 3 & 6 & 5
\end{array}
$$

Hence, the quotient polynomial is $q(x) = 2x^2 + 3x + 6$ and the remainder is $R = 5$.

2 $f(x) = 3x^4 - 2x^3 - 5x; g(x) = x + 2$

Solution   The dividend $f(x) = 3x^4 - 2x^3 + 0x^2 - 5x + 0$ has coefficients 3, -2, 0, -5, and 0, and the divisor has the form $g(x) = x - c$ with $c = -2$. By synthetic division:

$$
\begin{array}{r|rrrrr}
-2 & 3 & -2 & 0 & -5 & 0 \\
   &   & -6 & 16 & -32 & 74 \\
\hline
   & 3 & -8 & 16 & -37 & 74
\end{array}
$$

Hence, the quotient polynomial is $q(x) = 3x^3 - 8x^2 + 16x - 37$ and the remainder is $R = 74$.

A detailed proof that synthetic division always works is a bit tedious, so we shall not give it here. However, if you will work a simple example by both long division and synthetic division, you will see clearly that every bit of the arithmetic required in the long division is accounted for in the synthetic division.

PROBLEM SET 3

In problems 1–18, perform the indicated long division to find a quotient polynomial and a remainder polynomial. Be sure that the remainder is either the zero polynomial or a polynomial of lower degree than the divisor. Check your work by verifying the fact that dividend = (divisor)(quotient) + remainder.

1. $x - 5\,\overline{\smash{\big)}\,x^2 + 3x - 10}$

2. $3x^2 - 1\,\overline{\smash{\big)}\,6x^4 + 10x^2 + 7}$

3. $x - 1\,\overline{\smash{\big)}\,x^3 - 1}$

4. $4x - 3\,\overline{\smash{\big)}\,4x^6 + 5x^3 - 6}$

5. $2x^2 - 4x + 1\,\overline{\smash{\big)}\,6x^4 - 31x^2 + 26x - 6}$

6. $x + 1\,\overline{\smash{\big)}\,x^5 - 1}$

7. $2x - 3\,\overline{\smash{\big)}\,4x^4 - 12x^3 + 15x^2 - 17x}$

8. $2x^4 - 3x - 1\,\overline{\smash{\big)}\,-6x^4 - 7x^3 + 14x^2 - 5x + 1}$

9. $x^2 + 3\,\overline{\smash{\big)}\,x^3 + 3x^2 + 2x - 4}$

10. $1 - 4x - 2x^3\,\overline{\smash{\big)}\,x^2 - 3x^4 - x^3 + 5}$

11. $x^3 + x^2\,\overline{\smash{\big)}\,x^5 - 1}$

12. $x - c\,\overline{\smash{\big)}\,x^2 - 2cx + 2}$

13. $2x - 1\,\overline{\smash{\big)}\,x^2 + \tfrac{1}{2}x + 1}$

14. $t^2 - t + 1\,\overline{\smash{\big)}\,2t^4 + t^2 - 1}$

15. $3x^2 + 2x + 1\,\overline{\smash{\big)}\,x^3 + x^2 + 1}$

16. $x - c\,\overline{\smash{\big)}\,x^3 - c^3}$

17. $5t^2 - t + 4\,\overline{\smash{\big)}\,10t^3 + 13t^2 + 5t + 2}$

18. $x^2 + x + 1\,\overline{\smash{\big)}\,ax^2 + bx + c}$

In problems 19–22, use the quotient theorem (Theorem 2) to rewrite each rational expression as the sum of a polynomial and a proper rational expression.

19. $\dfrac{5x^3 + 3x^2 - x + 2}{x - 4}$

20. $\dfrac{x^2 - x + 1}{x^2 + x - 1}$

21. $\dfrac{5x^3 - 6x^2 - 68x - 16}{x^3 - 2x^2 - 8x}$

22. $\dfrac{ax + b}{cx + d}$

In problems 23–32, use synthetic division to obtain the quotient polynomial $q(x)$ and the remainder R upon division of the polynomial $f(x)$ by the first-degree polynomial $g(x)$.

23. $f(x) = 3x^3 - 2x^2 - x + 4; g(x) = x - 2$

24. $f(x) = x^6 - x^5 - x^2 - x - 1; g(x) = x - 2$

25. $f(x) = x^5 - 5x^3 + x - 16; g(x) = x + 2$

26. $f(x) = 5x^3 - 7x^2 + 3x - 2; g(x) = x - 1$

27. $f(x) = 3x^6 - 2x^4 + x^2; g(x) = x + 1$

28. $f(x) = 9x^2 - 3x + 1; g(x) = x - \tfrac{1}{3}$

29. $f(x) = -16x^3 - 12x^2 + 2x + 7; g(x) = x - \tfrac{1}{2}$

30. $f(x) = x^2 + 2x + 1; g(x) = x - c$

C 31. $f(x) = x^3 + x^2 + x + 1; g(x) = x - 1.1$

32. $f(x) = ax^2 + bx + c; g(x) = x - 1$

In problems 33–36, find the quotient and remainder (a) by long division and (b) by synthetic division.

33. $x - 3 \overline{\smash{\big)}\ 5x^3 - 11x^2 - 14x - 10}$

34. $x + 3 \overline{\smash{\big)}\ 2x^3 + 3x^2 - 5x + 12}$

35. $x + 4 \overline{\smash{\big)}\ -2x^3 - 6x^2 + 18x + 20}$

36. $x + \frac{1}{2} \overline{\smash{\big)}\ 5x^3 + x - 9}$

37. Find a value of k such that $x^3 + 2x^2 - 3kx - 10$ divided by $x + 3$ has a remainder of 8.

38. Find a value of k such that $x + 5$ is a factor of $x^3 + kx + 125$.

39. Find a value of k such that $x - \frac{1}{3}$ is a factor of $3x^3 - x^2 + kx - 5$.

40. Suppose that the polynomial $f(x)$ is divided by the nonzero polynomial $g(x)$ to produce a quotient polynomial $q(x)$ and a remainder polynomial $r(x)$. Now, suppose that $q(x)$ is divided by the nonzero polynomial $G(x)$ to produce a quotient polynomial $Q(x)$ and a remainder polynomial $R(x)$. Show that, if $f(x)$ is divided by the product $g(x)G(x)$, the quotient polynomial is $Q(x)$ and the remainder polynomial is $r(x) + g(x)R(x)$.

4   Values and Zeros of Polynomial Functions

In this section, we continue our study of polynomial functions and their graphs. We begin with the following important theorem, which shows that values of polynomial functions can be found by using the division algorithm.

Theorem 1   **The Remainder Theorem**

> If f is a polynomial function and c is a constant, then
>
> $$f(c) = R,$$
>
> where R is the remainder upon division of $f(x)$ by $x - c$.

Proof   By the division algorithm (Theorem 1, Section 3) with $g(x) = x - c$, we have

$$f(x) = (x - c)q(x) + r(x),$$

where either $r(x)$ is the zero polynomial or its degree is less than 1, the degree of $g(x)$. In either case, $r(x)$ is a constant polynomial, $r(x) = R$. Thus,

$$f(x) = (x - c)q(x) + R.$$

Substituting $x = c$ in the last equation, we find that

$$f(c) = (c - c)q(c) + R$$
$$= 0q(c) + R = R.$$

Example   Use the remainder theorem and synthetic division to find

$$f(-3) \quad \text{if} \quad f(x) = 4x^4 + 3x^3 - x^2 - 5x - 6.$$

Solution   By the remainder theorem, $f(-3)$ is the remainder when $f(x)$ is divided by $x - (-3)$. Using synthetic division, we have

$$
\begin{array}{r|rrrrr}
-3 & 4 & 3 & -1 & -5 & -6 \\
   &   & -12 & 27 & -78 & 249 \\
\hline
   & 4 & -9 & 26 & -83 & 243 = R.
\end{array}
$$

Therefore, $f(-3) = 243$.

One of the noteworthy consequences of the remainder theorem is the following.

Theorem 2   **The Factor Theorem**

Let f be a polynomial function and let c be a constant. Then $f(x) = 0$ if and only if $x - c$ is a factor of $f(x)$.

Proof   We have to prove that (a) if $f(c) = 0$, then $x - c$ is a factor of $f(x)$, and (b) if $x - c$ is a factor of $f(x)$, then $f(c) = 0$.

(a) Suppose that $f(c) = 0$. Then, we know by the remainder theorem that 0 is the remainder when $f(x)$ is divided by $x - c$; hence, $x - c$ is a factor of $f(x)$.

(b) Suppose that $x - c$ is a factor of $f(x)$. Then, we can write

$$f(x) = (x - c)q(x),$$

and it follows that

$$f(c) = (c - c)q(c) = 0q(c) = 0.$$

Example   Let $f(x) = x^3 + 2x^2 - 5x - 6$. Use the factor theorem to determine (a) whether $x + 1$ is a factor of $f(x)$ and (b) whether $x - 3$ is a factor of $f(x)$.

Solution   (a) We have $x + 1 = x - (-1)$, which has the form $x - c$ with $c = -1$. Since

$$f(-1) = (-1)^3 + 2(-1)^2 - 5(-1) - 6 = -1 + 2 + 5 - 6 = 0,$$

it follows from the factor theorem that $x + 1$ is a factor of $f(x)$.

(b) Here $x - 3$ has the form $x - c$ with $c = 3$. Since

$$f(3) = 3^3 + 2(3)^2 - 5(3) - 6 = 27 + 18 - 15 - 6 = 24 \neq 0,$$

it follows from the factor theorem that $x - 3$ is not a factor of $f(x)$.

If f is a function, then a root of the equation

$$f(x) = 0$$

is called a **zero of f**. By the factor theorem, each zero of a polynomial function f corresponds to a first-degree factor of $f(x)$. Because the polynomial $f(x)$ can't have more first-degree factors than its degree, we have the following result.

Theorem 3   **The Maximum Number of Zeros of a Polynomial Function**

> A polynomial function cannot have more zeros than its degree.

Note that a zero of a function f is the same as an x intercept of the graph of f. Therefore, as a consequence of Theorem 3, the graph of a polynomial function f cannot have more x intercepts than the degree of the polynomial $f(x)$.

The following theorem, which we state without proof, is very useful for finding zeros of polynomial functions.

Theorem 4   **The Rational Zeros Theorem**

> Let $f(x) = a_n x^n + a_{n-1} x^{n-1} + \cdots + a_1 x + a_0$ be a polynomial function of degree $n > 0$ with integers as coefficients. Then, if p/q is any rational zero of f and if the fraction p/q is reduced to lowest terms, p must be a factor of a_0 and q must be a factor of a_n.

You can use Theorem 4 to find the rational zeros of a polynomial function by carrying out the following procedure.

Procedure for Finding Rational Zeros of Polynomial Functions

Step 1. Check to see that all coefficients of $f(x)$ are integers. Find all factors of a_0 (the constant term) and of a_n (the coefficient of the highest degree term). The factors of a_0 are the possible values of p, and the factors of a_n are the possible values of q.

Step 2. Form all possible ratios p/q, where p is a factor of a_0 and q is a factor of a_n. These rational numbers are all possible rational zeros of f.

Step 3. Using synthetic division, check each of the possible rational zeros produced in step 2 to see if it is a zero of f. If none of them work, conclude that f has no rational zeros. If a rational zero p/q is found, proceed to step 4.

Step 4. If $c = p/q$ is the rational zero produced in step 3, then, by the factor theorem, $f(x) = (x - c)Q(x)$, where the coefficients of the quotient polynomial $Q(x)$ were determined when you performed the synthetic division. The remaining zeros of f are the zeros of Q.

Example   Find all rational zeros of $f(x) = x^3 + 8x^2 + 11x - 20$.

Solution   We follow the procedure above.

(1) All coefficients are integers, $a_0 = -20$, and $a_3 = 1$. The factors of a_0 are

$$p: \quad \pm 1, \pm 2, \pm 4, \pm 5, \pm 10, \pm 20$$

and the factors of a_3 are

$$q: \quad \pm 1.$$

(2) The possible quotients p/q are

$$\frac{p}{q}: \quad \pm 1, \ \pm 2, \ \pm 4, \ \pm 5, \ \pm 10, \ \pm 20.$$

(3) Of the twelve possible rational zeros in (2), three at most can be zeros (Theorem 3), but it may be that none of them are. We test them, one by one, using synthetic division. To test $p/q = 1$, we divide $x^3 + 8x^2 + 11x - 20$ by $x - 1$:

$$
\begin{array}{r|rrrr}
1 & 1 & 8 & 11 & -20 \\
  &   & 1 & 9 & 20 \\
\hline
  & 1 & 9 & 20 & 0 = R.
\end{array}
$$

Therefore, by the remainder theorem, 1 is a zero of f.

(4) The synthetic division just performed yields the coefficients of the quotient polynomial $Q(x) = x^2 + 9x + 20$. Thus,

$$f(x) = (x - 1)(x^2 + 9x + 20),$$

and the remaining zeros of f are the zeros of $x^2 + 9x + 20$. We could now apply the same procedure to the polynomial function $Q(x) = x^2 + 9x + 20$, but it's easier to factor:

$$x^2 + 9x + 20 = (x + 4)(x + 5).$$

We conclude that the remaining zeros are -4 and -5. Therefore, the rational zeros of f are 1, -4, and -5.

By using the rational zeros theorem and the factor theorem, you can often factor a polynomial function completely into prime factors. When this is possible, you can use the techniques discussed in Section 2 to sketch the graph of the polynomial function.

Example   Factor the polynomial function $f(x) = 4x^4 - 4x^3 - 25x^2 + x + 6$ completely into prime factors and c sketch its graph.

Solution   We begin by applying the procedure for finding rational zeros to the function f. Here $a_0 = 6$ and $a_4 = 4$. Thus

$$p: \quad \pm 1, \ \pm 2, \ \pm 3, \ \pm 6 \qquad \text{(factors of 6)}$$

$$q: \quad \pm 1, \ \pm 2, \ \pm 4 \qquad \text{(factors of 4)}$$

$$\frac{p}{q}: \quad \pm 1, \ \pm 2, \ \pm 3, \ \pm 6, \ \pm \tfrac{1}{2}, \ \pm \tfrac{3}{2}, \ \pm \tfrac{1}{4}, \ \pm \tfrac{3}{4}.$$

We test the possible rational zeros p/q, one by one, using synthetic division. The first one that works is -2:

$$
\begin{array}{r|rrrrr}
-2 & 4 & -4 & -25 & 1 & 6 \\
   &   & -8 & 24 & 2 & -6 \\
\hline
   & 4 & -12 & -1 & 3 & 0 = R.
\end{array}
$$

Therefore, -2 is a rational zero of f and the quotient polynomial is given by $Q(x) = 4x^3 - 12x^2 - x + 3$. Hence,

$$f(x) = (x + 2)(4x^3 - 12x^2 - x + 3).$$

Now, we repeat the whole procedure for $Q(x) = 4x^3 - 12x^2 - x + 3$. Here $a_0 = 3$ and $a_3 = 4$. Thus

$$p: \quad \pm 1, \pm 3 \qquad \text{(factors of 3)}$$

$$q: \quad \pm 1, \pm 2, \pm 4 \qquad \text{(factors of 4)}$$

$$\frac{p}{q}: \quad \pm 1, \pm 3, \pm\tfrac{1}{2}, \pm\tfrac{1}{4}, \pm\tfrac{3}{2}, \pm\tfrac{3}{4}.$$

We test the possible rational zeros p/q of $Q(x) = 4x^3 - 12x^2 - x + 3$, one by one, using synthetic division. The first one that works is 3:

$$
\begin{array}{r|rrrr}
3 & 4 & -12 & -1 & 3 \\
  &   & 12 & 0 & -3 \\
\hline
  & 4 & 0 & -1 & 0 = R.
\end{array}
$$

Therefore, 3 is a rational zero of Q and

$$4x^3 - 12x^2 - x + 3 = (x - 3)(4x^2 + 0x - 1)$$
$$= (x - 3)(4x^2 - 1).$$

It follows that

$$f(x) = (x + 2)(4x^3 - 12x^2 - x + 3)$$
$$= (x + 2)(x - 3)(4x^2 - 1)$$
$$= (x + 2)(x - 3)(2x - 1)(2x + 1).$$

Using the techniques discussed in Section 2, we can now sketch the graph of f (Figure 1). As usual accuracy is enhanced if a calculator is used to help determine points on the graph.

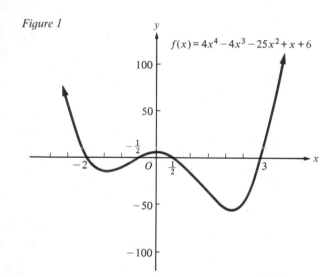

Figure 1

$f(x) = 4x^4 - 4x^3 - 25x^2 + x + 6$

A polynomial function with integer coefficients may have no rational zeros; for instance,

$$f(x) = x^2 + x + 1$$

has no real zeros, rational or irrational. (Why?) It may happen that some of the zeros of a polynomial function are rational and some are not; for instance,

$$g(x) = (x - 1)(2x^2 + 9x - 3)$$

has the rational zero 1, but the other two zeros are irrational. (Why?) Our discussion of polynomials continues in the final chapter, where we consider the question of complex zeros.

PROBLEM SET 4

In problems 1–10, use the remainder theorem and synthetic division to find the value of $f(c)$ for the given polynomial function f and the indicated value of c. [C] In each case, check your answer by a direct calculation of $f(c)$. (Use a calculator if you wish.)

1. $f(x) = x^3 - 9x^2 + 23x - 15,$ $\quad c = 1$
2. $f(x) = 2x^3 - 2x^2 - x - 2,$ $\quad c = -1$
3. $f(x) = 8x^3 - 25x^2 + 4x - 3,$ $\quad c = -3$
4. $f(x) = 3x^4 - x^3 + 2x - 10,$ $\quad c = 2$
5. $f(x) = x^5 - 2x^4 + x^3 - 3x^2 + 8,$ $\quad c = 2$
6. $f(x) = x^5 + 2x^3 - 4x + 5,$ $\quad c = -2$
7. $f(x) = 3x^4 + 2x^2 - 5,$ $\quad c = -2$
8. $f(x) = 16x^4 - 8x^2 + 12x - 1,$ $\quad c = \frac{1}{2}$
9. $f(x) = x^3 - 4x + 8,$ $\quad c = \frac{1}{3}$
10. $f(x) = 5x^3 - 20x^2 + 2x - 1,$ $\quad c = 0.2$

[C] In problems 11–18, use the factor theorem to determine (without actually dividing) whether or not the indicated binomial is a factor of the given polynomial. (You may want to use a calculator to find values of the polynomial functions.)

11. $f(x) = 4x^4 + 13x^3 - 13x^2 - 40x + 12;$ $\quad x + 2$
12. $g(x) = x^4 - 9x^3 + 18x^2 - 3;$ $\quad x + 1$
13. $H(x) = x^5 - 17x^3 + 75x + 9;$ $\quad x - 3$
14. $F(x) = 2x^4 - x^3 + x^2 + x - 3;$ $\quad x + 1$
15. $h(x) = 30x^3 - 20x^2 - 100x + 1000;$ $\quad x - 10$
16. $G(t) = t^4 + 2t^3 - 6t^2 - 14t - 7;$ $\quad t - 7$
17. $F(y) = 2y^3 + 4y - 2;$ $\quad y + \frac{1}{2}$
18. $f(z) = 81z^3 + 51z^2 - 153z + \frac{127}{3};$ $\quad z - \frac{1}{3}$

In problems 19–32, find all rational zeros of each polynomial function.

19. $f(x) = x^3 - 9x^2 + 23x - 15$
20. $g(x) = x^3 - 3x^2 - 4x + 12$
21. $h(x) = 8x^3 - 25x^2 + 4x - 3$
22. $p(x) = 2x^3 - 15x^2 + 27x - 10$
23. $F(x) = 2x^3 - 3x^2 + 2x + 2$
24. $G(x) = x^3 - 2x^2 - 13x - 10$
25. $f(x) = x^4 - 4x^3 + 7x^2 - 12x + 12$
26. $H(x) = 4x^4 + 4x^3 + 9x^2 + 8x + 2$
27. $f(x) = 4x^4 - 20x^3 + x^2 + 18x + 6$
28. $g(x) = x^5 - x^3 + 27x^2 - 27$
29. $Q(x) = x^5 - 17x^3 + 75x - 9$
30. $f(x) = \frac{1}{2}x^5 + 2x^4 + \frac{1}{2}x^2 - \frac{3}{2}x - 14$
31. $g(x) = 20x^5 - 9x^4 - 74x^3 + 30x^2 + 42x - 9$
32. $q(x) = x^5 + x^4 - \frac{5}{4}x^3 + \frac{25}{4}x^2 - 21x + 9$

In problems 33–36, factor each polynomial function completely into prime factors and [C] sketch its graph.

33. $f(x) = x^3 - 6x^2 - x + 6$
34. $g(x) = x^3 - 8x^2 + 2x - 16$
35. $Q(x) = 2x^4 - 3x^3 - 7x^2 + 12x - 4$
36. $G(x) = 16x^4 - 40x^2 + 9$

37. A box is to be constructed so that the length is twice the width and the depth exceeds the width by 2 inches. Find the dimensions of the box if its volume is 350 cubic inches. [Hint: Let x denote the width of the box, express the length and height in terms of x, and solve the equation length $\times$ height $\times$ width $= 350$.]
38. Suppose that $f(x) = a_n x^n + a_{n-1}x^{n-1} + \cdots + a_1 x + a_0$ is a polynomial function such that $f(x) = 0$ for all values of x. Prove that all of the coefficients $a_n, a_{n-1}, \ldots, a_1, a_0$ must be zero. [Hint: Use Theorem 3.]
39. A plot of land has the shape of a right triangle with a hypotenuse 1 kilometer longer than one of the sides. Find the lengths of the sides of the plot of land if its area is 6 square kilometers.
40. Suppose that f and g are polynomial functions such that $f(x) = g(x)$ for all values of x. Prove that the coefficients of f are the same as the coefficients of g. [Hint: Use problem 38.]

5   Rational Functions

Although the sum $f + g$, the difference $f - g$, and the product $f \cdot g$ of polynomial functions f and g are again polynomial functions, the quotient f/g can only be a polynomial function if g is a factor of f. A function of the form $R = f/g$, where f and g are polynomial functions, is called a **rational function.** In other words, a rational function is a function such as

$$R(x) = \frac{x^3 - 2x^2 + 5x - 4}{7x^2 - 3x + 2},$$

whose values are given by a rational expression in the independent variable. The domain of a rational function is the set of all real numbers *except the zeros of the denominator function.*

Example   Determine the domain of each rational function.

(a) $R(x) = \dfrac{3x^2 + 2}{x + 1}$ (b) $F(x) = \dfrac{x + 2}{x}$ (c) $H(x) = \dfrac{x^3 - 3x^2 - 2x - 1}{4}$

(d) $h(x) = \dfrac{1}{(x + 2)(x - 3)}$ (e) $T(x) = \dfrac{3x - 2}{x^2 + 1}$

Solution   (a) All real numbers except -1.   (b) All real numbers except 0.

(c) All real numbers.   (d) All real numbers except -2 and 3.

(e) All real numbers.

Figure 1

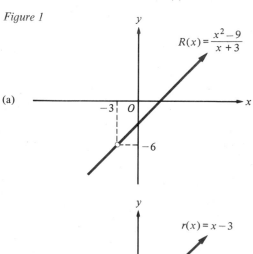

(a)

(b)

Recall that two rational expressions are said to be equivalent if one can be obtained from the other by canceling common factors or by multiplying numerator and denominator by the same nonzero polynomial. Strictly speaking, two rational functions defined by equivalent rational expressions may not be the same because their domains may be different. For instance, the rational function R defined by

$$R(x) = \frac{x^2 - 9}{x + 3} = \frac{(x - 3)(x + 3)}{x + 3}$$

isn't really the same as the rational function r defined by $r(x) = x - 3$ because

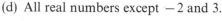

$$R(x) = \frac{(x - 3)(x + 3)}{x + 3} = x - 3 = r(x)$$

only if $x \neq -3$. Notice that -3 does not belong to the domain of R, but it does belong to the domain of r. The graph of R (Figure 1a) is the same as the graph of r (Figure 1b), except that the point $(-3, -6)$ does not belong to the graph of R.

5.1 Functions of the Form $R(x) = 1/x^n$

If n is a positive integer, the domain of the rational function $R(x) = 1/x^n$ consists of all real numbers except 0. Here we consider the graphs of such functions.

Example   Sketch the graph of each function.

(a) $F(x) = 1/x$ (b) $G(x) = 1/x^2$

Solution   (a) The function F is odd,

$$F(-x) = \frac{1}{(-x)} = -\frac{1}{x} = -F(x),$$

so its graph is symmetric about the origin. Therefore, we begin by sketching the portion of the graph for $x > 0$, and then we obtain the complete graph by using the symmetry. For small positive values of x, the reciprocal $1/x$ will be large. As x increases, $1/x$ decreases, coming closer and closer to 0 as x gets larger. This variation is shown in Table 1.

Table 1

x	$\frac{1}{100}$	$\frac{1}{10}$	$\frac{1}{3}$	$\frac{1}{2}$	1	2	3	10	100
$F(x) = \dfrac{1}{x}$	100	10	3	2	1	$\frac{1}{2}$	$\frac{1}{3}$	$\frac{1}{10}$	$\frac{1}{100}$

Figure 2

$F(x) = \dfrac{1}{x}$

$(\frac{1}{3},3)$

$(\frac{1}{2},2)$

$(1,1)$

$(2,\frac{1}{2})$ $(3,\frac{1}{3})$

By using this information and plotting a few points, we obtain the graph of F (Figure 2). Notice that although the graph comes closer and closer to the x and y axes, it never *touches* them; in other words, it has no x or y intercepts.

(b) The graph of $G(x) = 1/x^2$ is obtained in much the same way as the graph of $F(x) = 1/x$ in part (a). The main differences are that $G(x)$ is always positive and that G is an even function,

$$G(-x) = \frac{1}{(-x)^2} = \frac{1}{x^2} = G(x),$$

so its graph is symmetric about the y axis. Also, the graph of G is steeper than the graph of F near the origin, and it approaches the x axis more rapidly as x becomes large. You can see why by comparing Table 1 above with Table 2.

Table 2

x	$\frac{1}{100}$	$\frac{1}{10}$	$\frac{1}{3}$	$\frac{1}{2}$	1	2	3	10	100
$G(x) = \dfrac{1}{x^2}$	10,000	100	9	4	1	$\frac{1}{4}$	$\frac{1}{9}$	$\frac{1}{100}$	$\frac{1}{10,000}$

Figure 3

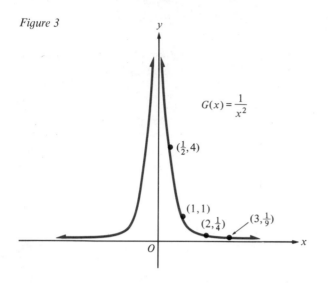

$$G(x) = \frac{1}{x^2}$$

$(\frac{1}{2}, 4)$

$(1, 1)$

$(2, \frac{1}{4})$ $(3, \frac{1}{9})$

By using this information and plotting a few points, we obtain the graph of G (Figure 3). Like the graph of F in part (a), the graph of G has no x or y intercepts.

The graphs in Figures 2 and 3 are typical of the graphs of functions of the form $R(x) = 1/x^n$, where n is a positive integer. If n is odd, the graph resembles that of $F(x) = 1/x$ (Figure 2); whereas if n is even, the graph resembles that of $G(x) = 1/x^2$ (Figure 3). For larger values of n, the graphs become steeper near the origin and they approach the x axis more rapidly as one moves away from the origin.

5.2 Horizontal and Vertical Asymptotes

For the graphs of the functions $R(x) = 1/x^n$, $n \geq 1$, the coordinate axes play a very special role. Indeed, they are **asymptotes** of these graphs in the following sense: If the values $f(x)$ of a function f approach a fixed number k as x (or $-x$) gets larger and larger without bound, we say that the line $y = k$ is a **horizontal asymptote** of the graph of f (Figure 4a). Similarly, if $|f(x)|$ gets larger and larger without bound as x approaches a fixed number h, the line $x = h$ is called a **vertical asymptote** of the graph of f (Figure 4b).

Figure 4

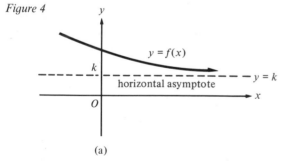

$y = f(x)$

k

horizontal asymptote $y = k$

(a)

From our discussion in Section 6 of Chapter 3, if a, h, and k are constants and n is a positive integer,

$$f(x) = \frac{a}{(x - h)^n} + k$$

is obtained by vertically "stretching" or "flattening" the graph of

$$R(x) = \frac{1}{x^n}$$

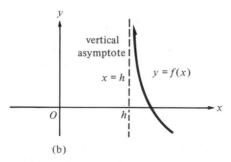

vertical asymptote

$x = h$ $y = f(x)$

h

(b)

by a factor of $|a|$, reflecting it across the x axis if $a < 0$, shifting it horizontally by $|h|$ units, and shifting it vertically by $|k|$ units. After the vertical stretching, flattening, or reflecting, the x and y axes will still be asymptotes. However, when you shift a graph, you will shift its asymptotes as well; hence, $x = h$ is a vertical asymptote and $y = k$ is a horizontal asymptote of the graph of f.

Examples   For each function, find the horizontal and vertical asymptotes of the graph, determine any y and x intercepts, and sketch the graph.

1 $f(x) = \dfrac{3}{x-2} + 4$

Solution   The graph will have the same general shape as the graph of $F(x) = 1/x$ in Figure 2, except that it is stretched vertically by a factor of 3 and then shifted 2 units to the right and 4 units upward (Figure 5). The asymptotes are $x = 2$ and $y = 4$. The y intercept is given by $f(0) = 3/(0-2) + 4 = \frac{5}{2}$ and the x intercept is obtained by solving the equation $f(x) = 0$, that is,

$$\frac{3}{x-2} + 4 = 0 \quad \text{or} \quad 3 + 4(x-2) = 0.$$

Thus, $4x - 5 = 0$, so $x = \frac{5}{4}$.

Figure 5

$$f(x) = \frac{3}{x-2} + 4$$

$(2,4)$

$(0, \frac{5}{2})$ $(\frac{5}{4}, 0)$

2 $H(x) = \dfrac{4x-5}{x-2}$

Solution   The function is not given in the form

$$H(x) = \frac{a}{(x-h)^n} + k,$$

but notice that the rational expression $(4x - 5)/(x - 2)$ isn't a proper fraction because its numerator has the same degree as its denominator. We should therefore divide and apply the quotient theorem (Section 3, Theorem 2):

$$\begin{array}{r} 4 \quad \longleftarrow \text{quotient} \\ x - 2 \overline{\smash{)}\ 4x - 5} \\ \underline{4x - 8} \\ 3 \quad \longleftarrow \text{remainder} \end{array}$$

Hence, by the quotient theorem,

$$H(x) = \frac{4x - 5}{x - 2} = 4 + \frac{3}{x - 2} = \frac{3}{x - 2} + 4,$$

and we see that H is the same as the function f whose graph has already been sketched in Figure 5.

3 $$F(x) = \frac{-1}{x^2 + 4x + 4}$$

Solution   Here the denominator is a perfect square, and we have

Figure 6

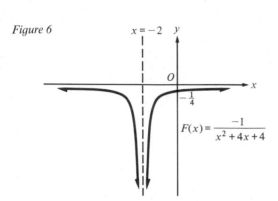

$$x = -2$$

$$F(x) = \frac{-1}{x^2 + 4x + 4}$$

$$F(x) = \frac{-1}{(x + 2)^2}.$$

The graph will have the same shape as the graph of $G(x) = 1/x^2$ (Figure 3), except that it is reflected across the x axis and then shifted 2 units to the left (Figure 6). The asymptotes are $x = -2$ and $y = 0$. The y intercept is given by $F(0) = \frac{-1}{(0 + 2)^2} = -\frac{1}{4}$. Since the equation $\frac{-1}{(x + 2)^2} = 0$ has no solution, there is no x intercept.

5.3   Graphs of General Rational Functions

Graphs of rational functions other than the very simple types already considered may be quite difficult to sketch without using calculus. However, in some cases, you can sketch an accurate graph using only the methods we have studied.

Before attempting to sketch the graph of a rational function $R(x) = f(x)/g(x)$, reduce the fraction by canceling all common factors of the numerator and denominator. (But, be careful—as we saw in Figure 1, such common factors might produce "holes" in the graph of R.) If the fraction is improper (that is, if the degree of the numerator isn't smaller than the degree of the denominator), perform a long division and use the quotient theorem (Section 3, Theorem 2) to write $R(x)$ as the sum of a polynomial and a proper fraction.

If the fraction $f(x)/g(x)$ is in reduced form, then the graph of $R(x) = f(x)/g(x)$ will have a vertical asymptote $x = h$ corresponding to each zero h of the denominator. Furthermore, these will be the only vertical asymptotes. Although the graph can have several vertical asymptotes, it can have at most one horizontal asymptote. To find it (if it exists), we recommend the following technique: Divide the numerator and denominator by the highest power of x that occurs in either, then study what happens as x becomes large. If the expression approaches a fixed number k as x becomes larger and larger without bound, then $y = k$ is a horizontal asymptote.

Example   Find all the horizontal and vertical asymptotes of the graph of $G(x) = \dfrac{1}{x^2 + 1}$, determine any x and y intercepts, and sketch the graph.

Solution   The graph of G has no vertical asymptote because the fraction is in reduced form and the denominator has no zeros. To find the horizontal asymptote, we divide the numerator and denominator of $\dfrac{1}{x^2 + 1}$ by x^2 to obtain

$$G(x) = \frac{1}{x^2 + 1} = \frac{\dfrac{1}{x^2}}{\dfrac{x^2}{x^2} + \dfrac{1}{x^2}} = \frac{\dfrac{1}{x^2}}{1 + \dfrac{1}{x^2}} \quad \text{for } x \neq 0.$$

As x becomes larger and larger without bound, $1/x^2$ approaches 0 and $G(x)$ approaches $\dfrac{0}{1 + 0} = 0$. Therefore, $y = 0$ (the x axis) is a horizontal asymptote of the graph. The y intercept is given by $G(0) = 1/(0^2 + 1) = 1$, and there is no x intercept. Because $G(-x) = G(x)$, the function G is even and its graph is symmetric about the y axis. Therefore, we can sketch the portion of the graph for $x \geq 0$ and reflect it across the y axis to obtain the complete graph. Notice that if we increase x, then we increase $x^2 + 1$, so we decrease $G(x) = 1/(x^2 + 1)$; in other words, on the interval $[0, \infty)$, G is a decreasing function. Using these facts and plotting a few points, we obtain a sketch of the graph (Figure 7).

Figure 7

PROBLEM SET 5

In problems 1–12, determine the domain of each rational function.

1. $f(x) = \dfrac{-5}{x^4}$

2. $r(x) = x - \dfrac{1}{x}$

3. $g(x) = \dfrac{-5}{x^2 + 5}$

4. $R(x) = \dfrac{2x^2 + 3x - 1}{(x - 1)(x + 2)}$

5. $G(x) = \dfrac{x^3 - 8}{40}$

6. $H(x) = \dfrac{40}{x^3 - 8}$

7. $P(x) = \dfrac{x^4 - 16}{2x^2 - x + 3}$

8. $T(x) = \dfrac{x^4 + 1}{2x^2 + 9x - 3}$

9. $F(x) = \dfrac{x^2 + x + 1}{x^2 + 8x + 15}$

10. $S(x) = \dfrac{1}{x} + \dfrac{3}{x^2 - 1}$

11. $p(x) = 1 + \dfrac{x^2 + x}{x^3 - x}$

12. $Q(x) = \dfrac{x^3 - 9x^2 + 32x + 13}{x^3 + 8x^2 + 11x - 20}$

In problems 13–40, the graph of each rational function can be obtained by stretching, flattening, reflecting, and/or shifting the graph of a function of the form $R(x) = 1/x^n$, for $n \geq 1$. In each case, find the horizontal and vertical asymptotes of the graph, determine any y and x intercepts, and $\boxed{\text{C}}$ sketch the graph (use a calculator, if you wish, to determine more points on the graph and thus improve the accuracy of your sketch).

13. $H(x) = \dfrac{3}{x}$

14. $g(x) = \dfrac{7}{x^2}$

15. $F(x) = -\dfrac{3}{x}$

16. $G(x) = -7x^{-2}$

17. $p(x) = \dfrac{4}{x^4}$

18. $T(x) = -5x^{-5}$

19. $P(x) = \dfrac{-4}{x^4}$

20. $Q(x) = -7x^{-8}$

21. $f(x) = \dfrac{1}{x} + 1$

22. $F(x) = 2 - \dfrac{1}{x}$

23. $g(x) = \dfrac{1}{x^2} - 4$

24. $G(x) = -1 - \dfrac{1}{x^2}$

25. $p(x) = \dfrac{1}{x - 3}$

26. $P(x) = \dfrac{2}{x + 2}$

27. $Q(x) = \dfrac{-2}{x + 1}$

28. $K(x) = \dfrac{-3}{x + 5}$

29. $f(x) = \dfrac{3}{x - 2} + 6$

30. $H(x) = \dfrac{2x + 1}{x - 4}$

31. $g(x) = \dfrac{-4x}{x - 1}$

32. $k(x) = \dfrac{-3}{(x + 2)^3} - 4$

33. $G(x) = \dfrac{5}{(x + 3)^2} - 4$

34. $p(x) = \dfrac{3x^2 + 4x + 7}{x^2 - 2x + 1}$

35. $F(x) = \dfrac{3}{x^2 - 2x + 1}$

36. $f(x) = \dfrac{2x^2 + 4x + 7}{x^2 + 2x + 1}$

37. $H(x) = \dfrac{-2}{x^2 - 4x + 4} + 5$

38. $g(x) = \dfrac{4x^2 + 16x + 15}{x^2 + 4x + 4}$

39. $P(x) = \dfrac{3x^2 - 18x + 28}{x^2 - 6x + 9}$

40. $T(x) = \dfrac{3x^2 - 1}{x^2} + 4$

In problems 41–50, find any horizontal and vertical asymptotes of the graph of each rational function, and $\boxed{\text{C}}$ sketch the graph.

41. $f(x) = \dfrac{x^2 - 4}{x + 2}$

42. $F(x) = \dfrac{x^2 - 2x + 1}{x - 1}$

43. $g(x) = \dfrac{1}{x^2 + 4}$

44. $H(x) = \dfrac{4}{x^2 + 9}$

45. $K(x) = \dfrac{-2}{x^2 + 9}$

46. $k(x) = \dfrac{3x^2 + 1}{x^2 - 1} + 1$

47. $Q(x) = \dfrac{1}{2x^2 + 1}$

48. $G(x) = \dfrac{x^2 + 5}{x^2 - 4}$

49. $H(x) = \dfrac{x^2}{x - 2} - 2$

50. $F(x) = \dfrac{-x^2}{x + 1} + 1$

In problems 51–56, find the horizontal asymptote (if it exists) of the graph of each rational function.

51. $f(x) = \dfrac{5x^2}{2x^2 - 3}$

52. $g(x) = \dfrac{x^3 + 1}{x^2 - 1}$

53. $F(x) = \dfrac{5x^2 - 7x + 3}{8x^2 + 5x + 1}$

54. $G(x) = \dfrac{7x^3 + 3x + 1}{x^3 - 2x + 3}$

55. $H(x) = \dfrac{x^2 - 1}{x^3 + 1}$

56. $K(x) = \dfrac{x^{99} + x^{98}}{x^{100} - x^{99}}$

$\boxed{\text{C}}$ **57.** According to Boyle's law, the pressure P and the volume V of a sample of gas maintained at constant temperature are related by an equation of the form $P = K/V$, where K is a numerical constant. If $K = 3.5$, (a) sketch a graph of P as a function of V, (b) determine the domain of this function, and (c) indicate whether P increases or decreases when V increases.

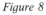 **58.** Sketch the graph of the rational function $f(x) = (2x^2 + 1)/(2x^2 - 3x)$. [Hint: The graph has a strange bend near $(-\frac{1}{3}, f(-\frac{1}{3}))$. Plot several points for values of x close to $-\frac{1}{3}$ to get an accurate sketch.]

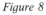 **59.** The electrical resistance R per unit length of a wire depends on the diameter d of the wire according to the formula $R = K/d^2$, where K is a constant. If $K = 0.1$, (a) sketch a graph of R as a function of d, (b) determine the domain of this function, and (c) indicate whether R increases or decreases when d decreases.

60. If a, b, c, and d are constants and c and d are not both zero, a function of the form $f(x) = (ax + b)/(cx + d)$ is called a *fractional linear function*. Discuss the graph of f with regard to (a) its x and y intercepts, (b) its horizontal and vertical asymptotes, and (c) its general shape.

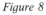 **61.** A person learning to type has an achievement record given by $N(t) = 60[(t - 2)/t]$ for $3 \le t \le 10$, where $N(t)$ is the number of words per minute at the end of t weeks. Sketch the graph of the rational function N.

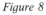 **62.** The electrical power P in watts consumed by a certain electrical circuit is related to the load resistance R in ohms by the equation $P = 100R(0.5 + R)^{-2}$ for $R \ge 0$. Sketch a graph of P as a function of R and use it to estimate the value of R for which P is maximum.

63. Explain why the graph in Figure 8 cannot be the graph of a rational function.

Figure 8

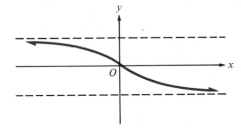

6   Ratio, Proportion, and Variation

When we describe the ways in which real-world quantities are related, we do not ordinarily write or speak in equations. Thus, in order to apply mathematical methods to problems in engineering, medicine, business, economics, the earth sciences, the life sciences, and so forth, we must be able to translate verbal statements about the relationships among variable quantities into mathematical formulas, and vice versa.

For instance, in economics, we might learn that the price of a certain commodity varies linearly as the demand for it; in engineering, that the power consumed by an electrical transmission line varies directly as the square of the current; or in physics, that the gravitational attraction between two homogeneous spheres varies inversely as the square of the distance between their centers. In medicine, we might be told that the rate at which bacteria increase in a bladder infection is proportional to the number of bacteria already present; in respiratory physiology, that alveolar ventilation is inversely proportional to carbon dioxide pressure; or in chemistry, that the rate of an autocatalytic reaction is jointly proportional to the amount of unreacted substance and the amount of the product. In this section we study the exact mathematical meaning of statements such as these.

6.1 Ratio and Proportion

Two quantities a and b are often compared by considering their *ratio* a/b. For instance, we might compare the fuel efficiency of two automobiles by forming the ratio of the gas mileage of one to that of the other. If $b \neq 0$, the **ratio** of a to b is written as

$$\frac{a}{b} \quad \text{or} \quad a \div b \quad \text{or} \quad a{:}b.$$

Although the notation $a{:}b$ isn't as popular as it once was, it is still used occasionally.

Example   If the ratio of a to b is 2/3, find the ratio of $5a + 3b$ to $7a - 4b$.

Solution   Here, $\dfrac{a}{b} = \dfrac{2}{3}$. To find the value of $\dfrac{5a + 3b}{7a - 4b}$, we divide numerator and denominator by b:

$$\frac{5a + 3b}{7a - 4b} = \frac{\dfrac{5a}{b} + \dfrac{3b}{b}}{\dfrac{7a}{b} - \dfrac{4b}{b}} = \frac{5\left(\dfrac{a}{b}\right) + 3}{7\left(\dfrac{a}{b}\right) - 4} = \frac{5\left(\dfrac{2}{3}\right) + 3}{7\left(\dfrac{2}{3}\right) - 4} = \frac{\dfrac{10}{3} + 3}{\dfrac{14}{3} - 4} = \frac{10 + 9}{14 - 12} = \frac{19}{2}.$$

An equation such as

$$\frac{a}{b} = \frac{c}{d} \quad \text{or} \quad a{:}b = c{:}d,$$

which expresses the equality of two ratios, is called a **proportion.** The proportion $a{:}b = c{:}d$ is often read, "a is to b as c is to d." If the proportion

$$a{:}b = c{:}d \quad \text{is written as} \quad \frac{a}{b} = \frac{c}{d}$$

and both sides of the equation are multiplied by bd, we have

$$\frac{a}{b}bd = \frac{c}{d}bd \quad \text{or} \quad ad = bc.$$

Conversely, if we divide both sides of the equation $ad = bc$ by bd, we obtain

$$\frac{ad}{bd} = \frac{bc}{bd},$$

which is just the original proportion $a{:}b = c{:}d$, so $a{:}b = c{:}d$ or $a/b = c/d$ is equivalent to the equation $ad = bc$. You may want to use the memory device

$$\frac{a}{b} \times \frac{c}{d},$$

indicating that a is multiplied by d and that b is multiplied by c when

$$\frac{a}{b} = \frac{c}{d} \quad \text{is rewritten as} \quad ad = bc.$$

This procedure is called **cross-multiplication.**

Example   Solve the proportion $\dfrac{5}{x-1} = \dfrac{3}{x+1}$ for x.

Solution   Cross-multiplying, we have

$$5(x+1) = 3(x-1) \quad \text{or} \quad 5x + 5 = 3x - 3.$$

Thus, $2x = -8$, so $x = -4$.

6.2 Variation

Suppose that x and y are variable quantities and that y is a function of x, that is, $y = f(x)$. As the quantity x varies, the quantity y, which depends on x, is subject to a corresponding variation. One of the simplest and also most important types of variation is defined as follows.

Definition 1   **Direct Variation**

> If there is a constant k such that
>
> $$y = kx$$
>
> holds for all values of x, we say that **y is directly proportional to x** or that **y varies directly as x** (or *with* x). The constant k is called the **constant of proportionality.**

If $y = kx$, then $y = 0$ when $x = 0$. Also, if $x \neq 0$, we have

$$\frac{y}{x} = k.$$

Therefore, if y is directly proportional to x, it follows that the ratio y/x maintains the constant value k as x varies through nonzero values. A graph of $y = kx$ is, of course, a straight line that has slope k and contains the origin (Figure 1).

Figure 1

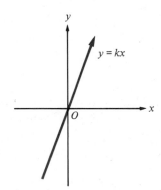

Example 1   Suppose that y varies directly as x and that $y = 6$ when $x = 2$. Express y as a function of x.

Solution   Let k be the constant of proportionality, so that $y = kx$. Substituting the given values $x = 2$ and $y = 6$, we find that

$$6 = k(2) \quad \text{or} \quad k = 3.$$

Therefore, $y = 3x$.

Example 2   Suppose that the rate r at which impulses are transmitted along a nerve fiber is directly proportional to the diameter d of the fiber. Given that $r = 20$ meters per second when $d = 6$ microns, express r as a function of d.

Solution   We have $r = kd$, and, in particular,

$$20 = k(6) \quad \text{or} \quad k = \frac{20}{6} = \frac{10}{3}.$$

Therefore, $r = \frac{10}{3}d$.

Figure 1 suggests the following natural extension of direct variation.

Definition 2   **Linear Variation**

Figure 2

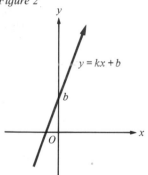

If there are constants k and b such that

$$y = kx + b,$$

then we say that y **varies linearly as** x (or *with x*).

Recall that the graph of the equation $y = kx + b$ is a straight line that has slope k and y intercept b (Figure 2). If you are given two distinct points on such a graph, you can use the methods in Section 3 of Chapter 3 to determine both the slope k and the y intercept b.

Example   In a certain income tax bracket, the amount y dollars of income tax that a family must pay to the Federal Government varies linearly with the amount x dollars of income after deductions. Suppose that $y = \$3260$ when $x = \$19,200$, and that $y = \$4380$ when $x = \$23,200$. Express y as a function of x and find y when $x = \$21,000$.

Solution   The graph of $y = f(x)$ is a straight line as in Figure 2. Its slope k is given by the slope formula (Theorem 1, Chapter 3, Section 2):

$$k = \frac{y_2 - y_1}{x_2 - x_1} = \frac{4380 - 3260}{23,200 - 19,200} = \frac{1120}{4000} = 0.28.$$

Thus, in point-slope form (Theorem 1, page 130), an equation of the graph is

$$y - 3260 = 0.28(x - 19{,}200) = 0.28x - 0.28(19{,}200) = 0.28x - 5376,$$

or

$$y = 0.28x - 5376 + 3260 = 0.28x - 2116.$$

In particular, when $x = \$21{,}000$, we have

$$y = 0.28(21{,}000) - 2116 = 5880 - 2116 = 3764 \text{ dollars.}$$

If we say that y *varies directly as the square of* x, we mean of course that $y = kx^2$ for some constant k. More generally, if n is a fixed rational number and $y = kx^n$ for some constant k, we say that **y varies directly as the nth power of x** or that **y is proportional to the nth power of x**.

Example C Suppose that the weight w of men of a given body type varies directly as the cube of their height h. If such a man is 1.85 meters tall and weighs 80 kilograms, and if a basketball player who has this same body type is 2 meters tall, what is the basketball player's weight?

Solution We have $w = kh^3$, where k is a constant. When $h = 1.85$, we have $w = 80$; that is,

$$80 = k(1.85)^3.$$

Therefore,

$$k = \frac{80}{(1.85)^3}.$$

Hence, when $h = 2$,

$$w = kh^3 = \frac{80}{(1.85)^3}(2)^3 \approx 101.08 \text{ kilograms.}$$

The situation in which y is proportional to the reciprocal of x arises so frequently that it is given a special name, **inverse variation**.

Definition 3 **Inverse Variation**

If there is a constant k such that

$$y = \frac{k}{x}$$

for all nonzero values of x, then we say that **y is inversely proportional to x** or that **y varies inversely as x** (or *with x*).

Naturally, if $y = k/x^n$ for some constant k and some rational number n, we say that **y is inversely proportional to the nth power of x**.

Example 1   Express y as a function of x if y is inversely proportional to the cube of x, and $y = 2$ when $x = 3$. Find the value of y when $x = \frac{1}{2}$.

Solution   We have $y = k/x^3$, where k is a constant. Substituting the values $x = 3$ and $y = 2$, we obtain

$$2 = \frac{k}{3^3} = \frac{k}{27},$$

so

$$k = 2(27) = 54$$

and

$$y = \frac{54}{x^3}.$$

When $x = \frac{1}{2}$, we have

$$y = \frac{54}{(1/2)^3} = \frac{54}{(1/8)}$$

$$= 54(8) = 432.$$

Example 2   A company is informed by its bookkeeper that the number S of units of a certain item sold per month seems to be inversely proportional to the quantity $p + 20$, where p is the selling price per unit in dollars. Suppose that 800 units per month are sold when the price is $10 per unit. How many units per month would be sold if the price were dropped to $5 per unit?

Solution   Since S is inversely proportional to $p + 20$, there is a constant k such that

$$S = \frac{k}{p + 20}.$$

We know that $S = 800$ when $p = 10$, so we have

$$800 = \frac{k}{10 + 20} = \frac{k}{30}$$

or

$$k = 30(800) = 24{,}000.$$

Therefore,

$$S = \frac{24{,}000}{p + 20}$$

gives S as a function of p. For $p = 5$ dollars, we have

$$S = \frac{24{,}000}{5 + 20} = \frac{24{,}000}{25} = 960,$$

so 960 units per month would be sold at a price of $5 per unit.

6.3 Joint and Combined Variation

Often the value of a variable quantity depends on the values of several other quantities; for instance, the amount of simple interest on an investment depends on the interest rate, the amount invested, and the period of time involved. For compound interest, the amount of interest depends on an additional variable: how often the compounding takes place. A situation in which one variable depends on several others is called **combined variation.** An important type of combined variation is defined as follows.

Definition 3   **Joint Variation**

> If a variable quantity y is proportional to the product of two or more variable quantities, we say that **y is jointly proportional** to these quantities, or that **y varies jointly as** (or *with*) these quantities.

Thus, if y is jointly proportional to u and v, then $y = kuv$ for some constant k. Sometimes the word "jointly" is omitted and we simply say that y is proportional to u and v.

Example   In chemistry, the absolute temperature T of a perfect gas varies jointly as its pressure P and its volume V. Given that $T = 500°$ Kelvin when $P = 50$ pounds per square inch and $V = 100$ cubic inches, find a formula for T in terms of P and V, and find T when $P = 100$ pounds per square inch and $V = 75$ cubic inches.

Solution   Since T varies jointly as P and V, there is a constant k such that $T = kPV$. Putting $T = 500$, $P = 50$, and $V = 100$, we find that

$$500 = k(50)(100) \quad \text{or} \quad k = \frac{500}{50(100)} = \frac{1}{10}.$$

Thus, the desired formula is

$$T = \frac{1}{10} PV.$$

When $P = 100$ and $V = 75$, we have

$$T = \frac{1}{10}(100)(75) = 750° \text{ Kelvin.}$$

We sometimes have joint variation together with inverse variation. Perhaps the most important historical discovery of this type of combined variation is *Newton's law of universal gravitation*, which states that the gravitational force of attraction F between two particles is jointly proportional to their masses m and M, and inversely proportional to the square of the distance d between them. In other words, $F = G\dfrac{mM}{d^2}$, where G is the constant of proportionality.

Example   © Careful measurements show that two 1-kilogram masses 1 meter apart exert a mutual gravitational attraction of 6.67×10^{-11} newton. (One pound of force is approximately 4.45 newtons.) The earth has a mass of 5.98×10^{24} kilograms. Find the earth's gravitational force on a space capsule that has a mass of 1000 kilograms and that is 10^8 meters from the center of the earth.

Solution   Putting $F = 6.67 \times 10^{-11}$ newton, $m = 1$ kilogram, $M = 1$ kilogram, and $d = 1$ meter in the formula $F = G(mM/d^2)$, we find that

$$6.67 \times 10^{-11} = G\frac{1(1)}{1^2} = G,$$

so that, for arbitrary values of m, M, and d,

$$F = (6.67 \times 10^{-11})\frac{mM}{d^2}.$$

Now we substitute $M = 5.98 \times 10^{24}$, $m = 10^3$, and $d = 10^8$ to obtain

$$F = (6.67 \times 10^{-11})\frac{10^3(5.98 \times 10^{24})}{(10^8)^2}$$

$$= (6.67 \times 10^{-11})(5.98 \times 10^{24+3-16})$$

$$= (6.67)(5.98) \times 10^{24+3-16-11} \approx 39.9 \times 10^0 = 39.9 \text{ newtons}$$

(or less than 9 pounds of gravitational force).

In solving the problem above, we used the fact—first proved by Newton himself using integral calculus—that in calculating the gravitational attraction of a homogeneous sphere, one can assume that all of the mass is concentrated at the center.

PROBLEM SET 6

In problems 1–6, find the ratio of the first quantity to the second. Write your answer in the form of a reduced fraction. If the quantities are measured in different units, you must first express both in the same units.

1. 51 pounds to 68 pounds
2. 700 grams to 21 kilograms
3. $\frac{3}{4}$ yard to 15 inches
4. 1 square foot to 1 square yard
5. 1500 cubic centimeters to $\frac{1}{10}$ cubic meter
6. 88 kilometers per hour to 11 miles per hour [Take one kilometer to be $\frac{5}{8}$ mile.]

In problems 7–10, find the ratio of each pair of expressions if the ratio of a to b is 4/5.

7. $6a - 7b$ to $3a + 4b$
8. $5a + 3b$ to $4a + 9b$
9. $3a + 11b$ to $2a - b$
10. $-7a$ to $3a + 13b$

In problems 11–16, solve each proportion for x.

11. $15:x = 10:7$
12. $\dfrac{x-1}{10} = \dfrac{17}{4}$
13. 8 is to $3x$ as 7 is to 5
14. $5 - 3x:-8 = 3:2x$
15. $25/(x+2) = 2/5$
16. $3x + 40:12 = 2x - 15:6$

In problems 17–22, assume that $a/b = c/d$ and show by an algebraic calculation that the indicated proportion follows from it. Assume that the values of a, b, c, and d are restricted so that denominators are not zero.

17. $\dfrac{c}{a} = \dfrac{d}{b}$

18. $b{:}a = d{:}c$

19. $c - a{:}a = d - b{:}b$

20. $\dfrac{b + a}{a} = \dfrac{d + c}{c}$

21. $\dfrac{b}{a + b} = \dfrac{d}{c + d}$

22. $\dfrac{a + b}{a - b} = \dfrac{c + d}{c - d}$

In problems 23–34, relate the quantities by writing an equation involving a constant of proportionality k.

23. y is directly proportional to t.

24. A varies directly with r^2.

25. The volume V of a sphere varies directly as the cube of its radius r.

26. The distance d through which a body falls is directly proportional to the square of the amount of time t during which it falls.

27. The power P provided by a wind generator is directly proportional to the cube of the wind speed v.

28. The kinetic energy K of a moving body is jointly proportional to its mass m and the square of its speed v.

29. The heat energy E conducted per hour through the wall of a house is jointly proportional to the area A of the wall and the difference T degrees between the inside and outside temperatures, and inversely proportional to the thickness d of the wall.

30. The maximum safe load W on a horizontal beam supported at both ends varies directly as the width w of the beam and the square of its depth d, and inversely as the length ℓ of the beam.

31. The force F of repulsion between two electrically charged particles is directly proportional to the charges Q_1 and Q_2 on the particles, and inversely proportional to the square of the distance d between them.

32. The intensity I of a sound wave is jointly proportional to the square of its amplitude A, the square of its frequency n, the speed of sound v, and the density d of the air.

33. The frequency n of a guitar string is directly proportional to the square root of the string tension F, and inversely proportional to the length ℓ of the string and the square root of its linear density (mass per unit length) d.

34. The rate r at which an infectious disease spreads is jointly proportional to the fraction p of the population that has the disease and the fraction $1 - p$ of the population that does not have the disease.

In problems 35–46, (a) find an equation relating the given quantities and (b) find the value of the indicated quantity under the given conditions.

35. y is directly proportional to x, and $y = 3$ when $x = 6$. Find y when $x = 12$.

36. y varies directly as x^3, and $y = 4$ when $x = 2$. Find y when $x = 4$.

37. w varies inversely as x, and $w = 3$ when $x = 4$. Find w when $x = 8$.

38. q varies inversely as $\sqrt{p}$, and $q = 2$ when $p = \frac{1}{4}$. Find q when $p = 9$.

39. u varies inversely as v^2, and $u = 2$ when $v = 10$. Find u when $v = 3$.

40. r is inversely proportional to $\sqrt[3]{t}$, and $r = 2$ when $t = \frac{1}{8}$. Find r when $t = 27$.

41. z is directly proportional to x and inversely proportional to y, and $z = 8$ when $x = 4$ and $y = 2$. Find z when $x = 8$ and $y = 4$.

42. z is directly proportional to x^3 and inversely proportional to y^2, and $z = 2$ when $x = 2$ and $y = 4$. Find z when $x = -2$ and $y = 2$.

43. H is directly proportional to t and inversely proportional to $\sqrt{s}$, and $H = 7$ when $t = 2$ and $s = 4$. Find H when $t = 4$ and $s = 9$.

44. s varies jointly as the product of x and $y^{2/3}$, and $s = 24$ when $x = 6$ and $y = 8$. Find s when $x = -3$ and $y = 1$.

45. V varies directly as T and inversely as P, and $V = 40$ when $T = 300$ and $P = 30$. Find V when $T = 324$ and $P = 24$.

46. y varies jointly as x^2 and z and inversely as w, and $y = 12$ when $x = 4$, $z = 3$ and $w = 2$. Find y when $x = -4$, $z = 5$, and $w = -1$.

47. In wildlife management, the population P of animals of a certain type that inhabit a particular area is often estimated by capturing M of these animals, banding or otherwise marking them, and releasing the marked animals. Later, after the marked animals have mixed with the rest of the animals in the area, a random sample consisting of S animals is captured and the number m of marked animals in the sample is counted. It is assumed that the ratio of marked animals in the population is the same as the ratio of marked animals in the sample, so that $M/P = m/S$. Solve for P in terms of M, S, and m.

48. In order to estimate the number of fish in a lake, ecologists catch 100 fish, band them, and release the banded fish to mix with the rest of the fish. Later, 50 fish are caught from the lake, and 2 of them are found to be banded. Estimate the number of fish in the lake. [See problem 47.]

49. At a constant speed v on level ground, the number N of gallons of gasoline used by an automobile is directly proportional to the amount of time T during which it travels. If the automobile uses 6 gallons in 1.5 hours, how many gallons will it use in 10 hours?

C 50. The distance d in which an automobile can stop when its brakes are applied varies directly as the square of its speed v. Skid marks from a car involved in an accident measured 173 feet. In a test, the same car going 40 miles per hour was braked to a panic stop in 88 feet. Was the driver exceeding the speed limit (55 miles per hour) when the accident occurred?

51. Fahrenheit temperature F varies linearly as Celsius temperature C. Find a formula for F in terms of C if $F = 212°$ when $C = 100°$, and $F = 32°$ when $C = 0°$.

C 52. A manufacturer finds that the cost C dollars of manufacturing N units of a certain product varies linearly with N. The manufacturer's *overhead* for the product is the cost if $N = 0$. Find the overhead, given that $C = \$4200$ when $N = 100$, and $C = \$13,000$ when $N = 500$.

C 53. A company estimates that its profit P dollars per month varies linearly with its sales revenue R. If $P = \$6000$ when $R = \$75,000$, and $P = \$16,000$ when $R = \$150,000$, find a formula for P in terms of R.

C 54. The shoe size of a normal man varies approximately as the 3/2 power of his height. If the average 6-foot man wears a size 11 shoe, what would you predict as the shoe size of a 7-foot basketball player?

55. The amount A of pollution entering the atmosphere in a certain region is found to be directly proportional to the 2/3 power of the number N of people in that region. If a population of $N = 8000$ people produces $A = 900$ tons of pollution per year, find a formula for A in terms of N.

C 56. In astronomy, Kepler's third law states that the time T required for a planet to make one revolution about the sun is directly proportional to the 3/2 power of the maximum radius of its orbit. If the maximum radius of the earth's orbit is 93 million miles and the maximum radius of Mars' orbit is 142 million miles, how many days are required for Mars to make one revolution about the sun?

57. The time t days required to pick up litter along a stretch of highway is inversely proportional to the number n of people on the work team. If 5 people can do the job in 3 days, how many days would be required for 8 people to do it?

C 58. The amount H of heat conducted per second through a cube is jointly proportional to the temperature difference T degrees between the opposite faces and the area A of one face, and inversely proportional to the length ℓ of an edge of the cube. The constant of proportionality for this combined variation is called the *coefficient of conductivity* for the material of the cube. Find the coefficient of conductivity for a cube with an edge that is

REVIEW PROBLEM SET 217

45 centimeters long, if 1.26 joules of heat per second are conducted through the cube when the temperature difference between opposite faces is 31.5 degrees Celsius.

59. The distance from the earth to the moon is approximately 3.8×10^8 meters, and the mass of the moon is approximately 7.3×10^{22} kilograms. Find the gravitational attraction of the moon on a person on earth whose mass is 70 kilograms. [See the example on page 214.]

60. Using the data in problem 59, find the ratio of a person's weight on the moon to the same person's weight on earth. Take the radius of the moon to be 1.7×10^6 meters and the radius of the earth to be 6.4×10^6 meters.

61. The frequency n hertz of a musical tone is directly proportional to the speed v of sound in air, and inversely proportional to the wavelength ℓ of the sound wave. The speed of sound in air varies linearly as the temperature T in degrees Celsius. Write a formula for the frequency n in terms of ℓ, T, and two constants.

62. Determine the numerical value of the two constants in problem 61, given the following data: $n = 440$ hertz when $T = 0°$ Celsius and $\ell = 0.75$ meter, and $n = 880$ hertz when $T = 20°$ Celsius and $\ell = 0.39$ meter.

63. Since ancient times, artists and artisans have recognized that figures whose proportions are in a certain ratio a/b, called the *golden ratio*, are especially pleasing. For instance, a rectangle whose width a and length b form the golden ratio is called a *golden rectangle*. The condition under which a/b is the golden ratio is that a is to b as b is to $a + b$. Calculate the golden ratio to three decimal places, and draw a golden rectangle.

REVIEW PROBLEM SET

In problems 1–12, find the vertex and the y and x intercepts of the graph of the quadratic function, determine whether the graph opens upward or downward, sketch the graph, and find the domain and range of the function.

1. $f(x) = 4x^2$
2. $g(x) = \frac{1}{4}x^2$
3. $h(x) = -\frac{1}{4}x^2$
4. $F(x) = 3x^2 + 2$
5. $G(x) = 3(x - 2)^2 + 1$
6. $H(x) = \frac{1}{3}[(x + 3)^2 + 2]$
7. $f(x) = x^2 - 3x + 2$
8. $g(x) = 6x^2 + 13x - 5$
9. $h(x) = -6x^2 - 7x + 20$
10. $G(x) = -2x^2 + x + 10$
11. $F(x) = 10x - 25 - x^2$
12. $H(x) = 7x + 2x^2 - 39$

13. Use the graph obtained in problem 7 to find the solution set of $x^2 - 3x + 2 \geq 0$.

14. Suppose that x_1 and x_2 are the zeros of the quadratic function f. Show that the x coordinate of the vertex of the graph of f is $\frac{1}{2}(x_1 + x_2)$.

15. Use the graph obtained in problem 9 to find the solution set of $-6x^2 - 7x + 20 \leq 0$.

16. Suppose that f is a quadratic function and that there is a number x_0 with $x_0 \neq 0$ and $f(x_0) = f(-x_0)$. Prove that f is an even function.

17. Find two numbers whose sum is 21 and whose product is maximum.

18. Find all values of K such that the graph of $g(x) = 3x^2 + Kx - 4K$ has no x intercepts.

19. The strength S of a new plastic is given by $S = 500 + 600T - 20T^2$, where T is the temperature in degrees Fahrenheit. At what temperature is the strength maximum?

20. The height of a projectile fired straight upward with an initial speed v_0 is given by the function $h = v_0 t - \frac{1}{2}gt^2$, where t is the elapsed time since the projectile was fired, and the constant g is the acceleration of gravity. At what time does the projectile reach its maximum height, and what is this height?

21. A manufacturer of sports trophies knows that the total cost C dollars of making x thousand trophies is given by $C = 600 + 60x$, and that the corresponding sales revenue R dollars is given by $R = 300x - 4x^2$. Find the number of trophies (in thousands) that will maximize the manufacturer's profit. [Hint: Profit = revenue − cost.]

22. A homeowner is planning a rectangular flower garden surrounded by an ornamental fence. The fencing for three sides of the garden costs $20 per meter, but the fencing for the fourth side, which faces the house, costs $30 per meter. If the homeowner has $1200 to spend on the fence, what dimensions of the garden will give it the maximum area?

In problems 23–26, decide which functions are polynomial functions and which are not.

23. $f(x) = \sqrt{3}x - 4x^4 + \frac{1}{2}$

24. $g(x) = 2x^{-2} + 3x^{-1} + x + 1$

25. $h(x) = 2\sqrt{x} + x - 4$

26. $H(x) = (x - 1)(x - 2)(x - 3)$

In each of problems 27–30, sketch the graphs of the functions f, g, F, and G on the same coordinate system.

27. $f(x) = \frac{1}{3}x^3$; $g(x) = -\frac{1}{3}x^3$; $F(x) = \frac{1}{3}x^3 + 1$; $G(x) = \frac{1}{3}(x + 1)^3 + 4$

28. $f(x) = \frac{1}{5}x^4$; $g(x) = -\frac{1}{5}x^4$; $F(x) = \frac{1}{5}x^4 - 1$; $G(x) = \frac{1}{5}(x - 1)^4 + 2$

29. $f(x) = \frac{1}{4}x^6$; $g(x) = -\frac{1}{4}x^6$; $F(x) = \frac{1}{4}x^6 + 4$; $G(x) = \frac{1}{4}(x + 1)^6 + 3$

30. $f(x) = \frac{1}{2}x^7$; $g(x) = -\frac{1}{2}x^7$; $F(x) = \frac{1}{2}x^7 + 2$; $G(x) = \frac{1}{2}(x + 2)^7 - 1$

In problems 31–34, find the y and x intercepts and sketch the graph of each function.

31. $f(x) = (x - 2)^3 + 8$

32. $g(x) = \frac{1}{3}(x + 3)^4 + 2$

33. $F(x) = \frac{1}{4}(x + 1)^5 - 8$

34. $G(x) = -\frac{2}{3}(x - 5)^6 - \frac{1}{2}$

In problems 35–42, (a) find the x intercepts of the graph of the function, (b) sketch the graph, and (c) use the graph to solve the given inequality. [C] (Use a calculator if you wish.)

35. $f(x) = x(x^2 - 9)$; $f(x) \le 0$

36. $g(x) = -(3x - 1)(x + 1)(x + 2)$; $g(x) \ge 0$

37. $F(x) = x(x^2 + 7x + 10)$; $F(x) > 0$

38. $H(x) = -x(3x^2 - 7x + 2)$; $H(x) < 0$

39. $G(x) = (x + 1)(10x^2 - 3x - 18)$; $G(x) \ge 0$

40. $f(x) = -(x^2 - 4)(x^2 - 9)$; $f(x) < 0$

41. $h(x) = -(x + 3)(2 - x)^2$; $h(x) > 0$

42. $g(x) = x^2(x - 1)^2(x + 2)^2$; $g(x) \le 0$

In problems 43–48, use long division to find the quotient polynomial $q(x)$ and the remainder polynomial $r(x)$ when $f(x)$ is divided by $g(x)$. Check your work by verifying that $r(x)$ is either the zero polynomial or a polynomial of degree less than $g(x)$, and that $f(x) = g(x)q(x) + r(x)$.

43. $f(x) = x^2 + 5x + 2$; $g(x) = x + 3$

44. $f(x) = 6x^4 + 38x^3 + 44x^2 - 96x + 27$; $g(x) = 2x + 6$

45. $f(x) = x^5 - 32$; $g(x) = x^3 - 8$

46. $f(x) = 8x^3 - 5x^2 - 51x - 18$; $g(x) = 2x^3 + x^2 + 1$

47. $f(x) = 6x^6 + 9x^5 + x^4 - 3x^3 + 3x^2 + 6x + 1$; $g(x) = 2x^2 + 3x + 1$

48. $f(x) = 3x^4 - 4x^3 + 5x^2 + x + 7$; $g(x) = 2x^2 + x + 2$

In problems 49–52, use the quotient theorem to rewrite each rational expression as the sum of a polynomial and a proper rational expression.

49. $\dfrac{4x^4 + x^3 - 2x + 3}{x + 1}$

50. $\dfrac{x^3 + 1}{x^2 + 1}$

51. $\dfrac{x^3 + 6x^2 + 10x}{x^2 + x + 2}$

52. $\dfrac{2x^4 + 3x^2 + 4x + 2}{x^4 + x^3 + x^2 + x + 1}$

In problems 53 and 54, divide $f(x)$ by $g(x)$ to obtain a quotient polynomial $q(x)$ and a constant remainder R, (a) using long division and (b) using synthetic division.

53. $f(x) = 3x^4 + 2x^3 + x^2 + x + 2$; $g(x) = x + 2$

54. $f(x) = x^5 - 3x^3 + 5x^2 - 12$; $g(x) = x - \frac{1}{2}$

In problems 55–60, (a) use synthetic division to obtain the quotient polynomial $q(x)$ and the constant remainder R upon division of the given polynomial by $x - c$ for the indicated value of c, and [C] (b) verify by direct substitution that R is the value of the polynomial when $x = c$.

55. $x^3 - 2x^2 + 3x - 5$; $c = -2$

56. $x^3 - 2x^2 + 7x - 1$; $c = -3$

57. $2x^3 - x^2 + 7$; $c = 4$

58. $5x^4 - 10x^3 - 12x - 7$; $c = -4$

59. $x^5 + 5x - 13$; $c = -1$

60. $3x^5 + 5x^4 - 2x^3 + x^2 - x + 1$; $c = 2$

C In problems 61 and 62, use synthetic division, the remainder theorem, and a calculator to find the indicated number rounded off to two decimal places.

61. $f(2.72)$ if $f(x) = 17.1x^4 + 33.3x^3 - 2.75x^2 + 11.1x + 21.8$
62. $g(3.14)$ if $g(x) = 13.5x^8 - 31.7x^6 + 22.1x^4 - 35.7x^2 + 21.2$

In problems 63–68, C (a) use the remainder theorem to find the remainder R when each division is performed, and (b) verify your result using synthetic division.

63. $x^3 - 2x^2 + 3x - 5$ divided by $x + 2$
64. $x^3 - 2x^2 + 7x - 1$ divided by $x + 3$
65. $2x^3 - x^2 + 7$ divided by $x + 4$
66. $5x^4 - 10x^3 - 12x - 7$ divided by $x - 4$
67. $x^5 + 5x - 13$ divided by $x - 1$
68. $3x^5 + 5x^4 - 2x^3 + x^2 - x + 1$ divided by $x - 2$

In problems 69–72, use the factor theorem to determine (without actually dividing) whether or not the indicated binomial is a factor of the given polynomial.

69. $x^4 - 4x - 69; x - 3$
70. $x^3 - 2x^2 + 1; x - 1$
71. $x^3 - 2x^2 + 3x + 4; x - 2$
72. $2x^4 + 3x - 26; x + 3$

In problems 73–78, find all rational zeros of each polynomial function.

73. $f(x) = x^3 - 8x^2 + 5x + 14$
74. $g(x) = x^3 - 4x^2 - 5x + 14$
75. $F(x) = x^4 - 4x^3 - 5x^2 + 36x - 36$
76. $G(x) = 4x^3 - 19x^2 + 32x - 15$
77. $h(x) = 4x^4 - 4x^3 - 7x^2 + 4x + 3$
78. $H(x) = 4x^4 - 2x^3 + 2x^2 + 10x + 3$

In problems 79–82, factor each polynomial function completely into prime factors, and C sketch the graph of the function.

79. $f(x) = x^3 + 2x^2 - x - 2$
80. $F(x) = x^3 - x^2 - 14x + 24$
81. $g(x) = x^4 - 4x^3 - 14x^2 + 36x + 45$
82. $G(x) = 2x^4 - x^3 - 14x^2 + 19x - 6$

C **83.** Find the original length of the edge of a cube if, after a slice 1 centimeter thick is cut from one side, the volume of the remaining solid is 448 cubic centimeters.

84. Corresponding to each polynomial function f is another polynomial function f' called its *derivative*, such that for each real number c, $f'(c) = q(c)$, where $q(x)$ is the quotient polynomial obtained by dividing $f(x)$ by $x - c$.
  (a) Find the derivative of $f(x) = 3x^2 + 2x + 7$.
  (b) Find the derivative of $f(x) = Ax^3 + Bx^2 + Cx + D$.

85. Let f be a polynomial function and suppose that a and b are constant real numbers with $a \neq b$. Show that the remainder upon division of $f(x)$ by $(x - a)(x - b)$ has the form

$$Ax + B, \quad \text{where} \quad A = \frac{f(b) - f(a)}{b - a} \quad \text{and} \quad B = \frac{bf(a) - af(b)}{b - a}.$$

86. Let f be a polynomial function and suppose that f' is the derivative of f, as in problem 84. If c is a constant, show that the remainder upon division of $f(x)$ by $(x - c)^2$ has the form

$$Ax + B, \quad \text{where} \quad A = f'(c) \quad \text{and} \quad B = f(c) - cf'(c).$$

In problems 87–96, indicate the domain of each rational function, find the y and x intercepts of its graph, determine the horizontal and vertical asymptotes of the graph, and C sketch it.

87. $f(x) = -\dfrac{2}{x}$
88. $g(x) = 1 - \dfrac{4}{x}$
89. $G(x) = 1 - \dfrac{3}{x^2}$
90. $P(x) = 2 + \dfrac{4}{x^2}$

91. $F(x) = \dfrac{2x}{x - 1}$
92. $H(x) = \dfrac{3x + 2}{2 - x}$
93. $p(x) = \dfrac{(x + 2)^2}{x^2 + 2x}$
94. $G(x) = \dfrac{4x^2}{x^2 - 4x}$

95. $T(x) = \dfrac{x^2 + 1}{x^2 - 3x}$
96. $k(x) = \dfrac{x^2 + 5x + 4}{x^2 + 5x}$

C **97.** If a resistor of resistance x ohms, $x \geq 0$, is connected in parallel with a resistor of resistance 1 ohm, the resulting net resistance y is given by $y = x/(1 + x)$ (Figure 1). (a) Sketch a graph of y as a function of x, (b) determine the domain and the range of this function, and (c) indicate whether y increases or decreases when x increases.

C **98.** According to the Doppler effect in physics, if a source of sound of frequency n is moving away from an observer with speed u, the frequency N of the sound heard by the observer

is given by $N = \dfrac{n}{1 + \dfrac{u}{v}}$, where v is the speed of sound in air. If $v = 768$ miles per hour and

$n = 440$ hertz, sketch the graph of N as a function of u for $0 \leq u \leq 100$ miles per hour.

Figure 1

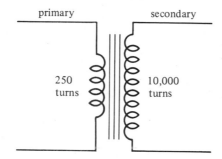

In problems 99–104, find the ratio of the first quantity to the second. Write your answer as a reduced fraction. If the quantities are measured in different units, you must first express both in the same units.

99. Twelve dollars to sixty-five cents.

100. 100 milliwatts to 25 microwatts.

101. 250 turns in the primary winding of a transformer to 10,000 turns in the secondary winding (Figure 2).

Figure 2

primary   secondary

250   10,000
turns   turns

102. 10 miles to 1000 feet.   **103.** 50 ounces to 2 pounds.   **104.** 1 kilometer to 1 mile.

In problems 105 and 106, find the ratio of each pair of expressions if the ratio of a to b is $\frac{2}{3}$.

105. $3a + 2b$ to $4a - 7b$ **106.** $5a - 3b$ to $3b + 7a$

In problems 107 and 108, solve each proportion for x.

107. $2:x + 1 = 4:5$ **108.** $2x - 3$ is to 3 as $1 - x$ is to 5

109. If $a:x = x:b$, x is called the *mean proportional* between a and b. Find the mean proportional between 2 and 8.

110. The ratio of the surface areas of two spheres is $\frac{9}{4}$. Find the ratio of their volumes. [Hint: The volume V and surface area A of a sphere of radius r are given by $V = \frac{4}{3}\pi r^3$ and $A = 4\pi r^2$.]

111. The *voltage gain* of an electronic amplifier is defined to be the ratio of the output voltage to the input voltage. Find the voltage gain of an amplifier if the input voltage is 0.005 volt and the output voltage is 25 volts.

112. One of the indicators of the quality of instruction at an educational institution is the student-to-teacher ratio. Is it desirable that this ratio be large or small?

In problems 113–122, relate the quantities by writing an equation involving at least one constant.

113. P is inversely proportional to V.

114. The surface area A of a sphere varies directly as the square of its diameter d.

115. For tax purposes, it is often assumed that the value V dollars of an article varies linearly with the time t years since it was purchased.

116. In a *voltage-controlled* electronic device, the change in output current, written ΔI, is directly proportional to the change in input voltage, written ΔV. [Δ is the capital Greek letter *delta*. It is often used to stand for "a change in."]

117. The force F of the wind on a blade of a wind-powered generator varies jointly with the area A of the blade and the square of the wind speed v.

118. The inductance L of a coil of wire is jointly proportional to the cross-sectional area A of the coil and the square of the number N of turns, and inversely proportional to the length ℓ of the coil.

119. The rate r at which a rumor is spreading in a population of size P is jointly proportional to the number N of people who have heard the rumor and the number $P - N$ of people who have not.

120. The collector current I_C of a transistor is jointly proportional to its current ratio β and its base current I_B. [β is the small Greek letter *beta*.]

121. The power P provided by a jet of water is jointly proportional to the cross-sectional area A of the jet and the cube of the speed v of the water in the jet.

122. In geology, it is found that the erosive force E of a swiftly flowing stream is directly proportional to the sixth power of the speed v of flow of the water.

In problems 123–130, (a) find an equation relating the given quantities and, (b) find the value of the indicated quantity under the specified conditions.

123. y is inversely proportional to x, and $y = 1$ when $x = 5$. Find y when $x = 25$.

124. y is jointly proportional to u and $\sqrt{v}$, and $y = 6$ when $u = 1$ and $v = 4$. Find y when $u = \frac{1}{3}$ and $v = 9$.

125. w is directly proportional to x and inversely proportional to y, and $w = 7$ when $x/y = 3$. Find w when $x = 24$ and $y = 6$.

126. y varies linearly with x, $y = -1$ when $x = 1$, and $y = 5$ when $x = -1$. Find y when $x = 0$.

127. y is directly proportional to x^2 and inversely proportional to $z + 3$, and $y = 4$ when $x = 2$ and $z = 1$. Find y when $x = 3$ and $z = 6$.

128. y is jointly proportional to $\sqrt[4]{x}$ and z^3, and $y = 32$ when $x = 16$ and $z = 2$. Find y when $x = 81$ and $z = \frac{1}{3}$.

129. y varies linearly as $\sqrt{x}$, $y = 3$ when $x = 1$, and $y = 5$ when $x = 4$. Find y when $x = 9$.

C **130.** y is directly proportional to x and inversely proportional to z^2, and $y = 1.422$ when $x = 0.4181$ and $z = 0.7135$. Find y when $x = 2.133$ and $z = 5.357$.

C **131.** The volume V of a sphere is directly proportional to the cube of its diameter. If a sphere of diameter 2 meters has a volume of approximately 4.19 cubic meters, find the approximate volume of a sphere of diameter 10 meters.

132. Kelvin temperature K varies linearly with Fahrenheit temperature F. Find a formula for K in terms of F if $K = 0$ when $F = -459$, and $K = 273$ when $F = 32$.

C **133.** The volume V of a block of iron varies linearly with its temperature T in degrees Celsius. Find a formula for the volume V of a block of iron if $V = 0.25$ cubic meter when $T = 0°$, and $V = 0.25018$ cubic meter when $T = 20°$.

C **134.** A company estimates that its profit P dollars per month varies linearly with the number of items n manufactured per month. If $P = \$180,000$ when $n = 5000$, and $P = \$300,000$ when $n = 8000$, find P as a function of x.

C **135.** A company's sales volume S per month varies directly as the number A of dollars per month spent for advertising, and inversely as the product of the selling price x dollars per unit and the inflation index I. If S is 20,000 units when $A = \$10,000$, $x = \$400$, and $I = 10\%$, find S when $A = \$15,000$, $x = \$500$, and $I = 15\%$.

5 Exponential and Logarithmic Functions

Until now, we have considered only functions defined by equations involving algebraic expressions. Although such *algebraic functions* are useful and important, they are not sufficient for all the requirements of applied mathematics. In this chapter we begin our study of the *transcendental functions*—that is, functions that transcend (go beyond) purely algebraic methods. Here we consider the exponential and logarithmic functions and give some of their many applications to the life sciences, finance, earth sciences, engineering, electronics, and other fields.

1 Exponential Functions

In Section 2 of Chapter 4, we studied power functions

$$p(x) = x^n$$

in which the base is the variable x and the exponent n is constant. In this section we shall study functions of the form

$$f(x) = b^x$$

in which the exponent is the variable x and the base b is constant. Such a function is called an **exponential function.**

If b is a positive real number, then b^x is defined as in Section 7 of Chapter 1 for all **rational** values of x. It is possible to extend the definition so that **irrational** numbers can also be used as exponents. Although the technical details of the extended definition depend on methods studied in calculus, the basic idea is quite simple: If $b > 0$ and x is an irrational number, then

$$b^x \approx b^r,$$

where r is a rational number obtained by rounding off x to a finite number of decimal places. Better and better approximations to b^x are obtained by rounding off x to more and more decimal places. For instance,

$$b^\pi \approx b^{3.14},$$

and a better approximation is given by

$$b^\pi \approx b^{3.14159}.$$

If b is a positive constant, and if you plot several points (x, b^x) for rational values of x, you will notice that these points seem to lie along a smooth curve. This curve is the graph of the exponential function $f(x) = b^x$ with base b. For instance, taking $b = 2$ and plotting several points $(x, 2^x)$ for rational values of x, we obtain Figure 1a. In Figure 1b, we have connected these points with a smooth curve to obtain the graph of $f(x) = 2^x$.

Figure 1

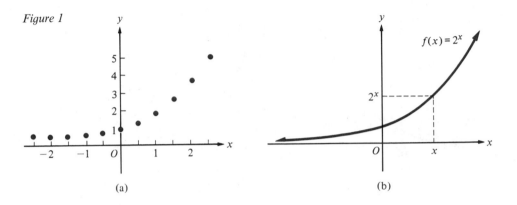

(a)   (b)

Example   Sketch the graph of (a) $f(x) = 3^x$ and (b) $g(x) = (\frac{1}{3})^x$.

Solution   We begin by calculating values of $f(x) = 3^x$ and of $g(x) = (\frac{1}{3})^x$ for several integer values of x, as shown in the table in Figure 2. Then we plot the corresponding points and connect them by smooth curves to obtain the graphs of $f(x) = 3^x$ (Figure 2a) and $g(x) = (\frac{1}{3})^x$ (Figure 2b). Because

$$g(x) = \left(\frac{1}{3}\right)^x = \frac{1}{3^x} = 3^{-x} = f(-x),$$

these curves are reflections of each other across the y axis.

Figure 2

x	3^x	$(\frac{1}{3})^x$
-2	$\frac{1}{9}$	9
-1	$\frac{1}{3}$	3
0	1	1
1	3	$\frac{1}{3}$
2	9	$\frac{1}{9}$

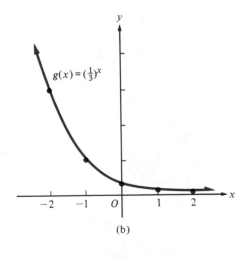

(a)   (b)

In general, graphs of exponential functions have shapes similar to the graphs in Figure 2. Thus, if $b > 1$, the graph of $f(x) = b^x$ is rising to the right (Figure 3a), while if $0 < b < 1$, the graph is falling to the right (Figure 3b). Of course, when $b = 1$, the graph is neither rising nor falling (Figure 3c). Notice that the graph of $f(x) = b^x$ always contains the point $(1, b)$ (because $b^1 = b$) and that its y intercept is always 1 (because $b^0 = 1$).

Figure 3

(a)

(b)

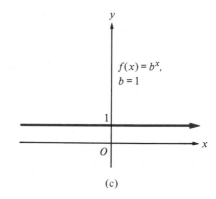

(c)

If $b > 0$, the domain of the exponential function $f(x) = b^x$ is $\mathbb{R}$. If $b > 1$ (Figure 3a), the graph of $f(x) = b^x$ comes as close to the x axis as we please if we move far enough to the left of the origin, but the curve never reaches the axis; in other words, the x axis is a horizontal asymptote. There is no vertical asymptote. As we move farther and farther to the right of the origin, the graph climbs higher and higher without bound; hence, the range of $f(x) = b^x$ is the interval $(0, \infty)$. Similar remarks apply to the graph of $f(x) = b^x$ for $0 < b < 1$ (Figure 3b).

Of course, you can sketch graphs of exponential functions more accurately if you plot more points. For this purpose, a calculator with a y^x key is a most useful tool.

Example   © Using a calculator with a y^x key, evaluate (a) $\sqrt{2}^{\sqrt{3}}$ and (b) $\pi^{-\sqrt{2}}$.

Solution   On a 10-digit calculator, we obtain
(a) $\sqrt{2}^{\sqrt{3}} = 1.414213562^{1.732050808} = 1.822634654$
(b) $\pi^{-\sqrt{2}} = 3.141592654^{-1.414213562} = 0.198117987$

The Properties of Rational Exponents (page 46) continue to hold for all real exponents, provided that all bases are positive. For instance, if a and b are positive, we have

$$a^x a^y = a^{x+y}, \quad (a^x)^y = a^{xy}, \quad \text{and} \quad (ab)^x = a^x b^x,$$

for all real values of x and y.

Example $\boxed{\text{C}}$ Using a calculator with a y^x key, verify that $\pi^{\sqrt{2}}\pi^{\sqrt{3}} = \pi^{\sqrt{2}+\sqrt{3}}$.

Solution   On a 10-digit calculator, we obtain

$$\pi^{\sqrt{2}}\pi^{\sqrt{3}} = (5.047497266)(7.262545040) = 36.65767623$$

and

$$\pi^{\sqrt{2}+\sqrt{3}} = 3.141592654^{3.146264370} = 36.65767624.$$

The discrepancy in the last decimal place is due to accumulated error caused by the rounding off.

The techniques of graph sketching presented in Section 6 of Chapter 3 and illustrated throughout Chapter 4 can be applied to exponential functions.

Examples   Sketch the graph of the given function, determine its domain, its range, and any horizontal or vertical asymptotes, and indicate whether the function is increasing or decreasing.

1 $h(x) = 3^{x+2}$

Solution   The graph of $f(x) = 3^x$ appears in Figure 2a. Since $f(x + 2) = 3^{x+2}$, we have

$$h(x) = f(x + 2).$$

Therefore, the graph of h is obtained by shifting the graph of f two units to the left (Figure 4). Evidently, the domain of h is $\mathbb{R}$, the range is the interval $(0, \infty)$, the x axis is a horizontal asymptote, there is no vertical asymptote, and h is increasing throughout its domain. Because

$$h(x) = 3^{x+2} = 3^x \cdot 3^2 = 9 \cdot 3^x = 9f(x),$$

the graph of h could also have been obtained from the graph of f by multiplying each ordinate by 9.

Figure 4

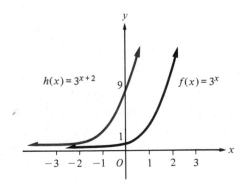

$h(x) = 3^{x+2}$ $f(x) = 3^x$

2    $k(x) = -2^x + 1$

Solution   The graph of $y = -2^x$ is obtained by reflecting the graph of $y = 2^x$ (Figure 1b) across the x axis (Figure 5), and the graph of $k(x) = -2^x + 1$ is obtained by shifting the graph of $y = -2^x$ one unit upward (Figure 5). Evidently, the domain of k is $\mathbb{R}$, the range is the interval $(-\infty, 1)$, the line $y = 1$ is a horizontal asymptote, there is no vertical asymptote, and k is decreasing throughout its domain.

Figure 5

The following examples illustrate the use of exponential functions in the world of business, investment, and finance.

Example 1   © Bankers use the **compound interest** formula

$$S = P\left(1 + \frac{r}{n}\right)^{nt}$$

for the *final value S* dollars of a *principal P* dollars invested for a *term* of t years at a *nominal annual interest rate r compounded n times per year*. If you invest $P = \$500$ at a nominal annual interest rate of 8% (that is, $r = 0.08$) compounded quarterly ($n = 4$), what is the final value S of your investment after a term of $t = 3$ years?

Solution   Using a calculator, we find that

$$S = P\left(1 + \frac{r}{n}\right)^{nt} = 500\left(1 + \frac{0.08}{4}\right)^{4(3)} = 500(1.02)^{12} = \$634.12.$$

Example 2   When a bank offers compound interest, it usually specifies not only the nominal annual interest rate r but also the *effective* simple annual interest rate R; that is, the rate of simple annual interest that would yield the same final value over a 1-year term as the compound interest. The formula

$$R = \left(1 + \frac{r}{n}\right)^{n} - 1$$

is used to calculate R in terms of r and n. Find the effective simple annual interest rate R corresponding to a nominal annual interest rate of 12% (that is, $r = 0.12$) compounded semiannually ($n = 2$).

Solution

$$R = \left(1 + \frac{r}{n}\right)^n - 1 = \left(1 + \frac{0.12}{2}\right)^2 - 1$$

$$= (1.06)^2 - 1 = 0.1236;$$

in other words, the effective simple annual interest rate is 12.36%.

Money that you will receive in the future is worth *less* to you than the same amount of money received now, because you miss out on the interest you could collect by investing the money now. For this reason, we use the idea of the *present value* of money to be received in the future. If you have an opportunity to invest P dollars at a nominal annual interest rate r compounded n times a year, this principal plus the interest it earns will amount to S dollars after t years, as given by

$$S = P\left(1 + \frac{r}{n}\right)^{nt}$$

Thus, P dollars in hand *right now* is worth S dollars to be received t years *in the future*. Solving the equation above for P in terms of S, we get an equation for the **present value** of an offer of S dollars to be received t years in the future:

$$P = S\left(1 + \frac{r}{n}\right)^{-nt}$$

You could do just as well by investing P dollars now and collecting S dollars from your investment after t years.

Example

[C] Find the present value of $500 to be paid to you 2 years in the future, if investments during this period are earning a nominal annual interest rate of 10% compounded monthly.

Solution

Here $S = 500$, $r = 0.10$, $n = 12$, $t = 2$, and

$$P = S\left(1 + \frac{r}{n}\right)^{-nt} = 500\left(1 + \frac{0.10}{12}\right)^{-12(2)}$$

$$= 500\left(1 + \frac{1}{120}\right)^{-24} = \$409.70.$$

The idea of present value helps people to make intelligent investment choices. It allows the investor to translate various complex arrangements for future payments into single figures that are easy to compare.

PROBLEM SET 1

1. By finding values of 4^x for $x = -2, -\frac{3}{2}, -1, -\frac{1}{2}, 0, \frac{1}{2}, 1, \frac{3}{2}$, and 2, plotting the resulting points $(x, 4^x)$, and drawing a smooth curve through these points, sketch the graph of $f(x) = 4^x$.

2. (a) Using the graph obtained in problem 1 and approximating $\sqrt{2}$ as 1.4, find the approximate value of $4^{\sqrt{2}}$. © (b) Using a calculator with a y^x key, find the value of $4^{\sqrt{2}}$ to as many decimal places as you can.

3. Sketch the graph of $g(x) = (\frac{1}{4})^x$.

© 4. Using a calculator with a y^x key, sketch the graph of $f(x) = (\frac{1}{2})^x$ for $-4 \leq x \leq 4$ as accurately as you can.

© In problems 5–12, use a calculator with a y^x key to find the value of each quantity to as many significant digits as you can.

5. $2^{\sqrt{2}}$
6. $2^{-\sqrt{2}}$
7. 2^π
8. $2^{-\pi} - \pi^{-2}$
9. $\sqrt{2}^{\sqrt{2}}$
10. π^π
11. $\sqrt{3}^{-\sqrt{5}}$
12. $3.0157^{2.7566}$

© In problems 13–18, use a calculator with a y^x key to verify each equation for the indicated values of the variables.

13. $a^x a^y = a^{x+y}$ for $a = 3.074$, $x = 2.183$, $y = 1.075$
14. $a^{x+y} = a^x a^y$ for $a = 2.471$, $x = 5.507$, $y = 0.012$
15. $(a^x)^y = a^{xy}$ for $a = 1.777$, $x = -2.058$, $y = 3.333$
16. $a^{x-y} = \dfrac{a^x}{a^y}$ for $a = \sqrt{2}$, $x = \sqrt{5}$, $y = \sqrt{3}$
17. $(ab)^x = a^x b^x$ for $a = \sqrt{7}$, $b = \pi$, $x = \sqrt{\pi}$
18. $\left(\dfrac{a}{b}\right)^x = \dfrac{a^x}{b^x}$ for $a = 2 + \pi$, $b = \sqrt{2} - 1$, $x = \sqrt{5} - \sqrt{3}$

In problems 19–28, sketch the graph of the given function, determine its domain, its range, and any horizontal or vertical asymptotes, and indicate whether the function is increasing or decreasing. (Just make a rough sketch—do not use a calculator or tables.)

19. $f(x) = 2^x + 1$
20. $g(x) = (\frac{2}{3})^x$
21. $h(x) = 4^x - 1$
22. $F(x) = (0.2)^x$
23. $G(x) = 3 \cdot 2^x$
24. $H(x) = 3 \cdot 2^{-x}$
25. $f(x) = 2^{-x} - 3$
26. $g(x) = \frac{3}{2}(\frac{1}{3})^{-x}$
27. $h(x) = 2^{x-3}$
28. $F(x) = 3 - 2^{x-1}$

© In problems 29–34, assume that you have invested a principal P dollars at a nominal annual interest rate r compounded n times per year for a term of t years. Calculate (a) the final value

$$S = P\left(1 + \frac{r}{n}\right)^{nt}$$ dollars of your investment and (b) the effective simple annual interest rate

$$R = \left(1 + \frac{r}{n}\right)^{n} - 1.$$

29. $P = \$1000$, $r = 0.07$ (7%), $n = 1$, $t = 13$ years
30. $P = \$1000$, $r = 0.07$ (7%), $n = 12$, $t = 13$ years
31. $P = \$1000$, $r = 0.12$ (12%), $n = 1$, $t = 13$ years
32. $P = \$1000$, $r = 0.12$ (12%), $n = 52$, $t = 13$ years
33. $P = \$50,000$, $r = 0.135$ (13.5%), $n = 12$, $t = \frac{1}{2}$ year
34. $P = \$25,000$, $r = 0.155$ (15.5%), $n = 52$, $t = \frac{1}{4}$ year

© 35. Suppose that a bank offers to pay a nominal annual interest rate of 0.08 (8%) on money left on deposit for 2 years. Assume that a principal $P = \$1000$ is deposited. Find the final value S after the 2-year term if the interest is compounded (a) annually, (b) semiannually, (c) quarterly, (d) monthly, (e) weekly, (f) daily, and (g) hourly.

© 36. Find out the nominal annual interest rate r offered by your local savings bank for regular savings accounts and the number of times n per year that the interest is compounded. Calculate the effective simple annual interest rate R.

 37. Find the present value of $1000 five years in the future at a nominal annual rate of 12% ($r = 0.12$) compounded weekly.

 38. A fund compounds interest quarterly. If a principal $P = \$14,000$ yields a final value $S = \$45,510$ after a 5-year term, (a) find the nominal annual interest rate r and (b) find the effective simple annual interest rate R.

 39. Suppose that someone owes you money, and that your local savings bank offers savings accounts at 8% nominal annual interest compounded monthly. Use this interest rate to determine the present value to you of money offered in the future. If your debtor offers to pay you $100 six months from now, what is the present value to you of this offer?

 40. On a boy's 16th birthday, his father promises to give him $25,000 when he turns 21 to help set him up in business. Local banks are offering savings accounts at a nominal annual interest rate of 8% compounded quarterly. The boy, who has studied the mathematics of finance, says, "Dad, I'll settle for _____ dollars right now!" Fill in the blank appropriately.

---

2   The Exponential Function with Base *e*

If we increase the base b, the graph of the exponential function $f(x) = b^x$ rises more rapidly. This is illustrated in Figure 1 for $b = 2$ and $b = 3$. Since the graphs of all exponential functions contain the point $(0, 1)$, a good indication of how rapidly such a graph rises is its "steepness" at this point. The steepness of the graph of $f(x) = b^x$ at the point $(0, 1)$ can be measured by the slope m of its **tangent line**—that is, the straight line that just grazes the curve at this point (Figure 2). Although the tangent line is easily sketched by eye, its precise determination requires the use of calculus.

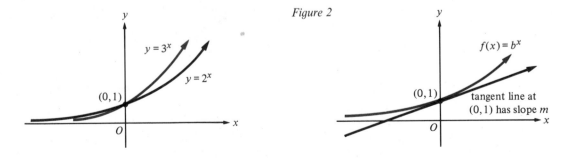

Figure 1

Figure 2

As we increase the base b of the exponential function $f(x) = b^x$, the graph of f in Figure 2 becomes steeper at the point $(0, 1)$ and the slope m of the tangent line increases. If you sketch accurate graphs for the values of b in Table I, draw the tangent lines at $(0, 1)$ by eye, and measure their slopes m, you will obtain approximately the values shown in the table (Problem 11).

Table 1

b	m
0.5	−0.69
1	0
2	0.69
3	1.1
4	1.4

From Table 1 we see that m is less than 1 when $b = 2$, and that m is greater than 1 when $b = 3$. As you might suspect, somewhere between 2 and 3 there is a value of b for which m is exactly 1. This particular value of the base is denoted by e, in honor of the great Swiss mathematician Leonhard Euler (1707–83) (pronounced "oiler"), who was one of the first to recognize its immense importance. Like π, the value of e is an irrational number. By using advanced mathematical methods and high-speed computers, the numerical value of e has been calculated to thousands of decimal places. Rounded off to three decimal places,

$$e \approx 2.718.$$

The graph of the exponential function

$$f(x) = e^x$$

rises at just the right rate so that the tangent line at $(0, 1)$ has slope $m = 1$. Using this fact, recalling the general shape of graphs of exponential functions with bases greater than 1, and plotting the points corresponding to

$$f(1) = e^1 = e \approx 2.718 \quad \text{and} \quad f(-1) = e^{-1} = \frac{1}{e} \approx \frac{1}{2.718} \approx 0.368,$$

we can sketch a reasonably accurate graph of $f(x) = e^x$ (Figure 3).

Figure 3

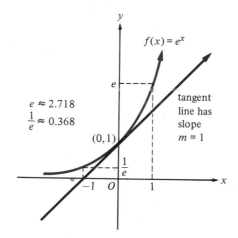

The exponential function $f(x) = e^x$ plays an important role in calculus and it is essential in the applications of mathematics to many fields ranging from engineering to public health. You will find a variety of these applications in the remainder of this chapter. Indeed, the function $f(x) = e^x$ is used so often that people simply call it the **exponential function.** Whenever anyone uses this term without specifying the base, you can be certain that the function $f(x) = e^x$ is intended. On some calculators and in many of the standard computer languages (such as BASIC), the exponential function is denoted by exp (or by EXP). Thus,

$$\exp(x) = e^x.$$

Example   © Using a calculator with an e^x (or exp) key, evaluate

(a) e^1 (b) $e^{\sqrt{2}}$ (c) $e^{-5.0321}$

Solution   Using a 10-digit calculator, we find that

(a) $e^1 = e = 2.718281828$
(b) $e^{\sqrt{2}} = e^{1.414213562} = 4.113250377$
(c) $e^{-5.0321} = 6.525093476 \times 10^{-3} = 0.006525093476$

Because the tangent line to the graph of $y = e^x$ at the point $(0, 1)$ has slope $m = 1$, the slope-intercept equation of the tangent line is $y = 1 + x$ (Figure 4).

Figure 4

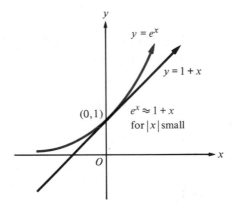

Obviously, the curve $y = e^x$ and the tangent line $y = 1 + x$ are very close together near the point $(0, 1)$; hence, for small values of $|x|$, we have

$$e^x \approx 1 + x,$$

and this approximation becomes more and more accurate as $|x|$ gets smaller and smaller.

Example   © How accurate is the approximation $e^x \approx 1 + x$ if $x = 0.01$?

Solution   Using a calculator, we find that

$$e^{0.01} = 1.010050167.$$

Since

$$1 + 0.01 = 1.010000000,$$

the discrepancy in the approximation $e^{0.01} \approx 1 + 0.01$ first occurs in the fifth decimal place.

Some banks offer savings accounts with interest compounded not quarterly, not weekly, not daily, not hourly, but *continuously*. The formula for continuously compounded interest involves the exponential function. Although the derivation of this formula requires methods studied in calculus, we can derive it informally as follows. We begin with the formula

$$S = P\left(1 + \frac{r}{n}\right)^{nt}$$

for the final value S dollars of a principal P dollars invested for t years at a nominal annual interest rate r compounded n times a year. We're interested in what happens as n gets larger and larger. Let $x = r/n$ and notice that the larger n is, the smaller x is. Using the approximation $e^x \approx 1 + x$ for small x and the fact that $xn = r$, we have

$$S = P\left(1 + \frac{r}{n}\right)^{nt} = P(1 + x)^{nt} \approx P(e^x)^{nt} = Pe^{xnt} = Pe^{rt}.$$

As n becomes larger and larger, $x = r/n$ becomes smaller and smaller, and the approximation

$$S \approx Pe^{rt}$$

becomes more and more accurate. Therefore, for **continuously compounded** interest at a nominal annual rate r, bankers use the formula

$$S = Pe^{rt}$$

for the final value S dollars of a principal P dollars invested for a term of t years.

Example   ⓒ The New Mattoon Savings Bank offers a savings account with continuously compounded interest at a nominal annual rate of 7% (that is, $r = 0.07$). (a) Sketch a graph showing the amount of money S dollars in such an account after t years, $0 \le t \le 20$, if a principal $P = \$100$ is deposited when $t = 0$. (b) What is the final value S dollars of an investment of $P = \$100$ for a term of $t = 20$ years?

Solution   (a) Here $r = 0.07$, $P = 100$, and

$$S = Pe^{rt} = 100e^{0.07t}.$$

The graph is sketched in Figure 5.

Figure 5

(b) When $t = 20$ years,

$$S = 100e^{0.07(20)}$$

$$= 100e^{1.4}$$

$$= \$405.52.$$

2.1 Exponential Growth and Decay

If x and y are variable quantities, we say that y **increases** or **grows exponentially** as a function of x if there are positive constants y_0 and k such that

$$y = y_0 e^{kx}.$$

Similarly, if

$$y = y_0 e^{-kx},$$

we say that y **decreases** or **decays exponentially** as a function of x. Graphs of y as a function of x for exponential growth and exponential decay are shown in Figure 6. Notice that y_0 is the y intercept in both of the graphs; that is, y_0 is the value of y when $x = 0$. (Why?) The constant k, which determines how rapidly the growth or decay takes place, is called the **growth constant** or the **decay constant.**

As we have seen, S dollars in a savings account with continuously compounded interest grows exponentially as a function of time. Since $S = Pe^{rt}$, the growth constant is equal to the nominal annual interest rate r. On the other hand, radioactive materials provide a good example of exponential decay.

Figure 6

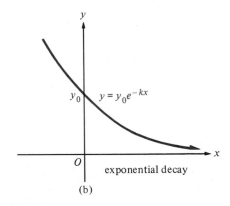

$y = y_0 e^{kx}$

exponential growth

(a)

$y = y_0 e^{-kx}$

exponential decay

(b)

Example   ⓒ Polonium, a radioactive element discovered by Marie Curie in 1898 and named after her native country Poland, decays exponentially. If y_0 grams of polonium are initially present, the number of grams y present after t days is given by

$$y = y_0 e^{-0.005t}.$$

(a) If $y_0 = 5$ grams, sketch a graph showing the amount y grams of polonium left after t days for $0 \leq t \leq 730$. (b) Of a 5-gram sample of polonium, how much is left after 2 years (730 days)?

Solution   (a) The graph of $y = 5e^{-0.005t}$ for $0 \leq t \leq 730$ is sketched in Figure 7.

Figure 7

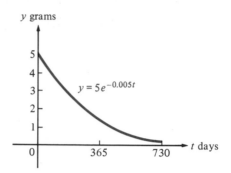

(b) When $t = 730$, we have

$$y = 5e^{-(0.005)(730)} = 5e^{-3.65} \approx 0.13 \text{ gram.}$$

PROBLEM SET 2

© In problems 1–10, use a calculator with an e^x (or exp) key to evaluate the given quantity to as many significant digits as you can.

1. e^{-1} 2. e^{-2} 3. e^3 4. $\exp(0.5)$ 5. $e^{\sqrt{5}}$
6. e^e 7. $e^{-3.11}$ 8. e^{-e} 9. e^π 10. $\exp(1 - \sqrt{2})$

© 11. Using a calculator with a y^x key, sketch accurate graphs of $y = b^x$ for the values of b in Table 1 on page 230. Draw tangent lines to these graphs by eye at the point $(0, 1)$, measure the slopes of the tangent lines, and compare your slopes to the entries in the table.

© 12. Using a calculator with an e^x (or exp) key and a sheet of graph paper (available at your college bookstore), sketch the graph of the exponential function $\exp x = e^x$ as accurately as you can for $-2 \leq x \leq 2$.

© In problems 13–16, use a calculator with an e^x (or exp) key to verify each equation for the indicated values of the variables.

13. $e^x e^y = e^{x+y}$ for $x = \sqrt{2}, \quad y = \sqrt{3}$ 14. $e^{x+y} = e^x e^y$ for $x = \sqrt{5}, \quad y = -\pi$

15. $(e^x)^y = e^{xy}$ for $x = \dfrac{\pi}{2}, \quad y = 1 - \sqrt{3}$ 16. $e^{x-y} = \dfrac{e^x}{e^y}$ for $x = 3.9, \quad y = 2.5$

In problems 17–24, use the graph of $y = e^x$ and the techniques of shifting, stretching, and reflecting to sketch the graph of each function. (Just make a rough sketch—do not use a calculator or tables.)

17. $f(x) = e^x + 1$ 18. $g(x) = e^{x+1}$ 19. $h(x) = -e^x$ 20. $F(x) = -3e^{x+1}$
21. $G(x) = e^{-x}$ 22. $H(x) = e^{1-x}$ 23. $f(x) = e^{-x} + 1$ 24. $g(x) = 2e^{1-x} + 2$

C **25.** How accurate is the approximation $e^x \approx 1 + x$ if:

(a) $x = 0.05$ (b) $x = 0.1$ (c) $x = 0.5$ (d) $x = 1$

26. If n is a large number, show that $\left(1 + \dfrac{1}{n}\right)^n \approx e$. [Hint: Let $x = 1/n$ and use the approximation $e^x \approx 1 + x$.]

C **27.** Suppose that you invest a principal of $P = \$1000$ at a nominal annual interest rate of 10% ($r = 0.1$) for a period of $t = 5$ years. Calculate the final value S of your investment if the interest is compounded (a) monthly and (b) continuously.

C **28.** The concentration C of a drug in a person's circulatory system decreases as the drug is eliminated by the liver and kidneys or absorbed by other organs. Medical researchers often use the equation $C = C_0 e^{-kt}$ to predict the concentration C at a time t hours after the drug is administered, where C_0 is the initial concentration when $t = 0$ and k is a constant depending on the type of drug. If $C_0 = 3$ milligrams per liter and $k = 0.173$, (a) sketch a graph of C as a function of t for $0 \le t \le 4$ hours and (b) find C when $t = 4$ hours.

C **29.** Ecologists have determined that the approximate population N of bears in a certain protected forest area is given by $N = 225e^{0.02t}$, where t is the elapsed time in years since 1977. (a) Sketch a graph showing the bear population N as a function of t for $0 \le t \le 10$ years and (b) estimate the number of bears that will inhabit the region in 1990.

30. If P dollars is invested for $t = 1$ year at a nominal annual interest rate r compounded continuously, the final value S dollars at the end of the year is given by $S = Pe^{r \cdot 1} = Pe^r$. Since P dollars invested for $t = 1$ year at a simple annual interest rate R yields a final value $S = P(1 + R)$ dollars, it follows that the effective simple annual interest rate R corresponding to the continuous nominal annual interest rate r satisfies the equation $P(1 + R) = Pe^r$. (a) Solve for R in terms of r. **C** (b) Find R if $r = 0.07$ (7%).

C **31.** Carbon 14 decays exponentially according to the equation $y = y_0 e^{-0.0001212t}$, where y grams is the amount left after t years and y_0 grams is the initial amount. (a) Sketch a graph of y as a function of t for $0 \le t \le 10{,}000$ years. (b) Of a 10-gram sample of carbon 14, how much will be left after 10,000 years?

C **32.** The electric current I in amperes flowing in a series circuit having an inductance L henrys, a resistance R ohms, and a constant electromotive force E volts (Figure 8) satisfies the equation

$$I = \frac{E}{R} - \frac{E}{R}\exp\left(-\frac{Rt}{L}\right),$$

where t is the time in seconds after the current begins to flow. If $E = 12$ volts, $R = 5$ ohms, and $L = 0.03$ henry, sketch the graph of I as a function of t.

Figure 8

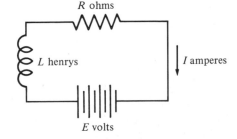

R ohms

L henrys

I amperes

E volts

C **33.** The population N of a small country after t years is given by $N = 2{,}000{,}000e^{0.03t}$. (a) What was the population when $t = 0$? (b) What is the projected population when $t = 20$ years?

34. Find a formula for the present value P of S dollars t years in the future at a nominal annual interest rate r compounded continuously.

C 35. A biologist finds that the number N of bacteria in a culture after t hours is given by $N = 2000e^{0.7t}$. (a) How many bacteria were present when $t = 0$? (b) How many bacteria will be present after $t = 12$ hours? (c) Sketch a graph of N as a function of t for $0 \le t \le 6$ hours. (d) Using the graph in part (c), estimate the time at which $N = 32,000$ bacteria.

C 36. In calculus, it is shown that $e^x \approx 1 + x + (x^2/2)$ gives a better estimate for e^x than does $e^x \approx 1 + x$. Compare the two estimates for $x = 0.01$.

C 37. In 1973, it was projected that in t years the annual consumption of gasoline in the United States would be A billion barrels, where $A = 2.4e^{0.0381t}$. Use this equation to estimate the gasoline consumption of the United States in the year 1990.

---

3   Exponential Equations and Logarithms

If you put \$100 in a savings account at 8% nominal annual interest compounded quarterly, the final value S dollars of your investment after t years is given by

$$S = 100\left(1 + \frac{0.08}{4}\right)^{4t} = 100(1.02)^{4t}.$$

It's natural to ask how long you'll have to wait until your money doubles. To find out, you have to solve the equation

$$200 = 100(1.02)^{4t} \quad \text{or} \quad 1.02^{4t} = 2$$

for t. We'll explain later (in Section 5.1) how to solve this equation, but you can verify using a calculator that the solution is (approximately) $t = 8.75$. In other words, you'll have to wait 8 years and 9 months for your money to double. An equation such as $1.02^{4t} = 2$, in which an unknown appears in an exponent, is called an **exponential equation.**

The simple exponential equations in the following examples can be solved by using the fact that

$$\boxed{\text{if} \quad b > 0, \quad b \neq 1, \quad \text{and} \quad b^x = b^y, \quad \text{then} \quad x = y}$$

(Problem 18).

Examples   Solve each exponential equation.

1 $2^x = 64$

Solution   We begin by expressing 64 as a power of 2 so that both sides of the equation will have the same base. Since $64 = 2^6$, we can rewrite $2^x = 64$ as $2^x = 2^6$, from which it follows that $x = 6$.

2 $36^t = 216^{2t-1}$

Solution   Since $36 = 6^2$ and $216 = 6^3$, we can rewrite the given equation as

$$(6^2)^t = (6^3)^{2t-1} \quad \text{or} \quad 6^{2t} = 6^{3(2t-1)},$$

from which it follows that

$$2t = 3(2t - 1), \quad 2t = 6t - 3, \quad 4t = 3, \quad \text{and} \quad t = \tfrac{3}{4}.$$

If $b > 0$, $b \neq 1$, and $c > 0$, the solution x of the exponential equation

$$b^x = c$$

is denoted by

$$x = \log_b c,$$

which is read "x equals the **logarithm to the base b of c**." In other words,

$\log_b c$ is the power to which you must raise b to obtain c.

Example   Find
(a) $\log_2 16$ (b) $\log_3 27$ (c) $\log_{10} \tfrac{1}{10}$ (d) $\log_e e^5$
(e) $\log_b 1$ for $b > 0$ and $b \neq 1$.

Solution   (a) We ask ourselves, "To what power must we raise 2 to obtain 16?" Since $2^4 = 16$, the answer is 4. Therefore, $\log_2 16 = 4$.
(b) Since $3^3 = 27$, it follows that $\log_3 27 = 3$.
(c) Since $10^{-1} = \tfrac{1}{10}$, it follows that $\log_{10} \tfrac{1}{10} = -1$.
(d) We ask ourselves, "To what power must we raise e to obtain e^5?" The answer is 5, so $\log_e e^5 = 5$.
(e) Since $b^0 = 1$, it follows that $\log_b 1 = 0$.

Using the fact that for $b > 0$, $b \neq 1$, and $c > 0$,

$$x = \log_b c \quad \text{if and only if} \quad b^x = c,$$

you can convert equations from logarithmic form to exponential form and vice versa. For instance:

Logarithmic Form	Exponential Form
$2 = \log_2 4$	$2^2 = 4$
$\log_{10} 10{,}000 = 4$	$10{,}000 = 10^4$
$-\tfrac{1}{2} = \log_{64} \tfrac{1}{8}$	$64^{-1/2} = \tfrac{1}{8}$
$\log_b x = y$	$b^y = x \quad (b > 0,\ b \neq 1,\ x > 0)$
$k = \log_x d$	$x^k = d \quad (x > 0,\ x \neq 1,\ d > 0)$

Notice that whenever you write $\log_b c$, you must make sure that c is positive and that b is positive and not equal to 1.

Examples Solve each equation.

1 $\log_3 x^2 = 4$

Solution The equation $\log_3 x^2 = 4$ is equivalent to $3^4 = x^2$; that is, $x^2 = 81$. The solutions are $x = 9$ and $x = -9$.

2 $\log_x 25 = 2$

Solution The equation $\log_x 25 = 2$ is equivalent to $x^2 = 25$, with the restriction that $x > 0$ and $x \neq 1$; hence, $x = 5$ is the solution.

Using the connection between logarithms and exponents, we can translate properties of exponents into properties of logarithms. Some of these properties are as follows.

Properties of Logarithms

Let M, N, and b be positive numbers, $b \neq 1$, and let y be any real number. Then:

(i) $b^{\log_b N} = N$

(ii) $\log_b b^y = y$

(iii) $\log_b(MN) = \log_b M + \log_b N$

(iv) $\log_b \dfrac{M}{N} = \log_b M - \log_b N$

(v) $\log_b N^y = y \log_b N$

(vi) $\log_b \dfrac{1}{N} = -\log_b N$

Properties (i) and (ii) are direct consequences of the definition of logarithms. We verify Property (iii) here and leave it to you to check Properties (iv), (v), and (vi) (Problem 60). To prove Property (iii), let

$$x = \log_b M \quad \text{and} \quad y = \log_b N,$$

so that

$$b^x = M \quad \text{and} \quad b^y = N.$$

Then,

$$\log_b(MN) = \log_b(b^x b^y) = \log_b(b^{x+y}) = x + y = \log_b M + \log_b N.$$

Examples Use the Properties of Logarithms to work the following.

1 Rewrite each expression as a sum or difference of multiples of logarithms.

(a) $\log_b \dfrac{z}{uv}$

(b) $\log_2 \dfrac{(x^2 + 5)(2x + 5)^{3/2}}{\sqrt[4]{3x + 1}}$

Solution   (a) Assuming that z, u, and v are positive, we have

$$\log_b \frac{z}{uv} = \log_b z - \log_b(uv) \qquad \text{[Property (iv)]}$$

$$= \log_b z - (\log_b u + \log_b v) \qquad \text{[Property (iii)]}$$

$$= \log_b z - \log_b u - \log_b v.$$

(b) Assuming that $2x + 5$ and $3x + 1$ are positive, we have

$$\log_2 \frac{(x^2 + 5)(2x + 5)^{3/2}}{\sqrt[4]{3x + 1}} = \log_2[(x^2 + 5)(2x + 5)^{3/2}] - \log_2 \sqrt[4]{3x + 1}$$

$$= \log_2(x^2 + 5) + \log_2(2x + 5)^{3/2} - \log_2(3x + 1)^{1/4}$$

$$= \log_2(x^2 + 5) + \tfrac{3}{2}\log_2(2x + 5) - \tfrac{1}{4}\log_2(3x + 1),$$

where we applied Property (v) in the last step.

2   Rewrite each expression as a single logarithm.

(a) $2\log_{10} x + 3\log_{10}(x + 1)$ (b) $\log_b\left(x + \dfrac{x}{y}\right) - \log_b\left(z + \dfrac{z}{y}\right)$

Solution   Assuming that all quantities whose logarithms are taken are positive, we have the following:

(a) $2\log_{10} x + 3\log_{10}(x + 1) = \log_{10} x^2 + \log_{10}(x + 1)^3 = \log_{10}[x^2(x + 1)^3]$

(b) $\log_b\left(x + \dfrac{x}{y}\right) - \log_b\left(z + \dfrac{z}{y}\right) = \log_b \dfrac{x + \dfrac{x}{y}}{z + \dfrac{z}{y}} = \log_b \dfrac{\left(1 + \dfrac{1}{y}\right)x}{\left(1 + \dfrac{1}{y}\right)z} = \log_b \dfrac{x}{z}$

3   Suppose that $\log_b 2 = 0.48$ and $\log_b 3 = 0.76$. Find
(a) $\log_b 6$ (b) $\log_b \tfrac{3}{2}$ (c) $\log_b \sqrt[4]{2}$.

Solution   (a) By Property (iii),

$$\log_b 6 = \log_b(2 \cdot 3) = \log_b 2 + \log_b 3 = 0.48 + 0.76 = 1.24.$$

(b) By Property (iv),

$$\log_b \tfrac{3}{2} = \log_b 3 - \log_b 2 = 0.76 - 0.48 = 0.28.$$

(c) By Property (v),

$$\log_b \sqrt[4]{2} = \log_b 2^{1/4} = \tfrac{1}{4}\log_b 2 = \tfrac{1}{4}(0.48) = 0.12.$$

4   Solve each equation:
(a) $\log_{10} x + \log_{10}(x + 21) = 2$
(b) $\log_7(3t^2 - 5t - 2) - \log_7(t - 2) = 1$

Solution   (a) We begin by noticing that x must be positive for $\log_{10} x$ to be defined. If x is positive, so is $x + 21$, and $\log_{10}(x + 21)$ is also defined. Applying Property (iii), we rewrite

$$\log_{10} x + \log_{10}(x + 21) = 2$$

as

$$\log_{10}[x(x + 21)] = 2.$$

The last equation can be rewritten in exponential form as

$$x(x + 21) = 10^2,$$

that is,

$$x^2 + 21x - 100 = 0.$$

Factoring, we have

$$(x + 25)(x - 4) = 0,$$

so $x = -25$ or $x = 4$. Since x must be positive, we can eliminate $x = -25$ as an extraneous root. Therefore, the solution is $x = 4$.

(b) Applying Property (iv), we can rewrite the given equation

$$\log_7(3t^2 - 5t - 2) - \log_7(t - 2) = 1$$

as

$$\log_7 \frac{3t^2 - 5t - 2}{t - 2} = 1,$$

provided that both $3t^2 - 5t - 2$ and $t - 2$ are positive. The last equation can be simplified by reducing the fraction,

$$\frac{3t^2 - 5t - 2}{t - 2} = \frac{(3t + 1)(t - 2)}{t - 2}$$
$$= 3t + 1,$$

so

$$\log_7(3t + 1) = 1,$$

that is,

$$3t + 1 = 7^1 = 7.$$

The solution of this equation is

$$t = (7 - 1)/3 = 2.$$

We must check this answer against the original restrictions on the variables. However, if $t = 2$, then $t - 2 = 0$ and $\log_7(t - 2)$ is undefined. Thus, $t = 2$ is an extraneous root, and the original equation has no solution.

5   Derive the **base-changing formula:** If $a > 0$, $b > 0$, $a \neq 1$, $b \neq 1$, and $c > 0$, then

$$\log_a c = \frac{\log_b c}{\log_b a}.$$

Solution   Let

$$x = \log_a c.$$

Then,

$$a^x = c,$$

and it follows that

$$\log_b a^x = \log_b c.$$

Using Property (v), the last equation can be rewritten as

$$x \log_b a = \log_b c,$$

or

$$x = \frac{\log_b c}{\log_b a}.$$

Therefore,

$$\log_a c = \frac{\log_b c}{\log_b a}.$$

PROBLEM SET 3

In problems 1–12, solve each exponential equation.

1. $2^x = 8$
2. $3^{x^2} = 81$
3. $25^x = 5$
4. $2^{x^3} = 256$
5. $3^{2x+1} = 27$
6. $(1/10)^{4x} = 1000$
7. $3^{2-8x} = 9^{3x+1}$
8. $8^{3t} = 32^{4t-1}$
9. $5^{x^2+x} = 25$
10. $7^{x^2+x} = 1$
11. $3^{2t} - 3^t - 6 = 0$ [Hint: Let $x = 3^t$.]
12. $2^{2x+1} + 2^x = 10$

13. Find:
   (a) $\log_2 4$
   (b) $\log_2 8$
   (c) $\log_3 81$
   (d) $\log_9 9^5$
   (e) $\log_3 \frac{1}{9}$
   (f) $\log_8 \frac{1}{64}$
   (g) $\log_{10} 100,000$

14. Find:
   (a) $\log_2 \frac{1}{4}$
   (b) $\log_3 \sqrt{3}$
   (c) $\log_9 1$
   (d) $\log_e e^\pi$
   (e) $\log_2 4^3$
   (f) $\log_3 9^{-0.5}$
   (g) $\log_{10} \dfrac{1}{100,000}$

15. Rewrite each logarithmic equation as an equivalent exponential equation:
   (a) $\log_2 32 = 5$
   (b) $\log_{16} 2 = \frac{1}{4}$
   (c) $\log_9 \frac{1}{3} = -\frac{1}{2}$
   (d) $\log_e e = 1$
   (e) $\log_{\sqrt{3}} 9 = 4$
   (f) $\log_{10} 10^n = n$
   (g) $\log_x x^5 = 5$

16. Give a geometric argument based on the graph of $f(x) = b^x$ to show that if $b > 0$, $b \neq 1$, and $c > 0$, then the exponential equation $b^x = c$ has exactly one solution.

17. Rewrite each exponential equation as an equivalent logarithmic equation:
   (a) $8^0 = 1$
   (b) $10^{-4} = 0.0001$
   (c) $4^4 = 256$
   (d) $27^{-1/3} = \frac{1}{3}$
   (e) $8^{2/3} = 4$
   (f) $a^c = y$

18. Using the result of problem 16, show that if $b > 0$, $b \neq 1$, and x and y are real numbers such that $b^x = b^y$, it follows that $x = y$.

In problems 19–42, solve each equation.

19. $x = \log_6 36$
20. $\log_5 x = 2$
21. $\log_x 125 = 3$
22. $x = \log_3 729$
23. $\log_7 x = 1$
24. $\log_3 x = 1$
25. $\log_x 16 = -\frac{4}{3}$
26. $\log_{\sqrt{3}} x = 6$
27. $\log_{\sqrt{2}} x = -6$

28. $\log_x \frac{27}{8} = -\frac{3}{2}$
29. $x = \log_{10} 10^{-7}$
30. $x = \log_e e^{-0.01}$
31. $x = \log_{27} \frac{1}{9}$
32. $x = \log_2(\log_4 256)$
33. $x = \log_5 \sqrt[4]{5}$
34. $x = \log_{3/4} \frac{4}{3}$
35. $\log_2(2x - 1) = 3$
36. $\log_5(2x - 3) = 2$
37. $\log_3(3x - 4) = 4$
38. $\log_7(2x - 7) = 0$
39. $\log_2(t^2 + 3t + 4) = 1$
40. $\log_5(y^2 - 4y) = 1$
41. $\log_4(9u^2 + 6u + 1) = 2$
42. $\log_3|3 - 2t| = 2$

In problems 43–50, rewrite each expression as a sum or difference of multiples of logarithms. (Make the necessary assumptions about the values of the variables.)

43. $\log_b[x(x + 1)]$
44. $\log_a(x^4\sqrt{y})$
45. $\log_{10}[x^2(x + 1)]$
46. $\log_c \sqrt{\dfrac{x}{x + 7}}$
47. $\log_3 \dfrac{x^3 y^2}{z}$
48. $\log_e \dfrac{t(t + 1)}{(t + 2)^3}$
49. $\log_e \sqrt{x(x + 3)}$
50. $\log_b \sqrt[3]{(x + 1)^2}\sqrt{x + 7}$

In problems 51–58, rewrite each expression as a single logarithm. (Make the necessary assumptions about the values of the variables.)

51. $2\log_3 x + 7\log_3 x$
52. $\log_{10} \dfrac{a^3}{b} + \log_{10} \dfrac{b^2}{5a}$
53. $\frac{1}{2}[\log_5 a - \log_5 3b]$
54. $\log_x \dfrac{y^5}{z^4} - \log_x \dfrac{y^3}{z^2}$
55. $\log_e \dfrac{x}{x - 1} + \log_e \dfrac{x^2 - 1}{x}$
56. $\log_b \dfrac{x + y}{z} - \log_b \dfrac{1}{x + y}$
57. $\log_3 \dfrac{x^2 + 14x - 15}{x^2 + 4x - 5} - \log_3 \dfrac{x^2 + 12x - 45}{x^2 + 6x - 27}$
58. $\log_e \dfrac{m^2 - 2m - 24}{m^2 - m - 30} + \log_e \dfrac{(m + 5)^2}{m^2 - 16}$

59. Suppose that $\log_b 2 = 0.53$, $\log_b 3 = 0.83$, $\log_b 5 = 1.22$, and $\log_b 7 = 1.48$. Find:
(a) $\log_b 21$ (b) $\log_b 35$ (c) $\log_b \frac{2}{7}$ (d) $\log_b \frac{35}{3}$
(e) $\log_b \sqrt{7}$ (f) $\log_b \sqrt[3]{42}$ (g) $\log_b 3\sqrt{8}$
Round off all answers to two decimal places.
60. Verify Properties (iv), (v), and (vi) on page 238.

In problems 61–68, solve each equation.

61. $\log_4 x + \log_4(x + 6) = 2$
62. $\log_{10} x + \log_{10}(x + 3) = 1$
63. $\log_7 x + \log_7(18x + 61) = 1$
64. $\log_2 x + \log_2(x - 2) = \log_2(9 - 2x)$
65. $\log_3(x^2 + x) - \log_3(x^2 - x) = 1$
66. $\log_5(4x^2 - 1) = 2 + \log_5(2x + 1)$
67. $\log_8(x^2 - 9) - \log_8(x + 3) = 2$
68. $2\log_2 x - \log_2(x - 1) = 2$

69. Suppose that $a > 0, b > 0, a \neq 1$, and $b \neq 1$. Using the base-changing formula (Example 5, page 240) and the fact that $\log_b b = 1$, show that

$$\log_a b = \frac{1}{\log_b a}.$$

70. Derive the following alternative base-changing formula: If $a > 0, b > 0, a \neq 1, b \neq 1$, and $c > 0$, then

$$\log_a c = (\log_a b)(\log_b c).$$

4 Logarithmic Functions

A function F of the form

$$F(x) = \log_b x,$$

where $b > 0$ and $b \neq 1$, is called a **logarithmic function with base b.** The domain of F is the interval $(0, \infty)$ of all positive real numbers. If we let

$$f(x) = b^x,$$

then, because of Properties (i) and (ii) on page 238,

$$b^{\log_b x} = x \quad \text{for } x > 0, \quad \text{and} \quad \log_b b^x = x \quad \text{for } x \text{ in } \mathbb{R}.$$

Therefore,

$$f[F(x)] = x \quad \text{for } x > 0, \quad \text{and} \quad F[f(x)] = x \quad \text{for } x \text{ in } \mathbb{R}.$$

In other words, the functions f and F are inverses of each other. (You may wish to review the idea of inverse functions in Section 8 of Chapter 3.) It follows that the graph of the logarithmic function $F(x) = \log_b x$ is the mirror image of the graph of the exponential function $f(x) = b^x$ across the line $y = x$.

Example   Sketch the graphs of (a) $F(x) = \log_3 x$ and (b) $G(x) = \log_{1/3} x$.

Solution   Graphs of the functions $f(x) = 3^x$ and $g(x) = (\frac{1}{3})^x$ were shown in Figure 2 of Section 1. By reflecting these graphs across the line $y = x$, we obtain the graphs of $F(x) = \log_3 x$ (Figure 1a) and $G(x) = \log_{1/3} x$ (Figure 1b). Notice that the graph of G can be obtained by reflecting the graph of F across the x axis. (For the reason why, see Problem 42.)

Figure 1

(a)

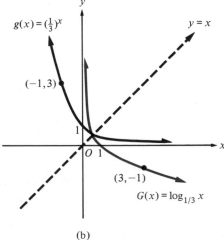

(b)

In general, graphs of logarithmic functions have the characteristic shapes shown in Figure 2. Thus, if $b > 1$, the graph of $F(x) = \log_b x$ rises to the right (Figure 2a), whereas if $0 < b < 1$, the graph falls to the right (Figure 2b).

Notice that the graph of $F(x) = \log_b x$ always contains the point $(b, 1)$ (because $\log_b b = 1$) and that its x intercept is always 1 (because $\log_b 1 = 0$). There is no y intercept; in fact, the y axis is a vertical asymptote of the graph. The range of $F(x) = \log_b x$ is $\mathbb{R}$.

Figure 2

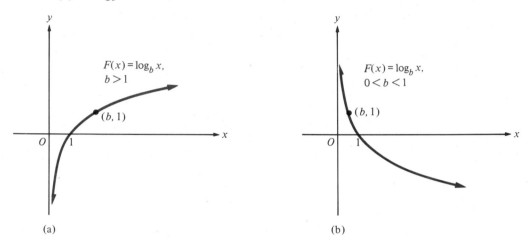

(a)   (b)

4.1 The Common and Natural Logarithm Functions

Before the development of electronic calculators and computers, logarithms were extensively used to facilitate numerical calculations. Because the usual positional system for writing numerals is based on 10, arithmetic calculation is easiest when logarithms with base 10 are used. Logarithms with base 10 are called *common logarithms*, and the symbol "log x" (with no subscript) is often used as an abbreviation for $\log_{10} x$. Thus, the **common logarithm function** is defined by

$$\log x = \log_{10} x \quad \text{for } x > 0.$$

These days, because of the wide availability of inexpensive and reliable electronic calculators, the common logarithm function is rarely used for purposes of numerical calculation. However, it still has many applications, ranging from the measurement of pH in chemistry to the measurement of sound pollution in the health sciences (see Section 5). For this reason, most scientific calculators have both a 10^x key and a log key.

Example 1   ⓒ Use a calculator with a log key to evaluate

(a) $\log 2110$ (b) $\log 0.004326$

Solution   On a 10-digit calculator, we obtain:

(a) $\log 2110 = 3.324282455$ (b) $\log 0.004326 = -2.363913485$

Example 2   ⓒ Use a calculator to verify that $10^{\log x} = x$ for $x = \pi$.

Solution   Rounded off to 9 decimal places,

$$\pi = 3.141592654$$

and

$$\log \pi = 0.497149873.$$

Now,

$$10^{0.497149873} = 3.141592654 = \pi,$$

confirming that

$$10^{\log \pi} = \pi.$$

If a calculator isn't available, you can use Appendix Table IB to find common logarithms. Appendix I explains how.

In advanced mathematics and its applications, many otherwise cumbersome formulas become much simpler if logarithmic and exponential functions with base $e \approx 2.718$ are used (see Section 2). Logarithms with base e are called *natural logarithms* and the symbol "$\ln x$" is often used as an abbreviation for $\log_e x$. Thus, the **natural logarithm function** is defined by

$$\ln x = \log_e x \quad \text{for } x > 0.$$

In other words, for $x > 0$,

$$y = \ln x \quad \text{if and only if} \quad e^y = x.$$

Of course, all scientific calculators have an ln key.

Example 1   ⓒ Use a calculator with an ln key to evaluate
(a) $\ln 7124$ (b) $\ln 0.05319$

Solution   On a 10-digit calculator, we obtain:
(a) $\ln 7124 = 8.871224644$ (b) $\ln 0.05319 = -2.933884870$

Example 2   ⓒ Use a calculator to verify that $\ln e^x = x$ for $x = \sqrt{5}$.

Solution   Rounded off to 9 decimal places,

$$\sqrt{5} = 2.236067977$$

and

$$e^{\sqrt{5}} = 9.356469012.$$

Now,

$$\ln 9.356469012 = 2.236067977 = \sqrt{5},$$

confirming that

$$\ln e^{\sqrt{5}} = \sqrt{5}.$$

If a calculator isn't available, you can use Appendix Table IA to find natural logarithms. Alternatively, you can use Appendix Table IB for common logarithms and the formula

$$\ln x = M \log x,$$

where $M = \ln 10 = 2.302585093$ (Problem 16).

Because the natural logarithm function is the inverse of the exponential function, the graph of

$$y = \ln x$$

can be obtained by reflecting the graph of $y = e^x$ across the line $y = x$ (Figure 3). Recall that the tangent line to the graph of $y = e^x$ at $(0,1)$ has slope $m = 1$ (Section 2, Figure 3). It follows that the tangent line to the graph of $y = \ln x$ at $(1,0)$ also has slope $m = 1$. If you keep this fact in mind whenever you sketch the graph of the natural logarithm function, you will obtain a more accurate graph.

Although common and natural logarithm functions are sufficient for most purposes, there are situations in which logarithms with bases other than 10 and e are useful. For instance, in communications engineering and computer science, the bases 2 and 8 are often used. Since scientific calculators ordinarily have keys only for log and ln, you must use the *base-changing formula*

$$\log_a c = \frac{\log c}{\log a}$$

Figure 3

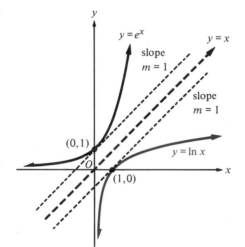

if you want to calculate logarithms to other bases. (See Example 5 on page 240.)

Example   © Use a calculator and the base-changing formula to find $\log_2 3$.

Solution
$$\log_2 3 = \frac{\log 3}{\log 2} = \frac{0.477121255}{0.301029996} = 1.584962500.$$

4.2   Graphs of Functions Involving Logarithms

In dealing with functions involving logarithms, you must keep in mind that logarithms of negative numbers and zero are undefined.

Example   Find the domain of (a) $h(x) = \log_2(x + 1)$ and (b) $g(x) = \log x^2$.

Solution   (a) $\log_2(x + 1)$ is defined if and only if $x + 1 > 0$; that is, $x > -1$. Therefore, the domain of h is the interval $(-1, \infty)$.

(b) $\log x^2$ is defined if and only if $x^2 > 0$, that is, $x \neq 0$. Therefore, the domain of g is the set of all nonzero real numbers.

The techniques of graph sketching presented in Section 6 of Chapter 3 can be applied to functions involving logarithms. As usual, accuracy is enhanced by plotting more points, and you may wish to use a calculator to find coordinates of such points quickly.

Examples   © Determine the domain and range of each function; find any y or x intercepts and any horizontal or vertical asymptotes of its graph, indicate where the function is increasing or decreasing, and sketch the graph.

1 $H(x) = \log_3(-x)$

Solution   The domain of H consists of all values of x for which $-x > 0$, that is, the interval $(-\infty, 0)$. Because $H(0)$ is undefined, there is no y intercept. The x intercept is the solution of the equation $H(x) = 0$; that is, $\log_3(-x) = 0$. Rewriting the last equation in exponential form, we obtain

$$-x = 3^0 = 1,$$

so the x intercept is $x = -1$. The graph of $H(x) = \log_3(-x)$ (Figure 4) is the mirror image of the graph of $F(x) = \log_3 x$ (Figure 1a) across the y axis. From the graph, we see that the range of H is the set $\mathbb{R}$ of all real numbers, and that the function H is decreasing over its entire domain. The y axis is a vertical asymptote, and there is no horizontal asymptote.

Figure 4

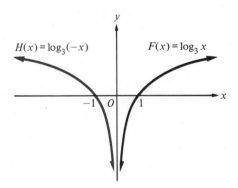

2 $f(x) = 1 + \ln(x - 2)$

Solution   The domain of f is the interval $(2, \infty)$; hence, $f(0)$ is undefined and there is no y intercept. The x intercept is the solution of the equation $f(x) = 0$, that is,

$$1 + \ln(x - 2) = 0 \quad \text{or} \quad \ln(x - 2) = -1.$$

Rewriting the last equation in exponential form, we obtain

$$x - 2 = e^{-1},$$

so the x intercept is $x = 2 + e^{-1} \approx 2.37$. The graph of $f(x) = 1 + \ln(x - 2)$

(Figure 5) is obtained by shifting the graph of $y = \ln x$ (Figure 3) 1 unit upward and 2 units to the right. From the graph, we see that the range of f is the set $\mathbb{R}$ of all real numbers, and that the function f is increasing over its entire domain. Because the y axis is a vertical asymptote of the graph of $y = \ln x$, it follows that the line $x = 2$ is a vertical asymptote of the graph of f. There is no horizontal asymptote.

Figure 5

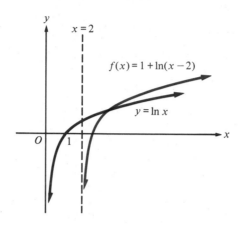

PROBLEM SET 4

In problems 1–6, sketch the graph of each logarithmic function by reflecting the graph of its inverse exponential function across the line $y = x$.

1. $f(x) = \log_2 x$ 2. $g(x) = \log_4 x$ 3. $F(x) = \log_{1/2} x$
4. $G(x) = \log_{1/4} x$ 5. $h(x) = \log_5 x$ 6. $H(x) = \log_{1/5} x$

[C] 7. Use a calculator with a log key to evaluate:
   (a) $\log 6.373$ (b) $\log 1230.4$ (c) $\log 0.03521$
   (d) $\log(3.047 \times 10^{11})$ (e) $\log(6.562 \times 10^{-9})$

8. Use Appendix Table IB to evaluate:
   (a) $\log 2.74$ (b) $\log 0.00333$ (c) $\log 3470$
   (d) $\log(9.09 \times 10^{21})$ (e) $\log(3.11 \times 10^{-13})$

[C] 9. Using a calculator, verify that $\log(xy) = \log x + \log y$ for (a) $x = 31.27$, $y = 5.246$ and
   (b) $x = \pi$, $y = \sqrt{2}$

[C] 10. Using a calculator, verify that (a) $\log 10^{\sqrt{7}} = \sqrt{7}$ and (b) $10^{\log \sqrt{7}} = \sqrt{7}$.

[C] 11. Use a calculator with an ln key to evaluate:
   (a) $\ln 4126$ (b) $\ln 2.704$ (c) $\ln 0.040404$
   (d) $\ln(7.321 \times 10^8)$ (e) $\ln(1.732 \times 10^{-7})$

12. Use Appendix Table IA to evaluate:
   (a) $\ln 2.68$ (b) $\ln 3.33$ (c) $\ln 25$ [Hint: $25 = 5^2$.]

[C] 13. Using a calculator, verify that (a) $\ln e^{\pi} = \pi$ and (b) $e^{\ln \pi} = \pi$.

[C] 14. Using a calculator, verify that $\ln y^x = x \ln y$ for $x = 77.01$ and $y = 3.352$.

[C] 15. Using a calculator and the base-changing formula, evaluate:
   (a) $\log_2 25$ (b) $\log_3 2$ (c) $\log_8 e$
   (d) $\log_{\pi} 5$ (e) $\log_{\sqrt{2}} 0.07301$

16. Show that, for $x > 0$, $\ln x = M \log x$, where $M = \ln 10$.

17. Find the domain of each function.

(a) $f(x) = \log(x - 2)$

(b) $g(x) = \log_2 \sqrt{x}$

(c) $h(x) = \log_8(4 - x)$

(d) $H(x) = \log|x|$

(e) $K(x) = \dfrac{1}{\ln x}$

(f) $G(x) = \log(x^2 - 5x + 7)$

(g) $k(x) = \ln(x^2 + 1)$

18. Find any y and x intercepts of the graph of each function in problem 17.

© In problems 19–32, determine the domain and range of each function; find any y or x intercepts and any horizontal or vertical asymptotes of its graph; indicate where the function is increasing or decreasing; and sketch the graph.

19. $f(x) = \log_3(x + 2)$

20. $g(x) = \log_2|x|$

21. $h(x) = (\ln x) - 1$

22. $F(x) = |\log_2 x|$

23. $G(x) = \ln(1 - x)$

24. $H(x) = \ln\sqrt{x}$

25. $f(x) = 2 + \ln(x - 1)$

26. $g(x) = (\ln x)^{-1}$

27. $h(x) = 2 - \log(1 - x)$

28. $F(x) = \log\dfrac{1}{x}$

29. $G(x) = \log x^2$

30. $H(x) = \log_{1/4} x^3$

31. $f(x) = \log_{1/3}(2 - x)$

32. $g(x) = \log_{1/5}|1 - x|$

In problems 33–40, justify each property of the natural logarithm function by using either its definition or the Properties of Logarithms given in Section 3.

33. $\ln 1 = 0$

34. $\ln\dfrac{1}{x} = -\ln x, \quad x > 0$

35. $\ln e = 1$

36. $\ln x = \dfrac{\log x}{\log e}, \quad x > 0$

37. $\ln xy = \ln x + \ln y, \quad x > 0, \quad y > 0$

38. $\ln x = \dfrac{1}{\log_x e}, \quad x > 0, \quad x \neq 1$

39. $\ln\dfrac{x}{y} = \ln x - \ln y, \quad x > 0, \quad y > 0$

40. $\ln y^x = x \ln y, \quad y > 0, \quad x \text{ in } \mathbb{R}$

41. Explain why every real number y can be written in the form $y = \ln x$ for a suitable value of x.

42. If $b > 0$ and $b \neq 1$, show that $\log_{1/b} x = -\log_b x$ holds for $x > 0$.

43. Sketch the graphs of $y = \log_2 x$, $y = \log_3 x$, and $y = \log_4 x$ on the same coordinate system. Describe the relationships among these graphs.

44. (a) Using the data in Table 1, Section 2, page 230, sketch a graph of m as a function of b.
(b) By looking at the graph in part (a), guess what function this is.

45. The function $F(x) = \ln|x|$ is used quite often in calculus. Sketch a graph of this function.

46. Solve the equation $\log_x(2x)^{3x} = 4^{\log_4 4x}$.

47. Find the value of b if the graph of $y = \log_b x$ contains the point $(\frac{1}{2}, -1)$.

48. If $a > 0$, $b > 0$, and $b \neq 1$, find a base c such that $a\log_b x = \log_c x$ holds for all $x > 0$.

49. Using the fact that the tangent line to the graph of $y = \ln x$ at $(1, 0)$ has slope $m = 1$, explain why $\ln x \approx x - 1$ for values of x close to 1, with the approximation becoming more and more accurate as x comes closer and closer to 1.

50. Show that $\ln(1 + x) \approx x$ if $|x|$ is small and that the approximation becomes more and more accurate as $|x|$ gets smaller and smaller. [Hint: Use the result of problem 49.]

51. The formula $y^x = e^{x \ln y}$ for $y > 0$ is used in calculus to write y^x in terms of the exponential and natural logarithm functions. Derive this formula. [Hint: $x \ln y = \ln y^x$.]

© **52.** Using a calculator with y^x, e^x, and $\ln$ keys, check the formula in problem 51 for $x = 1.59$ and $y = 7.47$.

53. Criticize the following statement: Since $\ln x^2 = 2 \ln x$, the graph of $y = \ln x^2$ can be found by doubling all ordinates on the graph of $y = \ln x$.

5   Applications of Exponential and Logarithmic Functions

In this and the next section, we present a small sample of the many and varied applications of exponential and logarithmic functions.

5.1   Solving Exponential Equations

You can often solve an exponential equation by taking the logarithm of both sides and using the Properties of Logarithms to simplify the resulting equation. For this purpose, you can use either common or natural logarithms.

Example   [c] Solve the exponential equation $7^{2x+1} = 3^{x-2}$.

Solution

$$7^{2x+1} = 3^{x-2}$$
$$\log 7^{2x+1} = \log 3^{x-2}$$
$$(2x+1)\log 7 = (x-2)\log 3$$
$$(2\log 7)x + \log 7 = (\log 3)x - 2\log 3$$
$$(2\log 7 - \log 3)x = -\log 7 - 2\log 3$$
$$(\log 7^2 - \log 3)x = -(\log 7 + \log 3^2)$$
$$(\log 49 - \log 3)x = -(\log 7 + \log 9)$$
$$(\log \tfrac{49}{3})x = -\log 63$$
$$x = \frac{-\log 63}{\log \tfrac{49}{3}}$$

Therefore, using a 10-digit calculator, we find that

$$x = \frac{-\log 63}{\log \tfrac{49}{3}} = \frac{-1.799340549}{1.213074825} = -1.483289004.$$

The following example shows how to answer the question that we raised in the introduction to Section 3.

Example   [c] If you put \$100 in a savings account at 8% nominal annual interest compounded quarterly, how long will it take for your money to double?

Solution   The final value S dollars of your investment after t years is given by

$$S = 100\left(1 + \frac{0.08}{4}\right)^{4t} = 100(1.02)^{4t}.$$

If t is the time required to double your money, then

$$200 = 100(1.02)^{4t} \quad \text{or} \quad 1.02^{4t} = 2.$$

Taking the logarithm of both sides of the last equation, we get

$$4t \log 1.02 = \log 2$$

so that

$$t = \frac{\log 2}{4 \log 1.02} \approx 8.75 \text{ years.}$$

5.2 Applications in Chemistry, Earth Sciences, Psychophysics, and Physics

Measuring pH in Chemistry   In chemistry, the **pH** of a substance is defined by

$$pH = -\log[H^+],$$

where $[H^+]$ is the concentration of hydrogen ions in the substance, measured in moles per liter. The pH of distilled water is 7. A substance with a pH of less than 7 is known as an *acid*, whereas a substance with a pH of greater than 7 is called a *base*.

Environmentalists constantly monitor the pH of rain and snow because of the destructive effects of "acid rain" caused largely by sulfur dioxide emissions from factories and coal-burning power plants. Because of dissolved carbon dioxide from the atmosphere, rain and snow have a natural concentration of $[H^+] = 2.5 \times 10^{-6}$ moles per liter.

Example   © Find the natural pH of rain and snow.

Solution
$$pH = -\log[H^+] = -\log(2.5 \times 10^{-6}) \approx -(-5.6) = 5.6.$$

Measuring Altitude: The Barometric Equation   The **barometric equation**

$$h = (30T + 8000) \ln \frac{P_0}{P}$$

relates the height h in meters above sea level, the air temperature T in degrees Celsius, the atmospheric pressure P_0 in centimeters of mercury at sea level, and the atmospheric pressure P in centimeters of mercury at height h. The altimeters most commonly used in aircraft measure the atmospheric pressure P and display the altitude by means of a scale calibrated according to the barometric equation.

Example   Atmospheric pressure at the summit of Pike's Peak in Colorado on a certain day measures 44.7 centimeters of mercury. If the average air temperature is 5° Celsius and the atmospheric pressure at sea level is 76 centimeters of mercury, find the height of Pike's Peak.

Solution   We use the barometric equation with $T = 5$, $P_0 = 76$, and $P = 44.7$ to obtain

$$h = [30(5) + 8000] \ln \frac{76}{44.7} = 8150 \ln \frac{76}{44.7}$$

$$\approx 4330 \text{ meters.}$$

Measuring Sensation   In 1860, the German physicist Gustav Fechner (1801–87) published a psychophysical law relating the intensity S of a sensation to the intensity P of the physical stimulus causing it. This law, which was based on experiments originally reported in 1829 by the German physiologist Ernst Weber (1795–1878), states that the change in S caused by a small change in P is proportional not to the change in P as one might suppose, but rather to the *percentage* of change in P. Using calculus, it can be shown, as a consequence, that S varies linearly as the natural logarithm of P, so that

$$S = A + B \ln P,$$

where A and B are suitable constants.

Suppose that P_0 denotes the **threshold intensity** of the physical stimulus; that is, the largest value of the intensity P for which there is no sensation. Then, substituting 0 for S and P_0 for P in the equation above, we find that

$$0 = A + B \ln P_0 \quad \text{or} \quad A = -B \ln P_0.$$

It follows that

$$S = -B \ln P_0 + B \ln P = B(\ln P - \ln P_0) = B \ln \frac{P}{P_0}.$$

If you prefer to write the relationship between S and P in terms of the common logarithm, use the equation

$$\ln \frac{P}{P_0} = M \log \frac{P}{P_0}, \qquad \text{where } M = \ln 10$$

(see Problem 16 on page 248) to rewrite $S = B \ln \dfrac{P}{P_0}$ as

$$S = MB \log \frac{P}{P_0}.$$

Finally, letting $C = MB$, you will obtain the **Weber–Fechner law** in the form

$$S = C \log \frac{P}{P_0}.$$

Choice of the constant C determines the units in which S is measured.

Early in the development of the telephone, it became necessary to have a unit to measure the loudness of telephone signals at various points in the network. The proposed unit, called the *bel* in honor of Alexander Graham Bell, the inventor of the telephone, is obtained by taking the constant $C = 1$ in the Weber–Fechner law. (The power P of the electrical signal is measured in watts.) It turns out that $\frac{1}{10}$ of a bel, called a **decibel** is roughly the smallest noticeable difference in the loudness of two sounds. For this reason, loudness is commonly measured in decibels rather than in bels. Thus, in electronics, an amplifier with an input of P_0 watts and an output of P watts is said to provide a *gain* of

$$S = 10 \log \frac{P}{P_0} \text{ decibels.}$$

Decibels are used to measure the loudness of sound produced by any source, not just a telephone receiver or an electronic amplifier. The intensity P of the air vibrations that produce the sensations of sound is commonly measured by the power (in watts) per square meter of wavefront. The threshold of human hearing is approximately

$$P_0 = 10^{-12} \text{ watt per square meter}$$

at the eardrum, and the formula

$$S = 10 \log \frac{P}{P_0}$$

thus gives the loudness in decibels produced by a sound wave of intensity P watts per square meter at the eardrum. The rustle of leaves in a gentle breeze corresponds to about 20 decibels, while loud thunder may reach 110 decibels or more. Sound in excess of 120 decibels can cause pain. In order to prevent "boilermaker's deafness," the American Academy of Ophthalmology and Otolaryngology recommends that no worker be exposed to a continuous sound level of 85 decibels or more for 5 hours or more per day without protective devices.

Example   © Suppose that an employee in a factory must work in the vicinity of a machine producing 60 decibels of sound during an 8-hour shift. A second machine producing 70 decibels is moved into the same vicinity. Does the worker now need ear protection?

Solution   We can add sound intensities, but not decibels; thus, a combination of 60 decibels and 70 decibels does *not* amount to 130 decibels. In fact, suppose that P_1 and P_2 are the sound-wave intensities in watts per square meter produced by the two machines separately. Then the sound-wave intensity P produced by the machines operating together is given by $P = P_1 + P_2$. We know that

$$10 \log \frac{P_1}{P_0} = 60 \quad \text{and} \quad 10 \log \frac{P_2}{P_0} = 70,$$

so that

$$\log \frac{P_1}{P_0} = 6 \quad \text{and} \quad \log \frac{P_2}{P_0} = 7;$$

that is,

$$\frac{P_1}{P_0} = 10^6 \quad \text{and} \quad \frac{P_2}{P_0} = 10^7$$

or

$$P_1 = 10^6 P_0 \quad \text{and} \quad P_2 = 10^7 P_0.$$

It follows that

$$P = P_1 + P_2 = 10^6 P_0 + 10^7 P_0,$$

and the loudness of the sound produced by the machines operating together is

$$S = 10 \log \frac{P}{P_0} = 10 \log \frac{10^6 P_0 + 10^7 P_0}{P_0} = 10 \log(10^6 + 10^7)$$

$$= 10 \log 11{,}000{,}000 \approx 70.4 \text{ decibels,}$$

well below the 85-decibel limit. Ear protection will not be needed.

Measuring the Rate of Radioactive Decay   As we mentioned in Section 2.1, radioactive materials decay according to the formula

$$y = y_0 e^{-kt},$$

where y is the mass of the material at time t and y_0 was the original mass when $t = 0$.

Let's consider how much the mass y changes when t changes by one unit of time. The change is

$$y_0 e^{-kt} - y_0 e^{-k(t+1)} = y_0(e^{-kt} - e^{-kt-k})$$

$$= y_0(e^{-kt} - e^{-kt}e^{-k})$$

$$= y_0 e^{-kt}(1 - e^{-k}).$$

The percentage of the change in mass in one unit of time is therefore given by

$$\frac{\text{change in mass}}{\text{original mass}} \times 100\% = \frac{y_0 e^{-kt}(1 - e^{-k})}{y_0 e^{-kt}} \times 100\%$$

$$= (1 - e^{-k}) \times 100\%.$$

In other words, if the percentage of the change in mass in one unit of time is expressed as a decimal K, we have

$$K = 1 - e^{-k} \quad \text{or} \quad e^{-k} = 1 - K,$$

so that

$$-k = \ln(1 - K) \quad \text{or} \quad k = -\ln(1 - K).$$

For small values of K, it can be shown (Problem 20) that $k \approx K$.

The rate of decay of a radioactive material can be measured not only in terms of k or K, but in terms of its **half-life,** which is defined to be the period of time T required for it to decay to half of its original mass. Thus,

$$\tfrac{1}{2}y_0 = y_0 e^{-kT} \quad \text{or} \quad \tfrac{1}{2} = e^{-kT};$$

that is,

$$e^{kT} = 2 \quad \text{or} \quad kT = \ln 2.$$

It follows that

$$T = \frac{\ln 2}{k}.$$

Example   [C] Potassium-42 is a radioactive element that is often used as a tracer in biological experiments. Its half-life is approximately 12.5 hours. (a) If y_0 milligrams of potassium-42 are initially present, write a formula for the number of milligrams y present after t hours. (b) What percent decrease in the amount of potassium-42 occurs in a sample during 1 hour? (c) Of a 1-milligram sample of potassium-42, how much is left after 5 hours?

Solution   (a) $T = 12.5$; hence, from the equation $T = \dfrac{\ln 2}{k}$, we have

$$k = \frac{\ln 2}{T} = \frac{\ln 2}{12.5} = 0.05545.$$

Therefore,

$$y = y_0 e^{-0.05545t}.$$

(b) Here $k = 0.05545$, so

$$K = 1 - e^{-k} = 1 - e^{-0.05545} \approx 0.0539.$$

Thus, the percentage of decrease per hour is about 5.39%.

(c) Putting $y_0 = 1$ milligram and $t = 5$ hours in the equation of part (a), we find that the amount left after 5 hours is

$$y = 1 \cdot e^{(-0.05545)(5)} = e^{-0.27725} \approx 0.758 \text{ milligram}.$$

PROBLEM SET 5

[C] In problems 1–6, use logarithms to solve each exponential equation.

1. $4^x = 3$
2. $2^{-x} = 5$
3. $7.07^x = 2001$
4. $8.97^x = 7.27^{x-1}$
5. $4^{2x+1} = 6^x$
6. $2^{3x} = 3^{2x-1}$

[C] 7. If you invest $1000 in a savings account at 6% nominal annual interest compounded weekly, how long will it take for your money to double?

8. If a sum of money is invested at a nominal annual interest rate r compounded continuously, show that it doubles in $(\ln 2)/r$ years.

[C] 9. Find the pH of each substance (rounded off to two significant digits).
   (a) eggs: $[H^+] = 1.6 \times 10^{-8}$ moles/liter
   (b) tomatoes: $[H^+] = 6.3 \times 10^{-5}$ moles/liter
   (c) milk: $[H^+] = 4 \times 10^{-7}$ moles/liter

[C] 10. Find the hydrogen ion concentration $[H^+]$ in moles/liter of each substance (rounded off to two significant digits).
   (a) vinegar: pH = 3.1
   (b) beer: pH = 4.3
   (c) lemon juice: pH = 2.3

[C] 11. Atmospheric pressure at the summit of Mt. Everest on a certain day measures 25.1 centimeters of mercury, and the air temperature is 0° Celsius. If the atmospheric pressure at sea level is 76 centimeters of mercury, use the barometric equation to find the approximate height of Mt. Everest.

[C] 12. Suppose that the pilot of a light aircraft has neglected to recalibrate the altimeter, which is set to a sea-level pressure of 76 centimeters of mercury, when the actual sea-level pressure has dropped to 75 centimeters because of weather conditions. Using $T = 0°$ Celsius in the barometric equation, find the amount of error in the pilot's altimeter reading.

[C] 13. Use the barometric equation with $T = 0°$ Celsius to find the elevation at which one-half of the atmosphere lies below and one-half lies above. [Hint: At this height, atmospheric pressure will have dropped to one-half its sea-level value.]

[C] 14. Using the barometric equation with $T = 0°$ Celsius and $P_0 = 76$ centimeters of mercury, sketch a graph of the height h as a function of atmospheric pressure P.

[C] 15. At takeoff, a certain supersonic jet produces a sound wave of intensity 0.2 watt per square meter. Taking P_0 to be 10^{-12} watt per square meter, find the loudness in decibels of the takeoff.

16. Show that a combination of two sounds with loudnesses S_1 and S_2 decibels produces a sound of loudness

$$S = 10 \log(10^{S_1/10} + 10^{S_2/10}) \text{ decibels.}$$

[C] 17. Find the ratio of the sound in watts per square meter at the threshold of pain (about 120 decibels) and at the threshold of hearing (0 decibels).

[C] 18. Suppose that the smallest weight you can perceive is $W_0 = 0.5$ gram, and that you can just barely notice the difference between 100 grams and 125 grams. Using the Weber–Fechner law, develop a scale of perceived heaviness. Call one unit on this scale a *heft* and write a formula for the number S of hefts corresponding to a weight of W grams.

[C] 19. The brightness of a star perceived by the naked eye is measured in units called *magnitudes*; the brightest stars are of magnitude 1 and the dimmest are of magnitude 6. If I is the actual intensity of light from a star, and I_0 is the intensity of light from a just-visible star, the magnitude M corresponding to I is given by

$$M = 6 - 2.5 \log \frac{I}{I_0}.$$

Calculate the ratio of light intensities from a star of magnitude 1 and a star of magnitude 5.

20. If k is a small positive number, show that $k \approx 1 - e^{-k}$. [Hint: Use the approximation $e^x \approx 1 + x$ for small $|x|$.]

[C] 21. The half-life of radium is 1656 years. (a) If y_0 grams of radium are initially present, write a formula for the number of grams y present after t years. (b) What percent decrease occurs in the amount of radium in a sample over 1 year? (c) How much of a 1-gram sample of radium is present after 20 years?

[C] 22. A manufacturing plant estimates that the value V dollars of a machine is decreasing exponentially according to the equation

$$V = 76,000e^{-0.14t},$$

where t is the number of years since the machine was placed in service. (a) Find the value of the machine after 12 years. (b) Find the "half-life" of the machine.

[C] 23. Sociologists in a certain country estimate that, of all couples married this year, 40% will be divorced within 15 years. Assuming that marriages fail exponentially, find the "half-life" of a marriage in that country.

[C] 24. The intensity I of light below the surface of the ocean decreases exponentially with the depth d, the decay constant k depending on the amount of dissolved organic material, the pollution level, and other factors. In the surface layer of the ocean, called the *photic zone*, there is sufficient light for plant growth. Nearly all life in the ocean is dependent on microscopic plants called *phytoplankton*, which can live only in the photic zone. In coastal waters, the photic zone has a depth of as little as 5 meters; while in clear oceanic waters this zone may extend to a depth of 150 meters. (a) Find the ratio of the decay constant k for the intensity of light in coastal waters to its value in clear oceanic waters. (b) Assuming that the light intensity at the bottom of the photic zone is 1% of the surface light intensity, find the value of k for clear oceanic waters. (c) Find the depth at which the light intensity in clear oceanic waters has dropped to half its value at the surface.

C 25. In psychological tests it is often found that if a group of people memorize a list of nonsense words, the fraction F of these people who remember all the words t hours later is given by $F = 1 - k\ln(t + 1)$, where k is a constant depending on the length of the list of words and other factors. A certain group was given such a memory test, and after 3 hours only half of the group's members could remember all the words. (a) Find the value of k for this experiment. (b) Predict the approximate fraction of group members who will remember all the words after 5 hours.

C 26. In engineering thermodynamics, it is shown that when n moles of a gas are compressed isothermally (that is, at a constant temperature $T°$ Celsius) from volume V_0 to volume V, the resulting work W joules done on the gas is given by

$$W = n(8.314)(T + 273)\ln\frac{V_0}{V}.$$

Calculate the work done in compressing $n = 5$ moles of carbon dioxide isothermally at a temperature of $T = 100°$ Celsius from an initial volume $V_0 = 0.5$ cubic meter to a final volume of $V = 0.1$ cubic meter.

---

6   Mathematical Models and Population Growth

In the later years of his life, the Italian scientist Galileo Galilei (1564–1642) wrote about his experiments with motion in a treatise called *Dialogues Concerning Two New Sciences*. Here he described his wonderful discovery that distances covered in consecutive equal time intervals by balls rolling down inclined planes are proportional to the successive odd positive integers. Thus, if $d(n)$ denotes the distance covered during the nth time interval, then

$$d(n) = k(2n - 1),$$

where k is the proportionality constant. Galileo determined that the constant k depends only on the incline and not on the mass of the ball or the material of which it is composed. He reasoned that the same result should hold for freely falling bodies if air resistance is neglected.

Galileo's discovery of a mathematical model for uniformly accelerated motion is considered to have been the beginning of the science of dynamics. A **mathematical model** is an equation or set of equations in which variables represent real-world quantities—for instance, distance and time intervals in the equation $d(n) = k(2n - 1)$. A mathematical model is often an *idealization* of the real-world situation it supposedly describes; for instance, Galileo's mathematical model assumes a perfectly smooth inclined plane, no friction, and no air resistance.

Figure 1

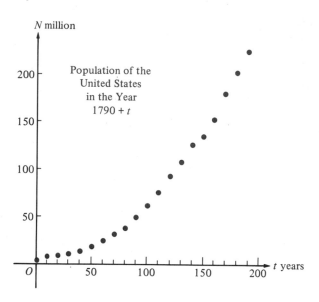

Population of the
United States
in the Year
$1790 + t$

In this section, we shall consider some of the mathematical models currently used in the life sciences to describe population growth and biological growth in general. The construction of realistic mathematical models nearly always requires the patient accumulation of experimental data, sometimes over a period of many years. In Figure 1, we have plotted points (t, N) showing the population N of the United States in the year $1790 + t$, according to the U.S. Census Bureau. These points seem to lie along a curve that is reminiscent of the graph of exponential growth (Figure 6a in Section 2.1) and they therefore suggest the mathematical model $N = N_0 e^{kt}$ for the growth of the population of the United States.

In 1798, the English economist Thomas Malthus made similar observations about the world population in his *Essay on the Principle of Population.* Because Malthus also proposed a linear model for the expansion of food resources, he forecast that the exponentially growing population would eventually be unable to feed itself. This dire prediction had such an impact on economic thought that the exponential model for population growth came to be known as the **Malthusian model.**

Consider a population growing according to the Malthusian model

$$N = N_0 e^{kt}.$$

Of course, N_0 is the population when $t = 0$; but what is the meaning of the growth constant k? To find out, let's consider how the population changes in 1 year. At the beginning of the $(t + 1)$st year, the population is $N_0 e^{kt}$, and at the end of this year it is $N_0 e^{k(t+1)}$. During the year, the population increase is

$$N_0 e^{k(t+1)} - N_0 e^{kt} = N_0(e^{kt+k} - e^{kt}) = N_0(e^{kt}e^k - e^{kt})$$

$$= N_0 e^{kt}(e^k - 1).$$

The percentage of the increase in population during the year is therefore given by

$$\frac{\text{increase during the year}}{\text{population at the beginning of the year}} \times 100\% = \frac{N_0 e^{kt}(e^k - 1)}{N_0 e^{kt}} \times 100\%$$

$$= (e^k - 1) \times 100\%.$$

In other words, if the yearly percentage increase in population is expressed as a decimal K, we have

$$K = e^k - 1 \quad \text{or} \quad e^k = 1 + K;$$

so that,

$$k = \ln(1 + K).$$

Example   © According to the U.S. Census Bureau, the population of the United States in 1980 was $N_0 = 226$ million. Suppose that the population grows according to the Malthusian model at 1.1% per year. (a) Write an equation for the population N million of the United States t years after 1980. (b) Predict N in the year 2000. (c) Predict N in the year 2020.

Solution   (a) 1.1% expressed as a decimal is $K = 0.011$. Therefore,

$$k = \ln(1 + K) = \ln 1.011 \approx 0.011,$$

and we have

$$N = N_0 e^{kt} = 226 e^{0.011t}.$$

(b) In the year 2000, $t = 20$ and

$$N = 226 e^{0.011(20)} = 226 e^{0.22} \approx 282 \text{ million}.$$

(c) In the year 2020, $t = 40$ and

$$N = 226 e^{0.011(40)} = 226 e^{0.44} \approx 351 \text{ million}.$$

In the example above, the growth constant k, rounded off to two significant digits, is the same as the yearly percentage of population increase expressed as a decimal K. This is no accident. Indeed, if K is small, then we can use the approximation $e^x \approx 1 + x$ (Section 2, page 231) with $x = K$ to get $e^K \approx 1 + K$, so that

$$k = \ln(1 + K) \approx \ln e^K = K.$$

If $K \leq 0.06$ (6% per year), the error in the approximation $k \approx K$ is less than 3%.

Whenever exponential growth is involved as in the Malthusian model, it is interesting to ask when the growing quantity doubles. If T is the **doubling time,** then

$$N_0 e^{k(t + T)} = 2N_0 e^{kt} \quad \text{or} \quad e^{kt + kT} = 2e^{kt};$$

that is,

$$e^{kt} e^{kT} = 2e^{kt} \quad \text{or} \quad e^{kT} = 2.$$

Thus,

$$kT = \ln 2,$$

and we have the formula

$$T = \frac{\ln 2}{k} \approx \frac{\ln 2}{K}$$

for the doubling time.

Example   © If the population of the United States grows according to the Malthusian model at 1.1% per year, in approximately how many years will it double?

Solution   Here $K = 0.011$, so $T \approx \dfrac{\ln 2}{K} \approx \dfrac{\ln 2}{0.011} \approx 63$ years.

When one-celled organisms reproduce by simple cell division in a culture containing an unlimited supply of nutrients, the Malthusian or exponential model $N = N_0 e^{kt}$ for the number N of organisms at time t is often quite accurate. In a natural environment, however, growth is often inhibited by various constraints that have a greater and greater effect as time goes on—depletion of the food supply, build-up of toxic wastes, physical crowding, and so on—and the Malthusian model may no longer apply.

If N_0 organisms are introduced into a habitat at time $t = 0$ and the population N of these organisms is plotted as a function of time t, the result is often an S-shaped curve (Figure 2). Typically, such a curve shows an increasing rate of growth up to a point P_I, called the **inflection point,** followed by a declining rate of growth as the population N levels off and approaches the maximum N_{max} that can be supported by the habitat. The horizontal line $N = N_{max}$ is an asymptote of the graph. Notice that the graph is bending upward to the left of the inflection point and bending downward to the right of it.

There are many equations which have graphs with the characteristic S-shape of Figure 2, and which are therefore used as mathematical models for population growth. Of these, one of the most popular is the **logistic model**

Figure 2

$$N = \frac{N_0 N_{max}}{N_0 + (N_{max} - N_0)e^{-kt}}$$

(Problem 8). Using calculus, it can be shown that the coordinates (t_I, N_I) of the inflection point P_I of the logistic model are

$$t_I = \frac{1}{k}\ln\frac{N_{max} - N_0}{N_0}, \qquad N_I = \frac{N_{max}}{2}.$$

Using logarithms, we can solve the logistic equation for t in terms of N; the result is

$$t = \frac{1}{k}\ln\frac{N(N_{max} - N_0)}{N_0(N_{max} - N)}$$

(Problem 10).

Example   © Suppose that the population N million of the United States grows according to the logistic model

$$N = \frac{N_0 N_{max}}{N_0 + (N_{max} - N_0)e^{-kt}},$$

where t is the time in years since 1780, and where $k = 0.03$. Use the following data:

The population in 1780 was 3 million and the population in 1880 was 50 million.
(a) Find N_{max}.
(b) Find N in the year 1980.
(c) Find N in the year 2000.
(d) Determine the year in which the inflection occurred.

Solution

(a) From the data given, $N_0 = 3$ million in 1780 when $t = 0$, while in 1880 when $t = 100$, $N = 50$ million; that is,

$$50 = \frac{N_0 N_{max}}{N_0 + (N_{max} - N_0)e^{-kt}} = \frac{3N_{max}}{3 + (N_{max} - 3)e^{-0.03(100)}}.$$

It follows that

$$50[3 + (N_{max} - 3)e^{-3}] = 3N_{max}$$

or

$$150 + 50e^{-3}N_{max} - 150e^{-3} = 3N_{max}.$$

Thus,

$$(3 - 50e^{-3})N_{max} = 150(1 - e^{-3}),$$

so, rounding off to two significant digits, we have

$$N_{max} = \frac{150(1 - e^{-3})}{3 - 50e^{-3}} \approx 280 \text{ million}.$$

(b) In 1980, we have $t = 200$ years and

$$N = \frac{N_0 N_{max}}{N_0 + (N_{max} - N_0)e^{-kt}}$$

$$= \frac{3(280)}{3 + (280 - 3)e^{-0.03(200)}} \approx 230 \text{ million}.$$

[Note: The correct value according to the 1980 census was 226 million, so our logistic model has predicted the correct value with an error of less than 2%.]

(c) In the year 2000, $t = 220$ years and

$$N = \frac{3(280)}{3 + (280 - 3)e^{-0.03(220)}} \approx 250 \text{ million}.$$

[Compare this with the prediction of 282 million according to the Malthusian model (page 259). Notice that the "leveling off" built into the logistic model has apparently taken hold.]

(d) $t_I = \frac{1}{k} \ln \frac{N_{max} - N_0}{N_0} = \frac{1}{0.03} \ln \frac{280 - 3}{3} = \frac{1}{0.03} \ln \frac{277}{3} \approx 150$ years.

Thus, according to the logistic model, the inflection would have taken place in the year $1780 + 150 = 1930$.

PROBLEM SET 6

C **1.** The population of a small country was 10 million in 1980 and it is growing according to the Malthusian model at 3% per year. (a) Write an equation for the population N million of the country t years after 1980. (b) Predict N in the year 2000. (c) Find the doubling time T for the population.

C **2.** The population of a certain city is growing according to the Malthusian model and it is expected to double in 35 years. Approximately what percent of growth will occur in this population over 1 year?

C **3.** The number of bacteria in an unrefrigerated chicken salad doubles in 3 hours. Assuming exponential growth, in how many hours will the number of bacteria be increased by a factor of 10?

4. Suppose that a population that is growing according to the Malthusian model increases by $100K\%$ per year. Write an *exact* formula (not an approximation) for the doubling time T in terms of K.

C **5.** The bacterium *Escherichia coli* is found in the human intestine. When *E. coli* is cultivated under ideal conditions in a biological laboratory, the population doubles in $T = 20$ minutes. Suppose that $N_0 = 10$ *E. coli* cells are placed in a nutrient broth medium extracted from yeast at time $t = 0$ minutes. (a) Write an equation for the number N of *E. coli* bacteria in the colony t minutes later. (b) Find N when $t = 60$ minutes.

6. Suppose that a population is growing according to the Malthusian model with doubling time T. If N_0 is the original size of the population when $t = 0$, show that the population t units of time later is given by $N = N_0 2^{t/T}$.

C **7.** The fruit fly *Drosophila melanogaster* is often used by biologists for genetic experiments because it breeds rapidly and has a short life cycle. Suppose a colony of *D. melanogaster* in a laboratory is observed to double in size in $T = 2$ days. Assuming a Malthusian model for growth of the colony, determine the approximate daily percentage increase in the size of the colony.

8. For the logistic model

$$N = \frac{N_0 N_{max}}{N_0 + (N_{max} - N_0)e^{-kt}},$$

show that (a) $N = N_0$ when $t = 0$, and (b) $N = N_{max}$ is a horizontal asymptote of the graph of N as a function of t. [Hint: As t gets larger and larger, e^{-kt} gets closer and closer to 0.]

C **9.** Suppose that a herd of 300 deer, newly introduced into a game preserve, grows according to the logistic model (problem 8) with $k = 0.1$. Assume that the herd has grown to 387 deer after 5 years. (a) Find N_{max}, the maximum possible size of the herd in this habitat. (b) Find the population of the herd after 7 years. (c) When does the inflection occur in the population of the herd?

10. Solve the logistic equation (problem 8) for t in terms of N.

C **11.** Sketch an accurate graph of the population N of the deer herd in problem 9 as a function of the time t in years since the herd was introduced into the habitat.

12. If a certain population N is thought to be growing according to the logistic model (problem 8), the value of the constant k is often found experimentally as follows: First, an estimate is made for N_{max}, the maximum possible size of the population in the given habitat. Then values of N are measured corresponding to several different values of t and

$$y = \ln \frac{N}{N_{max} - N}$$

is calculated for each value of N. The points (t, y) are plotted on a graph (Figure 3) and a straight line L that "best fits" these points is drawn. The constant k is taken to be the slope

Figure 3

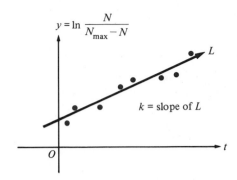

$$y = \ln \frac{N}{N_{\text{max}} - N}$$

L

$k = \text{slope of } L$

t

O

of the line L. Justify this procedure. [Hint: Use the result of problem 10 to show that $kt = y - b$, where b is the value of y when $t = 0$.]

[c] 13. Consider the alternative growth model

$$N = N_{\text{max}}\left[1 - \left(1 - \frac{N_0}{N_{\text{max}}}\right)e^{-ct}\right]$$

for the deer herd in problem 9, where $c = 0.05$, $N_0 = 300$, and $N_{\text{max}} = 700$. Sketch the graph of N as a function of t according to this model and compare it with the graph in problem 11.

14. Outline a procedure for determining the constant c for the alternative growth model in problem 13. The procedure should be similar to that in problem 12, but with

$$y = \ln \frac{N_{\text{max}}}{N_{\text{max}} - N}.$$

[c] 15. The **Gompertz growth model** is $N = N_{\text{max}} \exp(-Be^{-Ct})$, where $B = \ln \frac{N_{\text{max}}}{N_0}$ and C is a positive constant. (Recall that $\exp x = e^x$.) Take $N_0 = 300$, $N_{\text{max}} = 700$, and $C = 0.07$; sketch the resulting graph of N as a function of t; and compare it with the graph in problem 9.

16. For the Gompertz growth model (problem 15), show that (a) N_0 is the value of N when $t = 0$ and (b) $N = N_{\text{max}}$ is a horizontal asymptote of the graph.

REVIEW PROBLEM SET

1. Let $f(x) = 3^x$, $g(x) = (\frac{1}{5})^x$, and $h(x) = 2^{-x}$. Find:
   (a) $f(3)$ (b) $g(0)$ (c) $g(-1)$ (d) $h(2)$
   (e) $f[h(1)]$ (f) $g(\frac{1}{2})$

[c] 2. Using a calculator with a y^x key, find the value of each quantity to as many significant digits as you can.
   (a) $0.47308^{7.7703}$ (b) $32.273^{-0.35742}$

In problems 3–10, sketch the graph of the given function, determine its domain, its range, and any horizontal or vertical asymptotes, and indicate whether the function is increasing or decreasing. (Just make a rough sketch—do not use a calculator or tables.)

3. $f(x) = 2^{3+x}$ 4. $g(x) = (\frac{1}{3})^{2-x}$ 5. $F(x) = (\frac{1}{10})^{x+1}$ 6. $G(x) = 6^{-x} - 1$

7. $H(x) = 2^{-x} + 3$ 8. $h(x) = 3 + (\frac{1}{3})^{x-3}$ 9. $F(x) = 1 - 2(6^{-x})$ 10. $g(x) = 4(\frac{1}{10})^{1-x} + 2$

C 11. A savings account earns a nominal annual interest of 6% compounded quarterly. If there is $5000 in the account now, how much money will it contain after 5 years?

C 12. A certain bank advertises interest at a nominal annual rate of 6.5% compounded hourly. Find the effective simple annual interest rate.

C 13. Suppose that your local savings bank offers certificates of deposit paying nominal annual interest of 12% compounded semiannually. Use this interest rate to determine the present value to you of money in the future. If a debtor offers to pay you $1000 eighteen months from now, what is the present value to you of this offer?

14. True or false: If $f(x) = b^x$, then $f(x^2) = [f(x)]^2$. Justify your answer.

C 15. Use a calculator to find the value of each quantity to as many significant digits as you can.
(a) $e^{\sqrt{11}}$ (b) $e^{-\pi}$

C 16. Use a calculator to verify that $e^{x+y} = e^x e^y$ for $x = 7.7077$ and $y = 3.0965$.

In problems 17–22, sketch the graph of the given function, determine its domain, its range and any horizontal or vertical asymptotes, and indicate whether the function is increasing or decreasing. (Just make a rough sketch—do not use a calculator or tables.)

17. $f(x) = 3 + e^x$ 18. $g(x) = 1 - e^{-x}$ 19. $h(x) = 4e^{x-4}$

20. $F(x) = 3 - e^{2-x}$ 21. $G(x) = 3e^{x-1} + 2$ 22. $H(x) = 2 + \left(\dfrac{1}{e}\right)^x$

C 23. A savings bank with $28,000,000 in regular savings accounts is paying interest at a nominal rate of 5.5% compounded quarterly. The bank is contemplating offering the same rate of interest, but compounding continuously rather than quarterly. How much more interest will the bank have to pay out per year if the new plan is adopted?

24. Infusion of a glucose solution into the bloodstream is a standard medical technique. Medical technicians use the formula $y = A + (B - A)e^{-kt}$, where A, B, and k are positive constants, to determine the concentration y of glucose in the blood t minutes after the beginning of the infusion. (a) Interpret the meaning of the constant B. (b) Interpret the meaning of the constant A. (c) The constant k has to do with the rate at which the glucose is converted and removed from the bloodstream. If the glucose is being removed from the bloodstream rapidly, would this suggest that k is relatively large or small?

C 25. Assume that y grows or decays exponentially as a function of x, that y_0 is the value of y when $x = 0$, and that k is the growth or decay constant. Find the indicated quantity from the information given.
(a) Growth, $k = 5$, $y_0 = 3$. Find y when $x = 2$.
(b) Decay, $k = 0.12$, $y_0 = 5000$. Find y when $x = 10$.
(c) Growth, $y = 5$ when $x = 1$, $y = 7$ when $x = 3$. Find y_0.
(d) Decay, $k = 4$. Find the percentage of change in y when x changes from 10 to 11.

C 26. Using a calculator, determine the percent error in approximating e by $[1 + (1/n)]^n$ for $n = 1000$. [See problem 26 in Problem Set 2.]

In problems 27–38, (a) rewrite each logarithmic equation as an exponential equation and (b) solve the exponential equation without using a calculator or tables.

27. $x = \log_5 \frac{1}{125}$ 28. $\log_x 16 = -2$ 29. $\log_3(2 + x) = 1$

30. $\log_4 |2x| = 1$ 31. $\log_3 81 = 5x$ 32. $x = \log_3 9 \sqrt{3}$

33. $\log_x \frac{1}{49} = 2$ 34. $\log_5 |3x - 5| = 0$ 35. $\log_3 \frac{1}{81} = -2x$

36. $\ln(x^2 - 4) = 0$ 37. $1 + \log_7 49 = |x|$ 38. $\log |2x - 1| = 2$

39. Rewrite each exponential equation as an equivalent logarithmic equation.
(a) $3^6 = 729$ (b) $2^{-10} = \frac{1}{1024}$ (c) $64^{4/3} = 256$
(d) $x^a = w$ (e) $10^x = y$ (f) $e^x = y$

40. Solve each exponential equation without using a calculator or tables.
(a) $5^{2x} - 26(5^x) + 25 = 0$ (b) $2^x + 2^{-x} = 2$
(c) $7^{3x^2 - 2x} = 49(7^3)$ (d) $(2^{|x|+1} - 1)^{-1} = \frac{1}{3}$

41. Find each logarithm without using a calculator or tables.
(a) $\log_2 16$ (b) $\log_{16} 2$ (c) $\log_5 \sqrt{5}$ (d) $\log_7 1$
(e) $\log_3 \frac{1}{27}$ (f) $\ln e^{33}$ (g) $\log 100^{-0.7}$

42. Suppose that $\log_b 2 = 0.9345$, $\log_b 3 = 1.4812$, and $\log_b 7 = 2.6237$. Find each value.
(a) $\log_b 6$ (b) $\log_b \frac{7}{6}$ (c) $\log_b \frac{18}{7}$ (d) $\log_b 2^{64}$
(e) $\log_b 21$ (f) $\log_b 0.5$ (g) $\log_b 49$ (h) $\log_b \sqrt[5]{28}$
(i) $\log_b(36/\sqrt{56})$ (j) $\log_7 b$ [C] (k) b

[C] **43.** Use a calculator to find each value to as many significant digits as you can.
(a) $\log e$ (b) $\ln 100$ (c) $\ln 33.3$
(d) $2.07^{-15.32}$ (e) $\log(\log 14)$ (f) $\ln(\ln 200)$
(g) e^{e-1} (h) π^{π^π} (i) $\log e^{e^e}$

[C] **44.** Use a calculator with y^x, e^x, and $\ln$ keys to verify the formula $y^x = e^{x \ln y}$ for the indicated values of x and y.
(a) $x = 3.22, y = 2.03$ (b) $x = 4.71, y = 0.83$ (c) $x = -4.01, y = 8.11$

In problems 45–48, rewrite each expression as a sum or difference of multiples of logarithms. (Make the necessary assumptions about the values of the variables.)

45. $\log_b\left(\dfrac{\sqrt[n]{p}}{R^n}\right)$ **46.** $\ln(e^x q^n \sqrt{p})$ **47.** $\log\sqrt{\dfrac{4-x}{4+x}}$ **48.** $\log\dfrac{y^2\sqrt{4-y^2}}{16(3y+7)^{3/2}y^4}$

In problems 49–52, rewrite each expression as a single logarithm. (Make the necessary assumptions about the values of the variables.)

49. $a\log_b x + \dfrac{1}{c}\log_b y$ **50.** $4\log(x^2 - 1) - 2\log(x+1)$

51. $\ln\dfrac{x^2-4}{x^2-3x-4} - \ln\dfrac{x^2-3x-10}{2x^2+x-1}$ **52.** $\ln\sqrt{\dfrac{x}{x+2}} + \ln\dfrac{\sqrt{x^2-4}}{x^2}$

In problems 53–56, solve each equation.

53. $\log_3(x+1) + \log_3(x+3) = 1$ **54.** $\log_4(x+3) - \log_4 x = 2$
55. $\log_3(7-x) - \log_3(1-x) = 1$ **56.** $\log_6(x+3) + \log_6(x+4) = 1$

[C] **57.** Use a calculator and the base-changing formula to evaluate:
(a) $\log_2 11$ (b) $\log_8 10^{10}$ (c) $\log_{1/3} 0.707$ (d) $\log_{\sqrt{5}} \sqrt{2}$

58. Use Appendix Table IB (with linear interpolation when necessary) to find the common logarithm of each number rounded off to four decimal places.
(a) $\log 5.43$ (b) $\log 5430$ (c) $\log 0.00543$ (d) $\log 8.03$
(e) $\log 0.000803$ (f) $\log 903,200$ (g) $\log 7152$ (h) $\log 0.001347$

[C] In problems 59–64, use a calculator to verify each equation for the indicated values of the variables.

59. $\log\sqrt{x} = \frac{1}{2}\log x$ for $x = 33.20477$
60. $\log(xy) = \log x + \log y$ for $x = 72.1355$, $y = 0.774211$

61. $\ln x = \dfrac{\log x}{\log e}$ for $x = 77.0809$ **62.** $\log_a b = \dfrac{1}{\log_b a}$ for $a = 10$, $b = e$

63. $\log y^x = x\log y$ for $y = 0.001507$, $x = 10.1333$

64. $\log\dfrac{x}{y} = \log x - \log y$ for $x = 5.3204 \times 10^{-7}$, $y = 3.2211 \times 10^3$

C In problems 65–70, an error has been made. In each case, show that the equation is false by evaluating both sides with the aid of a calculator.

65. $\dfrac{\log \pi}{\log e} = \log \pi - \log e$?

66. $\log 6 = (\log 2)(\log 3)$?

67. $\dfrac{1}{\log \frac{2}{3}} = \log \frac{3}{2}$?

68. $-\log 5 = \log(-5)$?

69. $\log_2 3 = \log 8$?

70. $\dfrac{1}{\ln 2} + \dfrac{1}{\ln 3} = \dfrac{1}{\ln 5}$?

71. Find the domain of each function.
  (a) $f(x) = \log_2(4x - 3)$ (b) $g(x) = \ln |x + 1|$
  (c) $h(x) = \ln(x^2 - 4x - 4)$ (d) $F(x) = \log e^x$

72. Find the y and x intercepts of the graph of each function in problem 71.

C In problems 73–78, determine the domain and range of each function; find any y or x intercepts and any horizontal or vertical asymptotes of its graph; indicate where the function is increasing or decreasing; and sketch the graph.

73. $f(x) = 1 + \ln(x - 1)$

74. $g(x) = 1 - \log x^3$

75. $h(x) = \log \sqrt{4 - x}$

76. $F(x) = \dfrac{1}{\log_x 10}$

77. $G(x) = \ln |2 - x|$

78. $H(x) = \log_{1/e} x$

In problems 79–82, simplify each expression.

79. $\dfrac{10^{\log(x^2 - 3x + 2)}}{x - 2}$

80. $e^{(1/2)\ln(x^2 - 2x + 1)}$

81. $\dfrac{\log 10^{x^2 - 4x - 5}}{x + 1}$

82. $\dfrac{\log y^x}{10^{\log x}}$

C In problems 83–88, use logarithms and a calculator to solve each exponential equation.

83. $10^x = 20$

84. $e^{x+3} = 5^{-2}$

85. $10^{x+7} = 14^3$

86. $3^{x-1} = (15.4)^x$

87. $5^{2x} = 4(3^x)$

88. $3^{2x-1} = 4^{x+2}$

C **89.** As part of a psychological experiment, a group of students took an exam on material they had just studied in a physics course. At monthly intervals thereafter they were given equivalent exams. After t months, the average score S of the group was found to be given by $S = 78 - 55 \log(t + 1)$. What was the average score (a) when they took the original exam and (b) 5 months later? (c) When will the group have forgotten everything they learned?

C **90.** In 1935, the U.S. seismologist Charles Richter established the **Richter scale** for measuring the *magnitude M* of an earthquake in terms of the total energy E joules released by it. One form of Richter's equation is $M = \frac{2}{3}\log(E/E_0)$, where E_0 is a minimum amount of released energy used for comparison. An earthquake that measures 5.5 on the Richter scale, a magnitude corresponding to an energy release of 10^{13} joules, can cause local damage. An earthquake of magnitude 6 produces considerable destruction, and an earthquake of magnitude 7 or higher can result in catastrophic damage. (a) Find E_0. (b) The most powerful earthquake ever recorded occurred in Colombia on January 31, 1906, and measured 8.6 on the Richter scale. Approximately how many joules of energy were released?

C **91.** Suppose that a supersonic plane on takeoff produces 120 decibels of sound. How many decibels of sound would be produced by two such planes taking off side by side?

C **92.** An alternative method for measuring the loudness of sound is the *sone* scale. Whereas loudness S in decibels is given by $S = 10 \log(P/P_0)$, loudness s in sones is given by $s = 10^{2.4} P^{0.3}$, where P is the intensity of the sound wave in watts per square meter, and $P_0 = 10^{-12}$ watt per square meter. Show that a 10-decibel increase in loudness doubles the loudness on the sone scale.

C **93.** In 1921, President Warren G. Harding presented Marie Curie a gift of 1 gram of radium on behalf of the women of the United States. Using the fact that the half-life of radium is 1656 years, determine how much of the 1-gram gift was left in 1981.

94. A couple wishes to invest their life savings, P dollars, but finds that inflation is reducing the value of a dollar by $100K\%$ per year. They also find that they will have to pay state and local income taxes amounting to $100Q\%$ of their interest at the end of each year. Write a formula for the nominal annual interest rate r, compounded continuously, that the couple must receive on their investment just to break even at the end of the year.

© **95.** You have just won first prize in a state lottery and you have your choice of the following: (a) $30,000 will be placed in a savings account in your name, and the money will be compounded continuously at a nominal annual rate of 10%. (b) One penny will be placed in a fund in your name, and the amount in the fund will be doubled every 6 months over the next 12 years. Which plan do you choose and why?

© **96.** Suppose a bank offers savings accounts at a nominal annual rate r compounded n times per year. By regularly depositing a fixed sum of p dollars in such an account every $\frac{1}{n}$ th of a year, money can be accumulated for future needs. Such a plan is called a **sinking fund.** If it is required that such a sinking fund yield an amount S dollars after a term of t years, the periodic payment p dollars must be

$$p = \frac{Sr}{n\left[\left(1 + \frac{r}{n}\right)^{nt} - 1\right]}.$$

If a restaurant anticipates a capital expenditure of $80,000 for expanding in 5 years, how much money should be deposited quarterly in a sinking fund, at a nominal annual interest rate of 10% compounded quarterly, in order to provide the required amount for the expansion?

© **97.** An interest-bearing debt is said to be **amortized** if the principal P dollars and the interest I dollars are paid over a term of t years by regular payments of p dollars every $\frac{1}{n}$ th of a year. The amortization formulas are

$$p = \frac{Pr}{n\left[1 - \left(1 + \frac{r}{n}\right)^{-nt}\right]} \quad \text{and} \quad I = np\left[t - \frac{1 - \left(1 + \frac{r}{n}\right)^{-nt}}{r}\right].$$

Suppose you want to buy a car and need to borrow $P = \$6000$ from your local bank. The bank charges a nominal annual interest of 14% ($r = 0.14$) on automobile loans with monthly payments ($n = 12$) over a term of 36 months ($t = 3$ years). Use the amortization formulas to calculate (a) your monthly payment p dollars and (b) the total interest charge I dollars.

© **98.** An automobile dealer's advertisement reads: "No money down, $300 per month for 36 months, 12 percent annual interest rate compounded monthly on the unpaid balance, puts you in the driver's seat of a new Wildebeest." Use the amortization formulas (problem 97) to figure out how much of the $10,800 to be paid goes toward the car and how much is interest. [Take $p = 300$, $t = 3$, $r = 0.12$, and $n = 12$.]

© **99.** The amortization formulas (problem 97) with $n = 12$ apply if you borrow P dollars from a bank for a home mortgage at a nominal annual interest rate r payable in successive equal monthly payments of p dollars each over a term of t years. The balance due to the bank at the beginning of the kth month, P_k dollars, is customarily considered to be

$$P_k = P\frac{\left(1 + \frac{r}{12}\right)^{12t} - \left(1 + \frac{r}{12}\right)^{k-1}}{\left(1 + \frac{r}{12}\right)^{12t} - 1}.$$

Suppose that you purchase a new home for $65,000, paying $15,000 down, and taking out a 30-year mortgage on the remaining $50,000 at a nominal annual interest rate of 12%. (a) What is your monthly payment on the mortgage? (b) After 15 years, how much of the original $50,000 will be paid off?

© **100.** In problem 99, what part of your 180th payment goes toward interest and what part goes toward reduction of the principal?

© **101.** Suppose that the population N of cancer cells in an experimental culture grows exponentially as a function of time. If $N_0 = 3000$ cells are present when $t = 0$ days, and if the number of cells is increasing at a rate of 0.35% per day, (a) how many cells are present in the culture after 100 days and (b) what is the approximate doubling time?

102. Suppose that y is growing or decaying exponentially as a function of x and that k is the growth or decay constant. Assume that $y = y_1$ when $x = x_1$ and that $y = y_2$ when $x = x_2$. If $x_1 \neq x_2$, prove that

$$k = \frac{1}{x_2 - x_1} \ln \frac{y_2}{y_1}.$$

103. In 1960, the American scientist Willard Libby received the Nobel Prize in Physical Chemistry for his discovery of the technique of radiocarbon dating. When a plant or animal dies, it receives no more of the naturally occurring radioactive carbon C^{14} from the atmosphere. Libby developed methods for determining the fraction F of the original C^{14} that is left in a fossil and made an experimental determination of the half-life T of C^{14}. (a) If F is the fraction of the original C^{14} left in a fossil, show that the original plant or animal died t years ago, where $t = -T(\ln F/\ln 2)$.© (b) An ancient scroll is unearthed, and it is determined that it contains only 76% of its original C^{14}. If $T = 5580$ years, how old is the scroll?

© **104.** According to **Newton's law of cooling,** if a hot object with initial temperature T_0 is placed at time $t = 0$ in a surrounding medium with a lower temperature T_1, the object cools in such a way that its temperature T at time t is given by

$$T = T_1 + (T_0 - T_1)e^{-kt},$$

where the constant k depends on the materials involved. Suppose an iron ball is heated to 300°F and allowed to cool in a room where the air temperature is 80°F. If after 10 minutes the temperature of the ball has dropped to 250°F, what is its temperature after 20 minutes?

© **105.** A new species of plant is introduced on a small island. Assume that 500 plants were introduced initially, and that the number N of plants increases according to the logistic model (Section 6, page 260) with $k = 0.62$. Suppose that there are 1700 plants on the island 2 years after they were introduced. (a) Find N_{max}, the maximum possible number of plants the island will support. (b) Find N after 10 years. (c) When does the inflection occur? (d) Sketch a graph of N as a function of time t in years.

© **106.** A rumor started by a single individual is spreading among students at a university. Assume that the number N of students who have heard the rumor t days after it was started grows according to the logistic model, with $N_0 = 1$ and $N_{max} = 20,000$ (the entire student body). If 10,000 students have heard the rumor after 2 days, (a) find the value of k and (b) determine how long it takes until 15,000 students have heard the rumor.

6 Trigonometric Functions

The word "trigonometry" is derived from the Greek words "trigon" for triangle and "metra" for measurement. Trigonometry in the sense of "triangle measurement" is based on the relationship between the vertex angles of a right triangle and ratios of sides of the triangle. The six *trigonometric functions*, which are defined in terms of these ratios, are used routinely in calculations made by surveyors, engineers, and navigators. Trigonometric functions also have applications in the physical and life sciences, where they provide mathematical models for periodic phenomena —for instance, the ebb and flow of the tides, electromagnetic waves, seasonal variations in an animal's food supply, or the alpha rhythms of the human brain. We begin this chapter by considering angles and their measure. We define the trigonometric functions, first for acute angles in a right triangle and then for general angles. We study the evaluation of trigonometric functions, circular functions, graphs of trigonometric functions, and the simple harmonic model.

1 Angles and Their Measure

In order to define the trigonometric functions so that they can be used not only for triangle measurement but also for modeling periodic phenomena, we must give a definition of an angle that is somewhat more general than a vertex angle of a triangle.

If A and B are distinct points, the portion of the straight line that starts at A and continues indefinitely through B is called a **ray** with **endpoint** A (Figure 1a). An **angle** is determined by rotating a ray about its endpoint. This rotation can be indicated by a curved arrow as in Figure 1b. The endpoint of the rotated ray is called the **vertex** of the angle. The position of the ray before the rotation is called the **initial side** of the angle, and the position of the ray after the rotation is called the **terminal side.** Angles will often be denoted by small Greek letters, such as angle α in Figure 1b (α is the Greek letter *alpha*).

Figure 1

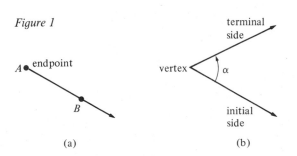

(a)

(b)

Angles determined by a *counterclockwise* rotation are said to be **positive** and angles determined by a *clockwise* rotation are said to be **negative.** In Figure 2a, angle β is positive (β is the Greek letter *beta*); in Figure 2b, angle γ is negative (γ is the Greek letter *gamma*).

Figure 2

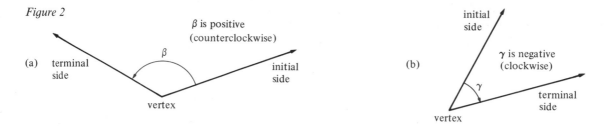

One full turn of a ray about its endpoint is called **one revolution** (Figure 3a); one-half of a revolution is called a **straight angle** (Figure 3b); and one-quarter of a revolution is called a **right angle** (Figure 3c). The initial and terminal sides of a right angle are perpendicular; hence, in drawing a right angle, we often replace the usual curved arrow by two perpendicular line segments (Figure 3c).

Figure 3

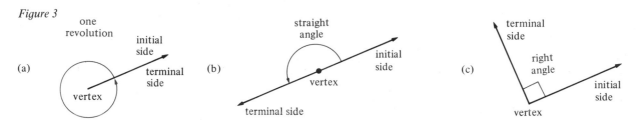

Angles that have the same initial sides and the same terminal sides are called **coterminal** angles. For instance, Figure 4 shows three coterminal angles α, β, and γ. Because different angles can be coterminal, an angle is not completely determined merely by specifying its initial and terminal sides. Nevertheless, angles are often named by using three letters such as *ABC* to denote a point *A* on the initial side, the vertex *B*, and a point *C* on the terminal side (Figure 5). Usually, the intended angle is the *smallest* rotation about the vertex that carries the initial side around to the terminal side. Sometimes, when it is perfectly clear which angle is intended, we refer to the angle by simply naming its vertex. For instance, the angle in Figure 5 could be called angle *B*.

Figure 4   *Figure 5*

There are various ways to assign numerical measures to angles. Of these, the most familiar is the *degree* measure: **One degree** (1°) is the measure of an angle formed by $\frac{1}{360}$ of a counterclockwise revolution. Negative angles are measured by a negative number of degrees; for instance, $-360°$ is the measure of one clockwise revolution. A positive right angle is $\frac{1}{4}$ of a counterclockwise revolution, and it therefore has a measure of $\frac{1}{4}(360°) = 90°$. Similarly, 180° represents 180°/360° or 1/2 of a counterclockwise revolution.

Example 1   What is the degree measure of one-eighth of a clockwise revolution?

Solution $\frac{1}{8}(-360°) = -45°.$

Example 2   What fraction of a revolution is an angle of 30°?

Solution $30°/360° = 1/12$ of a counterclockwise revolution.

Although fractions of a degree can be expressed as decimals, such fractions are sometimes given in "minutes" and "seconds." **One minute** (1′) is defined to be 1/60 of a degree and **one second** (1″) is defined to be 1/60 of a minute. Hence, one second is 1/3600 of a degree. Using the relationships

$$1' = \left(\frac{1}{60}\right)^{\circ}, \quad 1'' = \left(\frac{1}{3600}\right)^{\circ}, \quad 1° = 60', \quad \text{and} \quad 1' = 60'',$$

you can convert from degrees, minutes, and seconds to decimals and vice versa.

Example 1   [C] Express the angle measure 72°13′59″ as a decimal. Round off your answer to four decimal places.

Solution $72°13'59'' = 72° + \left(\frac{13}{60}\right)^{\circ} + \left(\frac{59}{3600}\right)^{\circ} \approx 72.2331°$

Example 2   Express the angle measure 173.372° in degrees, minutes, and seconds. Round off to the nearest second.

Solution $$0.372° = 0.372(1°) = 0.372(60') = 22.32'$$

and
$$0.32' = 0.32(1') = 0.32(60'') = 19.2'' \approx 19''.$$

Hence,
$$173.372° = 173° + 0.372° = 173° + 22.32' = 173° + 22' + 0.32'$$
$$\approx 173° + 22' + 19'' = 173°22'19''.$$

Because calculators work with angles expressed in decimal form, minutes and seconds are not as popular as they once were.

Figure 6

Figure 7

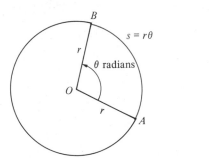

Although the degree measure of angles is used in most elementary applications of trigonometry, more advanced applications (especially those that involve calculus) require radian measure. **One radian** is the measure of an angle that has its vertex at the center of a circle (that is, a **central angle**) and intercepts an arc on the circle equal in length to the radius r (Figure 6).

A central angle of 2 radians in a circle of radius r intercepts an arc of length $2r$ on the circle, a central angle of $\frac{3}{4}$ radian intercepts an arc of length $\frac{3}{4}r$, and so forth. More generally, if a central angle AOB of θ radians (θ is the Greek letter *theta*) intercepts an arc $\overset{\frown}{AB}$ of length s on a circle of radius r (Figure 7), then we have

$$s = r\theta.$$

It follows that angle AOB has radian measure θ given by the formula

$$\theta = \frac{s}{r}.$$

Example 1   Find the length s of the arc intercepted on a circle of radius $r = 3$ meters by a central angle whose measure is $\theta = 4.75$ radians.

Solution $s = r\theta = 3(4.75) = 14.25$ meters.

Example 2 .   A central angle in a circle of radius 27 inches intercepts an arc of length 9 inches. Find the measure θ of the angle in radians.

Solution   Here $s = 9$ inches, $r = 27$ inches, and

$$\theta = \frac{s}{r} = \frac{9}{27} = \frac{1}{3} \text{ radian.}$$

A central angle of 360° corresponds to one revolution; hence it intercepts an arc $s = 2\pi r$ equal to the entire circumference of the circle (Figure 8). Therefore, if θ is the radian measure of the 360° angle,

Figure 8

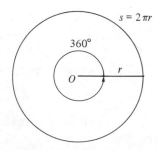

$$\theta = \frac{s}{r} = \frac{2\pi r}{r} = 2\pi \text{ radians;}$$

that is, $360° = 2\pi$ radians, or

$$180° = \pi \text{ radians.}$$

You can use this relationship to convert degrees into radians and vice versa. In particular,

$$1° = \frac{\pi}{180} \text{ radian} \quad \text{and} \quad 1 \text{ radian} = \left(\frac{180}{\pi}\right)°.$$

Thus,

multiply degrees by $\dfrac{\pi}{180°}$ to convert to radians and multiply radians by $\dfrac{180°}{\pi}$ to convert to degrees.

Example 1   Convert each degree measure to radian measure.

(a) $60°$

(b) $-22.5°$

Solution   (a) $60° = \dfrac{\pi}{180°}(60°) \text{ radians} = \dfrac{\pi}{3} \text{ radians}$

(b) $-22.5° = \dfrac{\pi}{180°}(-22.5°) \text{ radian} = -\dfrac{\pi}{8} \text{ radian}$

Example 2   Convert each radian measure to degree measure.

(a) $\dfrac{13\pi}{10}$ radians

(b) $-\dfrac{\pi}{4}$ radian

Solution   (a) $\dfrac{13\pi}{10} \text{ radians} = \dfrac{180°}{\pi}\left(\dfrac{13\pi}{10}\right) = 234°$ (b) $-\dfrac{\pi}{4} \text{ radian} = \dfrac{180°}{\pi}\left(-\dfrac{\pi}{4}\right) = -45°$

As the previous examples illustrate, when radian measures are expressed as rational multiples of π, we may leave them in that form rather than writing them as decimals. Also, when angles are measured in radians, the word "radian" is often omitted. For instance, rather than writing $60° = \pi/3$ radians, we write $60° = \pi/3$. Therefore, *when no unit of angular measure is indicated, it is always understood that radian measure is intended.*

Angles measured in radians are especially useful in studying the motion of a particle moving at constant speed around a circle of radius r with center O (Figure 9). Suppose that the moving particle starts at the point A and that t units of time later it is at the point P. If the arc $\overparen{AP}$ has length s, then the particle has moved s units of distance in t units of time, and we say that its *speed* v is given by

$$v = \frac{s}{t} \text{ units of distance per unit of time.}$$

Here we assume that v is a constant.

Figure 9

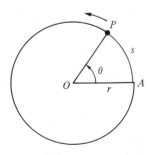

Even though the particle in Figure 9 is moving around the circle, we speak of $v = s/t$ as its **linear speed** (since we could imagine the circular path straightened into a line). Now suppose that the radial line segment $\overline{OP}$ sweeps through an angle of θ radians in t units of time. Then we say that the **angular speed** of the particle is θ/t radians per unit of time. Angular speed is usually denoted by the Greek letter ω (*omega*), so we have

$$\omega = \frac{\theta}{t} \text{ radians per unit of time.}$$

To find the relationship between linear speed v and angular speed ω, we begin by recalling the formula

$$s = r\theta.$$

Dividing both sides of this equation by t, we obtain

$$\frac{s}{t} = r\frac{\theta}{t}$$

or

$$v = r\omega.$$

In words, *linear speed is the product of the radius and the angular speed.*

Example   A stone is whirled around in a circle at the end of a string 70 centimeters long. If it makes 5 revolutions in 2 seconds, find (a) its angular speed ω and (b) its linear speed v.

Solution   (a) Since the stone travels five times around the circle in 2 seconds, the radial line segment (the string) sweeps through $5(2\pi) = 10\pi$ radians in 2 seconds. Therefore, the angular speed of the stone is

$$\omega = \frac{\theta}{t} = \frac{10\pi}{2} = 5\pi \text{ radians per second.}$$

(b) The linear speed of the stone is

$$v = r\omega = 70(5\pi) = 350\pi$$

$$\approx 1100 \text{ centimeters per second.}$$

Until now, we have been careful to distinguish between an *angle* (a rotation) and the *measure* of the angle in degrees or radians (a real number). However, continual use of such phrases as "the measure of the angle α is $30°$" can become tedious, so people often write "$\alpha = 30°$" as an abbreviation for the correct statement. We shall follow this practice whenever it seems convenient and harmless.

PROBLEM SET 1

In problems 1–4, find the degree measure of each angle.

1. One-sixth of a counterclockwise revolution.
2. Two-thirds of a clockwise revolution.
3. Eleven-fifths of a clockwise revolution.
4. The angle through which the small hand on a clock turns in 24 hours.

In problems 5–12, indicate the number of revolutions or the fraction of a revolution represented by the angle whose degree measure is given. Indicate whether the rotation is clockwise or counterclockwise.

5. $45°$	**6.** $-1080°$	**7.** $120°$	**8.** $800°$
9. $-330°$	**10.** $15°$	**11.** $-12°$	**12.** $-700°$

Ⓒ In problems 13–18, express each angle measure as a decimal.

13. $2'$	**14.** $5°6'0''$	**15.** $3''$
16. $20°45''$	**17.** $100°30'20''$	**18.** $-28°70'3''$

Ⓒ In problems 19–24, express each angle measure in degrees, minutes, and seconds.

19. $62.25°$	**20.** $-371.12°$	**21.** $-0.125°$	**22.** $880.23°$
23. $21.16°$	**24.** $1/17$ of a counterclockwise revolution		

In problems 25–30, s denotes the length of the arc intercepted on a circle of radius r by a central angle of θ radians. Find the missing quantity.

25. $r = 2$ meters, $\theta = 1.65$ radians, $s = ?$

26. $r = 1.8$ centimeters, $\theta = 8$ radians, $s = ?$

27. $r = 9$ feet, $s = 12$ feet, $\theta = ?$

28. $s = 4\pi$ kilometers, $\theta = \dfrac{\pi}{2}$ radians, $r = ?$

29. $r = 12$ inches, $\theta = \dfrac{5\pi}{18}$ radian, $s = ?$

30. $r = 5$ meters, $s = 13\pi$ meters, $\theta = ?$

31. Convert each degree measure to radians. Do not use a calculator. Write your answers as rational multiples of π.

  (a) $30°$ (b) $45°$ (c) $90°$ (d) $120°$ (e) $-150°$ (f) $520°$
  (g) $72°$ (h) $67.5°$ (i) $-330°$ (j) $450°$ (k) $21°$ (l) $-360°$

Ⓒ **32.** Use a calculator to convert each degree measure to an approximate radian measure expressed as a decimal. Round off all answers to five significant digits.

  (a) $7°$ (b) $33.333°$ (c) $-11.227°$ (d) $571°$ (e) $1229°$ (f) $0.017320°$

33. Convert each radian measure to degrees. Do not use a calculator.

  (a) $\dfrac{\pi}{2}$ (b) $\dfrac{\pi}{3}$ (c) $\dfrac{\pi}{4}$ (d) $\dfrac{\pi}{6}$ (e) $\dfrac{2\pi}{3}$ (f) $-\pi$

  (g) $\dfrac{3\pi}{5}$ (h) $-\dfrac{5\pi}{2}$ (i) $\dfrac{9\pi}{4}$ (j) $-\dfrac{3\pi}{8}$ (k) 7π (l) $-\dfrac{\pi}{14}$

Ⓒ **34.** Use a calculator to convert each radian measure to an approximate degree measure. Round off all answers to five significant digits.

  (a) $\dfrac{2}{3}$ (b) -2 (c) 200 (d) $\dfrac{7\pi}{12}$ (e) 2.7333 (f) 1.5708

35. Indicate both the degree measure and the radian measure of the angle formed by (a) three-eighths of a clockwise revolution, (b) four and one-sixth counterclockwise revolutions, and (c) the hands of a clock at 4:00.

36. (a) Find a formula for the length s of the arc intercepted by a central angle of α *degrees*.
   (b) Use part (a) to find s if $r = 10$ meters and $\alpha = 35°$.

37. As a monkey swings on a 10-meter rope from one platform to another, the rope turns through an angle of $\pi/3$. How far does the monkey travel along the arc of the swing?

C 38. What radius should be used for a circular monorail track if the track is to change its direction by $7°$ in a distance of 24 meters (measured along the arc of the track)?

C 39. A motorcycle is moving at a speed of 88 kilometers per hour. If the radius of its wheels is 0.38 meter, find the angle in *degrees* through which a spoke of a wheel turns in 3 seconds.

40. A wheel of radius r is turning at a constant speed of n revolutions per second. Find a formula for (a) the angular speed ω and (b) the linear speed v of a point on the rim of the wheel.

C 41. A satellite is orbiting a certain planet in a perfectly circular orbit of radius 7680 kilometers. If it makes two-thirds of a revolution every hour, find (a) its angular speed ω and (b) its linear speed v.

C 42. A **nautical mile** may be defined as the arc length intercepted on the surface of the earth by a central angle of measure 1 minute. The radius of the earth is 2.09×10^7 feet. How many feet are there in a nautical mile?

C 43. A girl rides her bicycle to school at a speed of 12 miles per hour. If the wheel diameter is 26 inches, what is the angular speed of the wheels?

C 44. To determine the approximate speed of the water in a river, a water wheel of radius 1.5 meters is used. If the wheel makes 13 revolutions per minute, what is the speed of the river in kilometers per hour?

C 45. A $33\frac{1}{3}$-rpm (revolutions per minute) phonograph record has a radius of 14.6 centimeters. What is the linear speed (in centimeters per second) of a point on the rim of the record?

46. A first pulley of radius r_1 and a second pulley of radius r_2 are connected by a belt (Figure 10). The first pulley makes n_1 revolutions per minute and the second makes n_2 revolutions per minute. A point on the rim of either pulley will be moving at the same linear speed v as the belt. (a) Find formulas for n_1 and n_2 in terms of v, r_1, and r_2. (b) Show that $n_1/n_2 = r_2/r_1$.

Figure 10   Figure 11

C 47. The earth, which is approximately 93,000,000 miles from the sun, revolves about the sun in a nearly circular orbit in approximately 365 days. Find the approximate linear speed (in miles per hour) of the earth in its orbit.

48. A **sector** OPQ of a circle is the region inside the circle bounded by the arc $\overset{\frown}{PQ}$ and the radial segments $\overline{OP}$ and $\overline{OQ}$ (Figure 11). If the central angle POQ has a measure of θ radians and r is the radius of the circle, show that the area A of the sector is given by $A = \frac{1}{2}r^2\theta$. [Hint: The area of the circle is πr^2. What fraction of this is the area of the sector?]

49. Use the formula $A = \frac{1}{2}r^2\theta$ (problem 48) to find the area A of a circular sector of radius r with a central angle of θ radians if (a) $r = 7$ centimeters and $\theta = 3\pi/14$ and (b) $r = 9$ inches and $\theta = 13\pi/9$.

50. A radar beam has an effective range of 70 miles and sweeps through an angle of $3\pi/4$ radians. What is the effective area in square miles swept by the radar beam?

2 Trigonometric Functions of Acute Angles in Right Triangles

In this section, we introduce the six functions—sine, cosine, tangent, cotangent, secant, and cosecant—upon which trigonometry is based. Here we define the values of these functions for acute angles in right triangles; in Section 3, we extend the definitions to general angles.

An angle with a measure greater than $0°$ but less than $90°$ is called an **acute angle.** For instance, in a right triangle ACB (Figure 1a) the vertex angle C is a right angle $(90°)$, and the other two vertex angles A and B are acute angles. Recall that the side $\overline{AB}$ opposite the right angle C is called the *hypotenuse.* If we denote one of the two acute angles by θ, we can abbreviate the lengths of the side **opposite** θ, the **hypotenuse,** and the side **adjacent** to θ as **opp, hyp,** and **adj,** respectively (Figure 1b). The six trigonometric functions of the acute angle θ are defined as follows:

Figure 1

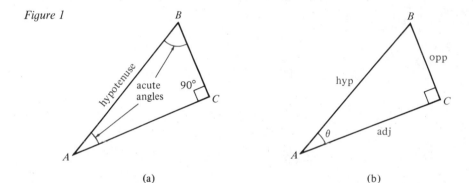

(a)   (b)

Name of Function	Abbreviation	Value at θ
sine	sin	$\sin \theta = \dfrac{\text{opp}}{\text{hyp}}$
cosine	cos	$\cos \theta = \dfrac{\text{adj}}{\text{hyp}}$
tangent	tan	$\tan \theta = \dfrac{\text{opp}}{\text{adj}}$
cotangent	cot	$\cot \theta = \dfrac{\text{adj}}{\text{opp}}$
secant	sec	$\sec \theta = \dfrac{\text{hyp}}{\text{adj}}$
cosecant	csc	$\csc \theta = \dfrac{\text{hyp}}{\text{opp}}$

The values of the six trigonometric functions depend only on the angle θ and not on the size of the right triangle ACB. Indeed, two right triangles with the same acute vertex angle θ are similar (Figure 2); hence, ratios of corresponding sides are equal.

Figure 2

similar triangles

Figure 3

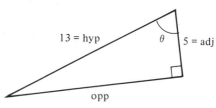

Example   Find the values of the six trigonometric functions of the acute angle θ in Figure 3.

Solution   For angle θ, we have

$$adj = 5 \quad and \quad hyp = 13,$$

but opp isn't given. However, by the Pythagorean theorem,

$$(adj)^2 + (opp)^2 = (hyp)^2,$$

so that

$$(opp)^2 = (hyp)^2 - (adj)^2 = 13^2 - 5^2 = 144.$$

Therefore,

$$opp = \sqrt{144} = 12.$$

It follows that

$$\sin \theta = \frac{opp}{hyp} = \frac{12}{13} \qquad \csc \theta = \frac{hyp}{opp} = \frac{13}{12}$$

$$\cos \theta = \frac{adj}{hyp} = \frac{5}{13} \qquad \sec \theta = \frac{hyp}{adj} = \frac{13}{5}$$

$$\tan \theta = \frac{opp}{adj} = \frac{12}{5} \qquad \cot \theta = \frac{adj}{opp} = \frac{5}{12}.$$

If you look at the ratios that define the six trigonometric functions, you will see that

$$\tan \theta = \frac{\sin \theta}{\cos \theta} \qquad \cot \theta = \frac{\cos \theta}{\sin \theta}$$

$$\sec \theta = \frac{1}{\cos \theta} \qquad \csc \theta = \frac{1}{\sin \theta}.$$

For instance,

$$\tan \theta = \frac{\text{opp}}{\text{adj}} = \frac{\text{opp/hyp}}{\text{adj/hyp}} = \frac{\sin \theta}{\cos \theta},$$

and the remaining equations are derived similarly (Problem 42). Perhaps the easiest way for you to become familiar with all six of the trigonometric functions is to learn the two basic definitions

$$\sin \theta = \frac{\text{opp}}{\text{hyp}} \quad \text{and} \quad \cos \theta = \frac{\text{adj}}{\text{hyp}}$$

and use the equations obtained above to write the values of the other four functions in terms of sines and cosines. Thus, in the example above, once you know that

$$\sin \theta = \frac{12}{13} \quad \text{and} \quad \cos \theta = \frac{5}{13},$$

you can immediately write

$$\tan \theta = \frac{\sin \theta}{\cos \theta} = \frac{12/13}{5/13} = \frac{12}{5} \qquad \cot \theta = \frac{\cos \theta}{\sin \theta} = \frac{5/13}{12/13} = \frac{5}{12}$$

$$\sec \theta = \frac{1}{\cos \theta} = \frac{1}{5/13} = \frac{13}{5} \qquad \csc \theta = \frac{1}{\sin \theta} = \frac{1}{12/13} = \frac{13}{12}.$$

Two angles are called **complementary** if their sum is a right angle. For instance, *the two acute angles of a right triangle are* **complementary angles**. (Why?) If θ is the degree measure of an angle, then the degree measure of the complementary angle is $90° - \theta$. For radian measure, the angle $(\pi/2) - \theta$ is complementary to θ.

Example Find the angle complementary to θ if (a) $\theta = 37°$ and (b) $\theta = \pi/5$.

Solution (a) $90° - \theta = 90° - 37° = 53°$ (b) $\dfrac{\pi}{2} - \dfrac{\pi}{5} = \dfrac{5\pi}{10} - \dfrac{2\pi}{10} = \dfrac{3\pi}{10}$

In Figure 4, the two acute angles α and β of right triangle ABC are complementary, a is the length of the side opposite α, b is the length of the side opposite β, and c is the length of the hypotenuse. Notice that the side opposite either one of the acute angles is adjacent to the other. Therefore,

Figure 4

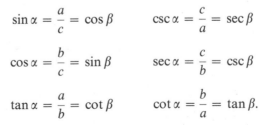

$$\sin \alpha = \frac{a}{c} = \cos \beta \qquad\qquad \csc \alpha = \frac{c}{a} = \sec \beta$$

$$\cos \alpha = \frac{b}{c} = \sin \beta \qquad\qquad \sec \alpha = \frac{c}{b} = \csc \beta$$

$$\tan \alpha = \frac{a}{b} = \cot \beta \qquad\qquad \cot \alpha = \frac{b}{a} = \tan \beta.$$

By using the idea of *cofunctions*, you can remember the equations above more easily. We say that sine and cosine are **cofunctions** of one another. Likewise, tangent and cotangent are said to be cofunctions, and so are secant and cosecant.

Thus, *the value of any trigonometric function of an acute angle is the same as the value of the cofunction of the complementary angle:*

> "Cosine" is "sine of the complementary angle."
> "Cotangent" is "tangent of the complementary angle."
> "Cosecant" is "secant of the complementary angle."

Example

Express the value of each trigonometric function as the value of the cofunction of the complementary angle.

(a) $\cos 70°$ (b) $\cot \dfrac{\pi}{12}$ (c) $\sin 10°$ (d) $\sec \dfrac{\pi}{4}$

Solution

(a) $\cos 70° = \sin(90° - 70°) = \sin 20°$ (b) $\cot \dfrac{\pi}{12} = \tan\left(\dfrac{\pi}{2} - \dfrac{\pi}{12}\right) = \tan \dfrac{5\pi}{12}$

(c) $\sin 10° = \cos(90° - 10°) = \cos 80°$ (d) $\sec \dfrac{\pi}{4} = \csc\left(\dfrac{\pi}{2} - \dfrac{\pi}{4}\right) = \csc \dfrac{\pi}{4}$

The special angles with measures of 30° ($\pi/6$ radian), 45° ($\pi/4$ radian), and 60° ($\pi/3$ radians) have a way of showing up over and over again in various applications of trigonometry. Notice that 30° and 60° angles are complementary, whereas a 45° angle is complementary to itself.

Using some elementary geometry, we can find the values of the six trigonometric functions of the special angles. To begin with, consider an equilateral triangle ABC having sides of length 2 units (Figure 5a). Recall from plane geometry that the vertex angles of an equilateral triangle all measure 60°, and that the bisector $\overline{AD}$ of vertex angle A is the perpendicular bisector of the opposite side $\overline{BC}$ (Figure 5b). Therefore,

$$|\overline{BD}| = \tfrac{1}{2}|\overline{BC}| = \tfrac{1}{2}(2) = 1 \quad\text{and}\quad \text{angle } DAB = \tfrac{1}{2}(60°) = 30°.$$

Applying the Pythagorean theorem to right triangle ADB, we have

$$|\overline{AD}|^2 + |\overline{BD}|^2 = |\overline{AB}|^2 \quad\text{or}\quad |\overline{AD}|^2 + 1^2 = 2^2,$$

so that

$$|\overline{AD}|^2 = 2^2 - 1^2 = 3 \quad\text{or}\quad |\overline{AD}| = \sqrt{3}.$$

Figure 5

(a)

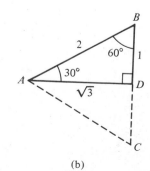

(b)

Thus, from triangle ADB in Figure 5b, we have:

$$\sin 30° = \cos 60° = 1/2 \qquad\qquad \csc 30° = \sec 60° = 2$$

$$\cos 30° = \sin 60° = \sqrt{3}/2 \qquad\qquad \sec 30° = \csc 60° = 2/\sqrt{3} = 2\sqrt{3}/3$$

$$\tan 30° = \cot 60° = 1/\sqrt{3} = \sqrt{3}/3 \qquad \cot 30° = \tan 60° = \sqrt{3}.$$

To find the values of the trigonometric functions of a 45° angle, we construct an isosceles right triangle with two equal sides of length 1 unit (Figure 6). We know from plane geometry that both of the acute angles in this right triangle measure 45°. Also, by the Pythagorean theorem,

$$1^2 + 1^2 = (\text{hyp})^2 \quad \text{or} \quad 2 = (\text{hyp})^2$$

so that

$$\text{hyp} = \sqrt{2}.$$

Figure 6

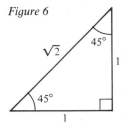

It follows that

$$\sin 45° = 1/\sqrt{2} = \sqrt{2}/2 \qquad \csc 45° = \sqrt{2}$$

$$\cos 45° = 1/\sqrt{2} = \sqrt{2}/2 \qquad \sec 45° = \sqrt{2}$$

$$\tan 45° = 1 \qquad\qquad\qquad \cot 45° = 1.$$

For convenience, the values of the trigonometric functions for the special angles are gathered in Table 1.

Table 1

θ degrees	θ radians	$\sin\theta$	$\cos\theta$	$\tan\theta$	$\cot\theta$	$\sec\theta$	$\csc\theta$
30°	$\dfrac{\pi}{6}$	$\dfrac{1}{2}$	$\dfrac{\sqrt{3}}{2}$	$\dfrac{\sqrt{3}}{3}$	$\sqrt{3}$	$\dfrac{2\sqrt{3}}{3}$	2
45°	$\dfrac{\pi}{4}$	$\dfrac{\sqrt{2}}{2}$	$\dfrac{\sqrt{2}}{2}$	1	1	$\sqrt{2}$	$\sqrt{2}$
60°	$\dfrac{\pi}{3}$	$\dfrac{\sqrt{3}}{2}$	$\dfrac{1}{2}$	$\sqrt{3}$	$\dfrac{\sqrt{3}}{3}$	2	$\dfrac{2\sqrt{3}}{3}$

You can find values of the trigonometric functions for angles other than the special angles in Table 1 by using a scientific calculator. If a calculator isn't available, you can use Appendix Tables IIA and IIB. Instructions for using these tables are included in Appendix II. Before using a calculator, you must decide whether you are going to measure angles in *degrees* or *radians* and set the calculator accordingly. (Refer to the instruction booklet furnished with your calculator.) Most calculators can deal with fractions of a degree only in decimal form. If angles are given in degrees, minutes, and seconds, you must first convert to decimal form before pressing a trigonometric key.

Examples   ⓒ Use a calculator to find the approximate value of each trigonometric function.

1 $\sin 41.3°$

Solution   We make sure the calculator is set in degree mode, enter 41.3, and press the sine (SIN) key. On a 10-digit calculator we get

$$\sin 41.3° = 0.660001668.$$

An 8-digit calculator gives this result rounded off as

$$\sin 41.3° = 0.6600017.$$

If a calculator isn't available, Appendix Table IIA can be used to obtain the result rounded off to four decimal places:

$$\sin 41.3° = 0.6600.$$

2 $\cos 19°21'17''$

Solution   We begin by using a calculator to convert to decimal form:

$$19°21'17'' = 19° + \left(\frac{21}{60}\right)° + \left(\frac{17}{3600}\right)° = 19.35472222°.$$

Now, we make sure the calculator is set in degree mode, and press the cosine (COS) key to get

$$\cos 19°21'17'' = \cos 19.35472222° = 0.943484853.$$

3 $\tan \dfrac{5\pi}{36}$

Solution   We set the calculator in *radian mode*, press the π key (or enter 3.141592654), multiply by 5, divide by 36, and press the tangent (TAN) key to obtain

$$\tan \frac{5\pi}{36} = 0.466307658.$$

4 $\sin 0.8432$

Solution   Because degrees are not indicated, we know that the angle is measured in radians. We set the calculator in radian mode, enter 0.8432, and press the sine (SIN) key to obtain

$$\sin 0.8432 = 0.746775185.$$

Most scientific calculators have keys only for sine, cosine, and tangent. To evaluate the cotangent function on such a calculator, you can use the fact that

$$\cot \theta = \frac{1}{\tan \theta}$$

(Problem 43). Similarly, to evaluate the secant or cosecant functions, you can use the relations

$$\sec \theta = \frac{1}{\cos \theta} \quad \text{and} \quad \csc \theta = \frac{1}{\sin \theta}.$$

Examples · © Use a calculator to find the approximate values of each trigonometric function.

1   cot 50°

Solution   With the calculator in degree mode, we enter 50, press the tangent (TAN) key, then we take the reciprocal of the result by pressing the 1/x key. (We're using the fact that cot 50° = 1/tan 50°.) The result is

$$\cot 50° = 0.839099631.$$

2   sec 1.098

Solution   With the calculator in radian mode, we enter 1.098, press the cosine (COS) key, and then press the 1/x key. The result is

$$\sec 1.098 = 2.195979642.$$

PROBLEM SET 2

In problems 1–8, find the values of the six trigonometric functions of the acute angle θ in the indicated figure.

1.

2.

3.

4.

5.

6.

7.

8.
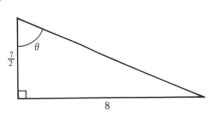

In problems 9–16, find the angle complementary to each of the given angles.

9. $35°$ **10.** $89°$ **11.** $\dfrac{\pi}{5}$ **12.** $\dfrac{\pi}{7}$ **13.** $\dfrac{5\pi}{12}$

14. $\dfrac{7\pi}{33}$ **15.** $22.34°$ **16.** one-sixth of a counterclockwise revolution

In problems 17–24, express the value of each trigonometric function as the value of the co-function of the complementary angle.

17. $\sin 31°$ **18.** $\cos 13°$ **19.** $\tan \dfrac{\pi}{9}$ **20.** $\csc \dfrac{\pi}{11}$

21. $\cot \dfrac{2\pi}{5}$ **22.** $\sin 9.03°$ **23.** $\csc 77.03°$ **24.** $\tan 0.774$

C In problems 25–34, use a calculator to find the approximate values of the six trigonometric functions of each angle.

25. $48°$ **26.** $7°$ **27.** $23°12'33''$ **28.** $82°55'55''$ **29.** $16.19°$

30. $\dfrac{9\pi}{10}$ **31.** $\dfrac{2\pi}{7}$ **32.** $\dfrac{5\pi}{21}$ **33.** 0.7764 **34.** 1.33

C In problems 35–38, use a calculator to verify that the given equation is true for the indicated value of the angle.

35. $\cos \theta = \sin(90° - \theta)$ for $\theta = 33°$ **36.** $\sin \theta = \cos(90° - \theta)$ for $\theta = 77.77°$

37. $\sin \theta = \cos\left(\dfrac{\pi}{2} - \theta\right)$ for $\theta = \dfrac{2\pi}{9}$ **38.** $\cos \theta = \sin\left(\dfrac{\pi}{2} - \theta\right)$ for $\theta = 1.0785$

39. How could you check to make sure that a calculator is set in degree mode? [Hint: We know that $\sin 30° = \frac{1}{2}$.]

40. Justify the following procedure: To convert from degrees, minutes, and seconds to degrees in decimal form, divide the number of seconds by 60, add the quotient to the number of minutes, divide this sum by 60, and add the result to the number of degrees.

C **41.** Using a calculator, verify the entries in Table 1 on page 281.

42. Show that

(a) $\cot \theta = \dfrac{\cos \theta}{\sin \theta}$ (b) $\sec \theta = \dfrac{1}{\cos \theta}$ (c) $\csc \theta = \dfrac{1}{\sin \theta}$

43. Show that $\cot \theta = \dfrac{1}{\tan \theta}$.

C **44.** Use a calculator to verify that $\tan \theta = \dfrac{\sin \theta}{\cos \theta}$ for

(a) $\theta = 10°$ (b) $\theta = \dfrac{3\pi}{13}$ (c) $\theta = 0.9474$

45. Using the Pythagorean theorem, show that, if θ is an acute angle in a right triangle, then

$$(\cos \theta)^2 + (\sin \theta)^2 = 1.$$

C **46.** Use a calculator to verify that $(\cos \theta)^2 + (\sin \theta)^2 = 1$ for

(a) $\theta = 20°$ (b) $\theta = \dfrac{4\pi}{17}$ (c) $\theta = 0.5678$

3 Trigonometric Functions of General Angles

In order to extend the definitions of the six trigonometric functions to general angles, we shall make use of the following ideas: In a cartesian coordinate system, an angle α is said to be in **standard position** if its vertex is at the origin O and its initial side coincides with the positive x axis (Figure 1). An angle is said to be *in* a certain quadrant if, when the angle is in standard position, the terminal side lies in that quadrant. For instance, a 65° angle is in quadrant I (Figure 2a). Slight variations of this terminology are often used; for instance, we might also say that a 65° angle **lies in quadrant I** or simply that it is a **quadrant I angle**. As Figure 2b shows, an angle of $-187°$ is a quadrant II angle.

If the terminal side of an angle in standard position lies along either the x axis or the y axis, then the angle is called **quadrantal**. For example, $-360°$, $-270°$, $-180°$, $-90°$, 0°, 90°, 180°, 270°, and 360° are quadrantal angles. Evidently, an angle is quadrantal if and only if its measure is an integer multiple of 90° ($\pi/2$ radians).

Figure 1

Figure 2

(a)

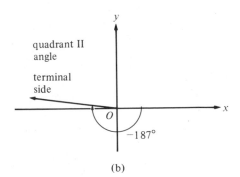

(b)

Example 1 For each point given, sketch two coterminal angles α and β in standard position whose terminal side contains the point. Arrange it so that α is positive, β is negative, and neither angle exceeds one revolution. In each case, name the quadrant in which the angle lies, or indicate that the angle is quadrantal.

(a) $(4, 3)$ (b) $(-4, 4)$ (c) $(-3, 0)$ (d) $(3, -4)$

Figure 3

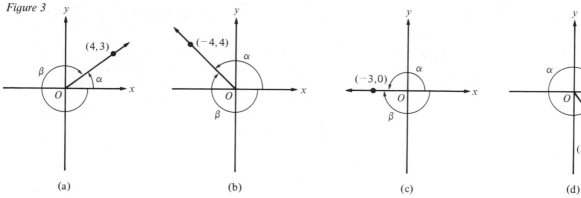

    (a)   (b)   (c)   (d)

Solution   (a) α and β are quadrant I angles (Figure 3a).

                (b) α and β are quadrant II angles (Figure 3b).

                (c) α and β are quadrantal angles (Figure 3c).

                (d) α and β are quadrant IV angles (Figure 3d).

Example 2   Specify and sketch three angles that are coterminal with a $30°$ angle in standard position.

Solution   The three angles shown in Figure 4 are obtained as follows:

$$30° + 360° = 390°,$$

$$30° + 2(360°) = 750°,$$

$$30° - 360° = -330°.$$

Of course, there are other possibilities, such as

$$30° + 5(360°) = 1830°, \quad 30° - 13(360°) = -4650°,$$

and so on.

Figure 4

Figure 5

Now, let θ be one of the acute angles of a right triangle OQP placed on a cartesian coordinate system with θ in standard position and the hypotenuse $\overline{OP}$ in quadrant I (Figure 5). Then $P = (x, y)$ is on the terminal side of θ, and we have

$$x = \text{adj} \quad \text{and} \quad y = \text{opp.}$$

If we let

$$r = \text{hyp} = \sqrt{x^2 + y^2},$$

then the six trigonometric functions of θ are given by the ratios shown in Figure 5.

Figure 5 suggests the following definition.

Definition 1   **Trigonometric Functions of a General Angle**

Let θ be an angle in standard position and suppose that (x, y) is any point other than $(0, 0)$ on the terminal side of θ (Figure 6). If $r = \sqrt{x^2 + y^2}$ is the distance between (x, y) and $(0, 0)$, then the six trigonometric functions of θ are defined by

$$\sin \theta = \frac{y}{r} \qquad \csc \theta = \frac{r}{y}$$

$$\cos \theta = \frac{x}{r} \qquad \sec \theta = \frac{r}{x}$$

$$\tan \theta = \frac{y}{x} \qquad \cot \theta = \frac{x}{y},$$

provided that the denominators are not zero.

Figure 6

By using similar triangles, you can see that the values of the six trigonometric functions in Definition 1 depend only on the angle θ and not on the choice of the point (x, y) on the terminal side of θ (Problem 30).

Example   Evaluate the six trigonometric functions of the angle θ in standard position if the terminal side of θ contains the point $(x, y) = (2, -1)$ (Figure 7).

Figure 7

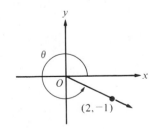

Solution   Here $x = 2$, $y = -1$, and

$$r = \sqrt{x^2 + y^2} = \sqrt{2^2 + (-1)^2} = \sqrt{5}.$$

Thus,

$$\sin \theta = \frac{y}{r} = \frac{-1}{\sqrt{5}} = -\frac{\sqrt{5}}{5} \qquad \csc \theta = \frac{r}{y} = \frac{\sqrt{5}}{-1} = -\sqrt{5}$$

$$\cos \theta = \frac{x}{r} = \frac{2}{\sqrt{5}} = \frac{2\sqrt{5}}{5} \qquad \sec \theta = \frac{r}{x} = \frac{\sqrt{5}}{2}$$

$$\tan \theta = \frac{y}{x} = \frac{-1}{2} = -\frac{1}{2} \qquad \cot \theta = \frac{x}{y} = \frac{2}{-1} = -2.$$

Figure 8

$x < 0, y > 0, r > 0$	$x > 0, y > 0, r > 0$
$\sin \theta > 0$	$\sin \theta > 0$
$\cos \theta < 0$	$\cos \theta > 0$
Q_{II} $\tan \theta < 0$	$\tan \theta > 0$
$\cot \theta < 0$	$\cot \theta > 0$ Q_I
$\sec \theta < 0$	$\sec \theta > 0$
$\csc \theta > 0$	$\csc \theta > 0$
$x < 0, y < 0, r > 0$	$x > 0, y < 0, r > 0$
$\sin \theta < 0$	$\sin \theta < 0$
$\cos \theta < 0$	$\cos \theta > 0$
Q_{III} $\tan \theta > 0$	$\tan \theta < 0$
$\cot \theta > 0$	$\cot \theta < 0$ Q_{IV}
$\sec \theta < 0$	$\sec \theta > 0$
$\csc \theta < 0$	$\csc \theta < 0$

If (x, y) is a point in quadrant I, then x, y, and $r = \sqrt{x^2 + y^2}$ are positive, so all six trigonometric functions have positive values for quadrant I angles. In quadrant II, x is negative while y and r are positive; so the values of the sine and cosecant are positive, whereas the values of the cosine, tangent, cotangent, and secant are negative. The algebraic signs of the trigonometric functions for angles in quadrants III and IV are easily found by observing the algebraic signs of x and y in these quadrants and keeping in mind that r is always positive (Figure 8).

Example   Find the quadrant in which θ lies if $\tan \theta > 0$ and $\sin \theta < 0$.

Solution   Let (x, y) be a point on the terminal side of θ (in standard position). Since $\tan \theta = \dfrac{y}{x} > 0$, it follows that x and y have the same algebraic sign. In addition, since $\sin \theta = \dfrac{y}{r} < 0$, it follows that $y < 0$; therefore, $x < 0$. Because $x < 0$ and $y < 0$, the point (x, y) is in quadrant III; hence, θ is a quadrant III angle.

The same basic relationships that we obtained in Section 2 for the trigonometric functions of acute angles continue to hold for general angles.

Theorem 1   **Reciprocal Identities**

If θ is an angle for which the functions are defined, then:

(i) $\csc \theta = \dfrac{1}{\sin \theta}$ (ii) $\sec \theta = \dfrac{1}{\cos \theta}$ (iii) $\cot \theta = \dfrac{1}{\tan \theta}$

Proof

Figure 9

Let (x, y) be a point on the terminal side of θ in standard position and let $r = \sqrt{x^2 + y^2}$ (Figure 9). Then, provided that the denominators are not zero:

$$\text{(i)} \quad \csc \theta = \frac{r}{y} = \frac{1}{y/r} = \frac{1}{\sin \theta}$$

$$\text{(ii)} \quad \sec \theta = \frac{r}{x} = \frac{1}{x/r} = \frac{1}{\cos \theta}$$

$$\text{(iii)} \quad \cot \theta = \frac{x}{y} = \frac{1}{y/x} = \frac{1}{\tan \theta}$$

In a similar way, you can prove the following theorem (Problem 35).

Theorem 2   **Quotient Identities**

If θ is an angle for which the functions are defined, then:

$$\tan \theta = \frac{\sin \theta}{\cos \theta} \quad \text{and} \quad \cot \theta = \frac{\cos \theta}{\sin \theta}.$$

Example   If $\sin \theta = 1/2$ and $\cos \theta = -\sqrt{3}/2$, find the values of the other four trigonometric functions of θ.

Solution   Using Theorems 1 and 2, we have

$$\tan \theta = \frac{\sin \theta}{\cos \theta} = \frac{1/2}{(-\sqrt{3}/2)} = -\frac{\sqrt{3}}{3} \qquad \cot \theta = \frac{1}{\tan \theta} = \frac{1}{(-\sqrt{3}/3)} = -\sqrt{3}$$

$$\sec \theta = \frac{1}{\cos \theta} = \frac{1}{(-\sqrt{3}/2)} = -\frac{2\sqrt{3}}{3} \qquad \csc \theta = \frac{1}{\sin \theta} = \frac{1}{1/2} = 2$$

Another important relationship is derived as follows: Again suppose that θ is an angle in standard position and that (x, y) is a point on the terminal side of θ (Figure 9). Because $r = \sqrt{x^2 + y^2}$, we have $x^2 + y^2 = r^2$, so

$$(\cos \theta)^2 + (\sin \theta)^2 = \left(\frac{x}{r}\right)^2 + \left(\frac{y}{r}\right)^2 = \frac{x^2}{r^2} + \frac{y^2}{r^2} = \frac{x^2 + y^2}{r^2} = \frac{r^2}{r^2} = 1.$$

The relationship

$$(\cos \theta)^2 + (\sin \theta)^2 = 1$$

is called the **fundamental Pythagorean identity** because its derivation involves the fact that $x^2 + y^2 = r^2$, which is a consequence of the Pythagorean theorem.

The fundamental Pythagorean identity is used quite often, and it would be bothersome to write the parentheses each time for $(\cos \theta)^2$ and $(\sin \theta)^2$; yet, if the parentheses were simply omitted, the resulting expressions would be misunderstood. (For instance, $\cos \theta^2$ is usually understood to mean the cosine of the square of θ.) Therefore, it is customary to write $\cos^2\theta$ and $\sin^2\theta$ to mean $(\cos\theta)^2$ and $(\sin\theta)^2$. Similar notation is used for the remaining trigonometric functions and for powers other than 2. Thus, $\cot^4\theta$ means $(\cot\theta)^4$, $\sec^n\theta$ means $(\sec\theta)^n$, and so forth. With this notation, the fundamental Pythagorean identity becomes

$$\cos^2\theta + \sin^2\theta = 1.$$

Actually, there are three Pythagorean identities—the fundamental identity and two others derived from it.

Theorem 3   **Pythagorean Identities**

If θ is an angle for which the functions are defined, then:

(i) $\cos^2\theta + \sin^2\theta = 1$

(ii) $1 + \tan^2\theta = \sec^2\theta$

(iii) $\cot^2\theta + 1 = \csc^2\theta$

Proof   We have already proved (i). To prove (ii), we divide both sides of (i) by $\cos^2\theta$ to obtain

$$1 + \frac{\sin^2\theta}{\cos^2\theta} = \frac{1}{\cos^2\theta}$$

or

$$1 + \left(\frac{\sin\theta}{\cos\theta}\right)^2 = \left(\frac{1}{\cos\theta}\right)^2,$$

provided that $\cos\theta \neq 0$. Since

$$\frac{\sin\theta}{\cos\theta} = \tan\theta \quad\text{and}\quad \frac{1}{\cos\theta} = \sec\theta,$$

we have

$$1 + \tan^2\theta = \sec^2\theta.$$

Identity (iii) is proved by dividing both sides of (i) by $\sin^2\theta$ (Problem 34).

Examples   In the following examples, the value of one of the trigonometric functions of an angle θ is given along with information about the quadrant in which θ lies. Find the values of the other five trigonometric functions of θ.

1 $\sin\theta = \dfrac{5}{13}$, θ in quadrant II.

Solution   By the fundamental Pythagorean identity, $\cos^2\theta + \sin^2\theta = 1$, so

$$\cos^2\theta = 1 - \sin^2\theta = 1 - \left(\frac{5}{13}\right)^2 = 1 - \frac{25}{169} = \frac{169 - 25}{169} = \frac{144}{169}.$$

Therefore,

$$\cos\theta = \pm\sqrt{\frac{144}{169}} = \pm\frac{12}{13}.$$

Because θ is in quadrant II, we know that $\cos\theta$ is negative; hence,

$$\cos\theta = -\frac{12}{13}.$$

It follows that

$$\tan\theta = \frac{\sin\theta}{\cos\theta} = \frac{5/13}{(-12/13)} = -\frac{5}{12} \qquad \cot\theta = \frac{1}{\tan\theta} = \frac{1}{(-5/12)} = -\frac{12}{5}$$

$$\sec\theta = \frac{1}{\cos\theta} = \frac{1}{(-12/13)} = -\frac{13}{12} \qquad \csc\theta = \frac{1}{\sin\theta} = \frac{1}{5/13} = \frac{13}{5}.$$

2 $\tan\theta = -8/7$ and $\sin\theta < 0$.

Solution   Because $\tan\theta < 0$ only in quadrants II and IV, and $\sin\theta < 0$ only in quadrants III and IV, it follows that θ must be in quadrant IV. By part (ii) of Theorem 3, $\sec^2\theta = 1 + \tan^2\theta$, so

$$\sec\theta = \pm\sqrt{1 + \tan^2\theta} = \pm\sqrt{1 + \left(-\frac{8}{7}\right)^2} = \pm\sqrt{1 + \frac{64}{49}} = \pm\sqrt{\frac{113}{49}} = \pm\frac{\sqrt{113}}{7}.$$

Since θ is in quadrant IV, $\sec\theta > 0$; hence,

$$\sec\theta = \frac{\sqrt{113}}{7}.$$

Because

$$\sec\theta = \frac{1}{\cos\theta},$$

it follows that

$$\cos\theta = \frac{1}{\sec\theta} = \frac{1}{\sqrt{113}/7} = \frac{7}{\sqrt{113}} = \frac{7\sqrt{113}}{113}.$$

Now,

$$\tan\theta = \frac{\sin\theta}{\cos\theta}$$

so

$$\sin\theta = (\tan\theta)(\cos\theta) = \left(-\frac{8}{7}\right)\left(\frac{7\sqrt{113}}{113}\right) = -\frac{8\sqrt{113}}{113}.$$

Finally,

$$\csc\theta = \frac{1}{\sin\theta} = \frac{1}{(-8\sqrt{113}/113)} = -\frac{\sqrt{113}}{8}$$

and

$$\cot\theta = \frac{1}{\tan\theta} = \frac{1}{(-8/7)} = -\frac{7}{8}.$$

PROBLEM SET 3

In problems 1–10, sketch two coterminal angles α and β in standard position whose terminal side contains the given point. Arrange it so that α is positive, β is negative, and neither angle exceeds one revolution. In each case, name the quadrant in which the angle lies, or indicate that the angle is quadrantal.

1. $(1,2)$ **2.** $(2,-\frac{4}{3})$ **3.** $(-5,0)$ **4.** $(-3,-4)$
5. $(5,-3)$ **6.** $(0,4)$ **7.** $(-1,1)$ **8.** $(\sqrt{2}/2,0)$
9. $(-\frac{3}{4},-\frac{3}{4})$ **10.** $(0,-3)$

In problems 11–18, specify and sketch three angles that are coterminal with the given angle in standard position.

11. $60°$ **12.** $-15°$ **13.** $-\dfrac{\pi}{4}$ **14.** $\dfrac{4\pi}{3}$

15. $-612°$ **16.** $-\dfrac{7\pi}{4}$ **17.** $\dfrac{5\pi}{6}$ **18.** $1440°$

In problems 19–28, evaluate the six trigonometric functions of the angle θ in standard position if the terminal side of θ contains the given point (x, y). [Do not use a calculator—leave all answers in the form of a fraction or an integer.] In each case, sketch one of the coterminal angles θ.

19. $(4,3)$ **20.** $(2,7)$ **21.** $(-5,12)$ **22.** $(-2,4)$
23. $(-3,-4)$ **24.** $(1,1)$ **25.** $(7,3)$ **26.** $(1,-3)$
27. $(-\sqrt{3},-\sqrt{2})$ **28.** $(20,21)$

29. Is there any angle θ for which $\sin\theta = \frac{5}{4}$? Explain.
30. Using similar triangles, show that the values of the six trigonometric functions in Definition 1 depend only on the angle θ and not on the choice of the point (x, y) on the terminal side of θ.
31. In each case, assume that θ is an angle in standard position and find the quadrant in which it lies.
 (a) $\tan\theta > 0$ and $\sec\theta > 0$ (b) $\sin\theta > 0$ and $\sec\theta < 0$
 (c) $\sin\theta > 0$ and $\cos\theta < 0$ (d) $\sec\theta > 0$ and $\tan\theta < 0$
 (e) $\tan\theta > 0$ and $\csc\theta < 0$ (f) $\cos\theta < 0$ and $\csc\theta < 0$
 (g) $\sec\theta > 0$ and $\cot\theta < 0$ (h) $\cot\theta > 0$ and $\sin\theta > 0$
32. Is there any angle θ for which $\sin\theta > 0$ and $\csc\theta < 0$? Explain.
33. Give the algebraic sign of each of the following.

 (a) $\cos 163°$ (b) $\sin 211°$ (c) $\sec(-355°)$ (d) $\tan\dfrac{7\pi}{4}$

 (e) $\cot\left(-\dfrac{28\pi}{45}\right)$ (f) $\csc\left(-\dfrac{5\pi}{12}\right)$ (g) $\sec\dfrac{38\pi}{15}$

34. Prove part (iii) of Theorem 3.
35. Prove Theorem 2.
36. If θ is an angle for which the functions are defined, show that $\sec\theta - (\sin\theta)(\tan\theta) = \cos\theta$.
37. If $\sin\theta = -\frac{5}{13}$ and $\cos\theta = \frac{12}{13}$, use the reciprocal and quotient identities to find (a) $\sec\theta$, (b) $\csc\theta$, (c) $\tan\theta$, and (d) $\cot\theta$.
38. If $\sec\theta = -\frac{5}{4}$ and $\csc\theta = -\frac{5}{3}$, use the reciprocal and quotient identities to find (a) $\sin\theta$, (b) $\cos\theta$, (c) $\tan\theta$, and (d) $\cot\theta$.

In problems 39–56, the value of one of the trigonometric functions of an angle θ is given along with information about the quadrant (Q) in which θ lies. Find the values of the other five trigonometric functions of θ.

39. $\sin \theta = \frac{4}{5}$, θ in Q_{I}

40. $\cos \theta = \frac{12}{13}$, θ in Q_{IV}

41. $\sin \theta = -\frac{3}{4}$, θ in Q_{III}

42. $\sin \theta = \frac{3}{4}$, θ not in Q_{I}

43. $\cos \theta = \frac{4}{7}$, $\sin \theta < 0$

44. $\cos \theta = \frac{1}{3}$, θ not in Q_{I}

45. $\csc \theta = \frac{3}{2}$, θ in Q_{I}

46. $\sec \theta = -\frac{5}{3}$, θ in Q_{III}

47. $\tan \theta = \frac{4}{3}$, θ in Q_{I}

48. $\tan \theta = 2$, $\sin \theta < 0$

49. $\cot \theta = -\frac{12}{5}$, $\csc \theta > 0$

50. $\csc \theta = -\frac{25}{7}$, $\sec \theta < 0$

51. $\sec \theta = -3$, $\csc \theta > 0$

52. $\cot \theta = -\sqrt{15}/7$, $\sin \theta > 0$

Ⓒ **53.** $\sin \theta = 0.4695$, θ in Q_{I}

Ⓒ **54.** $\tan \theta = -0.5095$, $\sin \theta < 0$

Ⓒ **55.** $\csc \theta = 2.613$, θ in Q_{II}

Ⓒ **56.** $\sec \theta = 2.924$, $\tan \theta > 0$

57. According to Theorem 1, the six trigonometric functions can be grouped in three reciprocal pairs. They can also be grouped in three pairs of cofunctions. Is any reciprocal pair also a pair of cofunctions?

4 Evaluation of Trigonometric Functions

In this section, we obtain values of the trigonometric functions for quadrantal angles, we introduce the idea of reference angles, and we discuss the use of a calculator to evaluate trigonometric functions of general angles.

In Definition 1, Section 3, the domain of each trigonometric function consists of all angles θ for which the denominator in the corresponding ratio is not zero. Because $r > 0$, it follows that $\sin \theta = y/r$ and $\cos \theta = x/r$ are defined for all angles θ. However, $\tan \theta = y/x$ and $\sec \theta = r/x$ are not defined when the terminal side of θ lies along the y axis (so that $x = 0$). Likewise, $\cot \theta = x/y$ and $\csc \theta = r/y$ are not defined when the terminal side of θ lies along the x axis (so that $y = 0$). Therefore, when you deal with a trigonometric function of a quadrantal angle, you must check to be sure that the function is actually defined for that angle.

Example Find the values (if they are defined) of the six trigonometric functions for the quadrantal angle $\theta = 90°$ (or $\theta = \pi/2$).

Solution In order to use Definition 1 in Section 3, we begin by choosing any point $(0, y)$, with $y > 0$, on the terminal side of the 90° angle (Figure 1). Because $x = 0$, it follows that $\tan 90°$ and $\sec 90°$ are undefined. Since $y > 0$, we have

Figure 1

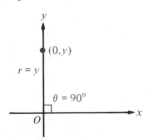

$$r = \sqrt{x^2 + y^2} = \sqrt{0^2 + y^2} = \sqrt{y^2} = |y| = y.$$

Therefore,

$$\sin 90° = \frac{y}{r} = \frac{y}{y} = 1 \qquad \cos 90° = \frac{x}{r} = \frac{0}{y} = 0$$

$$\csc 90° = \frac{r}{y} = \frac{y}{y} = 1 \qquad \cot 90° = \frac{x}{y} = \frac{0}{y} = 0.$$

The values of the trigonometric functions for other quadrantal angles are found in a similar manner (Problem 1). The results appear in Table 1. Dashes in the table indicate that the function is undefined for that angle.

Table 1

θ degrees	θ radians	$\sin\theta$	$\cos\theta$	$\tan\theta$	$\cot\theta$	$\sec\theta$	$\csc\theta$
0°	0	0	1	0	—	1	—
90°	$\dfrac{\pi}{2}$	1	0	—	0	—	1
180°	π	0	−1	0	—	−1	—
270°	$\dfrac{3\pi}{2}$	−1	0	—	0	—	−1
360°	2π	0	1	0	—	1	—

It follows from Definition 1 in Section 3 that the values of each of the six trigonometric functions remain unchanged if the angle is replaced by a coterminal angle. If an angle exceeds one revolution or is negative, you can change it to a nonnegative coterminal angle that is less than one revolution by adding or subtracting an integer multiple of 360° (or 2π radians). For instance,

$$\sin 450° = \sin(450° - 360°) = \sin 90° = 1$$

$$\sec 7\pi = \sec[7\pi - (3 \times 2\pi)] = \sec\pi = -1$$

$$\cos(-660°) = \cos[-660° + (2 \times 360°)] = \cos 60° = \tfrac{1}{2}.$$

Examples Replace each angle by a nonnegative coterminal angle that is less than one revolution and then find the values of the six trigonometric functions (if they are defined).

1 1110°

Solution By trial and error (or by dividing 1110 by 360), we find that the largest integer multiple of 360° that is less than 1110° is $3 \times 360° = 1080°$. Thus,

$$1110° - (3 \times 360°) = 1110° - 1080° = 30°.$$

(Or, we could have started with 1110° and repeatedly subtracted 360° until we obtained 30°.) It follows that

$$\sin 1110° = \sin 30° = 1/2 \qquad\qquad \csc 1110° = \csc 30° = 2$$

$$\cos 1110° = \cos 30° = \sqrt{3}/2 \qquad\qquad \sec 1110° = \sec 30° = 2\sqrt{3}/3$$

$$\tan 1110° - \tan 30° = \sqrt{3}/3 \qquad\qquad \cot 1110° = \cot 30° = \sqrt{3}.$$

2 $-5\pi/2$

Solution   We repeatedly add 2π to $-5\pi/2$ until we obtain a nonnegative coterminal angle:

$$(-5\pi/2) + 2\pi = -\pi/2 \quad \text{(still negative)}$$
$$(-\pi/2) + 2\pi = 3\pi/2.$$

Therefore, by Table 1 for quadrantal angles,

$$\sin(-5\pi/2) = \sin 3\pi/2 = -1 \qquad \cot(-5\pi/2) = \cot 3\pi/2 = 0$$
$$\cos(-5\pi/2) = \cos 3\pi/2 = 0 \qquad \csc(-5\pi/2) = \csc 3\pi/2 = -1,$$

and both $\tan(-5\pi/2)$ and $\sec(-5\pi/2)$ are undefined.

In Section 2 we found exact values of the six trigonometric functions for the special acute angle $30°$, $45°$, and $60°$ (see Table 1 on page 281). This information can be applied to certain angles in quadrants II, III, and IV by using the idea of a "reference angle." If θ is an angle in standard position and if θ is not a quadrantal angle, then the **reference angle** θ_R for θ is defined to be the positive acute angle formed by the terminal side of θ and the x axis. Figure 2 shows the reference angle θ_R for an angle θ in each of the four quadrants.

Figure 2

(a)

(b)

(c)

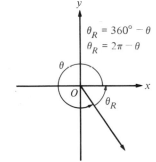

(d)

Example   Find the reference angle θ_R for each angle θ.

(a) $\theta = 60°$ (b) $\theta = \dfrac{3\pi}{4}$ (c) $\theta = 210°$ (d) $\theta = \dfrac{5\pi}{3}$

Solution   (a) By Figure 2a, $\theta_R = \theta = 60°$.

(b) By Figure 2b, $\theta_R = \pi - \theta = \pi - \dfrac{3\pi}{4} = \dfrac{\pi}{4}$.

(c) By Figure 2c, $\theta_R = \theta - 180° = 210° - 180° = 30°$.

(d) By Figure 2d, $\theta_R = 2\pi - \theta = 2\pi - \dfrac{5\pi}{3} = \dfrac{\pi}{3}$.

The value of any trigonometric function of an angle θ is the same as the value of the function for the reference angle θ_R, except possibly for a change of algebraic sign. Thus,

$$\sin\theta = \pm\sin\theta_R, \qquad \cos\theta = \pm\cos\theta_R,$$

and so forth. You can always determine the correct algebraic sign by considering which quadrant θ lies in.

Figure 3

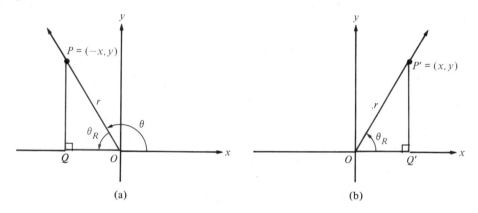

(a)   (b)

We shall verify the statement above for θ in quadrant II, and leave the remaining cases as an exercise (Problem 16). Figure 3a shows a quadrant II angle θ in standard position; the corresponding reference angle θ_R is shown in standard position in Figure 3b. In the two figures, we have chosen points P and P' as shown, both r units from the origin on the terminal sides of θ and θ_R. Dropping perpendiculars $\overline{PQ}$ and $\overline{P'Q'}$ from P and P' to the x axis, we find that the right triangles PQO and $P'Q'O$ are congruent. (Why?) Therefore, if the coordinates of P' are (x, y), it follows that the coordinates of P are $(-x, y)$. Applying Definition 1 in Section 3 to Figures 3a and 3b, we have:

$$\sin\theta = \frac{y}{r} = \sin\theta_R \qquad\qquad \csc\theta = \frac{r}{y} = \csc\theta_R$$

$$\cos\theta = \frac{-x}{r} = -\frac{x}{r} = -\cos\theta_R \qquad \sec\theta = \frac{r}{-x} = -\frac{r}{x} = -\sec\theta_R$$

$$\tan\theta = \frac{y}{-x} = -\frac{y}{x} = -\tan\theta_R \qquad \cot\theta = \frac{-x}{y} = -\frac{x}{y} = -\cot\theta_R.$$

Examples   Find the reference angle θ_R for the given angle θ, and then use the information in Table 1 on page 281 to find the values of the six trigonometric functions of θ.

1 $\theta = 135°$

Solution   The reference angle is $\theta_R = 180° - 135° = 45°$ (Figure 4). Since $135°$ is a quadrant II angle, we have

$$\sin 135° = \sin 45° = \frac{\sqrt{2}}{2} \qquad\qquad \csc 135° = \csc 45° = \sqrt{2}$$

$$\cos 135° = -\cos 45° = -\frac{\sqrt{2}}{2} \qquad \sec 135° = -\sec 45° = -\sqrt{2}$$

$$\tan 135° = -\tan 45° = -1 \qquad\qquad \cot 135° = -\cot 45° = -1.$$

Figure 4

2 $\theta = 210°$

Solution   The reference angle is $\theta_R = 210° - 180° = 30°$ (Figure 5). Since $210°$ is a quadrant III angle, we have

$$\sin 210° = -\sin 30° = -\frac{1}{2} \qquad\qquad \csc 210° = -\csc 30° = -2$$

$$\cos 210° = -\cos 30° = -\frac{\sqrt{3}}{2} \qquad \sec 210° = -\sec 30° = -\frac{2\sqrt{3}}{3}$$

$$\tan 210° = \tan 30° = \frac{\sqrt{3}}{3} \qquad\qquad \cot 210° = \cot 30° = \sqrt{3}.$$

Figure 5

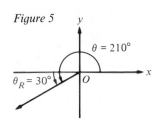

3 $\theta = \dfrac{5\pi}{3}$

Solution   The reference angle is $\theta_R = 2\pi - \dfrac{5\pi}{3} = \dfrac{\pi}{3}$ (Figure 6). Since $\dfrac{5\pi}{3}$ is a quadrant IV angle, we have

$$\sin \frac{5\pi}{3} = -\sin \frac{\pi}{3} = -\frac{\sqrt{3}}{2} \qquad\qquad \csc \frac{5\pi}{3} = -\csc \frac{\pi}{3} = -\frac{2\sqrt{3}}{3}$$

$$\cos \frac{5\pi}{3} = \cos \frac{\pi}{3} = \frac{1}{2} \qquad\qquad \sec \frac{5\pi}{3} = \sec \frac{\pi}{3} = 2$$

$$\tan \frac{5\pi}{3} = -\tan \frac{\pi}{3} = -\sqrt{3} \qquad\qquad \cot \frac{5\pi}{3} = -\cot \frac{\pi}{3} = -\frac{\sqrt{3}}{3}.$$

Figure 6

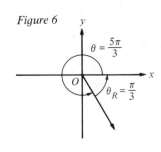

To find the reference angle θ_R for an angle θ in standard position when θ is negative or greater than one revolution, you can replace θ by a nonnegative coterminal angle that is less than one revolution, and then proceed as in the examples above.

Example   Find the reference angle θ_R for the given angle θ, and then use the information in Table 1 on page 281 to find the values of the six trigonometric functions of θ.

(a) $\theta = \dfrac{21\pi}{4}$ (b) $\theta = -300°$

Solution   (a) Here $\theta = \dfrac{21\pi}{4}$ is coterminal with $\dfrac{21\pi}{4} - 2(2\pi) = \dfrac{5\pi}{4}$, a quadrant III angle

(Figure 7a). The corresponding reference angle is $\theta_R = \dfrac{5\pi}{4} - \pi = \dfrac{\pi}{4}$, so

$$\sin \frac{21\pi}{4} = -\sin \frac{\pi}{4} = -\frac{\sqrt{2}}{2} \qquad \csc \frac{21\pi}{4} = -\csc \frac{\pi}{4} = -\sqrt{2}$$

$$\cos \frac{21\pi}{4} = -\cos \frac{\pi}{4} = -\frac{\sqrt{2}}{2} \qquad \sec \frac{21\pi}{4} = -\sec \frac{\pi}{4} = -\sqrt{2}$$

$$\tan \frac{21\pi}{4} = \tan \frac{\pi}{4} = 1 \qquad \cot \frac{21\pi}{4} = \cot \frac{\pi}{4} = 1.$$

Figure 7

(a)

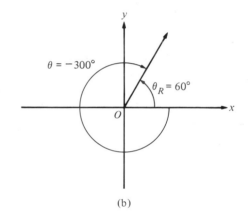

(b)

(b) The angle $\theta = -300°$ is coterminal with $-300° + 360° = 60°$, a quadrant I angle (Figure 7b). Here the reference angle is also $\theta_R = 60°$, and we have

$$\sin(-300°) = \sin 60° = \frac{\sqrt{3}}{2} \qquad \csc(-300°) = \csc 60° = \frac{2\sqrt{3}}{3}$$

$$\cos(-300°) = \cos 60° = \frac{1}{2} \qquad \sec(-300°) = \sec 60° = 2$$

$$\tan(-300°) = \tan 60° = \sqrt{3} \qquad \cot(-300°) = \cot 60° = \frac{\sqrt{3}}{3}.$$

In the examples above, we obtained *exact* values of the trigonometric functions for certain angles. Approximate values of these functions for all other angles can be found by using a scientific calculator. No special techniques are required beyond those already presented in Section 2 for evaluating the trigonometric functions of acute angles—correct algebraic signs are automatically provided. (If a calculator isn't available, you can use Appendix Tables IIA and IIB.)

Examples   © Use a calculator to find the approximate value of each trigonometric function.

1   cot 300.42°

Solution   With the calculator in degree mode, we enter 300.42, press the tangent (TAN) key, then press the 1/x key. (We're using the fact that $\cot 300.42° = \dfrac{1}{\tan 300.42°}$.) On a 10-digit calculator, the result is

$$\cot 300.42° = -0.587165830.$$

2 $\sec \dfrac{40\pi}{7}$

Solution   With the calculator in radian mode, we enter π, multiply by 40, divide by 7, press the cosine (COS) key, and then press the 1/x key. On a 10-digit calculator, we get

$$\sec \dfrac{40\pi}{7} = 1.603875452.$$

3 $\sin(-2.007)$

Solution   Because degrees are not indicated, we know that the angle is measured in radians. We set the calculator in radian mode, enter -2.007, and press the sine (SIN) key to obtain

$$\sin(-2.007) = -0.906362145.$$

PROBLEM SET 4

In problems 1 and 2, find the values (if they are defined) of the six trigonometric functions of the given quadrantal angles.

1. (a) 0°   (b) 180°   (c) 270°   (d) 360°
   [When you have finished, compare your answers with the results in Table 1.]

2. (a) 5π (b) 6π (c) -7π (d) $\dfrac{5\pi}{2}$ (e) $\dfrac{7\pi}{2}$

In problems 3–14, replace each angle by a nonnegative coterminal angle that is less than one revolution, and then find the exact values of the six trigonometric functions (if they are defined) for the angle.

3. 1440°   **4.** 810°   **5.** 900°   **6.** $-720°$ **7.** 750°

8. 1845°   **9.** $-675°$ **10.** $\dfrac{19\pi}{2}$ **11.** 5π **12.** $\dfrac{25\pi}{6}$

13. $-\dfrac{17\pi}{3}$ **14.** $-\dfrac{31\pi}{4}$

© **15.** What happens when you try to evaluate tan 90° on a calculator? [Try it.]
16. Let θ be a quadrant III angle in standard position and let θ_R be its reference angle. Show that the value of any trigonometric function of θ is the same as the value of the function for θ_R, except possibly for a change of algebraic sign. Repeat for θ in quadrant IV.

In problems 17–36, find the reference angle θ_R for each angle θ, and then find the exact values of the six trigonometric functions of θ.

17. $\theta = 150°$ **18.** $\theta = 120°$ **19.** $\theta = 240°$ **20.** $\theta = 225°$ **21.** $\theta = 315°$

22. $\theta = \dfrac{9\pi}{4}$ **23.** $\theta = -150°$ **24.** $\theta = 675°$ **25.** $\theta = -60°$ **26.** $\theta = -\dfrac{7\pi}{6}$

27. $\theta = -\dfrac{\pi}{4}$ **28.** $\theta = -\dfrac{13\pi}{6}$ **29.** $\theta = -\dfrac{2\pi}{3}$ **30.** $\theta = \dfrac{53\pi}{6}$ **31.** $\theta = \dfrac{7\pi}{4}$

32. $\theta = -\dfrac{50\pi}{3}$ **33.** $\theta = \dfrac{11\pi}{3}$ **34.** $\theta = -\dfrac{147\pi}{4}$ **35.** $\theta = -420°$ **36.** $\theta = -5370°$

37. Complete the following table. (Do not use a calculator.)

θ degrees	θ radians	$\sin\theta$	$\cos\theta$	$\tan\theta$	$\cot\theta$	$\sec\theta$	$\csc\theta$
210°	$7\pi/6$						
225°	$5\pi/4$						
240°	$4\pi/3$						
300°	$5\pi/3$						
315°	$7\pi/4$						
330°	$11\pi/6$						

38. A calculator is set in radian mode, π is entered, and the sine (SIN) key is pressed. The display shows -4.1×10^{-10}. But we know that $\sin\pi = 0$. Explain.

C In problems 39–58, use a calculator to find the approximate values of the six trigonometric functions of each angle.

39. $46°$ **40.** $162°$ **41.** $143°$ **42.** $-225.31°$ **43.** $-61.37°$

44. $852.48°$ **45.** $61°35'$ **46.** $-31°8'$ **47.** $-97°9'8''$ **48.** $\dfrac{12\pi}{5}$

49. $-\dfrac{2\pi}{7}$ **50.** $\dfrac{\sqrt{2\pi}}{3}$ **51.** 1.67 **52.** 3.725 **53.** -2.436

54. 16.57 **55.** -10.79 **56.** -6.203 **57.** 9.673 **58.** -3985.2

C In problems 59–62, use a calculator to verify that the equation is true for the indicated value of the angle θ.

59. $\tan\theta = \sin\theta/\cos\theta$ for $\theta = 35°$

60. $(\cos\theta)(\tan\theta) = \sin\theta$ for $\theta = 5\pi/7$

61. $\sin^2\theta + \cos^2\theta = 1$ for $\theta = 3\pi/5$

62. $1 + \tan^2\theta = \sec^2\theta$ for $\theta = 17.75°$

63. Verify that for $\theta = 0°, 30°, 45°, 60°$, and $90°$, we have $\sin\theta = \sqrt{0}/2, \sqrt{1}/2, \sqrt{2}/2, \sqrt{3}/2$, and $\sqrt{4}/2$, respectively. [Although there is no theoretical significance to this pattern, people often use it as a memory aid to help recall these values of $\sin\theta$.]

5   Graphs of the Sine and Cosine Functions

Until now, we have considered trigonometric functions of *angles*, measured either in degrees or radians. In advanced work, particularly in calculus and its applications, it is necessary to modify the trigonometric functions so that their domains consist of **real numbers.** For reasons that we shall soon explain, the resulting functions are sometimes called the **circular functions.**

If t is a real number, then the value of $\sin t$ is defined by

$$\sin t = \sin \theta,$$

where θ is an angle whose measure is t radians. Similarly, the values of $\cos t$, $\tan t$, $\cot t$, $\sec t$, $\csc t$ are defined to be the values of the corresponding trigonometric functions of an angle θ whose measure is t radians. We shall now explain just what these so-called circular functions have to do with circles.

Consider the **unit circle**; that is, the circle with center at the origin and radius $r = 1$ (Figure 1a). Let t be any real number. Starting at the point $(1,0)$ on the unit circle, we move $|t|$ units around the circle, counterclockwise if $t > 0$ and clockwise if $t < 0$, to the point (x, y) (Figure 1a). Now let θ be the angle in standard position whose terminal side contains the point (x, y). Recall from Section 1 (page 272) that $t = r\theta$. Since $r = 1$, it follows that angle θ has radian measure t. Hence,

$$\sin t = \sin \theta = \frac{y}{r} = \frac{y}{1} = y \quad \text{and} \quad \cos t = \cos \theta = \frac{x}{r} = \frac{x}{1} = x.$$

Therefore,

$$(x, y) = (\cos t, \sin t)$$

(Figure 1b).

In Figure 1b, the coordinates of $(\cos t, \sin t)$ repeat themselves each time the point goes completely around the circle; that is, each time t increases by 2π (the circumference of the unit circle). Thus, the sine and cosine functions are **periodic** in the sense that their values repeat themselves whenever the independent variable increases by 2π, that is,

$$\sin(t + 2\pi) = \sin t \quad \text{and} \quad \cos(t + 2\pi) = \cos t.$$

Figure 1

(a)

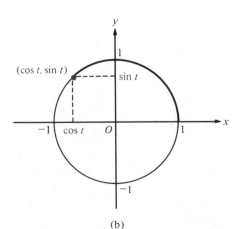

(b)

As t increases uniformly from 0 to 2π, the point $(\cos t, \sin t)$ (Figure 1b) moves once around the unit circle in a counterclockwise direction, and the coordinates $\cos t$ and $\sin t$ change as shown in Table 1.

Table 1

As t Increases from	$\sin t$	$\cos t$
0 to $\dfrac{\pi}{2}$	is positive and increases from 0 to 1	is positive and decreases from 1 to 0
$\dfrac{\pi}{2}$ to π	is positive and decreases from 1 to 0	is negative and decreases from 0 to -1
π to $\dfrac{3\pi}{2}$	is negative and decreases from 0 to -1	is negative and increases from -1 to 0
$\dfrac{3\pi}{2}$ to 2π	is negative and increases from -1 to 0	is positive and increases from 0 to 1

The idea of a circular function (that is, a trigonometric function of a real number rather than of an angle) is more a point of view than a new mathematical concept. For instance, put a calculator in radian mode, enter $\pi/3$, and press the sine key. The display will show the approximate value of $\sin \pi/3$, and it makes no difference whether you have in mind that $\pi/3$ is the measure of an angle in radians, the length of an arc on the unit circle, or simply a real number. In any case,

$$\sin \frac{\pi}{3} = \frac{\sqrt{3}}{2} \approx 0.866025404.$$

For this reason, people speak of "trigonometric functions" and "circular functions" more or less interchangeably.

On most scientific calculators, the number in the display register is denoted by x. Pressing the 1/x key gives you the reciprocal of x, pressing the x² key gives you the square of x, and, likewise, pressing the sine or cosine key gives you $\sin x$ or $\cos x$. This notation is consistent with the custom of denoting the independent variable of a function by x (unless there are good reasons for doing otherwise). Therefore, in what follows we shall sketch graphs of $y = \sin x$ and of $y = \cos x$ on the usual xy coordinate system. Notice that x, used in this way as the independent variable for a circular function, must not be confused with the abscissa of the point (x, y) in Figure 1.

We can make a rough sketch of the graph of $y = \sin x$ for $0 \le x \le 2\pi$ by using only the information in Table 1, if we replace t by x. However, we can make a more accurate sketch by using the values of $\sin x$ shown in Table 2. If even more accuracy is desired, additional points can be found with the aid of a calculator. The graph of $y = \sin x$ for $0 \le x \le 2\pi$ is sketched in Figure 2. This curve is called **one cycle** of the graph of $y = \sin x$.

Table 2

x	$y = \sin x$
0	0
$\pi/6$	$1/2$
$\pi/4$	$\sqrt{2}/2$
$\pi/3$	$\sqrt{3}/2$
$\pi/2$	1
$2\pi/3$	$\sqrt{3}/2$
$3\pi/4$	$\sqrt{2}/2$
$5\pi/6$	$1/2$
π	0
$7\pi/6$	$-1/2$
$5\pi/4$	$-\sqrt{2}/2$
$4\pi/3$	$-\sqrt{3}/2$
$3\pi/2$	-1
$5\pi/3$	$-\sqrt{3}/2$
$7\pi/4$	$-\sqrt{2}/2$
$11\pi/6$	$-1/2$
2π	0

Figure 2

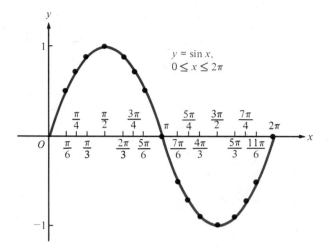

Because the sine is a periodic function, the complete graph of $y = \sin x$ consists of an endless sequence of copies of the cycle in Figure 2, repeated to the right and left over successive intervals of length 2π (Figure 3). The graph shows that the range of the sine function is the closed interval $[-1, 1]$. In particular,

$$-1 \leq \sin x \leq 1$$

holds for all real numbers x. Other features of the sine function are indicated by its graph. For instance, the graph appears to be symmetric about the origin, which suggests that the sine is an odd function,

$$\sin(-x) = -\sin x.$$

Figure 3

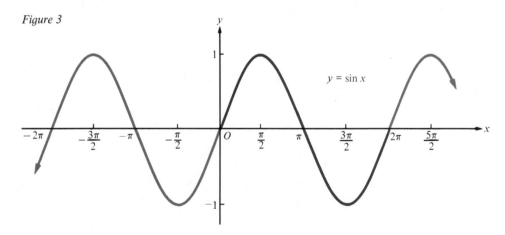

(We prove this in Section 1 of Chapter 7.) Another interesting feature suggested by the graph is that

$$\sin(x + \pi) = -\sin x$$

(Figure 4, page 304). This property is proved in Section 3 of Chapter 7.

Figure 4

Figure 5

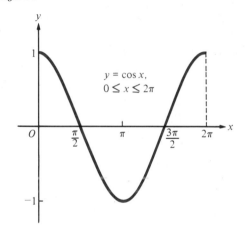

By using the information in Table 1 and the values of the cosine function for angles such as 0, $\pi/6$, $\pi/4$, $\pi/3$, $\pi/2$, and so on, you can sketch the graph of $y = \cos x$ for $0 \leq x \leq 2\pi$ (Problem 1). This curve (Figure 5) is called one *cycle* of the graph of $y = \cos x$. By extending the cycle to the right and to the left in a periodic fashion, we obtain the complete graph of $y = \cos x$ (Figure 6). Figure 6 also includes the graph of $y = \sin x$, which, as you can see, is the graph of $y = \cos x$ shifted $\pi/2$ units to the right. This indicates that

$$\sin x = \cos\left(x - \frac{\pi}{2}\right),$$

a fact which is proved in Section 3 of Chapter 7.

Like the sine function, the range of the cosine function is the interval $[-1, 1]$; hence,

$$-1 \leq \cos x \leq 1$$

holds for all real numbers x. The graph in Figure 6 appears to be symmetric about the y axis, indicating that the cosine is an even function,

$$\cos(-x) = \cos x.$$

(Again, this is proved in Section 1 of Chapter 7.) By drawing a figure for $y = \cos x$ similar to Figure 4, you can also see graphically that

$$\cos(x + \pi) = -\cos x$$

(Problem 2).

Figure 6

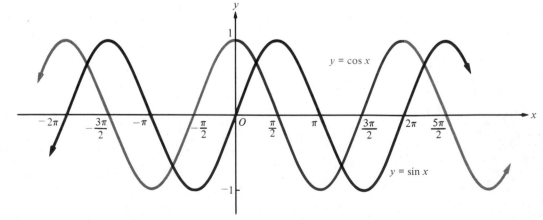

If the techniques of graphing discussed in Section 6 of Chapter 3 are used to stretch or shift the graphs of sine or cosine functions, the resulting curves always have the wavelike form shown in Figure 7. Such a curve is called a **simple harmonic curve,** a **sine wave,** or a **sinusoidal curve.** The highest points on a simple harmonic curve are called **crests** and the lowest points are called **troughs.** The crests lie along one straight line, the troughs lie along another, and the two lines are parallel. The straight line midway between the crest line and the trough line is called the **axis** of the simple harmonic curve, and the points where the curve crosses the axis are called **nodes.** The distance between the axis and the crest line (or trough line) is called the **amplitude** of the simple harmonic curve.

Figure 7

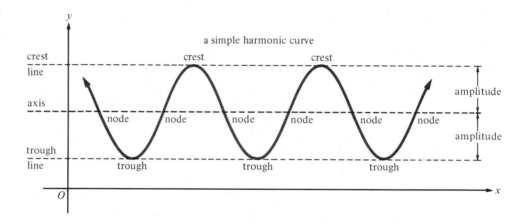

The portion of a simple harmonic curve from any one point P to the next point Q at which the curve starts to repeat itself is called a **cycle,** and the distance between P and Q is called the **period** of the simple harmonic curve (Figure 8). In sketching a cycle, we usually begin at a crest, node, or trough. If you are given (or have already sketched) one cycle, you can easily sketch as much of the rest of the curve as you please by repeating the cycle to the right and to the left over successive intervals, each of which has length equal to the period. It is often convenient to use different units of length along the x and y axes.

Figure 8

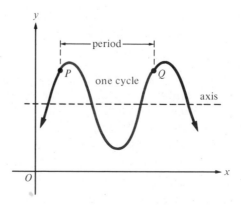

Examples

Sketch the graph of the function defined by each equation, find the amplitude and the period, and indicate one cycle (starting at a node) on your graph.

1   (a) $y = 3 \sin x$

(b) $y = -3 \sin x$

Solution

(a) We begin by sketching the graph of $y = \sin x$ (Figure 3); then we multiply each ordinate by 3 (Figure 9a). The x axis is the axis of the simple harmonic curve. The crests are 3 units above the axis, so the amplitude is 3. The period is 2π. One cycle, starting at $(0,0)$ and ending at $(2\pi, 0)$, is indicated on the graph.

Figure 9

(a)

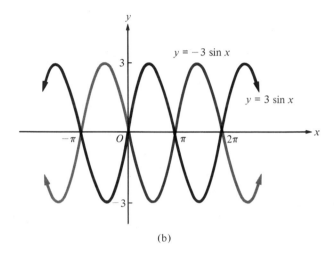

(b)

(b) We obtain the graph of $y = -3 \sin x$ by reflecting the graph of $y = 3 \sin x$ across the x axis (Figure 9b). The amplitude of the reflected graph is still 3, and its period is still 2π. One cycle, starting at $(0,0)$ and ending at $(2\pi, 0)$, is indicated on the graph.

2   (a) $y = \frac{1}{3} \cos x$

(b) $y = 1 + \frac{1}{3} \cos x$

Solution

(a) The graph of $y = \frac{1}{3} \cos x$ is obtained from the graph of $y = \cos x$ (Figure 6) by multiplying each ordinate by $\frac{1}{3}$ (Figure 10a). The amplitude is $\frac{1}{3}$, the period is 2π, and one cycle [starting at $x = \pi/2$ and ending at $x = (\pi/2) + 2\pi = 5\pi/2$] is indicated on the graph.

(b) The graph of $y = 1 + \frac{1}{3} \cos x$ is obtained by shifting the graph of $y = \frac{1}{3} \cos x$ one unit upward. This shift does not affect the amplitude, which is still $\frac{1}{3}$, or the period, which is still 2π. One cycle is indicated on the graph (Figure 10b).

Figure 10

(a)

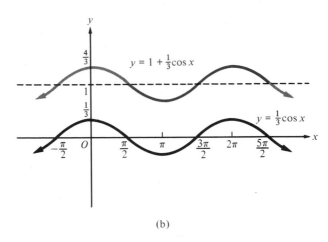

(b)

3 $y = 3 \sin 2x$

Solution   The graph of $y = 3 \sin 2x$ will cover one cycle as $2x$ varies from 0 to 2π. When $2x = 0$, we have $x = 0$; when $2x = 2\pi$, we have $x = \pi$. Therefore, the graph covers one cycle as x varies from 0 to π, and the period is π. As x varies from 0 to π, $\sin 2x$ will take on maximum and minimum values of $+1$ and -1; hence, $3 \sin 2x$ will take on maximum and minimum values of $+3$ and -3. Thus, the amplitude is 3. We use this information to sketch one cycle, and then repeat the cycle on either side (Figure 11). Figure 11 includes the graph of $y = 3 \sin x$ for comparison. Notice that the graph of $y = 3 \sin 2x$ oscillates up and down twice as fast as the graph of $y = 3 \sin x$.

Figure 11

4 $y = 3 \sin\left(2x - \dfrac{\pi}{2}\right)$

Solution   The graph will cover one cycle as $2x - \dfrac{\pi}{2}$ varies from 0 to 2π. When $2x - \dfrac{\pi}{2} = 0$, we have $x = \dfrac{\pi}{4}$; when $2x - \dfrac{\pi}{2} = 2\pi$, we have $2x = 2\pi + \dfrac{\pi}{2} = \dfrac{5\pi}{2}$, so $x = \dfrac{5\pi}{4}$.

Figure 12

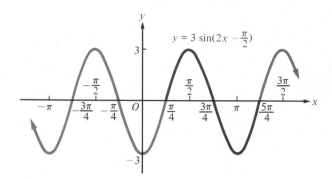

Therefore, the graph covers one cycle as x varies from $\dfrac{\pi}{4}$ to $\dfrac{5\pi}{4}$. It follows that the period is $\dfrac{5\pi}{4} - \dfrac{\pi}{4} = \pi$. The amplitude is 3 (Figure 12). Notice that the graph of $y = 3\sin\left(2x - \dfrac{\pi}{2}\right)$ is the graph of $y = 3\sin 2x$ (Figure 11) shifted $\dfrac{\pi}{4}$ unit to the right.

PROBLEM SET 5

1. Use the information in Table 1 (page 281) and the values of the cosine function for the angles $0, \dfrac{\pi}{6}, \dfrac{\pi}{4}, \dfrac{\pi}{3}, \dfrac{\pi}{2}, \dfrac{2\pi}{3}, \dfrac{3\pi}{4}, \dfrac{5\pi}{6}, \pi, \dfrac{7\pi}{6}, \dfrac{5\pi}{4}, \dfrac{4\pi}{3}, \dfrac{3\pi}{2}, \dfrac{5\pi}{3}, \dfrac{7\pi}{4}, \dfrac{11\pi}{6}$, and 2π to sketch an accurate graph of $y = \cos x$ for $0 \le x \le 2\pi$. When you have finished your sketch compare it with Figure 5. ⓒ (For enhanced accuracy, you may wish to use a calculator to obtain additional points on the graph.)

2. By sketching a figure for $y = \cos x$ similar to Figure 4, show graphical evidence for the fact that $\cos(x + \pi) = -\cos x$.

ⓒ In problems 3–6, use a calculator to verify that the equation is true for the indicated value of the variable.

3. $\sin(-x) = -\sin x$, $x = 4.203$

4. $\cos(-x) = \cos x$, $x = \dfrac{5\pi}{9}$

5. $\sin(x + \pi) = -\sin x$, $x = 2.771$

6. $\cos(x + \pi) = -\cos x$, $x = \dfrac{11\pi}{7}$

In problems 7–12, use the graphs of $y = \sin x$ (Figure 3) and $y = \cos x$ (Figure 6) to help answer each question.

7. If $-2\pi \le x \le 2\pi$, for what values of x does $\sin x$ reach its maximum value and what is this maximum value?

8. If $-2\pi \le x \le 2\pi$, for what values of x is $\cos x = \frac{1}{2}$?   9. If $-2\pi \le x \le 2\pi$, for what values of x is $\cos x = 0$?

10. If h is a constant with $0 < h < 1$ and if $-2\pi \le x \le 2\pi$, how many different values of x are there for which $\sin x = h$?

11. If $-2\pi \le x \le 2\pi$, for what values of x does $\sin x$ reach its minimum value and what is this minimum value?

12. Complete the following sentence: $\sin x$ reaches its maximum and minimum values at the same values of x for which the graph of $y = \cos x$ _____.

13. Suppose a friend who is just beginning to study trigonometry says, "I understand how to take the sine of an *angle*, but how do you take the sine of a *number*?" Answer your friend's question in your own words.

14. Suppose you are going to measure an acute angle with a protractor and then use a calculator to find the sine of the angle. Of course, if there is an error in your measurement, there will be an error in the calculated value of the sine. Will the error in the sine value be more pronounced when the acute angle is small or when it is large? Why?

In problems 15–34, sketch the graph of the function defined by each equation, find the amplitude and the period, and indicate one cycle on your graph. Start the cycle at a node for the sine functions and at a crest for the cosine functions.

15. $y = 2\sin x$

16. $y = -2\sin x$

17. $y = \frac{1}{2}\sin x$

18. $y = 1 + \cos x$

19. $y = 1 - \frac{2}{3}\cos x$

20. $y = 3\cos x - \frac{1}{2}$

21. $y = \sin 2x$

22. $y = 2\sin\frac{x}{3}$

23. $y = \cos 6x$

24. $y = 1 + \cos \pi x$

25. $y = 1 + \cos\frac{x}{2}$

26. $y = \cos\frac{\pi x}{2} + 1$

27. $y = 2 - \pi\cos \pi x$

28. $y = \frac{1}{3} - \frac{1}{3}\cos\frac{3\pi x}{2}$

29. $y = \cos\left(x - \frac{\pi}{6}\right)$

30. $y = 1 + \sin\left(x + \frac{\pi}{4}\right)$

31. $y = 2\cos\left(x - \frac{\pi}{3}\right)$

32. $y = -\cos\left(x - \frac{7\pi}{8}\right)$

33. $y = 1 - \cos\left(x - \frac{\pi}{2}\right)$

34. $y = -2 + 2\cos\left(2x - \frac{\pi}{2}\right)$

35. Suppose that a, b, and c are constants, $a > 0$, and $b > 0$. Show that the simple harmonic curve $y = a\sin(bx - c)$ has amplitude a and period $2\pi/b$.

---

6   Graphs of the Tangent, Cotangent, Secant, and Cosecant Functions

In the previous section, we studied the graphs of the sine and cosine functions. In this section, we consider the graphs of the remaining four trigonometric functions.

Example   © Sketch the graph of

$$y = \tan x \quad \text{for} \quad -\frac{\pi}{2} < x < \frac{\pi}{2}.$$

Solution   Table 1 shows values of tan x rounded off to two decimal places and obtained by using a calculator. Of course, $\tan x = \frac{\sin x}{\cos x}$ is undefined when cos x is zero; for instance, when $x = -\pi/2$ or when $x = \pi/2$. For values of x slightly smaller than $\pi/2$, the numerator sin x is close to 1, the denominator cos x is small, and thus,

$\tan x = \sin x / \cos x$ is very large. (Why?) Similarly, for values of x slightly larger than $-\pi/2$, $\tan x$ is negative with a very large absolute value. Thus the vertical lines

$$x = -\frac{\pi}{2} \quad \text{and} \quad x = \frac{\pi}{2}$$

are asymptotes of the graph. Plotting the points in Table 1, and using the information about the asymptotes, we obtain a sketch of the graph (Figure 1).

Table 1

x	$\tan x$
$-\pi/2$	undefined
$-5\pi/12$	-3.73
$-\pi/3$	-1.73
$-\pi/4$	-1
$-\pi/6$	-0.58
$-\pi/12$	-0.27
0	0
$\pi/12$	0.27
$\pi/6$	0.58
$\pi/4$	1
$\pi/3$	1.73
$5\pi/12$	3.73
$\pi/2$	undefined

Figure 1

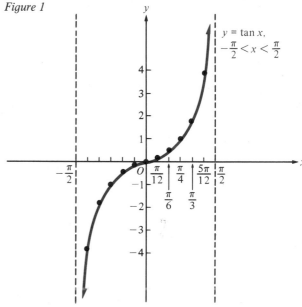

Like the sine and cosine functions, the tangent function is periodic; however, its values repeat themselves whenever the independent variable increases by π. In other words,

$$\tan(x + \pi) = \tan x$$

holds for every value of x in the domain of the tangent function. This can be seen by combining the identities

$$\sin(x + \pi) = -\sin x \quad \text{and} \quad \cos(x + \pi) = -\cos x,$$

which were mentioned in Section 5. Thus,

$$\tan(x + \pi) = \frac{\sin(x + \pi)}{\cos(x + \pi)} = \frac{-\sin x}{-\cos x} = \frac{\sin x}{\cos x} = \tan x.$$

Notice that the interval from $-\pi/2$ to $\pi/2$ in Figure 1 has length π units. Therefore, the complete graph of $y = \tan x$ consists of an endless sequence of copies of the curve in Figure 1, repeated to the right and left over successive intervals of length π (Figure 2).

Figure 2

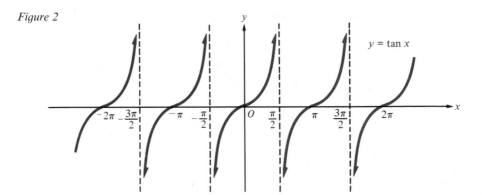

As you can see, the range of the tangent function is $\mathbb{R}$. Also, the graph appears to be symmetric about the origin, suggesting that the tangent is an odd function,

$$\tan(-x) = -\tan x$$

for all values of x in the domain.

Like the tangent function, the values of the cotangent function repeat themselves periodically whenever the independent variable increases by π; that is,

$$\cot(x + \pi) = \cot x$$

holds for every value of x in the domain of the cotangent function (Problem 4). Because $\cot x = 1/\tan x$, the graph of the cotangent function has vertical asymptotes where the graph of the tangent function has x intercepts and vice versa. To sketch the graph of $y = \cot x$, start by sketching the graph for $0 < x < \pi$ (Problem 1), then draw copies of this curve to the left and right over successive intervals of length π (Problem 2). The resulting graph is shown in Figure 3.

Figure 3

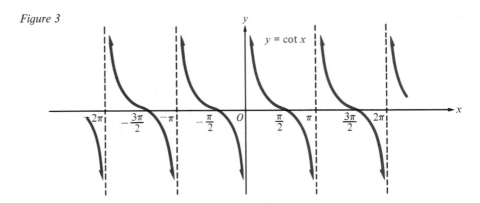

As you can see, the range of the cotangent function is $\mathbb{R}$. Also, the graph appears to be symmetric about the origin, suggesting that the cotangent is an odd function,

$$\cot(-x) = -\cot x$$

for all values of x in the domain.

Example $\boxed{\text{C}}$ Sketch the graph of $y = \sec x$.

Solution   Since $\sec x = 1/\cos x$, we obtain the graph of the secant function by using a calculator to find the reciprocals of nonzero ordinates of points on the graph of the cosine function (Figure 4). The graph of the secant function has vertical asymptotes where the graph of the cosine function has x intercepts.

Figure 4

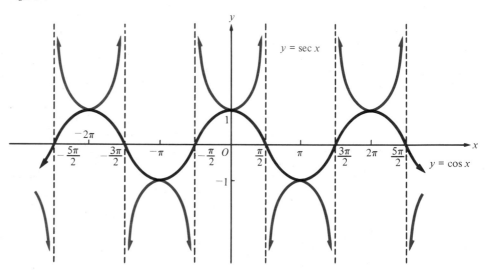

As you can see, the range of the secant function consists of the intervals $(-\infty, -1]$ and $[1, \infty)$. In particular,

$$|\sec x| \geq 1$$

holds for all values of x in the domain of the secant function.

You can sketch the graph of $y = \csc x$ (Figure 5) by using a calculator to find the reciprocals of nonzero ordinates of points on the graph of $y = \sin x$ (Problem 3). Like the secant function, the range of the cosecant function consists of the two intervals $(-\infty, -1]$ and $[1, \infty)$, and

$$|\csc x| \geq 1$$

holds for all values of x in the domain of the function.

You can use the techniques of graphing discussed in Section 6 of Chapter 3 to stretch, shift, or reflect the graphs of the tangent, cotangent, secant, and cosecant functions.

Figure 5

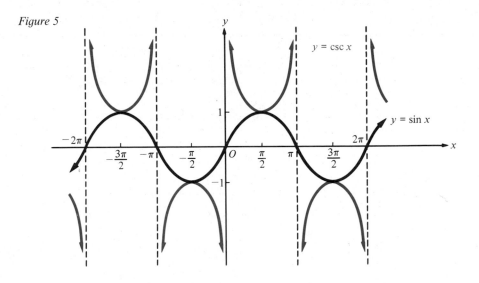

Example   Sketch the graph of $y = 3 \sec(x - \pi)$.

Solution   The graph is obtained by multiplying ordinates of points on the graph of $y = \sec x$ (Figure 4) by 3 and then shifting the resulting graph π units to the right (Figure 6).

Figure 6

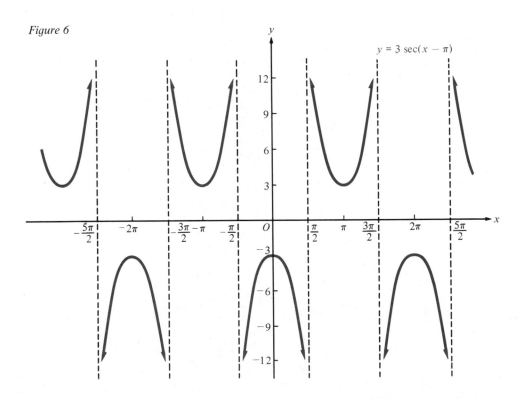

PROBLEM SET 6

C **1.** (a) Make a table showing the values of $\cot x$, rounded off to two decimal places, for the following values of x: $\dfrac{\pi}{12}, \dfrac{\pi}{6}, \dfrac{\pi}{4}, \dfrac{\pi}{3}, \dfrac{5\pi}{12}, \dfrac{\pi}{2}, \dfrac{7\pi}{12}, \dfrac{2\pi}{3}, \dfrac{3\pi}{4}, \dfrac{5\pi}{6}$, and $\dfrac{11\pi}{12}$. (b) Use the table in part (a) to sketch an accurate graph of $y = \cot x$ for $0 < x < \pi$.

2. Using the result of problem 1, sketch the graph of $y = \cot x$. When you have finished your sketch, compare it with Figure 3.

C **3.** With the aid of a calculator, sketch an accurate graph of $y = \csc x$. When you have finished, compare your graph with Figure 5.

4. Assuming that $\sin(x + \pi) = -\sin x$ and $\cos(x + \pi) = -\cos x$, show that $\cot(x + \pi) = \cot x$ holds for all values of x in the domain of the cotangent function.

C In problems 5–10, use a calculator to verify that the equation is true for the indicated value of the variable.

5. $\tan(x + \pi) = \tan x$, $\quad x = \dfrac{2\pi}{3}$

6. $\cot(x + \pi) = \cot x$, $\quad x = -\dfrac{5\pi}{12}$

7. $\tan(-x) = -\tan x$, $\quad x = 1.334$

8. $\cot(-x) = -\cot x$, $\quad x = 0.7075$

9. $\sec(-x) = \sec x$, $\quad x = \dfrac{5\pi}{12}$

10. $\csc(-x) = -\csc x$, $\quad x = 0.3544$

In problems 11–16, use the graphs of $y = \tan x$, $y = \cot x$, $y = \sec x$, and $y = \csc x$ to help answer each question.

11. If $-2\pi \le x \le 2\pi$, for what values of x is $\tan x = 0$?

12. If h is a positive constant and if $-2\pi \le x \le 2\pi$, how many different values of x are there for which $\tan x = h$?

13. If $-2\pi \le x \le 2\pi$, for what values of x does $\sec x$ reach its smallest positive value and what is this value?

14. How is the graph of $y = -\tan x$ related to the graph of $y = \cot x$?

15. For the graph of $y = \sec x$, what is the distance from one vertical asymptote to the next?

16. How is the graph of $y = \csc x$ related to the graph of $y = \sec x$?

17. Assuming that the cosine is an even function, show that the secant is also an even function.

18. The portion of the graph of

$$y = \sec x \qquad \text{for } (-\pi/2) < x < (\pi/2)$$

looks vaguely like a parabola. Could it possibly be a parabola? Why?

In problems 19–32, use the graphs of $y = \tan x$, $y = \cot x$, $y = \sec x$, and $y = \csc x$ and the techniques of shifting, stretching, and reflecting to sketch the graph of each equation.

19. $y = 2\tan x$

20. $y = -2\tan x$

21. $y = 1 - \tfrac{1}{2}\cot x$

22. $y = 3 - \sec x$

23. $y = 1 + \dfrac{1}{2}\sec x$

24. $y = \dfrac{2}{3}\csc\left(x - \dfrac{\pi}{3}\right)$

25. $y = \tan\left(x - \dfrac{\pi}{4}\right)$

26. $y = \tan\left(x + \dfrac{\pi}{4}\right)$

27. $y = \dfrac{2}{3}\cot\left(x - \dfrac{\pi}{2}\right)$

28. $y = 5\sec\left(x - \dfrac{\pi}{2}\right)$

29. $y = \sec\left(x + \dfrac{\pi}{6}\right)$

30. $y = 1 - \csc\left(x + \dfrac{5\pi}{6}\right)$

31. $y = 2\csc\left(x - \dfrac{\pi}{3}\right)$

32. $y = 2 + \dfrac{1}{2}\csc\left(x + \dfrac{\pi}{3}\right)$

7   The Simple Harmonic Model

In the real world there are a number of quantities that oscillate or vibrate in a uniform manner, repeating themselves periodically in definite intervals of time. Examples include alternating electrical currents; sound waves; light waves, radio waves, and other electromagnetic waves; pendulums, mass-spring systems, and other mechanical oscillators; tides, geysers, seismic waves, the seasons, climatic cycles, and other periodic phenomena of interest in the earth sciences; and biological phenomena ranging from a human heartbeat to periodic variation of the population of a plant or animal species. Simple harmonic curves often provide useful mathematical models for such oscillations. It is only necessary to replace the variable x in the equation of a simple harmonic curve by the variable t representing units of time.

The mathematical model for a quantity y that is oscillating in a **simple harmonic** manner is the equation

$$y = a \cos(\omega t - \phi) + k$$

where a and ω are *positive* constants and ϕ and k are constants. Here a is the **amplitude**, ω (the Greek letter omega) is called the **angular frequency** (see Problem 36), ϕ (the Greek letter phi) is called the **phase angle**, k is the **vertical shift,** and the independent variable t is the **time.**

To sketch a graph of

$$y = a \cos(\omega t - \phi) + k,$$

we begin by noticing that y will go through one cycle as $\omega t - \phi$ goes from 0 to 2π. When $\omega t - \phi = 0$, we have $t = \phi/\omega$; when $\omega t - \phi = 2\pi$, we have $t = (\phi/\omega) + (2\pi/\omega)$. Therefore, one cycle of the graph of $y = a \cos(\omega t - \phi) + k$ covers the interval from $t = \phi/\omega$ to $t = (\phi/\omega) + (2\pi/\omega)$ (Figure 1). The vertical shift k moves the axis of the simple harmonic curve from $y = 0$ to $y = k$ (Figure 1). The **period** of the oscillation is the amount of time T required for y to proceed through one cycle. From Figure 1, you can see that

$$T = \left(\frac{\phi}{\omega} + \frac{2\pi}{\omega}\right) - \frac{\phi}{\omega} = \frac{2\pi}{\omega}.$$

The equation

$$T = \frac{2\pi}{\omega}$$

is often written in the alternative form

$$\omega = \frac{2\pi}{T}.$$

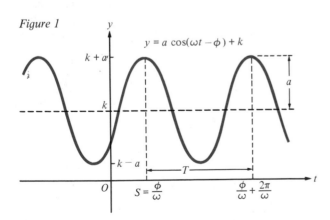

Figure 1

Notice that one cycle (from crest to crest) of the graph of $y = a\cos(\omega t - \phi)$ has the same general shape as one cycle (from crest to crest) of the cosine function (Figure 5 in Section 5), but shifted horizontally by an amount ϕ/ω. For this reason, the quantity ϕ/ω is called the **phase shift**. We denote the phase shift by S, so that

$$S = \frac{\phi}{\omega}.$$

The last equation is often written in the alternative form

$$\phi = \omega S.$$

Example 1   If y is oscillating according to the simple harmonic model

$$y = 2\cos\left(\frac{\pi}{4}t - \frac{\pi}{2}\right) - 1,$$

find (a) the amplitude a, (b) the angular frequency ω, (c) the period T, (d) the phase angle ϕ, (e) the phase shift S, and (f) the vertical shift k. Then (g) sketch the graph of one cycle.

Solution   (a) $a = 2$ (b) $\omega = \dfrac{\pi}{4}$

(c) $T = \dfrac{2\pi}{\omega} = \dfrac{2\pi}{(\pi/4)} = 8$ (d) $\phi = \dfrac{\pi}{2}$

(e) $S = \dfrac{\phi}{\omega} = \dfrac{\pi/2}{\pi/4} = 2$ (f) $k = -1$

(g) See Figure 2.

Example 2   The curve in Figure 3 is one cycle of the graph of a simple harmonic model. Determine (a) the amplitude a, (b) the period T, (c) the phase shift S, (d) the angular frequency ω, (e) the phase angle ϕ, and (f) the vertical shift k. Then (g) write an equation for y in terms of t.

Figure 2

Figure 3

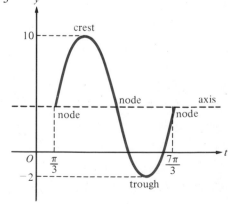

Solution   (a) The vertical distance between the crest and the trough is $10 - (-2) = 12$ units, so $a = \frac{12}{2} = 6$.

(b) $T = \dfrac{7\pi}{3} - \dfrac{\pi}{3} = 2\pi$

(c) We need to calculate the distance S from the y axis to the crest. The distance from the y axis to the first node is $\pi/3$ units. The horizontal distance from this node to the crest is one-fourth of the period T, or $T/4 = 2\pi/4 = \pi/2$. Hence,

$$S = \frac{\pi}{3} + \frac{\pi}{2} = \frac{5\pi}{6}.$$

(d) $\omega = \dfrac{2\pi}{T} = \dfrac{2\pi}{2\pi} = 1$ (e) $\phi = \omega S = 1\left(\dfrac{5\pi}{6}\right) = \dfrac{5\pi}{6}$

(f) The axis is a units below the crest, so $k = 10 - a = 10 - 6 = 4$. (The axis is the horizontal line $y = 4$.)

(g) $y = a\cos(\omega t - \phi) + k = 6\cos\left(t - \dfrac{5\pi}{6}\right) + 4$

In the simple harmonic model

$$y = a\cos(\omega t - \phi) + k,$$

one cycle requires $T = 2\pi/\omega$ units of time, so y **oscillates** *through* $1/T = \omega/(2\pi)$ *cycles in one unit of time.* We call $1/T$ the **frequency of the oscillation.** The Greek letter v (nu) is often used to denote frequency:

$$v = \frac{1}{T} = \frac{\omega}{2\pi}.$$

For instance, a quantity oscillating with a period of $T = 1/60$ second will oscillate through $v = 1/T = 60$ cycles per second.

One cycle per second is called a **hertz** (abbreviated Hz) in honor of the German physicist Heinrich Hertz (1857–94), who discovered radio waves in the late 1880s. Broadcast AM radio waves have frequencies of thousands of hertz (kilohertz), whereas FM and TV waves have frequencies of millions of hertz (megahertz). Alternating current generated by electrical utilities in the United States has a standard frequency of 60 Hz (that is, 60 cycles per second).

The idea of simple harmonic oscillation is nicely illustrated by a **simple pendulum,** which is an idealized object consisting of a point mass m suspended by a weightless string of length ℓ (Figure 4). When pulled to one side of its vertical position and released, it moves periodically to and fro. Let y denote the displacement of the mass from its vertical position, measured along the arc of the swing (positive to the right and negative to the left) at time t. Suppose that $y = a$

Figure 4

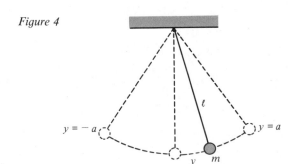

when $t = 0$, the instant of release. Then, if a is not too large, the quantity y will oscillate (approximately) according to a simple harmonic model

$$y = a \cos \omega t,$$

with amplitude a, phase angle $\phi = 0$, vertical shift $k = 0$, and period

$$T = 2\pi \sqrt{\frac{\ell}{g}},$$

where g is the acceleration of gravity. (At the surface of the earth $g \approx 32$ feet/sec² or $g \approx 9.8$ meters/sec².)

Example   ⓒ A pendulum of length $\ell = 1.2$ meters is pulled to the right through an arc of $a = 0.05$ meter and released at $t = 0$ seconds. Find (a) the period T, (b) the frequency v, (c) the angular frequency ω, and (d) the equation for y as a function of t. Round off your answers in parts (a), (b), and (c) to 2 significant digits.

Solution   (a) $T = 2\pi \sqrt{\dfrac{\ell}{g}} = 2\pi \sqrt{\dfrac{1.2}{9.8}} \approx 2.2$ seconds

(b) $v = \dfrac{1}{T} \approx \dfrac{1}{2.2} \approx 0.45$ Hz

(c) $\omega = \dfrac{2\pi}{T} \approx \dfrac{2\pi}{2.2} \approx 2.9$

(d) $y = a \cos \omega t = 0.05 \cos(2.9t)$

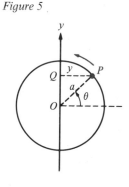

Figure 5

The simple harmonic model does not accurately represent the motion of a pendulum swinging with a large amplitude a. But if a is no larger than $\ell/4$, the formula $T = 2\pi \sqrt{\ell/g}$ gives the true period with an error of less than 1%.

There is an important relationship between simple harmonic motion and circular motion with constant speed (called **uniform circular motion**) which was discussed in Section 1. *Simple harmonic motion is the perpendicular projection along a diameter of uniform circular motion.* Indeed, in Figure 5, suppose that the point P is moving counterclockwise with constant angular speed ω around a circle of radius $r = a$ with center at the origin O of the y axis. Let Q be the perpendicular projection of P on the y axis, and let y denote the y coordinate of Q. Then y oscillates according to the simple harmonic model

$$y = a \cos(\omega t - \phi)$$

with amplitude a, angular frequency ω, and vertical shift $k = 0$ (Problem 36). If the central angle θ (Figure 5) has the value $\theta = \theta_0$ when $t = 0$, then the phase angle ϕ is given by

$$\phi = \frac{\pi}{2} - \theta_0$$

(Problem 36).

PROBLEM SET 7

In problems 1–10, suppose that y is oscillating according to the given simple harmonic model. Find (a) the amplitude a, (b) the angular frequency ω, (c) the period T, (d) the phase angle ϕ, (e) the phase shift S, and (f) the vertical shift k. Then (g) sketch the graph of one cycle.

1. $y = 2\cos\left(t - \dfrac{\pi}{3}\right) + 1$

2. $y = 3\cos\left(t + \dfrac{\pi}{6}\right) - 5$

3. $y = 4\cos\left(t + \dfrac{\pi}{4}\right) + 2$

4. $y = 2\cos(2t - \pi) + 2$

5. $y = 3\cos\left(3t + \dfrac{5\pi}{2}\right)$

6. $y = \dfrac{1}{2}\cos\left(2t - \dfrac{\pi}{2}\right) + 1$

7. $y = \dfrac{3}{4}\cos(4t + 12) - \dfrac{3}{4}$

8. $y = \dfrac{1}{3}\cos\left(\dfrac{\pi t}{4} - \dfrac{7\pi}{8}\right) + \dfrac{2}{3}$

9. $y = 110\cos\left(120\pi t + \dfrac{3\pi}{2}\right)$

10. $y = \sqrt{3}\cos(t - \sqrt{10}) - \dfrac{\sqrt{3}}{3}$

11. In the odd problems from 1 to 9, find the frequency v.

12. For the simple harmonic model $y = a\cos(\omega t - \phi) + k$, show that $\phi = 2\pi S/T$, where S is the phase shift and T is the period.

In problems 13–18, the figure shows one cycle of the graph of a simple harmonic model. Determine (a) the amplitude a, (b) the period T, (c) the phase shift S, (d) the angular frequency ω, (e) the phase angle ϕ, and (f) the vertical shift k. Then (g) write an equation for y in terms of t.

13.

14.

15.

16.

17.

18.

19. In problems 13, 15, and 17, find the frequency v.

20. For the graph of a simple harmonic model with period T, complete the following sentences:
(a) The distance between two successive nodes is _____ units. (b) The horizontal distance between a node and the next crest or trough is _____ units.

21. Figure 6 shows one cycle, from crest to crest, of the graph of a simple harmonic model

$$y = a\cos(\omega t - \phi) + k.$$

Show that

Figure 6

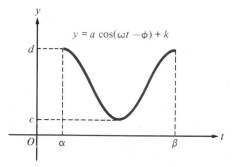

(a) $a = \dfrac{d - c}{2}$

(b) $\omega = \dfrac{2\pi}{\beta - \alpha}$

(c) $\phi = \dfrac{2\pi\alpha}{\beta - \alpha}$

(d) $k = \dfrac{d + c}{2}$

22. Assuming that $\sin x = \cos[x - (\pi/2)]$, show that the graph of $y = a\sin(\omega t - \mu) + k$ is the same as that of the simple harmonic model $y = a\cos(\omega t - \phi) + k$, where $\phi = \mu + (\pi/2)$. [μ is the Greek letter mu.]

23. Figure 7 shows one cycle, starting at a node, of the graph of a simple harmonic model

$$y = a\cos(\omega t - \phi) + k.$$

Show that

Figure 7

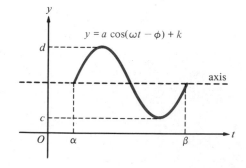

(a) $a = \dfrac{d - c}{2}$

(b) $\omega = \dfrac{2\pi}{\beta - \alpha}$

(c) $\phi = \dfrac{(3\alpha + \beta)\pi}{2(\beta - \alpha)}$

(d) $k = \dfrac{d + c}{2}$

24. What does *amplitude modulation* (AM) mean? What does *frequency modulation* (FM) mean? [If you don't know, use a dictionary.]

In problems 25–30, sketch one cycle of the graph of a quantity y oscillating according to a simple harmonic model $y = a\cos(\omega t - \phi) + k$, with amplitude a, frequency v, phase angle ϕ, and vertical shift k.

25. $a = 1, v = 50$ Hz, $\phi = \dfrac{\pi}{2}, k = 0$

26. $a = 220, v = 60$ Hz, $\phi = 0, k = 0$

27. $a = \dfrac{2}{3}, v = \dfrac{1}{4\pi}$ Hz, $\phi = \dfrac{\pi}{4}, k = \dfrac{2}{3}$

28. $a = 5, v = \dfrac{1}{28}$ cycle per day, $\phi = \dfrac{-3\pi}{14}, k = 0$

29. $a = 8.4, v = 0.18$ cycle per week, $\phi = 0, k = 0$

30. $a = 5 \times 10^{-7}, v = 88.5$ megahertz, $\phi = 0.8, k = 0$

31. A pendulum of length $\ell = 0.5$ foot is pulled to the right through an arc of $a = 0.1$ foot and released at $t = 0$ seconds. Find the period T, the frequency v, the angular frequency ω, and the equation giving the displacement y at time t.

© **32.** An astronaut on the surface of the moon finds that a pendulum has a period 2.45 times as long as it does on the surface of the earth. From this information, she is able to calculate the acceleration of gravity on the moon. How does she do this, and what is her answer?

33. You are lost on a desert island and your watch is broken. You have managed to salvage a toolbox containing, among other things, a tape measure and a ball of string. Explain how you go about constructing a crude "clock" that will measure time in seconds.

© **34.** In the northern sky, the constellation Cepheus includes a number of stars that pulsate in brightness or magnitude. One of these stars, Beta Cephei, has a magnitude y that varies from a minimum of 3.141 to a maximum of 3.159 with a period of 4.5 hours. Using a simple harmonic model with a phase angle of $\phi = 6.33$, (a) sketch a graph showing the variation of y over the time interval $t = 0$ to $t = 8$ hours, and (b) determine the magnitude of Beta Cephei at $t = 4$ hours.

Figure 8

© **35.** A typical mass-spring system consists of a mass m suspended on a spring (Figure 8). A vertical y axis is set up, with the equilibrium level of the mass at $y = k$. Suppose that the mass is raised to a level $y = k + a$ and released at time $t = 0$. It is shown in physics that, neglecting friction and air resistance, the mass will bob up and down according to a simple harmonic model $y = a\cos(\omega t - \phi) + k$ with phase angle $\phi = 0$ and frequency

$$v = \frac{1}{2\pi}\sqrt{\frac{K}{m}}.$$

Here K, a constant that measures the stiffness of the spring, is the amount of force required to stretch it by one unit of distance. If $m = 0.05$ kilogram, $a = 0.04$ meter, and $K = 1.25$ newtons per meter, sketch two cycles of the graph of y.

Figure 9

36. In Figure 5, suppose that the point P is moving counterclockwise around the circle with angular speed ω and that $\theta = \theta_0$ when $t = 0$. Show that $y = a\cos(\omega t - \phi)$, where $\phi = (\pi/2) - \theta_0$. Hint: Show that $\theta = \omega t + \theta_0$, then use $\sin\theta = \cos[\theta - (\pi/2)]$.

37. An electrical circuit consisting of a coil with inductance L henries and a capacitor with a capacitance of C farads is called an LC-circuit (Figure 9). If the capacitor is charged and the switch is closed at time $t = 0$ seconds, then the potential difference E volts across the plates of the capacitor will oscillate in a simple harmonic manner with amplitude E_0, angular frequency $\omega = 1/\sqrt{LC}$, and phase angle $\phi = 0$. If $E_0 = 100$ volts, $L = 10$ henries, and $C = 0.0005$ farad, find the period T, the frequency v, and the equation giving the voltage E at time t seconds.

ⓒ **38.** A **sound wave** in air consists of a continuous train of alternate compressions and rarefactions traveling at a speed of approximately 331 meters per second. The distance between any two successive compressions is called the **wavelength.** If the air pressure at any one fixed point varies in a simple harmonic manner, the sound heard by a human ear at that point is perceived as a **pure tone.** The amplitude of this variation determines the loudness of the tone, and its frequency determines the pitch. The pure tone middle C has a frequency of 256 Hz. (a) Find the wavelength of a sound wave that produces the pure tone middle C. (b) Suppose the sound wave in (a) causes an air pressure variation of amplitude 0.05 newton/meter^2 at a certain point. Using a phase angle $\phi = \pi/2$, write the equation for the air pressure y at this point at time t, if normal undisturbed atmospheric pressure is 10^5 newtons/meter^2.

ⓒ **39.** Some people believe that we are all subject to periodic variations, called **biorhythms,** in our physical stamina, emotional well-being, and intellectual ability. The physical cycle is supposed to have a period of 23 days, the emotional cycle a period of 28 days, and the intellectual cycle a period of 33 days. Three graphs plotted on the same coordinate system (using, say, three different colors) showing the variations of these three factors for a particular individual over a period of a month or more make up a **biorhythm chart.** Most practitioners use simple harmonic curves all of the same (arbitrary) amplitude a. Taking $a = 1$, plot a biorhythm chart over a 62-day period for an individual with the following phase angles: physical phase angle = 5.804, emotional phase angle = 3.030, and intellectual phase angle = 6.045.

ⓒ **40.** Believers in biorhythms (see problem 39) refer to the nodes of any one of the three biorhythm curves as **critical points,** and to the corresponding days as **critical days.** Contrary to what one might at first think, it is believed that the critical points are more significant than the crests or the troughs of the biorhythm curves. For the biorhythm chart plotted in problem 39, locate the physical, emotional, and intellectual critical days. Are there any doubly critical days—that is, days on which two curves pass through a node simultaneously? What about triply critical days?

REVIEW PROBLEM SET

In problems 1–4, find the degree measure of the given angle.

1. One-tenth of a counterclockwise revolution.
2. Five-sixths of a counterclockwise revolution.
3. Four-fifths of a clockwise revolution.
4. The angle through which the minute hand on a clock turns in 5 hours.

In problems 5–10, indicate the number of revolutions or the fraction of a revolution represented by the given angle. Indicate whether the rotation is clockwise or counterclockwise.

5. $60°$ **6.** $140°$ **7.** $-1440°$ **8.** $-310°$
9. $930°$ **10.** $-610°$

ⓒ In problems 11–14, express each angle measure as a decimal. Round off to four decimal places.

11. $2°3'$ **12.** $23°19'13''$ **13.** $55°45'35''$ **14.** $5''$

In problems 15–18, express each angle measure in degrees, minutes, and seconds.

15. $87.35°$ **16.** $-62.45°$ ⓒ **17.** $-24.53°$ ⓒ **18.** $165.37°$

In problems 19–24, s denotes the length of the arc intercepted on a circle of radius r by a central angle of θ radians. Find the missing quantity.

19. $r = 5$ meters, $\theta = 0.57$ radian, $s = ?$

20. $r = 40$ centimeters, $s = 4$ centimeters, $\theta = ?$

21. $s = 3\pi$ feet, $\theta = \pi$ radians, $r = ?$

22. $r = 13$ kilometers, $\theta = \dfrac{3\pi}{7}$ radians, $s = ?$

23. $r = 2$ meters, $s = \pi$ meters, $\theta = ?$

24. $s = 17\pi$ microns, $\theta = \dfrac{5\pi}{6}$ radians, $r = ?$

In problems 25–30, convert each degree measure to radian measure. Do not use a calculator. Write your answer as a rational multiple of π.

25. $80°$ **26.** $570°$ **27.** $-355°$ **28.** $-810°$ **29.** $-310°$ **30.** $765°$

In problems 31–36, convert each radian measure to degree measure. Do not use a calculator.

31. $\dfrac{2\pi}{5}$ **32.** $-\dfrac{13\pi}{4}$ **33.** $-\dfrac{7\pi}{8}$ **34.** $\dfrac{35\pi}{3}$ **35.** $\dfrac{51\pi}{4}$ **36.** $\dfrac{18\pi}{5}$

C **37.** Use a calculator to convert each degree measure to an approximate radian measure rounded off to four decimal places.
(a) $5°$ (b) $27.7533°$ (c) $-17.173°$ (d) $35°16'55''$

C **38.** Use a calculator to convert each radian measure to an approximate degree measure rounded off to four decimal places.
(a) 5 (b) 3.9 (c) -7.63 (d) -21.403

C In problems 39–42, s denotes the length of the arc intercepted on a circle of radius r by a central angle θ measured in *degrees*. Find the missing quantity. Round off your answers to three significant digits.

39. $r = 10$ feet, $\theta = 36°$, $s = ?$

40. $s = 111$ centimeters, $\theta = 135°$, $r = ?$

41. $r = 12$ meters, $s = 3\pi$ meters, $\theta = ?$

42. $r = 5$ kilometers, $\theta = 65°$, $s = ?$

C In problems 43 and 44, find the area $A = \frac{1}{2}r^2\theta$ of a sector of a circle of radius r with central angle θ. (See problem 48 in Problem Set 1.)

43. $r = 25$ centimeters, $\theta = \dfrac{\pi}{6}$

44. $r = 3.5$ meters, $\theta = 60°$

C **45.** The minute hand on a tower clock is 0.6 meter long. How far does the tip of the hand travel in 4 minutes?

C **46.** In problem 45, what is the area of the sector swept out by the minute hand in 4 minutes?

C **47.** A wheel with radius 2 feet makes 30 revolutions per minute. Find the linear speed of a point on the rim of the wheel.

C **48.** Find the approximate diameter of the moon if its disk subtends an angle of $30'$ at a point on the earth 240,000 miles away (Figure 1). [Hint: Approximate the diameter $|DE|$ by the length of the arc $\overset{\frown}{BC}$.]

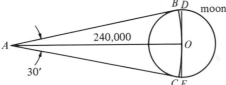

Figure 1

C **49.** A satellite in a circular orbit above the earth is known to have a linear speed of 9.92 kilometers per second. In 10 seconds it moves along an arc that subtends an angle of $0.75°$ at the center of the earth. If the radius of the earth is 6371 kilometers, how high is the satellite above the surface of the earth?

C **50.** A laser beam is projected from a point on the surface of the earth to a reflector on the surface of the moon 384,000 kilometers away. If the laser beam diverges at an angle of $1''$, find the approximate diameter of the beam when it strikes the surface of the moon.

C **51.** A phonograph record is rotating at 33.33 revolutions per minute. When the needle is in a groove 10 centimeters from the center, what is the linear speed of the groove with respect to the needle?

C **52.** A small sprocket (toothed wheel) of radius 7 centimeters is connected by a chain to a large sprocket of radius 11 centimeters. The small sprocket is turning at 60 revolutions per minute. (a) Find the linear speed of the chain connecting the sprockets. (b) Find the rate of rotation of the large sprocket.

Figure 2

53. Find the values of the six trigonometric functions of the acute angle θ in Figure 2.

54. Given that $\sin 67° = 0.9205$, $\tan 67° = 2.356$, and $\sec 67° = 2.559$, find the values of
(a) $\cos 23°$ (b) $\cot 23°$ (c) $\csc 23°$

55. Express the value of each trigonometric function as the value of the cofunction of the complementary angle.

(a) $\sin 50°$ (b) $\cos \dfrac{\pi}{7}$ (c) $\sec 89°$ (d) $\cot \dfrac{3\pi}{28}$

C **56.** Suppose that α and β are complementary acute angles and that $\sin \alpha = 0.9063$. Find:
(a) $\cos \beta$ (b) $\csc \alpha$ (c) $\sec \beta$.

57. Sketch two coterminal angles α and β in standard position whose terminal side contains each point. Arrange it so that α is positive, β is negative, and neither angle exceeds one revolution. In each case, name the quadrant in which the angle lies or indicate that it is quadrantal.
(a) $(5, 12)$ (b) $(-3, 5)$ (c) $(-7, -6)$ (d) $(0, -4)$
(e) $(\sqrt{2}, -\sqrt{3})$ (f) $(\sqrt{5}, 0)$

58. Sketch each angle in standard position and name the quadrant in which it lies or specify that it is quadrantal.
(a) $60°$ (b) $-210°$ (c) $110°$ (d) $-2160°$ (e) $-340°$ (f) $-750°$

59. In each case, specify and sketch three different angles that are coterminal with the given angle in standard position.

(a) $-15°$ (b) $460°$ (c) $170°$ (d) $-980°$ (e) $\dfrac{5\pi}{3}$

60. Indicate which of the six trigonometric functions are *not* defined for each angle.

(a) $1260°$ (b) 37π (c) 38π (d) $\dfrac{19\pi}{2}$

61. For each point (x, y), evaluate the six trigonometric functions of an angle θ in standard position if the terminal side of θ contains (x, y). In each case, sketch one of the coterminal angles θ thus formed. Do not use a calculator.
(a) $(-3, 5)$ (b) $(2, -3)$ (c) $(-6, -8)$ (d) $(\sqrt{3}, -1)$

62. Without using a calculator, give the algebraic sign of:

(a) $\sin 183°$ (b) $\tan \dfrac{37\pi}{39}$ (c) $\cos(-269°)$ (d) $\cot\left(-\dfrac{27\pi}{17}\right)$.

63. Indicate the quadrant in which θ lies if
(a) $\sin \theta > 0$ and $\cos \theta < 0$ (b) $\tan \theta > 0$ and $\sec \theta < 0$
(c) $\csc \theta < 0$ and $\tan \theta < 0$ (d) $\sin \theta < 0$ and $\sec \theta > 0$

64. If n is an integer, explain why $\cos(n\pi) = (-1)^n$.

65. If $\sin \theta = -\frac{4}{5}$ and $\cos \theta = \frac{3}{5}$, use the reciprocal and quotient identities to find the values of the other four trigonometric functions.

66. If $\sin\theta = 0.3145$ and $\cos\theta = -0.9493$, use the reciprocal and quotient identities to find the values of the other four trigonometric functions.

In problems 67–74, the value of one of the trigonometric functions is given along with information about the quadrant (Q) in which θ lies. Find the values of the other five trigonometric functions of θ.

67. $\sin\theta = -\frac{5}{13}$, θ in Q_{IV}
69. $\csc\theta = \frac{13}{12}$, θ in Q_{II}
71. $\sin\theta = \frac{3}{5}$, $\cos\theta < 0$
73. $\cot\theta = 0.6249$, $\sin\theta < 0$

68. $\tan\theta = \frac{3}{2}$, θ not in Q_I
70. $\sec\theta = -\frac{5}{4}$, θ in Q_{III}
72. $\sec\theta = -5$, $\csc\theta < 0$
74. $\tan\theta = 0.5543$, $\sec\theta < 0$

75. Without using a calculator, find the values (if they are defined) of the six trigonometric functions of

(a) $-180°$ (b) $540°$ (c) $990°$ (d) $-360°$ (e) $\dfrac{7\pi}{2}$ (f) -5π

(g) 17π (h) 18π

76. If n is an integer, explain why $\sin\left[\dfrac{(2n-1)\pi}{2}\right] = (-1)^{n+1}$.

77. Using reference angles and the values of the six trigonometric functions for $30°$, $45°$, and $60°$ (or $\pi/6$, $\pi/4$, and $\pi/3$), find the values of the six trigonometric functions of:

(a) $-150°$ (b) $-315°$ (c) $780°$ (d) $\dfrac{13\pi}{3}$ (e) $-\dfrac{15\pi}{4}$

78. Trigonometric tables ordinarily give approximate values of the trigonometric functions for acute angles only. Explain in your own words how to use such a table to find values of these functions for angles that are not acute.

In problems 79–96, use a calculator to find the approximate values of the indicated trigonometric functions.

79. $\sin 27°20'$
82. $\cos(-113.81°)$
85. $\sec 16.43°$
88. $\csc 5.132$
91. $\cos(-19.213)$
94. $\cot\sqrt{3}$

80. $\sin(421°15')$
83. $\tan(-117°15'30'')$
86. $\sec(-248.2°)$
89. $\cot(-3.18)$
92. $\csc 18.113$
95. $\cos\dfrac{\sqrt{2\pi}}{4}$

81. $\cos 53.47°$
84. $\tan(-281°31'25'')$
87. $\csc\dfrac{4\pi}{5}$
90. $\cot(-7.167)$
93. $\sin 5.015$
96. $\sin[\tan(-71.32)]$

In problems 97–106, use the graphs of the six trigonometric functions (Sections 5 and 6) and the techniques of graphing (Section 6 of Chapter 3) to sketch the graph of each equation.

97. $y = 1 + \frac{1}{3}\sin x$
99. $y = 2 + \dfrac{1}{2}\cos\left(x - \dfrac{\pi}{2}\right)$
101. $y = -\tan\left(x - \dfrac{\pi}{6}\right)$
103. $y = \sec(x - \pi)$
105. $y = \dfrac{2}{3}\cot\left(x - \dfrac{\pi}{2}\right)$

98. $y = 1 - \frac{2}{5}\sin(x + \pi)$
100. $y = -0.1\cos\left(x - \dfrac{\pi}{4}\right)$
102. $y = \sec\left(x - \dfrac{\pi}{4}\right) + 1$
104. $y = 1 - 4\csc\left(x - \dfrac{\pi}{3}\right)$
106. $y = 2 - \tan(x + \pi)$

107. If $-2\pi \le x \le 2\pi$, for what values of x does $\cos x$ reach its maximum and minimum values and what are these values?

108. If $-2\pi \le x \le 2\pi$, for what values of x is $\sin x = \sqrt{3}/2$?

109. Sketch the graphs of

$$y = \sin\left(\frac{3x}{2} - \frac{\pi}{4}\right) \quad \text{and} \quad y = \cos\left(\frac{3x}{2} - \frac{\pi}{4}\right)$$

on the same coordinate system.

110. Sketch the graph of $y = |\sin x|$.

In problems 111–116, suppose that y is a quantity that is oscillating according to the given simple harmonic model. Find (a) the amplitude a, (b) the angular frequency ω, (c) the period T, (d) the phase angle ϕ, (e) the phase shift S, (f) the vertical shift k, and (g) the frequency v. Then (h) sketch the graph of one cycle.

111. $y = 2\cos(\pi t + \pi) + 1$

112. $y = 3\cos(0.4t - 1) + 3$

113. $y = 3\cos\left(\frac{t}{2} - \frac{\pi}{2}\right) - 3$

114. $y = 5\cos\left(\frac{t}{3} + \frac{5\pi}{6}\right) - 2$

115. $y = 0.2\cos(0.25t - \pi)$

116. $y = 0.1\cos\left[3\left(t - \frac{\pi}{3}\right)\right]$

C **117.** A pendulum of length $\ell = 5.2$ meters is pulled to the right through an arc of $a = 0.2$ meter and released at $t = 0$ seconds. Find the period T, the frequency v, the angular frequency ω, and the equation giving the displacement y at time t. Sketch a graph showing two cycles of the motion starting from $t = 0$.

C **118.** The long-period variable star T Cephei has a magnitude y that varies from a minimum of 5.13 to a maximum of 10.87 with a period of 390 days. Using a simple harmonic model with a phase angle of $\phi = 0.16$, (a) sketch a graph showing the variation of y over the time interval $t = 0$ to $t = 600$ days, and (b) determine the magnitude of T Cephei at $t = 24$ days.

119. A sound wave in air with a frequency of 440 Hz is perceived as the pure tone "concert A." Suppose that this sound wave causes a sinusoidal air-pressure variation of amplitude 0.03 newton/meter^2 at a certain point. Using a phase angle of $\phi = \pi/2$, write the equation for the air pressure y at this point at time t if normal undisturbed atmospheric pressure is 10^5 newtons/meter.

C **120.** An LC-circuit (problem 37, Section 7) contains an inductance of $L = 5 \times 10^{-4}$ henry and a variable capacitor that allows the circuit to be tuned from a frequency of 5.5×10^5 Hz to a frequency of 1.6×10^6 Hz. Sketch a graph showing the capacitance C as a function of the frequency v over this range of values.

121. People who believe in biorhythms (see problem 39, Section 7) claim that the physical, emotional, and intellectual curves start with a common node at birth, $t = 0$ days. Assuming this, determine the number of years that will elapse before the three biorhythms again reach a common node.

C **122.** In engineering and physics, an oscillation with an amplitude that decreases exponentially with the passage of time is said to be *damped*. The mathematical model for **damped simple harmonic oscillation** is

$$y = ae^{-Kt}\cos(\omega t - \phi) + k,$$

where the positive constant K is called the *damping constant*. Sketch a graph of a damped harmonic oscillation for $K = \frac{1}{2}$, $a = 2$, $\omega = \pi/4$, $\phi = \pi/2$, and $k = 0$.

7 Trigonometric Identities and Equations

In Section 3 of Chapter 6, we derived the fundamental *reciprocal*, *quotient*, and *Pythagorean* identities. In this chapter, we establish the remaining standard trigonometric identities, including the important formulas for *sums*, *differences*, and *multiples* of angles. (For convenience, all of the standard trigonometric identities are listed inside the back cover of this book.) In many applications, ranging from calculus, engineering, and physics to economics and the life sciences, these identities and formulas are routinely used to simplify complicated expressions and to help solve equations involving trigonometric functions.

1 The Fundamental Trigonometric Identities

A **trigonometric equation** is, by definition, an equation that involves at least one trigonometric function of a variable. Such an equation is called a **trigonometric identity** if it is true for all values of the variable for which both sides of the equation are defined. An equation that is not an identity is called a **conditional equation.**

For instance, the trigonometric equation

$$\csc t = \frac{1}{\sin t}$$

is an identity, since it is true for all values of t (except, of course, for those values for which $\csc t$ or $1/\sin t$ is undefined). On the other hand, the trigonometric equation

$$\sin t = \cos t$$

is a conditional equation, since there are values of t (for instance, $t = 0$) for which it isn't true.

Now we are going to derive the trigonometric identities

$$\sin(-\theta) = -\sin \theta \quad \text{and} \quad \cos(-\theta) = \cos \theta$$

suggested by the graphs of the sine and cosine functions in Section 5 of Chapter 6. Figure 1 shows an angle θ and the corresponding angle $-\theta$, both in standard position. Evidently, the points P and Q,

Figure 1

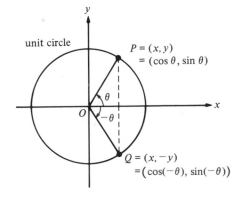

327

where the terminal sides of these angles intersect the unit circle, are mirror images of each other across the x axis. Therefore, if $P = (x, y)$ it follows that $Q = (x, -y)$. In Section 5 of Chapter 6, we showed that

$$P = (x, y) = (\cos\theta, \sin\theta).$$

Likewise,

$$Q = (x, -y) = (\cos(-\theta), \sin(-\theta)).$$

Therefore,

$$\sin(-\theta) = -y = -\sin\theta$$

and

$$\cos(-\theta) = x = \cos\theta.$$

If we now combine the identities obtained above with the quotient identity, $\tan\theta = \sin\theta/\cos\theta$, we find that

$$\tan(-\theta) = \frac{\sin(-\theta)}{\cos(-\theta)} = \frac{-\sin\theta}{\cos\theta} = -\tan\theta.$$

Similar arguments apply to $\cot(-\theta)$, $\sec(-\theta)$, and $\csc(-\theta)$ (Problem 51). The results are summarized in the following theorem.

Theorem 1   **Even–Odd Identities**

For all values of θ in the domains of the functions:

(i) $\sin(-\theta) = -\sin\theta$ (ii) $\cos(-\theta) = \cos\theta$ (iii) $\tan(-\theta) = -\tan\theta$

(iv) $\cot(-\theta) = -\cot\theta$ (v) $\sec(-\theta) = \sec\theta$ (vi) $\csc(-\theta) = -\csc\theta$

Notice that only the cosine and its reciprocal the secant are even functions—the remaining four trigonometric functions are odd. The even–odd identities are often used to simplify expressions as the following example illustrates.

Example   Use the even–odd identities to simplify each expression.

(a) $\dfrac{\sin(-\theta) + \cos(-\theta)}{\sin(-\theta) - \cos(-\theta)}$ (b) $1 + \tan^2(-t)$

Solution   (a) $\dfrac{\sin(-\theta) + \cos(-\theta)}{\sin(-\theta) - \cos(-\theta)} = \dfrac{-\sin\theta + \cos\theta}{-\sin\theta - \cos\theta}$

$$= \frac{-(\sin\theta - \cos\theta)}{-(\sin\theta + \cos\theta)} = \frac{\sin\theta - \cos\theta}{\sin\theta + \cos\theta}$$

(b) $1 + \tan^2(-t) = 1 + [\tan(-t)]^2 = 1 + (-\tan t)^2 = 1 + \tan^2 t = \sec^2 t$

The reciprocal, quotient, and Pythagorean identities obtained in Chapter 6, Section 3, together with the even–odd identities, are often called the *fundamental trigonometric identities*. For convenience, we summarize these identities here.

Fundamental Trigonometric Identities

1. $\csc\theta = \dfrac{1}{\sin\theta}$
2. $\sec\theta = \dfrac{1}{\cos\theta}$
3. $\cot\theta = \dfrac{1}{\tan\theta}$

4. $\tan\theta = \dfrac{\sin\theta}{\cos\theta}$
5. $\cot\theta = \dfrac{\cos\theta}{\sin\theta}$
6. $\cos^2\theta + \sin^2\theta = 1$

7. $1 + \tan^2\theta = \sec^2\theta$
8. $\cot^2\theta + 1 = \csc^2\theta$
9. $\sin(-\theta) = -\sin\theta$

10. $\cos(-\theta) = \cos\theta$
11. $\tan(-\theta) = -\tan\theta$
12. $\cot(-\theta) = -\cot\theta$

13. $\sec(-\theta) = \sec\theta$
14. $\csc(-\theta) = -\csc\theta$

Not only should you memorize these fourteen fundamental identities, but they should become so familiar to you that you can recognize them quickly even when they are written in equivalent forms. For instance, $\csc\theta = 1/\sin\theta$ can also be written as

$$(\sin\theta)(\csc\theta) = 1 \quad \text{or} \quad \sin\theta = \frac{1}{\csc\theta}.$$

Incidentally, a product of values of trigonometric functions such as $(\sin\theta)(\csc\theta)$ is usually written simply as $\sin\theta\csc\theta$, unless the parentheses are necessary to prevent confusion.

Examples   Simplify each trigonometric expression by using the fundamental identities.

1 $\csc\theta\cos\theta$

Solution
$$\csc\theta\cos\theta = \frac{1}{\sin\theta}\cos\theta = \frac{\cos\theta}{\sin\theta} = \cot\theta$$

2 $\tan^2 t - \sec^2 t$

Solution   Because $1 + \tan^2 t = \sec^2 t$, it follows that
$$\tan^2 t - \sec^2 t = -1.$$

3 $\csc^4 x - 2\csc^2 x\cot^2 x + \cot^4 x$

Solution   The given expression is the square of $\csc^2 x - \cot^2 x$. Because $\cot^2 x + 1 = \csc^2 x$, we have $\csc^2 x - \cot^2 x = 1$. Therefore,
$$\csc^4 x - 2\csc^2 x\cot^2 x + \cot^4 x = (\csc^2 x - \cot^2 x)^2 = 1^2 = 1.$$

The reciprocal and quotient identities enable us to write $\csc\theta$, $\sec\theta$, $\tan\theta$, and $\cot\theta$ in terms of $\sin\theta$ and $\cos\theta$. Therefore, *any trigonometric expression can be rewritten in terms of sines and cosines only*. This fact and the Pythagorean identity $\cos^2\theta + \sin^2\theta = 1$ can often be used to simplify trigonometric expressions.

Examples   Rewrite each trigonometric expression in terms of sines and cosines, and then simplify the result.

1 $\csc t - \dfrac{\cot t}{\sec t}$

Solution

$$\csc t - \frac{\cot t}{\sec t} = \frac{1}{\sin t} - \frac{\dfrac{\cos t}{\sin t}}{\dfrac{1}{\cos t}} = \frac{1}{\sin t} - \frac{\cos t}{\sin t} \cos t$$

$$= \frac{1}{\sin t} - \frac{\cos^2 t}{\sin t} = \frac{1 - \cos^2 t}{\sin t} = \frac{\sin^2 t}{\sin t} = \sin t$$

2 $\dfrac{\csc^2 x \sec^2 x}{\csc^2 x + \sec^2 x}$

Solution

$$\frac{\csc^2 x \sec^2 x}{\csc^2 x + \sec^2 x} = \frac{\left(\dfrac{1}{\sin^2 x}\right)\left(\dfrac{1}{\cos^2 x}\right)}{\dfrac{1}{\sin^2 x} + \dfrac{1}{\cos^2 x}} = \frac{\sin^2 x \cos^2 x \left(\dfrac{1}{\sin^2 x}\right)\left(\dfrac{1}{\cos^2 x}\right)}{\sin^2 x \cos^2 x \left(\dfrac{1}{\sin^2 x} + \dfrac{1}{\cos^2 x}\right)}$$

$$= \frac{1}{\cos^2 x + \sin^2 x} = \frac{1}{1} = 1$$

The Pythagorean identity $\cos^2 \theta + \sin^2 \theta = 1$ can be rewritten as

$$\sin^2 \theta = 1 - \cos^2 \theta \quad \text{or} \quad \cos^2 \theta = 1 - \sin^2 \theta.$$

Therefore, we have

$$\sin \theta = \pm\sqrt{1 - \cos^2 \theta}$$

and

$$\cos \theta = \pm\sqrt{1 - \sin^2 \theta}.$$

In either case, the correct algebraic sign is determined by the quadrant or coordinate axis containing the terminal side of the angle θ in standard position. After you have rewritten a trigonometric expression in terms of sines and cosines, you can use these equations to bring the expression into a form involving *only the sine* or *only the cosine*.

Example   Rewrite the expression $\cot \theta \csc^2 \theta$ in terms of $\sin \theta$ only.

Solution

$$\cot \theta \csc^2 \theta = \frac{\cos \theta}{\sin \theta} \cdot \frac{1}{\sin^2 \theta} = \frac{\cos \theta}{\sin^3 \theta}$$

$$= \frac{\pm\sqrt{1 - \sin^2 \theta}}{\sin^3 \theta}$$

Algebraic expressions not originally containing trigonometric functions can often be simplified by substituting trigonometric expressions for the variable. This technique, called **trigonometric substitution,** is routinely used in calculus to rewrite radical expressions as trigonometric expressions containing no radicals.

Example If a is a positive constant, rewrite the radical expression $\sqrt{a^2 - u^2}$ as a trigono- metric expression containing no radical by using the trigonometric substitution $u = a \sin \theta$. Assume that $-\dfrac{\pi}{2} \le \theta \le \dfrac{\pi}{2}$ so that $\cos \theta \ge 0$.

Solution
$$\sqrt{a^2 - u^2} = \sqrt{a^2 - (a \sin \theta)^2} = \sqrt{a^2 - a^2 \sin^2 \theta}$$
$$= \sqrt{a^2(1 - \sin^2 \theta)} = \sqrt{a^2 \cos^2 \theta}$$
$$= a \cos \theta$$

PROBLEM SET 1

In problems 1–6, use the even–odd identities to simplify each expression.

1. $\sin(-\theta)\cos(-\theta)$

2. $\cot^2(-u) + 1$

3. $\tan t + \tan(-t)$

4. $\cos(-x)\sec x$

5. $\dfrac{1 + \csc(-\alpha)}{1 - \cot(-\beta)}$

6. $[1 + \sin \gamma][1 + \sin(-\gamma)]$

In problems 7–28, use the fundamental identities to simplify each expression.

7. $\sec \theta \sin \theta$

8. $1 + \dfrac{\tan \alpha}{\cot \alpha}$

9. $\cot v \sec v$

10. $\dfrac{\csc^2 u}{1 + \tan^2 u}$

11. $\dfrac{\csc \beta}{\sec \beta}$

12. $\dfrac{\sin^2 \theta - 1}{\sec \theta}$

13. $\cot^2 \alpha - \csc^2 \alpha$

14. $\dfrac{\sec^2 t - 1}{\sec^2 t}$

15. $(\csc u - 1)(\csc u + 1)$

16. $\dfrac{1 + \cot^2 y}{1 + \tan^2 y}$

17. $\dfrac{1}{\sec^2 x} + \dfrac{1}{\csc^2 x}$

18. $\dfrac{(\sec \gamma - 1)(\sec \gamma + 1)}{\tan \gamma}$

19. $\sin^4 t + 2 \cos^2 t \sin^2 t + \cos^4 t$

20. $\sin^4 u + 2 \cos^2 u - \cos^4 u$

21. $\tan^4 \alpha - 2 \tan^2 \alpha \sec^2 \alpha + \sec^4 \alpha$

22. $(1 + \tan^2 \theta)(1 - \sin^2 \theta)$

23. $\cos x \sin^3 x + \sin x \cos^3 x$

24. $(1 - \cos^2 \beta)(1 + \cot^2 \beta)$

25. $\dfrac{1}{\sin t \cos t} - \dfrac{\cos t}{\sin t}$

26. $\dfrac{\cos \gamma}{1 - \sin \gamma} + \dfrac{\cos \gamma}{1 + \sin \gamma}$

27. $\dfrac{\sin t}{1 + \cos t} + \dfrac{1 + \cos t}{\sin t}$

28. $\dfrac{\sin \alpha + \sin \beta}{\cos \alpha + \cos \beta} + \dfrac{\cos \alpha - \cos \beta}{\sin \alpha - \sin \beta}$

In problems 29–38, rewrite each trigonometric expression in terms of sines and cosines, and then simplify the result.

29. $\dfrac{\tan x}{\sec x}$

30. $(\cos \theta + \tan \theta \sin \theta)\cot \theta$

31. $\dfrac{\csc(-t)}{\sec(-t)\cot t}$

32. $\dfrac{\csc^2 x + \sec^2 x}{\csc^2 x \sec x}$

33. $\dfrac{\sec \alpha}{\csc \alpha(\tan \alpha + \cot \alpha)}$

34. $\dfrac{\sin y + \tan y}{1 + \sec y}$

35. $\dfrac{1 + \tan\theta}{\sec\theta}$

36. $\dfrac{\cot(-t) - 1}{1 - \tan(-t)}$

37. $\dfrac{\tan u + \sin u}{\cot u + \csc u}$

38. $\dfrac{\csc\beta}{\csc\beta + \tan\beta} + \dfrac{\csc\beta}{\csc\beta - \tan\beta}$

In problems 39–44, rewrite each expression in terms of the indicated function only.

39. $\sec^2\theta \tan\theta$ in terms of $\cos\theta$

40. $\dfrac{\sin t + \cot t \cos t}{\cot t}$ in terms of $\sec t$

41. $\dfrac{1 + \cot^2 x}{\cot^2 x}$ in terms of $\cos x$

42. $\dfrac{\csc^2 y + \sec^2 y}{\csc y \sec y}$ in terms of $\tan y$

43. $\dfrac{\sin(-\alpha) + \tan(-\alpha)}{1 + \sec(-\alpha)}$ in terms of $\sin\alpha$

44. $(\cot u + \csc u)(\tan u - \sin u)$ in terms of $\sec u$

In problems 45–48, rewrite each radical expression as a trigonometric expression containing no radical, by making the indicated trigonometric substitution. Assume that a is a positive constant.

45. $\sqrt{a^2 + u^2}$, $u = a\tan\theta$, $-\pi/2 < \theta < \pi/2$

46. $\sqrt{u^2 - a^2}$, $u = a\sec\theta$, $0 \le \theta < \pi/2$

47. $\sqrt{(4 - x^2)^3}$, $x = 2\cos t$, $0 \le t \le \pi$

48. $\sqrt{9 - 25x^2}$, $x = (3/5)\sin u$, $-\pi/2 \le u \le \pi/2$

49. Show that every trigonometric expression can be rewritten in terms of each of the six trigonometric functions alone, provided that ambiguous $\pm$ signs are allowed.

50. Let (x, y) be a point 1 unit from the origin (that is, on the unit circle) on the terminal side of an angle θ in standard position. Sketch figures to show that $(x, -y)$ is the point 1 unit from the origin on the terminal side of $-\theta$ for the following cases: (a) θ lies in quadrants II, III, or IV; (b) θ is a quadrantal angle; (c) θ is a negative angle.

51. Prove parts (iv), (v), and (vi) of Theorem 1.

2   Verifying Trigonometric Identities

By combining the fundamental trigonometric identities, it is possible to derive a large number of related identities. It isn't necessary to memorize these derived identities—when you need one of them, you can obtain it by manipulating the fundamental identities.

Although there is no universal step-by-step procedure for proving that a trigonometric equation is an identity, there are some useful guidelines. Perhaps the most important one concerns what *not* to do: *Do not treat a trigonometric equation as if it were an identity until after you have proved that it really is one.* The easiest way to avoid this pitfall is to *treat both sides of the equation separately*. To see the necessity for being careful, consider the following simple example. It's clear that

$$\sin x = -\sin x$$

is *not* an identity. Yet, if we square both sides, we get

$$\sin^2 x = \sin^2 x,$$

which *is* an identity.

Here are some additional suggestions.

Suggestion 1. Take one side of the equation and try to reduce it to the other side by a sequence of manipulations. It usually pays to start with the more complicated side and try to reduce it to the simpler side.

Suggestion 2. If Suggestion 1 doesn't seem to work, try to simplify each side of the equation separately. If you can reduce each side to the *same* expression, you can conclude that the original equation is an identity.

Suggestion 3. In carrying out Suggestion 1 or 2, try the following:

(a) Perform indicated additions or subtractions of fractions. (First obtain common denominators, of course.)

(b) Perform indicated multiplications or divisions of expressions.

(c) Simplify fractions by canceling common factors in the numerator and denominator.

(d) See whether you can come closer to your goal by factoring combinations of terms.

(e) Try multiplying both numerator and denominator of a fraction by the same expression.

(f) Try rewriting all trigonometric expressions in terms of sines and cosines.

Examples   Prove that each equation is an identity.

1 $(\sin^2 \theta)(1 + \cot^2 \theta) = 1$

Solution   We follow Suggestion 1, starting with the left side of the equation (the more complicated side):

$$(\sin^2 \theta)(1 + \cot^2 \theta) = \sin^2 \theta \csc^2 \theta = (\sin \theta \csc \theta)^2 = 1^2 = 1.$$

2 $\cos^4 t - \sin^4 t = 1 - 2\sin^2 t$

Solution   Again, we try Suggestion 1. The left side of the equation involves fourth powers, so it is perhaps a bit more complicated than the right side. Therefore, we start with the left side. Trying parts (a)–(f) of Suggestion 3, one at a time, we find that (d) is the first reasonable idea. So we begin by factoring:

$$\cos^4 t - \sin^4 t = (\cos^2 t)^2 - (\sin^2 t)^2 = (\cos^2 t + \sin^2 t)(\cos^2 t - \sin^2 t)$$
$$= 1(\cos^2 t - \sin^2 t) = \cos^2 t - \sin^2 t = (1 - \sin^2 t) - \sin^2 t$$
$$= 1 - 2\sin^2 t.$$

3 $2\sec^2 x = \dfrac{1}{1 - \sin x} + \dfrac{1}{1 + \sin x}$

Solution   Here we start with the right side, which seems more complicated than the left side, and we apply part (a) of Suggestion 3. Thus,

$$\frac{1}{1 - \sin x} + \frac{1}{1 + \sin x} = \frac{1(1 + \sin x)}{(1 - \sin x)(1 + \sin x)} + \frac{1(1 - \sin x)}{(1 + \sin x)(1 - \sin x)}$$

$$= \frac{1 + \sin x}{1 - \sin^2 x} + \frac{1 - \sin x}{1 - \sin^2 x} = \frac{1 + \sin x + 1 - \sin x}{1 - \sin^2 x}$$

$$= \frac{2}{1 - \sin^2 x} = \frac{2}{\cos^2 x} = 2 \sec^2 x.$$

4 $$\frac{1 + \sin(-t)}{\cos(-t)} = \frac{\cos(-t)}{1 - \sin(-t)}$$

Solution   Since $\sin(-t) = -\sin t$ and $\cos(-t) = \cos t$, the left side of the equation is equal to $(1 - \sin t)/\cos t$ and the right side is equal to $\cos t/(1 + \sin t)$. Therefore, it will be enough to prove that

$$\frac{1 - \sin t}{\cos t} = \frac{\cos t}{1 + \sin t}.$$

Notice that it would *not* be correct to "cross multiply" here, since we don't know that the equation is true—indeed, our job is to show that it is true. Thus, we work with both sides separately. Neither side appears to be more complicated than the other, so let's start with the left side. We try part (e) of Suggestion 3. Thus,

$$\frac{1 - \sin t}{\cos t} = \frac{(1 - \sin t)(1 + \sin t)}{\cos t(1 + \sin t)} = \frac{1 - \sin^2 t}{\cos t(1 + \sin t)} = \frac{\cos^2 t}{\cos t(1 + \sin t)}$$

$$= \frac{\cos t}{1 + \sin t}.$$

Example   Show that the trigonometric equation

$$\sin x + \cos x = \tan x + 1$$

is not an identity.

Solution   If we let $x = 0$, we find that both sides of the equation are defined and that the equation becomes $1 = 1$, which is true. From this, however, we can't conclude that the equation is an identity, since there may be other values of x for which it is false. If we let $x = \pi/2$, the left side of the equation is $\sin(\pi/2) + \cos(\pi/2) = 1 + 0 = 1$, but the right side of the equation is undefined. Again, this is inconclusive, since an identity is only required to be true for values of the variable for which both sides are defined. Suppose we try $x = \pi/4$. Then the left side of the equation is $\sin(\pi/4) + \cos(\pi/4) = (\sqrt{2}/2) + (\sqrt{2}/2) = \sqrt{2}$, whereas the right side of the equation is $\tan(\pi/4) + 1 = 1 + 1 = 2$. Therefore, the equation is not an identity.

PROBLEM SET 2

In problems 1–52, show that each trigonometric equation is an identity.

1. $\sin\theta\sec\theta = \tan\theta$

2. $\cos\alpha\tan\alpha\csc\alpha = 1$

3. $\tan x\cos x = \sin x$

4. $\sin\beta\cot\beta\sec\beta = 1$

5. $\csc(-t)\tan(-t) = \sec t$

6. $\sin(-u) = \sin^2 u\csc(-u)$

7. $\tan\alpha\sin\alpha + \cos\alpha = \sec\alpha$

8. $\dfrac{\sec x\csc x}{\tan x + \cot x} = 1$

9. $\dfrac{\sin\beta}{\csc\beta} + \dfrac{\cos\beta}{\sec\beta} = 1$

10. $2 - \sin^2\theta = 1 + \cos^2\theta$

11. $\cos^2 t(1 + \tan^2 t) = 1$

12. $\sec^2 v(1 - \sin^2 v) = 1$

13. $\sec^2 w\cot^2 w - \cos^2 w\csc^2 w = 1$

14. $\tan^4 u - \sec^4 u = 1 - 2\sec^2 u$

15. $\sin^2\theta\cot^2\theta + \cos^2\theta\tan^2\theta = 1$

16. $\cot^2\gamma - \cos^2\gamma = \cot^2\gamma\cos^2\gamma$

17. $\sin^2 v + \tan^2 v + \cos^2 v = \sec^2 v$

18. $2\csc\beta - \cot\beta\cos\beta = \sin\beta + \csc\beta$

19. $\sin^2 x + \cos^2 x(1 - \tan^2 x) = \cos^2 x$

20. $\sin^4 t - \cos^4 t + 2\sin^2 t\cot^2 t = 1$

21. $\dfrac{\tan\theta}{1 + \tan^2\theta} = \dfrac{\sin\theta}{\sec\theta}$

22. $\dfrac{\cos^2 s}{\sin s} + \dfrac{1}{\csc s} = \csc s$

23. $\dfrac{\sin^2 t}{\cos t} + \cos t = \sec t$

24. $\dfrac{1}{\tan\theta + \cot\theta} = \sin\theta\cos\theta$

25. $\dfrac{\sin^3 t}{\cos t} + \sin t\cos t = \tan t$

26. $\dfrac{\csc x - \sec x}{\csc x + \sec x} = \dfrac{\cot x - 1}{\cot x + 1}$

27. $\dfrac{\sin\beta + \cos\beta}{\sin\beta - \cos\beta} = \dfrac{\sec\beta + \csc\beta}{\sec\beta - \csc\beta}$

28. $\left(\dfrac{1 + \csc t}{\csc t}\right)^2\sec^2 t = \dfrac{1 + \sin t}{1 - \sin t}$

29. $\dfrac{\sin x\cos x}{1 - 2\sin^2 x} = \dfrac{\tan x}{1 - \tan^2 x}$

30. $\dfrac{(1 - \cot y)^2}{\csc^2 y} + 2\sin y\cos y = 1$

31. $\dfrac{\tan u\sin u}{\tan u - \sin u} = \dfrac{\sin u}{1 - \cos u}$

32. $(\sec\gamma - \tan\gamma)^2 = \dfrac{1 - \sin\gamma}{1 + \sin\gamma}$

33. $\dfrac{1 - \cot(-\alpha)}{1 - \tan(-\alpha)} = \cot\alpha$

34. $\dfrac{1}{\csc x - \cot x} = \dfrac{2}{\sin x} - \dfrac{1}{\csc x + \cot x}$

35. $(\cot\beta + \csc\beta)^2 = \dfrac{\sec\beta + 1}{\sec\beta - 1}$

36. $\dfrac{\csc^2 t + \sec^2 t}{\csc t\sec t} = \cot t + \tan t$

37. $\dfrac{\cos\alpha}{1 + \cos\alpha\tan\alpha} = \dfrac{1 - \cos\alpha\tan\alpha}{\cos\alpha}$

38. $\dfrac{\cot\beta - \csc\beta + 1}{\cot\beta + \csc\beta - 1} = \dfrac{\sin\beta}{1 + \cos\beta}$

39. $\dfrac{\sin\theta}{\cot\theta + \csc\theta} - \dfrac{\sin\theta}{\cot\theta - \csc\theta} = 2$

40. $\dfrac{\tan x - \tan y}{1 + \tan x\tan y} = \dfrac{\cot y - \cot x}{1 + \cot x\cot y}$

41. $\dfrac{1}{\sin^2 t} + \dfrac{1}{\cos^2 t} = \dfrac{1}{\sin^2 t - \sin^4 t}$

42. $\cos^6\theta - \sin^6\theta = (2\cos^2\theta - 1)(1 - \sin^2\theta\cos^2\theta)$

43. $\dfrac{\cos(-\alpha)}{1 + \tan(-\alpha)} - \dfrac{\sin(-\alpha)}{1 + \cot(-\alpha)} = \sin\alpha + \cos\alpha$

44. $(1 + \tan\beta + \sec\beta)(1 + \cot\beta - \csc\beta) = 2$

45. $\dfrac{\sec u}{\csc u(1 + \sec u)} + \dfrac{1 + \cos u}{\sin u} = 2\csc u$

46. $\dfrac{\sec^4 x + \tan^4 x}{\sec^2 x\tan^2 x} = \dfrac{\cos^4 x}{\sin^2 x} + 2$

47. $(1 + \tan\beta + \cot\beta)(\cos\beta - \sin\beta) = \dfrac{\csc\beta}{\sec^2\beta} - \dfrac{\sec\beta}{\csc^2\beta}$

48. $\sqrt{\dfrac{\sec\gamma - \tan\gamma}{\sec\gamma + \tan\gamma}} = \dfrac{|\cos\gamma|}{1 + \sin\gamma}$

49. $(1 - \cot w)^2(1 + \cot w)^2 + 4\cot^2 w = \csc^4 w$

50. $\dfrac{\cot\alpha + \csc\alpha}{\sin\alpha + \cot(-\alpha) + \csc(-\alpha)} + \sec\alpha = 0$

51. $(1 + \sin\omega t + \cos\omega t)^2 = 2(1 + \sin\omega t)(1 + \cos\omega t)$

52. $\dfrac{\cot x}{1 - \tan x} + \dfrac{\tan x}{1 - \cot x} = 1 + \dfrac{\sec x}{\sin x}$

In problems 53–59, show that the given trigonometric equation is *not* a trigonometric identity.

53. $\sin\theta - \sec\theta = \tan\theta - 1$

54. $(\sin x + \cos x)^2 = 1$

55. $\dfrac{\sin t + \tan t}{\cos t + \tan t} = \tan t$

56. $\cos(\gamma + \pi) = \cos\gamma$

57. $\sqrt{1 + \sin^2 u} = 1 + \sin u$

58. $\ln(\sin x) = \sin(\ln x)$

59. $\sin(t^2) = \sin^2 t$

60. Give an example of a trigonometric equation that is true for three different values of the variable, but isn't an identity.

3 Trigonometric Formulas for Sums and Differences

The fundamental trigonometric identities considered in Section 1 express relationships among trigonometric functions of a single variable. In this section we develop trigonometric identities involving sums or differences of *two* variables.

Since $75° = 30° + 45°$, you might ask whether $\cos 75°$ can be found by calculating $\cos 30° + \cos 45°$. The answer is *no;* $\cos 75° < 1$, whereas

$$\cos 30° + \cos 45° = \frac{\sqrt{3}}{2} + \frac{\sqrt{2}}{2} = \frac{\sqrt{3} + \sqrt{2}}{2} > 1.$$

The correct formula for the cosine of the sum of two angles is given by the following theorem.

Theorem 1 **Cosine of a Sum**

$$\cos(\alpha + \beta) = \cos\alpha\cos\beta - \sin\alpha\sin\beta$$

To see why the formula in Theorem 1 is correct, consider the unit circles in Figures 1 (a) and (b). In Figure 1(a), angle AOB is $\alpha + \beta$. Triangle COD in Figure 1(b) is obtained by rotating triangle AOB through the angle $-\alpha$. In Figure 1(a), we have $A = (\cos(\alpha + \beta), \sin(\alpha + \beta))$ and $B = (1, 0)$, so, by the distance formula,

$$|\overline{AB}|^2 = [\cos(\alpha + \beta) - 1]^2 + [\sin(\alpha + \beta) - 0]^2$$
$$= \cos^2(\alpha + \beta) - 2\cos(\alpha + \beta) + 1 + \sin^2(\alpha + \beta)$$
$$= [\cos^2(\alpha + \beta) + \sin^2(\alpha + \beta)] - 2\cos(\alpha + \beta) + 1$$
$$= 1 - 2\cos(\alpha + \beta) + 1 = 2 - 2\cos(\alpha + \beta).$$

Figure 1

(a)

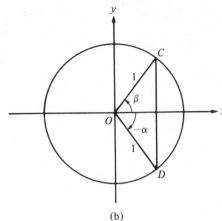

(b)

In Figure 1(b),

$$C = (\cos\beta, \sin\beta) \quad \text{and} \quad D = (\cos(-\alpha), \sin(-\alpha)) = (\cos\alpha, -\sin\alpha),$$

so, by the distance formula again,

$$|\overline{CD}|^2 = (\cos\beta - \cos\alpha)^2 + [\sin\beta - (-\sin\alpha)]^2 = (\cos\beta - \cos\alpha)^2 + (\sin\beta + \sin\alpha)^2$$
$$= \cos^2\beta - 2\cos\beta\cos\alpha + \cos^2\alpha + \sin^2\beta + 2\sin\beta\sin\alpha + \sin^2\alpha$$
$$= (\cos^2\beta + \sin^2\beta) + (\cos^2\alpha + \sin^2\alpha) - 2\cos\beta\cos\alpha + 2\sin\beta\sin\alpha$$
$$= 1 + 1 - 2(\cos\beta\cos\alpha - \sin\beta\sin\alpha)$$
$$= 2 - 2(\cos\alpha\cos\beta - \sin\alpha\sin\beta).$$

Because triangle *AOB* is congruent to triangle *COD*, it follows that

$$|\overline{AB}|^2 = |\overline{CD}|^2.$$

Hence,

$$2 - 2\cos(\alpha + \beta) = 2 - 2(\cos\alpha\cos\beta - \sin\alpha\sin\beta)$$
$$-2\cos(\alpha + \beta) = -2(\cos\alpha\cos\beta - \sin\alpha\sin\beta)$$
$$\cos(\alpha + \beta) = \cos\alpha\cos\beta - \sin\alpha\sin\beta,$$

which is the formula given in Theorem 1.

Example   Find the exact value of $\cos 75°$.

Solution   By Theorem 1,

$$\cos 75° = \cos(30° + 45°) = \cos 30° \cos 45° - \sin 30° \sin 45°$$
$$= \frac{\sqrt{3}}{2} \cdot \frac{\sqrt{2}}{2} - \frac{1}{2} \cdot \frac{\sqrt{2}}{2} = \frac{(\sqrt{3} - 1)\sqrt{2}}{4}.$$

Theorem 2   **Cosine of a Difference**

$$\cos(\alpha - \beta) = \cos\alpha\cos\beta + \sin\alpha\sin\beta$$

Proof   In Theorem 1, replace β by $-\beta$ to obtain

$$\cos(\alpha - \beta) = \cos[\alpha + (-\beta)] = \cos\alpha\cos(-\beta) - \sin\alpha\sin(-\beta)$$

$$= \cos\alpha\cos\beta - \sin\alpha(-\sin\beta) = \cos\alpha\cos\beta + \sin\alpha\sin\beta.$$

Example   Find the exact value of $\cos 15°$.

Solution   By Theorem 2,

$$\cos 15° = \cos(45° - 30°) = \cos 45°\cos 30° + \sin 45°\sin 30°$$

$$= \frac{\sqrt{2}}{2}\cdot\frac{\sqrt{3}}{2} + \frac{\sqrt{2}}{2}\cdot\frac{1}{2} = \frac{\sqrt{2}(\sqrt{3} + 1)}{4}.$$

In Section 2 of Chapter 6, we discussed the function-cofunction relationships for trigonometric functions of *acute angles*. The following theorem shows that these relationships hold for arbitrary angles.

Theorem 3   **Cofunction Theorem**

If θ is a real number or an angle measured in radians, then

(i) $\cos\left(\dfrac{\pi}{2} - \theta\right) = \sin\theta$ (ii) $\sin\left(\dfrac{\pi}{2} - \theta\right) = \cos\theta$

(iii) $\cot\left(\dfrac{\pi}{2} - \theta\right) = \tan\theta$ (iv) $\tan\left(\dfrac{\pi}{2} - \theta\right) = \cot\theta$

(v) $\csc\left(\dfrac{\pi}{2} - \theta\right) = \sec\theta$ (vi) $\sec\left(\dfrac{\pi}{2} - \theta\right) = \csc\theta$

Proof   We prove (i), (ii), and (iii) here and leave the remaining proofs as an exercise (Problem 49).

(i) By Theorem 2,

$$\cos\left(\frac{\pi}{2} - \theta\right) = \cos\frac{\pi}{2}\cos\theta + \sin\frac{\pi}{2}\sin\theta = 0\cdot\cos\theta + 1\cdot\sin\theta = \sin\theta.$$

(ii) In part (i), written from right to left, we replace θ by $(\pi/2) - \theta$ to obtain

$$\sin\left(\frac{\pi}{2} - \theta\right) = \cos\left[\frac{\pi}{2} - \left(\frac{\pi}{2} - \theta\right)\right] = \cos\theta.$$

(iii) By parts (i) and (ii),

$$\cot\left(\frac{\pi}{2} - \theta\right) = \frac{\cos\left(\dfrac{\pi}{2} - \theta\right)}{\sin\left(\dfrac{\pi}{2} - \theta\right)} = \frac{\sin\theta}{\cos\theta} = \tan\theta.$$

Naturally, if we measure an angle α in degrees instead of radians, the relationships expressed in Theorem 3 are written as

$$\cos(90° - \alpha) = \sin\alpha, \quad \sin(90° - \alpha) = \cos\alpha,$$

and so forth.

Now, by combining Theorem 2 and Theorem 3, we can prove the following important theorem.

Theorem 4   **Sine of a Sum**

$$\sin(\alpha + \beta) = \sin\alpha\cos\beta + \sin\beta\cos\alpha$$

Proof   By parts (i) and (ii) of Theorem 3,

$$\sin(\alpha + \beta) = \cos\left[\frac{\pi}{2} - (\alpha + \beta)\right] = \cos\left(\frac{\pi}{2} - \alpha - \beta\right) = \cos\left[\left(\frac{\pi}{2} - \alpha\right) - \beta\right]$$

$$= \cos\left(\frac{\pi}{2} - \alpha\right)\cos\beta + \sin\left(\frac{\pi}{2} - \alpha\right)\sin\beta = \sin\alpha\cos\beta + \cos\alpha\sin\beta$$

$$= \sin\alpha\cos\beta + \sin\beta\cos\alpha.$$

Example   Find the exact value of $\sin\dfrac{7\pi}{12}$.

Solution   By Theorem 4, we have

$$\sin\frac{7\pi}{12} = \sin\left(\frac{\pi}{3} + \frac{\pi}{4}\right) = \sin\frac{\pi}{3}\cos\frac{\pi}{4} + \sin\frac{\pi}{4}\cos\frac{\pi}{3}$$

$$= \frac{\sqrt{3}}{2}\cdot\frac{\sqrt{2}}{2} + \frac{\sqrt{2}}{2}\cdot\frac{1}{2} = \frac{\sqrt{2}(\sqrt{3} + 1)}{4}.$$

The following theorem follows directly from Theorem 4 and the even–odd identities.

Theorem 5   **Sine of a Difference**

$$\sin(\alpha - \beta) = \sin\alpha\cos\beta - \sin\beta\cos\alpha$$

Proof   Replacing β by $-\beta$ in Theorem 4, we have

$$\sin(\alpha - \beta) = \sin[\alpha + (-\beta)] = \sin\alpha\cos(-\beta) + \sin(-\beta)\cos\alpha$$
$$= \sin\alpha\cos\beta - \sin\beta\cos\alpha.$$

The four identities

$$\sin(\alpha + \beta) = \sin\alpha\cos\beta + \sin\beta\cos\alpha \qquad \sin(\alpha - \beta) = \sin\alpha\cos\beta - \sin\beta\cos\alpha$$
$$\cos(\alpha + \beta) = \cos\alpha\cos\beta - \sin\alpha\sin\beta \qquad \cos(\alpha - \beta) = \cos\alpha\cos\beta + \sin\alpha\sin\beta$$

are used so often that they should be memorized. We recommend that you memorize the first identity in the form: *The sine of the sum of two angles is equal to the sine of the first times the cosine of the second plus the sine of the second times the cosine of the first.* The remaining three identities can be stated in a similar manner.

Example 1   Simplify each expression:

(a) $\sin(90° + \theta)$

(b) $\cos\left(\dfrac{\pi}{2} + x\right)$

Solution   (a) $\sin(90° + \theta) = \sin 90° \cos\theta + \sin\theta\cos 90° = 1 \cdot \cos\theta + (\sin\theta)\cdot 0 = \cos\theta$

(b) $\cos\left(\dfrac{\pi}{2} + x\right) = \cos\dfrac{\pi}{2}\cos x - \sin\dfrac{\pi}{2}\sin x = 0 \cdot \cos x - 1 \cdot \sin x = -\sin x$

Example 2   Suppose that α and β are angles in standard position; α is in quadrant I, $\sin\alpha = \frac{3}{5}$, β is in quadrant II, and $\cos\beta = -\frac{5}{13}$. Find (a) $\cos\alpha$, (b) $\sin\beta$, (c) $\sin(\alpha + \beta)$, (d) $\cos(\alpha + \beta)$, and (e) the quadrant containing angle $\alpha + \beta$ in standard position.

Solution   (a) Since α is in quadrant I, $\cos\alpha$ is positive, and

$$\cos\alpha = \sqrt{1 - \sin^2\alpha} = \sqrt{1 - (\tfrac{3}{5})^2} = \sqrt{\tfrac{16}{25}} = \tfrac{4}{5}$$

(b) Since β is in quadrant II, $\sin\beta$ is positive and

$$\sin\beta = \sqrt{1 - \cos^2\beta} = \sqrt{1 - (-\tfrac{5}{13})^2} = \sqrt{\tfrac{144}{169}} = \tfrac{12}{13}$$

(c) $\sin(\alpha + \beta) = \sin\alpha\cos\beta + \sin\beta\cos\alpha = \frac{3}{5}(-\frac{5}{13}) + \frac{12}{13}\cdot\frac{4}{5} = \frac{33}{65}$

(d) $\cos(\alpha + \beta) = \cos\alpha\cos\beta - \sin\alpha\sin\beta = \frac{4}{5}(-\frac{5}{13}) - \frac{3}{5}\cdot\frac{12}{13} = -\frac{56}{65}$

(e) Since $\sin(\alpha + \beta)$ is positive and $\cos(\alpha + \beta)$ is negative, it follows that $\alpha + \beta$ is a quadrant II angle.

Example 3   Simplify each expression:

(a) $\cos 25° \cos 20° - \sin 25° \sin 20°$

(b) $\sin\dfrac{2\pi}{5}\cos\dfrac{3\pi}{5} + \sin\dfrac{3\pi}{5}\cos\dfrac{2\pi}{5}$

Solution   (a) Reading the formula for the cosine of a sum from right to left, we have

$$\cos 25° \cos 20° - \sin 25° \sin 20° = \cos(25° + 20°) = \cos 45° = \sqrt{2}/2.$$

(b) Reading the formula for the sine of a sum from right to left, we have

$$\sin\frac{2\pi}{5}\cos\frac{3\pi}{5} + \sin\frac{3\pi}{5}\cos\frac{2\pi}{5} = \sin\left(\frac{2\pi}{5} + \frac{3\pi}{5}\right) = \sin\pi = 0.$$

Example 4   As we noticed in Section 5 of Chapter 6, the graphs of the sine and cosine functions suggest that

$$\sin(x + \pi) = -\sin x \quad\text{and}\quad \cos(x + \pi) = -\cos x.$$

Prove these identities.

Solution
$$\sin(x + \pi) = \sin x \cos\pi + \sin\pi \cos x = (\sin x)(-1) + 0 \cdot \cos x$$
$$= -\sin x$$
$$\cos(x + \pi) = \cos x \cos\pi - \sin x \sin\pi = (\cos x)(-1) - (\sin x)\cdot 0$$
$$= -\cos x$$

Example 5   Prove the trigonometric identity

$$\cos(x + y)\cos(x - y) = \cos^2 x - \sin^2 y.$$

Solution   Beginning with the left side and applying the formulas for the cosine of a sum and for the cosine of a difference, we have

$$\cos(x + y)\cos(x - y) = (\cos x \cos y - \sin x \sin y)(\cos x \cos y + \sin x \sin y)$$
$$= (\cos x \cos y)^2 - (\sin x \sin y)^2$$
$$= \cos^2 x \cos^2 y - \sin^2 x \sin^2 y$$
$$= \cos^2 x(1 - \sin^2 y) - (1 - \cos^2 x)\sin^2 y$$
$$= \cos^2 x - \cos^2 x \sin^2 y - \sin^2 y + \cos^2 x \sin^2 y$$
$$= \cos^2 x - \sin^2 y.$$

Do you see what gave us the idea to substitute $1 - \sin^2 y$ for $\cos^2 y$ and $1 - \cos^2 x$ for $\sin^2 x$ in the fourth step of the calculation above? [Hint: We wanted the expression $\cos^2 x - \sin^2 y$ to show up.]

Using the formulas for the sine and cosine of a sum, we can derive a useful formula for the tangent of a sum.

Theorem 6   **Tangent of a Sum**

If $\cos\alpha \neq 0$, $\cos\beta \neq 0$, and $\cos(\alpha + \beta) \neq 0$, then

$$\tan(\alpha + \beta) = \frac{\tan\alpha + \tan\beta}{1 - \tan\alpha\tan\beta}.$$

Proof
$$\tan(\alpha + \beta) = \frac{\sin(\alpha + \beta)}{\cos(\alpha + \beta)} = \frac{\sin\alpha\cos\beta + \sin\beta\cos\alpha}{\cos\alpha\cos\beta - \sin\alpha\sin\beta}$$

Dividing numerator and denominator of the fraction on the right side of the equation above by $\cos\alpha\cos\beta$, we get

$$\tan(\alpha + \beta) = \frac{\dfrac{\sin\alpha\cos\beta}{\cos\alpha\cos\beta} + \dfrac{\sin\beta\cos\alpha}{\cos\alpha\cos\beta}}{\dfrac{\cos\alpha\cos\beta}{\cos\alpha\cos\beta} - \dfrac{\sin\alpha\sin\beta}{\cos\alpha\cos\beta}} = \frac{\dfrac{\sin\alpha}{\cos\alpha} + \dfrac{\sin\beta}{\cos\beta}}{1 - \left(\dfrac{\sin\alpha}{\cos\alpha}\right)\left(\dfrac{\sin\beta}{\cos\beta}\right)}$$

$$= \frac{\tan\alpha + \tan\beta}{1 - \tan\alpha\tan\beta}.$$

Example 1   Assuming that $\cos\alpha \neq 0$, $\cos\beta \neq 0$, and $\cos(\alpha - \beta) \neq 0$, show that

$$\tan(\alpha - \beta) = \frac{\tan\alpha - \tan\beta}{1 + \tan\alpha\tan\beta}.$$

Solution   In Theorem 6, we replace β by $-\beta$ to obtain

$$\tan(\alpha - \beta) = \tan[\alpha + (-\beta)] = \frac{\tan\alpha + \tan(-\beta)}{1 - \tan\alpha\tan(-\beta)} = \frac{\tan\alpha - \tan\beta}{1 + \tan\alpha\tan\beta}.$$

Example 2   Simplify the expression $\dfrac{\tan 43° + \tan 17°}{1 - \tan 43° \tan 17}$.

Solution   Reading the formula in Theorem 6 from right to left, we have

$$\frac{\tan 43° + \tan 17°}{1 - \tan 43° \tan 17°} = \tan(43° + 17°) = \tan 60° = \sqrt{3}.$$

PROBLEM SET 3

In problems 1–12, find the exact value of the indicated trigonometric function.

1. $\sin 105°$ [Hint: $105° = 60° + 45°$.]   2. $\cos 105°$
3. $\sin 195°$ [Hint: $195° = 150° + 45°$.]   4. $\tan 195°$

5. $\cos\dfrac{17\pi}{12}$ $\left[\text{Hint: } \dfrac{17\pi}{12} = \dfrac{5\pi}{4} + \dfrac{\pi}{6}.\right]$ 6. $\cot\dfrac{17\pi}{12}$

7. $\sin\dfrac{11\pi}{12}$ $\left[\text{Hint: } \dfrac{11\pi}{12} = \dfrac{7\pi}{6} - \dfrac{\pi}{4}.\right]$ 8. $\sec\dfrac{11\pi}{12}$

9. $\tan\dfrac{13\pi}{12}$ $\left[\text{Hint: } \dfrac{13\pi}{12} = \dfrac{3\pi}{4} + \dfrac{\pi}{3}.\right]$ 10. $\cot\dfrac{13\pi}{12}$

11. $\tan\dfrac{19\pi}{12}$ $\left[\text{Hint: } \dfrac{19\pi}{12} = \dfrac{11\pi}{6} - \dfrac{\pi}{4}.\right]$ 12. $\csc\dfrac{19\pi}{12}$

In problems 13–34, simplify each expression.

13. $\sin(180° - \theta)$

14. $\cos(180° + t)$

15. $\cos(\pi - \alpha)$

16. $\sin(270° - x)$

17. $\sin(360° - s)$

18. $\cos(2\pi - \gamma)$

19. $\tan(180° - t)$

20. $\sec(\pi + u)$

21. $\csc\left(\dfrac{3\pi}{2} + \beta\right)$

22. $\tan(270° + \theta)$ [Caution: Theorem 6 doesn't work here.]

23. $\sin 17° \cos 13° + \sin 13° \cos 17°$

24. $\cos 43° \cos 17° - \sin 43° \sin 17°$

25. $\sin\dfrac{17\pi}{36}\cos\dfrac{11\pi}{36} - \sin\dfrac{11\pi}{36}\cos\dfrac{17\pi}{36}$

26. $\cos\dfrac{5\pi}{7}\sin\dfrac{2\pi}{7} + \cos\dfrac{2\pi}{7}\sin\dfrac{5\pi}{7}$

27. $\cos 2x \cos x + \sin 2x \sin x$

28. $\sin(\alpha - \beta)\cos\beta + \cos(\alpha - \beta)\sin\beta$

29. $\cos(\alpha + \beta)\cos\beta + \sin(\alpha + \beta)\sin\beta$

30. $\cos(x + y) - \cos(x - y)$

31. $\dfrac{\tan 42° + \tan 18°}{1 - \tan 42° \tan 18°}$

32. $\dfrac{\tan 50° - \tan 20°}{1 + \tan 50° \tan 20°}$

33. $\dfrac{\tan 8x - \tan 7x}{1 + \tan 8x \tan 7x}$

34. $\tan\left(\dfrac{\pi}{2} + \theta\right)$

35. Suppose that α and β are angles in standard position; α is in quadrant I, $\sin\alpha = \frac{4}{5}$, β is in quadrant II, and $\sin\beta = \frac{12}{13}$. Find (a) $\cos\alpha$, (b) $\cos\beta$, (c) $\sin(\alpha + \beta)$, (d) $\cos(\alpha + \beta)$, (e) $\sin(\alpha - \beta)$, (f) $\tan(\alpha + \beta)$, and (g) the quadrant containing the angle $\alpha + \beta$ in standard position.

36. Suppose that $\dfrac{\pi}{2} < s < \pi$, $\dfrac{\pi}{2} < t < \pi$, $\sin s = \frac{5}{13}$, and $\tan t = -\frac{3}{4}$. Find (a) $\sin(s - t)$, (b) $\cos(s - t)$, (c) $\sin(s + t)$, (d) $\sec(s + t)$, and (e) the quadrant containing the angle θ in standard position if the radian measure of θ is $s + t$.

37. If α and β are positive acute angles, $\sin(\alpha + \beta) = \frac{56}{65}$, and $\sin\beta = \frac{5}{13}$, find $\sin\alpha$. [Hint: $\sin\alpha = \sin((\alpha + \beta) - \beta)$.]

38. Derive a formula for $\tan(\alpha + \beta)$ for the case in which $\cos\alpha = 0$ and $\sin\beta \neq 0$.

In problems 39–48, show that each equation is an identity.

39. $\dfrac{\sin(\alpha + \beta)}{\sin\alpha\cos\beta} = 1 + \cot\alpha\tan\beta$

40. $\tan(\theta + \pi) = \tan\theta$

41. $\sin\left(t - \dfrac{\pi}{3}\right) + \cos\left(t - \dfrac{\pi}{6}\right) = \sin t$

42. $\cos(x + y)\cos(x - y) = 1 - \sin^2 x - \sin^2 y$

43. $\tan\left(x + \dfrac{\pi}{2}\right) = \tan\left(x - \dfrac{\pi}{2}\right)$

44. $\tan\alpha - \tan\beta = \dfrac{\sin(\alpha - \beta)}{\cos\alpha\cos\beta}$

45. $\cot(x + y) = \dfrac{\cot x \cot y - 1}{\cot x + \cot y}$

46. $\dfrac{\sin(s + t)}{\sin(s - t)} = \dfrac{\tan s + \tan t}{\tan s - \tan t}$

47. $\dfrac{1 - \tan^2\left(\dfrac{\pi}{4} - \theta\right)}{1 + \tan^2\left(\dfrac{\pi}{4} - \theta\right)} = 2\sin\theta\cos\theta$

48. $\tan^2\left(\dfrac{\pi}{2} - t\right)\sec^2 t - \sin^2\left(\dfrac{\pi}{2} - t\right)\csc^2 t = 1$

49. Prove parts (iv), (v), and (vi) of Theorem 3.

50. If $x + y = \pi/4$, show that

$$\tan y = \dfrac{1 - \tan x}{1 + \tan x}.$$

© In problems 51–56, use a calculator to verify the equation for the indicated values of the variables.

51. $\cos(\alpha - \beta) = \cos\alpha\cos\beta + \sin\alpha\sin\beta$ for $\alpha = \dfrac{3\pi}{5}$, $\beta = \dfrac{\pi}{5}$

52. $\sin(x + y) = \sin x\cos y + \sin y\cos x$ for $x = 2.31$, $y = 1.07$

53. $\tan(\alpha + \beta) = \dfrac{\tan\alpha + \tan\beta}{1 - \tan\alpha\tan\beta}$ for $\alpha = 22.4°$, $\beta = 17.3°$

54. $\cos(s + t) = \cos s\cos t - \sin s\sin t$ for $s = \dfrac{5\pi}{11}$, $t = \dfrac{7\pi}{9}$

55. $\sin(s - t) = \sin s\cos t - \cos s\sin t$ for $s = 0.791$, $t = 1.234$

56. $\tan(\alpha - \beta) = \dfrac{\tan\alpha - \tan\beta}{1 + \tan\alpha\tan\beta}$ for $\alpha = 91.7°$, $\beta = 87.2°$

4   Double-Angle and Half-Angle Formulas

In this section we study the important double-angle and half-angle formulas. These identities, which are obtained as special cases of the formulas derived in the last section, enable us to evaluate trigonometric functions of 2θ or $\theta/2$ in terms of trigonometric functions of θ.

The sine of twice an angle is *not* in general the same as twice the sine of the angle. For instance, $\sin 30° = \frac{1}{2}$; $\sin 2(30°) = \sin 60° = \sqrt{3}/2 \neq 2\sin 30°$. The correct formula for the sine of twice an angle is given by the following theorem.

Theorem 1   **Double-Angle Formula for Sine**

$$\sin 2\theta = 2\sin\theta\cos\theta$$

Proof   In the formula

$$\sin(\alpha + \beta) = \sin\alpha\cos\beta + \sin\beta\cos\alpha,$$

let $\alpha = \beta = \theta$, so that

$$\sin(\theta + \theta) = \sin\theta\cos\theta + \sin\theta\cos\theta$$

or

$$\sin 2\theta = 2\sin\theta\cos\theta.$$

Example 1   Verify Theorem 1 for $\theta = 30°$.

Solution   We know that $\sin 60° = \sqrt{3}/2$. Using Theorem 1 with $\theta = 30°$, we also get

$$\sin 60° = \sin 2(30°) = 2\sin 30°\cos 30° = 2\left(\frac{1}{2}\right)\left(\frac{\sqrt{3}}{2}\right) = \frac{\sqrt{3}}{2}.$$

Example 2   Write $\sin 100°$ in terms of trigonometric functions of $50°$.

Solution $$\sin 100° = \sin 2(50°) = 2 \sin 50° \cos 50°$$

Example 3   Reduce the expression $2 \sin 8x \cos 8x$ to the value of a function of $16x$.

Solution   Reading Theorem 1 from right to left with $\theta = 8x$, we have

$$2 \sin 8x \cos 8x = \sin[2(8x)] = \sin 16x.$$

Example 4   Suppose that α is a quadrant I angle and that $\sin \alpha = \frac{3}{5}$. Find $\sin 2\alpha$.

Solution   We are going to use the formula $\sin 2\alpha = 2 \sin \alpha \cos \alpha$, but first we must find $\cos \alpha$. Because α is a quadrant I angle, we know that $\cos \alpha$ is positive, so

$$\cos \alpha = \sqrt{1 - \sin^2 \alpha} = \sqrt{1 - (\tfrac{3}{5})^2} = \sqrt{\tfrac{16}{25}} = \tfrac{4}{5}.$$

Therefore,

$$\sin 2\alpha = 2 \sin \alpha \cos \alpha = 2(\tfrac{3}{5})(\tfrac{4}{5}) = \tfrac{24}{25}.$$

There are three double-angle formulas for the cosine—a basic formula and two alternative versions of it.

Theorem 2   **Double-Angle Formulas for Cosine**

> (i) $\cos 2\theta = \cos^2 \theta - \sin^2 \theta$
> (ii) $\cos 2\theta = 2 \cos^2 \theta - 1$
> (iii) $\cos 2\theta = 1 - 2 \sin^2 \theta$

Proof   (i) In the formula

$$\cos(\alpha + \beta) = \cos \alpha \cos \beta - \sin \alpha \sin \beta,$$

let $\alpha = \beta = \theta$, so that

$$\cos(\theta + \theta) = \cos \theta \cos \theta - \sin \theta \sin \theta$$

or

$$\cos 2\theta = \cos^2 \theta - \sin^2 \theta.$$

(ii) In the basic formula just derived, substitute $1 - \cos^2 \theta$ for $\sin^2 \theta$ to obtain

$$\cos 2\theta = \cos^2 \theta - (1 - \cos^2 \theta) = 2 \cos^2 \theta - 1.$$

(iii) In the basic formula $\cos 2\theta = \cos^2 \theta - \sin^2 \theta$, substitute $1 - \sin^2 \theta$ for $\cos^2 \theta$ to obtain

$$\cos 2\theta = 1 - \sin^2 \theta - \sin^2 \theta = 1 - 2 \sin^2 \theta.$$

Example 1   Write $\cos(\pi/4)$ in terms of trigonometric functions of $\pi/8$.

Solution   By Theorem 2, with $\theta = \pi/8$,

$$\cos\frac{\pi}{4} = \cos 2\left(\frac{\pi}{8}\right) = \cos^2\frac{\pi}{8} - \sin^2\frac{\pi}{8}.$$

Alternatively, we could have written

$$\cos\frac{\pi}{4} = 2\cos^2\frac{\pi}{8} - 1 \quad \text{or} \quad \cos\frac{\pi}{4} = 1 - 2\sin^2\frac{\pi}{8}.$$

Example 2   Reduce the expression $1 - 2\sin^2 5\theta$ to the value of a function of 10θ.

Solution   Reading part (iii) of Theorem 2 from right to left and replacing θ by 5θ, we have

$$1 - 2\sin^2 5\theta = \cos 10\theta.$$

Example 3   If $\sin\beta = -\frac{7}{10}$, find $\cos 2\beta$.

Solution   By part (iii) of Theorem 2 with $\theta = \beta$, we have
$$\cos 2\beta = 1 - 2\sin^2\beta = 1 - 2(-\tfrac{7}{10})^2 = \tfrac{50}{50} - \tfrac{49}{50} = \tfrac{1}{50}.$$

Example 4   Express $\cos 4x$ in terms of $\sin x$.

Solution   By part (ii) of Theorem 2 with $\theta = 2x$,
$$\cos 4x = \cos 2(2x) = 2\cos^2 2x - 1 = 2(\cos 2x)^2 - 1.$$

Therefore, by part (iii) of Theorem 2 with $\theta = x$,
$$\cos 4x = 2(1 - 2\sin^2 x)^2 - 1 = 2(1 - 4\sin^2 x + 4\sin^4 x) - 1$$
$$= 1 - 8\sin^2 x + 8\sin^4 x.$$

If we solve the equation $\cos 2\theta = 2\cos^2\theta - 1$ in part (ii) of Theorem 2 for $\cos^2\theta$, we obtain the identity

$$\cos^2\theta = \frac{1 + \cos 2\theta}{2}.$$

Similarly, the equation $\cos 2\theta = 1 - 2\sin^2\theta$ in part (iii) of Theorem 2 can be solved for $\sin^2\theta$, and we have

$$\sin^2\theta = \frac{1 - \cos 2\theta}{2}.$$

The formulas above for $\cos^2\theta$ and $\sin^2\theta$ are used extensively in calculus, and they provide the basis for the half-angle formulas to be derived shortly. We record them for future use in the following theorem.

Theorem 3   **Formulas for $\cos^2 \theta$ and $\sin^2 \theta$**

$$\cos^2 \theta = \frac{1 + \cos 2\theta}{2} \quad \text{and} \quad \sin^2 \theta = \frac{1 - \cos 2\theta}{2}$$

Example 1   Prove the identity
$$\sin^4 \theta = \tfrac{3}{8} - \tfrac{1}{2}\cos 2\theta + \tfrac{1}{8}\cos 4\theta.$$

Solution   Using the second equation of Theorem 3, we can write

$$\sin^4 \theta = (\sin^2 \theta)^2 = \left(\frac{1 - \cos 2\theta}{2}\right)^2 = \tfrac{1}{4}(1 - 2\cos 2\theta + \cos^2 2\theta).$$

Now, using the first equation of Theorem 3 with θ replaced by 2θ, we have

$$\sin^4 \theta = \tfrac{1}{4}(1 - 2\cos 2\theta + \cos^2 2\theta) = \tfrac{1}{4}\left[1 - 2\cos 2\theta + \left(\frac{1 + \cos 4\theta}{2}\right)\right]$$

$$= \tfrac{1}{4} - \tfrac{1}{2}\cos 2\theta + \tfrac{1}{8} + \tfrac{1}{8}\cos 4\theta = \tfrac{3}{8} - \tfrac{1}{2}\cos 2\theta + \tfrac{1}{8}\cos 4\theta.$$

If we put $\alpha = \beta = \theta$ in Theorem 6 of Section 3, we obtain the double-angle formula for the tangent:

$$\tan 2\theta = \frac{2\tan\theta}{1 - \tan^2\theta}.$$

A formula for $\tan^2 \theta$ can be obtained as follows:

$$\tan^2 \theta = \frac{\sin^2\theta}{\cos^2\theta} = \frac{\left(\dfrac{1 - \cos 2\theta}{2}\right)}{\left(\dfrac{1 + \cos 2\theta}{2}\right)}$$

so that

$$\tan^2 \theta = \frac{1 - \cos 2\theta}{1 + \cos 2\theta}.$$

If you let $\theta = \alpha/2$ in Theorem 3 and in the preceding formula for $\tan^2 \theta$, you will obtain the formulas in the following theorem (Problem 34).

Theorem 4   **Half-Angle Formulas**

(i) $\cos^2 \dfrac{\alpha}{2} = \dfrac{1 + \cos\alpha}{2}$

(ii) $\sin^2 \dfrac{\alpha}{2} = \dfrac{1 - \cos\alpha}{2}$

(iii) $\tan^2 \dfrac{\alpha}{2} = \dfrac{1 - \cos\alpha}{1 + \cos\alpha},$ for $\cos\alpha \neq -1$.

The equations in Theorem 4 can be written in the form

$$\cos\frac{\alpha}{2} = \pm\sqrt{\frac{1+\cos\alpha}{2}}, \quad \sin\frac{\alpha}{2} = \pm\sqrt{\frac{1-\cos\alpha}{2}}, \quad \text{and} \quad \tan\frac{\alpha}{2} = \pm\sqrt{\frac{1-\cos\alpha}{1+\cos\alpha}}.$$

In using these formulas, you can determine the algebraic sign (+ or −) by considering which quadrant contains the angle $\alpha/2$ in standard position.

Example 1   Find the exact values of (a) $\sin 22.5°$ and (b) $\cos(\pi/12)$.

Solution   (a) Because $22.5°$ is a quadrant I angle, we have

$$\sin 22.5° = \sin\frac{45°}{2} = \sqrt{\frac{1-\cos 45°}{2}} = \sqrt{\frac{1-(\sqrt{2}/2)}{2}} = \sqrt{\frac{2-\sqrt{2}}{4}} = \frac{\sqrt{2-\sqrt{2}}}{2}$$

(b) Because $\pi/12$ is a quadrant I angle, we have

$$\cos\frac{\pi}{12} = \cos\frac{\pi/6}{2} = \sqrt{\frac{1+\cos(\pi/6)}{2}} = \sqrt{\frac{1+(\sqrt{3}/2)}{2}} = \sqrt{\frac{2+\sqrt{3}}{4}} = \frac{\sqrt{2+\sqrt{3}}}{2}$$

Example 2   Suppose that $\frac{3\pi}{2} < \theta < 2\pi$ and $\cos\theta = \frac{5}{13}$. Find (a) $\sin\frac{\theta}{2}$ and (b) $\tan\frac{\theta}{2}$.

Solution   Because $3\pi/2 < \theta < 2\pi$, we have $\frac{1}{2}(3\pi/2) < \frac{1}{2}\theta < \frac{1}{2}(2\pi)$, or $3\pi/4 < \theta/2 < \pi$. Therefore, $\theta/2$ is a quadrant II angle, and so $\sin(\theta/2) > 0$ and $\tan(\theta/2) < 0$.

(a) $\sin\dfrac{\theta}{2} = \sqrt{\dfrac{1-\cos\theta}{2}} = \sqrt{\dfrac{1-(5/13)}{2}} = \sqrt{\dfrac{1}{2}\left(\dfrac{8}{13}\right)} = \sqrt{\dfrac{4}{13}} = \dfrac{2\sqrt{13}}{13}$

(b) $\tan\dfrac{\theta}{2} = -\sqrt{\dfrac{1-\cos\theta}{1+\cos\theta}} = -\sqrt{\dfrac{1-(5/13)}{1+(5/13)}} = -\sqrt{\dfrac{13-5}{13+5}}$

$\qquad = -\sqrt{\dfrac{8}{18}} = -\sqrt{\dfrac{4}{9}} = -\dfrac{2}{3}$

Example 3   Let $z = \tan(\theta/2)$. Show that:

(a) $\cos\theta = \dfrac{1-z^2}{1+z^2}$ \qquad (b) $\sin\theta = \dfrac{2z}{1+z^2}$ \qquad (c) $\tan\dfrac{\theta}{2} = \dfrac{1-\cos\theta}{\sin\theta}$

Solution   (a) Using part (iii) of Theorem 4, we have

$$\frac{1-z^2}{1+z^2} = \frac{1-\tan^2\dfrac{\theta}{2}}{1+\tan^2\dfrac{\theta}{2}} = \frac{1-\left(\dfrac{1-\cos\theta}{1+\cos\theta}\right)}{1+\left(\dfrac{1-\cos\theta}{1+\cos\theta}\right)}$$

$$= \frac{1+\cos\theta-(1-\cos\theta)}{1+\cos\theta+(1-\cos\theta)} = \frac{2\cos\theta}{2} = \cos\theta.$$

(b) Using the fundamental identities, we have

$$\frac{2z}{1+z^2} = \frac{2\tan\dfrac{\theta}{2}}{1+\tan^2\dfrac{\theta}{2}} = \frac{2\tan\dfrac{\theta}{2}}{\sec^2\dfrac{\theta}{2}} = 2\tan\frac{\theta}{2}\cos^2\frac{\theta}{2}$$

$$= 2\frac{\sin\dfrac{\theta}{2}}{\cos\dfrac{\theta}{2}}\cos^2\frac{\theta}{2} = 2\sin\frac{\theta}{2}\cos\frac{\theta}{2}.$$

But, by Theorem 1,

$$2\sin\frac{\theta}{2}\cos\frac{\theta}{2} = \sin\left[2\left(\frac{\theta}{2}\right)\right] = \sin\theta.$$

Therefore,

$$\frac{2z}{1+z^2} = \sin\theta.$$

(c) Using the results of (a) and (b), we have

$$\frac{1-\cos\theta}{\sin\theta} = \frac{1-\left(\dfrac{1-z^2}{1+z^2}\right)}{\left(\dfrac{2z}{1+z^2}\right)} = \frac{(1+z^2)-(1-z^2)}{2z} = \frac{2z^2}{2z} = z = \tan\frac{\theta}{2}.$$

Example 4   Find the exact value of $\tan\dfrac{\pi}{8}$.

Solution   By part (c) of Example 3 above,

$$\tan\frac{\pi}{8} = \tan\left[\frac{1}{2}\left(\frac{\pi}{4}\right)\right] = \frac{1-\cos\dfrac{\pi}{4}}{\sin\dfrac{\pi}{4}} = \frac{1-\dfrac{\sqrt{2}}{2}}{\dfrac{\sqrt{2}}{2}} = \frac{2-\sqrt{2}}{\sqrt{2}} = \frac{2}{\sqrt{2}} - 1 = \sqrt{2} - 1.$$

Example 5   If $z = \tan\dfrac{\theta}{2}$, write $\dfrac{1}{1-\cos\theta}$ in terms of z.

Solution   By part (a) of Example 3,

$$\frac{1}{1-\cos\theta} = \frac{1}{1-\left(\dfrac{1-z^2}{1+z^2}\right)} = \frac{1+z^2}{(1+z^2)-(1-z^2)} = \frac{1+z^2}{2z^2}.$$

PROBLEM SET 4

In problems 1–6, use the double-angle formulas to write the given expression in terms of the values of trigonometric functions of an angle half as large.

1. $\sin 76°$ **2.** $\tan 62°$ **3.** $\cos 144°$ **4.** $\cos 4x$ **5.** $\sin \dfrac{2\pi}{9}$ **6.** $\tan \dfrac{\pi}{7}$

In problems 7–12, use the double-angle formulas to reduce the given expression to the value of a single trigonometric function of an angle twice as large.

7. $2\sin 29° \cos 29°$

8. $\cos^2 4\alpha - \sin^2 4\alpha$

9. $2\cos^2 \dfrac{5\theta}{2} - 1$

10. $\dfrac{\tan 5t}{1 - \tan^2 5t}$

11. $1 - 2\sin^2 \dfrac{\pi}{17}$

12. $2\sin\left(\dfrac{x + y}{2}\right)\cos\left(\dfrac{x + y}{2}\right)$

In problems 13–18, use the given information to find (a) $\sin 2\theta$, (b) $\cos 2\theta$, and (c) $\tan 2\theta$.

13. $\sin \theta = \frac{4}{5}$, θ in quadrant I   **14.** $\csc \theta = \frac{5}{3}$, θ in quadrant II
15. $\cos \theta = -\frac{5}{13}$, θ in quadrant III   **16.** $\sec \theta = -\frac{17}{8}$, θ not in quadrant II
17. $\tan \theta = -\frac{12}{5}$, θ in quadrant II   **18.** $\cot \theta = \frac{1}{2}$, θ in quadrant III

In problems 19–26, rewrite each expression in a form involving only a single trigonometric function.

19. $\dfrac{\sin 2x}{2\sin x}$

20. $(\sin 2\alpha + \cos 2\alpha)^2 - 2\sin 4\alpha$

21. $(\sin t - \cos t)^2 + 3\sin 2t$

22. $\dfrac{1 - \cos 2u}{\sin 2u}$

23. $\sin x \cos^3 x + \sin^3 x \cos x$

24. $\dfrac{\sin 2\beta}{1 + \cos 2\beta}$

25. $\dfrac{\tan 2\theta - \sin 2\theta}{2\tan 2\theta}$

26. $\dfrac{\sin 2x}{1 - \cos 2x}$

In problems 27–33, (a) use the half-angle formulas to write the given expression in terms of the value of a trigonometric function of an angle twice as large, and (b) find the exact numerical value of each expression.

27. $\cos 15°$

28. $\sin 67.5°$

29. $\cos \dfrac{5\pi}{8}$

30. $\tan \dfrac{\pi}{12}$

31. $\tan 157.5°$

32. $\cos \dfrac{11\pi}{12}$

33. $\sin 202.5°$

34. Prove Theorem 4.

In problems 35–38, use the given information to find (a) $\sin(\theta/2)$ (b) $\cos(\theta/2)$, and (c) $\tan(\theta/2)$.

35. $\cos \theta = \dfrac{7}{25}$, $0 < \theta < 90°$ **36.** $\cos \theta = -\dfrac{4}{5}$, θ in quadrant II

37. $\cos \theta = -\dfrac{5}{13}$, $\pi < \theta < \dfrac{3\pi}{2}$ **38.** $\cot \theta = -\dfrac{15}{8}$, θ in quadrant IV

In problems 39–44, reduce the given expression to the value of a single trigonometric function of an angle half as large.

39. $-\sqrt{\dfrac{1 + \cos 250°}{2}}$

40. $-\sqrt{\dfrac{1 - \cos 400°}{2}}$

41. $\dfrac{1 - \cos 6x}{\sin 6x}$

42. $\dfrac{\sin 10\theta}{1 - \cos 10\theta}$

43. $\sqrt{\dfrac{1 + \cos(2\pi/5)}{2}}$

44. $\sqrt{\dfrac{1 - \cos(11\pi/15)}{2}}$

In problems 45–55, prove that each trigonometric equation is an identity.

45. $\cos^4\theta - \sin^4\theta = \cos 2\theta$

46. $2\csc 2x = \sec x \csc x$

47. $\sin 4t = 2(\sin 2t)(1 - 2\sin^2 t)$

48. $\dfrac{\sin^2 2u}{\sin^2 u} = 4\cos^2 u$

49. $\sin 3\theta = (\sin\theta)(3 - 4\sin^2\theta)$ [Hint: $3\theta = \theta + 2\theta$.]

50. $\cos 3t = 4\cos^3 t - 3\cos t$

51. $\dfrac{\sec^2 x}{2 - \sec^2 x} = \sec 2x$

52. $\sin^2 t \cos^2 t = \frac{1}{8}(1 - \cos 4t)$

53. $\dfrac{\cot\alpha - 1}{\cot\alpha + 1} = \dfrac{\cos 2\alpha}{1 + \sin 2\alpha}$

54. $\cos^4 x = \frac{3}{8} + \frac{1}{2}\cos 2x + \frac{1}{8}\cos 4x$

55. $\csc\theta - \cot\theta = \tan\dfrac{\theta}{2}$

56. Show that $\cos 20°$ is a root of the equation $8x^3 - 6x - 1 = 0$. [Hint: Use problem 50.]

In problems 57–60, write the given expression in terms of z, where $z = \tan(\theta/2)$. (Use the results of Example 3, page 348.)

57. $\sec\theta$

58. $\csc\theta$

59. $\cot\theta$

60. $\dfrac{\sin\theta}{2 - \cos\theta}$

61. Using Theorem 3, show that, if b and c are positive constants, the quantity $y = b\cos^2 ct$ oscillates in accordance with the simple harmonic model $y = a\cos(\omega t - \phi) + k$, with $a = b/2$, $\omega = 2c$, $\phi = 0$, and $k = b/2$.

---

5 Product Formulas, Sum Formulas, and Graphs of Combinations of Sine and Cosine Functions

In applied mathematics, it is often necessary to rewrite a product of sine or cosine functions as a sum or difference of such functions, or vice versa. This can be done by using the identities in Theorems 1 and 2.

Theorem 1   **Product Formulas**

(i) $\sin\alpha\cos\beta = \frac{1}{2}[\sin(\alpha + \beta) + \sin(\alpha - \beta)]$

(ii) $\cos\alpha\sin\beta = \frac{1}{2}[\sin(\alpha + \beta) - \sin(\alpha - \beta)]$

(iii) $\cos\alpha\cos\beta = \frac{1}{2}[\cos(\alpha + \beta) + \cos(\alpha - \beta)]$

(iv) $\sin\alpha\sin\beta = \frac{1}{2}[\cos(\alpha - \beta) - \cos(\alpha + \beta)]$

Proof   We prove (i) and (iv) here and leave (ii) and (iii) as exercises (Problem 22).

(i) Using the formulas

$$\sin(\alpha + \beta) = \sin\alpha\cos\beta + \sin\beta\cos\alpha$$
$$\sin(\alpha - \beta) = \sin\alpha\cos\beta - \sin\beta\cos\alpha,$$

and adding the second equation to the first, we obtain

$$\sin(\alpha + \beta) + \sin(\alpha - \beta) = 2\sin\alpha\cos\beta.$$

It follows that

$$\sin\alpha\cos\beta = \tfrac{1}{2}[\sin(\alpha + \beta) + \sin(\alpha - \beta)].$$

(iv) Using the formulas

$$\cos(\alpha - \beta) = \cos\alpha\cos\beta + \sin\alpha\sin\beta$$
$$\cos(\alpha + \beta) = \cos\alpha\cos\beta - \sin\alpha\sin\beta,$$

and subtracting the second equation from the first, we obtain

$$\cos(\alpha - \beta) - \cos(\alpha + \beta) = 2\sin\alpha\sin\beta.$$

It follows that

$$\sin\alpha\sin\beta = \tfrac{1}{2}[\cos(\alpha - \beta) - \cos(\alpha + \beta)].$$

Example   Express each product as a sum or difference:

(a) $\sin 5\theta \cos 3\theta$ (b) $\sin 3t \sin t$

Solution   (a) By part (i) of Theorem 1 with $\alpha = 5\theta$ and $\beta = 3\theta$,

$$\sin 5\theta \cos 3\theta = \tfrac{1}{2}[\sin(5\theta + 3\theta) + \sin(5\theta - 3\theta)] = \tfrac{1}{2}\sin 8\theta + \tfrac{1}{2}\sin 2\theta.$$

(b) By part (iv) of Theorem 1 with $\alpha = 3t$ and $\beta = t$,

$$\sin 3t \sin t = \tfrac{1}{2}[\cos(3t - t) - \cos(3t + t)] = \tfrac{1}{2}\cos 2t - \tfrac{1}{2}\cos 4t.$$

Theorem 2   **Sum Formulas**

$$\text{(i)} \quad \sin\gamma + \sin\theta = 2\sin\left(\frac{\gamma + \theta}{2}\right)\cos\left(\frac{\gamma - \theta}{2}\right)$$

$$\text{(ii)} \quad \sin\gamma - \sin\theta = 2\cos\left(\frac{\gamma + \theta}{2}\right)\sin\left(\frac{\gamma - \theta}{2}\right)$$

$$\text{(iii)} \quad \cos\gamma + \cos\theta = 2\cos\left(\frac{\gamma + \theta}{2}\right)\cos\left(\frac{\gamma - \theta}{2}\right)$$

$$\text{(iv)} \quad \cos\gamma - \cos\theta = -2\sin\left(\frac{\gamma + \theta}{2}\right)\sin\left(\frac{\gamma - \theta}{2}\right)$$

Proof   We prove (i) here and leave (ii), (iii), and (iv) as exercises (Problem 24). In part (i) of Theorem 1, let

$$\alpha = \frac{\gamma + \theta}{2} \quad \text{and} \quad \beta = \frac{\gamma - \theta}{2}.$$

Thus,

$$\alpha + \beta = \frac{\gamma + \theta}{2} + \frac{\gamma - \theta}{2} = \frac{2\gamma}{2} = \gamma$$

and

$$\alpha - \beta = \frac{\gamma + \theta}{2} - \frac{\gamma - \theta}{2} = \frac{2\theta}{2} = \theta.$$

It follows that

$$\sin\left(\frac{\gamma + \theta}{2}\right)\cos\left(\frac{\gamma - \theta}{2}\right) = \tfrac{1}{2}[\sin \gamma + \sin \theta],$$

or

$$\sin \gamma + \sin \theta = 2\sin\left(\frac{\gamma + \theta}{2}\right)\cos\left(\frac{\gamma - \theta}{2}\right).$$

Example 1   Express $\sin 10x + \sin 4x$ as a product.

Solution   Using part (i) of Theorem 2 with $\gamma = 10x$ and $\theta = 4x$, we have

$$\sin 10x + \sin 4x = 2\sin\left(\frac{10x + 4x}{2}\right)\cos\left(\frac{10x - 4x}{2}\right) = 2\sin 7x \cos 3x.$$

Example 2   Prove that the equation

$$\frac{\sin(x + h) - \sin x}{h} = \cos\left(x + \frac{h}{2}\right)\frac{\sin(h/2)}{h/2}$$

is an identity.

Solution   Using part (ii) of Theorem 2 with $\gamma = x + h$ and $\theta = x$, we have

$$\frac{\sin(x + h) - \sin x}{h} = \frac{2\cos\left[\dfrac{(x + h) + x}{2}\right]\sin\left[\dfrac{(x + h) - x}{2}\right]}{h}$$

$$= \frac{2\cos\left(\dfrac{2x + h}{2}\right)\sin \dfrac{h}{2}}{h}$$

$$= \cos\left(x + \frac{h}{2}\right)\frac{\sin(h/2)}{h/2}.$$

5.1 Graphs of Combinations of Sine and Cosine Functions

We have already graphed the sum of algebraic functions in Section 7 of Chapter 3 by adding ordinates. The same method works for graphing sums of trigonometric functions.

Example   © Sketch $y = \sin x + 2 \cos 2x$ by adding ordinates.

Solution   We begin by sketching the graphs of

$$y_1 = \sin x \quad \text{and} \quad y_2 = 2 \cos 2x$$

on the same coordinate system (Figure 1). Then we add the ordinates y_1 and y_2 for selected values of x to obtain values of

$$y = \sin x + 2 \cos 2x.$$

(A calculator is useful here.) By plotting the resulting points and connecting them with a smooth curve, we obtain the desired graph (Figure 1).

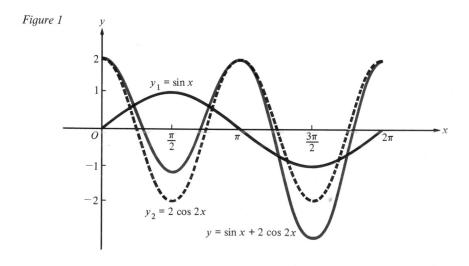

Figure 1

In sketching the graphs of certain combinations of sines and cosines, you will find the following theorem very helpful.

Theorem 3   **Reduction of $A \cos \theta + B \sin \theta$ to the Form $a \cos(\theta - \phi)$**

Suppose that A and B are constants; let $a = \sqrt{A^2 + B^2}$, and let ϕ be an angle in standard position with the point (A, B) on its terminal side. Then $A = a \cos \phi$, $B = a \sin \phi$, and, for all values of θ,

$$A \cos \theta + B \sin \theta = a \cos(\theta - \phi).$$

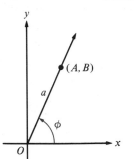

Figure 2

Proof Notice that $a = \sqrt{A^2 + B^2}$ is the distance between (A, B) and $(0,0)$ (Figure 2). Hence, by the definitions of $\sin \phi$ and $\cos \phi$,

$$\sin \phi = \frac{B}{a} \quad \text{and} \quad \cos \phi = \frac{A}{a}.$$

It follows that

$$B = a \sin \phi \quad \text{and} \quad A = a \cos \phi.$$

Therefore,

$$A \cos \theta + B \sin \theta = (a \cos \phi) \cos \theta + (a \sin \phi) \sin \theta$$
$$= a(\cos \theta \cos \phi + \sin \theta \sin \phi) = a \cos(\theta - \phi).$$

Example Sketch the graph of $y = \sqrt{3} \cos \pi t + \sin \pi t$.

Solution We could use the method of adding ordinates; but, since the same quantity πt appears in both terms, we prefer to use Theorem 3 with $\theta = \pi t$, $A = \sqrt{3}$, and $B = 1$. Then $a = \sqrt{A^2 + B^2} = \sqrt{4} = 2$ and the equations $A = a \cos \phi$ and $B = a \sin \phi$ become $\sqrt{3} = 2 \cos \phi$ and $1 = 2 \sin \phi$, or

$$\cos \phi = \frac{\sqrt{3}}{2} \quad \text{and} \quad \sin \phi = \frac{1}{2}.$$

Because both $\cos \phi$ and $\sin \phi$ are positive, ϕ is a quadrant I angle. Evidently, $\phi = \pi/6$. Therefore, by Theorem 3, we can rewrite the original equation in the form $y = a \cos(\theta - \phi)$, that is,

$$y = 2 \cos\left(\pi t - \frac{\pi}{6}\right).$$

According to the discussion in Section 7 of Chapter 6, the graph is a simple harmonic curve with amplitude $a = 2$, angular frequency $\omega = \pi$, phase angle $\phi = \pi/6$, phase shift $S = \phi/\omega = (\pi/6)/\pi = \frac{1}{6}$, period $T = 2\pi/\omega = 2\pi/\pi = 2$, and vertical shift $k = 0$ (Figure 3).

Figure 3

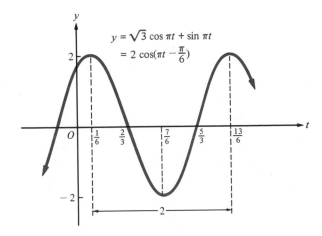

To sketch the graph of a product of sine and cosine functions, you can use Theorem 1 to rewrite the product as a sum or difference, and then use the method of adding or subtracting ordinates.

Example   [C] Sketch the graph of

$$y = 2 \sin \frac{3x}{2} \sin \frac{x}{2}.$$

Solution   Using the product formula

$$\sin \alpha \sin \beta = \tfrac{1}{2}[\cos(\alpha - \beta) - \cos(\alpha + \beta)]$$

in part (iv) of Theorem 1 with $\alpha = 3x/2$ and $\beta = x/2$, we find that

$$2 \sin \frac{3x}{2} \sin \frac{x}{2}$$

$$= \cos\left(\frac{3x}{2} - \frac{x}{2}\right) - \cos\left(\frac{3x}{2} + \frac{x}{2}\right)$$

$$= \cos x - \cos 2x.$$

Therefore, the original equation can be rewritten as

$$y = \cos x - \cos 2x.$$

In Figure 4, we have sketched graphs of $y_1 = \cos x$ and $y_2 = \cos 2x$ on the same coordinate system. Then, subtracting the ordinate y_2 from the ordinate y_1 for selected values of x, we obtain corresponding values of $y = \cos x - \cos 2x$. (Here a calculator proves to be quite useful.) By plotting the resulting points and connecting them with a smooth curve, we obtain the required graph (Figure 4).

Figure 4

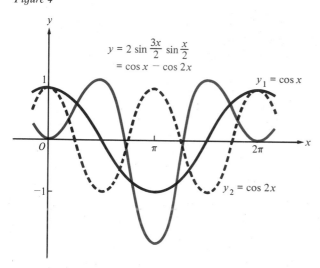

Many natural variations result from combinations of two or more simple harmonic oscillations with different frequencies, amplitudes, and phase angles. These variations may no longer be simple harmonic oscillations. For instance, pulsations in the brightness or magnitude of a variable star may be caused by a combination of factors, such as periodic expansion and contraction, variable temperature, and periodic eclipses by a companion star. The resulting pulsations, although they may occur in a regular pattern such as that shown in Figure 5, may not be simple harmonic oscillations.

Figure 5

When two simple harmonic oscillations (for instance, musical tones) combine by addition, the resulting oscillation is called a **superposition** of the original oscillations. Thus, a superposition of the oscillations represented by

$$y_1 = a_1 \cos(\omega_1 t - \phi_1) + k_1 \quad \text{and} \quad y_2 = a_2 \cos(\omega_2 t - \phi_2) + k_2$$

is represented by

$$y = y_1 + y_2.$$

The graph of such a superposition can be sketched by adding ordinates. This graph is often **periodic,** in that it repeats itself over intervals of length T (called the **period**).

It can be shown that the superposition of two simple harmonic oscillations with periods T_1 and T_2 will be a periodic oscillation only if T_1/T_2 is a rational number. If

$$T_1/T_2 = n/m,$$

where n and m are positive integers and the fraction n/m is reduced to lowest terms, then the period T of the superposition is given by

$$T = mT_1 = nT_2.$$

Example   Find the period T of the oscillation represented by

$$y = 2\cos\left(3\pi t - \frac{\pi}{3}\right) + \frac{1}{2}\cos\left(\frac{4}{5}\pi t + \frac{\pi}{12}\right).$$

Solution   The oscillation is a superposition of the simple harmonic oscillations

$$y_1 = 2\cos\left(3\pi t - \frac{\pi}{3}\right) \quad \text{and} \quad y_2 = \frac{1}{2}\cos\left(\frac{4}{5}\pi t + \frac{\pi}{12}\right),$$

which have periods

$$T_1 = \frac{2\pi}{\omega_1} = \frac{2\pi}{3\pi} = \frac{2}{3} \quad \text{and} \quad T_2 = \frac{2\pi}{\omega_2} = \frac{2\pi}{(4/5)\pi} = \frac{5}{2}.$$

Here,

$$\frac{T_1}{T_2} = \frac{2/3}{5/2} = \frac{4}{15},$$

so the period of the superposition is given by

$$T = 15T_1 = 15(\tfrac{2}{3}) = 10$$

[and also by $T = 4T_2 = 4(\tfrac{5}{2}) = 10$].

Even if the superposition of two simple harmonic oscillations is periodic, it usually isn't a simple harmonic oscillation (for instance, see Figure 4). However, *the superposition of simple harmonic oscillations*

$$y_1 = a_1 \cos(\omega t - \phi_1) + k_1 \quad \text{and} \quad y_2 = a_2 \cos(\omega t - \phi_2) + k_2,$$

which have the same period $T = 2\pi/\omega$, is also a simple harmonic oscillation of period T (for instance, see Figure 3). We leave the proof as an exercise (Problem 60).

PROBLEM SET 5

In problems 1–10, use Theorem 1 to express each product as a sum or difference.

1. $\sin 105° \cos 40°$

2. $\cos \dfrac{5\pi}{12} \sin \dfrac{\pi}{12}$

3. $2 \cos \dfrac{5\pi}{8} \cos \dfrac{\pi}{8}$

4. $10 \sin 160° \sin 42°$

5. $\sin 3\theta \cos 5\theta$

6. $\cos 7y \sin 3y$

7. $\sin 7s \sin 5t$

8. $\cos 3t \cos 5t$

9. $\sin x \cos 3x$

10. $a\cos(\omega_1 t + \phi_1)\cos(\omega_2 t + \phi_2)$

In problems 11–20, use Theorem 2 to express each sum or difference as a product.

11. $\sin 80° + \sin 20°$

12. $\sin 50° - \sin 20°$

13. $\cos \dfrac{3\pi}{8} + \cos \dfrac{\pi}{8}$

14. $\cos \dfrac{7\pi}{12} - \cos \dfrac{5\pi}{12}$

15. $\sin 4\theta + \sin 2\theta$

16. $\sin 4x - \sin 6x$

17. $\cos 2t - \cos 4t$

18. $\cos 3\gamma + \cos 8\gamma$

19. $\sin(\alpha + \beta) - \sin \alpha$

20. $a\cos(\omega_1 t + \phi_1) + a\cos(\omega_2 t + \phi_2)$

21. Use Theorem 2 to simplify each expression; then find its exact numerical value.

   (a) $\dfrac{\cos 75° - \cos 15°}{\sin 75° + \sin 15°}$

   (b) $\dfrac{\sin 80° + \sin 10°}{\cos 80° + \cos 10°}$

22. Prove parts (ii) and (iii) of Theorem 1.

23. Show that $\sin 50° + \sin 10° = \sin 70°$.

24. Prove parts (ii), (iii), and (iv) of Theorem 2.

In problems 25–38, prove that each equation is a trigonometric identity.

25. $\dfrac{\sin 5\alpha + \sin \alpha}{\cos \alpha - \cos 5\alpha} = \cot 2\alpha$

26. $\dfrac{\cos \theta + \cos 9\theta}{\sin \theta + \sin 9\theta} = \cot 5\theta$

27. $2 \sin 3s \cos 2t = \sin(3s + 2t) + \sin(3s - 2t)$

28. $\sin(x + y)\sin(x - y) = \cos^2 y - \cos^2 x$

29. $2 \sin\left(\dfrac{\pi}{4} + x\right)\sin\left(\dfrac{\pi}{4} - x\right) = \cos 2x$

30. $(\sin \alpha)(\sin 3\alpha + \sin \alpha) = \cos^2 \alpha - \cos \alpha \cos 3\alpha$

31. $\dfrac{\sin 4t - \sin 2t}{\sin 4t + \sin 2t} = \dfrac{\tan t}{\tan 3t}$

32. $\dfrac{2 \sin 3x}{\cos 2x} - \dfrac{1}{\sin x} = -\cos 4x \sec 2x \csc x$

33. $\dfrac{\cos 5s + \cos 2s}{\sin 5s + \sin 2s} = \cot \dfrac{7s}{2}$

34. $\dfrac{\sin x + \sin y}{\sin x - \sin y} = \dfrac{\tan \frac{1}{2}(x + y)}{\tan \frac{1}{2}(x - y)}$

35. $\dfrac{\sin x - \sin y}{\cos x - \cos y} = -\cot\left(\dfrac{x + y}{2}\right)$

36. $\dfrac{\cos x + \cos y}{\cos y - \cos x} = \cot\left(\dfrac{x + y}{2}\right)\cot\left(\dfrac{x - y}{2}\right)$

37. $\dfrac{\cos(x + h) - \cos x}{h} = -\sin\left(x + \dfrac{h}{2}\right)\dfrac{\sin(h/2)}{h/2}$

38. $\dfrac{\cos 4t + \cos 3t + \cos 2t}{\sin 4t + \sin 3t + \sin 2t} = \cot 3t$

C In problems 39–44, use addition or subtraction of ordinates to sketch each graph.

39. $y = \sin x + \cos x$

40. $y = \sin 2x + 2 \cos x$

41. $y = \sin x - \cos 2x$

42. $y = 3 \sin 2\pi x + 2 \sin 4\pi x$

43. $y = \sin 2t + \sin 3t$

44. $y = 2 \cos t - \sin \frac{1}{2} t$

In problems 45–48, use Theorem 3 to rewrite each equation in the form $y = a\cos(\omega t - \phi)$, and then sketch the graph.

45. $y = \sin 2t + \cos 2t$

46. $y = -\sqrt{3}\cos\left(\dfrac{\pi}{2} t\right) + \sin\left(\dfrac{\pi}{2} t\right)$

47. $y = \cos 2\pi t - \sqrt{3}\sin 2\pi t$

48. $y = 3\cos(2t - 1) - 3\sin(2t - 1)$

Ⓒ In problems 49–54, use Theorem 1 to rewrite the product as a sum or difference, and then sketch the graph by adding or subtracting ordinates.

49. $y = \cos 4x \sin 2x$

50. $y = 3 \cos 3t \cos 7t$

51. $y = 2 \sin 3t \sin t$

52. $y = -\sin \dfrac{3x}{2} \cos \dfrac{x}{2}$

53. $y = -\cos \pi x \cos 2\pi x$

54. $y = 5 \sin \dfrac{4t}{3} \cos \dfrac{2t}{3}$

In problems 55–58, find the period T of the oscillation represented by each equation.

55. $y = 3 \cos(2\pi t - \pi) + 2 \cos\left(\dfrac{\pi}{2} t - \dfrac{\pi}{2}\right) + 4$

56. $y = 2 \cos\left(\dfrac{3}{2} t + 1\right) + 4 \cos\left(\dfrac{2}{3} t - 2\right) - 5$

57. $y = \cos \dfrac{\sqrt{2}}{2} t + \cos(3\sqrt{2} t - 1)$

58. $y = \sqrt{3} \cos\left(\dfrac{4\pi}{3} t - \dfrac{\pi}{7}\right) + 2 \sin\left(\dfrac{3\pi}{2} t + \dfrac{\pi}{5}\right)$

Ⓒ **59.** Suppose you sketch the graph of

$$y = 19 \sin x - 10 \sin 2x$$

on the interval $0 \le x \le \pi$ by plotting points corresponding to

$$x = 0, \frac{\pi}{6}, \frac{\pi}{4}, \frac{\pi}{3}, \frac{\pi}{2}, \frac{2\pi}{3}, \frac{3\pi}{4}, \frac{5\pi}{6}, \text{ and } \pi.$$

If you connect these points with a smooth curve, you will obtain the graph shown in Figure 6. However, this is *not* the correct graph. Find the error, correct it, and redraw the graph.

Figure 6

60. Assume $y_1 = a_1 \cos(\omega t - \phi_1) + k_1$ and $y_2 = a_2 \cos(\omega t - \phi_2) + k_2$. Let $\theta = \omega t - \phi_1$, $A = a_1 + a_2 \cos(\phi_2 - \phi_1)$, $B = a_2 \sin(\phi_2 - \phi_1)$, $k = k_1 + k_2$, and $y = y_1 + y_2$.
(a) Show that $\omega t - \phi_2 = \theta - (\phi_2 - \phi_1)$.
(b) Use the result in part (a) to show that

$$y_2 = a_2 \cos(\phi_2 - \phi_1) \cos \theta + a_2 \sin(\phi_2 - \phi_1) \sin \theta + k_2.$$

(c) Use the result in part (b) to show that

$$y = A \cos \theta + B \sin \theta + k.$$

(d) Use the result in part (c) together with Theorem 3 to rewrite y in the form

$$y = a \cos(\theta - \phi) + k.$$

(e) Conclude that $y = a \cos(\omega t - \phi_0) + k$, where $\phi_0 = \phi_1 + \phi$.

Ⓒ **61.** Even when sketched carefully and correctly, the graph of $y = 19 \sin x - 10 \sin 2x$ (see problem 59) appears to have a maximum at $x = 2\pi/3$. Use a calculator to check values of y for values of x close to $2\pi/3$ to see if this is really so.

62. A 50 kilohertz radio frequency oscillation with the equation $y_1 = 10 \sin(\pi \times 10^5)t$ is amplitude modulated by an audio frequency tone with the equation $y_2 = \sin(2\pi \times 10^4)t$ to produce a signal with the equation $y = y_2 y_1$. Sketch three graphs over the interval $0 \le t \le \frac{1}{5000}$ to show (a) the radio frequency oscillation, (b) the audio frequency tone, and (c) the amplitude modulated signal.

6   Inverse Trigonometric Functions

In Sections 2, 3, and 4 of Chapter 6, we learned how to find the values (or the approximate values) of the trigonometric functions of known angles. In many applications of trigonometry, the problem is the other way around—to find unknown angles when information is given about the values of trigonometric functions of the angles. The **inverse trigonometric functions** can be used to solve such problems.

Recall from Section 8 of Chapter 3 that by reflecting the graph of an invertible function f across the line $y = x$ we obtain the graph of the inverse function f^{-1} (Figure 1). However, in order for f to be invertible, no horizontal line can intersect its graph more than once.

Figure 1

Figure 2

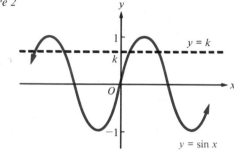

Because the sine function is periodic, any horizontal line $y = k$ with $-1 \leq k \leq 1$ intersects the graph of $y = \sin x$ repeatedly (Figure 2), and therefore the sine function is *not* invertible. But consider the portion of the graph of $y = \sin x$ between $x = -\pi/2$ and $x = \pi/2$ (Figure 3a). No horizontal line intersects this curve more than once, so its reflection across the line $y = x$ (Figure 3b) is the graph of a function. This function is often called the "inverse sine" and denoted by $\sin^{-1}$.

Figure 3

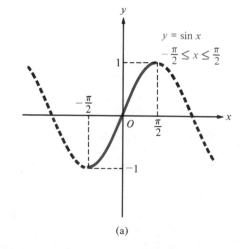

(a)   (b)

Of course, the terminology "inverse sine function" and the notation $\sin^{-1}$ aren't really correct because the sine function (defined on the set of all real numbers) has no inverse. Because of this, some people use the notation Sine (with a *capital* S) for the function in Figure 3a and Sin^{-1} for its inverse in Figure 3b. Another source of difficulty is the possible confusion of $\sin^{-1} x$ with $(\sin x)^{-1} = 1/\sin x = \csc x$. To avoid such confusion, many people use the notation "$\arcsin x$," meaning the arc (or angle) whose sine is x, rather than $\sin^{-1} x$. Because both the notations $\sin^{-1} x$ and $\arcsin x$ are in common use in mathematics and its applications, we shall use them interchangeably. Thus, we make the following definition.

Definition 1   **The Inverse Sine or Arcsine**

> The *inverse sine* or *arcsine* function, denoted by $\sin^{-1}$ or arcsin, is defined by
> $$\sin^{-1} x = y \quad \text{if and only if} \quad x = \sin y \quad \text{and} \quad -\frac{\pi}{2} \le y \le \frac{\pi}{2}.$$

In words, $\sin^{-1} x$ is the angle (or number) between $-\pi/2$ and $\pi/2$ whose sine is x. The graph of
$$y = \sin^{-1} x \quad \text{or} \quad y = \arcsin x$$

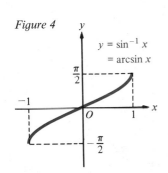

Figure 4

(Figure 4) shows that $\sin^{-1}$ is an increasing function with domain $[-1,1]$ and range $[-\pi/2, \pi/2]$. The graph appears to be symmetric about the origin, indicating that $\sin^{-1}$ is an odd function; that is,
$$\sin^{-1}(-x) = -\sin^{-1} x.$$

A scientific calculator can be used to find approximate values of $\sin^{-1} x$ (or $\arcsin x$). (If a calculator isn't available, you can find these values by reading Appendix Table IIB "backwards.") Some calculators have a $\sin^{-1}$ key; some have a key marked arcsin; and some have an INV key which must be pressed before the SIN key to give the inverse sine. Again, you must be careful to put the calculator in radian mode if you want $\sin^{-1} x$ in radians—otherwise, you will get the angle in *degrees* (between $-90°$ and $90°$) whose sine is x.

Example   Find each value:

(a) $\sin^{-1}\frac{1}{2}$ (b) $\arcsin(-\sqrt{3}/2)$ C (c) $\sin^{-1} 0.7321$

Solution   (a) Here a calculator isn't needed because we know that $\pi/6$ radian ($30°$) is an angle whose sine is $\frac{1}{2}$. Thus, because $-\pi/2 \le \pi/6 \le \pi/2$, we have $\sin^{-1}\frac{1}{2} = \pi/6$.

(b) Again we don't need a calculator because we know that $-\pi/3$ radians ($-60°$) is an angle whose sine is $-\sqrt{3}/2$. Thus, because $-\pi/2 \le -\pi/3 \le \pi/2$, we have $\arcsin(-\sqrt{3}/2) = -\pi/3$.

(c) Using a 10-digit calculator in radian mode, we obtain
$$\sin^{-1} 0.7321 = 0.821399673.$$

(In degree mode, we get $\sin^{-1} 0.7321 = 47.06273457°$.)

The remaining five trigonometric functions are also periodic, and therefore, like the sine function, they are not invertible. However, by restricting these functions to suitable intervals we can define corresponding inverses, just as we did for the sine function. The appropriate portions of the graphs of the cosine and tangent functions are shown in Figure 5. Thus, we make the following definitions.

Figure 5

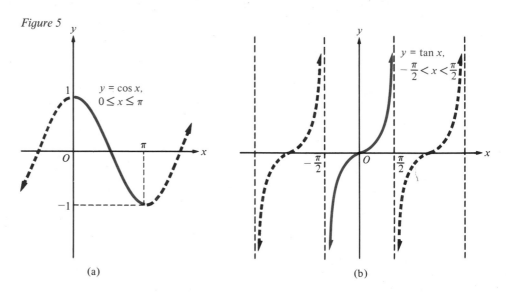

(a)   (b)

Definition 2   **The Inverse Cosine or Arccosine**

> The *inverse cosine* or *arccosine* function, denoted by $\cos^{-1}$ or arccos, is defined by
>
> $$\cos^{-1} x = y \quad \text{if and only if} \quad x = \cos y \quad \text{and} \quad 0 \le y \le \pi.$$

Definition 3   **The Inverse Tangent or Arctangent**

> The inverse tangent or arctangent function, denoted by $\tan^{-1}$ or arctan, is defined by
>
> $$\tan^{-1} x = y \quad \text{if and only if} \quad x = \tan y \quad \text{and} \quad -\frac{\pi}{2} < y < \frac{\pi}{2}.$$

Figure 6

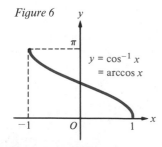

In words, $\cos^{-1} x$ is the angle (or number) between 0 and π whose cosine is x. Likewise, $\tan^{-1} x$ is the angle (or number) between $-\pi/2$ and $\pi/2$ whose tangent is x.

The graph of $y = \cos^{-1} x$, obtained by reflecting the curve in Figure 5a across the line $y = x$, shows that $\cos^{-1}$ (or arccos) is a decreasing function with domain $[-1, 1]$ and range $[0, \pi]$ (Figure 6). Notice that the graph of $\cos^{-1}$ is *not* symmetric about the origin nor about the y axis. Thus, in spite of the fact that cosine is an even function, $\cos^{-1}$ is not.

Figure 7

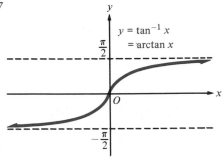

The graph of $y = \tan^{-1} x$, obtained by reflecting the curve in Figure 5b across the line $y = x$, shows that $\tan^{-1}$ (or arctan) is an increasing function with domain $\mathbb{R}$ and range $(-\pi/2, \pi/2)$ (Figure 7). Notice that the lines $y = -\pi/2$ and $y = \pi/2$ are horizontal asymptotes. The graph appears to be symmetric about the origin, indicating that $\tan^{-1}$ is an odd function; that is,

$$\tan^{-1}(-x) = -\tan^{-1} x$$

(Problem 39).

Example 1   Find each value:

(a) $\cos^{-1}(-\sqrt{2}/2)$ © (b) arccos 0.6675

(c) arctan $\sqrt{3}$ © (d) $\tan^{-1}(-2.498)$

Solution   (a) If $\cos^{-1}(-\sqrt{2}/2) = y$, then $-\sqrt{2}/2 = \cos y$ and $0 \le y \le \pi$. Therefore, $y = 3\pi/4$.

(b) Using a 10-digit calculator in radian mode, we find that

$$\text{arccos } 0.6675 = 0.839950077.$$

(c) If arctan $\sqrt{3} = y$, then $\sqrt{3} = \tan y$ and $-\pi/2 < y < \pi/2$. Therefore, $y = \pi/3$.

(d) Using a 10-digit calculator in radian mode, we find that

$$\tan^{-1}(-2.498) = -1.190013897.$$

If you enter a number x between -1 and 1 in a calculator, take $\sin^{-1} x$, and then take the sine of the result, you will get x back again. Thus,

$$x \xrightarrow{\;\sin^{-1}\;} y \xrightarrow{\;\sin\;} x.$$

Do you see why? Therefore, we have the identity

$$\sin(\sin^{-1} x) = x \quad \text{for} \quad -1 \le x \le 1.$$

However, if you try the same thing the other way around, first taking the sine and then taking the inverse sine, you might not get your original number back. For instance, starting with $x = 2$, we get (on a 10-digit calculator in radian mode)

$$2 \xrightarrow{\;\sin\;} 0.909297427 \xrightarrow{\;\sin^{-1}\;} 1.141592654.$$

The reason is simply that $\sin^{-1} 0.909297427$ is the number *between $-\pi/2$ and $\pi/2$* whose sine is 0.909297427, and 2 does not lie between $-\pi/2$ and $\pi/2$. However, if you start with a number *between $-\pi/2$ and $\pi/2$*, take the sine, and then take the inverse sine, you will get your original number back. In other words,

$$\sin^{-1}(\sin x) = x \quad \text{for} \quad -\pi/2 \le x \le \pi/2.$$

Similar rules apply to cos and $\cos^{-1}$, and to tan and $\tan^{-1}$ (Problems 45 and 47).

In calculus, it's sometimes necessary to find exact values of expressions such as $\sin(\tan^{-1}\frac{2}{3})$. The following example shows how this can be done.

Example Find the exact value of $\sin(\tan^{-1}\frac{2}{3})$.

Solution We begin by sketching an acute angle θ in a right triangle such that $\tan^{-1}\frac{2}{3} = \theta$; that is, $\frac{2}{3} = \tan\theta = \text{opp/adj}$. This is accomplished simply by letting opp $= 2$ units and adj $= 3$ units (Figure 8). By the Pythagorean theorem,

$$\text{hyp} = \sqrt{\text{opp}^2 + \text{adj}^2} = \sqrt{2^2 + 3^2} = \sqrt{13}.$$

Figure 8

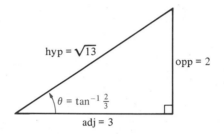

Therefore,

$$\sin(\tan^{-1}\tfrac{2}{3}) = \sin\theta = \frac{\text{opp}}{\text{hyp}} = \frac{2}{\sqrt{13}} = \frac{2\sqrt{13}}{13}.$$

Another method for finding exact values of expressions involving inverse trigonometric functions is to use identities such as the one developed in the following example.

Example Show that $\cos(\tan^{-1} x) = \dfrac{1}{\sqrt{1 + x^2}}$.

Solution Let $\tan^{-1} x = y$, so that $x = \tan y$ and $-\pi/2 < y < \pi/2$. Then

$$1 + x^2 = 1 + \tan^2 y = \sec^2 y.$$

Because $-\pi/2 < y < \pi/2$, it follows that $\sec y > 0$; hence,

$$\sqrt{1 + x^2} = \sec y.$$

Therefore,

$$\frac{1}{\sqrt{1 + x^2}} = \frac{1}{\sec y} = \cos y = \cos(\tan^{-1} x).$$

The remaining inverse trigonometric functions $\cot^{-1}$, $\sec^{-1}$, and $\csc^{-1}$ (that is, arccot, arcsec, and arccsc, respectively) aren't used as often as $\sin^{-1}$, $\cos^{-1}$, and $\tan^{-1}$, so you won't find them on the keys of many scientific calculators. (See Problems 48–51.)

PROBLEM SET 6

In problems 1–12, evaluate each expression without using a calculator or tables.

1. $\sin^{-1} 1$

2. $\arcsin \dfrac{\sqrt{3}}{2}$

3. $\arcsin\left(-\dfrac{\sqrt{2}}{2}\right)$

4. $\cos^{-1}\left(-\dfrac{1}{2}\right)$

5. $\arccos 1$

6. $\cos^{-1} \dfrac{\sqrt{3}}{2}$

7. $\sin^{-1} \dfrac{\sqrt{2}}{2}$

8. $\cos^{-1} 0$

9. $\arccos \dfrac{1}{2}$

10. $\arctan 1$

11. $\tan^{-1}(-1)$

12. $\tan^{-1} \dfrac{\sqrt{3}}{3}$

[c] In problems 13–24, use a calculator (or Appendix Table IIB) to evaluate each expression. Give answers in *radians*.

13. $\arcsin 0.6442$

14. $\arccos 0.6675$

15. $\cos^{-1} 0.9051$

16. $\tan^{-1} 0.2500$

17. $\arctan 2$

18. $\sin^{-1}(-0.5495)$

19. $\tan^{-1}(-3.224)$

20. $\cos^{-1}(-\frac{1}{8})$

21. $\arcsin(-0.5505)$

22. $\arccos\left(-\dfrac{5}{11}\right)$

23. $\sin^{-1} \dfrac{\sqrt{5}}{4}$

24. $\tan^{-1}\left(-\dfrac{\sqrt{7}}{3}\right)$

In problems 25–36, find the exact value of each expression without using a calculator or tables.

25. $\sin\left(\sin^{-1}\dfrac{3}{4}\right)$

26. $\tan(\tan^{-1} 3)$

27. $\sin^{-1}\left(\sin\dfrac{\pi}{6}\right)$

28. $\tan^{-1}\left(\tan\dfrac{5\pi}{4}\right)$

29. $\sin\left(\tan^{-1}\dfrac{4}{3}\right)$

30. $\tan\left(\sin^{-1}\dfrac{\sqrt{5}}{5}\right)$

31. $\cos\left(\sin^{-1}\dfrac{\sqrt{10}}{10}\right)$

32. $\cos[\arctan(-2)]$

33. $\tan\left(\sin^{-1}\dfrac{4}{5}\right)$

34. $\sin(2\arcsin \frac{2}{3})$

35. $\sec(\cos^{-1}\frac{7}{10})$

36. $\cos(2\sin^{-1}\frac{1}{8})$

In problems 37–44, show that the given equation is an identity.

37. $\sin^{-1}(-x) = -\sin^{-1} x$

38. $\cos^{-1} x = \dfrac{\pi}{2} - \sin^{-1} x$

39. $\arctan(-x) = -\arctan x$

40. $\cos^{-1}(-x) = \pi - \cos^{-1} x$

41. $\tan(\sin^{-1} x) = \dfrac{x}{\sqrt{1 - x^2}}$

42. $\sin(\arctan x) = \dfrac{x}{\sqrt{1 + x^2}}$

43. $\sin(2\arcsin x) = 2x\sqrt{1 - x^2}$

44. $\tan(\frac{1}{2}\arccos x) = \sqrt{\dfrac{1 - x}{1 + x}}$

45. For what values of x is it true that
(a) $\cos(\cos^{-1} x) = x$?
(b) $\cos^{-1}(\cos x) = x$?

[c] **46.** To the nearest hundredth of a degree, find the two vertex angles α and β of a 3–4–5 right triangle (Figure 9).

47. For what values of x is it true that
(a) $\tan(\arctan x) = x$?
(b) $\arctan(\tan x) = x$?

48. The **inverse cotangent** function, $\cot^{-1}$ or arccot, is defined by $\cot^{-1} x = y$ if and only if $x = \cot y$ and $0 < y < \pi$. (a) Sketch the graph of $y = \cot^{-1} x$. (b) Show that

$$\cot^{-1} x = \dfrac{\pi}{2} - \tan^{-1} x.$$

Figure 9

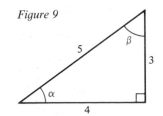

49. One popular definition of the **inverse secant** function, $\sec^{-1}$ or arcsec, is $\sec^{-1} x = y$ if and only if $x = \sec y$ and $0 \leq y \leq \pi$ with $y \neq \pi/2$. (a) Sketch the graph of $y = \sec^{-1} x$. (b) Show that

$$\sec^{-1} x = \cos^{-1}\left(\frac{1}{x}\right) \qquad \text{for} \quad |x| \geq 1.$$

50. An alternative definition of the inverse secant function (see problem 49) used in some calculus textbooks is $\sec^{-1} x = y$ if and only if $x = \sec y$ and either $0 \leq y < \pi/2$ or else $\pi \leq y < 3\pi/2$. Sketch the graph of $\sec^{-1}$ according to this definition.

51. One popular definition of the **inverse cosecant** function, $\csc^{-1}$ or arccsc, is $\csc^{-1} x = y$ if and only if $x = \csc y$ and $-\pi/2 \leq y \leq \pi/2$ with $y \neq 0$. (a) Sketch the graph of $y = \csc^{-1} x$. (b) Show that

$$\csc^{-1} x = \sin^{-1}\left(\frac{1}{x}\right) \qquad \text{for} \quad |x| \geq 1.$$

7 Trigonometric Equations

In Section 1, we mentioned that a trigonometric equation that is not an identity is called a **conditional** equation. To **solve** a conditional trigonometric equation means to find all values of the unknown for which the equation is true; these values are called the **solutions** of the equation. If a side condition such as $0° \leq \theta < 360°$ is given along with a trigonometric equation, we understand that the solutions consist of all values of θ that satisfy *both* the equation and the side condition.

An equation, such as $\sin \theta = \sqrt{3}/2$, whose left side is a trigonometric function of the unknown and whose right side is a constant, is called a **simple** trigonometric equation. A side condition such as $0° \leq \theta < 360°$ may or may not be involved. Even if there is no side condition, we usually *begin* by solving the equation for nonnegative values of the unknown that are less than $360°$ or 2π radians. The complete set of solutions is then obtained by adding all integer multiples of $360°$ or 2π radians to the values thus obtained. (Why?)

Examples   Solve each simple trigonometric equation. Use degree or radian measure as indicated.

1 $\sin \theta = \dfrac{\sqrt{3}}{2}, \qquad 0° \leq \theta < 360°$

Solution   Recall that $\sin 60° = \sqrt{3}/2$. Therefore, $\theta = 60°$ is the only solution in quadrant I. (Why?) Because the values of $\sin \theta$ are negative in quadrants III and IV, the only possible remaining solution between $0°$ and $360°$ must lie in quadrant II, and its reference angle must be $\theta_R = 60°$ (Figure 1). Hence, $\theta = 180° - 60° = 120°$. Therefore, the solutions are

$$\theta = 60° \quad \text{and} \quad \theta = 120°.$$

Figure 1

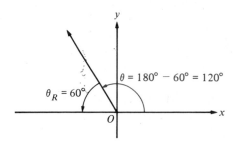

2 $\sec \alpha = -\sqrt{2}, \quad 0° \le \alpha < 360°$

Solution   Recall that the values of $\sec \alpha$ are negative only in quadrants II and III, and that $\sec 45° = \sqrt{2}$. Therefore, by constructing angles in quadrants II and III with 45° reference angles (Figure 2), we find the solutions

$$\alpha = 180° - 45° = 135° \quad \text{and} \quad \alpha = 180° + 45° = 225°.$$

Figure 2

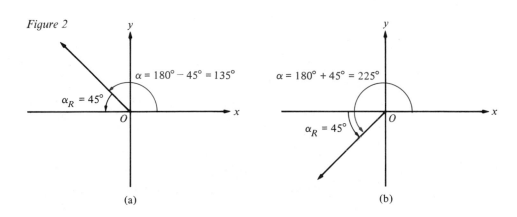

(a)   (b)

3 $\sec t = -\sqrt{2}, \quad t$ in radians

Solution   Here there is no side condition. Nevertheless, we *begin* by solving the equation with the side condition $0 \le t < 2\pi$. Proceeding as in Example 2, but measuring angles in radians, we find the two solutions

$$t = \pi - \frac{\pi}{4} = \frac{3\pi}{4} \quad \text{and} \quad t = \pi + \frac{\pi}{4} = \frac{5\pi}{4}.$$

Therefore, the complete set of solutions consists of all real numbers

$$t = \frac{3\pi}{4} + 2\pi k \quad \text{and} \quad t = \frac{5\pi}{4} + 2\pi k,$$

where k denotes an arbitrary integer.

© 4 $\cot x = -0.8333, \qquad 0 \le x < 2\pi$

Solution   We begin by rewriting the equation as

$$\tan x = 1/(-0.8333)$$

so that we can use the $\tan^{-1}$ (or arctan) key on a scientific calculator. If we simply calculate $\tan^{-1}[1/(-0.8333)]$, we get -0.876077723, which is *not* a solution because it doesn't satisfy the condition $0 \le x < 2\pi$. Recall that the values of $\tan x$ are negative in quadrants II and IV. With the calculator in radian mode, we enter 0.8333, take the reciprocal, and then take the inverse tangent to obtain the reference angle

$$\theta_R = \tan^{-1}(1/0.8333) = 0.876077723$$

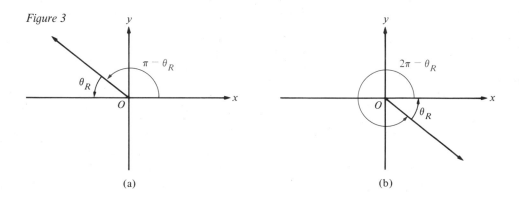

Figure 3

(a)

(b)

(Figure 3). Therefore, the solution in quadrant II is $x = \pi - \theta_R$ and the solution in quadrant IV is $x = 2\pi - \theta_R$; that is,

$$x = \pi - 0.876077723 = 2.265514931$$

and

$$x = 2\pi - 0.876077723 = 5.407107585.$$

Snell's law of refraction, which was discovered around 1620 by the Dutch physicist Willebrord Snell (1591–1626), says that a light ray is bent (refracted) as it passes from a first medium into a second medium according to the equation

$$\frac{\sin \alpha}{\sin \beta} = \mu,$$

where α is the **angle of incidence** (Figure 4), β is the **angle of refraction** (Figure 4), and μ (the Greek letter mu) is a constant called the **index of refraction** of the second medium with respect to the first.

Figure 4

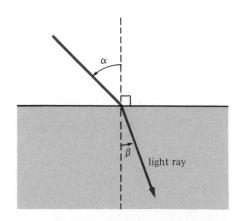

Example   C The index of refraction of flint glass with respect to air is $\mu = 1.650$. Determine the angle of refraction β of a ray of light that strikes a block of flint glass with an angle of incidence $\alpha = 35°$.

Solution   From Snell's law $\dfrac{\sin \alpha}{\sin \beta} = \mu$, we obtain the simple trigonometric equation

$$\sin \beta = \frac{\sin \alpha}{\mu} = \frac{\sin 35°}{1.650}.$$

From the geometry of the problem (Figure 4), we have the side condition $0 < \beta < 90°$. Using a calculator in degree mode, we find that

$$\beta = \sin^{-1}\left(\frac{\sin 35°}{1.650}\right) = 20.34°$$

to the nearest hundredth of a degree.

Many of the techniques used for solving trigonometric equations are similar to those used for solving algebraic equations. The basic idea is to reduce the equation to one or more simple trigonometric equations, and then use the method illustrated above.

Examples   Solve each trigonometric equation subject to the side condition $0° \leq \theta < 360°$ or $0 \leq t < 2\pi$.

1 $\cos 3t = \sqrt{2}/2$

Solution   Let $x = 3t$, and rewrite the equation as the simple trigonometric equation $\cos x = \sqrt{2}/2$. The side condition $0 \leq t < 2\pi$ is equivalent to $0 \leq 3t < 6\pi$ or $0 \leq x < 6\pi$. The solutions of $\cos x = \sqrt{2}/2$ for $0 \leq x < 2\pi$, obtained using the method described previously, are $x = \pi/4$ and $x = 7\pi/4$. Adding 2π to these, we obtain two more solutions,

$$x = \frac{\pi}{4} + 2\pi = \frac{9\pi}{4} \quad \text{and} \quad x = \frac{7\pi}{4} + 2\pi = \frac{15\pi}{4},$$

both of which satisfy the condition $2\pi \leq x < 4\pi$. Finally, adding 2π again, we obtain two more solutions,

$$x = \frac{9\pi}{4} + 2\pi = \frac{17\pi}{4} \quad \text{and} \quad x = \frac{15\pi}{4} + 2\pi = \frac{23\pi}{4},$$

both of which satisfy the condition $4\pi \leq x < 6\pi$. Thus, the solutions for $0 \leq x < 6\pi$ are

$$3t = x = \frac{\pi}{4}, \frac{7\pi}{4}, \frac{9\pi}{4}, \frac{15\pi}{4}, \frac{17\pi}{4}, \text{and } \frac{23\pi}{4},$$

and so the values of t are

$$t = \frac{x}{3} = \frac{\pi}{12}, \frac{7\pi}{12}, \frac{3\pi}{4}, \frac{5\pi}{4}, \frac{17\pi}{12}, \text{and } \frac{23\pi}{12}.$$

2 $\qquad \cos\theta\cot\theta = \cos\theta$

Solution $\qquad$ We rewrite the equation as

$$\cos\theta\cot\theta - \cos\theta = 0 \quad\text{or}\quad \cos\theta(\cot\theta - 1) = 0,$$

so that, with the side condition $0° \le \theta < 360°$, we have:

$\cos\theta = 0$	$\cot\theta - 1 = 0$
$\theta = 90°, 270°$	$\cot\theta = 1$
	$\theta = 45°, 225°$

Therefore, the solutions are

$$\theta = 45°, 90°, 225°, \text{ and } 270°.$$

(Note: We were careful *not* to divide both sides of the original equation by the variable factor $\cos\theta$. Had we done so, we would have lost the solutions $\theta = 90°$ and $\theta = 270°$.)

3 $\qquad 2\cos^2 t - \sin t - 1 = 0.$

Solution $\qquad$ Using the Pythagorean identity $\cos^2 t + \sin^2 t = 1$, we rewrite the equation as

$$2(1 - \sin^2 t) - \sin t - 1 = 0 \quad\text{or}\quad -2\sin^2 t - \sin t + 1 = 0,$$

that is,

$$2\sin^2 t + \sin t - 1 = 0 \quad\text{or}\quad (2\sin t - 1)(\sin t + 1) = 0.$$

Setting each factor equal to zero and keeping in mind the side condition $0 \le t < 2\pi$, we have:

$2\sin t - 1 = 0$	$\sin t + 1 = 0$
$\sin t = \tfrac{1}{2}$	$\sin t = -1$
$t = \dfrac{\pi}{6}, \dfrac{5\pi}{6}$	$t = \dfrac{3\pi}{2}$

Therefore, the solutions are

$$t = \frac{\pi}{6}, \frac{5\pi}{6}, \text{ and } \frac{3\pi}{2}.$$

4 $\qquad \tan\theta - \sec\theta = 1$

Solution $\qquad$ We rewrite the equation as $\sec\theta = \tan\theta - 1$, and square both sides. (Note: When proving a trigonometric *identity*, we are not permitted to square both sides. However, in solving a *conditional* equation, we may square both sides *provided that we later check for extraneous roots*.) We have

$$\sec^2\theta = \tan^2\theta - 2\tan\theta + 1.$$

Using the Pythagorean identity $1 + \tan^2\theta = \sec^2\theta$, we can rewrite the last equation as

$$1 + \tan^2\theta = \tan^2\theta - 2\tan\theta + 1 \quad \text{or} \quad 0 = -2\tan\theta,$$

that is,

$$\tan\theta = 0.$$

Therefore, with the side condition $0° \leq \theta < 360°$, we have

$$\theta = 0° \quad \text{or} \quad \theta = 180°.$$

We must now check our solutions. Substituting $\theta = 0°$ in the original equation, we get

$$\tan 0° - \sec 0° = 1 \quad \text{or} \quad 0 - 1 = 1,$$

which is *false.* Therefore, $\theta = 0°$ isn't a solution. Substituting $\theta = 180°$ in the original equation, we get

$$\tan 180° - \sec 180° = 1 \quad \text{or} \quad 0 - (-1) = 1,$$

which is *true.* Thus, $\theta = 180°$ is the only solution.

5 © $5\tan^2 t + 2\tan t - 7 = 0$ (Round off to four decimal places.)

Solution   Factoring the left side of the equation, we have

$$(5\tan t + 7)(\tan t - 1) = 0.$$

Setting the first factor equal to zero, we obtain the simple trigonometric equation

$$\tan t = -\tfrac{7}{5} = -1.4.$$

Using a calculator (as in Example 4 on page 368), we find that

$$t \approx 2.1910 \quad \text{or} \quad t \approx 5.3326.$$

Setting the second factor equal to zero, we obtain the simple trigonometric equation

$$\tan t = 1,$$

which has the solutions

$$t = \frac{\pi}{4} \approx 0.7854 \quad \text{and} \quad t = \frac{5\pi}{4} \approx 3.9270.$$

Hence, rounded off to four decimal places, the solutions are

$$t = 0.7854, 2.1910, 3.9270, \text{ and } 5.3326.$$

Ecologists often use the following equations as a **simple predator–prey model:**

$$x = a_1 \cos\omega t + k_1$$
$$y = a_2 \cos(\omega t - \phi) + k_2,$$

where x is the number of prey at time t and y is the number of predators at time t.

Here ω, a_1, a_2, k_1, k_2, and ϕ are constants determined by the particular predator–prey relationship. Notice that when $t = 0$, the prey population has its maximum value $x = a_1 + k_1$.

That x and y oscillate in a simple harmonic fashion is explained as follows: As the predators begin consuming the prey, the predator population increases and the prey population decreases. After a while, there aren't enough prey left to support the increased number of predators, and the predators begin to die off. Fewer predators means that more prey survive, and the prey population begins to increase. These oscillations continue indefinitely, unless something disturbs the ecological balance.

Example

In a certain habitat, the simple predator–prey model for owls (the predator) and field mice (the prey) is determined by the values $\omega = 2$, $a_1 = 500$, $a_2 = 10$, $k_1 = 2000$, $k_2 = 50$, and $\phi = \pi/4$, where t is the time in years measured from an instant when the prey population had its maximum value $x = 2500$. (a) Find the values of $t > 0$ for which the owl population is $y = 55$. [c] (b) Find the field mouse population x when $y = 55$.

Solution

(a) Substitute $y = 55$, $a_2 = 10$, $\omega = 2$, $\phi = \pi/4$, and $k_2 = 50$ in the equation $y = a_2 \cos(\omega t - \phi) + k_2$ to obtain

$$55 = 10 \cos\left(2t - \frac{\pi}{4}\right) + 50;$$

that is,

$$\cos\left(2t - \frac{\pi}{4}\right) = \frac{1}{2}.$$

It follows that

$$2t - \frac{\pi}{4} = \frac{\pi}{3} + 2\pi n \quad \text{or} \quad 2t - \frac{\pi}{4} = \frac{5\pi}{3} + 2\pi n,$$

where n is an integer. Thus,

$$t = \frac{7\pi}{24} + \pi n \quad \text{or} \quad t = \frac{23\pi}{24} + \pi n.$$

Since $t > 0$, then n must be a *nonnegative* integer.

(b) When $t = \dfrac{7\pi}{24} + \pi n$,

$$x = 500 \cos 2\left(\frac{7\pi}{24} + \pi n\right) + 2000$$

$$= 500 \cos\left(\frac{7\pi}{12} + 2\pi n\right) + 2000$$

$$= 500 \cos \frac{7\pi}{12} + 2000 \approx 1871.$$

When $t = \dfrac{23\pi}{24} + \pi n$,

$$x = 500\cos 2\left(\dfrac{23\pi}{24} + \pi n\right) + 2000 = 500\cos\left(\dfrac{23\pi}{12} + 2\pi n\right) + 2000$$

$$= 500\cos\dfrac{23\pi}{12} + 2000 \approx 2483.$$

Thus, when there are 55 owls in the habitat, the model predicts that there are either 1871 or 2483 field mice.

PROBLEM SET 7

In problems 1–16, solve each simple trigonometric equation. Use degree or radian measure as indicated. Do not use a calculator or tables.

1. $\cos\theta = 0$, $\quad 0° \le \theta < 360°$
2. $\cos x = 1$, $\quad 0 \le x < 2\pi$
3. $\cot\theta = 1$, $\quad 0° \le \theta < 360°$
4. $\tan s = 1$, $\quad 0 \le s < 2\pi$
5. $\sin t = \frac{1}{2}$, $\;0 \le t < 2\pi$
6. $\sec\alpha = 2$, $\quad 0° \le \alpha < 360°$
7. $\tan x = -\sqrt{3}$, $\quad 0 \le x < 2\pi$
8. $\csc t = -2$, $\quad 0 \le t < 2\pi$
9. $\sin t = -1$, $\quad 0 \le t < 2\pi$
10. $\sec\theta = -2\sqrt{3}/3$, $\quad 0° \le \theta < 720°$
11. $\tan\theta = 0$, $\quad \theta$ in degrees
12. $\sin\theta = 0$, $\quad 0° \le \theta < 720°$
13. $\sec t = \sqrt{2}$, $\quad t$ in radians
14. $\sec x = 2$, $\quad x$ in radians
15. $\tan x = \sqrt{3}$, $\quad x$ in radians
16. $\csc\beta = 2\sqrt{3}/3$, $\quad \beta$ in degrees

[c] In problems 17–22, use a calculator to solve each simple trigonometric equation. Use degree or radian measure as indicated and round off all answers to four decimal places.

17. $\sin\theta = \frac{3}{4}$, $\quad 0° \le \theta < 360°$
18. $\csc x = 4.5201$, $\quad x > 0$ in radians
19. $\cos t = \frac{2}{3}$, $\quad 0 \le t < 2\pi$
20. $\tan\beta = \frac{5}{4}$, $\quad 0° \le \beta < 720°$
21. $\cot x = 0.2884$, $\quad x$ in radians
22. $\sec\alpha = 1.5763$, $\quad \alpha$ in degrees

In problems 23–46, solve each trigonometric equation with the side condition $0° \le \theta < 360°$ or $0 \le t < 2\pi$. Do not use a calculator or tables.

23. $\sin 2\theta = \frac{1}{2}$
24. $\sec 2t = 2$
25. $\tan 3\theta = \sqrt{3}$
26. $\csc 3t = -2$
27. $\sin\frac{1}{2}t = \sqrt{3}/2$
28. $\sec\frac{1}{2}\theta = 2\sqrt{3}/3$
29. $\cos\frac{1}{3}t = 0$
30. $\tan\frac{1}{3}\theta = 1$
31. $2\sin\theta + 1 = 0$
32. $\sin t\cos t = 0$
33. $\sin^2 t = \frac{1}{2}\sin t$
34. $\sin(\theta + 40°) = 1$

35. $3\tan^2\theta - 1 = 0$
36. $2\cos\left(t - \dfrac{\pi}{6}\right) = -\sqrt{2}$
37. $\sec^2 t - 2 = 0$

38. $4\sin t\cos t = 1$
39. $\tan t\sin t = \sqrt{3}\sin t$
40. $\cos\theta + \sqrt{3}\sin\theta = -1$
41. $\tan^2\theta - \tan\theta = 0$
42. $\sin\theta + \cos\theta = 1$
43. $\cot^2 t + \csc^2 t = 3$
44. $\tan^4 t - 2\sec^2 t + 3 = 0$
45. $(\sin\theta - 1)(2\cos\theta + \sqrt{3}) = 0$
46. $\sec^2 t + \csc^2 t = \sec^2 t\csc^2 t$

[c] In problems 47–50, solve each trigonometric equation with the side condition $0° \le \theta < 360°$ or $0 \le t < 2\pi$. Use a calculator and round off all answers to four decimal places.

47. $7\sin^2 t - 10\sin t + 3 = 0$
48. $\cot^2\theta - 2\cot\theta - 3 = 0$
49. $3\sin^2\theta + \sqrt{3}\cos\theta = 3$
50. $8\sec^2 t + 2\tan t - 9 = 0$

51. An oscillating signal voltage is given by the equation $E = 75 \cos(\omega t - \phi)$ millivolts, with angular frequency $\omega = 120\pi$, phase angle $\phi = \pi/2$, and time t in seconds. This voltage is applied to an oscilloscope, and the triggering circuit of the oscilloscope starts the sweep when E reaches the value 35 millivolts. Find the smallest positive value of t for which triggering occurs.

52. An irrigation ditch has a cross section in the shape of an isosceles trapezoid that is wider at the top than at the bottom (Figure 5). The bottom and the equal sides of the trapezoid are each 2 meters long. If θ is the acute angle between the horizontal and the side of the ditch, (a) show that $A = 4 \sin \theta (1 + \cos \theta)$ gives the cross-sectional area of the ditch, and (b) solve for θ if $A = 3\sqrt{3}$ square meters.

Figure 5

53. The index of refraction of water with respect to air is $\mu = 1.333$. Use Snell's law (page 368) to determine the angle of refraction β of a ray of light that strikes the surface of a tank of water at an angle of incidence $\alpha = 15°$.

54. In optics, it is shown that if the index of refraction of a first medium with respect to a second is μ, then the index of refraction of the second medium with respect to the first is $1/\mu$. The index of refraction of flint glass with respect to air is $\mu = 1.650$. Figure 6 shows a light ray initiating in a block of flint glass and emerging into the air. The *critical angle* α_c for flint glass is the value of α for which $\beta = 90°$. (For $\alpha \geq \alpha_c$ the ray is totally reflected back into the glass when it strikes the surface.) Find the critical angle for flint glass.

Figure 6

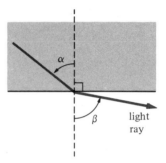

55. Suppose that in a certain tropical habitat, the simple predator–prey model (page 371) for boa constrictors (the predator) and wild pigs (the prey) is determined by the values $\omega = \frac{1}{2}$, $a_1 = 10$, $a_2 = 4$, $k_1 = 40$, $k_2 = 10$, and $\phi = \pi/3$, where t is the time in years measured from an instant when the prey population had its maximum value $x = 50$. (a) Find the values of t for which the boa population is $y = 8$. (b) Find the wild pig population x when $y = 8$.

REVIEW PROBLEM SET

In problems 1–10, simplify each expression.

1. $\dfrac{\sin(-\theta)}{\cos(-\theta)}$

2. $\dfrac{-\sin(-\alpha)}{-\cos(-\alpha)}$

3. $\csc x - \cos x \cot x$

4. $\sec \theta - \sin \theta \tan \theta$

5. $\csc^2 t \tan^2 t - 1$

6. $(\cot x + 1)^2 - \csc^2 x$

7. $\dfrac{\sec^2 u + 2 \tan u}{1 + \tan u}$

8. $\dfrac{\sec \beta}{\cot \beta + \tan \beta}$

9. $\dfrac{\sin^2 \theta + 2 \cos^2 \theta}{\sin \theta \cos \theta} - 2 \cot \theta$

10. $\dfrac{1}{\csc y - \cot y} - \dfrac{1}{\csc y + \cot y}$

In problems 11 and 12, rewrite each radical expression as a trigonometric expression containing no radical by making the indicated trigonometric substitution.

11. $\sqrt{(25 - x^2)^3}$, $x = 5 \sin \theta$, $-\dfrac{\pi}{2} \leq \theta \leq \dfrac{\pi}{2}$

12. $\dfrac{x}{\sqrt{x^2 - 4}}$, $x = 2 \sec t$, $0 \leq t < \dfrac{\pi}{2}$

In problems 13–26, prove that each equation is an identity.

13. $\dfrac{\sin \theta}{\csc \theta} - 1 = \dfrac{-1}{\sec^2 \theta}$

14. $\dfrac{\csc \alpha}{\cot \alpha + \tan \alpha} = \cos \alpha$

15. $\dfrac{1 - \cot^2 t}{\tan^2 t - 1} = \cot^2 t$

16. $\dfrac{1}{\tan \beta} + \dfrac{\sin \beta}{1 + \cos \beta} = \csc \beta$

17. $\dfrac{1 + \tan x}{1 + \cot x} = \dfrac{\sec x}{\csc x}$

18. $\dfrac{\sin y + \tan y}{\cot y + \csc y} = \dfrac{\sin y}{\cot y}$

19. $\dfrac{\sin \beta + \cos \beta}{\sec \beta + \csc \beta} = \dfrac{\sin \beta}{\sec \beta}$

20. $\dfrac{1 - (\sin t - \cos t)^2}{\sin t} = 2 \cos t$

21. $\dfrac{1 - \tan \theta}{1 + \tan \theta} = \dfrac{\cot \theta - 1}{\cot \theta + 1}$

22. $(\sec u - \tan u)^2 = \dfrac{1 - \sin u}{1 + \sin u}$

23. $\dfrac{\cos s}{\sec s - \tan s} = 1 + \sin s$

24. $\sqrt{\dfrac{\sec x - 1}{\sec x + 1}} = \dfrac{|\sin x|}{1 + \cos x}$

25. $\dfrac{\sin t}{\cot t} - \dfrac{\cos t}{\sec t} = \dfrac{\tan t - \cos t \cot t}{\csc t}$

26. $(\csc \omega + \sec \omega)^2 = \dfrac{\sec^2 \omega + 2 \tan \omega}{\sin^2 \omega}$

In problems 27–36, simplify each expression.

27. $\cos(360° - \theta)$

28. $\tan(2\pi - \beta)$

29. $\sin(270° + \alpha)$

30. $\cos(270° - \phi)$

31. $\sin(2\pi + t)$

32. $\cot\left(\dfrac{3\pi}{2} + x\right)$

33. $\sin 37° \cos 23° + \cos 37° \sin 23°$

34. $\dfrac{\tan(\pi/5) + \tan(\pi/20)}{1 - \tan(\pi/5)\tan(\pi/20)}$

35. $\sin x \cos y - \sin\left(x + \dfrac{\pi}{2}\right)\sin(-y)$

36. $\cos(\pi - t) - \tan t \cos\left(\dfrac{\pi}{2} - t\right)$

37. Use the fact that $\dfrac{7\pi}{12} = \dfrac{\pi}{4} + \dfrac{\pi}{3}$ to find the exact numerical value of

(a) $\sin \dfrac{7\pi}{12}$ (b) $\cos \dfrac{7\pi}{12}$ (c) $\tan \dfrac{7\pi}{12}$.

38. Use the fact that $75° = 120° - 45°$ to find the exact numerical value of
(a) $\sin 75°$ (b) $\cos 75°$ (c) $\sec 75°$.

In problems 39–48, assume that α is in quadrant IV, $\cos \alpha = \frac{3}{5}$, β is in quadrant I, $\sin \beta = \frac{8}{17}$, γ is in quadrant II, $\cos \gamma = -\frac{24}{25}$, θ is in quadrant II, and $\sin \theta = \frac{5}{13}$. Find the exact numerical value of each expression.

39. $\sin(\alpha + \beta)$

40. $\cos(\gamma + \theta)$

41. $\sin(\beta + \theta)$

42. $\sin(\alpha - \gamma)$

43. $\cos(\beta - \gamma)$

44. $\sin(\beta - \gamma)$

45. $\tan(\beta - \gamma)$

46. $\sec(\beta - \gamma)$

47. $\sin(\theta - \gamma)$

48. $\cos(\beta - \theta)$

In problems 49–56, prove that each equation is an identity.

49. $\sin(90° + \theta) = \sin(90° - \theta)$

50. $\sin\left(t + \dfrac{\pi}{6}\right) + \cos\left(t + \dfrac{\pi}{3}\right) = \cos t$

51. $\dfrac{\sin(s + t)}{\cos s \cos t} = \tan s + \tan t$

52. $\sin(\beta + 30°) + \cos(60° - \beta) = 2 \sin(\beta + 30°)$

53. $\dfrac{\cos(x - y)}{\cos x \sin y} = \tan x + \cot y$

54. $\tan(\alpha + 135°) = \dfrac{\tan \alpha - 1}{\tan \alpha + 1}$

55. $\sin(\alpha - \beta)\cos \beta + \cos(\alpha - \beta)\sin \beta = \sin \alpha$

56. $\tan(\alpha + \beta)\tan(\alpha - \beta) = \dfrac{\tan^2 \alpha - \tan^2 \beta}{1 - \tan^2 \alpha \tan^2 \beta}$

57. If $\sec x = \frac{25}{7}$ and $0 < x < \pi/2$, find the exact numerical value of
(a) $\sin 2x$ (b) $\cos 2x$ (c) $\cos(x/2)$ (d) $\tan(x/2)$.

58. If $\tan\theta = \frac{5}{12}$ and $180° < \theta < 270°$, find the exact numerical value of
(a) $\sin(\theta/2)$ (b) $\cos(\theta/2)$ (c) $\cos 2\theta$, (d) $\tan 2\theta$.

In problems 59–68, simplify each expression.

59. $\cos^2 2x - \sin^2 2x$

60. $1 - 2\sin^2\dfrac{t}{2}$

61. $2\sin\dfrac{t}{2}\cos\dfrac{t}{2}$

62. $\cos^4 2\theta - \sin^4 2\theta$

63. $2\sin^2\dfrac{\theta}{2} + \cos\theta$

64. $\dfrac{\sin 4\pi t}{4\sin\pi t\cos\pi t}$

65. $\dfrac{\tan\omega t}{1 - \tan^2\omega t}$

66. $\dfrac{2\tan(t/2)}{1 - \tan^2(t/2)}$

67. $\dfrac{\cos^2(v/2) - \cos v}{\sin^2(v/2)}$

68. $2\sin 2x\cos^3 2x + 2\sin^3 2x\cos 2x$

In problems 69–84, prove that each equation is an identity.

69. $\dfrac{\sin 2\theta + \sin\theta}{\cos 2\theta + \cos\theta + 1} = \tan\theta$

70. $\dfrac{\sec x - 1}{2\sec x} = \sin^2\dfrac{x}{2}$

71. $\tan t + \cot t = 2\csc 2t$

72. $\csc w - \cot w = \tan\dfrac{w}{2}$

73. $\dfrac{\cos^2\dfrac{x}{2} - \cos x}{\sin^2\dfrac{x}{2}} = 1$

74. $\dfrac{\tan\theta - \sin\theta}{\sin\theta\sec\theta} = \tan\dfrac{\theta}{2}\sin\theta$

75. $8\sin^2\dfrac{\theta}{2}\cos^2\dfrac{\theta}{2} = 1 - \cos 2\theta$

76. $\cos^4\dfrac{x}{2} + \sin^4\dfrac{x}{2} = 1 - \dfrac{1}{2}\sin^2 x$

77. $\dfrac{\tan x + \sin x}{2\tan x} = \cos^2\dfrac{x}{2}$

78. $\dfrac{\tan\alpha - \sin\alpha}{2\tan\alpha} = \sin^2\dfrac{\alpha}{2}$

79. $\sin 4t = 4\sin t\cos t(1 - 2\sin^2 t)$

80. $\cos 4w = 8\cos^4 w - 8\cos^2 w + 1$

81. $4\sin\beta\cos^2\dfrac{\beta}{2} = \sin 2\beta + 2\sin\beta$

82. $\dfrac{1 - \cos x - \tan^2\dfrac{x}{2}}{\sin^2\dfrac{x}{2}} = \dfrac{2\cos x}{1 + \cos x}$

83. $\dfrac{\sin 3x}{\sin x} + \dfrac{\cos 3x}{\cos x} = \sin 4x\sec x\csc x$

84. $\dfrac{\cos 3u}{\sin u} - \dfrac{\sin 3u}{\cos u} = \dfrac{\cos 4u}{\sin u\cos u}$

In problems 85–88, express each product as a sum or difference.

85. $\sin\dfrac{3x}{2}\cos\dfrac{x}{2}$

86. $\cos 4\beta\sin 2\beta$

87. $\sin 37.5°\sin 7.5°$

88. $\sin 75°\cos 15°$

In problems 89–92, rewrite each expression as a product.

89. $\sin 55° + \sin 5°$

90. $\cos\dfrac{5\pi}{3} + \cos\dfrac{\pi}{12}$

91. $\sin 4\beta - \sin\beta$

92. $\sin 11t + \sin 5t$

In problems 93–98, prove that each equation is an identity.

93. $\dfrac{\cos 5x + \cos 3x}{\sin 5x - \sin 3x} = \cot x$

94. $\dfrac{2\sin 2\theta\cos\theta - \sin\theta}{\cos\theta - 2\sin 2\theta\sin\theta} = \tan 3\theta$

95. $4\sin 3t\cos 3t\sin t = \cos 5t - \cos 7t$

96. $4\sin 4\alpha\cos 2\alpha\sin\alpha = \cos\alpha - \cos 3\alpha + \cos 5\alpha - \cos 7\alpha$

97. $\dfrac{\sin\theta + \sin 3\theta + \sin 5\theta}{\cos\theta + \cos 3\theta + \cos 5\theta} = \tan 3\theta$

98. $2\sin\left(\dfrac{\alpha - \beta}{2} + \dfrac{\pi}{4}\right)\cos\left(\dfrac{\alpha + \beta}{2} - \dfrac{\pi}{4}\right) = \sin\alpha + \cos\beta$

C In problems 99–102, use addition or subtraction of ordinates to sketch each graph. (A calculator will be helpful.)

99. $y = 2\sin 3x + 3\cos 2x$

100. $y = 0.12\sin\pi t + 0.6\sin 2\pi t$

101. $y = \sin t - \sin\dfrac{t}{3}$

102. $y = \sin x + \tan x$

C **103.** The variations in air pressure caused at a certain point by simultaneously playing the pure tone "concert A" and the pure tone one octave higher are given by

$$y = 0.12\sin 880\pi t + 0.06\sin 1760\pi t,$$

where t is the time in seconds. Sketch the graph showing these air-pressure variations.

C **104.** Believers in biorhythms sometimes plot a *composite* biorhythm curve by adding all three ordinates (physical, emotional, and intellectual) on a biorhythm chart. Using the data in problem 39 on page 322, sketch a composite biorhythm curve over a 62-day period.

C **105.** Use the product formula to rewrite the quantity y given by $y = 4\sin t\cos(4t/3)$ as a superposition of two simple harmonic oscillations. Then plot a graph of y as a function of t using the method of addition of ordinates.

106. The equation for a radio frequency carrier wave with frequency v_r, which is amplitude modulated by a pure audio tone with frequency v_a, is

$$y = a(1 + m\sin 2\pi v_a t)\sin 2\pi v_r t,$$

where a and m are constants. Rewrite this equation to show that y is a superposition of simple harmonic oscillations.

In problems 107–110, use Theorem 3 on page 354 to rewrite the given equation in the form $y = a\cos(\omega t - \phi)$ and then sketch the graph.

107. $y = 2\sin\left(\dfrac{\pi}{3}t - \pi\right) - 2\cos\left(\dfrac{\pi}{3}t - \pi\right)$

C **108.** $y = 4\cos\left(\dfrac{\pi}{2}t - \dfrac{\pi}{3}\right) - 3\sin\left(\dfrac{\pi}{2}t - \dfrac{\pi}{3}\right)$

109. $y = \sqrt{3}\cos(2t - 3) - \sin(2t - 3)$

C **110.** $y = 2\cos(3t - 1) - 3\sin(3t - 1)$

In problems 111 and 112, find the period T of the oscillation represented by each equation.

111. $y = 2\cos\left(\dfrac{\pi}{3}t - \dfrac{\pi}{4}\right) + \sqrt{3}\cos\left(\dfrac{\pi}{5}t - \dfrac{\pi}{3}\right) + 2$

112. $y = \sqrt{2}\cos\left(\dfrac{\sqrt{3}}{2}t - \sqrt{5}\right) - \sin\left(\dfrac{\sqrt{3}}{3}t - \sqrt{3}\right) - 1$

113. Write the oscillation represented by $y = 2\cos\pi t\cos 3\pi t$ as a superposition of two simple harmonic oscillations and determine its period T.

114. Figure 1 illustrates the phenomenon of *beats* or *hetrodyning* which occurs when two sounds of slightly different frequencies v_1 and v_2 are superposed. If the sounds are represented by $y_1 = a\cos(2\pi v_1 t)$ and $y_2 = a\cos(2\pi v_2 t)$, show that the superposition $y = y_1 + y_2$ can be regarded as a sound of frequency $(v_1 + v_2)/2$ amplitude modulated by an oscillation of frequency $|v_1 - v_2|/2$. Explain why one hears "beats" with a frequency of $|v_1 - v_2|$.

Figure 1

In problems 115–118, evaluate each expression without the use of a calculator or tables.

115. $\sin^{-1}(-\tfrac{1}{2})$ **116.** $\arccos(-\sqrt{3}/2)$ **117.** $\arctan\sqrt{3}$ **118.** $\operatorname{arcsec}\sqrt{2}$

© In problems 119–128, use a calculator (or Appendix Table IIB) to evaluate each expression. Give answers in radians.

119. $\sin^{-1}0.3750$ **120.** $\arccos(-0.3901)$ **121.** $\cos^{-1}0.9273$ **122.** $\tan^{-1}57.29$
123. $\arctan 1.425$ **124.** $\cos^{-1}\tfrac{1}{8}$ **125.** $\arcsin(-\tfrac{3}{4})$ **126.** $\tan^{-1}8$
127. $\cos^{-1}(-\tfrac{10}{11})$ **128.** $\arctan 1.007$

In problems 129–134, find the exact value of each expression without using a calculator or tables.

129. $\cos(\tan^{-1}\tfrac{4}{3})$ **130.** $\sin[\arctan(-\tfrac{5}{12})]$ **131.** $\sin[\sin^{-1}(-\tfrac{12}{13})]$
132. $\tan[\arccos(-\tfrac{3}{5})]$ **133.** $\arcsin[\sin(19\pi/14)]$ **134.** $\sin(\sin^{-1}\tfrac{2}{3}+\sin^{-1}\tfrac{3}{4})$

In problems 135–138, show that the given equation is an identity.

135. $\sin(\cos^{-1}x)=\sqrt{1-x^2}$ **136.** $\sin(\tfrac{1}{2}\arccos x)=\sqrt{\dfrac{1-x}{2}}$

137. $\tan(\arccos x)=\dfrac{\sqrt{1-x^2}}{x}$ **138.** $\tan(\tan^{-1}x+\tan^{-1}y)=\dfrac{x+y}{1-xy}$

139. A picture a meters high hangs on a wall so that its bottom is b meters above the eye level of an observer. If the observer stands x meters from the wall (Figure 2), show that the angle θ subtended by the picture is given by

$$\theta = \arctan\frac{a+b}{x} - \arctan\frac{b}{x}.$$

Figure 2

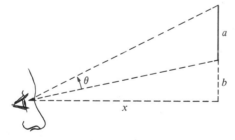

140. Two parallel lines, each at a distance a from the center of a circle of radius r, cut off the region of area A shown in Figure 3. Show that

$$A = 2a\sqrt{r^2-a^2} + 2r^2\arcsin\frac{a}{r}.$$

Figure 3

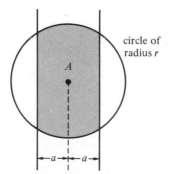

circle of radius r

A

In problems 141–152, solve each trigonometric equation with the side condition $0° \leq \theta < 360°$ or $0 \leq t < 2\pi$. Do not use a calculator or tables.

141. $\tan \theta = \dfrac{\sqrt{3}}{3}$

142. $4 \sin^2 t - 3 = 0$

143. $4 \cos^2 t - 3 = 0$

144. $\tan^2 \theta + (1 - \sqrt{3}) \tan \theta - \sqrt{3} = 0$

145. $2 \sin^2 \theta + \sqrt{2} \sin \theta = 0$

146. $\tan t = \sec t + 1$

147. $\cot^2 t + 3 \csc t + 3 = 0$

148. $\cot \theta = \csc \theta - 1$

149. $\tan 4\theta = \sqrt{3}$

150. $\sin 3t \cos t = \sin t \cos 3t + 1$

151. $\sin 3\theta + \sin \theta = 0$

152. $\cos \dfrac{t}{4} = \dfrac{\sqrt{3}}{2}$

© In problems 153–156, solve each trigonometric equation with the side condition $0° \leq \theta < 360°$ or $0 \leq t < 2\pi$. Use a calculator and round off all answers to four decimal places.

153. $\cos 2\theta = \frac{3}{5}$

154. $\cot \dfrac{\theta}{2} = -2.017$

155. $6 \sin^2 t + \sin t - 2 = 0$

156. $2 \sec^2 t + \tan t - 3 = 0$

157. The index of refraction of a diamond with respect to air is $\mu = 2.417$. Use Snell's law (page 368) to determine the angle of refraction β of a ray of light that strikes the surface of a diamond at an angle of incidence $\alpha = 20°$.

158. A ray of light passing through a plate of material with parallel faces is displaced, but not deviated (Figure 4); that is, the emerging ray is parallel to the ingoing ray. If d is the amount of displacement, t the thickness of the plate, α the angle of incidence, and β the angle of refraction, show that

$$\beta = \arctan\left(\tan \alpha - \frac{d}{t} \sec \alpha\right).$$

Figure 4

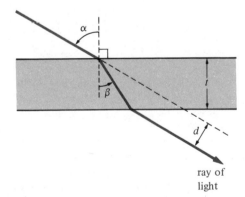

ray of light

8 Applications of Trigonometry

In this chapter, we study some of the applications of the trigonometric functions to geometry. We begin by "solving triangles," that is, finding certain parts of triangles when other parts are known. To do this, we derive and apply the *law of sines* and the *law of cosines*. The chapter also includes an introduction to vectors and a discussion of polar coordinates.

1 Right Triangles

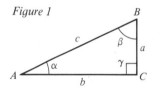

Figure 1

In the right triangle ACB shown in Figure 1, the angles are denoted by α at vertex A, β at vertex B, and γ at vertex C. The lengths of the sides opposite angles α, β, and γ are denoted by a, b, and c. Note that angles α and β are complementary acute angles, that angle γ is a right angle, and that c is the hypotenuse of right triangle ACB. Therefore,

$$\sin \alpha = \cos \beta = \frac{a}{c} \qquad \csc \alpha = \sec \beta = \frac{c}{a}$$

$$\cos \alpha = \sin \beta = \frac{b}{c} \qquad \sec \alpha = \csc \beta = \frac{c}{b}$$

$$\tan \alpha = \cot \beta = \frac{a}{b} \qquad \cot \alpha = \tan \beta = \frac{b}{a}.$$

If the lengths of two sides of a right triangle are given, or if one side and an acute angle are given, then these formulas can be used to solve for the remaining angles and sides of the triangle. This procedure is called **solving the right triangle.**

In solving a right triangle, it is necessary to use a calculator or a table of trigonometric functions, unless the special angles 30°, 45°, or 60° are involved. You should always keep in mind that solutions obtained using either a calculator or tables are *approximations*. In this section, we round off all angles to the nearest hundredth of a degree, and all side lengths to four significant digits.

Examples   © Assume that right triangle ACB is labeled as in Figure 1. In each case, solve the triangle and sketch it.

1 $b = 31, \quad \alpha = 43.33°$

Solution   We must find a, c, and β. Because α and β are complementary,

$$\beta = 90° - \alpha = 90° - 43.33° = 46.67°.$$

To find a, we notice that

$$\tan \alpha = \frac{a}{b}, \quad \text{so} \quad a = b \tan \alpha.$$

Using a calculator, we find that

$$\tan \alpha = \tan 43.33° = 0.943341386.$$

Therefore, rounded off to four significant digits,

$$a = b \tan \alpha = 31 \tan 43.33° = 29.24.$$

(Note that it isn't necessary to write down the intermediate result $\tan 43.33° = 0.943341386$. We have done it here so you can check your calculator work.) To find c, we notice that

$$\cos \alpha = \frac{b}{c}, \quad \text{so} \quad c = \frac{b}{\cos \alpha}.$$

Therefore, rounded off to four significant digits,

$$c = \frac{b}{\cos \alpha} = \frac{31}{\cos 43.33°} = \frac{31}{0.727413564} = 42.62.$$

(Again, if you are using your calculator efficiently, it won't be necessary to write down the fraction $31/0.727413564$. You should be able to calculate $31/\cos 43.33°$ directly with only a few key strokes.) Notice that we could have used the Pythagorean theorem $c^2 = a^2 + b^2$ to solve for c:

$$c = \sqrt{a^2 + b^2} = \sqrt{(29.24)^2 + 31^2}$$
$$= 42.61.$$

(The discrepancy in the last decimal place was caused by using the rounded off value for a. The result $c = 42.62$ is actually more accurate.) The triangle is shown in Figure 2.

Figure 2

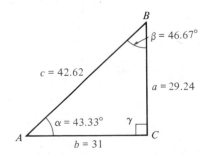

2 $a = 4, \quad b = 3$

Solution   By the Pythagorean theorem,

$$c = \sqrt{a^2 + b^2} = \sqrt{16 + 9} = 5.$$

Now we have

$$\sin \alpha = \frac{a}{c} = \frac{4}{5} = 0.8,$$

so

$$\alpha = \sin^{-1} 0.8 = 53.13°.$$

It follows that

$$\beta = 90° - \alpha = 90° - 53.13° = 36.87°.$$

The triangle is shown in Figure 3.

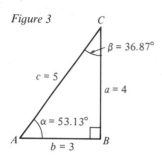

Figure 3

In many applications of trigonometry, we represent the relevant features of a real-world situation with triangles, and then solve the triangles for their unknown parts.

Example   © A railway track rises 300 feet per mile (5280 feet). Find the angle α at which the track is inclined from the horizontal (Figure 4).

Solution   From Figure 4,

$$\tan \alpha = \tfrac{300}{5280},$$

so

$$\alpha = \tan^{-1} \tfrac{300}{5280} = 3.25°.$$

Figure 4

The acute angle formed by a horizontal line and an observer's line of sight to any object above the horizontal is called an **angle of elevation** (Figure 5a). Similarly, the acute angle formed by a horizontal line and an observer's line of sight to any object below the horizontal is called an **angle of depression** (Figure 5b).

Figure 5

(a)

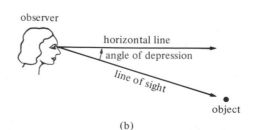

(b)

Example 1 © From a point on level ground 75 meters from the base of a television transmitting tower, the angle of elevation of the top of the tower is 68.17°. Find the height h of the tower (Figure 6).

Solution In Figure 6, $\tan 68.17° = \dfrac{h}{75}$, so,

Figure 6

$$h = 75 \tan 68.17° = 187.2 \text{ meters.}$$

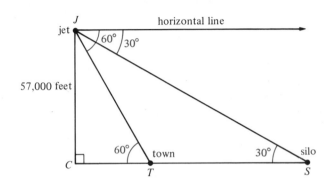

Figure 7

Example 2 © A high-altitude military-reconnaissance jet photographs a missile silo under construction near a small town. The jet is at an altitude of 57,000 feet, and the angles of depression of the town and silo are 60° and 30°, respectively. Assuming that the jet, the silo, and the town lie in the same vertical plane, find the distance between the town and the silo (Figure 7).

Solution In Figure 7, we want to find $|\overline{TS}|$, which is $|\overline{CS}| - |\overline{CT}|$. From elementary geometry, we know that alternate interior angles are equal, so angle $CTJ = 60°$ and angle $CSJ = 30°$. In right triangle TCJ,

$$\tan 60° = \frac{|\overline{CJ}|}{|\overline{CT}|} = \frac{57{,}000}{|\overline{CT}|},$$

so

$$|\overline{CT}| = \frac{57{,}000}{\tan 60°} = \frac{57{,}000}{\sqrt{3}} = \frac{57{,}000\sqrt{3}}{3} = 19{,}000\sqrt{3}.$$

In right triangle SCJ,

$$\tan 30° = \frac{|\overline{CJ}|}{|\overline{CS}|} = \frac{57{,}000}{|\overline{CS}|},$$

so

$$|\overline{CS}| = \frac{57{,}000}{\tan 30°} = \frac{57{,}000}{\sqrt{3}/3} = 57{,}000\sqrt{3}.$$

Hence,

$$|\overline{TS}| = |\overline{CS}| - |\overline{CT}| = 57{,}000\sqrt{3} - 19{,}000\sqrt{3} = 38{,}000\sqrt{3} \approx 65{,}820 \text{ feet.}$$

In some applications of trigonometry, especially in surveying and navigation, the **direction** or **bearing** of a point Q as viewed from a point P is defined to be the positive angle the ray from P through Q makes with a north–south line through P. Such an angle is specified as being measured east or west from north or south. For instance, in Figure 8, the bearing from P to Q is 61° east of north, or N61°E. Similarly, the bearing from P to R is N28°W; the bearing from P to T is S15°W; and the bearing from P to V is S40°E.

Figure 8

Figure 9

Example   C From an observation point P, a forest ranger sights a fire F directly to the north. Another ranger at point Q, 5 kilometers due east of P, sights the same fire at a bearing N21.15°W (Figure 9). Find the distance d between P and the fire.

Solution   In Figure 9, angle PQF is complementary to the 21.15° angle; therefore,

$$\text{angle } PQF = 90° - 21.15°$$
$$= 68.85°.$$

Also,

$$\tan 68.85° = \frac{d}{5},$$

so

$$d = 5 \tan 68.85° = 12.92.$$

Therefore, the fire is approximately 12.92 kilometers due north of observation point P.

© **PROBLEM SET 1**

In problems 1–18, assume that the right triangle ACB is labeled as in Figure 10. In each case solve the triangle and sketch it. Round off all angles to the nearest hundredth of a degree and all side lengths to four significant digits.

1. $a = 5$, $b = 4$
3. $a = 10$, $c = 15$
5. $b = 1700$, $\alpha = 31.23°$
7. $a = 7.132$, $c = 9.209$
9. $a = 3.32$, $b = 4331$
11. $b = 2570$, $\beta = 15.45°$
13. $b = 5.673 \times 10^{-3}$ meter, $\alpha = 67°45'$
15. $b = 30.73$ miles, $c = 77.17$ miles
17. $a = 4.932 \times 10^{-6}$ meter, $b = 4.101 \times 10^{-6}$ meter

2. $b = 91$, $\beta = 30°$
4. $c = 100$, $\beta = 41°$
6. $c = 10^4$, $\beta = 17.81°$
8. $b = 0.01523$, $\beta = 11.3°$
10. $a = 113.5$, $c = 217.0$
12. $a = 50$, $b = 120$
14. $a = 8.141 \times 10^5$ meters, $\alpha = 1.10°$
16. $a = 9.200 \times 10^7$ kilometers, $\alpha = 51°33'$
18. $c = 1.410 \times 10^{15}$ kilometers, $\alpha = 62°13'$

Figure 10

19. A 30-foot ladder leaning against a vertical wall just reaches a window sill. If the ladder makes an angle of 47° with the level ground, how high is the window sill? (Round off your answer to the nearest foot.)
20. A guy wire 8 meters long helps support a CB base antenna mounted on top of a flat roof. If the wire makes an angle of 49.5° with the horizontal roof, how far above the roof is it attached to the antenna? (Round off your answer to two decimal places.)
21. A kite string makes an angle of 28° with the level ground, and 73 meters of string is out. How high is the kite? (Round off your answer to the nearest meter.)
22. An engineer is designing an access ramp for an elevated expressway that is 40 feet above level ground. The ramp must be straight and cannot be inclined more than 15° from the horizontal. Because of existing structures near the expressway, the horizontal distance between the beginning of the ramp and the expressway cannot exceed 150 feet. Can the engineer design such a ramp?
23. A jetliner is climbing so that its path is a straight line that makes an angle of 8.5° with the horizontal. How many meters does the jetliner rise while traveling 300 meters along its path? (Round off your answer to the nearest meter.)
24. A portion of a tunnel under a river is straight for 150 meters and descends 10 meters in this distance. (a) To the nearest hundredth of a degree, what angle does this part of the tunnel make with the horizontal? (b) What is the *horizontal* distance, to the nearest 0.1 meter, between the ends of this part of the tunnel?
25. A monument 22 meters high casts a shadow 31 meters long. Find the angle of elevation of the sun to the nearest tenth of a degree.
26. To measure the height of a cloud cover at night, a spotlight is aimed straight upward from the ground. The resulting spot of light on the clouds is viewed from a point on the level ground 850 meters from the spotlight, and the angle of elevation is measured at 61.8°. Find the height of the cloud cover to the nearest meter.
27. A wrecking company has contracted to knock down an old brick chimney. In order to determine in advance just where the chimney will fall, it is necessary to find its height. The top of the chimney is viewed from a point 150 meters from its base, and the angle of elevation is measured at 33.25°. Find the height of the chimney, rounded off to the nearest 0.1 meter.
28. A rectangular panel to collect solar energy rests on flat ground and is tilted toward the sun. The edge resting on the ground is 3.217 meters long, and the upper edge is 1.574 meters above the ground. The panel is located near Chicago and its latitude is 41°50'. Solar engineers recommend that the angle between the panel and the ground be equal to the latitude of its location. Assuming that this recommendation has been followed, find the surface area of the panel. (Round off your answer to three decimal places.)

29. A lifeguard is seated on a high platform so that her eyes are 7 meters above sea level. Suddenly she spots the dorsal fin of a great white shark at a 4° angle of depression. Estimate, to the nearest meter, the horizontal distance between the platform and the shark.

30. A customs officer located on a straight shoreline observes a smuggler's motorboat making directly for the closest point on shore, an abandoned lighthouse. The angle between the shoreline and the officer's line of sight to the motorboat is 34.6°, the motorboat is traveling at 18 knots (nautical miles per hour), and it is 6 nautical miles from the lighthouse. The officer immediately departs by car on a straight road along the shoreline, hoping to reach the lighthouse 10 minutes before the smugglers in order to apprehend them as they land. How fast must the officer drive? (One nautical mile is 6076 feet, whereas one statute, or ordinary, mile is 5280 feet.)

31. Biologists studying the migration of birds are following a migrating flock in a light plane. The birds are flying at a constant altitude of 1200 feet and the plane is following at a constant altitude of 1700 feet. The biologists must maintain a distance of at least 600 feet between the plane and the flock in order to avoid disturbing the birds; therefore, they must monitor the angle of depression of the flock from the plane. Find the maximum allowable angle of depression, rounded off to the nearest degree.

32. An observer on a bluff 100 meters above the surface of a Scottish loch sees the head of an aquatic monster at an angle of depression of 18.45°. The monster, swimming directly away from the observer, immediately submerges. Five minutes later, the monster's head reappears, now at an angle of depression of 14.05°. How fast is the monster swimming?

33. A nature photographer using a telephoto lens photographs a rare bird roosting on a high branch of a tree at an angle of elevation of 22.5°. The distance between the lens and the bird is 330 feet. In order to obtain a more detailed photograph of the bird, the photographer cautiously moves closer to the base of the tree. The angle of elevation of the bird is now 51.25°. Find the new distance between the photographer's lens and the bird.

34. The angle of elevation of the top of a tower from a point 100 meters from the top is measured to be α degrees. Thus, the height h of the tower is calculated to be $h = 100 \sin \alpha$ meters. Suppose, however, that a 0.1° error has been made, and that the true angle of elevation is $\alpha + 0.1°$. Then the true height of the tower is $100 \sin(\alpha + 0.1°)$, and the error E in the calculated value of h is given by $E = 100 \sin(\alpha + 0.1°) - 100 \sin \alpha$, the true value minus the calculated value. Using a calculator, find E if (a) $\alpha = 20°$, (b) $\alpha = 40°$, and (c) $\alpha = 60°$. (Round off your answer to two decimal places.)

35. A lifeguard at station A sights a swimmer directly south of him. Another lifeguard at station B, 230 feet directly east of A, sights the same swimmer at a bearing of S49.5°W. How far is the swimmer from station A?

36. Refer to problem 34. (a) Sketch a graph of E as a function of α for $0 < \alpha < 90°$. (b) Does the error E increase or decrease as α increases? (c) If you know that your measurement of α is subject to an error of as much as 0.1°, to how many decimal places should you round off your calculated value of h?

37. An airplane A is in the air at a position due west of an airport, and another airplane B is 15 miles south of A. From B, the bearing of the airport is N58.5°E. How far is airplane A from the airport?

38. To find the width $|\overline{PQ}|$ of a lake, a surveyor measures 3000 feet from P in the direction of N41°40′W to locate point R. The surveyor then determines that the bearing of Q from R is N71°30′E. Find the width of the lake if the point Q is located so that angle QPR is a right angle.

39. At 1:00 P.M. a ship is 24 nautical miles directly east of a lighthouse. The ship is sailing due north at 32 knots (nautical miles per hour). What is the bearing of the lighthouse from the ship at 3:00 P.M.?

2   The Law of Sines

In this section and the next, we shall be studying relationships among the three angles α, β, and γ and the opposite sides a, b, and c of a *general* triangle ABC (Figure 1). The following theorem, called the *law of sines*, relates the lengths of the three sides to the sines of the three vertex angles.

Figure 1

Theorem 1   **The Law of Sines**

> In the general triangle ABC in Figure 1,
>
> $$\frac{\sin \alpha}{a} = \frac{\sin \beta}{b} = \frac{\sin \gamma}{c}$$

Proof   We show that $\dfrac{\sin \alpha}{a} = \dfrac{\sin \beta}{b}$ and leave the similar proof that $\dfrac{\sin \beta}{b} = \dfrac{\sin \gamma}{c}$ as an exercise (Problem 38). Drop a perpendicular $\overline{CD}$ from vertex C to the straight line ℓ containing the vertices A and B. Figure 2 shows the case in which both angles α and β are acute, so that D lies *between* A and B on ℓ. [If either α or β is not acute, D falls outside of the segment $\overline{AB}$ on ℓ, and the following argument has to be modified slightly (Problem 37).] In right triangles ADC and BDC, we have

$$\sin \alpha = \frac{|\overline{CD}|}{b} \quad \text{and} \quad \sin \beta = \frac{|\overline{CD}|}{a},$$

so

$$b \sin \alpha = |\overline{CD}| \quad \text{and} \quad a \sin \beta = |\overline{CD}|.$$

Consequently,

$$b \sin \alpha = a \sin \beta.$$

Dividing both sides of the last equation by ab, we obtain

$$\frac{\sin \alpha}{a} = \frac{\sin \beta}{b}.$$

Figure 2

Notice that the law of sines (Theorem 1) can be written in the alternative form

$$\frac{a}{\sin \alpha} = \frac{b}{\sin \beta} = \frac{c}{\sin \gamma}.$$

If two angles of triangle ABC are given, then the third angle can be found by using the relationship

$$\alpha + \beta + \gamma = 180°;$$

hence, the three denominators $\sin \alpha$, $\sin \beta$, and $\sin \gamma$ can be found by using a calculator (or a table of sines). Now, if any one of the sides a, b, or c is also given, then the equations

$$\frac{a}{\sin \alpha} = \frac{b}{\sin \beta} = \frac{c}{\sin \gamma}$$

can be solved for the remaining two sides.

The following examples illustrate the procedure for solving a triangle when *two angles and one side* are given (or can be determined from information provided). Unless otherwise indicated, we shall round off all angles to the nearest hundredth of a degree, and all side lengths to four significant digits.

Example 1   In triangle ABC (Figure 3), suppose that $\alpha = 41°$, $\beta = 77°$, and $a = 74$. Solve for γ, b, and c.

Figure 3

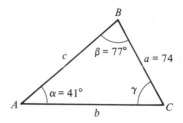

Solution

$$\gamma = 180° - \alpha - \beta = 180° - 41° - 77° = 62°$$

By the law of sines,

$$\frac{a}{\sin 41°} = \frac{b}{\sin 77°} = \frac{c}{\sin 62°};$$

hence, since $a = 74$,

$$b = \frac{a \sin 77°}{\sin 41°} = \frac{74 \sin 77°}{\sin 41°} = 109.9$$

and

$$c = \frac{a \sin 62°}{\sin 41°} = \frac{74 \sin 62°}{\sin 41°} = 99.59.$$

Example 2   A statue of height 70 meters ($\overline{BC}$ in Figure 4) stands atop a hill of height h. From a point A at ground level, the angle of elevation of the base B of the statue is $20.75°$ and the angle of elevation of the top C of the statue is $28.30°$. Find the height h.

Figure 4

Solution   In right triangle ADB, we have

$$\sin 20.75° = h/c$$

so

$$h = c \sin 20.75°.$$

Our plan is to find c and then use the last equation to find h. We know the length of one side $\overline{BC}$ in triangle ABC, so we can use the law of sines to find $c = |\overline{AB}|$, provided that we can find two of the vertex angles in this triangle. Angle α is easily found:

$$\alpha = 28.30° - 20.75° = 7.55°.$$

In right triangle ADC, angle γ is complementary to angle DAC; hence,

$$\gamma = 90° - \text{angle } DAC = 90° - 28.30° = 61.70°.$$

Now we can apply the law of sines to triangle ABC to obtain

$$\frac{c}{\sin \gamma} = \frac{70}{\sin \alpha}$$

so that

$$c = \frac{70 \sin \gamma}{\sin \alpha} = \frac{70 \sin 61.70°}{\sin 7.55°},$$

and it follows that

$$h = c \sin 20.75° = \left(\frac{70 \sin 61.70°}{\sin 7.55°}\right) \sin 20.75° = 166.2 \text{ meters}.$$

Example 3   Solve the triangle ABC if $\alpha = 82.17°$, $\gamma = 103.50°$, and $b = 615$.

Solution   We have

$$\beta = 180° - (\alpha + \gamma) = 180° - 185.67° = -5.67°,$$

which is impossible because we cannot have a triangle with a negative vertex angle. We conclude that there is no triangle satisfying the given conditions.

More generally, if the specifications for a triangle require that the sum of two vertex angles exceeds 180°, then no such triangle will exist.

2.1 The Ambiguous Case

Because there are several possibilities, the situation in which you are given the *lengths of two sides* of a triangle and the *angle opposite one of them* is called the **ambiguous case.** For instance, suppose you are given side a, side b, and angle α in triangle ABC. You might try to construct triangle ABC from this information by drawing a line segment $\overline{AC}$ of length b and a ray ℓ that starts at A and makes an angle α with $\overline{AC}$ (Figure 5). To find the remaining vertex B, you could use a compass to draw an arc of a circle of radius a with center C. If the arc intersects the ray ℓ at point B, then ABC is the desired triangle. (Why?)

Figure 5

As Figure 6 illustrates, there are actually *three* possibilities if you try to construct triangle ABC by the method above: The circle does not intersect the ray ℓ at all and there is **no triangle ABC** (Figure 6a); it intersects the ray ℓ in exactly one point B and there is just **one triangle ABC** (Figures 5 and 6b); or it intersects the ray ℓ in two points B_1 and B_2 and there are **two triangles AB_1C and AB_2C** that satisfy the given conditions (Figure 6c).

Figure 6

(a)

(b)

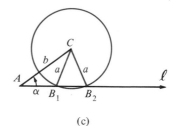

(c)

In the ambiguous case, you can use a calculator (or Appendix Table IIA) to solve the triangle ABC without bothering to draw any diagrams at all. Just use the law of sines,

$$\frac{\sin \alpha}{a} = \frac{\sin \beta}{b},$$

to evaluate $\sin \beta$:

$$\sin \beta = b\,\frac{\sin \alpha}{a}, \qquad 0 < \beta < 180°.$$

Recall that the sine of an angle is never greater than 1; hence, if $b\,\dfrac{\sin \alpha}{a} > 1$, this trigonometric equation has no solution, and *no triangle satisfies the given conditions.* If $b\,\dfrac{\sin \alpha}{a} = 1$, the equation has only one solution, $\beta = 90°$. If $b\,\dfrac{\sin \alpha}{a} < 1$, the trigonometric equation has two solutions (Problem 39), namely

$$\beta_1 = \sin^{-1}\!\left(b\,\frac{\sin \alpha}{a}\right) \quad \text{and} \quad \beta_2 = 180° - \beta_1.$$

Once you have determined β (or β_1 and β_2), you know two angles and two sides of the triangle (or triangles), and you can solve the triangle by using the methods previously explained. Even if there are two solutions β_1 and β_2 of the trigonometric equation for β, it is possible that only one of these solutions corresponds to an actual triangle satisfying the given conditions (see Figure 6b).

Examples   Solve each triangle by using the method for the ambiguous case.

1   © $a = 95$, $b = 117$, $\alpha = 65°$

Solution   Here,

$$b\,\frac{\sin\alpha}{a} = 117\,\frac{\sin 65°}{95} = 1.116 > 1,$$

so there is *no triangle* satisfying the given conditions.

2 $a = 10$, $b = 20$, $\alpha = 30°$

Solution   Because

$$\sin\beta = b\,\frac{\sin\alpha}{a} = 20\,\frac{\sin 30°}{10} = 20\,\frac{1/2}{10} = 1,$$

it follows that $\beta = 90°$, and the triangle is a right triangle with hypotenuse $b = 20$ (Figure 7). Therefore,

$$\gamma = 90° - \alpha = 90° - 30° = 60°.$$

Also, by the Pythagorean theorem,

$$c^2 + a^2 = b^2,$$

so

$$c = \sqrt{b^2 - a^2}$$
$$= \sqrt{20^2 - 10^2}$$
$$= \sqrt{300} = 10\sqrt{3}.$$

Figure 7

3   © $a = 10$, $b = 11$, $\alpha = 57°$

Solution   Here

$$b\,\frac{\sin\alpha}{a} = 11\,\frac{\sin 57°}{10} = 0.9225 < 1.$$

Thus,

$$\beta_1 = \sin^{-1}\left(b\,\frac{\sin\alpha}{a}\right) = \sin^{-1}\left(11\,\frac{\sin 57°}{10}\right) = 67.30°$$

and

$$\beta_2 = 180° - \beta_1 = 180° - 67.30° = 112.70°.$$

Figure 8

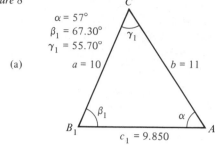

(a)

$\alpha = 57°$
$\beta_1 = 67.30°$
$\gamma_1 = 55.70°$
$a = 10$
$b = 11$
$c_1 = 9.850$

(b)

$\alpha = 57°$
$\beta_2 = 112.7°$
$\gamma_2 = 10.3°$
$a = 10$
$b = 11$
$c_2 = 2.132$

The solution $\beta_1 = 67.30°$ leads to a triangle AB_1C with

$$\gamma_1 = 180° - (\alpha + \beta_1)$$
$$= 180° - (57° + 67.30°) = 55.70°$$

and because $\dfrac{c_1}{\sin \gamma_1} = \dfrac{a}{\sin \alpha}$

$$c_1 = \frac{a \sin \gamma_1}{\sin \alpha} = \frac{10 \sin 55.70°}{\sin 57°} = 9.850$$

(Figure 8a). The solution $\beta_2 = 112.70°$ leads to a second triangle AB_2C with

$$\gamma_2 = 180° - (\alpha + \beta_2)$$
$$= 180° - (57° + 112.70°) = 10.30°$$

and because $\dfrac{c_2}{\sin \gamma_2} = \dfrac{a}{\sin \alpha}$

$$c_2 = \frac{a \sin \gamma_2}{\sin \alpha} = \frac{10 \sin 10.3°}{\sin 57°} = 2.132$$

(Figure 8b).

4 © $a = 70$, $b = 65$, $\alpha = 40°$

Solution   Here

$$b \frac{\sin \alpha}{a} = 65 \frac{\sin 40°}{70} = 0.5969 < 1.$$

Thus,

$$\beta_1 = \sin^{-1}\left(b \frac{\sin \alpha}{a}\right) = \sin^{-1}\left(65 \frac{\sin 40°}{70}\right) = 36.65°$$

and

$$\beta_2 = 180° - \beta_1 = 143.35°.$$

Because it is impossible to have an angle $\beta_2 = 143.35°$ in a triangle that already has an angle of 40° (143.35° + 40° > 180°), the solution $\beta_2 = 143.35°$ does not lead to an actual triangle. However, the solution $\beta = \beta_1 = 36.65°$ leads to

$$\gamma = 180° - (\alpha + \beta) = 180° - (40° + 36.65°) = 103.35°$$

and because $\dfrac{c}{\sin \gamma} = \dfrac{a}{\sin \alpha}$

$$c = \frac{a \sin \gamma}{\sin \alpha} = \frac{70 \sin 103.35°}{\sin 40°} = 106.0$$

(Figure 9).

Figure 9

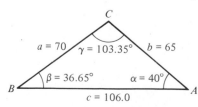

$a = 70$
$\gamma = 103.35°$
$b = 65$
$\beta = 36.65°$
$\alpha = 40°$
$c = 106.0$

PROBLEM SET 2

Ⓒ In problems 1–8, use the law of sines to solve each triangle *ABC*. Round off angles to the nearest hundredth of a degree and side lengths to four significant digits.

1. $a = 32, \beta = 57°, \gamma = 38°$
2. $b = 16, \alpha = 120°, \gamma = 30°$
3. $b = 24, \beta = 38°, \gamma = 21°$
4. $a = 42.9, \alpha = 32°, \beta = 81.5°$
5. $b = 47.3, \alpha = 48.31°, \gamma = 57.82°$
6. $a = 120, \alpha = 132°, \beta = 61°$
7. $a = 95, \alpha = 132°, \beta = 24.45°$
8. $b = 67.5, \alpha = 61.25°, \beta = 65.75°$

Ⓒ In problems 9–16, solve each triangle *ABC* by using the method for the ambiguous case. Round off angles to the nearest hundredth of a degree and side lengths to four significant digits. Be certain to find all possible triangles that satisfy the given conditions.

9. $a = 50, b = 30, \alpha = 45°$
10. $a = 40, b = 70, \alpha = 30°$
11. $a = 31, b = 33, \alpha = 60°$
12. $a = 10, b = 26, \beta = 110°$
13. $a = 27, b = 52, \alpha = 70°$
14. $b = 4.5, c = 9, \gamma = 60°$
15. $a = 65.52, b = 55.51, \alpha = 111.5°$
16. $a = 12.41, b = 81.69, \beta = 36.67°$

Ⓒ In problems 17–24, solve each triangle *ABC*. Round off angles to the nearest hundredth of a degree and side lengths to four significant digits. In the ambiguous cases, be certain to find all possible triangles that satisfy the given conditions.

17. $b = 52.05$ feet, $\alpha = 42.85°, \beta = 61.50°$
18. $b = 194.5$ meters, $\alpha = 82.25°, \beta = 69.45°$
19. $a = 110$ kilometers, $b = 100$ kilometers, $\alpha = 59.33°$
20. $a = 8.143 \times 10^6$ meters, $b = 1.271 \times 10^7$ meters, $\alpha = 34.65°$
21. $a = 2090$ meters, $b = 5579$ meters, $\alpha = 22°$
22. $a = 2.011 \times 10^{-10}$ meter, $b = 1.509 \times 10^{-10}$ meter, $\beta = 30.2°$
23. $a = 10.07$ kilometers, $b = 15.15$ kilometers, $\alpha = 67.67°$
24. $c = 0.4351$ meter, $\beta = 48.50°, \gamma = 96.25°$

Ⓒ 25. A surveyor lays out a baseline segment $\overline{AB}$ of length 346 meters. An inaccessible point *C* forms the third vertex of a triangle *ABC*. If the surveyor determines that angle $CAB = 62°40'$ and that angle $CBA = 54°30'$, find $|\overline{AC}|$ and $|\overline{BC}|$.

Ⓒ 26. The crankshaft $\overline{OA}$ of an engine is 2 inches long and the connecting rod $\overline{AP}$ is 8 inches long (Figure 10). If angle *APO* is 10° at a certain instant, find angle *AOP* at this instant. [Caution: There may be more than one solution.]

Figure 10

Figure 11

Ⓒ 27. An observation tower stands on the edge of a vertical cliff rising directly above a river (Figure 11). From the top of the tower, the angle of depression of a point on the opposite shore of the river is 24°; from the base of the tower, the angle of depression of the same point is 18°. If the top of the tower is 11 meters above its base, how wide is the river?

ⓒ **28.** Plans for an oil pipeline require that it be diverted, as shown in Figure 12, around the mating grounds of a caribou herd. Suppose that angle $BAC = 22°$, angle $ABC = 48°$, and $|\overline{AB}| = 7$ kilometers. If the pipeline costs \$200,000 per kilometer, find the additional cost of the pipeline because of the diversion.

Figure 12

ⓒ **29.** A nature photographer spots a rare northern shrike perched in a tree at an angle of elevation of 14°. Using the range finder on her camera, she determines that the shrike is 103 meters from her lens. She creeps toward the bird until its angle of elevation is 20°. Find the distance between the lens and the bird in the photographer's new position.

ⓒ **30.** A commercial jet is flying at constant speed on a level course that will carry it directly over Albuquerque, New Mexico. Through a hole in the cloud cover, the captain observes the lights of the city at an angle of depression of 7°. Three minutes later, the captain catches a second glimpse of the city, now at an angle of depression of 13°. In how many *more* minutes will the jet be directly over the city?

ⓒ **31.** A sports parachutist is sighted simultaneously by two observers who are 5 kilometers apart. The observed angles of elevation are 26.5° and 18.2°. Assuming that the parachutist and the two observers lie in the same vertical plane, how high above the ground is the parachutist?

ⓒ **32.** Riding at constant speed across the desert on his faithful camel Clyde, Lawrence sights the top of a tall palm tree. He assumes that the palm tree is located at an oasis, and heads directly toward it. When first sighted, the top of the palm tree was at an angle of elevation of 4°. Twenty minutes later, the angle of elevation has increased to 9°. In how many *more* minutes will Lawrence reach the palm tree?

ⓒ **33.** A CB transmitter is illegally transmitting at excessive power from point C, and it is under surveillance by two FCC agents situated 3.8 kilometers apart at points A and B. Using direction finders, the agents find that angle $CAB = 40.1°$ and angle $CBA = 31.8°$. How far is the illegal transmitter from the agent at point A?

ⓒ **34.** Two Coast Guard stations A and B are located along a straight coastline. Station A is 20 miles due north of station B. A supertanker in distress is observed from station A at a bearing of S9°10′E, and from station B at a bearing of S37°40′E. Find the distance from the supertanker to the nearest point on the coastline.

ⓒ **35.** Two joggers running at the same constant speed set out from the same point C. They run along separate straight paths toward a straight highway. The first jogger arrives at point A along the highway after running for 70 minutes. Some time later, the second jogger arrives at point B along the highway. If angle CAB is 105° and angle CBA is 50°, how many minutes were required by the second jogger to run from C to B?

36. In triangle ABC, use the law of sines and the fact that $\sin \gamma = \sin[180° - (\alpha + \beta)]$ to prove that $c = b\cos\alpha + a\cos\beta$.

37. In the proof of Theorem 1, suppose that angle α is not acute. Redraw Figure 2 to illustrate this situation and then show that $\dfrac{\sin\alpha}{a} = \dfrac{\sin\beta}{b}$ is still true. [Hint: $\sin(180° - \alpha) = \sin\alpha$.]

38. Complete the proof of Theorem 1 by showing that $\dfrac{\sin\beta}{b} = \dfrac{\sin\gamma}{c}$.

39. If $0 < \beta_1 < 90°$ and $\beta_2 = 180° - \beta_1$, show that $90° < \beta_2 < 180°$ and $\sin \beta_1 = \sin \beta_2$.

40. Use the law of sines to establish the **law of tangents:** For any triangle ABC,

$$\frac{a - b}{a + b} = \frac{\tan\left(\dfrac{\alpha - \beta}{2}\right)}{\tan\left(\dfrac{\alpha + \beta}{2}\right)}.$$

Begin by proving each of the following:

(a) $\dfrac{a - b}{b} = \dfrac{\sin \alpha - \sin \beta}{\sin \beta}$ (b) $\dfrac{a + b}{b} = \dfrac{\sin \alpha + \sin \beta}{\sin \beta}$ (c) $\dfrac{a - b}{a + b} = \dfrac{\sin \alpha - \sin \beta}{\sin \alpha + \sin \beta}$

© In problems 41–43, use the law of tangents in problem 40 to solve triangle ABC.

41. $a = 9, \beta = 73°, \alpha = 69°$ **42.** $b = 27, \beta = 33°, \gamma = 77°$ **43.** $a = 15, \gamma = 67°, \alpha = 81°$

44. The identity $(a - b)\cos\dfrac{\gamma}{2} = c \sin\left(\dfrac{\alpha - \beta}{2}\right)$ is called **Mollweide's formula.** It is often used to check the solution of a triangle ABC because it contains all six parts of the triangle. If, when values are substituted in Mollweide's formula, the result is not a true statement, then an error has been made in solving the triangle. Prove Mollweide's formula.

© **45.** Use Mollweide's formula to check your solutions of the triangles in problems 41 and 43.

46. (a) Show that the area $\mathscr{A}$ of triangle ABC is given by the formula $\mathscr{A} = \frac{1}{2}ab \sin \gamma$. ©(b) Use the formula in part (a) to find the area of triangle ABC if $a = 12.7$ meters, $b = 8.91$ meters, and $\gamma = 46.5°$.

3   The Law of Cosines

The law of sines is not effective for solving a triangle when *two sides and the angle between them are given*, or when *all three sides are given*. However, the following *law of cosines* can be used in these cases.

Theorem 1   **The Law of Cosines**

In triangle ABC (Figure 1):

   (i) $c^2 = a^2 + b^2 - 2ab \cos \gamma$

   (ii) $a^2 = b^2 + c^2 - 2bc \cos \alpha$

   (iii) $b^2 = a^2 + c^2 - 2ac \cos \beta$

Figure 1

Proof   We prove (i) here; (ii) and (iii) are proved in exactly the same way. To begin with, we place triangle ABC on a cartesian coordinate system with vertex C at the origin and vertex A on the positive x axis (Figure 2). Then vertex B has coordinates $B = (a\cos\gamma, a\sin\gamma)$. (Why?) Because $A = (b, 0)$, it follows from the distance formula that

Figure 2

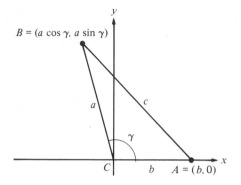

$$c^2 = (a\cos\gamma - b)^2 + (a\sin\gamma - 0)^2$$
$$= a^2\cos^2\gamma - 2ab\cos\gamma + b^2 + a^2\sin^2\gamma$$
$$= a^2(\cos^2\gamma + \sin^2\gamma) + b^2 - 2ab\cos\gamma$$
$$= a^2 + b^2 - 2ab\cos\gamma.$$

Rather than memorizing the three formulas in Theorem 1, we recommend that you learn the following statement, which summarizes all three formulas.

The Law of Cosines

> The square of the length of any side of a triangle is equal to the sum of the squares of the lengths of the other two sides minus twice the product of the lengths of these two sides and the cosine of the angle between them.

It's interesting to notice that, when the angle mentioned above is $90°$, the law of cosines reduces to the Pythagorean theorem (Problem 35).

Example 1   ⓒ In triangle ABC (Figure 3), find c to four significant digits if $a = 5$, $b = 10$, and $\gamma = 37.85°$.

Solution   By the law of cosines,

$$c^2 = a^2 + b^2 - 2ab\cos\gamma = 5^2 + 10^2 - 2(5)(10)\cos 37.85°$$
$$= 25 + 100 - 100\cos 37.85° = 46.04.$$

Therefore,

$$c = \sqrt{46.04} = 6.785.$$

Figure 3

Example 2 $\boxed{\text{c}}$ In triangle ABC (Figure 4), find β to the nearest hundredth of a degree if $a = 3$, $b = 4$, and $c = 2$.

Figure 4

Solution   We begin by writing the law of cosines in the form that contains $\cos \beta$:

$$b^2 = a^2 + c^2 - 2ac \cos \beta.$$

Solving this equation for $\cos \beta$, we obtain

$$\cos \beta = \frac{a^2 + c^2 - b^2}{2ac}.$$

Therefore,

$$\beta = \cos^{-1} \frac{a^2 + c^2 - b^2}{2ac} = \cos^{-1} \frac{3^2 + 2^2 - 4^2}{2(3)(2)}$$

$$= \cos^{-1} \frac{-3}{12} = \cos^{-1}\left(-\frac{1}{4}\right) = 104.48°.$$

Example 3 $\boxed{\text{c}}$ In order to find the horizontal distance between two towers A and B that are separated by a grove of trees, a point C that is readily accessible from both A and B is chosen (Figure 5), and the following measurements are made: $|\overline{CA}| = 341$ meters, $|\overline{CB}| = 549$ meters, angle $ACB = 115.5°$. Find $|\overline{AB}|$ to four significant digits.

Figure 5

Solution   By the law of cosines,

$$|\overline{AB}|^2 = |\overline{CA}|^2 + |\overline{CB}|^2 - 2|\overline{CA}||\overline{CB}| \cos (\text{angle } ACB)$$

$$= 341^2 + 549^2 - 2(341)(549) \cos 115.5°;$$

hence,

$$|\overline{AB}| = \sqrt{341^2 + 549^2 - 2(341)(549) \cos 115.5°} = 760.8 \text{ meters.}$$

The law of cosines can be used to establish a number of interesting formulas involving the geometry of triangles (see Problems 36–42). One of the most useful of these is **Hero's** (or Heron's) **area formula:**

$$\mathscr{A} = \sqrt{s(s-a)(s-b)(s-c)} \qquad \text{where } s = \tfrac{1}{2}(a+b+c),$$

which gives the area $\mathscr{A}$ of a triangle with sides of lengths a, b, and c (Problem 40). The quantity s, which is one-half of the perimeter $a+b+c$ of the triangle, is called the *semiperimeter*.

Example ⒸA cornfield that is bounded by three straight roads has side lengths $a = 600$ feet, $b = 400$ feet, and $c = 500$ feet (Figure 6). Find the area $\mathscr{A}$ of the cornfield.

Solution We use Hero's formula. The semiperimeter of the cornfield is

Figure 6

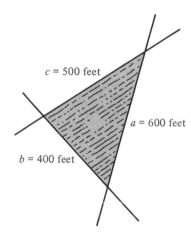

$c = 500$ feet
$a = 600$ feet
$b = 400$ feet

$$s = \tfrac{1}{2}(a+b+c)$$
$$= \tfrac{1}{2}(600+400+500)$$
$$= 750.$$

Therefore,

$$\mathscr{A} = \sqrt{s(s-a)(s-b)(s-c)}$$
$$= \sqrt{750(750-600)(750-400)(750-500)}$$
$$= \sqrt{750(150)(350)(250)} = \sqrt{9,843,750,000}$$
$$\approx 99,216 \text{ square feet.}$$

PROBLEM SET 3

Ⓒ In problems 1–12, use the law of cosines to find the specified unknown part of the triangle ABC (Figure 1). Round off angles to the nearest hundredth of a degree and side lengths to four significant digits.

1. Find c if $a = 3, b = 10$, and $\gamma = 60°$.
2. Find a if $b = 68, c = 14$, and $\alpha = 24.5°$.
3. Find b if $a = 23, c = 31$, and $\beta = 112°$.
4. Find a if $b = 3.2, c = 2.4$, and $\alpha = 117°$.
5. Find a if $c = 5.78, b = 4.78$, and $\alpha = 35.25°$.
6. Find β if $a = 200, b = 50$, and $c = 177$.
7. Find γ if $a = 2, b = 3$, and $c = 4$.
8. Find α if $a = 2\sqrt{61}, b = 8$, and $c = 10$.
9. Find β if $a = 1240, b = 876$, and $c = 918$.
10. Find γ if $a = 0.64, b = 0.27$, and $c = 0.49$.
11. Find α if $a = 189, b = 214$, and $c = 325$.
12. Find β if $a = 189, b = 214$, and $c = 325$.

Ⓒ In problems 13–18, use the law of cosines or the law of sines (or both) to solve each triangle ABC. Round off angles to the nearest hundredth of a degree and side lengths to four significant digits. In the ambiguous cases, be certain to find all possible triangles that satisfy the given conditions.

13. $a = 327, b = 251, \gamma = 72.45°$
14. $a = 100, b = 150, c = 300$
15. $a = 312, b = 490, \alpha = 33.75°$
16. $a = 87, c = 124, \alpha = 35.33°$
17. $a = 321, b = 456, c = 654$
18. $a = 59, b = 22, \beta = 117°20'$

© **19.** To find the distance between two points A and B, a surveyor chooses a point C that is 52 meters from A and 64 meters from B. If angle ACB is 72°20′, find the distance between A and B. Round off your answer to four significant digits.

© **20.** A tracking antenna A on the surface of the earth is following a spacecraft B (Figure 7). The tracking antenna is aimed 36° above the horizon, and the distance from the antenna to the spacecraft is 6521 miles. If the radius of the earth is 3959 miles, how high is the spacecraft above the surface of the earth?

Figure 7

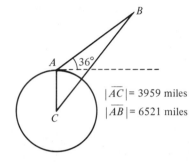

$|\overline{AC}| = 3959$ miles
$|\overline{AB}| = 6521$ miles

© **21.** Engineers are planning to construct a straight tunnel from point A on one side of a hill to point B on the other side. From a third point C, the distances $|\overline{CA}| = 837$ meters and $|\overline{CB}| = 1164$ meters are measured. If angle ACB is measured as 44.5°, find the length of the tunnel when it is completed. Round off your answer to four significant digits.

© **22.** A molecule of ammonia (NH_3) consists of three hydrogen atoms and one nitrogen atom, which form the vertices of a pyramid with four triangular faces. In one of the faces containing the nitrogen atom (N) and two hydrogen atoms (H), the distance between the two H atoms is 1.628 angstroms and the distance from the N atom to either H atom is 1.014 angstroms (Figure 8). [One angstrom (abbreviated Å) is 10^{-10} meter.] Find the angle formed by the two line segments joining the H atoms to the N atom.

© **23.** Two jet fighters leave an air base at the same time and fly at the same speed along straight courses, forming an angle of 135.65° with each other. After the jets have each flown 402 kilometers, how far apart are they?

© **24.** A ship leaves a harbor at noon and sails S52°W at 22 knots (nautical miles per hour) until 2:30 P.M. At that time it changes its course and sails N22°W at a reduced speed of 18 knots until 5:00 P.M. At 5:00 P.M., how far is the ship from the harbor and what is its bearing from the harbor?

© **25.** Two straight roads cross at an angle of 75°. A bus on one road is 5 miles from the intersection and moving away from it at 50 miles per hour. At the same instant, a truck on the other road is 10 miles from the intersection and moving away from it at the rate of 55 miles per hour (Figure 9). If the bus and truck maintain constant speeds, what is the distance between them after 45 minutes?

© **26.** The navigator of an oil tanker plots a straight-line course from point A to point B. Because of an error, the tanker proceeds from point A along a slightly different straight-line course. After sailing for 5 hours on the wrong course, the error is discovered and the tanker is turned through an angle of 23° so that it is now heading directly toward point B. After 3 more hours, the tanker reaches point B. Assuming that the tanker was proceeding at a constant speed, calculate the time lost because of the error.

© In problems 27–30, use Hero's area formula to find the area of a triangle with the given side lengths a, b, and c. Round off all answers to four significant digits.

27. $a = 5$ centimeters, $b = 8$ centimeters, $c = 7$ centimeters
28. $a = 7.301$ kilometers, $b = 11.15$ kilometers, $c = 15.22$ kilometers

Figure 8

Figure 9

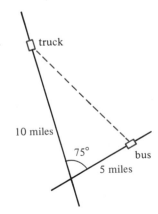

29. $a = 13.03$ feet, $b = 15.77$ feet, $c = 17.20$ feet

30. $a = 1.110 \times 10^5$ meters, $b = 9.812 \times 10^4$ meters, $c = 1.728 \times 10^5$ meters

© **31.** Offshore drilling rights are granted to RIPCO oil company in a triangular region bounded by sides of lengths 33 kilometers, 28 kilometers, and 7 kilometers. Find the area of this region.

© **32.** A bird sanctuary along the straight shoreline of a lake is bounded by straight line segments $\overline{AB}$, $\overline{BC}$, and $\overline{CD}$ as shown in Figure 10. Suppose that $|\overline{AB}| = 320$ meters, $|\overline{BC}| = 200$ meters, $|\overline{CD}| = 360$ meters, angle $ABC = 146°$, and angle $BCD = 77°$. Find (a) the area of the bird sanctuary and (b) the distance $|\overline{AD}|$ along the shoreline. [Hint: Break the quadrilateral $ABCD$ into two triangles—for instance, ABC and ACD—then solve these triangles.]

Figure 10

shoreline

© **33.** The Great Pyramid of Cheops at Al Giza, Egypt, has four congruent triangular faces. Each face is an isosceles triangle with equal sides of length 219 meters and base 230 meters (Figure 11). Find the *total* surface area of the pyramid.

Figure 11

© **34.** In problem 33, find (a) the height h of the pyramid and (b) the volume V of the pyramid. [Hint: The volume of a pyramid is one-third the height times the area of the base.]

35. Show that the Pythagorean theorem is a special case of the law of cosines.

36. Let $s = \frac{1}{2}(a + b + c)$ be the semiperimeter of triangle ABC (Figure 12). Combine the law of cosines

$$a^2 = b^2 + c^2 - 2bc \cos \alpha$$

and the half-angle formula

$$\cos^2 \frac{\alpha}{2} = \frac{1 + \cos \alpha}{2}$$

to obtain the formula

$$\cos \frac{\alpha}{2} = \sqrt{\frac{s(s - a)}{bc}}.$$

Figure 12

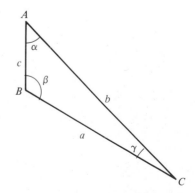

37. For every triangle ABC (Figure 12), show that

$$\frac{\cos \alpha}{a} + \frac{\cos \beta}{b} + \frac{\cos \gamma}{c} = \frac{a^2 + b^2 + c^2}{2abc}.$$

38. Proceeding as in problem 36, but using the half-angle formula

$$\sin^2 \frac{\alpha}{2} = \frac{1 - \cos \alpha}{2},$$

show that

$$\sin \frac{\alpha}{2} = \sqrt{\frac{(s - b)(s - c)}{bc}}.$$

39. For every triangle ABC (Figure 12), show that γ is an acute angle if and only if $a^2 + b^2 > c^2$.

40. In Figure 13, the area $\mathscr{A}$ of triangle ABC is given by $\mathscr{A} = \frac{1}{2}hb$.
   (a) Show that $\mathscr{A} = \frac{1}{2}bc \sin \alpha$.
   (b) Using the double-angle formula, show that

$$\sin \alpha = 2 \sin \frac{\alpha}{2} \cos \frac{\alpha}{2}.$$

   (c) Combine (a) and (b) to obtain

$$\mathscr{A} = bc \sin \frac{\alpha}{2} \cos \frac{\alpha}{2}.$$

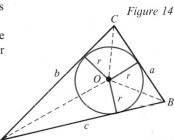

Figure 13

$\mathscr{A} = \frac{1}{2}hb$

   (d) Combine (c), problem 36, and problem 38 to obtain Hero's formula

$$\mathscr{A} = \sqrt{s(s - a)(s - b)(s - c)}.$$

41. Suppose you are given three positive numbers a, b, and c. Under what conditions will there exist a triangle ABC with side lengths a, b, and c? [Hint: Under what conditions could you form a triangle with three sticks of wood of lengths a, b, and c?]

42. The center O of the circle inscribed in a triangle ABC always lies at the intersection of the bisectors of the three vertex angles (Figure 14). If $s = \frac{1}{2}(a + b + c)$ is the semiperimeter of triangle ABC, show that the radius r of the inscribed circle is given by

$$r = \sqrt{\frac{(s - a)(s - b)(s - c)}{s}}.$$

Figure 14

[Hint: Use Hero's formula and the fact that the area of triangle ABC is the sum of the areas of triangles AOB, BOC, and COA.]

4 Vectors

Many applications of mathematics involve force, velocity, acceleration, or other quantities that have both a *magnitude* and a *direction*. For instance, if you push on a stone, the force you exert has both a magnitude (say, 5 pounds) and a direction (say, due east). You can represent this force by an arrow that has a length (5 units) corresponding to the magnitude and that points in the appropriate direction (due east) (Figure 1).

Figure 1

force: 5 pounds due east

Figure 2

More generally, a **vector** can be thought of as an arrow, that is, a line segment with a direction specified by an arrowhead. We use letters with bars, such as $\bar{A}$, $\bar{B}$, $\bar{C}$, and so forth, to denote vectors. If $\bar{V}$ is a vector, we denote the **length** or **magnitude** of $\bar{V}$ by $|\bar{V}|$ (Figure 2). A vector $\bar{V}$ with tail end or **initial point** A and head end or **terminal point** B may be written as

$$\bar{V} = \overrightarrow{AB}$$

Figure 3

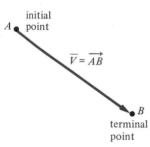

(Figure 3). This notation tells us that we are dealing with a *vector* $\overrightarrow{AB}$ rather than a line segment $\overline{AB}$, and it shows the direction of the vector—from the initial point A to the terminal point B. Notice that the magnitude

$$|\bar{V}| = |\overrightarrow{AB}| = |\overline{AB}|$$

is just the distance between A and B.

Let's agree to say that two vectors $\overrightarrow{AB}$ and $\overrightarrow{CD}$ are **equal** and to write

$$\overrightarrow{AB} = \overrightarrow{CD}$$

if they are parallel and have the same direction and magnitude (Figure 4). In other words, you may move a vector around freely if you keep it parallel to its original position. In particular, you can always move a vector $\bar{R}$ in the xy plane so that its tail end (initial point) is at the origin O (Figure 5). Then $\bar{R}$ will extend from O to its head end (terminal point) $P = (x, y)$, and we have

$$\bar{R} = \overrightarrow{OP}.$$

Figure 4

Figure 5

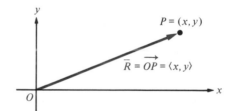

We say that $\bar{R} = \overrightarrow{OP}$ is the **position vector** of the point $P = (x, y)$; we call x and y the **components** of the vector $\bar{R}$; and we write

$$\bar{R} = \langle x, y \rangle.$$

The special symbol $\langle x, y \rangle$ is used to avoid confusing the *vector* $\langle x, y \rangle$ and the *point* $P = (x, y)$.

The **zero vector,** denoted by $\bar{0}$, is defined by $\bar{0} = \langle 0, 0 \rangle$. You can think of $\bar{0}$ as an arrow whose length has shrunk to 0, so that its initial and terminal points coincide. Not only is $\bar{0}$ the only vector whose magnitude is 0, but it is the only vector whose direction is indeterminate.

Let $\bar{R}$ be a nonzero vector in the xy plane. Move $\bar{R}$, if necessary, so that its initial point is at the origin O. If $\bar{R}$ falls along the terminal side of an angle θ in standard position, we say that θ is a **direction angle** for $\bar{R}$ (Figure 6). Notice that any angle coterminal with θ is also a direction angle for $\bar{R}$. The vector $\bar{R}$ can be

Figure 6

Figure 7

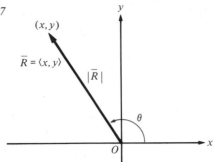

specified either by giving its components $\bar{R} = \langle x, y \rangle$ or by giving its magnitude (length) $|\bar{R}|$ and its direction angle θ (Figure 7). The quantities x, y, $|\bar{R}|$, and θ are related as follows:

> (i) $|\bar{R}| = \sqrt{x^2 + y^2}$
>
> (ii) $\cos \theta = \dfrac{x}{|\bar{R}|}$ or $x = |\bar{R}| \cos \theta$
>
> (iii) $\sin \theta = \dfrac{y}{|\bar{R}|}$ or $y = |\bar{R}| \sin \theta$
>
> (iv) $\tan \theta = \dfrac{y}{x}$ if $x \neq 0$.

(See Definition 1 on page 287.) These equations can be used to find $|\bar{R}|$ and θ in terms of x and y or vice versa.

Example 1   © Sketch each vector, find its magnitude, and find its smallest positive direction angle θ.

(a) $\bar{A} = \langle 4, 3 \rangle$ (b) $\bar{B} = \langle 2\sqrt{3}, -1 \rangle$

Solution   The two vectors, which extend from the origin to the points $(4, 3)$ and $(2\sqrt{3}, -1)$, are shown in Figure 8.

(a) By equation (i) above,
$$|\bar{A}| = \sqrt{4^2 + 3^2} = \sqrt{25} = 5 \text{ units.}$$

Figure 8

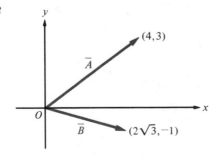

By equation (ii) above,
$$\cos \theta = \frac{4}{|\bar{A}|} = \frac{4}{5} = 0.8.$$

Because the point $(4, 3)$ is in quadrant I, it follows that θ is an acute angle; hence,
$$\theta = \cos^{-1} 0.8 \approx 36.87°.$$

(b) $|\bar{B}| = \sqrt{(2\sqrt{3})^2 + (-1)^2} = \sqrt{13}$ units. Because the point $(2\sqrt{3}, -1)$ is in quadrant IV, it follows that $270° < \theta < 360°$. Also,

$$\cos\theta = \frac{2\sqrt{3}}{\sqrt{13}}$$
$$= 2\sqrt{3/13},$$

so the reference angle θ_R corresponding to θ is given by

$$\theta_R = \cos^{-1}2\sqrt{3/13} \approx 16.10°.$$

Therefore,

$$\theta = 360° - \theta_R \approx 360° - 16.10° = 343.90°.$$

Example 2   Find the components of the vector $\bar{V}$ if $|\bar{V}| = 3$ and a direction angle for $\bar{V}$ is $\theta = 150°$.

Solution   We use equations (ii) and (iii) in the forms $x = |\bar{V}|\cos\theta$ and $y = |\bar{V}|\sin\theta$. Thus, we have

$$x = |\bar{V}|\cos\theta = 3\cos 150° = 3\left(-\frac{\sqrt{3}}{2}\right) = -\frac{3\sqrt{3}}{2}$$

and

$$y = |\bar{V}|\sin\theta = 3\sin 150° = 3\left(\frac{1}{2}\right) = \frac{3}{2}$$

so

$$\bar{V} = \left\langle -\frac{3\sqrt{3}}{2}, \frac{3}{2}\right\rangle.$$

As we mentioned earlier, forces can be represented by vectors: the length $|\bar{F}|$ of a force vector $\bar{F}$ is the magnitude of the force, while the direction of $\bar{F}$ indicates the direction of the force. Laboratory experiments have revealed the following: If two forces $\bar{F}_1$ and $\bar{F}_2$ act simultaneously on a particle, the effect is the same as if a single force $\bar{F}$, called the **resultant,** were acting on the particle. The resultant $\bar{F}$ forms a diagonal of the parallelogram in which $\bar{F}_1$ and $\bar{F}_2$ are adjacent edges (Figure 9). The resultant vector $\bar{F}$ is also called the **sum** of the two vectors $\bar{F}_1$ and $\bar{F}_2$, and we write

$$\bar{F} = \bar{F}_1 + \bar{F}_2.$$

Figure 9

Figure 10

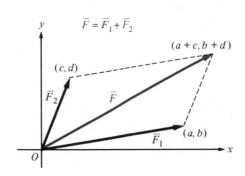

By applying the **parallelogram rule,** illustrated in Figure 9, you can form the sum $\bar{F} = \bar{F}_1 + \bar{F}_2$ of any two vectors, whether they represent forces or not. (Also, see Problem 31.) By examining Figure 10, you can see that *the components of the sum of two vectors are the sums of their corresponding components,* that is,

$$\langle a, b \rangle + \langle c, d \rangle = \langle a + c, b + d \rangle.$$

(See Problem 35).

Example   If $\bar{A} = \langle 2, -7 \rangle$ and $\bar{B} = \langle \sqrt{3}, 5 \rangle$, find the components of $\bar{A} + \bar{B}$.

Solution
$$\bar{A} + \bar{B} = \langle 2, -7 \rangle + \langle \sqrt{3}, 5 \rangle = \langle 2 + \sqrt{3}, -7 + 5 \rangle = \langle 2 + \sqrt{3}, -2 \rangle.$$

To *double* a force $\bar{F}$ means to double its magnitude and leave its direction the same. Thus, $2\bar{F}$ is a vector twice as long as $\bar{F}$, and pointing in the same direction (Figure 11).

Figure 11

More generally, if s is a real number and $\bar{F}$ is a vector, we define $s\bar{F}$ to be a vector $|s|$ times as long as $\bar{F}$, and in the same direction if s is positive, or in the opposite direction if s is negative. Naturally, if either $s = 0$ or $\bar{F} = \bar{0}$ (or both), we define $s\bar{F} = \bar{0}$. In a product $s\bar{F}$, the real number s is often called a **scalar** (because it corresponds to a point on a number scale). Thus, the vector $s\bar{F}$ is referred to as a **scalar multiple** of the vector $\bar{F}$. Notice that

$$|s\bar{F}| = |s|\,|\bar{F}|.$$

When a vector is multiplied by a scalar, the components of the vector are multiplied by the scalar. Thus, if $\bar{F} = \langle a, b \rangle$, then $s\bar{F} = \langle sa, sb \rangle$, that is,

$$s\langle a, b \rangle = \langle sa, sb \rangle.$$

Figure 12

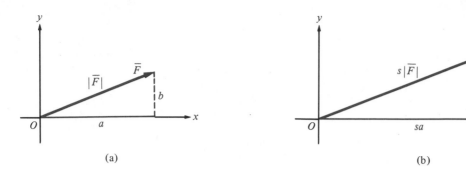

(a)

(b)

You can confirm this for the case in which s is positive and $\bar{F} = \langle a, b \rangle$ lies in quadrant I by considering the similar triangles in Figure 12a and b. The remaining cases are left as an exercise (Problem 34).

Example   Let $A = \langle -2, \sqrt{3} \rangle$ and $\bar{B} = \langle 1, 4 \rangle$. Find the components of
(a) $5\bar{A}$ (b) $-3\bar{B}$ (c) $2\bar{A} + 3\bar{B}$.

Solution   (a) $5\bar{A} = 5\langle -2, \sqrt{3} \rangle = \langle 5(-2), 5\sqrt{3} \rangle = \langle -10, 5\sqrt{3} \rangle$
(b) $-3\bar{B} = -3\langle 1, 4 \rangle = \langle -3(1), -3(4) \rangle = \langle -3, -12 \rangle$
(c) $2\bar{A} + 3\bar{B} = 2\langle -2, \sqrt{3} \rangle + 3\langle 1, 4 \rangle = \langle 2(-2), 2\sqrt{3} \rangle + \langle 3(1), 3(4) \rangle$
$\qquad\qquad = \langle -4, 2\sqrt{3} \rangle + \langle 3, 12 \rangle = \langle -4 + 3, 2\sqrt{3} + 12 \rangle$
$\qquad\qquad = \langle -1, 2\sqrt{3} + 12 \rangle$

By the definition of multiplication of a vector by a scalar, the vector $(-1)\bar{F}$ has the same magnitude as $\bar{F}$, but points in the opposite direction. For simplicity, $(-1)\bar{F}$ is often written as $-\bar{F}$ (Figure 13). If $\bar{F} = \langle a, b \rangle$, then $-\bar{F} = \langle -a, -b \rangle$, that is,

$$-\langle a, b \rangle = \langle -a, -b \rangle$$

(Problem 36). The **difference** of two vectors $\bar{A}$ and $\bar{B}$ is defined by

$$\bar{A} - \bar{B} = \bar{A} + (-\bar{B}).$$

In other words, to **subtract** a vector $\bar{B}$ from a vector $\bar{A}$, you add $-\bar{B}$ to $\bar{A}$ (see Problem 33.) Evidently, if $\bar{A} = \langle a, b \rangle$ and $\bar{B} = \langle c, d \rangle$, then $\bar{A} - \bar{B} = \langle a - c, b - d \rangle$, that is,

$$\langle a, b \rangle - \langle c, d \rangle = \langle a - c, b - d \rangle$$

(Problem 37).

Figure 13

Example   If $\bar{A} = \langle 4, -1 \rangle$ and $\bar{B} = \langle 7, \frac{2}{3} \rangle$, find the components of

(a) $-\bar{A}$

(b) $\bar{A} - \bar{B}$

(c) $2\bar{A} - 3\bar{B}$.

Solution   (a) $-\bar{A} = -\langle 4, -1 \rangle = \langle -4, -(-1) \rangle = \langle -4, 1 \rangle$

(b) $\bar{A} - \bar{B} = \langle 4, -1 \rangle - \langle 7, \frac{2}{3} \rangle = \langle 4 - 7, -1 - \frac{2}{3} \rangle = \langle -3, -\frac{5}{3} \rangle$

(c) $2\bar{A} - 3\bar{B} = 2\langle 4, -1 \rangle - 3\langle 7, \frac{2}{3} \rangle = \langle 8, -2 \rangle - \langle 21, 2 \rangle$
$= \langle 8 - 21, -2 - 2 \rangle = \langle -13, -4 \rangle$

Figure 14

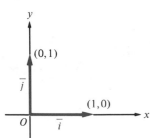

A popular alternative way to represent vectors in the xy plane is based on the two special vectors $\bar{i} = \langle 1, 0 \rangle$ and $\bar{j} = \langle 0, 1 \rangle$ (Figure 14). Notice that $\bar{i}$ and $\bar{j}$, which are called the **standard unit basis vectors,** have length 1 and point in the directions of the positive x and y axes, respectively. If $\bar{R} = \langle x, y \rangle$, then

$$\bar{R} = \langle x, y \rangle = \langle x, 0 \rangle + \langle 0, y \rangle = x\langle 1, 0 \rangle + y\langle 0, 1 \rangle = x\bar{i} + y\bar{j}.$$

In other words, $\bar{R} = \langle x, y \rangle$ and $\bar{R} = x\bar{i} + y\bar{j}$ are just two different ways of saying that $\bar{R}$ is the vector with components x and y. Using the standard unit basis vectors, we can write the following rules for addition, subtraction, and multiplication of vectors by scalars:

(i) $(a\bar{i} + b\bar{j}) + (c\bar{i} + d\bar{j}) = (a + c)\bar{i} + (b + d)\bar{j}$

(ii) $(a\bar{i} + b\bar{j}) - (c\bar{i} + d\bar{j}) = (a - c)\bar{i} + (b - d)\bar{j}$

(iii) $s(a\bar{i} + b\bar{j}) = (sa)\bar{i} + (sb)\bar{j}$.

Example   If $\bar{A} = 2\bar{i} + 5\bar{j}$ and $\bar{B} = 3\bar{i} - \bar{j}$, find

(a) $\bar{A} + \bar{B}$ (b) $\bar{A} - \bar{B}$ (c) $\frac{4}{5}\bar{A}$ (d) $3\bar{A} - 2\bar{B}$

Solution   (a) $\bar{A} + \bar{B} = (2\bar{i} + 5\bar{j}) + (3\bar{i} - \bar{j}) = (2 + 3)\bar{i} + (5 - 1)\bar{j} = 5\bar{i} + 4\bar{j}$

(b) $\bar{A} - \bar{B} = (2\bar{i} + 5\bar{j}) - (3\bar{i} - \bar{j}) = (2 - 3)\bar{i} + (5 + 1)\bar{j} = -\bar{i} + 6\bar{j}$

(c) $\frac{4}{5}\bar{A} = \frac{4}{5}(2\bar{i} + 5\bar{j}) = [\frac{4}{5}(2)]\bar{i} + [\frac{4}{5}(5)]\bar{j} = \frac{8}{5}\bar{i} + 4\bar{j}$

(d) $3\bar{A} - 2\bar{B} = 3(2\bar{i} + 5\bar{j}) - 2(3\bar{i} - \bar{j}) = (6\bar{i} + 15\bar{j}) - (6\bar{i} - 2\bar{j})$
$= (6 - 6)\bar{i} + [15 - (-2)]\bar{j} = 17\bar{j}$

PROBLEM SET 4

In problems 1–12, sketch each vector, find its magnitude, and find its smallest nonnegative direction angle θ.

1. $\bar{R} = \langle -4, 4 \rangle$
C 2. $\bar{A} = \langle 6, 8 \rangle$
C 3. $\bar{B} = \langle 4, 3 \rangle$
4. $\bar{D} = \langle 5, 0 \rangle$

5. $\bar{C} = \langle 1, -\sqrt{3} \rangle$
6. $\bar{F} = \langle 0, 5 \rangle$
C 7. $\bar{H} = \langle \frac{3}{8}, \frac{1}{2} \rangle$
C 8. $\bar{K} = \langle 3.71, -4.88 \rangle$

C 9. $\bar{M} = \langle -\sqrt{5}, \sqrt{11} \rangle$
10. $\bar{N} = \langle \pi, \pi\sqrt{3} \rangle$
C 11. $\bar{P} = 3\bar{i} + 4\bar{j}$
C 12. $\bar{Q} = 2\sqrt{5}\bar{i} - 4\bar{j}$

In problems 13–20, the magnitude $|\bar{V}|$ and a direction angle θ of a vector $\bar{V}$ are given. In each case, find the components of $\bar{V}$.

13. $|\bar{V}| = 5, \theta = 45°$

14. $|\bar{V}| = 3, \theta = 180°$

15. $|\bar{V}| = 4\sqrt{2}, \theta = 135°$

16. $|\bar{V}| = 9, \theta = -\dfrac{5\pi}{3}$

17. $|\bar{V}| = 2, \theta = -\dfrac{5\pi}{6}$

© 18. $|\bar{V}| = 4, \theta = 82°$

© 19. $|\bar{V}| = 7.03, \theta = 75.25°$

© 20. $|\bar{V}| = 8.12, \theta = 247°10'$

In problems 21–30, find (a) $\bar{A} + \bar{B}$, (b) $\bar{A} - \bar{B}$, (c) $3\bar{A} + 4\bar{B}$, and (d) $2\bar{A} - 3\bar{B}$.

21. $\bar{A} = \langle 3,4 \rangle, \bar{B} = \langle -1,4 \rangle$

22. $\bar{A} = \langle -2,6 \rangle, \bar{B} = \langle 3, -5 \rangle$

23. $\bar{A} = \langle -3,2 \rangle, \bar{B} = \langle \frac{5}{2}, \frac{7}{3} \rangle$

24. $\bar{A} = \langle 3, -3 \rangle, \bar{B} = \langle -5,0 \rangle$

25. $\bar{A} = 2\langle 1, -3 \rangle, \bar{B} = 3\langle 3, -5 \rangle$

26. $\bar{A} = -\bar{i} - \bar{j}, \bar{B} = 9\bar{j}$

27. $\bar{A} = 2\bar{i} + 5\bar{j}, \bar{B} = -3\bar{i} + 7\bar{j}$

28. $\bar{A} = \frac{7}{4}\bar{i} - \bar{j}, \bar{B} = \bar{i} - \frac{11}{3}\bar{j}$

29. $\bar{A} = 6\bar{i} - 5\bar{j}, \bar{B} = 2(2\bar{i} - 3\bar{j})$

30. $\bar{A} = 3.07\bar{i} + 0.33\bar{j}, \bar{B} = 7.65\bar{i} - 3.44\bar{j}$

31. The sum $\bar{A} + \bar{B}$ of two vectors can be obtained geometrically by applying the following **head-to-tail rule:** Move $\bar{B}$ if necessary so that its tail end coincides with the head end of $\bar{A}$ (Figure 15). Then $\bar{A} + \bar{B}$ is the vector whose tail end is the tail end of $\bar{A}$ and whose head end is the head end of $\bar{B}$. Show that the head-to-tail rule follows from the parallelogram rule.

Figure 15

32. Using the head-to-tail rule (problem 31), give a geometric justification of the following equation, which is known as the **associative law of vector addition:**

$$\bar{A} + (\bar{B} + \bar{C}) = (\bar{A} + \bar{B}) + \bar{C}.$$

33. Justify the **triangle rule** for subtraction of vectors: Move $\bar{B}$ if necessary so that its tail end coincides with the tail end of $\bar{A}$ (Figure 16). Then $\bar{A} - \bar{B}$ is the vector whose tail end is the head end of $\bar{B}$ and whose head end is the head end of $\bar{A}$.

Figure 16

34. Complete the argument begun in the text (Figure 12) to show that if $\bar{F} = \langle a,b \rangle$, then $s\bar{F} = \langle sa, sb \rangle$. Consider the cases in which $\bar{F}$ lies in quadrants II, III, and IV as well as the situations in which s is 0 or negative.

35. Figure 10 shows a special situation in which both vectors $\bar{F}_1$ and $\bar{F}_2$ lie in quadrant I. Check several other configurations to convince yourself that the components of $\bar{F}_1 + \bar{F}_2$ are always equal to the sum of the corresponding components of $\bar{F}_1$ and $\bar{F}_2$.

36. Prove that if $\bar{F} = \langle a,b \rangle$, then $-\bar{F} = \langle -a, -b \rangle$.

37. Prove that if $\bar{A} = \langle a,b \rangle$ and $\bar{B} = \langle c,d \rangle$, then $\bar{A} - \bar{B} = \langle a - c, b - d \rangle$.

38. If $\bar{A} + \bar{X} = \bar{B}$, prove that $\bar{X} = \bar{B} - \bar{A}$.

39. If $\bar{A} + \bar{X} = \bar{A}$, prove that $\bar{X} = \bar{0}$.

40. If $s\bar{F} = \bar{0}$, prove that either $s = 0$ or $\bar{F} = \bar{0}$.

41. Prove that $s(\bar{A} + \bar{B}) = s\bar{A} + s\bar{B}$.

42. If $\bar{A}$ and $\bar{B}$ are two vectors and α is the angle between them, then the **dot product** of $\bar{A}$ and $\bar{B}$ is defined to be the number

$$\bar{A} \cdot \bar{B} = |\bar{A}||\bar{B}| \cos \alpha.$$

Use the law of cosines to show that if $\bar{A} = \langle a,b \rangle$ and $\bar{B} = \langle c,d \rangle$, then

$$\bar{A} \cdot \bar{B} = ac + bd.$$

43. Use the formula obtained in problem 42 to calculate $\bar{A} \cdot \bar{B}$ if (a) $\bar{A} = \langle 5,1 \rangle$ and $\bar{B} = \langle 4, -7 \rangle$; (b) $\bar{A} = \langle -3,2 \rangle$ and $\bar{B} = \langle -4,3 \rangle$; (c) $\bar{A} = \bar{i} + 2\bar{j}$ and $\bar{B} = 3\bar{i} - 5\bar{j}$; (d) $\bar{A} = \bar{i}$ and $\bar{B} = \bar{j}$; and (e) $\bar{A} = \bar{B} = \bar{i}$.

44. Use the result of problem 42 to show that for any three vectors $\bar{A}$, $\bar{B}$, and $\bar{C}$,

$$\bar{A} \cdot (\bar{B} + \bar{C}) = \bar{A} \cdot \bar{B} + \bar{A} \cdot \bar{C}.$$

5 Polar Coordinates

Until now we have specified the position of points in the plane by means of cartesian coordinates; however, in some situations it is simpler and more natural to use a different system called *polar coordinates*. To set up a polar coordinate system in the plane, we choose a fixed point O called the **pole** and a fixed ray with endpoint O. The ray is called the **polar axis** (Figure 1a).

Figure 1

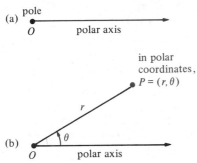

Now let P be any point in the plane and denote by r the distance between P and the pole O, so that $r = |\overline{OP}|$ (Figure 1b). If $P \neq O$, then the ray containing P with endpoint O forms the terminal side of an angle θ with the polar axis as its initial side (Figure 1b). We shall refer to the ordered pair (r, θ) as the **polar coordinates** of the point P, and we write

$$P = (r, \theta).$$

The angle θ can be measured either in degrees or radians.

The polar coordinates (r, θ) locate the point P on a grid formed by concentric circles with center O and rays with endpoint O (Figure 2). The value of r determines a particular circle of radius r; the value of θ designates a specific ray making an angle θ with the polar axis; and P lies at the intersection of the circle and the ray. For instance, the point

$$P = (4, 30°)$$

lies at the intersection of the circle $r = 4$ and the ray $\theta = 30°$.

Figure 2

If $r = 0$ in the polar coordinate system, we understand that the point $(r, \theta) = (0, \theta)$ is at the pole O no matter what the angle θ may be. Thus,

$$O = (0, \theta).$$

Also, it is convenient to allow r to be negative by making the following definition: The point $(-r, \theta°)$ is located $|r|$ units from the pole O, but on the ray opposite to the $\theta°$ ray, that is, on the $\theta° + 180°$ ray (Figure 3). Therefore,

Figure 3

$$(-r, \theta°) = (r, \theta° + 180°),$$

or if θ is measured in radians,

$$(-r, \theta) = (r, \theta + \pi).$$

To "plot the polar point (r, θ)," means to draw a diagram showing the pole O, the polar axis, and the point P whose polar coordinates are (r, θ). You will find it easier to plot polar points if you use polar graph paper.

Example   Plot each point in the polar coordinate system.

(a) $(2, 60°)$ (b) $(4, 135°)$ (c) $\left(-\frac{5}{2}, 30°\right)$ (d) $\left(2, -\frac{\pi}{4}\right)$

(e) $(2, 0°)$ (f) $(0, 0°)$ (g) $(-4, -45°)$

Solution   (a) To plot the polar point $(2, 60°)$, we measure an angle of $60°$ from the polar axis and then locate the point 2 units from the pole on the terminal side of this angle. The polar points in parts (b) through (g) are plotted in a similar way (Figure 4).

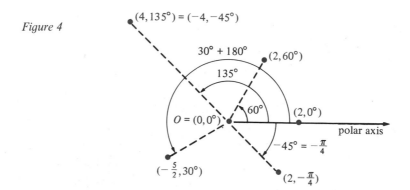

Figure 4

Unlike the cartesian coordinate system, in which a point P has only one representation, a point P has many different representations in the polar coordinate system. Not only do we have $(r, \theta°) = (-r, \theta° + 180°)$ but, because $\pm 360°$ corresponds to one revolution about the pole, we also have

$$(r, \theta°) = (r, \theta° + 360°) = (r, \theta° - 360°).$$

Indeed, if n is any integer,

$$(r, \theta°) = (r, \theta° + 360° \cdot n),$$

or, if angles are measured in radians,

$$(r, \theta) = (r, \theta + 2\pi n).$$

Example   Give other polar representations $P = (r, \theta°)$ of the polar point $P = (2, 30°)$ that satisfy each of the following conditions:

(a) $r < 0$ and $0° \leq \theta° < 360°$ (b) $r > 0$ and $-360° < \theta° \leq 0°$

(c) $r < 0$ and $-360° < \theta° \leq 0°$

Solution   (a) $P = (2, 30°) = (-2, 30° + 180°) = (-2, 210°)$
(b) $P = (2, 30°) = (2, 30° - 360°) = (2, -330°)$
(c) $P = (-2, 210°) = (-2, 210° - 360°) = (-2, -150°)$

It is sometimes necessary to convert from cartesian to polar coordinates or vice versa. To do this, we place the pole for the polar coordinate system at the origin of the cartesian coordinate system, and position the polar axis along the positive x axis (Figure 5). Now suppose that the point P has cartesian coordinates (x, y) and polar coordinates (r, θ) with $r > 0$. Then $\cos\theta = x/r$ and $\sin\theta = y/r$; hence,

Figure 5

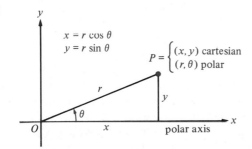

$$x = r\cos\theta \quad \text{and} \quad y = r\sin\theta.$$

It can be shown (Problem 51) that these equations hold in all cases, even when $r \leq 0$.

Example   Convert the following polar coordinates to cartesian coordinates:

(a) $(4, 30°)$ (b) $\left(6, -\dfrac{3\pi}{4}\right)$ $\boxed{\text{C}}$ (c) $(-2, 133.5°)$

Solution   We use the conversion equations $x = r\cos\theta$ and $y = r\sin\theta$.

(a) $x = r\cos\theta = 4\cos 30° = 4\left(\dfrac{\sqrt{3}}{2}\right) = 2\sqrt{3}$

$y = r\sin\theta = 4\sin 30° = 4(\tfrac{1}{2}) = 2$

Hence, the cartesian coordinates are $(2\sqrt{3}, 2)$.

(b) $x = r\cos\theta = 6\cos\left(-\dfrac{3\pi}{4}\right) = 6\left(-\dfrac{\sqrt{2}}{2}\right) = -3\sqrt{2}$

$y = r\sin\theta = 6\sin\left(-\dfrac{3\pi}{4}\right) = 6\left(-\dfrac{\sqrt{2}}{2}\right) = -3\sqrt{2}$

Hence, the cartesian coordinates are $(-3\sqrt{2}, -3\sqrt{2})$.

(c) Using a calculator, and rounding off to four significant digits, we have

$$x = r \cos \theta = -2 \cos 133.5° = 1.377$$
$$y = r \sin \theta = -2 \sin 133.5° = -1.451$$

Hence, the approximate cartesian coordinates are $(1.377, -1.451)$.

From the conversion equations $x = r \cos \theta$ and $y = r \sin \theta$, it follows that

$$x^2 + y^2 = r^2 \cos^2 \theta + r^2 \sin^2 \theta$$
$$= r^2(\cos^2 \theta + \sin^2 \theta)$$
$$= r^2,$$

so

$$r = \pm\sqrt{x^2 + y^2}.$$

Also, if $x \neq 0$,

$$\frac{y}{x} = \frac{r \sin \theta}{r \cos \theta} = \frac{\sin \theta}{\cos \theta} = \tan \theta,$$

so

$$\tan \theta = \frac{y}{x} \quad \text{for} \quad x \neq 0.$$

The equations $r = \pm\sqrt{x^2 + y^2}$ and $\tan \theta = y/x$ do not determine r and θ uniquely in terms of x and y, simply because the point P whose cartesian coordinates are (x, y) has an unlimited number of different representations in the polar coordinate system. Thus, in using these equations to find the polar coordinates of P, you must pay attention to the quadrant in which P lies, because this will help you to determine an appropriate value of θ.

Example 1   Convert the following cartesian coordinates to polar coordinates with $r \geq 0$ and $-180° < \theta° \leq 180°$:

(a) $(2, 2)$

C (b) $(5.031, -2.758)$

Solution   (a) $r = \sqrt{2^2 + 2^2} = \sqrt{8} = 2\sqrt{2}$, and $\tan \theta = 2/2 = 1$. Because the point lies in quadrant I, it follows that $0° < \theta° < 90°$, and so $\theta = 45°$. Hence, the polar coordinates are $(2\sqrt{2}, 45°)$.

(b) Using a calculator, and rounding off to four significant digits, we have

$$r = \sqrt{(5.031)^2 + (-2.758)^2} = 5.737.$$

Here $\tan \theta = -2.758/5.031$, and, because the point lies in quadrant IV, we have $-90° < \theta° < 0°$. Therefore, rounded off to the nearest hundredth of a degree,

$$\theta = \tan^{-1} \frac{-2.758}{5.031} = -28.73°.$$

Hence, the approximate polar coordinates are $(5.737, -28.73°)$.

Example 2 Convert the following cartesian coordinates to polar coordinates with $r \geq 0$ and $-\pi < \theta \leq \pi$.

(a) $(0, -7)$ © (b) $(-33.01, 29.77)$

Solution (a) $r = \sqrt{0^2 + (-7)^2} = \sqrt{49} = 7$. Because the point lies on the negative y axis, $\theta = -\pi/2$, and the polar coordinates are $(7, -\pi/2)$.

(b) Using a calculator, and rounding off to four significant digits, we have

$$r = \sqrt{(-33.01)^2 + (29.77)^2} = 44.45.$$

Because the point lies in quadrant II, we have $\pi/2 < \theta < \pi$, and because $\tan\theta = 29.77/(-33.01)$, the corresponding reference angle is

$$\theta_R = \tan^{-1}\frac{29.77}{33.01} = 0.7338$$

(rounded off to four significant digits). Hence,

$$\theta = \pi - \theta_R = 2.408,$$

and the approximate polar coordinates are $(44.45, 2.408)$.

PROBLEM SET 5

In problems 1–20, plot each point in the polar coordinate system.

1. $(6, 30°)$
2. $\left(10, \frac{\pi}{3}\right)$
3. $(3, 120°)$
4. $(0, 25°)$

5. $\left(4, -\frac{\pi}{6}\right)$
6. $(-4, 90°)$
7. $(-5, 135°)$
8. $\left(\frac{5}{2}, -270°\right)$

9. $(0, -45°)$
10. $\left(-\frac{7}{3}, \frac{\pi}{6}\right)$
11. $\left(\frac{17}{2}, -45°\right)$
12. $(5, -315°)$

13. $\left(\frac{5}{3}, -210°\right)$
14. $(3.75, -330°)$
15. $\left(-4.5, -\frac{\pi}{3}\right)$
16. $(-4.75, -30°)$

17. $\left(-2, -\frac{3\pi}{4}\right)$
18. $\left(-\frac{1}{2}, -\frac{3\pi}{2}\right)$
19. $(4.8, 0°)$
20. $(-3.25, 0)$

21. Plot the point $(5, 45°)$ in the polar coordinate system, and then give other polar representations $(r, \theta°)$ of the same point that satisfy each of the following conditions:
(a) $r < 0$ and $0° \leq \theta° < 360°$
(b) $r > 0$ and $-360° < \theta° \leq 0°$
(c) $r < 0$ and $-360° < \theta° \leq 0°$

22. Plot the point $\left(-3, \frac{13\pi}{6}\right)$ in the polar coordinate system, and then give other polar representations (r, θ) of the same point that satisfy each of the following conditions:
(a) $r < 0$ and $0 \leq \theta < 2\pi$
(b) $r > 0$ and $-2\pi < \theta \leq 0$
(c) $r < 0$ and $-2\pi < \theta \leq 0$

In problems 23–34, convert the polar coordinates to cartesian coordinates. If use of a calculator is indicated, round off your answers to four significant digits.

23. $(4, 45°)$

24. $(3, 60°)$

25. $\left(-5, \dfrac{\pi}{4}\right)$

26. $\left(0, \dfrac{\pi}{7}\right)$

27. $\left(1, -\dfrac{\pi}{4}\right)$

28. $\left(-5, \dfrac{5\pi}{6}\right)$

29. $(-4, 150°)$

© **30.** $(-3, 25°)$

© **31.** $(2.765, -122.73°)$

© **32.** $(-6.772, 133.33°)$

© **33.** $(17.76, -37.07°)$

© **34.** $(3.271 \times 10^7, 44.33°)$

In problems 35–44, convert the cartesian coordinates to polar coordinates with $r \geq 0$ and $-180° < \theta° \leq 180°$. If use of a calculator is indicated, round off r to four significant digits, and θ to the nearest hundredth of a degree.

35. $(\sqrt{2}, \sqrt{2})$

36. $(-1, \sqrt{3})$

37. $(1, \sqrt{3})$

38. $(-2, 2)$

39. $(-3, 0)$

© **40.** $(4, 3)$

© **41.** $(-5, 12)$

© **42.** $(-2, -3)$

© **43.** $(13.01, -15.57)$

© **44.** $(3.001 \times 10^{-5}, 2.774 \times 10^{-5})$

In problems 45–50, convert the cartesian coordinates to polar coordinates with $r \geq 0$ and $-\pi < \theta \leq \pi$. If use of a calculator is indicated, round off r and θ to four significant digits.

45. $(2\sqrt{3}, -2)$

46. $(-3, -3\sqrt{3})$

47. $(0, -2)$

48. $(4\sqrt{3}, -4)$

© **49.** $(-72.02, -91.33)$

© **50.** $(-2.102 \times 10^7, 9.901 \times 10^6)$

51. Show that the equations

$$x = r \cos \theta \quad \text{and} \quad y = r \sin \theta$$

work in all cases, even when $r \leq 0$, to convert the polar coordinates (r, θ) of a point P into the cartesian coordinates (x, y) of P.

---

6 Applications of Vectors and Polar Coordinates

In this section, we present a few of the many applications of vectors and polar coordinates. Notice that there is a particularly close connection between vectors and polar coordinates: If $\bar{R}$ is the position vector of a point P, and if θ is a direction angle for $\bar{R}$ (Figure 1), then $(|\bar{R}|, \theta)$ are polar coordinates for P and

$$\bar{R} = \langle |\bar{R}| \cos \theta, |\bar{R}| \sin \theta \rangle.$$

Figure 1

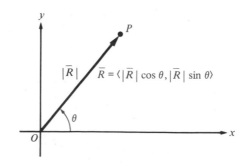

6.1   Applications of Vectors

The following example illustrates how vectors may be used to solve problems involving forces.

Example   © Two forces $\bar{F}_1$ and $\bar{F}_2$ act on a point. If $|\bar{F}_1| = 30$ pounds, $|\bar{F}_2| = 50$ pounds, and the angle α from $\bar{F}_1$ to $\bar{F}_2$ is 60°, find (a) the magnitude $|\bar{F}|$ of the resultant force $\bar{F}$, and (b) the angle θ from $\bar{F}_1$ to $\bar{F}$.

Solution   We begin by setting up an xy coordinate system with the origin O at the point on which both $\bar{F}_1$ and $\bar{F}_2$ act, and with the positive x axis in the direction of $\bar{F}_1$ (Figure 2). Then $\alpha = 60°$ is a direction angle for $\bar{F}_2$; hence, the components of $\bar{F}_2$ are given by

$$\bar{F}_2 = \langle |\bar{F}_2| \cos \alpha, |\bar{F}_2| \sin \alpha \rangle$$

$$= \langle 50 \cos 60°, 50 \sin 60° \rangle$$

$$= \left\langle \frac{50}{2}, \frac{50\sqrt{3}}{2} \right\rangle = \langle 25, 25\sqrt{3} \rangle.$$

Figure 2

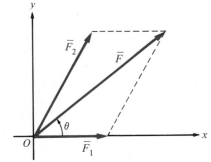

Since $\bar{F}_1$ lies along the positive x axis and $|\bar{F}_1| = 30$, $\bar{F}_1 = \langle 30, 0 \rangle$. As a consequence,

$$\bar{F} = \bar{F}_1 + \bar{F}_2 = \langle 30, 0 \rangle + \langle 25, 25\sqrt{3} \rangle$$

$$= \langle 30 + 25, 0 + 25\sqrt{3} \rangle = \langle 55, 25\sqrt{3} \rangle.$$

(a) $|\bar{F}| = \sqrt{55^2 + (25\sqrt{3})^2} = \sqrt{3025 + 1875} = \sqrt{4900} = 70$ pounds

Figure 3

(b) The angle θ from $\bar{F}_1$ to $\bar{F}$ is the direction angle of $\bar{F}$ (Figure 3). Because the components of $\bar{F}$ are both positive, $\bar{F}$ lies in quadrant I, so θ is a quadrant I angle. Therefore, because

$$\cos \theta = \frac{55}{|\bar{F}|} = \frac{55}{70},$$

it follows that

$$\theta = \cos^{-1} \frac{55}{70} \approx 38.21°.$$

Another important application of vectors is in navigation, where the following terminology is commonly used:

1. The **heading** of an airplane is the direction in which it is pointed, and its **air speed** is its speed relative to the air. The vector $\bar{V}_1$ whose magnitude is the air speed and whose direction is the heading represents the **velocity** of the airplane relative to the air (Figure 4).

(a) Applying the law of cosines to triangle ABC, we have

$$|\bar{V}|^2 = |\overline{AC}|^2 = |\overline{AB}|^2 + |\overline{BC}|^2 - 2|\overline{AB}||\overline{BC}|\cos 55°$$
$$= 550^2 + 50^2 - 2(550)(50)\cos 55°$$

so

$$|\bar{V}| = \sqrt{550^2 + 50^2 - 2(550)(50)\cos 55°}$$
$$\approx 522.9 \text{ miles per hour.}$$

(b) Applying the law of sines to triangle ABC, we have

$$\frac{\sin \alpha}{50} = \frac{\sin 55°}{|\bar{V}|},$$

so

$$\sin \alpha = \frac{50 \sin 55°}{|\bar{V}|} = \frac{50 \sin 55°}{522.9}.$$

Because α is a quadrant I angle, it follows that

$$\alpha = \sin^{-1} \frac{50 \sin 55°}{522.9} \approx 4.49°.$$

Hence, rounded off to the nearest hundredth of a degree, the drift angle is 4.49°.

(c) From Figure 5, the vector $\bar{V}$ is $30° + \alpha = 34.49°$ east of due north; hence, the course of the jet is N34.49°E.

6.2 Graphs of Polar Equations

In calculus, polar coordinates are often used to write equations of curves in the plane. An equation, such as $r^2 = 9 \cos 2\theta$, that relates the polar coordinates r and θ is called a **polar equation.** Because each point in the plane has a multitude of different polar representations, it is necessary to define the graph of a polar equation with some care: The **graph** of a polar equation consists of all points P in the plane that have at least one polar representation (r, θ) that satisfies the equation.

Example   Sketch the graph of each polar equation:

(a) $r = 4$ (b) $r^2 = 16$ (c) $\theta = \dfrac{\pi}{6}$

Figure 6   **Solution**   (a) The graph of $r = 4$ is a circle of radius 4 with center at the pole (Figure 6). Notice, for instance, that the point $P = (4, -\pi)$ belongs to the graph in spite of the fact that not all of its representations, such as $P = (-4, 0)$ or $P = (-4, 2\pi)$, satisfy the equation $r = 4$.

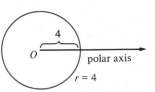

(b) The equation $r^2 = 16$ is equivalent to $|r| = 4$, and its graph is the same as the graph of $r = 4$ (Figure 6).

(c) The graph of $\theta = \pi/6$ consists of the entire line through O, making an angle of $\pi/6$ radian (30°) with the polar axis—not just the *ray* as one might at first think (Figure 7). Points of the form $P = \left(r, \dfrac{\pi}{6} + \pi\right)$ belong to the graph because they can be rewritten as $P = (-r, \pi/6)$.

Figure 7

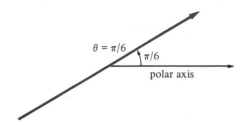

By making the substitutions $x = r\cos\theta$ and $y = r\sin\theta$, you can convert a cartesian equation into a polar equation.

Example   Find a polar equation corresponding to the cartesian equation $x^2 + (y + 4)^2 = 16$, and sketch the graph.

Solution   In Section 1.2 of Chapter 3, we learned that the graph of an equation of the form $(x - h)^2 + (y - k)^2 = r^2$ is a circle with center (h, k) and radius r. Hence, the graph of $x^2 + (y + 4)^2 = 16$ is a circle of radius 4 with center at the point with cartesian coordinates $(0, -4)$ (Figure 8). Rewriting the equation as

$$x^2 + y^2 + 8y + 16 = 16 \quad \text{or} \quad x^2 + y^2 + 8y = 0,$$

substituting $x = r\cos\theta$ and $y = r\sin\theta$, and using the fact that $x^2 + y^2 = r^2$, we obtain the polar equation

$$r^2 + 8r\sin\theta = 0$$

or

$$r + 8\sin\theta = 0.$$

Figure 8

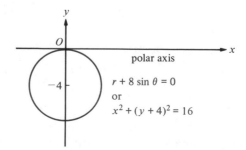

In the last step of the example above, we have divided by r; hence, we may have lost the solution $r = 0$. In this particular case, we have not lost the solution $r = 0$ because, if $r + 8\sin\theta = 0$, then $r = 0$ when $\theta = 0$. More generally, if you multiply a polar equation by r, you *may* introduce an extraneous solution $r = 0$ that doesn't really belong and, conversely, if you divide a polar equation by r, you *may* lose a solution $r = 0$ that really does belong. You must check each case.

Example   Find a cartesian equation corresponding to the polar equation $r = 4\tan\theta\sec\theta$, and sketch the graph.

Solution   We have

$$r = 4\tan\theta\sec\theta = 4\,\frac{\sin\theta}{\cos\theta}\cdot\frac{1}{\cos\theta}$$

or

$$r\cos^2\theta = 4\sin\theta.$$

Multiplication by r gives

$$r^2\cos^2\theta = 4r\sin\theta$$

or

$$(r\cos\theta)^2 = 4r\sin\theta.$$

Because $x = r\cos\theta$ and $y = r\sin\theta$, the last equation can be rewritten as

$$x^2 = 4y \quad\text{or}\quad y = \tfrac{1}{4}x^2.$$

By Section 1 of Chapter 4, the graph of $y = \tfrac{1}{4}x^2$ is a parabola with a vertical axis of symmetry and vertex at $(0,0)$ (Figure 9). [Notice that multiplication by r did not introduce an extraneous solution $r = 0$ because, if $r\cos^2\theta = 4\sin\theta$, then $r = 0$ when $\theta = 0$.]

Figure 9

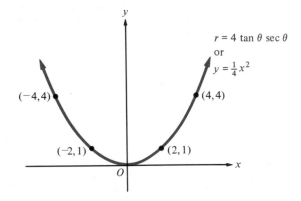

You can also sketch the graph of a polar equation by plotting polar points and connecting them with a smooth curve, in much the same way that you sketch the graph of a cartesian equation.

Example   ⒸSketch the graph of $r = 2(1 - \cos\theta)$.

Solution   Because $\cos(-\theta) = \cos\theta$, it follows that if the polar point $P = (r, \theta)$ belongs to the graph, so does the polar point $Q = (r, -\theta)$. Therefore, the graph is **symmetric about the polar axis.** We sketch the top half of the graph for $0 \le \theta \le \pi$ using the data in the table below, and then we reflect the top half across the polar axis to obtain the bottom half (Figure 10).

Figure 10

θ	$r = 2(1 - \cos\theta)$
0	0
$\pi/6$	0.27 (approximately)
$\pi/4$	0.59 (approximately)
$\pi/3$	1
$\pi/2$	2
$2\pi/3$	3
$5\pi/6$	3.73 (approximately)
π	4

$r = 2(1 - \cos\theta)$

The curve in Figure 10 is called a *cardioid*.

PROBLEM SET 6

Ⓒ In problems 1–4, assume that the two forces $\bar{F}_1$ and $\bar{F}_2$ act on a point and that α is the angle from $\bar{F}_1$ to $\bar{F}_2$. If $\bar{F}$ is the resultant force, find (a) $|\bar{F}|$ and (b) the angle θ from $\bar{F}_1$ to $\bar{F}$.

1. $|\bar{F}_1| = 25$ pounds, $|\bar{F}_2| = 40$ pounds, $\alpha = 45°$
3. $|\bar{F}_1| = 22$ newtons, $|\bar{F}_2| = 33$ newtons, $\alpha = 120°$

2. $|\bar{F}_1| = 41$ newtons, $|\bar{F}_2| = 7$ newtons, $\alpha = 27°$
4. $|\bar{F}_1| = 3.4$ pounds, $|\bar{F}_2| = 7.3$ pounds, $\alpha = 133°$

Ⓒ 5. Two children are pulling a third child across the ice on a sled. The first child pulls on her rope with a force of 10 pounds, and the second child pulls on his rope with a force of 7 pounds. If the angle between the ropes is 30°, (a) what is the magnitude of the resultant force, and (b) what angle does the resultant force vector make with the first child's rope?

Ⓒ 6. Two tugboats are pulling an ocean freighter (Figure 11). The first tugboat exerts a force of 1300 kilograms on a cable making an angle of 25° with the axis of the ship. The second tugboat pulls on a cable making an angle of 32° with the axis of the ship. If the resultant force vector lies directly along the axis of the ship, (a) what force is the second tugboat exerting on the ship, and (b) what is the magnitude of the resultant force vector?

Figure 11

C 7. An airplane is headed N95°E with an air speed of 800 kilometers per hour. The wind is blowing N25°E at a speed of 45 kilometers per hour. Find (a) the ground speed, (b) the drift angle, and (c) the course of the plane.

C 8. A long-distance swimmer can swim 3 kilometers per hour in still water. She is swimming away from and at right angles to a straight shoreline. If a current of 5 kilometers per hour is flowing parallel to the shoreline, find her speed relative to the land and the angle θ between the shoreline and her actual velocity vector.

C 9. A balloon is rising 3 meters per second, while a wind is blowing horizontally at 2 meters per second. Find (a) the speed of the balloon, and (b) the angle its track makes with the horizontal.

10. Figure 12 shows the sail $\overline{AB}$ of an iceboat and a particle P of air moving with velocity $\overline{V}_1$ perpendicular to the iceboat's track. Suppose that $\overline{V}_2$ is the velocity of the iceboat and that the sail $\overline{AB}$ makes an angle θ with the iceboat's track. Assume that P is moving at such a speed that it just maintains contact with the sail. (a) Show that $|\overline{V}_1| = |\overline{V}_2| \tan \theta$. (b) Explain why an iceboat can be propelled by a wind at a speed greater than the speed of the wind.

Figure 12

C 11. Three children located at points A, B, and C tug on ropes attached to a ring at point O (Figure 13). The child at A pulls with a force of 4 kilograms in the direction S71°W; the child at B pulls with a force of 12 kilograms in the direction S21°E; and the child at C pulls with a force of 7 kilograms in the direction N82°E. In what direction and with what force should a fourth child pull on a rope attached to the ring in order to counterbalance the other three forces?

Figure 13

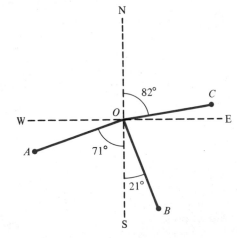

12. Suppose that two forces $\overline{F}_1$ and $\overline{F}_2$ act on a point and that α is the angle from $\overline{F}_1$ to $\overline{F}_2$. Let $\overline{F}$ be the resultant of $\overline{F}_1$ and $\overline{F}_2$ and let θ be the angle from $\overline{F}_1$ to $\overline{F}$.

(a) Show that $|\overline{F}| = (|\overline{F}_1|^2 + |\overline{F}_2|^2 + 2|\overline{F}_1||\overline{F}_2| \cos \alpha)^{1/2}$.

(b) Show that $\theta = \sin^{-1} \dfrac{|\overline{F}_2| \sin \alpha}{|\overline{F}|}$.

In problems 13–18, sketch the graph of each polar equation.

13. $r = 1$

14. $r^2 = 9$

15. $\theta = \dfrac{\pi}{3}$

16. $\theta^2 = \dfrac{\pi^2}{16}$

17. $\theta + \dfrac{\pi}{2} = 0$

18. $\theta^2 - \dfrac{4\pi}{3}\theta + \dfrac{7\pi^2}{36} = 0$

In problems 19–24, find a polar equation corresponding to each cartesian equation and sketch the graph. Simplify your answer if possible.

19. $x^2 + y^2 = 25$

20. $y = 6x^2$

21. $3x - 2y = 6$

22. $xy = 1$

23. $(x - 1)^2 + y^2 = 1$

24. $y = \sin x$

In problems 25–30, find a cartesian equation corresponding to each polar equation and sketch the graph. Simplify your answer if possible.

25. $r = 1$

26. $r = 3\cos\theta$

27. $r = \dfrac{1}{\sin\theta - 2\cos\theta}$

28. $\theta = \dfrac{\pi}{12}$

29. $r\cos\theta = \tan\theta$

30. $r + \dfrac{1}{r} = 2(\cos\theta + \sin\theta)$

In problems 31–34, use the point-plotting method to sketch the graph of each polar equation.

31. $r = 2(1 - \sin\theta)$ (This is called a *cardioid*.)

32. $r = 3\sin 3\theta$ (This is called a *three-leaved rose*.)

33. $r = \dfrac{6\theta}{\pi}$ (This is called an *Archimedean spiral*.)

34. $r = 3 - 2\cos\theta$ (This is called a *limaçon*.)

REVIEW PROBLEM SET

[c] In problems 1–4, assume that the right triangle ACB is labeled as in Figure 1. In each case, solve the triangle and sketch it. Round off all angles to the nearest hundredth of a degree and all side lengths to four significant digits.

1. $c = 200$ meters, $\alpha = 31.25°$

2. $a = 92.70$ kilometers, $\alpha = 34.45°$

3. $b = 47.33$ microns, $\beta = 56.15°$

4. $b = 1.151 \times 10^4$ meters, $c = 4.703 \times 10^4$ meters

Figure 1

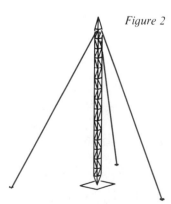

Figure 2

[c] **5.** A straight sidewalk is inclined at an angle of 4.5° from the horizontal. How far must a person walk along this sidewalk to change his or her elevation by 2 meters?

[c] **6.** A ship in distress at night fires a signal rocket straight upward. At an altitude of 0.2 kilometer, the rocket explodes with a brilliant flash that is observed at an angle of elevation of 15.75° by an approaching rescue ship. How far is the rescue ship from the ship in distress?

[c] **7.** A broadcasting antenna tower 200 meters high is to be held vertical by three cables running from a point 10 meters below the top of the tower to concrete anchors sunk in the ground (Figure 2). If the cables are to make angles of 60° with the horizontal, how many meters of cable will be required?

8. Let θ be the angle of depression of the horizon as seen by an astronaut A in a space vehicle h meters above the surface of a planet of radius r meters (Figure 3). Show that $h = r(\sec\theta - 1)$.

Figure 3

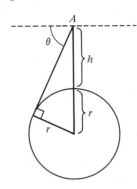

C 9. A certain species of bird is known to fly at an average altitude of 260 meters during migration. An ornithologist observes a migrating flock of these birds flying directly away from her at an angle of elevation of 30°. One minute later, she observes that the angle of elevation of the flock has changed to 20°. Approximately how fast are the birds flying?

C 10. In dealing with alternating current circuits, electrical engineers measure the opposition to current flow by a quantity called **impedance,** whose magnitude Z is measured in ohms. In a circuit driven by an alternating electromotive force (emf) and containing only a resistor and an inductor connected in series, the resistor and inductor both contribute to the impedance. If the resistance is R ohms, the inductance is L henrys, and the driving emf has a frequency of v hertz, the relationships among Z, R, L, and v is shown by an **impedance triangle** as in Figure 4. The angle ϕ is called the **phase** and the quantity $2\pi vL$ is called the **inductive reactance.** If $R = 4000$ ohms, $L = 2$ henrys, and $v = 60$ hertz, find Z and ϕ.

C 11. An electronic echo locator on a commercial fishing boat indicates a school of fish at a slant distance of 575 meters from the boat with an angle of depression of 33.60°. What is the depth of the school of fish?

Figure 4

C 12. A jetliner is flying at an altitude of 10 kilometers. The captain is preparing for a descent to an airport along a straight flight path making an angle of 5° with the horizontal. If the descent will take 12 minutes, what will be the average airspeed of the plane during the descent?

C 13. A ship leaves port and sails S48°10′E for 41 nautical miles. At this point, it turns and sails S41°50′W for 75 nautical miles. Find the bearing of the ship from the port and its distance from the port.

C 14. Two buildings A and B are 2175 feet apart; the angle of depression from the top of B to the top of A is 12.17°; and the angle of depression from the top of B to the bottom of A is 53.35°. Find the heights of the buildings.

C In problems 15–20, use the law of sines to find the specified part of each triangle ABC (Figure 5). Round off angles to the nearest hundredth of a degree and side lengths to four significant digits.

15. Find a if $b = 13$ feet, $\alpha = 47°$, and $\gamma = 118°$.
16. Find b if $a = 8.12$ centimeters, $\alpha = 62.45°$, and $\beta = 79.30°$.
17. Find c if $a = 23$ miles, $\alpha = 65°$, and $\beta = 70°$.
18. Find a if $c = 13.70$ kilometers, $\alpha = 62°50′$, and $\beta = 57°30′$.
19. Find γ if $b = 24$ meters, $c = 14$ meters, and $\beta = 38°$.
20. Find α if $a = 5.88$ microns, $c = 12.35$ microns, and $\gamma = 106.55°$.

Figure 5

C In problems 21–26, solve each triangle ABC (Figure 5) by using the method for the ambiguous case. Round off angles to the nearest hundredth of a degree and side lengths to four significant digits. Be certain to find all possible triangles that satisfy the given conditions.

21. $b = 50\sqrt{3}$ feet, $c = 150$ feet, $\beta = 30°$
22. $a = 4.50$ kilometers, $b = 5.30$ kilometers, $\alpha = 60.33°$
23. $b = 50$ centimeters, $c = 58$ centimeters, $\gamma = 57.25°$

24. $b = 5.94$ angstroms, $c = 7.23$ angstroms, $\beta = 38°$
25. $a = 8.00$ miles, $b = 10.00$ miles, $\alpha = 54°$
26. $a = 40.33$ nautical miles, $b = 42.01$ nautical miles, $\alpha = 110.05°$

© In problems 27–32, use the law of cosines to find the specified part of each triangle ABC (Figure 5). Round off angles to the nearest hundredth of a degree and side lengths to four significant digits.

27. Find α if $a = 7$ feet, $b = 8$ feet, and $c = 3$ feet.
28. Find a if $b = 11$ meters, $c = 12$ meters, and $\alpha = 81°$.
29. Find c if $a = 14$ kilometers, $b = 8$ kilometers, and $\gamma = 37°$.
30. Find b if $a = 7$ centimeters, $c = 11$ centimeters, and $\beta = 53.35°$.
31. Find a if $b = 12.30$ inches, $c = 4.85$ inches, and $\alpha = 161.15°$.
32. Find γ if $a = 48.31$ meters, $b = 35.11$ meters, and $c = 63.27$ meters.

© In problems 33–38, use an appropriate method to solve each triangle ABC (Figure 5). Round off angles to the nearest hundredth of a degree and side lengths to four significant digits.

33. $a = 49.7$ kilometers, $b = 111$ kilometers, $\gamma = 41.05°$
34. $\alpha = 37.17°$, $\beta = 82.25°$, $a = 7777$ meters
35. $a = 1.12$ microns, $c = 0.98$ micron, $\gamma = 55°$
36. $a = 33.39$ centimeters, $b = 72.27$ centimeters, $c = 106.2$ centimeters
37. $\alpha = 42.45°$, $\beta = 32.15°$, $b = 1.41$ nautical miles
38. $a = 1.80$ millimeters, $b = 1.20$ millimeters, $\alpha = 47°$

© 39. The three sides of a triangle have lengths $a = 27$ meters, $b = 44$ meters, and $c = 65$ meters. Use Hero's formula (page 398) to find the area of the triangle.

© 40. A section of West Germany is called the brown-coal triangle because it contains vast reserves of lignite (brown coal). It is a triangular area with vertices at the industrial cities of Aachen, Cologne, and Düsseldorf. The distance from Aachen to Cologne is 67 kilometers; the distance from Cologne to Düsseldorf is 37 kilometers; and the distance from Düsseldorf to Aachen is 75 kilometers. Find the approximate area of the brown-coal triangle in square kilometers.

© 41. In the western North Atlantic there is a mysterious region, shaped roughly like a triangle with sides of length 925 miles, 850 miles, and 1300 miles, that has been the scene of a number of unexplained disappearances and disasters involving both ships and aircraft. Find the approximate area of this region in square miles.

© 42. Engineers excavating a new subway tunnel in a large city encounter a region of very hard rock (Figure 6). They can either drill the tunnel straight through the rock along path $\overline{AC}$ at a cost of \$5500 per foot, or they can divert the tunnel along path $\overline{AB}$ and then path $\overline{BC}$ at a cost of \$4700 per foot. If angle $BAC = 20°$ and angle $ACB = 10°$, which path would be the cheaper?

Figure 6

© 43. Two observers are situated on level ground on opposite sides of a tall building. The top of the building is 1600 meters from the first observer at an angle of elevation of 15. The top of the building is 650 meters from the second observer. How far apart are the observers?

C 44. A vertical utility pole is supported on an embankment by a guy wire from 1 meter below the top of the pole to a point 20 meters up the embankment from the foot of the pole (Figure 7). If the embankment makes an angle of 17° with the horizontal and if the guy wire makes an angle of 21° with the horizontal, how tall is the utility pole?

Figure 7

C 45. A CB radio in a person's car can communicate with a base station located at the person's home over a maximum distance of 12 miles. The car leaves the home and travels due west for 5 miles to an interstate highway. The car moves onto the straight highway at precisely 1:00 P.M. traveling N30°E at 55 miles per hour. At what time (to the nearest minute) is communication lost between the car and the base station?

C 46. Apollo objects in the solar system are asteroidlike masses whose orbits intersect the orbit of the earth. Geologists believe that several craterlike formations in Canada and elsewhere were caused by collisions of such objects with the earth. An astronomer observes through a telescope that the angle formed by the lines of sight to two Apollo objects is 43.33°, and that the distances to the two objects are 15,000,000 and 43,000,000 kilometers. How far apart are the two Apollo objects?

C 47. A natural-gas pipeline is to be constructed through a swamp. An engineer obtains the distance across the swamp by establishing points A and B at both ends of it along the path the pipeline is to follow, and finding a third point C outside the swamp at a location that can be seen from both A and B. If $|\overline{CA}| = 1100$ meters, $|\overline{CB}| = 990$ meters, and angle ACB is 75°, what is the distance $|\overline{AB}|$ across the swamp?

C 48. A group of antinuclear activists is conducting a protest march to the construction site of a nuclear power plant. They are marching at constant speed along a flat plane directly toward a huge concrete cooling tower that has already been finished. At 7:00 A.M. the leader of the march observes that the top of the cooling tower is at an angle of elevation of 2°. At 7:30 A.M. the angle of elevation has increased to 5°. Estimate to the nearest minute the time of arrival of the protest group at the construction site.

C 49. A subatomic particle that has been created at point C in a bubble chamber travels in a straight line for 3.5 cm to point B, where it collides with another particle and its path is deflected. It then travels in a straight line from B for 2.3 cm to point A, where it is annihilated. If $|\overline{CA}| = 4.4$ cm, find the deflection angle $\theta = 180° -$ angle CBA.

50. Two small ships are traveling at constant speeds on straight-line courses that intersect. The captain of the first ship measures the bearing of the second ship at the beginning and at the end of a 2-minute interval. The bearing has changed during this interval, so the captain concludes that the ships will not collide. Is this reasoning correct? Why or why not?

In problems 51–62, let $\bar{A} = \langle 2, -1 \rangle$, $\bar{B} = \langle 3, 3 \rangle$, and $\bar{C} = \langle -5, 4 \rangle$. Evaluate each expression.

51. $5\bar{A}$

52. $-4\bar{B}$

53. $\bar{A} + \bar{B}$

54. $\bar{A} - \bar{B}$

55. $2\bar{A} + 3\bar{B}$

56. $3\bar{A} + \bar{C}$

57. $2\bar{C} - \bar{B}$

58. $3\bar{B} + 4\bar{C} - \bar{A}$

59. $|4\bar{A}|$

60. $|\bar{A} - \bar{B}|$

61. $|\bar{A} + \bar{B} - \bar{C}|$

62. $|2\bar{C} - 3\bar{B} + \bar{A}|$

63. Find the components of the vector $\bar{A}$ from the given information about its length $|\bar{A}|$ and its direction angle θ:

(a) $|\bar{A}| = 50, \theta = 30°$ **C** (b) $|\bar{A}| = 70, \theta = 61°$ **C** (c) $|\bar{A}| = 25, \theta = 310°$ (d) $|\bar{A}| = 250, \theta = -\dfrac{\pi}{6}$

64. Give an example to show that in general $|\bar{A} + \bar{B}| \neq |\bar{A}| + |\bar{B}|$. Are there any cases in which equality does hold?

65. Find the magnitude and the smallest nonnegative direction angle of each vector:
(a) $\langle 1, \sqrt{3} \rangle$ C (b) $\langle 12, 15 \rangle$ C (c) $\langle -30, 40 \rangle$ C (d) $\langle \frac{3}{16}, \frac{1}{4} \rangle$ C (e) $3\bar{i} + \sqrt{7}\bar{j}$

66. Give an example to show that the direction angle of the sum of two vectors is not necessarily the sum of their direction angles.

C **67.** Suppose that $\bar{F}_1$ and $\bar{F}_2$ represent forces acting on a point, and that α is the angle from $\bar{F}_1$ to $\bar{F}_2$. If $\bar{F}$ denotes the resultant force, find the magnitude $|\bar{F}|$ and the angle θ from $\bar{F}_1$ to $\bar{F}$ in each of the following cases:
(a) $|\bar{F}_1| = 20$ pounds, $|\bar{F}_2| = 18$ pounds, $\alpha = 71°$
(b) $|\bar{F}_1| = 139$ newtons, $|\bar{F}_2| = 156$ newtons, $\alpha = 83°$
(c) $|\bar{F}_1| = 7$ tonnes, $|\bar{F}_2| = 9$ tonnes, $\alpha = 125°$
(d) $|\bar{F}_1| = 300$ dynes, $|\bar{F}_2| = 450$ dynes, $\alpha = 270°$

C **68.** Find the angle θ from the position vector $\bar{R}_1 = \langle 5, 3 \rangle$ to the position vector $\bar{R}_2 = \langle -7, 2 \rangle$.

C **69.** A cargo jet is headed N45°W with an air speed of 500 knots. The wind is blowing N60°E at a speed of 70 knots. Find (a) the ground speed, (b) the drift angle, and (c) the course of the jet.

C **70.** Using ropes, two workers are gently lowering a stone weighing 60 kilograms as shown in Figure 8. The ropes make angles of 60° and 45° with the horizontal. Make a vector diagram showing all three forces acting on the stone, including the force of gravity. Using the fact that all three forces are in equilibrium—that is, their vector sum is $\bar{0}$—find the magnitude of the force exerted by each of the workers on their ropes.

Figure 8

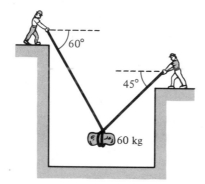

C **71.** An oceanographic research vessel leaves Miami on a course for a small island whose bearing is S43°20′E. The vessel is moving through the water at a constant speed of 20 knots, and the captain has charted the course to compensate for the fact that a Gulf Stream current of 7 knots is running due north. Determine the heading of the vessel.

72. If a force represented by a vector $\bar{F}$ acts on a particle and produces a displacement represented by a vector $\bar{D}$, the *work* done is given by $|\bar{F}||\bar{D}| \cos \alpha$, where α is the angle between $\bar{F}$ and $\bar{D}$. Calculate the work done by the force $\bar{F} = \langle -3, 4 \rangle$ in producing the displacement $\bar{D} = \langle 0, 7 \rangle$. [If you use the law of cosines, you won't need a calculator to solve this problem.]

73. Convert the polar coordinates to cartesian coordinates:
(a) $(2, 45°)$ (b) $\left(-2, \dfrac{\pi}{4}\right)$ C (c) $(-1, 280°)$
(d) $(1.5, 0°)$ (e) $(-4, -210°)$ (f) $(6, 315°)$

74. Plot the point $P = (-5, \pi/6)$ in the polar coordinate system and give other polar representations $P = (r, \theta)$ of P that satisfy each of the following conditions:

(a) $r < 0$ and $0 \le \theta < 2\pi$ (b) $r > 0$ and $-2\pi < \theta \le 0$

(c) $r < 0$ and $-2\pi < \theta \le 0$

75. Convert the cartesian coordinates to polar coordinates with $r \ge 0$ and $-\pi < \theta \le \pi$. If an approximation is necessary, round off the angle θ to the nearest hundredth of a radian.

(a) $(-1, 0)$ (b) $(-7, -7)$ Ⓒ (c) $(-5, 12)$

Ⓒ (d) $(3, -4)$ Ⓒ (e) $(-8, 15)$ Ⓒ (f) $(-30, -16)$

76. Show that the distance d between the points (r_1, θ_1) and (r_2, θ_2) in the polar coordinate system is given by the formula

$$d = \sqrt{r_1^2 - 2r_1 r_2 \cos(\theta_1 - \theta_2) + r_2^2}.$$

In problems 77–84, convert each cartesian equation into a corresponding polar equation by making the substitutions $x = r \cos\theta$, $y = r \sin\theta$, and $x^2 + y^2 = r^2$. Simplify your answer if possible.

77. $y = 3x + 1$ **78.** $(x - 1)^2 + (y - 2)^2 = 4$ **79.** $x^2 + y^2 = 4y$

80. $y^2(2 - x) = x^3$ **81.** $y^2 = 6x$ **82.** $(x^2 + y^2)^2 = x^2 - y^2$

83. $4x^2 + y^2 = 4$ **84.** $x^2 - 4y^2 = 4$

In problems 85–88, convert each polar equation into a corresponding cartesian equation.

85. $r = \cos\theta + \sin\theta$ **86.** $r = \sin 2\theta$ **87.** $r^2 \cos 2\theta = 2$ **88.** $\theta = \dfrac{5\pi}{4}$

89. Using the result of problem 87, sketch the graph of $r^2 \cos 2\theta = 2$.

90. Using the result in problem 76, write the polar form of the equation of a circle of radius a with center at (r_0, θ_0).

Ⓒ In problems 91–94, sketch the graph of each polar equation by the point-plotting method.

91. $r = 4(1 + \cos\theta)$ (This is called a *cardioid*.)

92. $r = 2 - 3\sin\theta$ (This is called a *limaçon with a loop*.)

93. $r = \cos 2\theta$ (This is called a *four-leaved rose*.)

94. $r^2 = \cos 2\theta$ (This is called a *lemniscate*.)

9 Systems of Equations and Inequalities

In Chapter 2, we studied methods for solving equations and inequalities in *one unknown*. Now we turn our attention to equations and inequalities containing *two or more unknowns*. Because applications of mathematics frequently involve many unknown quantities, it is very important to be able to deal with more than one unknown. In this chapter, we study systems of equations, determinants, systems of linear inequalities, and linear programming.

1 Systems of Linear Equations

A collection of two or more equations is called a **system** of equations. The equations in such a system are customarily written in a column enclosed by a brace on the left; for instance,

$$\begin{cases} 2x + y = -4 \\ x + 2y = 1 \end{cases}$$

is a system of two linear equations in the two unknowns x and y. Such a system is usually written as shown, with first-degree polynomials on the left and constants on the right of the equal signs.

If every equation in a system is true when we substitute particular numbers for the unknowns, we say that the substitution is a **solution** of the system. For instance, the substitution $x = -3$, $y = 2$ is a solution of the system

$$\begin{cases} 2x + y = -4 \\ x + 2y = 1. \end{cases}$$

(Why?) Sometimes we write such a solution as an ordered pair $(-3, 2)$, with the value of x first and the value of y second. Then the solution $(-3, 2)$ can be interpreted geometrically as the point where the graph of $2x + y = -4$ intersects the graph of $x + 2y = 1$ (Figure 1).

Figure 1

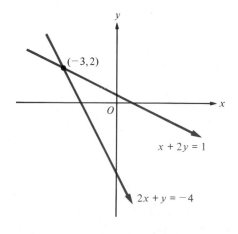

$(-3, 2)$

$x + 2y = 1$

$2x + y = -4$

In this section, we concentrate on solving systems of linear equations—later (in Section 7) we study the more general case in which nonlinear equations may be involved. The most general system of two linear equations in x and y can be written as

$$\begin{cases} ax + by = h \\ cx + dy = k, \end{cases}$$

where a, b, c, d, h, and k are constants. Unless both a and b, or both c and d, are zero, the graphs of the equations in this system are straight lines. If you draw these lines on the same coordinate system, one of the following cases will occur:

Case 1. The two lines intersect at exactly one point and there is exactly one solution (corresponding to this point). In this case, we say that the equations in the system are **consistent.**

Case 2. The two lines are parallel, and therefore do not intersect. In this case, there is no solution, and we say that the equations in the system are **inconsistent.**

Case 3. The two lines coincide. In this case, every point on the common line corresponds to a solution, and we say that the equations in the system are **dependent.**

Example Use graphs to determine whether the equations in each system are consistent, inconsistent, or dependent, and indicate the solutions (if any) on the graphs.

(a) $\begin{cases} x - 2y = 2 \\ x + y = 5 \end{cases}$ (b) $\begin{cases} x + 2y = 4 \\ 2x + 4y = -3 \end{cases}$ (c) $\begin{cases} 2x + 4y = 6 \\ x + 2y = 3 \end{cases}$

Solution The graphs of the equations in systems (a), (b), and (c) are shown in Figure 2.

(a) In Figure 2a, the graphs intersect at a point, so there is one solution, and the equations are consistent.

(b) In Figure 2b, the two lines have the same slope ($m = -\frac{1}{2}$), so they are parallel, there is no solution, and the equations are inconsistent.

(c) In Figure 2c, the two equations have the same graph, so there are infinitely many solutions (one for each point on the graph) and the equations are dependent.

Figure 2

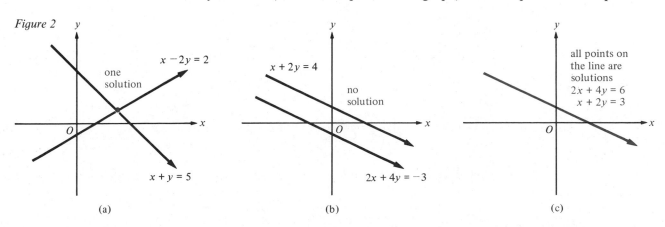

(a)   (b)   (c)

1.1 The Substitution Method

There are various algebraic methods for solving systems of equations. The **substitution method** works as follows:

> Choose one of the equations and solve it for *one* of the unknowns in terms of the remaining ones. Then substitute this solution into the remaining equation or equations. You will then have a system involving one fewer equation and one fewer unknown.

If necessary, you can repeat this procedure until the resulting system becomes simple enough to be solved.

Examples   Use the substitution method to solve each system of equations.

1 $\begin{cases} 2x + 3y = 1 \\ 3x - \ y = 7 \end{cases}$

Solution   The second equation is easily solved for y in terms of x to obtain

$$y = 3x - 7.$$

We now substitute $3x - 7$ for y in the first equation $2x + 3y = 1$ to get

$$2x + 3(3x - 7) = 1$$

or

$$2x + 9x - 21 = 1,$$

that is,

$$11x = 22 \quad \text{or} \quad x = 2.$$

To obtain the corresponding value of y, we go back to the equation $y = 3x - 7$ and substitute $x = 2$; hence,

$$y = 3(2) - 7 = -1.$$

Therefore, the solution is $x = 2$ and $y = -1$, or $(2, -1)$.

2 $\begin{cases} 2x + \ y - \ z = \ \ \ 3 \\ 2x - 2y + 8z = -24 \\ \ x + 3y + 5z = \ \ -2 \end{cases}$

Solution   We begin by solving the first equation for z in terms of x and y:

$$z = 2x + y - 3.$$

Now, we substitute $2x + y - 3$ for z in the second and third equations to obtain

$$\begin{cases} 2x - 2y + 8(2x + y - 3) = -24 \\ \ x + 3y + 5(2x + y - 3) = \ \ -2. \end{cases}$$

Simplifying these two equations, we have

$$\begin{cases} 18x + 6y = 0 \\ 11x + 8y = 13. \end{cases}$$

This simpler system can now be solved by another use of the substitution procedure. Solving the equation $18x + 6y = 0$ for y in terms of x, we get

$$y = -3x.$$

Substituting $y = -3x$ in the equation $11x + 8y = 13$, we have

$$11x + 8(-3x) = 13 \quad \text{or} \quad -13x = 13;$$

hence,

$$x = -1.$$

Now, we substitute $x = -1$ in the previous equation $y = -3x$ to get

$$y = -3(-1) = 3.$$

Having found that $x = -1$ and $y = 3$, we need only substitute these values back into the equation $z = 2x + y - 3$ to find that

$$z = 2(-1) + 3 - 3 = -2.$$

Hence, our solution is $x = -1, y = 3, z = -2$. This solution can also be written as an ordered *triple* $(-1, 3, -2)$.

1.2 The Elimination Method

Although the method of substitution can be quite efficient for solving a system of two or three equations, it tends to become cumbersome when more than three equations are involved. A more serious drawback of the substitution method is that it does not lend itself to being programmed on a computer. An alternative method, called the **method of elimination,** is similar in spirit to the method of substitution, but is much more systematic. It leads directly to **matrix methods** of solution, which are easily performed by computers.

The method of elimination is based on the idea of *equivalent* systems of equations. Two systems of equations (linear or not) are said to be **equivalent** if they have exactly the same solutions. You can change a system of equations into an equivalent system by any of the following three **elementary operations:**

1. Interchange the position of two equations in the system.

2. Multiply or divide an equation by a nonzero constant.

3. Add a constant multiple of one equation to another equation.

To solve a system of linear equations by the method of elimination, use a sequence of elementary operations to reduce the given system to a simple equivalent system whose solution is obvious. You do this by using the elementary operations to eliminate unknowns from the equations (which accounts for the name of the method). The following example illustrates the method of elimination.

Example   Use the method of elimination to solve the system of linear equations

$$\begin{cases} \frac{1}{3}x + y = 3 \\ -2x + 5y = 4. \end{cases}$$

Solution   We begin by multiplying the first equation by 3 in order to remove the fraction $\frac{1}{3}$. The result is the equivalent system

$$\begin{cases} x + 3y = 9 \\ -2x + 5y = 4. \end{cases}$$

Next we multiply the first equation by 2 and add it to the second equation in order to eliminate x from the latter. [The actual addition of

$$2x + 6y = 18 \quad \text{to} \quad -2x + 5y = 4 \quad \text{to obtain} \quad 11y = 22$$

is best done separately to avoid messing up the system with arithmetic calculations.] The resulting equivalent system is

$$\begin{cases} x + 3y = 9 \\ 11y = 22. \end{cases}$$

[Note that the first equation was *not changed* by this operation—it was used to eliminate x from the second equation, but *only* the second equation was changed.] Now we divide the second equation by 11 to obtain the equivalent system

$$\begin{cases} x + 3y = 9 \\ y = 2. \end{cases}$$

The resulting system is so simple that its solution is at hand. We just substitute $y = 2$ from the second equation back into the first equation to get

$$x + 3(2) = 9 \quad \text{or} \quad x = 3.$$

Thus, the solution is

$$x = 3 \text{ and } y = 2 \quad \text{or} \quad (3,2).$$

In the solution above, the final step in which the value of y is substituted into a previous equation is called *back substitution*.

PROBLEM SET 1

In problems 1–10, use graphs to determine whether the equations in each system are consistent, inconsistent, or dependent, and indicate the solutions (if any) on the graphs.

1. $\begin{cases} 4x + y = 5 \\ 3x - y = 2 \end{cases}$

2. $\begin{cases} y = 5x - 2 \\ y = 2x + 1 \end{cases}$

3. $\begin{cases} 3x - y = 4 \\ -6x + 2y = -8 \end{cases}$

4. $\begin{cases} x - y = 4 \\ -3x - 3y = -12 \end{cases}$

5. $\begin{cases} 2x + y = 3 \\ 4x + 2y = 7 \end{cases}$

6. $\begin{cases} 2u - v = 5 \\ u + 3v = -1 \end{cases}$

7. $\begin{cases} x + \frac{1}{3}y = 2 \\ 3x + y = -2 \end{cases}$

8. $\begin{cases} x = 2 \\ y = x \end{cases}$

9. $\begin{cases} \frac{1}{6}x + \frac{1}{4}y = \frac{1}{6} \\ \frac{1}{4}x - \frac{1}{2}y = 2 \end{cases}$

10. $\begin{cases} 0.5x - 1.2y = 0.3 \\ 0.7x + 1.5y = 3.6 \end{cases}$

11. Solve the systems in problems 1, 3, and 5 by the substitution method.

12. Solve the systems in problems 2, 4, and 6 by the substitution method.

In problems 13–20, use the substitution method to solve each system of linear equations.

13. $\begin{cases} x - y = -2 \\ 2x - 3y = -7 \end{cases}$

14. $\begin{cases} x = 3 - y \\ 5y = 12 - 2x \end{cases}$

15. $\begin{cases} 2u + 3v = 5 \\ u - 2v = 6 \end{cases}$

16. $\begin{cases} 6s + 5t = 7 \\ 3s - 7t = 13 \end{cases}$

17. $\begin{cases} \frac{1}{2}x - \frac{3}{4}y = 1 \\ 3x + \phantom{\frac{3}{4}}y = 1 \end{cases}$

18. $\begin{cases} 13x + 11y = 21 \\ 7x + 6y = -3 \end{cases}$

19. $\begin{cases} x - 2y + 3z = -3 \\ 2x - 3y - z = 7 \\ 3x + y - 2z = 6 \end{cases}$

20. $\begin{cases} s - 5t + 4u = 8 \\ 3s + t - 2u = 7 \\ -9s - 3t + 6u = 5 \end{cases}$

In problems 21–28, use the method of elimination to solve each system of linear equations.

21. $\begin{cases} 2x + y = 10 \\ 3x - y = 5 \end{cases}$

22. $\begin{cases} 5u + v = 14 \\ 2u + v = 5 \end{cases}$

23. $\begin{cases} 2x + 3y = 18 \\ -7x + 9y = 15 \end{cases}$

24. $\begin{cases} x + \frac{1}{2}y = 2 \\ 3x - \phantom{\frac{1}{2}}y = 1 \end{cases}$

25. $\begin{cases} x + y + z = 6 \\ 2x - y + z = 3 \\ x + 2y - 3z = -4 \end{cases}$

26. $\begin{cases} x - 5y + 4z = 8 \\ 3x + y - 2z = 7 \\ 9x + 3y - 6z = -5 \end{cases}$

27. $\begin{cases} 2x + y + z = 20 \\ x + 2y + 2z - 16 \\ x + y + 2z = 12 \end{cases}$

28. $\begin{cases} 2x_1 + 3x_2 - 2x_3 = 3 \\ 8x_1 + x_2 + x_3 = 2 \\ 2x_1 + 2x_2 + x_3 = 1 \end{cases}$

29. An appliance store sells driers for $280 each and washing machines for $315 each. On a certain day, the store sells a total of 39 washers and driers, and its total receipts for them are $11,375. Let x denote the number of driers sold on this day and let y denote the number of washing machines sold. (a) Write two linear equations that must be satisfied by x and y. (b) Solve the resulting system.

30. Solve the system

$$\begin{cases} \dfrac{1}{s} + \dfrac{4}{t} - \dfrac{3}{u} = 4 \\[2mm] \dfrac{2}{s} - \dfrac{3}{t} + \dfrac{1}{u} = 1 \\[2mm] -\dfrac{3}{s} + \dfrac{2}{t} + \dfrac{2}{u} = -3 \end{cases}$$

for s, t, and u. $\left[\text{Hint: Let } x = \dfrac{1}{s},\, y = \dfrac{1}{t},\, \text{and } z = \dfrac{1}{u}.\right]$

2   The Elimination Method Using Matrices

When you solve a system of linear equations by the elimination method, your arithmetic involves only the numerical coefficients and constants in the equations—the unknowns just "go along for the ride." This suggests abbreviating a system of linear equations by writing only the coefficients and constants. For instance, to abbreviate the system

$$\begin{cases} \frac{1}{3}x + y = 3 \\ -2x + 5y = 4, \end{cases}$$

we write only the coefficients on the left and the constants on the right:

$$\begin{array}{cc|c} \frac{1}{3} & 1 & 3 \\ -2 & 5 & 4. \end{array}$$

(The dashed line separating the coefficients and the constants is optional.) It is customary to enclose the resulting array of numbers in square brackets,

$$\begin{bmatrix} \frac{1}{3} & 1 & 3 \\ -2 & 5 & 4 \end{bmatrix},$$

and to refer to it as the **matrix** of the system.

Example 1   Write the matrix of the system

$$\begin{cases} y - 2x + \frac{2}{3}z = 7 \\ z + 5x = 3 \\ -y - z = 4. \end{cases}$$

Solution   We begin by rewriting the equations so that the unknowns on the left appear in the order x, y, z. Missing unknowns are written with zero coefficients:

$$\begin{cases} -2x + y + \frac{2}{3}z = 7 \\ 5x + 0y + \phantom{\frac{2}{3}}z = 3 \\ 0x - y - \phantom{\frac{2}{3}}z = 4. \end{cases}$$

The corresponding matrix is

$$\begin{bmatrix} -2 & 1 & \frac{2}{3} & 7 \\ 5 & 0 & 1 & 3 \\ 0 & -1 & -1 & 4 \end{bmatrix}.$$

Example 2   Write the system of linear equations in x, y, and z, corresponding to the matrix

$$\begin{bmatrix} 0 & 5 & -1 & \frac{2}{3} \\ 1 & 2 & 0 & 0 \end{bmatrix}.$$

Solution   Since there are only two horizontal rows in the matrix, there are only two equations in the corresponding system:

$$\begin{cases} 0x + 5y - z = \frac{2}{3} \\ x + 2y + 0z = 0 \end{cases} \quad \text{or} \quad \begin{cases} 5y - z = \frac{2}{3} \\ x + 2y = 0. \end{cases}$$

The horizontal rows in a matrix are called simply the **rows** and the vertical columns are called simply the **columns.** The numbers that appear in the matrix are called the **entries** or **elements** of the matrix. To specify a particular entry in a matrix, you can give its "address" by indicating the row and column to which it belongs. For instance, in the matrix

$$\begin{bmatrix} 0 & 5 & -1 & \frac{2}{3} \\ 1 & 2 & 0 & 0 \end{bmatrix},$$

the entry in the first row and third column is -1; the entry in the second row and second column is 2; and so forth.

The three elementary operations on the equations of a system (Section 1) are represented by the following **elementary row operations** on the corresponding matrix:

---

1. Interchange two rows.

2. Multiply or divide all elements of a row by a nonzero constant.

3. Add a constant multiple of the elements of one row to the corresponding elements of another row.

We denote an interchange of the ith and jth rows of a matrix by $R_i \rightleftarrows R_j$; the multiplication of the ith row by a constant c by cR_i; and the addition of such a constant multiple to the jth row by $cR_i + R_j$. By performing these elementary row operations on a matrix, you can solve the corresponding system of linear equations by the elimination method. When an unknown is eliminated from a particular equation, a zero will appear in the corresponding row and column of the matrix.

For instance, the solution of the example on page 432 in Section 1 is abbreviated as follows:

$$\begin{bmatrix} \frac{1}{3} & 1 & | & 3 \\ -2 & 5 & | & 4 \end{bmatrix} \xrightarrow{3R_1} \begin{bmatrix} 1 & 3 & | & 9 \\ -2 & 5 & | & 4 \end{bmatrix} \xrightarrow{2R_1 + R_2} \begin{bmatrix} 1 & 3 & | & 9 \\ 0 & 11 & | & 22 \end{bmatrix}$$

$$\xrightarrow{\frac{1}{11}R_2} \begin{bmatrix} 1 & 3 & | & 9 \\ 0 & 1 & | & 2 \end{bmatrix};$$

that is,

$$\begin{cases} x + 3y = 9 \\ y = 2, \end{cases}$$

from which the solution $x = 3$, $y = 2$ is found, as before, by back substitution. In this calculation, the last matrix is in **echelon form** in the sense that the following conditions are satisfied:

> 1. The **leading entry**—that is, the first nonzero entry, reading from left to right—in each row is 1.
>
> 2. The leading entry in each row after the first is to the right of the leading entry in the previous row.
>
> 3. If there are any rows with no leading entry—that is, rows consisting entirely of zeros—they are placed at the bottom of the matrix.

By a sequence of elementary row operations, the matrix corresponding to any system of linear equations can be brought into echelon form, and then the solution can be obtained by back substitution.

Example   Solve the system of equations

$$\begin{cases} 2x + y - z = 5 \\ 2x - 2y + 8z = -10 \\ 4y + z = 7. \end{cases}$$

Solution   We write the matrix of the system and then reduce it to echelon form by a sequence of elementary row operations as follows:

$$\begin{bmatrix} 2 & 1 & -1 & \vdots & 5 \\ 2 & -2 & 8 & \vdots & -10 \\ 0 & 4 & 1 & \vdots & 7 \end{bmatrix} \xrightarrow{(-1)R_1 + R_2} \begin{bmatrix} 2 & 1 & -1 & \vdots & 5 \\ 0 & -3 & 9 & \vdots & -15 \\ 0 & 4 & 1 & \vdots & 7 \end{bmatrix} \xrightarrow{(-\frac{1}{3})R_2}$$

$$\begin{bmatrix} 2 & 1 & -1 & \vdots & 5 \\ 0 & 1 & -3 & \vdots & 5 \\ 0 & 4 & 1 & \vdots & 7 \end{bmatrix} \xrightarrow{(-4)R_2 + R_3} \begin{bmatrix} 2 & 1 & -1 & \vdots & 5 \\ 0 & 1 & -3 & \vdots & 5 \\ 0 & 0 & 13 & \vdots & -13 \end{bmatrix} \xrightarrow{\frac{1}{13}R_3}$$

$$\begin{bmatrix} 2 & 1 & -1 & \vdots & 5 \\ 0 & 1 & -3 & \vdots & 5 \\ 0 & 0 & 1 & \vdots & -1 \end{bmatrix} \xrightarrow{\frac{1}{2}R_1} \begin{bmatrix} 1 & \frac{1}{2} & -\frac{1}{2} & \vdots & \frac{5}{2} \\ 0 & 1 & -3 & \vdots & 5 \\ 0 & 0 & 1 & \vdots & -1 \end{bmatrix}.$$

The last matrix is in echelon form and corresponds to the system

$$\begin{cases} x + \frac{1}{2}y - \frac{1}{2}z = \frac{5}{2} \\ y - 3z = 5 \\ z = -1. \end{cases}$$

Now we back substitute $z = -1$ from the third equation into the second equation $y - 3z = 5$ to obtain

$$y - 3(-1) = 5 \quad \text{or} \quad y = 2.$$

Finally, we back substitute $z = -1$ and $y = 2$ into the first equation $x + \frac{1}{2}y - \frac{1}{2}z = \frac{5}{2}$ to obtain

$$x + \tfrac{1}{2}(2) - \tfrac{1}{2}(-1) = \tfrac{5}{2} \quad \text{or} \quad x = 1.$$

Therefore, the solution is

$$x = 1, y = 2, \text{ and } z = -1 \quad \text{or} \quad (1, 2, -1).$$

For the relatively simple systems of linear equations considered here and in the problems at the end of this section, you can find, by trial and error, a suitable sequence of elementary row operations that reduces the matrix to echelon form. More advanced textbooks on linear algebra describe step-by-step procedures (eliminating all guesswork) for doing this. In these textbooks, you can also find a proof that, for any system of linear equations, there are just three possibilities:

Case 1. There is exactly one solution, in which case we say that the equations in the system are **consistent.**

Case 2. There is no solution, in which case we say that the equations in the system are **inconsistent.**

Case 3. There is an infinite number of solutions, in which case we say that the equations in the system are **dependent.**

In working with systems of linear equations, you can use whatever letters you please to denote the unknowns. Sometimes it is convenient to use just one letter with different subscripts (for instance, $x_1, x_2, x_3, \ldots, x_m$) to represent the different unknowns. In any case, before you can form the corresponding matrix, you must decide in what order the unknowns are to be written in the equations. When subscripts are used, the unknowns are usually written in the numerical order of the subscripts.

PROBLEM SET 2

In problems 1–4, consider the matrix $\begin{bmatrix} 0 & 5 & -3 \\ 2 & -1 & 7 \end{bmatrix}$.

1. What are the elements in the first row?
2. What are the elements in the second column?
3. What is the element in the second row and first column?
4. What is the element in the first row and third column?

In problems 5–8, write the matrix of each system of linear equations.

5. $\begin{cases} \frac{3}{4}x - \frac{2}{3}y = \frac{1}{7} \\ -x + 5y = 6 \end{cases}$

6. $\begin{cases} 0.5x_1 + 3.2x_2 = 7.1 \\ 5.3x_1 - 3.0x_2 = -6.5 \end{cases}$

7. $\begin{cases} 40x - z + 22y = -17 \\ y + z = 0 \\ 17y - 13x + 12z = 5 \end{cases}$

8. $\begin{cases} 3x_3 - 2x_2 = x_1 \\ 2x_1 - 5x_3 = -x_2 \\ x_2 + x_3 = 6 \end{cases}$

In problems 9–12, write the system of linear equations in x and y or in x, y, and z corresponding to each matrix.

9. $\begin{bmatrix} 1 & 3 & | & 0 \\ 2 & -4 & | & 1 \end{bmatrix}$

10. $\begin{bmatrix} 0.1 & 3.2 & | & -1.7 \\ -4.4 & 0 & | & 0 \end{bmatrix}$

11. $\begin{bmatrix} 2 & 5 & 3 & | & 1 \\ -3 & 7 & \frac{1}{2} & | & \frac{3}{4} \\ 0 & \frac{2}{3} & 0 & | & -\frac{4}{5} \end{bmatrix}$

12. $\begin{bmatrix} 3 & 2 & | & 3 \\ 1 & \frac{2}{3} & | & 1 \\ 5 & 0 & | & -2 \end{bmatrix}$

In problems 13–36, write the matrix of the system of linear equations, reduce the matrix to echelon form by a sequence of elementary row operations, then use back substitution to solve the system.

13. $\begin{cases} x + y = 4 \\ x - 4y = 8 \end{cases}$

14. $\begin{cases} x + 6y = 7 \\ 11x - 7y = -10 \end{cases}$

15. $\begin{cases} 3x + y = 15 \\ 3x - 7y = 15 \end{cases}$

16. $\begin{cases} -6x_1 + 2x_2 = 3 \\ 2x_1 + 5x_2 = 3 \end{cases}$

17. $\begin{cases} 4s + 3t = 17 \\ 2s + 3t = 13 \end{cases}$

18. $\begin{cases} \frac{1}{2}x + \frac{2}{3}y = 6 \\ -\frac{3}{2}x + \frac{1}{2}y = -3 \end{cases}$

19. $\begin{cases} 2x + y = 6 \\ 3x + 4y = 4 \end{cases}$

20. $\begin{cases} \frac{1}{7}u + \frac{3}{7}v = 1 \\ -\frac{1}{7}u + \frac{4}{7}v = 1 \end{cases}$

21. $\begin{cases} x + y = 1 \\ 2x + 2y = 0 \end{cases}$

22. $\begin{cases} 5x - 2y = y - 1 \\ 4x - 5y = 3 - 2x \end{cases}$

23. $\begin{cases} \frac{1}{3}x_1 + \frac{1}{6}x_2 = 1 \\ x_1 - x_2 = 3 \end{cases}$

24. $\begin{cases} x + by = 2 \\ bx + y = 3 \end{cases}$

25. $\begin{cases} x + 5y - z = -7 \\ 3x + 4y - 2z = 2 \\ 2x - 3y + 5z = 19 \end{cases}$

26. $\begin{cases} 3x + 2y + 5z = 7 \\ 2x - 3y - 2z = -3 \\ x + 2y + 3z = 5 \end{cases}$

27. $\begin{cases} 2x + y - 3z = 11 \\ x - 2y + 4z = -3 \\ 3x + y - 2z = 12 \end{cases}$

28. $\begin{cases} u + 5v - w = 2 \\ 2u + v + w = 7 \\ u - v + 2w = 11 \end{cases}$

29. $\begin{cases} 8r + 3s - 18t = 1 \\ 16r + 6s - 6t = 7 \\ 4r + 9s + 12t = 9 \end{cases}$

30. $\begin{cases} \frac{2}{3}x_1 + \frac{1}{4}x_2 - \frac{1}{3}x_3 = 3 \\ -\frac{3}{2}x_1 + \frac{1}{8}x_2 + x_3 = 1 \\ \frac{1}{2}x_1 - x_2 + x_3 = 4 \end{cases}$

31. $\begin{cases} x + 3y + z = 0 \\ 2y + 4z = 1 \\ -x + 3z = 2 \end{cases}$

32. $\begin{cases} x + 2y - z = 0 \\ 2x - y + 3z = 1 \\ 3x - 2y = -1 \end{cases}$

33. $\begin{cases} x_1 + 2x_2 - x_3 = 0 \\ 2x_1 - 2x_2 + x_3 = 0 \\ 6x_1 + 4x_2 + 3x_3 = 0 \end{cases}$

34. $\begin{cases} 3y - 2z = 2 \\ 4x + 5z = -1 \\ 5x + y = 0 \end{cases}$

35. $\begin{cases} 2x + 3z = 1 \\ x - y + 2z = 0 \\ 3x - y + 5z = -1 \end{cases}$

36. $\begin{cases} 3x_1 + 5x_2 - x_3 = 2 \\ x_2 + x_3 = 3 \\ 2x_1 + 3x_2 - x_3 = 4 \end{cases}$

3   Matrix Algebra

In Section 2, we used matrices merely as abbreviations for simultaneous systems of linear equations. Matrices, however, have a life of their own and, since their invention in 1858 by the English mathematician Arthur Cayley (1821–95), they have played an ever increasing role in applications ranging from economics to quantum mechanics.

Under suitable conditions, matrices can be added, subtracted, multiplied, and (in a sense) divided. In this section, we take a brief look at the resulting "algebra" of matrices. Here we make no attempt to be complete, and we recommend that you look into a textbook on linear algebra or matrix theory for more details.

We use capital letters A, B, C, and so on to denote matrices. Here we consider only matrices whose elements are real numbers. These elements are arranged in a rectangular pattern of horizontal rows and vertical columns; for instance, the matrix

$$A = \begin{bmatrix} -1 & 0 & 5 & 7 & 2 \\ 3 & -2 & \frac{2}{3} & 0 & \frac{1}{2} \\ \frac{5}{4} & 0 & -4 & 5 & 1 \end{bmatrix}$$

has 3 rows and 5 columns. If a matrix has n rows and m columns, we call it an **n by m matrix.** Thus, A is a 3 by 5 matrix. By a **square matrix,** we mean a matrix with the same number of rows as columns. For instance, the matrix

$$\begin{bmatrix} 1 & -5 \\ -3 & 2 \end{bmatrix}$$

is a square 2 by 2 matrix. Notice that an n by m matrix has nm elements.

If A and B are n by m matrices and if each element of A is equal to the corresponding element of B, we say that the matrices A and B are **equal** and we write

$$A = B.$$

Such a **matrix equation** represents nm ordinary equations in a highly compact form. (Why?)

If C and D are two n by m matrices, then their **sum** $C + D$ is defined to be the n by m matrix obtained by adding the corresponding elements of C and D. Likewise, the **difference** $C - D$ is defined to be the n by m matrix obtained by subtracting the elements of D from the corresponding elements of C.

Example   Let $C = \begin{bmatrix} 2 & 1 & 4 \\ 3 & -5 & 8 \end{bmatrix}$ and $D = \begin{bmatrix} -3 & 2 & -1 \\ 4 & -1 & 5 \end{bmatrix}$.

(a) Find $C + D$.   (b) Find $C - D$.

Solution   (a) $C + D = \begin{bmatrix} 2 & 1 & 4 \\ 3 & -5 & 8 \end{bmatrix} + \begin{bmatrix} -3 & 2 & -1 \\ 4 & -1 & 5 \end{bmatrix}$

$$= \begin{bmatrix} 2 + (-3) & 1 + 2 & 4 + (-1) \\ 3 + 4 & -5 + (-1) & 8 + 5 \end{bmatrix}$$

$$= \begin{bmatrix} -1 & 3 & 3 \\ 7 & -6 & 13 \end{bmatrix}$$

(b) $C - D = \begin{bmatrix} 2 & 1 & 4 \\ 3 & -5 & 8 \end{bmatrix} - \begin{bmatrix} -3 & 2 & -1 \\ 4 & -1 & 5 \end{bmatrix}$

$$= \begin{bmatrix} 2 - (-3) & 1 - 2 & 4 - (-1) \\ 3 - 4 & -5 - (-1) & 8 - 5 \end{bmatrix} = \begin{bmatrix} 5 & -1 & 5 \\ -1 & -4 & 3 \end{bmatrix}$$

Notice that you can only add or subtract matrices of the same shape—that is, with the same number of rows and the same number of columns. Because matrices are added by adding their corresponding elements, it follows from the commutative and associative properties of real numbers (Section 1 of Chapter 1) that matrix addition is also commutative and associative. Thus, if A, B, and C are matrices of the same shape, we have

$$A + B = B + A \qquad \text{(commutative property of addition)}$$

and

$$A + (B + C) = (A + B) + C \qquad \text{(associative property of addition).}$$

A matrix all of whose elements are zero is called a **zero matrix.** The zero matrix with n rows and m columns is denoted by $0_{n,m}$. For instance,

$$0_{1,3} = \begin{bmatrix} 0 & 0 & 0 \end{bmatrix} \qquad 0_{2,2} = \begin{bmatrix} 0 & 0 \\ 0 & 0 \end{bmatrix} \qquad 0_{4,3} = \begin{bmatrix} 0 & 0 & 0 \\ 0 & 0 & 0 \\ 0 & 0 & 0 \\ 0 & 0 & 0 \end{bmatrix}$$

and so forth. The subscripts indicating the shape of a zero matrix are often omitted because you can tell from the context how many rows and how many columns are involved. For instance, we have the property

$$A + 0 = A \qquad \text{(additive identity property),}$$

where it is understood that 0 denotes the zero matrix with the same shape as the matrix A.

If A is a matrix, we define $-A$ to be the matrix obtained by multiplying each element of A by -1. For instance,

$$-\begin{bmatrix} 2 & -\frac{1}{2} & 4 & 0 \\ -3 & 2 & 0 & -5 \end{bmatrix} = \begin{bmatrix} -2 & \frac{1}{2} & -4 & 0 \\ 3 & -2 & 0 & 5 \end{bmatrix}.$$

More generally, if k is any real number, we define kA to be the matrix obtained by multiplying each element of A by k. In particular,

$$(-1)A = -A.$$

Notice that $-A$ is an **additive inverse** of A in the sense that

$$A + (-A) = 0 \qquad \text{(additive inverse property).}$$

Example   If $A = \begin{bmatrix} 3 & -1 & 2 \\ 0 & 2 & -4 \end{bmatrix}$, find (a) $-A$ and (b) $\frac{1}{2}A$.

Solution   (a) $-A = -\begin{bmatrix} 3 & -1 & 2 \\ 0 & 2 & -4 \end{bmatrix}$

$$= \begin{bmatrix} -3 & 1 & -2 \\ 0 & -2 & 4 \end{bmatrix}$$

(b) $\frac{1}{2}A = \frac{1}{2}\begin{bmatrix} 3 & -1 & 2 \\ 0 & 2 & -4 \end{bmatrix}$

$$= \begin{bmatrix} \frac{3}{2} & -\frac{1}{2} & 1 \\ 0 & 1 & -2 \end{bmatrix}$$

The product of matrices is defined in a somewhat unexpected way. In order to introduce matrix multiplication, we begin by considering special matrices having only one row or column. A 1 by n matrix

$$R = \begin{bmatrix} x_1 & x_2 & x_3 & \cdots & x_n \end{bmatrix}$$

is called a **row vector,** and an n by 1 matrix

$$C = \begin{bmatrix} y_1 \\ y_2 \\ y_3 \\ \vdots \\ y_n \end{bmatrix}$$

is called a **column vector.** If the number of elements in R is the same as the number of entries in C, we define the **product** RC of R and C to be the number obtained by pairing each element of R with the corresponding element of C, multiplying these pairs, and adding the resulting products. Thus,

$$RC = x_1 y_1 + x_2 y_2 + x_3 y_3 + \cdots + x_n y_n.$$

Example   Find RC if $R = \begin{bmatrix} 2 & 8 & 3 \end{bmatrix}$ and $C = \begin{bmatrix} 60 \\ 20 \\ 300 \end{bmatrix}$.

Solution

$$RC = \begin{bmatrix} 2 & 8 & 3 \end{bmatrix} \begin{bmatrix} 60 \\ 20 \\ 300 \end{bmatrix} = 2(60) + 8(20) + 3(300) = 1180$$

Although the "row by column" product may seem contrived at first, it has many practical uses. For instance, if a furniture store sells 2 tables, 8 chairs, and 3 sofas for \$60 per table, \$20 per chair, and \$300 per sofa, the product of

the **demand vector** $\begin{bmatrix} 2 & 8 & 3 \end{bmatrix}$

and

the **revenue vector** $\begin{bmatrix} 60 \\ 20 \\ 300 \end{bmatrix}$

gives the **total revenue**

$$\begin{bmatrix} 2 & 8 & 3 \end{bmatrix} \begin{bmatrix} 60 \\ 20 \\ 300 \end{bmatrix} = 2(60) + 8(20) + 3(300) = 1180 \text{ dollars.}$$

As another example, notice that the linear equation

$$a_1 x_1 + a_2 x_2 + a_3 x_3 + \cdots + a_n x_n = k$$

can be written in matrix form as

$$\begin{bmatrix} a_1 & a_2 & a_3 \cdots a_n \end{bmatrix} \begin{bmatrix} x_1 \\ x_2 \\ x_3 \\ \vdots \\ x_n \end{bmatrix} = k.$$

The rows of an n by m matrix can be regarded as row vectors, and its columns as column vectors. This permits us to give the following definition.

Definition 1   **The Product of Matrices**

> Let A be an n by m matrix and let B be an m by p matrix. We define the **product** AB to be the n by p matrix determined by the following procedure: To find the element in the ith row and jth column of AB, we multiply the ith row of A by the jth column of B.

Example   Find AB if $A = \begin{bmatrix} 3 & -2 & 4 \\ 5 & 1 & -3 \end{bmatrix}$ and $B = \begin{bmatrix} 3 & 6 \\ -4 & 5 \\ 2 & -2 \end{bmatrix}.$

Solution   Here A is a 2 by 3 matrix and B is a 3 by 2 matrix, so AB will be a 2 by 2 matrix.

$$AB = \begin{bmatrix} 3 & -2 & 4 \\ 5 & 1 & -3 \end{bmatrix} \begin{bmatrix} 3 & 6 \\ -4 & 5 \\ 2 & -2 \end{bmatrix} = \begin{bmatrix} \text{1st row of } A \text{ times 1st column of } B & \text{1st row of } A \text{ times 2nd column of } B \\ \text{2nd row of } A \text{ times 1st column of } B & \text{2nd row of } A \text{ times 2nd column of } B \end{bmatrix}$$

$$= \begin{bmatrix} 3(3) + (-2)(-4) + 4(2) & 3(6) + (-2)(5) + 4(-2) \\ 5(3) + 1(-4) + (-3)(2) & 5(6) + 1(5) + (-3)(-2) \end{bmatrix} = \begin{bmatrix} 25 & 0 \\ 5 & 41 \end{bmatrix}.$$

Notice that two matrices can be multiplied only if they fit together in the sense that the first matrix has as many columns as the second matrix has rows. If they do fit together in this way, the product has as many rows as the first matrix and as many columns as the second. If A is an n by m matrix, B is an m by p matrix, and C is a p by q matrix, it can be shown that

$$A(BC) = (AB)C \quad \text{(associative property of multiplication)}.$$

Distributive properties can also be proved, so that, for matrices of the appropriate shapes

$$A(B + C) = AB + AC \quad \text{(distributive property)}$$

and

$$(D + E)F = DF + EF \quad \text{(distributive property)}.$$

If A and B are square matrices of the same size, we can form the product AB and also the product BA; however, in general, $AB \neq BA$, so *the commutative property of multiplication fails for matrices.*

Example Let $A = \begin{bmatrix} 1 & -1 \\ 2 & 3 \end{bmatrix}$ and $B = \begin{bmatrix} 5 & -2 \\ 4 & 1 \end{bmatrix}$. Find (a) AB and (b) BA.

Solution (a) $AB = \begin{bmatrix} 1 & -1 \\ 2 & 3 \end{bmatrix}\begin{bmatrix} 5 & -2 \\ 4 & 1 \end{bmatrix} = \begin{bmatrix} 1(5)+(-1)4 & 1(-2)+(-1)1 \\ 2(5)+3(4) & 2(-2)+3(1) \end{bmatrix} = \begin{bmatrix} 1 & -3 \\ 22 & -1 \end{bmatrix}$

(b) $BA = \begin{bmatrix} 5 & -2 \\ 4 & 1 \end{bmatrix}\begin{bmatrix} 1 & -1 \\ 2 & 3 \end{bmatrix} = \begin{bmatrix} 5(1)+(-2)2 & 5(-1)+(-2)3 \\ 4(1)+1(2) & 4(-1)+1(3) \end{bmatrix} = \begin{bmatrix} 1 & -11 \\ 6 & -1 \end{bmatrix}$

Notice that $AB \neq BA$.

Using the idea of matrix multiplication, you can write a system of linear equations such as

$$\begin{cases} a_1x_1 + a_2x_2 + a_3x_3 = k_1 \\ b_1x_1 + b_2x_2 + b_3x_3 = k_2 \\ c_1x_1 + c_2x_2 + c_3x_3 = k_3 \end{cases}$$

in the matrix form

$$\begin{bmatrix} a_1 & a_2 & a_3 \\ b_1 & b_2 & b_3 \\ c_1 & c_2 & c_3 \end{bmatrix}\begin{bmatrix} x_1 \\ x_2 \\ x_3 \end{bmatrix} = \begin{bmatrix} k_1 \\ k_2 \\ k_3 \end{bmatrix}.$$

Thus,

$$AX = K,$$

where A represents the matrix of coefficients

$$A = \begin{bmatrix} a_1 & a_2 & a_3 \\ b_1 & b_2 & b_3 \\ c_1 & c_2 & c_3 \end{bmatrix},$$

X represents the column vector of unknowns, and K represents the column vector of constants:

$$X = \begin{bmatrix} x_1 \\ x_2 \\ x_3 \end{bmatrix}, \quad K = \begin{bmatrix} k_1 \\ k_2 \\ k_3 \end{bmatrix}.$$

If $AX = K$ were an ordinary equation, you could multiply both sides by the reciprocal of A to obtain the solution

$$X = A^{-1}K.$$

As we shall see, it is often possible to solve the matrix equation in much the same way by using the **multiplicative inverse** A^{-1} of the matrix A.

A square n by n matrix with 1 in each position on the diagonal running from upper left to lower right, and zeros elsewhere, is called the *n* **by** *n* **identity matrix** and is denoted by I_n. For instance,

$$I_2 = \begin{bmatrix} 1 & 0 \\ 0 & 1 \end{bmatrix} \quad \text{and} \quad I_3 = \begin{bmatrix} 1 & 0 & 0 \\ 0 & 1 & 0 \\ 0 & 0 & 1 \end{bmatrix}.$$

The subscript indicating the size of the identity matrix is often omitted because you can tell from the context how many rows and columns are involved. An identity matrix I plays a role in matrix algebra similar to the role played by 1 in ordinary algebra. In particular, for matrices C and D of the appropriate shape,

$$IC = C \quad \text{and} \quad DI = D \qquad \text{(multiplicative identity property)}.$$

By analogy, with the reciprocal a^{-1} of a nonzero number in ordinary algebra, we have the following definition in matrix algebra.

Definition 2   **The Inverse of a Square Matrix**

Let A be a square n by n matrix. We say that A is **nonsingular** if there exists an n by n matrix A^{-1} such that

$$AA^{-1} = A^{-1}A = I.$$

If A^{-1} exists, it is called the **inverse** of the matrix A.

It can be shown that a nonsingular matrix A has a unique inverse A^{-1}, which you can find by carrying out the following procedure:

Form the n by $2n$ matrix

$$[A \mid I]$$

in which the first n columns are the columns of A and the last n columns are the columns of the identity matrix I_n. Then, using the elementary row operations (Section 2), reduce this matrix to the form

$$[I \mid B],$$

so that the first n columns are the columns of I_n. Then the matrix B formed by the last n columns is the inverse of A, that is,

$$B = A^{-1}.$$

Example   Find A^{-1} if $A = \begin{bmatrix} 1 & -1 & 1 \\ 0 & 2 & -1 \\ 2 & 3 & 0 \end{bmatrix}.$

Solution We start by forming the matrix $[A \mid I]$:

$$\begin{bmatrix} 1 & -1 & 1 & \vdots & 1 & 0 & 0 \\ 0 & 2 & -1 & \vdots & 0 & 1 & 0 \\ 2 & 3 & 0 & \vdots & 0 & 0 & 1 \end{bmatrix}.$$

Now, we execute elementary row operations on the entire matrix until the left half is transformed into the identity matrix:

$$\begin{bmatrix} 1 & -1 & 1 & \vdots & 1 & 0 & 0 \\ 0 & 2 & -1 & \vdots & 0 & 1 & 0 \\ 2 & 3 & 0 & \vdots & 0 & 0 & 1 \end{bmatrix} \xrightarrow{-2R_1 + R_3} \begin{bmatrix} 1 & -1 & 1 & \vdots & 1 & 0 & 0 \\ 0 & 2 & -1 & \vdots & 0 & 1 & 0 \\ 0 & 5 & -2 & \vdots & -2 & 0 & 1 \end{bmatrix}$$

$$\xrightarrow{\frac{1}{2}R_2} \begin{bmatrix} 1 & -1 & 1 & \vdots & 1 & 0 & 0 \\ 0 & 1 & -\frac{1}{2} & \vdots & 0 & \frac{1}{2} & 0 \\ 0 & 5 & -2 & \vdots & -2 & 0 & 1 \end{bmatrix}$$

$$\xrightarrow{R_2 + R_1} \begin{bmatrix} 1 & 0 & \frac{1}{2} & \vdots & 1 & \frac{1}{2} & 0 \\ 0 & 1 & -\frac{1}{2} & \vdots & 0 & \frac{1}{2} & 0 \\ 0 & 5 & -2 & \vdots & -2 & 0 & 1 \end{bmatrix}$$

$$\xrightarrow{-5R_2 + R_3} \begin{bmatrix} 1 & 0 & \frac{1}{2} & \vdots & 1 & \frac{1}{2} & 0 \\ 0 & 1 & -\frac{1}{2} & \vdots & 0 & \frac{1}{2} & 0 \\ 0 & 0 & \frac{1}{2} & \vdots & -2 & -\frac{5}{2} & 1 \end{bmatrix}$$

$$\xrightarrow{R_3 + R_2} \begin{bmatrix} 1 & 0 & \frac{1}{2} & \vdots & 1 & \frac{1}{2} & 0 \\ 0 & 1 & 0 & \vdots & -2 & -2 & 1 \\ 0 & 0 & \frac{1}{2} & \vdots & -2 & -\frac{5}{2} & 1 \end{bmatrix}$$

$$\xrightarrow{-1R_3 + R_1} \begin{bmatrix} 1 & 0 & 0 & \vdots & 3 & 3 & -1 \\ 0 & 1 & 0 & \vdots & -2 & -2 & 1 \\ 0 & 0 & \frac{1}{2} & \vdots & -2 & -\frac{5}{2} & 1 \end{bmatrix}$$

$$\xrightarrow{2R_3} \begin{bmatrix} 1 & 0 & 0 & \vdots & 3 & 3 & -1 \\ 0 & 1 & 0 & \vdots & -2 & -2 & 1 \\ 0 & 0 & 1 & \vdots & -4 & -5 & 2 \end{bmatrix}.$$

Therefore,

$$A^{-1} = \begin{bmatrix} 3 & 3 & -1 \\ -2 & -2 & 1 \\ -4 & -5 & 2 \end{bmatrix}.$$

By using matrix multiplication, you can check that $AA^{-1} = I$ and $A^{-1}A = I$ (Problem 43).

The following example illustrates the use of the inverse of a matrix to solve a system of linear equations.

Example   Use the inverse of the matrix of coefficients to solve the system of linear equations

$$\begin{cases} x - y + z = 8 \\ 2y - z = -7 \\ 2x + 3y = 1 \end{cases}$$

Solution   The matrix of coefficients is

$$A = \begin{bmatrix} 1 & -1 & 1 \\ 0 & 2 & -1 \\ 2 & 3 & 0 \end{bmatrix}.$$

If we let

$$X = \begin{bmatrix} x \\ y \\ z \end{bmatrix} \quad \text{and} \quad K = \begin{bmatrix} 8 \\ -7 \\ 1 \end{bmatrix},$$

we can write the system of linear equations as the matrix equation

$$AX = K.$$

In the previous example, we showed that A is nonsingular with

$$A^{-1} = \begin{bmatrix} 3 & 3 & -1 \\ -2 & -2 & 1 \\ -4 & -5 & 2 \end{bmatrix}.$$

If we multiply $AX = K$ on the left by A^{-1}, we obtain

$$A^{-1}(AX) = A^{-1}K$$

$$(A^{-1}A)X = A^{-1}K$$

$$IX = A^{-1}K$$

$$X = A^{-1}K.$$

Therefore,

$$\begin{bmatrix} x \\ y \\ z \end{bmatrix} = X = A^{-1}K$$

$$= \begin{bmatrix} 3 & 3 & -1 \\ -2 & -2 & 1 \\ -4 & -5 & 2 \end{bmatrix}\begin{bmatrix} 8 \\ -7 \\ 1 \end{bmatrix}$$

$$= \begin{bmatrix} 3(8) + 3(-7) + (-1)1 \\ (-2)8 + (-2)(-7) + 1(1) \\ (-4)8 + (-5)(-7) + 2(1) \end{bmatrix} = \begin{bmatrix} 2 \\ -1 \\ 5 \end{bmatrix}.$$

It follows that $x = 2$, $y = -1$, and $z = 5$.

PROBLEM SET 3

In problems 1–8, find (a) $A + B$, (b) $A - B$, (c) $-3A$, and (d) $-3A + 2B$.

1. $A = \begin{bmatrix} 2 & -3 \\ 5 & 1 \end{bmatrix}$, $B = \begin{bmatrix} 6 & 4 \\ 3 & -2 \end{bmatrix}$

2. $A = \begin{bmatrix} 3 & 2 & -4 & 1 \\ 0 & 3 & -5 & 6 \end{bmatrix}$, $B = \begin{bmatrix} -7 & 3 & 0 & 4 \\ 1 & 0 & -1 & -3 \end{bmatrix}$

3. $A = \begin{bmatrix} 3 & 2 \\ -2 & 5 \\ 2 & 1 \\ -4 & 4 \end{bmatrix}$, $B = \begin{bmatrix} -2 & 3 \\ 3 & 1 \\ 4 & -2 \\ 1 & 0 \end{bmatrix}$

4. $A = \begin{bmatrix} 1 & \frac{1}{3} & 3 \\ 2 & 0 & -\frac{4}{3} \\ 1 & \sqrt{3} & -2 \end{bmatrix}$, $B = \begin{bmatrix} 1 & \frac{1}{2} & -1 \\ 3 & -1 & 0 \\ 2 & 0 & -\frac{3}{2} \end{bmatrix}$

5. $A = \begin{bmatrix} 2 & -3 & 2 & -3 \\ -3 & 2 & 1 & 1 \\ 4 & 1 & -3 & 4 \end{bmatrix}$, $B = \begin{bmatrix} 2 & -3 & 0 & 2 \\ 3 & 2 & -1 & 5 \\ 0 & -2 & 1 & 0 \end{bmatrix}$

6. $A = \begin{bmatrix} 0 \\ 1 \\ 2 \\ 3 \\ 4 \end{bmatrix}$, $B = \begin{bmatrix} -1 \\ 3 \\ -4 \\ 2 \\ 0 \end{bmatrix}$

7. $A = \begin{bmatrix} 1 & \frac{1}{6} & 0 \\ \frac{4}{3} & \pi & -2 \\ 1 & 0 & \frac{5}{3} \end{bmatrix}$, $B = \begin{bmatrix} 1 & -\frac{5}{6} & \sqrt{2} \\ \frac{3}{2} & 1 & 0 \\ 3 & \frac{5}{2} & 0 \end{bmatrix}$

8. $A = \begin{bmatrix} 3.1 & 2.5 \\ 6.8 & 1.1 \\ 4.7 & -8.2 \end{bmatrix}$, $B = \begin{bmatrix} 1.9 & 0 \\ 7.4 & 1 \\ -1 & 2 \end{bmatrix}$

In problems 9–20, let

$$A = \begin{bmatrix} a_1 & a_2 \\ a_3 & a_4 \end{bmatrix}, \quad B = \begin{bmatrix} b_1 & b_2 \\ b_3 & b_4 \end{bmatrix}, \quad \text{and} \quad C = \begin{bmatrix} c_1 & c_2 \\ c_3 & c_4 \end{bmatrix},$$

and let p and q denote arbitrary numbers. Verify each equation by direct calculation.

9. $A + (B + C) = (A + B) + C$

10. $p(B + C) = pB + pC$

11. $(p + q)A = pA + qA$

12. $(pq)A = p(qA)$

13. $A + (-A) = 0$

14. $A(BC) = (AB)C$

15. $A + 0 = A$

16. $A(B + C) = AB + AC$

17. $AI = A$

18. $(A + B)C = AC + BC$

19. $IA = A$

20. $0A = 0$

In problems 21–42, let

$$A = \begin{bmatrix} 1 & -1 & 3 \\ 2 & 0 & 4 \\ 2 & -3 & 6 \end{bmatrix}, \quad B = \begin{bmatrix} -1 & 2 & -1 \\ -3 & 4 & 3 \\ 0 & -1 & 2 \end{bmatrix}, \quad C = \begin{bmatrix} 2 & -1 \\ 0 & 2 \\ -3 & 1 \end{bmatrix},$$

$$D = \begin{bmatrix} 1 & 3 & 2 \\ 4 & -1 & -2 \end{bmatrix}, \quad E = \begin{bmatrix} 1 & 3 \\ -1 & 2 \end{bmatrix}, \quad F = \begin{bmatrix} 2 & 0 \\ 4 & -1 \end{bmatrix}.$$

Find each product, if it exists.

21. EF

22. CE

23. FE

24. EC

25. EE

26. BC

27. FF

28. CB

29. AB

30. BA

31. CD

32. AC

33. AA

34. ABC

35. EFE

36. ACB

37. $(A + B)C$

38. $A(A + B)$

39. $E(F - I)$

40. $EDCF$

41. DE

42. $(A - B)(A + B)$

43. Check the solution to the example on page 445 by verifying that $AA^{-1} = I$ and that $A^{-1}A = I$.

44. Let

$$A = \begin{bmatrix} a & b \\ c & d \end{bmatrix}$$

and suppose that $ad - bc \neq 0$. Prove that A is nonsingular with

$$A^{-1} = (ad - bc)^{-1} \begin{bmatrix} d & -b \\ -c & a \end{bmatrix}.$$

In problems 45–58, find the inverse of each matrix if it exists. Check your answers (see problem 43).

45. $\begin{bmatrix} 1 & 1 \\ 1 & -4 \end{bmatrix}$ **46.** $\begin{bmatrix} 1 & 6 \\ 11 & -7 \end{bmatrix}$ **47.** $\begin{bmatrix} -6 & 2 \\ 2 & 5 \end{bmatrix}$ **48.** $\begin{bmatrix} 3 & 6 \\ 1 & 2 \end{bmatrix}$ **49.** $\begin{bmatrix} 1 & -1 \\ 9 & 3 \end{bmatrix}$

50. $\begin{bmatrix} 1 & b \\ b & 1 \end{bmatrix}$ **51.** $\begin{bmatrix} 1 & 1 \\ 1 & 1 \end{bmatrix}$ **52.** $\begin{bmatrix} 0 & 1 \\ 1 & 0 \end{bmatrix}$ **53.** $\begin{bmatrix} 1 & 2 & -1 \\ 2 & -1 & 3 \\ 3 & -2 & 3 \end{bmatrix}$ **54.** $\begin{bmatrix} 1 & 5 & -1 \\ 2 & 1 & 1 \\ 1 & -1 & 2 \end{bmatrix}$

55. $\begin{bmatrix} 1 & 2 & -1 \\ 2 & -2 & 1 \\ 6 & 4 & 3 \end{bmatrix}$ **56.** $\begin{bmatrix} 1 & 2 & -1 \\ 1 & 3 & 2 \\ 2 & 5 & 1 \end{bmatrix}$ **57.** $\begin{bmatrix} 1 & 3 & 1 \\ 0 & 2 & 4 \\ -1 & 0 & 3 \end{bmatrix}$ **58.** $\begin{bmatrix} 0 & 0 & 1 \\ 0 & 1 & 0 \\ 1 & 0 & 0 \end{bmatrix}$

In problems 59–68, use the inverse of the matrix of coefficients to solve each system of linear equations.

59. $\begin{cases} x + y = 4 \\ x - 4y = 8 \end{cases}$ (See problem 45.)

60. $\begin{cases} x + 6y = 7 \\ 11x - 7y = -10 \end{cases}$ (See problem 46.)

61. $\begin{cases} -6x + 2y = 3 \\ 2x + 5y = 3 \end{cases}$ (See problem 47.)

62. $\begin{cases} x_1 + 6x_2 = a \\ 11x_1 - 7x_2 = b \end{cases}$ (See problem 46.)

63. $\begin{cases} r - s = 6 \\ 9r + 3s = 14 \end{cases}$ (See problem 49.)

64. $\begin{cases} x + by = 2 \\ bx + y = 3 \end{cases}$ (See problem 50.)

65. $\begin{cases} x + 2y - z = 6 \\ 2x - y + 3z = -13 \\ 3x - 2y + 3z = -16 \end{cases}$ (See problem 53.)

66. $\begin{cases} u + 5v - w = 2 \\ 2u + v + w = 7 \\ u - v + 2w = 11 \end{cases}$ (See problem 54.)

67. $\begin{cases} x + 3y + z = 0 \\ 2y + 4z = 1 \\ -x + 3z = 2 \end{cases}$ (See problem 57.)

68. $\begin{cases} x + 2y - z = 0 \\ 2x - 2y + z = 0 \\ 6x + 4y + 3z = 0 \end{cases}$ (See problem 55.)

69. In economics, a square matrix in which the element in the ith row and jth column indicates the number of units of commodity number i used to produce one unit of commodity number j is called a **technology matrix.** Let $T = \begin{bmatrix} a & b \\ c & d \end{bmatrix}$ be the technology matrix for a simple economic model involving only two commodities. The column vector $X = \begin{bmatrix} x_1 \\ x_2 \end{bmatrix}$ in which x_1 (respectively, x_2) represents the number of units of commodity number 1 (respectively, commodity number 2) produced in unit time is called the **intensity vector.** (a) Give the economic interpretation of the product TX. (b) Give the economic interpretation of $X - TX$.

70. In problem 69, let d_1 (respectively, d_2) denote the surplus number of units of commodity number 1 (respectively, commodity number 2) required per unit time for export. If $I - T$ is a nonsingular matrix and $D = \begin{bmatrix} d_1 \\ d_2 \end{bmatrix}$, give the economic interpretation of $(I - T)^{-1}D$.

4   Determinants and Cramer's Rule

In this section, we consider an alternative method, called Cramer's rule, for solving systems of linear equations. Although Cramer's rule applies to systems of n linear equations in n unknowns for any positive integer n, its practical use is usually limited to the cases $n = 2$ or $n = 3$. For larger values of n, the elimination method using matrices (Section 2) is usually more efficient.

Cramer's rule is based on the idea of a *determinant*. If a, b, c, and d are any four numbers, the symbol

$$\begin{vmatrix} a & b \\ c & d \end{vmatrix}$$

is called a 2 by 2 **determinant** with *entries* or *elements* a, b, c, and d. Its **value** is defined to be the number $ad - cb$, that is,

$$\begin{vmatrix} a & b \\ c & d \end{vmatrix} = ad - cb.$$

The memory aid

$$\begin{vmatrix} a & b \\ c & d \end{vmatrix} \quad = ad - cb$$

is often helpful.

Example   Evaluate the determinant $\begin{vmatrix} 4 & -3 \\ 2 & 1 \end{vmatrix}$.

Solution
$$\begin{vmatrix} 4 & -3 \\ 2 & 1 \end{vmatrix} = 4(1) - 2(-3) = 10$$

Now, let's see how determinants can be used to solve systems of linear equations. Consider the system

$$\begin{cases} ax + by = h \\ cx + dy = k. \end{cases}$$

If we multiply the first equation by d and the second equation by b, we obtain

$$\begin{cases} adx + bdy = hd \\ bcx + bdy = bk. \end{cases}$$

So, subtracting the second equation from the first, we eliminate the terms involving y and get

$$adx - bcx = hd - bk \quad \text{or} \quad (ad - bc)x = hd - bk.$$

Using determinants, we can rewrite the last equation as

$$\begin{vmatrix} a & b \\ c & d \end{vmatrix} x = \begin{vmatrix} h & b \\ k & d \end{vmatrix}.$$

Therefore,

$$x = \frac{\begin{vmatrix} h & b \\ k & d \end{vmatrix}}{\begin{vmatrix} a & b \\ c & d \end{vmatrix}},$$

provided that $\begin{vmatrix} a & b \\ c & d \end{vmatrix} \neq 0$. A similar calculation (Problem 51) yields

$$y = \frac{\begin{vmatrix} a & h \\ c & k \end{vmatrix}}{\begin{vmatrix} a & b \\ c & d \end{vmatrix}}.$$

Thus, if

$$D = \begin{vmatrix} a & b \\ c & d \end{vmatrix}, \quad D_x = \begin{vmatrix} h & b \\ k & d \end{vmatrix}, \quad D_y = \begin{vmatrix} a & h \\ c & k \end{vmatrix},$$

and $D \neq 0$, then the system

$$\begin{cases} ax + by = h \\ cx + dy = k \end{cases}$$

has one and only one solution:

$$x = \frac{D_x}{D}, \quad y = \frac{D_y}{D}.$$

This is **Cramer's rule** for two linear equations in two unknowns.

In Cramer's rule, D is called the **coefficient determinant** because its entries are the coefficients of the unknowns in the system:

$$\begin{cases} ax + by = h \\ cx + dy = k, \end{cases} \qquad D = \begin{vmatrix} a & b \\ c & d \end{vmatrix}.$$

Notice that D_x is obtained by replacing the *first* column of D (the coefficients of x), and that D_y is obtained by replacing the *second* column of D (the coefficients of y) by the constants on the right in the system of equations:

$$\begin{cases} ax + by = h \\ cx + dy = k, \end{cases} \qquad D_x = \begin{vmatrix} h & b \\ k & d \end{vmatrix}, \quad D_y = \begin{vmatrix} a & h \\ c & k \end{vmatrix}.$$

Keep in mind that Cramer's rule can be applied only when $D \neq 0$. If $D = 0$, it can be shown (Problem 53) that the equations in the system are either inconsistent or dependent.

Example   Use Cramer's rule (if applicable) to solve each system:

(a) $\begin{cases} 2x - y = 7 \\ x + 3y = 14 \end{cases}$ (b) $\begin{cases} 2x - y = 7 \\ 4x - 2y = 3 \end{cases}$

Solution   (a) $D = \begin{vmatrix} 2 & -1 \\ 1 & 3 \end{vmatrix} = 2(3) - 1(-1) = 7, \quad D_x = \begin{vmatrix} 7 & -1 \\ 14 & 3 \end{vmatrix} = 7(3) - 14(-1) = 35,$

$D_y = \begin{vmatrix} 2 & 7 \\ 1 & 14 \end{vmatrix} = 2(14) - 1(7) = 21;$

hence, by Cramer's rule

$$x = \frac{D_x}{D} = \frac{35}{7} = 5 \quad \text{and} \quad y = \frac{D_y}{D} = \frac{21}{7} = 3.$$

(b) Here,

$$D = \begin{vmatrix} 2 & -1 \\ 4 & -2 \end{vmatrix} = 2(-2) - 4(-1) = 0,$$

so Cramer's rule does not apply. [Actually, the system of equations is inconsistent—it has no solution.]

To extend Cramer's rule to systems of three linear equations in three unknowns, we begin by extending the definition of a determinant. The **value** of a 3 by 3 determinant is defined in terms of 2 by 2 determinants as follows:

$$\begin{vmatrix} a_1 & a_2 & a_3 \\ b_1 & b_2 & b_3 \\ c_1 & c_2 & c_3 \end{vmatrix} = a_1 \begin{vmatrix} b_2 & b_3 \\ c_2 & c_3 \end{vmatrix} - a_2 \begin{vmatrix} b_1 & b_3 \\ c_1 & c_3 \end{vmatrix} + a_3 \begin{vmatrix} b_1 & b_2 \\ c_1 & c_2 \end{vmatrix}.$$

Notice the *negative sign* on the middle term. We refer to this as the **expansion formula** for 3 by 3 determinants.

In the expansion formula, notice that each entry in the first row is multiplied by the 2 by 2 determinant that remains when the row and column containing the multiplier are (mentally) crossed out. Thus:

a_1 is multiplied by $\begin{vmatrix} \cancel{a_1} & \cancel{a_2} & \cancel{a_3} \\ \cancel{b_1} & b_2 & b_3 \\ \cancel{c_1} & c_2 & c_3 \end{vmatrix} = \begin{vmatrix} b_2 & b_3 \\ c_2 & c_3 \end{vmatrix}$

a_2 is multiplied by $\begin{vmatrix} \cancel{a_1} & \cancel{a_2} & \cancel{a_3} \\ b_1 & \cancel{b_2} & b_3 \\ c_1 & \cancel{c_2} & c_3 \end{vmatrix} = \begin{vmatrix} b_1 & b_3 \\ c_1 & c_3 \end{vmatrix}$, and

a_3 is multiplied by $\begin{vmatrix} \cancel{a_1} & \cancel{a_2} & \cancel{a_3} \\ b_1 & b_2 & \cancel{b_3} \\ c_1 & c_2 & \cancel{c_3} \end{vmatrix} = \begin{vmatrix} b_1 & b_2 \\ c_1 & c_2 \end{vmatrix}.$

To **expand** a 3 by 3 determinant means to find its value by using the expansion formula. Again, we emphasize: When expanding a 3 by 3 determinant, *don't forget the negative sign on the middle term.*

Example   Expand the determinant $\begin{vmatrix} 3 & 1 & 2 \\ -4 & 2 & 4 \\ 1 & 0 & 5 \end{vmatrix}.$

Solution

$$\begin{vmatrix} 3 & 1 & 2 \\ -4 & 2 & 4 \\ 1 & 0 & 5 \end{vmatrix} = 3\begin{vmatrix} 2 & 4 \\ 0 & 5 \end{vmatrix} - 1\begin{vmatrix} -4 & 4 \\ 1 & 5 \end{vmatrix} + 2\begin{vmatrix} -4 & 2 \\ 1 & 0 \end{vmatrix}$$

$$= 3[2(5) - 0(4)] - 1[(-4)5 - 1(4)] + 2[(-4)0 - 1(2)]$$

$$= 3(10) - 1(-24) + 2(-2) = 50$$

Now we can state **Cramer's rule** for solving a system

$$\begin{cases} a_1 x + a_2 y + a_3 z = k_1 \\ b_1 x + b_2 y + b_3 z = k_2 \\ c_1 x + c_2 y + c_3 z = k_3 \end{cases}$$

of three linear equations in three unknowns: Form the *coefficient determinant*

$$D = \begin{vmatrix} a_1 & a_2 & a_3 \\ b_1 & b_2 & b_3 \\ c_1 & c_2 & c_3 \end{vmatrix}.$$

If $D \neq 0$, form the determinants

$$D_x = \begin{vmatrix} k_1 & a_2 & a_3 \\ k_2 & b_2 & b_3 \\ k_3 & c_2 & c_3 \end{vmatrix}, \quad D_y = \begin{vmatrix} a_1 & k_1 & a_3 \\ b_1 & k_2 & b_3 \\ c_1 & k_3 & c_3 \end{vmatrix}, \quad D_z = \begin{vmatrix} a_1 & a_2 & k_1 \\ b_1 & b_2 & k_2 \\ c_1 & c_2 & k_3 \end{vmatrix}.$$

Then the solution of the system of linear equations is

$$x = \frac{D_x}{D}, \quad y = \frac{D_y}{D}, \quad z = \frac{D_z}{D}.$$

If $D \neq 0$, this is the only solution of the system; that is, the equations in the system are consistent (see page 437). If $D = 0$, the equations are either inconsistent or dependent, and Cramer's rule is not applicable. You can find a proof of Cramer's rule in a textbook on linear algebra.

Example   Use Cramer's rule to solve the system

$$\begin{cases} 3x - 2y + z = -9 \\ x + 2y - z = 5 \\ 2x - y + 3z = -10. \end{cases}$$

Solution $$D = \begin{vmatrix} 3 & -2 & 1 \\ 1 & 2 & -1 \\ 2 & -1 & 3 \end{vmatrix} = 3\begin{vmatrix} 2 & -1 \\ -1 & 3 \end{vmatrix} - (-2)\begin{vmatrix} 1 & -1 \\ 2 & 3 \end{vmatrix} + 1\begin{vmatrix} 1 & 2 \\ 2 & -1 \end{vmatrix}$$

$$= 3[2(3) - (-1)(-1)] + 2[1(3) - 2(-1)] + 1[1(-1) - 2(2)]$$

$$= 3(5) + 2(5) + 1(-5) = 20$$

Because $D \neq 0$, the system is consistent, and we can solve it by applying Cramer's rule:

$$D_x = \begin{vmatrix} -9 & -2 & 1 \\ 5 & 2 & -1 \\ -10 & -1 & 3 \end{vmatrix} = -9 \begin{vmatrix} 2 & -1 \\ -1 & 3 \end{vmatrix} - (-2) \begin{vmatrix} 5 & -1 \\ -10 & 3 \end{vmatrix} + 1 \begin{vmatrix} 5 & 2 \\ -10 & -1 \end{vmatrix} = -20$$

$$D_y = \begin{vmatrix} 3 & -9 & 1 \\ 1 & 5 & -1 \\ 2 & -10 & 3 \end{vmatrix} = 3 \begin{vmatrix} 5 & -1 \\ -10 & 3 \end{vmatrix} - (-9) \begin{vmatrix} 1 & -1 \\ 2 & 3 \end{vmatrix} + 1 \begin{vmatrix} 1 & 5 \\ 2 & -10 \end{vmatrix} = 40$$

$$D_z = \begin{vmatrix} 3 & -2 & -9 \\ 1 & 2 & 5 \\ 2 & -1 & -10 \end{vmatrix} = 3 \begin{vmatrix} 2 & 5 \\ -1 & -10 \end{vmatrix} - (-2) \begin{vmatrix} 1 & 5 \\ 2 & -10 \end{vmatrix} + (-9) \begin{vmatrix} 1 & 2 \\ 2 & -1 \end{vmatrix} = -40;$$

hence,

$$x = \frac{D_x}{D} = \frac{-20}{20} = -1, \quad y = \frac{D_y}{D} = \frac{40}{20} = 2, \quad z = \frac{D_z}{D} = \frac{-40}{20} = -2.$$

4.1 Properties of Determinants

Determinants have a number of useful properties, some of which we now state (without proof).

Property 1. If you interchange any two rows or any two columns of a determinant, you change its algebraic sign.

For instance,

$$\begin{vmatrix} 2 & 3 \\ 5 & 6 \end{vmatrix} = - \begin{vmatrix} 5 & 6 \\ 2 & 3 \end{vmatrix}.$$

(Check this yourself.)

Property 2. If you multiply every entry in one row or one column of a determinant by a constant k, the effect is to multiply the value of the determinant by k.

For instance,

$$\begin{vmatrix} 2k & 3 \\ 5k & 6 \end{vmatrix} = k \begin{vmatrix} 2 & 3 \\ 5 & 6 \end{vmatrix}.$$

(Check this yourself.)

Property 2 allows you to "factor out" a common factor of all the elements of a single row or column of a determinant. For instance,

$$\begin{vmatrix} 8 & 28 & 3 \\ 3 & -14 & 2 \\ 5 & 42 & 4 \end{vmatrix} = 7\begin{vmatrix} 8 & 4 & 3 \\ 3 & -2 & 2 \\ 5 & 6 & 4 \end{vmatrix}.$$

> **Property 3.** If you add a constant multiple of the entries in any one row of a determinant to the corresponding entries in any other row, the value of the determinant will not change. Likewise for columns.

For instance, in the determinant

$$\begin{vmatrix} -3 & 5 \\ 6 & -4 \end{vmatrix},$$

if you add 2 times the first row to the second row, the value of the determinant won't change, that is,

$$\begin{vmatrix} -3 & 5 \\ 6 & -4 \end{vmatrix} = \begin{vmatrix} -3 & 5 \\ 6 + 2(-3) & -4 + 2(5) \end{vmatrix} = \begin{vmatrix} -3 & 5 \\ 0 & 6 \end{vmatrix}.$$

(Check this yourself.)

> **Property 4.** If any two rows or any two columns of a determinant are the same, its value is zero. More generally, if the corresponding entries in any two rows or in any two columns are proportional, the value of the determinant is zero.

For instance,

$$\begin{vmatrix} 1 & 5 & -7 \\ 2 & 3 & 1 \\ 1 & 5 & -7 \end{vmatrix} = 0 \quad \text{and} \quad \begin{vmatrix} 2 & -1 & 10 \\ 3 & 7 & 15 \\ 4 & 2 & 20 \end{vmatrix} = 0.$$

The **main diagonal** of a determinant is the diagonal running from upper left to lower right. For instance, in the determinant

$$\begin{vmatrix} a & b & c \\ u & v & w \\ x & y & z \end{vmatrix}$$

the entries on the main diagonal are a, v, and z. A determinant is said to be in **triangular form** if all entries below the main diagonal are zero. For instance,

$$\begin{vmatrix} 3 & 5 & -7 \\ 0 & 2 & 4 \\ 0 & 0 & 6 \end{vmatrix}$$

is in triangular form.

> **Property 5.** If a determinant is in triangular form, its value is the product of the entries on its main diagonal.

For instance,

$$\begin{vmatrix} 3 & 5 & -7 \\ 0 & 2 & 4 \\ 0 & 0 & 6 \end{vmatrix} = 3(2)(6) = 36.$$

(Check this yourself.)

By using Properties 1–5, you can often evaluate a determinant more easily than by applying the expansion formula. The usual idea is to use Properties 1–3 to bring the determinant into triangular form, and then to apply Property 5.

Example   Use the properties of determinants to evaluate:

$$\text{(a)} \begin{vmatrix} 3 & -1 & 2 \\ 6 & -2 & 4 \\ 7 & 0 & 3 \end{vmatrix} \qquad\qquad \text{(b)} \begin{vmatrix} 4 & 3 & 3 \\ 1 & 0 & 2 \\ 6 & 6 & 7 \end{vmatrix}$$

Solution   (a) The second row is proportional to the first, so, by Property 4,

$$\begin{vmatrix} 3 & -1 & 2 \\ 6 & -2 & 4 \\ 7 & 0 & 3 \end{vmatrix} = 0.$$

(b)
$$\begin{vmatrix} 4 & 3 & 3 \\ 1 & 0 & 2 \\ 6 & 6 & 7 \end{vmatrix} = 3\begin{vmatrix} 4 & 1 & 3 \\ 1 & 0 & 2 \\ 6 & 2 & 7 \end{vmatrix}$$

(We factored out 3 from the second column—Property 2.)

$$= -3\begin{vmatrix} 1 & 4 & 3 \\ 0 & 1 & 2 \\ 2 & 6 & 7 \end{vmatrix}$$

(We interchanged the first and second columns—Property 1.)

$$= -3\begin{vmatrix} 1 & 4 & 3 \\ 0 & 1 & 2 \\ 0 & -2 & 1 \end{vmatrix}$$

(We added -2 times the first row to the third row—Property 3.)

$$= -3\begin{vmatrix} 1 & 4 & 3 \\ 0 & 1 & 2 \\ 0 & 0 & 5 \end{vmatrix}$$

(We added 2 times the second row to the third row—Property 3.)

$$= -3(1)(1)(5)$$ (We used Property 5.)
$$= -15.$$

It should come as no surprise to you to learn that 4 by 4, 5 by 5, and, in general, n by n determinants can be defined, and Properties 1–5 continue to hold for them as well. You can learn more about determinants from a textbook on linear algebra. Determinants have a wide variety of uses (other than for solving systems of linear equations) in linear algebra, calculus, and other branches of mathematics.

PROBLEM SET 4

In problems 1–8, evaluate each determinant.

1. $\begin{vmatrix} 2 & 3 \\ 1 & 4 \end{vmatrix}$

2. $\begin{vmatrix} e & \pi \\ \sqrt{3} & \sqrt{2} \end{vmatrix}$

3. $\begin{vmatrix} 6 & -4 \\ 3 & 7 \end{vmatrix}$

4. $\begin{vmatrix} x & -y \\ y & x \end{vmatrix}$

5. $\begin{vmatrix} \sqrt{6} & -2\sqrt{5} \\ 3\sqrt{5} & 4\sqrt{6} \end{vmatrix}$

6. $\begin{vmatrix} x+y & x+y \\ x-y & x+y \end{vmatrix}$

7. $\begin{vmatrix} \log 100 & 2 \\ \log 10 & 3 \end{vmatrix}$

8. $\begin{vmatrix} x & -x \\ y & -y \end{vmatrix}$

In problems 9–14, use Cramer's rule (when applicable) to solve each system of linear equations.

9. $\begin{cases} 5x + 7y = -2 \\ 3x + 4y = -1 \end{cases}$

10. $\begin{cases} \frac{1}{2}x - \frac{2}{3}y = \frac{3}{4} \\ \frac{1}{3}x + 2y = \frac{5}{6} \end{cases}$

11. $\begin{cases} 2x_1 + x_2 = 5 \\ x_1 - 2x_2 = 0 \end{cases}$

12. $\begin{cases} 3u - 4v = 1 \\ -4u + \frac{16}{3}v = 2 \end{cases}$

13. $\begin{cases} 8u + 3v = 9 \\ 4u - 6v = 7 \end{cases}$

14. $\begin{cases} ax + y = 0 \\ x + ay = 0 \end{cases}$

In problems 15–22, expand each determinant.

15. $\begin{vmatrix} 2 & 3 & -1 \\ 5 & 7 & 0 \\ 2 & -3 & 1 \end{vmatrix}$

16. $\begin{vmatrix} 1 & 5 & -7 \\ 3 & 0 & 2 \\ -1 & 4 & 1 \end{vmatrix}$

17. $\begin{vmatrix} 2 & 0 & 4 \\ 1 & 5 & 0 \\ 0 & 7 & 1 \end{vmatrix}$

18. $\begin{vmatrix} a & b & c \\ x & y & z \\ 1 & 1 & 1 \end{vmatrix}$

19. $\begin{vmatrix} 2 & -3 & 1 \\ 1 & 2 & 3 \\ 0 & 1 & 2 \end{vmatrix}$

20. $\begin{vmatrix} e & \sqrt{2} & \sqrt{3} \\ \pi & 0 & 1 \\ -1 & 2 & 0 \end{vmatrix}$

21. $\begin{vmatrix} \frac{1}{2} & 1 & -\frac{2}{3} \\ \frac{5}{2} & -\frac{4}{3} & 1 \\ \frac{3}{2} & 0 & \frac{3}{4} \end{vmatrix}$

22. $\begin{vmatrix} 0 & a & b \\ a & 0 & c \\ b & c & 0 \end{vmatrix}$

In problems 23–28, use Cramer's rule (when applicable) to solve each system of linear equations.

23. $\begin{cases} x + 2y - z = -3 \\ 2x - y + z = 5 \\ 3x + 2y - 2z = -3 \end{cases}$

24. $\begin{cases} -u + 2v + w = -1 \\ 4u - 2v - w = 3 \\ 4u + 2v - w = 5 \end{cases}$

25. $\begin{cases} -3x + 4y + 6z = 30 \\ x + 2z = 6 \\ -x - 2y + 3z = 8 \end{cases}$

26. $\begin{cases} 2y - 3x = 1 \\ 3z - 2y = 5 \\ x + z = 4 \end{cases}$

27. $\begin{cases} 2x + 5y - z = 3 \\ -3x - 2y + 7z = 4 \\ -x + 3y + 6z = 0 \end{cases}$

28. $\begin{cases} 3x + y - z = 14 \\ x + 3y - z = 16 \\ x + y - 3z = -10 \end{cases}$

In problems 29–36, use Properties 1–6 (Section 4.1) to evaluate each determinant.

29. $\begin{vmatrix} 1 & -2 & 3 \\ 2 & 3 & -2 \\ 3 & 1 & -1 \end{vmatrix}$

30. $\begin{vmatrix} 5 & 2 & 3 \\ 4 & -5 & -6 \\ 7 & -8 & -9 \end{vmatrix}$

31. $\begin{vmatrix} 1 & -5 & 2 \\ -4 & -1 & 5 \\ 3 & -4 & 3 \end{vmatrix}$

32. $\begin{vmatrix} 2 & 4 & 3 \\ -6 & 0 & 4 \\ -1 & 1 & 3 \end{vmatrix}$

33. $\begin{vmatrix} -1 & 1 & 1 \\ 4 & 2 & 3 \\ 1 & 3 & 0 \end{vmatrix}$

34. $\begin{vmatrix} 4 & 5 & 6 \\ 2 & 2 & 2 \\ 7 & 2 & -7 \end{vmatrix}$

35. $\begin{vmatrix} 3 & 1 & 4 \\ 1 & 7 & 3 \\ 5 & -10 & 5 \end{vmatrix}$

36. $\begin{vmatrix} 9 & 3 & 3 \\ -2 & 0 & 6 \\ 2 & 1 & 1 \end{vmatrix}$

© In problems 37 and 38, evaluate each determinant with the aid of a calculator.

37. $\begin{vmatrix} 2.03 & -7.07 & 1.55 \\ 3.71 & 2.22 & 5.77 \\ 6.65 & -8.56 & 3.65 \end{vmatrix}$

38. $\begin{vmatrix} 0.071 & 0.029 & -0.095 \\ 0.101 & 0.210 & 0.055 \\ 0.077 & -0.101 & 0.039 \end{vmatrix}$

In problems 39–44, evaluate each determinant mentally. Indicate which property or properties you use.

39. $\begin{vmatrix} 3 & \sqrt{2} & 19 \\ 0 & 1 & \frac{5}{2} \\ 0 & 0 & -4 \end{vmatrix}$ **40.** $\begin{vmatrix} 3 & 1 & 4 \\ -6 & -2 & -8 \\ 1 & 5 & -7 \end{vmatrix}$ **41.** $\begin{vmatrix} 1 & 1 & 1 \\ -1 & -1 & -1 \\ a & b & c \end{vmatrix}$

42. $\begin{vmatrix} a & b & c \\ 0 & d & h \\ 0 & 0 & \sqrt{5} \end{vmatrix}$ **43.** $\begin{vmatrix} 0 & 1 & 0 \\ 1 & 0 & 0 \\ 0 & 0 & 1 \end{vmatrix}$ **44.** $\begin{vmatrix} 1 & 2 & 3 \\ 3 & 2 & 1 \\ 4 & 4 & 4 \end{vmatrix}$

In problems 45–50, assume that $\begin{vmatrix} a & b & c \\ u & v & w \\ x & y & z \end{vmatrix} = -3$. Find the value of each determinant.

45. $\begin{vmatrix} u & v & w \\ a & b & c \\ x & y & z \end{vmatrix}$ **46.** $\begin{vmatrix} u & v & w \\ x & y & z \\ a & b & c \end{vmatrix}$ **47.** $\begin{vmatrix} c & a & b \\ w & u & v \\ z & x & y \end{vmatrix}$ **48.** $\begin{vmatrix} -a & -b & -c \\ 3u & 3v & 3w \\ 4x & 4y & 4z \end{vmatrix}$

49. $\begin{vmatrix} a+u & b+v & c+w \\ u & v & w \\ x+u & y+v & z+w \end{vmatrix}$ **50.** $\begin{vmatrix} a & b & c \\ u-2a & v-2b & w-2c \\ 3x & 3y & 3z \end{vmatrix}$

51. Complete the proof of Cramer's rule for two linear equations in two unknowns (page 450) by showing that if $D \neq 0$, then $y = D_y/D$. [Hint: Multiply the first equation by c and the second equation by a; then subtract the first equation from the second.]

52. Show that Property 4 (Section 4.1) can be derived from Property 2 and Property 3.

53. If $\begin{vmatrix} a & b \\ c & d \end{vmatrix} = 0$, show that the equations in the system $\begin{cases} ax + by = h \\ cx + dy = k \end{cases}$ are either inconsistent or dependent. [Hint: The equations $Dx = D_x$ and $Dy = D_y$ are true even if $D = 0$.]

54. Solve each equation for x:

(a) $\begin{vmatrix} 4 & -1 \\ x & 3 \end{vmatrix} = 2$ (b) $\begin{vmatrix} -1 & 5 & -2 \\ 2 & -2 & x \\ 3 & 1 & 0 \end{vmatrix} = -3$

55. Show that in the xy plane, an equation of the line that contains the two points (a,b) and (c,d) is

$$\begin{vmatrix} x & y & 1 \\ a & b & 1 \\ c & d & 1 \end{vmatrix} = 0.$$

56. Show that

$$\begin{vmatrix} a_1+a_2 & b_1+b_2 & c_1+c_2 \\ u & v & w \\ x & y & z \end{vmatrix} = \begin{vmatrix} a_1 & b_1 & c_1 \\ u & v & w \\ x & y & z \end{vmatrix} + \begin{vmatrix} a_2 & b_2 & c_2 \\ u & v & w \\ x & y & z \end{vmatrix}.$$

57. Solve the equation

$$\begin{vmatrix} 1 & x-2 & 2 \\ -2 & x-1 & 3 \\ 1 & 2 & x \end{vmatrix} = 0.$$

58. Show that $x - a$ and $x - b$ are factors of

$$\begin{vmatrix} 1 & 1 & 1 \\ x & a & b \\ x^2 & a^2 & b^2 \end{vmatrix}.$$

59. Show that the equation

$$\begin{vmatrix} a - x & b \\ b & c - x \end{vmatrix} = 0,$$

in which a, b, and c are real numbers and x is the unknown, always has real roots.

---

5   Applications of Systems of Linear Equations

In this section, we present several problems that can be worked by setting up and solving systems of linear equations.

5.1 Partial Fractions

If you add the two fractions $\dfrac{2}{x - 2}$ and $\dfrac{3}{x + 1}$, you obtain

$$\frac{2}{x - 2} + \frac{3}{x + 1} = \frac{2(x + 1) + 3(x - 2)}{(x - 2)(x + 1)} = \frac{5x - 4}{(x - 2)(x + 1)}.$$

The reverse process of "taking the fraction $\dfrac{5x - 4}{(x - 2)(x + 1)}$ apart" into the sum of simpler fractions,

$$\frac{5x - 4}{(x - 2)(x + 1)} = \frac{2}{x - 2} + \frac{3}{x + 1},$$

is called **decomposing** $\dfrac{5x - 4}{(x - 2)(x + 1)}$ into the **partial fractions** $\dfrac{2}{x - 2}$ and $\dfrac{3}{x + 1}$.
The process of decomposing fractions into simpler partial fractions is used routinely in calculus and other branches of mathematics. Here we shall give you a brief introduction to the subject—you can find further details in calculus textbooks.
    It is easiest to decompose a rational expression into partial fractions if the degree of the numerator is less than the degree of the denominator, and if the denominator is (or can be) factored into linear factors that are all different from one another. In this case, you simply provide a partial fraction of the form

$$\frac{\text{constant}}{\text{linear factor}}$$

for each linear factor in the denominator.

Example   Decompose $\dfrac{6x^2 + 2x + 2}{x(x - 2)(2x + 1)}$ into partial fractions.

Solution   The linear factors in the denominator are x, $x - 2$, and $2x + 1$. Since these factors are different from each other, we must provide partial fractions

$$\frac{\text{constant}}{x}, \quad \frac{\text{constant}}{x - 2}, \quad \text{and} \quad \frac{\text{constant}}{2x + 1}.$$

We denote the three constants by A, B, and C, so that

$$\frac{6x^2 + 2x + 2}{x(x - 2)(2x + 1)} = \frac{A}{x} + \frac{B}{x - 2} + \frac{C}{2x + 1}.$$

To determine the values of A, B, and C, we begin by multiplying both sides of this equation by $x(x - 2)(2x + 1)$ to clear the fractions. Thus, we have

$$6x^2 + 2x + 2 = A(x - 2)(2x + 1) + Bx(2x + 1) + Cx(x - 2)$$

or

$$6x^2 + 2x + 2 = A(2x^2 - 3x - 2) + B(2x^2 + x) + C(x^2 - 2x).$$

Collecting like powers on the right, we obtain

$$6x^2 + 2x + 2 = (2A + 2B + C)x^2 + (-3A + B - 2C)x - 2A.$$

Now we equate the coefficients of like powers of x on both sides of the equation:

$$6 = 2A + 2B + C, \quad 2 = -3A + B - 2C, \quad \text{and} \quad 2 = -2A.$$

In other words, the constants A, B, and C satisfy the system of linear equations

$$\begin{cases} 2A + 2B + C = 6 \\ -3A + B - 2C = 2 \\ -2A = 2. \end{cases}$$

Solving this system by one of the methods given in Sections 1, 2, and 4, we find that

$$A = -1, \quad B = 3, \quad \text{and} \quad C = 2.$$

Therefore,

$$\frac{6x^2 + 2x + 2}{x(x - 2)(2x + 1)} = \frac{-1}{x} + \frac{3}{x - 2} + \frac{2}{2x + 1}.$$

A factor of the form $(ax + b)^2$ in the denominator requires two partial fractions:

$$\frac{B}{ax + b} + \frac{C}{(ax + b)^2}.$$

Example   Decompose $\dfrac{3x^2 + 4x + 2}{x(x + 1)^2}$ into partial fractions.

Solution   We begin by writing

$$\frac{3x^2 + 4x + 2}{x(x + 1)^2} = \frac{A}{x} + \frac{B}{x + 1} + \frac{C}{(x + 1)^2}.$$

Multiplying both sides of the equation by $x(x + 1)^2$, we have

$$3x^2 + 4x + 2 = A(x + 1)^2 + Bx(x + 1) + Cx$$
$$= A(x^2 + 2x + 1) + B(x^2 + x) + Cx$$
$$= (A + B)x^2 + (2A + B + C)x + A.$$

Equating coefficients of like powers of x on both sides of the last equation, we obtain the system

$$\begin{cases} A + B & = 3 \\ 2A + B + C & = 4 \\ A & = 2. \end{cases}$$

Solving this system, we find that $A = 2$, $B = 1$, and $C = -1$. Hence,

$$\frac{3x^2 + 4x + 2}{x(x + 1)^2} = \frac{2}{x} + \frac{1}{x + 1} + \frac{-1}{(x + 1)^2}.$$

A prime quadratic factor of the form $ax^2 + bx + c$ in the denominator requires a partial fraction

$$\frac{Bx + C}{ax^2 + bx + c}.$$

Example   Decompose $\dfrac{8x^2 + 3x + 20}{(x + 1)(x^2 + 4)}$ into partial fractions.

Solution   We begin by writing

$$\frac{8x^2 + 3x + 20}{(x + 1)(x^2 + 4)} = \frac{A}{x + 1} + \frac{Bx + C}{x^2 + 4}.$$

Multiplying both sides by $(x + 1)(x^2 + 4)$, we obtain

$$8x^2 + 3x + 20 = A(x^2 + 4) + (Bx + C)(x + 1)$$
$$= Ax^2 + 4A + Bx^2 + Bx + Cx + C$$
$$= (A + B)x^2 + (B + C)x + 4A + C.$$

Equating coefficients, we get the system

$$\begin{cases} A + B & = 8 \\ B + C & = 3 \\ 4A + C & = 20. \end{cases}$$

The solution of this system is $A = 5$, $B = 3$, and $C = 0$; hence,

$$\frac{8x^2 + 3x + 20}{(x + 1)(x^2 + 4)} = \frac{5}{x + 1} + \frac{3x}{x^2 + 4}.$$

5.2 Other Applications of Systems of Linear Equations

Word problems that lead to a system of linear equations can be solved by using a slight variation of the procedure given in Chapter 2 (pages 68–69). We only need to modify step 2 of this procedure by introducing as many letters as may be necessary to represent all of the unknown quantities in the problem. Step 3 will then produce a *system* of equations to be solved for these unknowns.

Example   The price of admission to a play was $2 for adults, $1 for senior citizens, and $0.50 for children. Altogether, 270 tickets were sold and the total revenue was $360. Twice as many children as senior citizens attended the play. How many adults, how many children, and how many senior citizens attended the play?

Solution   Let x = the number of adults, y = the number of children, and z = the number of senior citizens who attended the play. From the fact that 270 tickets were sold, we have

$$x + y + z = 270.$$

Because the total revenue was $360, it follows that

$$2x + \tfrac{1}{2}y + z = 360.$$

Since twice as many children as senior citizens attended,

$$2z = y \quad \text{or} \quad y - 2z = 0.$$

Thus, we have the system of linear equations

$$\begin{cases} x + y + z = 270 \\ 2x + \tfrac{1}{2}y + z = 360 \\ y - 2z = 0. \end{cases}$$

Solving this system (say, by the elimination method) we find that $x = 135$, $y = 90$, and $z = 45$.

PROBLEM SET 5

In problems 1–26, decompose each fraction into partial fractions.

1. $\dfrac{3}{(x - 3)(x - 2)}$

2. $\dfrac{x}{(x - 1)(x - 4)}$

3. $\dfrac{x + 2}{(x + 5)(x - 1)}$

4. $\dfrac{3x + 7}{x^2 - 2x - 3}$

5. $\dfrac{x^2 - 5x - 3}{x(x - 2)(x + 2)}$

6. $\dfrac{x + 12}{x^3 - x^2 - 6x}$

7. $\dfrac{8x + 2}{x^3 - x}$

8. $\dfrac{x^2 - 16x - 12}{x^3 - 3x^2 - 4x}$

9. $\dfrac{x^2 + x - 1}{x^3 + x^2 - 6x}$

10. $\dfrac{2x^2 + 5x - 4}{x^3 + x^2 - 2x}$

11. $\dfrac{-2x^2 + x - 1}{(x - 3)(x - 1)^2}$

12. $\dfrac{13x - 12}{x^2(x - 3)}$

13. $\dfrac{1}{x^2(x - 1)}$

14. $\dfrac{3x + 4}{(x + 2)^2(x - 6)}$

15. $\dfrac{3x^2 + 18x + 15}{(x - 1)(x + 2)^2}$

16. $\dfrac{x + 4}{(x + 1)^2(x - 1)^2}$

17. $\dfrac{4x^2 - 7x + 10}{(x + 2)(3x - 2)^2}$

18. $\dfrac{1}{x^4 - 2x^3 + x^2}$

19. $\dfrac{x + 10}{(x + 1)(x^2 + 1)}$

20. $\dfrac{x^2 + 2x + 3}{(x - 2)(x^2 + 2x + 2)}$

21. $\dfrac{x^5 + 9x^3 + 1}{x^3 + 9x}$ **22.** $\dfrac{4x + 3}{(x^2 + 1)(x^2 + 2)}$ **23.** $\dfrac{t + 3}{t(t^2 + 1)}$ **24.** $\dfrac{x}{(x + 1)^2(x^2 + 1)}$

25. $\dfrac{u}{u^4 - 1}$ **26.** $\dfrac{3s + 1}{s^2(s^2 + 1)}$

In problems 27–30, find the values of the constants in each decomposition into partial fractions.

27. $\dfrac{3x^2 - 2x - 4}{x^3(x + 2)} = \dfrac{A}{x} + \dfrac{B}{x^2} + \dfrac{C}{x^3} + \dfrac{D}{x + 2}$ **28.** $\dfrac{x^3 + 3x^2 + 1}{(x^2 + 1)^2} = \dfrac{Ax + B}{x^2 + 1} + \dfrac{Cx + D}{(x^2 + 1)^2}$

29. $\dfrac{t^3 - t^2}{(t^2 + 3)^2} = \dfrac{At + B}{t^2 + 3} + \dfrac{Ct + D}{(t^2 + 3)^2}$ **30.** $\dfrac{x^5 - 2x^4 + 2x^3 + x - 2}{x^2(x^2 + 1)^2} = \dfrac{A}{x} + \dfrac{B}{x^2} + \dfrac{Cx + D}{x^2 + 1} + \dfrac{Ex + G}{(x^2 + 1)^2}$

31. The price of admission for a sporting event was $2 for adults and $1 for children. Altogether, 925 tickets were sold, and the resulting revenue was $1150. How many adults and how many children attended the game?

32. A veterinarian has put certain animals on a diet. Each animal receives, among other things, exactly 25 grams of protein and 9.5 grams of fat for each feeding. If the veterinarian buys two food mixes, the first containing 10% protein and 8% fat, and the second containing 20% protein and 2% fat, how many grams of each should be combined to provide the right diet for a single feeding of 10 animals?

33. One angle x of a triangle is $10°$ greater than a second angle y, and $40°$ less than the third angle z. Find x, y, and z.

34. A certain three-digit number is 56 times the sum of its digits. The unit's digit is 4 more than the ten's digit. If the unit's digit and the hundred's digit were interchanged, the resulting number would be 99 less than the original number. Find the number.

35. A department store has sold 80 men's suits of three different types at a discount. If the suits had been sold at their original prices—type I suits for $80, type II suits for $90, and type III suits for $95—the total receipts would have been $6825. However, the suits were sold for $75, $80, and $85, respectively, and the total receipts amounted to $6250. Determine the number of suits of each type sold during the sale.

36. A collection of dimes and quarters amounts to $2.70. If the total number of coins is 15, how many coins of each type are in the collection?

37. Suppose that the demand and supply equations for coal in a certain marketing area are $q = -2p + 150$ and $q = 3p$, respectively, where p is the price per ton in dollars and q is the quantity of coal in thousands of tons. In economics, **market equilibrium** is said to occur when these equations hold simultaneously. Solve the system for market equilibrium.

38. A chemist has two solutions, the first containing 20% acid and the second containing 50% acid. She wishes to mix the two solutions to obtain 9 liters of a 30% acid solution. How many liters of each solution should she use?

39. Suppose that x dollars is invested at a simple annual interest rate of 8.5%, and that y dollars is invested at 9.5%. If the total amount invested is $17,000 and the total interest from the two investments at the end of the year is $1535, find x and y.

40. A company has two hydraulic presses, an old one and an improved model. With both presses working together, a certain job is done in 2 hours and 24 minutes. On another job of the same kind, the old press is operated alone for 3 hours, then the new press is also put into operation and the two presses together finish the job in an additional 1 hour and 12 minutes. How long would it take each press operating alone to do this job?

41. On a certain date, 3 pounds of coffee, 4 quarts of milk, and 2 cans of tuna fish cost $13.20. One can of tuna fish cost twice as much as 1 quart of milk. Six months later, because of inflation, the price of the coffee had increased by 15%, the price of milk by 5%, and the price of tuna fish by 10%; so the same grocery order cost $14.82. Find the original prices of a pound of coffee, a quart of milk, and a can of tuna fish.

42. Professor Grumbles, who teaches morning and afternoon statistics classes of equal size, is accused of male chauvinism because in the two classes taken together, 80% of the male students passed, but only 20% of the female students did. However, the professor contends that the accusation is false because in the morning class, 10% of the men and 10% of the women passed; whereas in the afternoon class, 90% of the men and 90% of the women passed. Furthermore, the total number of men in the two classes is the same as the total number of women. Is this possible, and if so, how?

6   Systems of Linear Inequalities and Linear Programming

Many real-world applications of mathematics—especially those that involve the allocation of limited resources—give rise to systems of linear inequalities. In this section we shall consider problems involving systems of linear inequalities in *two* unknowns. The solution of such a system can be represented by a set of points in the *xy* plane, and the problem can be solved geometrically. The section includes a brief discussion of *linear programming*.

6.1   The Graph of a Linear Inequality

By a *linear inequality* in the two unknowns x and y, we mean an inequality having one of the forms

$$ax + by + c > 0, \quad ax + by + c < 0, \quad ax + by + c \geq 0, \quad \text{or} \quad ax + by + c \leq 0,$$

where a, b, and c are constants and a and b are not both zero. The first two inequalities are called *strict*; the second two, *nonstrict*. The **graph** of such an inequality is defined to be the set of all points (x, y) in the *xy* plane whose coordinates satisfy the inequality.

In order to study the graph of a linear inequality in x and y, we begin by considering the graph of the linear *equation*

$$ax + by + c = 0$$

obtained by (temporarily) replacing the inequality sign with an equal sign. This graph is a straight line which divides the *xy* plane into two regions called **half-planes,** one on each side of the line (Figure 1). A half-plane is called **closed** if the points on the boundary line belong to it; a half-plane is said to be **open** if the points on the boundary line do not belong to it.

In Figure 1, the expression $ax + by + c$ is zero only for points on the line separating the two

Figure 1

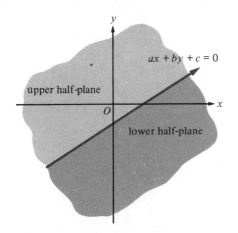

half-planes, so it is either positive or negative for points (x, y) in the open half-planes. Actually, $ax + by + c$ is positive on one of the open half-planes and negative on the other (Problems 42 and 43). Thus, we have the following procedure.

Procedure for Sketching the Graph of a Linear Inequality

Step 1. Draw the graph of the linear equation obtained by (temporarily) replacing the inequality sign with an equal sign. If the inequality is strict ($>$ or $<$), draw a dashed line; if the inequality is nonstrict ($\geq$ or $\leq$), draw a solid line.

Step 2. Select any convenient test point in one of the two open half-planes determined by the line in step 1. If the coordinates of the test point satisfy the inequality, shade the half-plane that contains it; otherwise, shade the other half-plane. The shaded half-plane is the graph of the inequality—a dashed boundary line indicates an open half-plane, and a solid boundary line indicates a closed one.

Examples   Sketch the graph of each linear inequality.

1 $4x + 3y - 12 > 0$.

Solution   We follow the procedure above.

(1) Since the inequality is strict, we draw the graph of $4x + 3y - 12 = 0$ as a dashed line (Figure 2).

(2) We test the inequality at the origin $O = (0, 0)$ by substituting $x = 0$ and $y = 0$ to obtain $-12 > 0$, which is *false*. Therefore, we shade the open half-plane *not* containing the origin (Figure 2).

2 $-2x + y \leq 0$.

Solution   Again, we follow the procedure.

(1) Since the inequality is nonstrict, we draw the graph of $-2x + y = 0$ as a solid line (Figure 3).

Figure 2

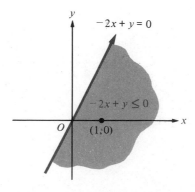

Figure 3

(2) We test the inequality at the point $(1,0)$ by substituting $x = 1$ and $y = 0$ to obtain $-2 \leq 0$, which is *true*. Therefore, we shade the closed half-plane containing the point $(1,0)$ (Figure 3).

6.2 The Graph of a System of Linear Inequalities

The **graph** of a system of linear inequalities is defined to be the set of all points (x, y) in the xy plane whose coordinates satisfy every inequality in the system. Such a graph is obtained by sketching the graphs of all the inequalities on the same coordinate system. The region where all of these graphs overlap is the graph of the system of linear inequalities.

Examples   Sketch the graph of each system of linear inequalities.

1
$$\begin{cases} x + y \leq 2 \\ -x + 3y \geq 4 \end{cases}$$

Solution   Using the procedure given in Section 6.1, we sketch the graphs of $x + y \leq 2$ and $-x + 3y \geq 4$ on the same coordinate system (Figure 4). The graph of $x + y \leq 2$ is the closed half-plane below the line $x + y = 2$, and the graph of $-x + 3y \geq 4$ is the closed half-plane above the line $-x + 3y = 4$. These two half-planes overlap in the region shaded in Figure 4; hence, this shaded region is the graph of the system of equations.

Figure 4

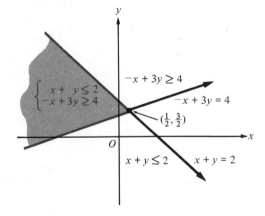

2
$$\begin{cases} 2x + y \geq 2 \\ x - 2y \leq 3 \\ x \geq 0 \\ y \geq 0 \end{cases}$$

Solution   Again, we begin by sketching the graphs of the four linear inequalities on the same coordinate system (Figure 5). Notice that the graph of $x \geq 0$ is the closed half-plane

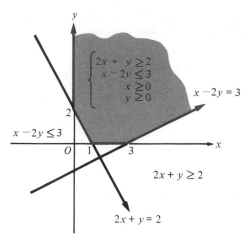

Figure 5

to the right of the y axis, and that the graph of $y \geq 0$ is the closed half-plane above the x axis. These two closed half-planes overlap in the region consisting of the first quadrant together with the positive x and y axes and the origin. We intersect this region with the closed half-plane above the line $2x + y = 2$ [the graph of $2x + y \geq 2$] and the closed half-plane above the line $x - 2y = 3$ [the graph of $x - 2y \leq 3$] to obtain the graph of the system of inequalities (Figure 5).

If the graph of a system of linear inequalities in x and y is a nonempty set of points, it will be bounded by straight line segments or rays meeting at "corner points" called **vertices.** After you have sketched the graph, you can see these vertices and you can find their coordinates by solving appropriate pairs of linear equations. For instance, in Figure 4 on page 465, the coordinates $(\frac{1}{2}, \frac{3}{2})$ of the vertex are found by solving the system

$$\begin{cases} x + y = 2 \\ -x + 3y = 4. \end{cases}$$

The graph in Figure 5 has three vertices: $(0, 2)$, $(1, 0)$, and $(3, 0)$.

Because the graphs in Figures 4 and 5 extend indefinitely in certain directions, we say that they are **unbounded.** In Section 6.3, we shall discuss systems of linear inequalities whose graphs are **bounded** in the sense that they are cut off in every direction by line segments. (See Figures 6 and 7.)

6.3 Linear Programming

Suppose that an oil company's profit depends on what portion of a limited allocation of crude oil the company refines into gasoline and what portion it converts into heating oil. Assume that there is a governmental restriction requiring that at least a certain fraction of the crude oil be converted into heating oil. In order to achieve the greatest possible profit under the governmental constraint, the company will have to plan or "program" its activities.

A problem in which it is necessary to find the maximum (largest) or minimum (smallest) value of a certain quantity (such as profit, cost, revenue, distance, or time) under a given set of constraints (restrictions) is sometimes called a **programming** problem. Here the word "programming" is intended to suggest planning. The quantity whose maximum or minimum value is desired is called the **objective function.** The objective function depends on one or more variables, and the constraints are expressed as conditions on the possible values of these variables. When the objective function is a linear (first-degree) polynomial in two or more variables and the constraints are expressed as a system of nonstrict linear inequalities, we have a **linear programming** problem.

Here we consider only linear programming problems in which the objective function F depends on two variables x and y, and the graph G of the system of linear inequalities that express the constraints is a *bounded* region in the xy plane. Thus, F has the form

$$F = ax + by + c,$$

where a, b, and c are constants, and G is bounded by a finite number of line segments meeting at a finite number of vertices. Under these circumstances, it can be shown that *F has a maximum and a minimum value on G, and these values occur at certain vertices of G.* Thus, to find these maximum and minimum values, you merely *list all the vertices of G and calculate the value of F at each one.* The largest and smallest of these values are the maximum and minimum values of F on the region G.

Example   Find the maximum and minimum values of the objective function

$$F = 3x + 4y + 1,$$

subject to the constraints

$$\begin{cases} x + 2y \le 8 \\ x + y \le 5 \\ x \ge 0 \\ y \ge 0. \end{cases}$$

Solution   We begin by using the method described in Section 6.2 to sketch the graph G of the system representing the constraints (Figure 6). By solving appropriate pairs of linear equations, we find the coordinates of the vertices of G. These coordinates are listed in Table 1 along with the corresponding values of $F = 3x + 4y + 1$ at each vertex. From Table 1, the minimum value of F, which is 1, occurs at $(0,0)$; and the maximum value of F, which is 19, occurs at $(2,3)$.

Figure 6

Table 1

Vertex	Value of $F = 3x + 4y + 1$	
$(0,0)$	1	minimum
$(5,0)$	16	
$(2,3)$	19	maximum
$(0,4)$	17	

The following example illustrates a typical application of linear programming.

Example   A large school system wants to design a lunch menu containing two food items X and Y. Each ounce of X supplies 1 unit of protein, 2 units of carbohydrates, and 1 unit of fat. Each ounce of Y supplies 1 unit of protein, 1 unit of carbohydrates, and 1 unit of fat. The two items together must provide at least 7 units of protein, at least 10 units of carbohydrates, and no more than 8 units of fat per serving. If each ounce of X costs 12 cents and each ounce of Y costs 8 cents, how many ounces of each item should each serving contain to meet the dietary requirements at the lowest cost?

Solution   Let x and y denote the number of ounces per serving of X and Y, respectively. Since each ounce of X costs 12 cents and each ounce of Y costs 8 cents, the cost per serving is

$$F = 12x + 8y \text{ cents.}$$

Since each ounce of X or of Y supplies 1 unit of protein, each serving will supply $x + y$ units of protein. To meet the protein requirement, we must have

$$x + y \geq 7.$$

Similarly, to meet the carbohydrate requirement, we must have

$$2x + y \geq 10,$$

and to meet the fat requirement,

$$x + y \leq 8.$$

Because the number of ounces per serving cannot be negative, we also have the conditions $x \geq 0$ and $y \geq 0$. Therefore, the problem is to minimize the objective function

$$F = 12x + 8y$$

subject to the constraints

$$\begin{cases} x + y \geq 7 \\ 2x + y \geq 10 \\ x + y \leq 8 \\ x \geq 0 \\ y \geq 0. \end{cases}$$

The graph G of the system representing these constraints is sketched in Figure 7, from which we find the vertices listed in Table 2. Thus, if each serving contains 3 ounces of item X and 4 ounces of item Y, all dietary requirements will be met at the minimum cost of 68 cents per serving.

Figure 7

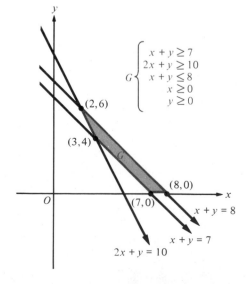

Table 2

Vertex	Value of $F = 12x + 8y$	
(7,0)	84	
(8,0)	96	
(2,6)	72	
(3,4)	68	minimum

PROBLEM SET 6

In problems 1–10, sketch the graph of each linear inequality.

1. $x \geq 0$ 2. $y < 1$ 3. $2x + y - 3 \leq 0$ 4. $5x - 2y \leq 2$
5. $3x + 2y \geq 6$ 6. $2x \geq 2 - y$ 7. $3x + 5y < 15$ 8. $4x > 3 - 2y$
9. $x < y$ 10. $1 < x - y \leq 5$

In problems 11–23, (a) sketch the graph of the system of linear inequalities, (b) determine whether the graph is bounded or unbounded, and (c) find all vertices of the graph.

11. $\begin{cases} 2x + 3y \leq 7 \\ 3x - y \leq 5 \end{cases}$
12. $\begin{cases} 2x + y < 3 \\ x + 3y > 4 \end{cases}$
13. $\begin{cases} x + y \leq 4 \\ y > 2x - 4 \end{cases}$
14. $\begin{cases} x + y < 1 \\ -x + 2y \geq 4 \\ y > 0 \end{cases}$

15. $\begin{cases} y + 1 < 3x \\ y - x > 3 \end{cases}$
16. $\begin{cases} 3x + y < 4 \\ y - 2x \geq -1 \\ x \leq 0 \end{cases}$
17. $\begin{cases} 4x + 7y \leq 28 \\ 2x - 3y \geq -6 \\ y \geq -2 \end{cases}$
18. $\begin{cases} x + 3y \leq 6 \\ 3x - 2y \leq 4 \\ y \geq 0 \end{cases}$

19. $\begin{cases} -2x + y \leq 2 \\ x + 2y \leq 8 \\ x \geq 0 \\ y \geq 0 \end{cases}$
20. $\begin{cases} 2x - 3y \geq 2 \\ y \geq 6 - x \\ x \geq 4 \end{cases}$
21. $\begin{cases} x + 2y \leq 6 \\ 3x + y \leq 9 \\ x \geq 0 \\ y \geq 0 \end{cases}$
22. $\begin{cases} \dfrac{x}{2} - 1 \leq y \leq 3 + 3x \\ 0 \leq 4x \leq 12 - 3y \end{cases}$

23. $\begin{cases} 0 \leq x \leq 7 - 2y \\ 0 \leq 8y \leq 5x + 3 \end{cases}$ [Hint: Rewrite as a system of four inequalities.]

24. A hardware store has display space for at most 40 spray cans of rustproofing paint, x cans of red and y cans of gray. If there are to be at least 10 cans of each type in the display, (a) sketch a graph showing the possible numbers of red and gray cans in the display, and (b) find all vertices of the graph.

In problems 25–34, find the maximum and minimum values of the objective function F subject to the indicated constraints.

25. $F = 2x - y + 3$
$\begin{cases} x + y \leq 1 \\ x \geq 0 \\ y \geq 0 \end{cases}$
26. $F = 3x + 2y - 1$
$\begin{cases} x + 2y \leq 7 \\ 5x - 8y \leq -3 \\ x \geq 0 \\ y \geq 0 \end{cases}$
27. $F = 5x + 4y$
$\begin{cases} x + 2y \geq 3 \\ x + 2y \leq 5 \\ x \geq 0 \\ y \geq 0 \end{cases}$
28. $F = 4x - y + 7$
$\begin{cases} 3x + 8y \leq 120 \\ 3x + y \leq 36 \\ x \geq 0 \\ y \geq 0 \end{cases}$

29. $F = 3x - 5y + 2$
$\begin{cases} x + y \leq 10 \\ x - 3y \geq -18 \\ x \geq 0 \\ y \geq 0 \end{cases}$
30. $F = 10x + 12y$
$\begin{cases} 0.2x + 0.4y \leq 30 \\ 0.2x + 0.2y \leq 20 \\ x \geq 0 \\ y \geq 0 \end{cases}$
31. $F = \frac{3}{2}x + y$
$\begin{cases} \frac{1}{2}x + y \geq 2 \\ \frac{1}{2}x - y \geq 1 \\ x \leq 6 \\ y \geq 0 \end{cases}$
32. $F = -10x + 5y + 3$
$\begin{cases} x + \frac{3}{2}y \geq -60 \\ x + y \geq -50 \\ x \leq 0 \\ y \leq 0 \end{cases}$

33. $F = \frac{1}{2}x + \frac{3}{4}y$
$\begin{cases} \frac{1}{3}x + y \leq 30 \\ x + \frac{1}{2}y \leq 40 \\ x + y \geq 10 \end{cases}$
34. $F = 0.15x + 0.1y$
$\begin{cases} 4x + 5y \leq 2000 \\ 12x + 5y \leq 3000 \\ x + y \geq 100 \\ x \geq 0 \\ y \geq 0 \end{cases}$

35. An electronics company manufactures two models of household smoke detectors. Model A requires 1 unit of labor and 4 units of parts; model B requires 1 unit of labor and 3 units

of parts. If 90 units of labor and 320 units of parts are available, and if the company makes a profit of $5 on each model A detector and $4 on each model B detector, how many of each model should it manufacture to maximize its profit?

C **36.** A family owns and operates a 312-acre farm on which it grows cotton and peanuts. The task of planting, picking, ginning, and baling the cotton requires 35 person-hours of labor per acre; the task of planting, harvesting, and bagging the peanuts requires 27 person-hours of labor per acre. The family is able to devote 9500 person-hours to these activities. If the profit for each acre of cotton grown is $173 and the profit for each acre of peanuts grown is $152, how many acres should be planted in cotton and how many in peanuts to maximize the family's profit?

37. In problem 35, suppose that the company raises its prices so that its profit on each model A detector is $7 and on each model B detector is $5. Now how many detectors of each type should it manufacture to maximize its profit?

38. A supplier has 105 pounds of leftover beef which must be sold before it spoils. Of this beef, 15 pounds is prime grade, 40 pounds is grade A, and the rest is utility grade. A local restaurant will buy ground beef consisting of 20% prime grade, 40% grade A, and 40% utility grade for 75 cents per pound. A hamburger stand will buy ground beef consisting of 40% grade A and 60% utility grade for 55 cents per pound. How much ground beef should the supplier sell to the restaurant and how much to the hamburger stand in order to maximize its revenue?

39. A town operates two recycling centers. Each day that Center I is open, 300 pounds of glass, 200 pounds of paper, and 100 pounds of aluminum are deposited there. Each day that Center II is open, 100 pounds of glass, 600 pounds of paper, and 100 pounds of aluminum are deposited there. The town has contracted to supply at least 1200 pounds of glass, 2400 pounds of paper, and 800 pounds of aluminum per week to a salvage company. Supervision and maintenance at Center I costs the town $40 each day it is open, and Center II costs $50 each day it is open. How many days a week should each center remain open so as to minimize the total weekly cost for supervision and maintenance, yet allow the town to fulfill its contract with the salvage company?

40. In problem 38, suppose that the owner of the restaurant learns that the meat is in danger of spoiling, and decides to pay only 50 cents per pound for it. If the hamburger stand will still pay 55 cents per pound, how much ground beef should the supplier now sell to the restaurant and how much to the hamburger stand?

41. Find a condition on the positive constants a and b such that the objective function

$$F = ax + by,$$

subject to the constraints $2x + 3y \leq 9$, $x - y \leq 2$, $x \geq 0$, and $y \geq 0$, will take on the *same* maximum value at the vertex $(0, 3)$ and the vertex $(3, 1)$.

42. Suppose that a, b, and c are constants and $b \neq 0$. Let L be the line

$$ax + by + c = 0.$$

(a) If $b > 0$, show that $ax + by + c > 0$ for (x, y) above L, and that $ax + by + c < 0$ for (x, y) below L.

(b) If $b < 0$, show that $ax + by + c < 0$ for (x, y) above L, and that $ax + by + c > 0$ for (x, y) below L.

43. Suppose that a and c are constants and $a \neq 0$. Let L be the vertical line

$$ax + c = 0.$$

(a) If $a > 0$, show that $ax + c > 0$ for (x, y) to the right of L, and that $ax + c < 0$ for (x, y) to the left of L.

(b) If $a < 0$, show that $ax + c < 0$ for (x, y) to the right of L, and that $ax + c > 0$ for (x, y) to the left of L.

7   Systems Containing Nonlinear Equations

In this section, we consider the solution of systems containing nonlinear equations. To avoid technical difficulties, we shall study only the relatively simple case of two equations in two unknowns. The solution of such a system can be found (at least approximately) by sketching graphs of the two equations on the same coordinate system, and determining the points where the two graphs intersect.

The substitution and elimination methods, introduced for systems of linear equations in Sections 1.1 and 1.2, can often be used (with minor modifications) for systems containing nonlinear equations. Even then, graphs can be sketched to determine the number of solutions and as a rough check on the calculations.

Examples   In each case, sketch graphs to determine the number of solutions of each system, and then solve the system.

1
$$\begin{cases} x^2 - 2y = 0 \\ x + 2y = 6 \end{cases}$$

Solution   The graph of $x^2 - 2y = 0$, or $y = \frac{1}{2}x^2$, is a parabola opening upward with vertex at the origin; the graph of $x + 2y = 6$ is a line that intersects the parabola at two points (Figure 1). Thus, there are two solutions of the system. To find these solu-

Figure 1

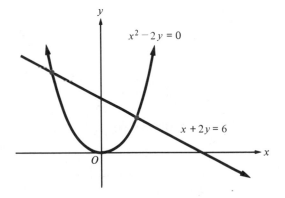

tions algebraically, we use the method of elimination. Adding the second equation to the first (to eliminate y), we obtain the equivalent system

$$\begin{cases} x^2 + x = 6 \\ x + 2y = 6. \end{cases}$$

Now the first equation is quadratic in x and can be solved by factoring:

$$x^2 + x - 6 = 0 \quad \text{or} \quad (x + 3)(x - 2) = 0,$$

so $x = -3$ or $x = 2$. Substituting these values, one at a time, into the second equation $x + 2y = 6$, we find that

$$y = \tfrac{9}{2} \quad \text{when} \quad x = -3 \quad \text{and} \quad y = 2 \quad \text{when} \quad x = 2.$$

Hence, the two solutions are $(-3, \tfrac{9}{2})$ and $(2, 2)$.

2 $$\begin{cases} x^2 + y^2 = 25 \\ x^2 + y\ \ = 13 \end{cases}$$

Solution   Sketching the graphs of the two equations (a circle and a parabola) on the same coordinate system, we see four points of intersection (Figure 2). Thus, there are four solutions of the system. Again, we can use the method of elimination. Subtracting the second equation from the first (to eliminate x^2), we obtain the equivalent system

$$\begin{cases} y^2 - y = 12 \\ x^2 + y = 13. \end{cases}$$

Figure 2

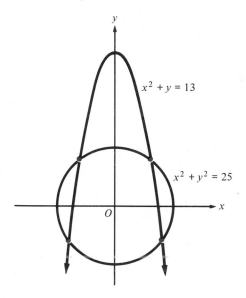

Now the first equation is quadratic in y and can be solved by factoring:

$$y^2 - y - 12 = 0 \quad \text{or} \quad (y + 3)(y - 4) = 0,$$

so $y = -3$ or $y = 4$. Substituting $y = -3$ into the second equation, $x^2 + y = 13$, we obtain

$$x^2 - 3 = 13 \quad \text{or} \quad x^2 = 16;$$

hence, $x = 4$ or $x = -4$. Similarly, substituting $y = 4$ into the second equation, we obtain

$$x^2 + 4 = 13 \quad \text{or} \quad x^2 = 9;$$

hence, $x = 3$ or $x = -3$. Therefore, the solutions are $(-4, -3)$, $(4, -3)$, $(-3, 4)$, and $(3, 4)$.

When the elimination method can be used, as in the examples above, it is usually the most efficient way to solve the system. Otherwise, try the substitution method.

Example   Solve the system $\begin{cases} x^2 + 2y^2 = 18 \\ xy = 4. \end{cases}$

Solution   Here there doesn't seem to be a simple way to eliminate variables, so we try the substitution method. Because the second equation is easily solved for y in terms of x, we begin there. If $xy = 4$, then $x \neq 0$ and

$$y = \frac{4}{x}.$$

Substituting $4/x$ for y in the first equation, we obtain

$$x^2 + 2\left(\frac{4}{x}\right)^2 = 18 \quad \text{or} \quad x^2 + \frac{32}{x^2} = 18.$$

Multiplying both sides of the last equation by x^2, we get

$$x^4 + 32 = 18x^2 \quad \text{or} \quad x^4 - 18x^2 + 32 = 0.$$

Factoring, we have

$$(x^2 - 16)(x^2 - 2) = 0 \quad \text{or} \quad (x - 4)(x + 4)(x - \sqrt{2})(x + \sqrt{2}) = 0.$$

Setting each factor equal to zero gives us

$$x = 4, \quad x = -4, \quad x = \sqrt{2}, \quad \text{or} \quad x = -\sqrt{2}.$$

For each of these values of x, there is a corresponding value of y given by $y = 4/x$. Therefore, the solutions are

$$(4, 1), \quad (-4, -1), \quad (\sqrt{2}, 2\sqrt{2}), \quad \text{and} \quad (-\sqrt{2}, -2\sqrt{2}).$$

If a system of equations contains exponential or logarithmic functions, the properties of these functions may be used to help find the solutions.

Example   Solve the system $\begin{cases} y - \log_4(6x + 10) = 1 \\ y + \log_4 x = 2. \end{cases}$

Solution   We use the method of elimination. Subtracting the first equation from the second, we obtain

$$\log_4 x + \log_4(6x + 10) = 1.$$

Now we recall a basic property of logarithms [Property (iii), page 238], and the definition of logarithms to rewrite the last equation as

$$\log_4[x(6x + 10)] = 1 \quad \text{or} \quad x(6x + 10) = 4^1;$$

that is,

$$6x^2 + 10x - 4 = 0 \quad \text{or} \quad 3x^2 + 5x - 2 = 0.$$

Factoring, we get

$$(3x - 1)(x + 2) = 0; \quad \text{hence,} \quad x = \tfrac{1}{3} \quad \text{or} \quad x = -2.$$

Because $\log_4 x$ is undefined when x is negative, $x = -2$ is an extraneous root, and we must reject it. Substituting $x = \frac{1}{3}$ into the equation $y = 2 - \log_4 x$, we find that

$$y = 2 - \log_4 \tfrac{1}{3} = 2 - \log_4 3^{-1} = 2 - (-1)\log_4 3 = 2 + \log_4 3.$$

Hence, the solution is $(\frac{1}{3}, 2 + \log_4 3)$.

We close this section with a brief indication of the way in which systems of equations, not all of which need be linear, arise in practical situations. We choose economics as our area of application. If p denotes the price per unit of a commodity, and q denotes the number of units of the commodity demanded in the marketplace at price p, a graph of q as a function of p produces a **demand curve** (Figure 3a). Note that q will ordinarily be a decreasing function of p. (Why?) On the other hand, if p represents the price per unit of a commodity in the marketplace, and q denotes the number of units that manufacturers are willing to supply at that price, a graph of q as a function of p produces a **supply curve** (Figure 3b). In this case, q will ordinarily be an increasing function of p. (Why?) If the supply and demand curves are plotted on the same coordinate system, the point where they intersect is called the **market equilibrium point** (Figure 3c). At the equilibrium price (the p coordinate of the market equilibrium point), the quantity supplied will be equal to the quantity demanded. Under the usual interpretations of price, supply, and demand, only the portions of the supply and demand curves that fall in the first quadrant are economically meaningful.

Figure 3

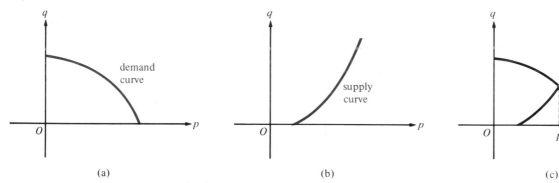

(a)   (b)   (c)

Example   Suppose that the weekly demand q millions of gallons of synthetic fuel made from oil shale is related to the price p in dollars per gallon by the demand equation $8p^2 + 5q = 100$. If the market price is p dollars per gallon, assume that producers are willing to supply q million gallons of the synthetic fuel per week according to the supply equation $6p^2 - p - 3q = 5$. Find (a) the market equilibrium point and (b) the equilibrium price.

Solution   (a) The market equilibrium point is found by solving the system

$$\begin{cases} 8p^2 + 5q = 100 \\ 6p^2 - p - 3q = 5. \end{cases}$$

Multiplying the second equation by 5 and adding 3 times the first equation to the result, we obtain the equivalent system

$$\begin{cases} 8p^2 + 5q = 100 \\ 54p^2 - 5p = 325. \end{cases}$$

The second equation can be solved by factoring:

$$54p^2 - 5p - 325 = 0 \quad \text{or} \quad (2p - 5)(27p + 65) = 0;$$

hence, $p = \frac{5}{2}$ or $p = -\frac{65}{27}$. Because of the economic interpretation, p cannot be negative, so we reject the second solution and retain only the solution $p = \frac{5}{2}$. Substituting $p = \frac{5}{2}$ into the first equation $8p^2 + 5q = 100$, we obtain

$$8(\tfrac{5}{2})^2 + 5q = 100 \quad \text{or} \quad 5q = 50,$$

so

$$q = 10.$$

Therefore, the market equilibrium point is $(p, q) = (\frac{5}{2}, 10)$.

(b) The equilibrium price is

$$p = \tfrac{5}{2} = \$2.50 \text{ per gallon.}$$

PROBLEM SET 7

In problems 1–6, sketch graphs to determine the number of solutions of each system of equations, and then solve the system.

1. $\begin{cases} x^2 - 2y = 0 \\ 3x + 2y = 10 \end{cases}$
2. $\begin{cases} x^2 - 2y = 3 \\ x - y = 1 \end{cases}$
3. $\begin{cases} y^2 - 3x = 0 \\ 2x - y - 3 \end{cases}$
4. $\begin{cases} x^2 + y^2 = 25 \\ x^2 + y = 19 \end{cases}$
5. $\begin{cases} x^2 + y^2 = 4 \\ x - 2y = 4 \end{cases}$
6. $\begin{cases} 2x^2 + y = 9 \\ y - x^2 - 5x = 1 \end{cases}$

In problems 7–24, solve each system of equations by the elimination or substitution methods.

7. $\begin{cases} 2x^2 + y = 4 \\ 2x - y = 1 \end{cases}$
8. $\begin{cases} x^2 + y^2 = 1 \\ x^2 - y = 3 \end{cases}$
9. $\begin{cases} x^2 - y^2 = 3 \\ -2x + y = 1 \end{cases}$
10. $\begin{cases} x^2 + y^2 = 1 \\ -x + 2y = -2 \end{cases}$
11. $\begin{cases} x^2 + 2y^2 = 22 \\ 2x^2 + y^2 = 17 \end{cases}$
12. $\begin{cases} x^2 + y^2 = 625 \\ x + y = 35 \end{cases}$
13. $\begin{cases} 2s^2 - 4t^2 = 8 \\ s^2 + 2t^2 = 10 \end{cases}$
14. $\begin{cases} 4r^2 + 7s^2 = 23 \\ -3r^2 + 11s^2 = -1 \end{cases}$
15. $\begin{cases} 4h^2 + 7k^2 = 32 \\ -3h^2 + 11k^2 = 41 \end{cases}$
16. $\begin{cases} 2a^2 - 5b^2 + 8 = 0 \\ a^2 - 7b^2 + 4 = 0 \end{cases}$
17. $\begin{cases} x^2 - y = 2 \\ 2x^2 + y = 6x + 7 \end{cases}$
18. $\begin{cases} x^2 + y^2 - 8y = -7 \\ y - x^2 = 1 \end{cases}$
19. $\begin{cases} x^2 - xy + 2y^2 = 8 \\ xy = 4 \end{cases}$
20. $\begin{cases} 4x^2 - 6xy + 9y^2 = 63 \\ 2x - 3y + 3 = 0 \end{cases}$
21. $\begin{cases} x - y = 21 \\ \sqrt{x} + \sqrt{y} = 7 \end{cases}$
22. $\begin{cases} \dfrac{3}{x^2} - \dfrac{2}{y^2} = 1 \\ \dfrac{7}{x^2} - \dfrac{6}{y^2} = 2 \end{cases}$
23. $\begin{cases} y - \sqrt[4]{x} = 0 \\ y^2 - \sqrt[4]{x} = 2 \end{cases}$
24. $\begin{cases} y = \sqrt[4]{x} - 2 \\ y^2 = \sqrt[4]{x} - 2 + 12 \end{cases}$

25. $\begin{cases} (x - y)^2 + (x + y)^2 = 17 \\ (x - y)^2 - 4(x + y)^2 = 12 \end{cases}$

26. $\begin{cases} \dfrac{2}{(x + 1)^2} - \dfrac{5}{(y - 1)^2} = 3 \\ \dfrac{1}{(x + 1)^2} + \dfrac{3}{(y - 1)^2} = 7 \end{cases}$

In problems 27–32, use appropriate properties of the exponential and logarithmic functions to solve each system.

27. $\begin{cases} y - \log_6(x + 3) = 1 \\ y + \log_6(x + 4) = 2 \end{cases}$

28. $\begin{cases} y + \log_3(x + 1) = 2 \\ y - \log_3(x + 3) = 1 \end{cases}$

29. $\begin{cases} y = 5 + \log_{10} x \\ y - \log_{10}(x + 6) = 7 \end{cases}$

30. $\begin{cases} \log_4(x^2 + y^2) = 2 \\ 2y - x = 4 \end{cases}$

31. $\begin{cases} x - 5^y = 0 \\ x - 25^y = -20 \end{cases}$

32. $\begin{cases} y + 15 = 10^x \\ y - 10^{2-x} = 0 \end{cases}$

33. The ratio of two positive numbers is 3 to 8, and their product is 864. Find the numbers.

34. A sporting goods store sells two circular targets for archery. The radius of the larger target is 10 centimeters more than the radius of the smaller target, and the difference between the areas of the two targets is 2300 square centimeters. Find the radii of the targets.

35. A commercial artist is using a triangular template whose altitude exceeds its base by 4 inches, and whose area is 30 square inches. Find the altitude and base of the template.

36. The sum of the squares of the digits of a certain two-digit number is 61. The product of the number and the number with the digits reversed is 3640. What is the number?

37. A company determines that its total monthly production cost C in thousands of dollars satisfies the equation $C^2 = 8x + 4$ and that its monthly revenue R in thousands of dollars satisfies the equation $8R - 3x^2 = 0$, where x is the number of thousands of units of its product manufactured and sold per month. When the cost of manufacturing the product equals the revenue obtained from selling it, the company **breaks even.** How many units must the company produce in order to break even?

38. A commercial jet was delayed for 15 minutes before takeoff because of a problem with baggage loading. To make up for the delay, the pilot increased the jet's air speed and took advantage of a favorable tail wind. This increased the jet's average speed by 36 miles per hour, with the result that it landed on time, 3 hours and 45 minutes after takeoff. Find the usual average ground speed and the distance between the two airports.

39. Suppose that the annual demand, q million cars, for a front-wheel-drive economy car is related to its sticker price, p thousand dollars, by the demand equation $2q^2 = 9 - p$. At a sticker price of p thousand dollars, the manufacturers are willing to build q million cars a year according to the supply equation $q^2 + 5q = p - 1$. Find (a) the market equilibrium point and (b) the equilibrium price of a car.

40. Find formulas for the length l and the width w of a rectangle in terms of its area A and its perimeter P.

REVIEW PROBLEM SET

In problems 1–6, use graphs to determine whether the equations in each system are consistent, inconsistent, or dependent, and indicate the solutions (if any) on the graphs.

1. $\begin{cases} y = 2x + 7 \\ 2x - y = 5 \end{cases}$

2. $\begin{cases} 4x + 2y = 3 \\ 2x + y = 5 \end{cases}$

3. $\begin{cases} 4x - y = 3 \\ -2x + 3y = 1 \end{cases}$

4. $\begin{cases} -2x + 6y = -8 \\ x - 3y = 4 \end{cases}$

5. $\begin{cases} 5x + 5y = 10 \\ -3x - 3y = -6 \end{cases}$

6. $\begin{cases} x - 3y = 4 \\ 2x + y = 15 \end{cases}$

In problems 7–12, use the substitution method to solve each system of linear equations.

7. $\begin{cases} x - y = 1 \\ 2x + y = 5 \end{cases}$

8. $\begin{cases} s - t = 1 \\ 2s + t = -4 \end{cases}$

9. $\begin{cases} 3u + 2v = 7 \\ -u + 4v = 3 \end{cases}$

10. $\begin{cases} 6x_1 = 15 + 9x_2 \\ x_1 = 7 - \frac{3}{2}x_2 \end{cases}$

11. $\begin{cases} 2x - y - z = 0 \\ 2x + 3y = 1 \\ 8x - 3z = 4 \end{cases}$

12. $\begin{cases} x_1 + x_2 - x_3 = 0 \\ x_1 - x_2 + x_3 = 2 \\ 2x_1 + x_2 - 4x_3 = -8 \end{cases}$

In problems 13–16, use the method of elimination (without matrices) to find the solution of each system of linear equations.

13. $\begin{cases} 6x - 9y = -3 \\ 2x + y = 3 \end{cases}$

14. $\begin{cases} -4x_1 + 5x_2 = 1 \\ x_1 - 2x_2 = -1 \end{cases}$

15. $\begin{cases} 2x - z = 12 \\ x + y = 7 \\ 5x + 4z = -9 \end{cases}$

16. $\begin{cases} x_1 + x_2 + x_3 = 6 \\ 2x_1 - x_2 + x_3 = 3 \\ 3x_1 + x_2 - x_3 = 2 \end{cases}$

In problems 17–24, solve the system of linear equations by the elimination method using matrices.

17. $\begin{cases} 5x + 2y = 3 \\ 2x - 3y = 5 \end{cases}$

18. $\begin{cases} 2x_1 + 3y_1 = 4 \\ 3x_1 + 5y_1 = 5 \end{cases}$

19. $\begin{cases} 5u - v = 19 \\ -2u + 3v = 8 \end{cases}$

20. $\begin{cases} s + 26 = -4t \\ 3s + 8 = 2t \end{cases}$

21. $\begin{cases} 2x + y - z = -3 \\ 3x - 2y + 2z = 13 \\ x + 3y + 4z = 0 \end{cases}$

22. $\begin{cases} 2x + y = 11 \\ x + 2y - 4z = 17 \\ 3x - y + 3z = 1 \end{cases}$

23. $\begin{cases} 2x_1 - x_2 + 2x_3 = -8 \\ 3x_1 - x_2 - 4x_3 = 3 \\ x_1 + 2x_2 - 3x_3 = 9 \end{cases}$

24. $\begin{cases} \frac{1}{2}x + \frac{1}{3}y - \frac{1}{4}z = 2 \\ \frac{1}{3}x + \frac{1}{4}y - \frac{1}{2}z = \frac{1}{6} \\ -\frac{3}{2}x + 3y + 2z = 23 \end{cases}$

In problems 25–36, perform the indicated matrix operations.

25. $A + 3B$ and $-2A + 4B$ if $A = \begin{bmatrix} -1 & 1 \\ 2 & 3 \end{bmatrix}$ and $B = \begin{bmatrix} -5 & 1 \\ 2 & 4 \end{bmatrix}$

26. $2A - B$ and $3A + 3B$ if $A = \begin{bmatrix} 2 & -1 & 3 \\ 1 & 2 & 4 \end{bmatrix}$ and $B = \begin{bmatrix} -3 & 2 & 1 \\ 5 & -2 & 4 \end{bmatrix}$

27. $4A + 2B$ and $B - 2A$ if $A = \begin{bmatrix} 2 & 4 \\ 5 & -2 \\ -4 & 3 \end{bmatrix}$ and $B = \begin{bmatrix} -1 & 0 \\ 3 & 4 \\ 0 & -5 \end{bmatrix}$

28. $\frac{1}{2}A - 2B + C$ if $A = \begin{bmatrix} 3 & 1 \\ 1 & 0 \\ 4 & -1 \\ 0 & 3 \end{bmatrix}$, $B = \begin{bmatrix} -2 & 2 \\ 1 & 4 \\ 3 & 1 \\ 1 & -2 \end{bmatrix}$, and $C = \begin{bmatrix} 3 & -1 \\ 5 & -1 \\ 0 & 2 \\ 8 & 0 \end{bmatrix}$

29. AB and BA if $A = \begin{bmatrix} 2 & 1 & -1 \\ 3 & 1 & 2 \end{bmatrix}$ and $B = \begin{bmatrix} -1 & 2 \\ 2 & 4 \\ 0 & 5 \end{bmatrix}$

30. $A^2, B^2,$ and A^2B^2 if $A = \begin{bmatrix} 1 & -1 \\ 2 & 3 \end{bmatrix}$ and $B = \begin{bmatrix} 1 & 3 \\ -2 & 1 \end{bmatrix}$

31. AB if $A = \begin{bmatrix} 1 & -1 & 2 \\ 3 & 0 & -2 \\ 1 & 4 & 0 \end{bmatrix}$ and $B = \begin{bmatrix} 1 & -2 \\ 2 & 3 \\ 4 & -1 \end{bmatrix}$

32. $(A - 2B)B$ if $A = \begin{bmatrix} -2 & 1 \\ 3 & 4 \end{bmatrix}$ and $B = \begin{bmatrix} 4 & 2 \\ 3 & -1 \end{bmatrix}$

33. $A^{-1}, AA^{-1},$ and $A^{-1}A$ if $A = \begin{bmatrix} 2 & -4 \\ 3 & 7 \end{bmatrix}$

34. $(AB)^{-1} - A^{-1}B^{-1}$ if $A = \begin{bmatrix} 1 & 2 \\ 3 & -1 \end{bmatrix}$ and $B = \begin{bmatrix} -2 & 4 \\ 1 & 3 \end{bmatrix}$

35. A^{-1} if $A = \begin{bmatrix} 2 & -1 & 0 \\ 1 & 0 & 1 \\ 1 & -2 & 0 \end{bmatrix}$

36. $A^{-1}, B^{-1},$ and $(AB)^{-1} - B^{-1}A^{-1}$ if $A = \begin{bmatrix} 1 & 3 \\ 4 & -1 \end{bmatrix}$ and $B = \begin{bmatrix} 2 & 3 \\ 4 & 1 \end{bmatrix}$

37. Find a value of x for which $AB = BA$ if $A = \begin{bmatrix} 1 & x \\ 0 & -1 \end{bmatrix}$ and $B = \begin{bmatrix} 2 & 3 \\ 0 & 1 \end{bmatrix}$

38. If $A = \begin{bmatrix} a & b \\ c & d \end{bmatrix}$, the *determinant* of A, in symbols, det A, is defined by det $A = \begin{vmatrix} a & b \\ c & d \end{vmatrix}$.

If A and B are 2 by 2 matrices, prove that $\det(AB) = (\det A)(\det B)$.

39. If A and B are square n by n nonsingular matrices, show that AB is nonsingular.

40. Let A be a 2 by 2 matrix and let $d = \det A$ (see problem 38). If A is nonsingular, prove that $d \neq 0$ and that $\det(A^{-1}) = d^{-1}$.

In problems 41–46, (a) write the system of linear equations in the form $AX = K$ for suitable matrices A, X, and K; and (b) solve each system with the aid of A^{-1}.

41. $\begin{cases} 3x + 2y = 11 \\ 4x - 3y = 9 \end{cases}$

42. $\begin{cases} 7x_1 + 10x_2 = -3 \\ 5x_1 + 2x_2 = 3 \end{cases}$

43. $\begin{cases} 3x + 2y + z = 7 \\ 2x + 3z = 10 \\ 5x - y = -8 \end{cases}$

44. $\begin{cases} 7x - 4z = k_1 \\ 5y + 3z = k_2 \\ -3x + 7y = k_3 \end{cases}$

45. $\begin{cases} 2x + z = a \\ x + 6y + 4z = b \\ -x - y = c \end{cases}$

46. $\begin{cases} 3x + 2y = 8 - 2z \\ x + 6z = 8 + 5y \\ 6x = 4 + 8z \end{cases}$

In problems 47 and 48, evaluate each determinant.

47. $\begin{vmatrix} 1/2 & 4/3 \\ -5/2 & 2/3 \end{vmatrix}$

Ⓒ **48.** $\begin{vmatrix} 5.007 & 13.142 \\ -3.733 & 2.501 \end{vmatrix}$

In problems 49 and 50, use the expansion formula to evaluate each determinant.

49. $\begin{vmatrix} 7 & 4 & 5 \\ 6 & -5 & 1 \\ 3 & 2 & 0 \end{vmatrix}$

Ⓒ **50.** $\begin{vmatrix} 4.21 & 3.72 & -2.02 \\ -1.59 & 7.07 & -8.83 \\ 2.22 & 3.14 & 2.03 \end{vmatrix}$

In problems 51–58, evaluate each determinant by any method you wish.

51. $\begin{vmatrix} 1 & -1 & 2 \\ 2 & -2 & 4 \\ \sqrt{2} & -\sqrt{2} & 2\sqrt{2} \end{vmatrix}$

52. $\begin{vmatrix} 1 & -1 & 2 \\ 2 & -2 & 1 \\ \sqrt{2} & -\sqrt{2} & 2\sqrt{2} \end{vmatrix}$

53. $\begin{vmatrix} x & 0 & 0 \\ 2 & x & 1 \\ 0 & 3 & 2 \end{vmatrix}$

54. $\begin{vmatrix} 25 & -6 & 0 \\ 0 & 3 & -4 \\ 35 & 9 & 2 \end{vmatrix}$

55. $\begin{vmatrix} 4 & \sqrt{2} & \pi \\ 0 & 5 & e \\ 0 & 0 & -1 \end{vmatrix}$

56. $\begin{vmatrix} 1 & 1 & 1 \\ x & a & b \\ x^2 & a^2 & b^2 \end{vmatrix}$

57. $\begin{vmatrix} 2 & 1 & -3 \\ 4 & 3 & -2 \\ 3 & -1 & 4 \end{vmatrix}$

58. $\begin{vmatrix} 1 & 3 & 4 \\ 2 & 2 & 4 \\ 3 & 1 & 4 \end{vmatrix}$

In problems 59–64, assume that

$$\begin{vmatrix} a & b & c \\ u & v & w \\ x & y & z \end{vmatrix} = 2.$$

Find the value of each determinant.

59. $\begin{vmatrix} c & a & b \\ w & u & v \\ z & x & y \end{vmatrix}$

60. $\begin{vmatrix} -a & -b & -c \\ -u & -v & -w \\ -x & -y & -z \end{vmatrix}$

61. $\begin{vmatrix} 2a & 2b & 2c \\ 2u & 2v & 2w \\ 2x & 2y & 2z \end{vmatrix}$

62. $\begin{vmatrix} a & b+a & 2c \\ u & v+u & 2w \\ x & y+x & 2z \end{vmatrix}$

63. $\begin{vmatrix} a+u & b+v & c+w \\ u+x & v+y & w+z \\ -x & -y & -z \end{vmatrix}$

64. $\begin{vmatrix} a & u & x \\ b & v & y \\ c & w & z \end{vmatrix}$

In problems 65–70, use Cramer's rule to solve each system.

65. $\begin{cases} 3x - 4y = -2 \\ 4x + y = 7 \end{cases}$

66. $\begin{cases} 3x_1 + 10x_2 = 3 \\ 6x_1 - 5x_2 = 16 \end{cases}$

67. $\begin{cases} x + y + 2z = 10 \\ 5x + 3y - z = 1 \\ 3x - y - 3z = -3 \end{cases}$

68. $\begin{cases} 3r + 6s - 5t = -1 \\ r - 2s + 4t = 4 \\ 5r + 6s - 7t = -5 \end{cases}$

69. $\begin{cases} x_1 + x_2 = 0 \\ 2x_1 - 3x_2 + x_3 = 7 \\ 3x_2 - 7x_3 = -17 \end{cases}$

70. $\begin{cases} s + t - u = 0 \\ 2s + 2u = 1 \\ 3t + u = 3 \end{cases}$

71. If $P = (a, b)$ and $Q = (c, d)$ are points in the xy plane, and $P \neq (0,0)$, show that the perpendicular distance from Q to the line through $O = (0,0)$ and P is the absolute value of

$$\begin{vmatrix} a & b \\ c & d \end{vmatrix}(a^2 + b^2)^{-1/2}.$$

72. For the points P and Q in problem 71, show that the absolute value of $\begin{vmatrix} a & b \\ c & d \end{vmatrix}$ is the area of the parallelogram that has $\overline{OP}$ and $\overline{OQ}$ as two adjacent sides.

In problems 73–78, decompose each fraction into partial fractions.

73. $\dfrac{2x}{(x-1)(x+1)}$

74. $\dfrac{4x+7}{x^2+5x+4}$

75. $\dfrac{4x^2+13x-9}{x(x+3)(x-1)}$

76. $\dfrac{-x^2+13x-26}{(x+1)^2(x-4)}$

77. $\dfrac{x^2+4}{x(x^2+1)}$

78. $\dfrac{6x^3+5x^2+21x+12}{x(x+1)(x^2+4)}$

79. If the supermarket price of a certain cut of beef is p dollars per pound, then q million pounds will be sold according to the demand equation $p + q = 4$. When the supermarket price is p dollars per pound, a large packing company will supply q million pounds of this meat according to the supply equation $p - 4q + 2 = 0$. Find the point of market equilibrium.

80. Two cars start from points 400 kilometers apart and travel toward each other at constant speeds. If the first car travels 20 kilometers per hour faster than the second, and if they both meet after 4 hours, find the speeds of both vehicles.

Ⓒ **81.** A person invested a total of $40,000, part of it in a conservative investment at 8.5% simple annual interest, and the rest in a riskier investment at 11.2% simple annual interest. At the end of the year, both investments paid off at the stipulated rates for a total interest of $4129. How much was invested at each rate of interest?

82. An industrial chemist, in preparing a batch of fertilizer, mixes 10 tons of nitrogen compounds with 12 tons of phosphates. The total cost of the ingredients is $5400. A second batch of fertilizer, prepared for special soil conditions, is mixed from 15 tons of nitrogen compounds and 8 tons of phosphates, for a total cost of $6100. Find the cost per ton of the nitrogen compounds and the cost per ton of the phosphates.

83. Byron, Jason, and Adrian receive a total weekly allowance of $12, which is split three ways. If Jason's allowance plus twice the sum of Byron's and Adrian's is $20, and if Adrian's allowance plus twice the sum of Byron's and Jason's is $22, find each boy's weekly allowance.

84. Joe is 4 years younger than twice Gus's age, and Jamal is 3 years older than Gus. Six years from now, Joe will be $\frac{4}{3}$ times as old as Gus will be. Find their present ages.

In problems 85–91, (a) sketch the graph of each system of linear inequalities, (b) determine whether the graph is bounded or unbounded, and (c) find all of its vertices.

85. $\begin{cases} 2y - x > 10 \\ y + x \geq 5 \end{cases}$

86. $\begin{cases} 4y > x - 16 \\ y + x < 0 \end{cases}$

87. $\begin{cases} 3y + x \leq 2 \\ y > x + 1 \end{cases}$

88. $\begin{cases} y \leq x + 1 \\ x + y \leq 4 \\ x \geq 0 \end{cases}$

89. $\begin{cases} y - 3x \leq 2 \\ y + 2x \leq 4 \\ y \geq 0 \end{cases}$

90. $\begin{cases} x + y \leq 500 \\ 3y \leq x \\ x \leq 400 \\ y \geq 60 \end{cases}$

91. $\begin{cases} x + y \geq 5 \\ 2y \geq 8 - x \\ y \leq 5 \\ x \leq 10 \end{cases}$

92. A psychology professor is planning a 50-minute class. The professor will devote x minutes to a review of old material, and y minutes to a lecture on new material. The remaining time will be used for classroom discussion of the new material. At least 20 minutes must be reserved for the lecture on new material; at least 10 minutes will be needed for the classroom discussion; and no more than 15 minutes will be required for the review of old material. Sketch a graph in the xy plane illustrating the professor's options, and find all vertices of the graph.

In problems 93–96, find the maximum and minimum values of the objective function F subject to the indicated constraints.

93. $F = 4x + 7y + 1$
$\begin{cases} y - x + 1 \geq 0 \\ 2y + x - 10 \geq 0 \\ x \geq 0 \\ y \leq 5 \end{cases}$

94. $F = 3x + 5y$
$\begin{cases} y \leq 4 - x \\ y \geq \frac{1}{2}x - 2 \\ x \geq 0 \\ y \geq 0 \end{cases}$

95. $F = x + 5y$
$\begin{cases} x + y \leq 4 \\ 4x + y \leq 7 \\ x \geq 0 \\ y \geq 0 \end{cases}$

96. $F = x$
$\begin{cases} 2y + x \leq 16 \\ x - y \leq 10 \\ x \geq 0 \end{cases}$

97. A firm manufactures two products, A and B. For each product, two different machines, M_1 and M_2, are used. Each machine can operate up to 20 hours per day. Product A requires 1 hour of time on machine M_1 and 3 hours of time on machine M_2 per 100 units. Product B requires 2 hours of time on machine M_1 and 1 hour of time on machine M_2 per 100 units. The firm makes a profit of $20 per unit on product A and $10 per unit on product B. Determine how many units of each product should be manufactured per day in order to maximize the profit.

c **98.** A woman has $20,000 to invest and two investment opportunities, one conservative and one somewhat riskier. The conservative investment pays 8.5% simple annual interest, and the risky investment pays 9.5% simple annual interest. She wants to invest at most three times as much in the risky investment as in the conservative one. Find her maximum possible dividend after 1 year if she decides to invest no more than $16,000 in the conservative investment and at least $2400 in the risky investment.

In problems 99–108, solve each system of equations.

99. $\begin{cases} y^2 - 12x = 0 \\ y^2 + 9x^2 = 9 \end{cases}$

100. $\begin{cases} x^2 - y = 0 \\ x^2 + 4y^2 = 4 \end{cases}$

101. $\begin{cases} 2x + y = 1 \\ 4x^2 + y^2 = 13 \end{cases}$

102. $\begin{cases} 4x^2 - y^2 = 7 \\ 2x^2 + 5y^2 = 8 \end{cases}$

103. $\begin{cases} y^2 - x^2 = 3 \\ xy = 2 \end{cases}$

104. $\begin{cases} 9x^2 - 4y^2 = 36 \\ x^2 + y^2 = 43 \end{cases}$

105. $\begin{cases} \log_4(x^2 - 4y^2) = 1 \\ x - y = 1 \end{cases}$

106. $\begin{cases} x - y = 2 \\ 2^{x^2 - y^2} = 256 \end{cases}$

107. $\begin{cases} \log_2(2x^2 + y) = 2 \\ \log_2(2x - y) = 0 \end{cases}$

108. $\begin{cases} \dfrac{3}{x^2} - \dfrac{2}{y^2} = \dfrac{1}{2} \\ \dfrac{6}{x^2} + \dfrac{1}{y} = \dfrac{5}{2} \end{cases}$

109. The length of the hypotenuse of a right triangle is 25 centimeters, and the perimeter is 56 centimeters. Find the lengths of the legs.

110. A manufacturer of drafting supplies makes right triangles from plastic sheets. The perimeters of the triangles are 60 centimeters, and each triangle has a hypotenuse of 25 centimeters. Find the lengths of the sides of the triangles.

111. A town parking lot has an area of 7500 square meters. The town engineer suggests enlarging the lot by increasing both its length and its width by 10 meters. She points out that this will increase the area of the lot by 1850 square meters. Find the present dimensions of the lot.

112. In three years, Joshua will be twice as old as Miriam. At present, twice his age is the same as the product of Rebecca's and Miriam's ages. If Rebecca is now twice as old as Miriam, find the ages of all three children.

10 Additional Topics in Algebra and Analytic Geometry

In this chapter, we give a brief introduction to several topics in algebra and analytic geometry that extend, supplement, and round out the ideas presented in earlier chapters. These topics include the complex number system, mathematical induction, the binomial theorem, sequences, series, permutations and combinations, probability, and conic sections. Here we can only give you the "flavor" of these topics—we urge you to consult more advanced textbooks for a more detailed and systematic presentation.

1 Complex Numbers

In Section 3.4 of Chapter 2, we introduced the so-called "imaginary number" $i = \sqrt{-1}$ and used it to form "complex numbers" $a + bi$, where a and b are real numbers. Using such complex numbers, we were able to obtain roots of quadratic equations with negative discriminants.

Leaving aside for now the question of just what i is, let's work with complex numbers a bit and see what happens. If we suppose that the basic algebraic properties of the real numbers continue to operate for complex numbers, then, for any real numbers a, b, c, and d, we have:

1. $(a + bi) + (c + di) = a + c + bi + di = (a + c) + (b + d)i$

2. $(a + bi) - (c + di) = a - c + bi - di = (a - c) + (b - d)i$

3. $(a + bi)(c + di) = ac + adi + bic + bidi = ac + bdi^2 + (ad + bc)i$
$$= ac + bd(-1) + (ad + bc)i = (ac - bd) + (ad + bc)i.$$

We take 1, 2, and 3 above as *definitions* of addition, subtraction, and multiplication of complex numbers. Notice that the sum, difference, and product of complex numbers is again a complex number. It is understood that $i^2 = -1$ and that the real number 0 has its usual additive and multiplicative properties, so that

$$0 + bi = bi$$

and

$$0i = 0.$$

Finally, let's agree that

$$a + bi = c + di \quad \text{means that} \quad a = c \quad \text{and} \quad b = d.$$

The real numbers a and b are called the **real part** and the **imaginary part,** respectively, of the complex number $a + bi$.

The set of all complex numbers, equipped with the algebraic operations of addition, subtraction, and multiplication, is called the **complex number system** and is denoted by the symbol $\mathbb{C}$. Note that a real number a can be regarded as a complex number whose imaginary part is zero: $a = a + 0i$; therefore, the real number system $\mathbb{R}$ forms part of the complex number system $\mathbb{C}$. It isn't difficult to show that the complex numbers have the same basic algebraic properties—commutative, associative, distributive, identity, and inverse—as the real numbers (Chapter 1, page 3). Of these, the most intriguing is certainly the multiplicative inverse property—the fact that every nonzero complex number has a reciprocal or multiplicative inverse.

If $a + bi \neq 0$, you can obtain the reciprocal $\dfrac{1}{a + bi}$ by a trick similar to that for rationalizing the denominator of a fraction: Just *multiply numerator and denominator by the* **complex conjugate** $a - bi$ *of* $a + bi$. Thus,

$$\frac{1}{a + bi} = \frac{a - bi}{(a + bi)(a - bi)} = \frac{a - bi}{a^2 - (bi)^2} = \frac{a - bi}{a^2 - b^2 i^2} = \frac{a - bi}{a^2 - b^2(-1)}$$

$$= \frac{a - bi}{a^2 + b^2} = \left(\frac{a}{a^2 + b^2}\right) + \left(\frac{-b}{a^2 + b^2}\right)i.$$

In other words, $\dfrac{a}{a^2 + b^2}$ and $\dfrac{-b}{a^2 + b^2}$ are the real and imaginary parts of $\dfrac{1}{a + bi}$. Notice that the denominator $a^2 + b^2$ is a positive real number because of the fact that $a + bi \neq 0$, so a and b cannot both be zero. You can verify by direct calculation that the complex number $\left(\dfrac{a}{a^2 + b^2}\right) + \left(\dfrac{-b}{a^2 + b^2}\right)i$ is indeed the multiplicative inverse $(a + bi)^{-1}$ of $a + bi$ (Problem 66).

If $a + bi \neq 0$, we define the **quotient** $\dfrac{c + di}{a + bi}$ to be the product of $c + di$ and $\dfrac{1}{a + bi}$. If you multiply the numerator and denominator of a quotient by the complex conjugate of the denominator, you will obtain a fraction with a positive real number in the denominator. This allows you to find the real and imaginary parts of the original quotient.

Example   Express each complex number in the form $a + bi$, where a and b are real numbers:

(a) $(3 + 5i) + (6 + 5i)$

(b) $(2 - 3i) - (6 + 4i)$

(c) $(4 + 3i)(2 + 4i)$

(d) $\dfrac{1}{4 + 3i}$

(e) $\dfrac{2 - 3i}{1 - 4i}$

Solution

(a) $(3 + 5i) + (6 + 5i) = 3 + 6 + 5i + 5i = 9 + 10i$

(b) $(2 - 3i) - (6 + 4i) = 2 - 6 - 3i - 4i = -4 - 7i$

(c) $(4 + 3i)(2 + 4i) = 8 + 16i + 6i + 12i^2 = 8 + 22i + 12(-1) = -4 + 22i$

(d) Multiplying numerator and denominator by $4 - 3i$, the complex conjugate of $4 + 3i$, we obtain

$$\frac{1}{4 + 3i} = \frac{4 - 3i}{(4 + 3i)(4 - 3i)} = \frac{4 - 3i}{16 - 9i^2} = \frac{4 - 3i}{25} = \frac{4}{25} - \frac{3}{25}i.$$

(e) Multiplying numerator and denominator by $1 + 4i$, the complex conjugate of $1 - 4i$, we obtain

$$\frac{2 - 3i}{1 - 4i} = \frac{(2 - 3i)(1 + 4i)}{(1 - 4i)(1 + 4i)} = \frac{2 + 8i - 3i - 12i^2}{1 - 16i^2} = \frac{2 + 5i - 12(-1)}{1 - 16(-1)}$$

$$= \frac{14 + 5i}{17} = \frac{14}{17} + \frac{5}{17}i.$$

Complex numbers, like real numbers, can be denoted by letters of the alphabet and treated as variables or unknowns. The letters z and w are special favorites for this purpose. If $z = a + bi$, where a and b are real numbers, the complex conjugate of z is often written as $\bar{z} = a - bi$. Notice that

$$z\bar{z} = (a + bi)(a - bi) = a^2 - (bi)^2 = a^2 - b^2i^2 = a^2 - b^2(-1) = a^2 + b^2.$$

Thus, $z\bar{z}$ *is always a nonnegative real number.* Its principal square root is called the **absolute value** or **modulus** of z and is written $|z|$. Thus, by definition,

$$|z| = \sqrt{z\bar{z}} = \sqrt{a^2 + b^2}.$$

Therefore,

$$z\bar{z} = |z|^2.$$

Example   If $z = 3 + 4i$, find (a) $\bar{z}$, (b) $z + \bar{z}$, (c) $z - \bar{z}$, (d) $z\bar{z}$, and (e) $|z|$.

Solution

(a) $\bar{z} = 3 - 4i$

(b) $z + \bar{z} = (3 + 4i) + (3 - 4i) = 6$

(c) $z - \bar{z} = (3 + 4i) - (3 - 4i) = 8i$

(d) $z\bar{z} = (3 + 4i)(3 - 4i) = 9 + 16 = 25$

(e) $|z| = \sqrt{z\bar{z}} = \sqrt{25} = 5$

As we have mentioned, a real number a can be regarded as a complex number $a + 0i$ with imaginary part zero. Thus, $\bar{a} = a - 0i = a$, so *each real number is its own complex conjugate.* In particular,

$$|a| = \sqrt{a\bar{a}} = \sqrt{a^2}.$$

Complex conjugation "preserves" all of the algebraic operations; that is, if z and w are complex numbers, then:

(i) $\overline{z + w} = \bar{z} + \bar{w}$ (ii) $\overline{z - w} = \bar{z} - \bar{w}$

(iii) $\overline{zw} = \bar{z}\bar{w}$ (iv) $\overline{\left(\dfrac{z}{w}\right)} = \dfrac{\bar{z}}{\bar{w}}$, $w \neq 0.$

We leave the verification of (i)–(iv) as an exercise (Problem 65). Properties (i) and (iii) can be extended to more than two complex numbers. For instance, if u, v, and w are three complex numbers, we can apply (i) twice to obtain

$$\overline{u + v + w} = \overline{(u + v) + w} = \overline{(u + v)} + \bar{w} = (\bar{u} + \bar{v}) + \bar{w} = \bar{u} + \bar{v} + \bar{w}.$$

Thus, for complex numbers, *the conjugate of a sum is the sum of the conjugates*, and a similar result holds for products. Using property (iii), we have

$$|zw| = \sqrt{(zw)(\overline{zw})} = \sqrt{zw\,\bar{z}\,\bar{w}} = \sqrt{z\,\bar{z}\,w\,\bar{w}} = \sqrt{|z|^2|w|^2} = |z|\,|w|;$$

that is, *the absolute value of a product of complex numbers is the product of their absolute values*. A similar calculation (Problem 69) shows that for $w \neq 0$,

$$\left|\frac{z}{w}\right| = \frac{|z|}{|w|}.$$

Of course, integer powers of complex numbers are defined just as they are for real numbers: $z^2 = zz$, $z^3 = zzz$, and so on; and, if $z \neq 0$, $z^0 = 1$, $z^{-1} = \dfrac{1}{z}$, $z^{-2} = \dfrac{1}{z^2}$, and so on. Notice that

$$i^1 = i \qquad i^5 = i \qquad i^9 = i$$
$$i^2 = -1 \qquad i^6 = -1 \qquad i^{10} = -1$$
$$i^3 = -i \qquad i^7 = -i \qquad i^{11} = -i$$
$$i^4 = 1 \qquad i^8 = 1 \qquad i^{12} = 1$$

and so on. Thus, the positive integer powers of i endlessly repeat the pattern of the first four.

Example   Find i^{59}.

Solution $i^{59} = i^{56+3} = i^{56}i^3 = (i^4)^{14}i^3 = 1^{14}(-i) = -i.$

Although complex numbers were originally introduced to provide roots for certain polynomial equations (such as quadratic equations with negative discriminants), they now have a wide variety of important applications in physics and engineering. For example, in 1893, Charles P. Steinmetz (1865–1923), an American electrical engineer born in Germany, developed a theory of alternating currents based on the complex numbers.

In direct current theory, **Ohm's law**

$$E = IR$$

relates the electromotive force (voltage) E, the current I, and the resistance R. Because of inductive and capacitative effects, voltage and current may be out of phase in alternating current circuits, and the equation $E = IR$ may no longer hold. Steinmetz saw that, by representing voltage and current with *complex numbers E and I*, he could deal algebraically with phase differences. Furthermore, he combined the resistance R, the inductive effect X_L (called **inductive reactance**), and the capacitative effect X_C (called **capacitative reactance**) in a single complex number

$$Z = R + (X_L - X_C)i,$$

called **complex impedance.** Using the complex numbers E, I, and Z, Steinmetz showed that Ohm's law for alternating currents takes the form

$$E = IZ.$$

Today, these ideas of Steinmetz are used routinely by electrical engineers all over the world. It has been said that Steinmetz "generated electricity with the square root of minus one."

PROBLEM SET 1

In problems 1–52, express each complex number in the form $a + bi$, where a and b are real numbers.

1. $(2 + 3i) + (7 - 2i)$
2. $(-1 + 2i) + (3 + 4i)$
3. $(4 + i) + 2(3 - i)$
4. $(4 + 2i) + 3(2 - 5i)$
5. $(5 - 4i) + 14$
6. $(\frac{1}{2} - \frac{2}{3}i) + (\frac{3}{4} + \frac{1}{6}i)$
7. $(3 + 2i) - (5 + 4i)$
8. $(2 - 3i) - i$
9. $3(1 + 2i) - 4(2 + i)$
10. $(\frac{1}{2} - \frac{4}{3}i) - \frac{1}{6}(5 + 7i)$
11. $2(5 + 4i) - 3(7 + 4i)$
12. $i - \frac{1}{2}(1 + 5i)$

13. $-(-3 + 5i) - (4 + 9i)$
14. $(\sqrt{2} + \sqrt{3}i) - \left(\dfrac{\sqrt{2}}{2} - \dfrac{\sqrt{3}}{3}i\right)$
15. $(2 + i)(1 + 5i)$

16. $(4 + 3i)(-1 + 2i)$
17. $(7 + 4i)(3 + 6i)$
18. $(7 - 6i)(-5 - i)$
19. $(3 - 2i)(-3 + i)$
20. $i(3 + 7i)$
21. $(-7 + 3i)(-3 + 2i)$
22. $(\frac{2}{3} + \frac{3}{5}i)(\frac{1}{2} - \frac{1}{3}i)$
23. $-8i(5 + 8i)$
24. $(\sqrt{2} + \sqrt{3}i)(\sqrt{2} - \sqrt{3}i)$
25. $(-4i)(-5i)$
26. $(-2i)(3i)(-4i)$
27. $(\frac{1}{2} + \frac{1}{3}i)(\frac{1}{2} - \frac{1}{3}i)$
28. $(\sqrt{2} + \sqrt{3}i)^2$
29. i^{21}
30. i^{41}
31. i^{201}
32. $(1 + i)^3$
33. $(4 + 2i)^2$

34. $(1 + i)^4$
35. $\left(\dfrac{1}{2} + \dfrac{\sqrt{3}}{2}i\right)^2$
36. $(1 - i)^4$

37. $\dfrac{1}{2 + 3i}$
38. $\dfrac{1}{3 - 4i}$
39. $\dfrac{3}{7 + 2i}$

40. $\dfrac{-4i}{6 - i}$
41. $\dfrac{3 - 4i}{4 + 2i}$
42. $\dfrac{\pi + 4i}{2 - i}$

43. $\dfrac{7 + 2i}{3 - 5i}$
44. $\dfrac{1}{i}$
45. $\dfrac{4 + i}{(3 - 2i) + (4 - 3i)}$

46. $\dfrac{2 + 3i}{4 - 3i} + \dfrac{3 + 5i}{1 - 2i}$

47. $\dfrac{2 - 6i}{3 + i} - \dfrac{4 + i}{3 + i}$

48. $\dfrac{(1 - i)(2 + i)}{(2 - 3i)(3 - 4i)}$

49. $\dfrac{3 - 2i}{3 + i} + \dfrac{4i}{3 - 7i}$

50. $\dfrac{3i^3 - 5}{1 + i^5}$

51. $\dfrac{3 - 2i}{(2 + i)(5 + 2i)}$

52. $\left(\dfrac{3 - i^7}{i^9 - 3}\right)^2$

In problems 53–64, calculate (a) $\bar{z}$, (b) $z + \bar{z}$, (c) $z - \bar{z}$, (d) $z\bar{z}$, and (e) $|z|$.

53. $z = 2 + i$

54. $z = i$

55. $z = -i$

56. $z = (1 + i)^2$

57. $z = \dfrac{1 + i}{1 - i}$

58. $z = -3i^5$

59. $z = -12 + 5i$

60. $z = 7i^{101}$

61. $z = \dfrac{4 - 3i}{2 + 4i}$

62. $z = (2 + i)^{-1}$

63. $z = 5$

64. $z = 1 + i + i^2 + i^3$

65. Show that (a) $\overline{z + w} = \bar{z} + \bar{w}$, (b) $\overline{z - w} = \bar{z} - \bar{w}$, (c) $\overline{zw} = \bar{z}\bar{w}$, and (d) if $w \neq 0$, then $\overline{z/w} = \bar{z}/\bar{w}$.

66. Assume that $a + bi \neq 0$. By multiplying, show that

$$(a + bi)\left[\left(\dfrac{a}{a^2 + b^2}\right) + \left(\dfrac{-b}{a^2 + b^2}\right)i\right] = 1.$$

67. Assume that z is a nonzero complex number.
 (a) Show that $z^{-1} = \bar{z}/|z|^2$.   (b) Show that $\overline{z^{-1}} = (\bar{z})^{-1}$.

68. If z is a complex number, show that (a) $\frac{1}{2}(z + \bar{z})$ is the real part of z, and that (b) $\dfrac{1}{2i}(z - \bar{z})$

 is the imaginary part of z.

69. If $w \neq 0$, show that $\left|\dfrac{z}{w}\right| = \dfrac{|z|}{|w|}$.

70. Prove the **triangle inequality** for complex numbers:

$$|z + w| \leq |z| + |w|.$$

 [Hint: Prove that $|z + w|^2 \leq (|z| + |w|)^2$. Use the fact that $|z + w|^2 = (z + w)(\overline{z + w})$.]

2 Polar Form and nth Roots of Complex Numbers

Figure 1

We begin this section by giving an answer to the question (temporarily put aside in Section 1) of just what complex numbers are. Although there are several possible representations of the complex numbers, their interpretation as geometric points in the plane has the most intuitive appeal. According to this scheme, if a and b are real numbers, then the complex number $a + bi$ is represented by the point (a, b) in the xy plane (Figure 1). When each point in a coordinate plane is made to correspond to a complex number in this way, we refer to the plane as the **complex plane**. This correspondence is so compelling that it is natural to *identify* a complex number with its corresponding geometric point, and we shall do so in what follows. Under this identification, there's nothing at all imaginary about i or any other complex number; nevertheless, the word "imaginary" continues to be used for historical reasons.

Figure 2

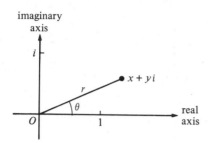

In the complex plane, the x axis is called the **real axis** because its points correspond to real numbers, and the y axis is called the **imaginary axis** because its points correspond to multiples of i by real numbers (Figure 2). Notice that the absolute value

$$|x + yi| = \sqrt{x^2 + y^2}$$

is just the distance $r = \sqrt{x^2 + y^2}$ between the point $x + yi$ and the origin (Figure 2). An angle θ in standard position that contains the point $x + yi$ on its terminal side is called an **argument** of $x + yi$. Thus, the absolute value $r = |x + yi|$ and an argument θ of $x + yi$ form polar coordinates (r, θ) of the point $x + yi$. It follows that

$$x = r\cos\theta \quad \text{and} \quad y = r\sin\theta$$

(Chapter 8, Section 5, page 411).

If z is a complex number with real part x, imaginary part y, absolute value $r = |z|$, and argument θ, then you can write z either in **cartesian form** (also called *rectangular form*)

$$z = x + yi$$

or in **polar form**

$$z = r\cos\theta + (r\sin\theta)i = r(\cos\theta + i\sin\theta).$$

Example 1   Plot the point $z = 4(\cos 60° + i\sin 60°)$ in the complex plane and rewrite z in cartesian form.

Solution   Here $r = 4$ and $\theta = 60°$. The point z is plotted in Figure 3. In cartesian form,

$$z = 4\cos 60° + (4\sin 60°)i$$

$$= 4\left(\frac{1}{2}\right) + \left[4\left(\frac{\sqrt{3}}{2}\right)\right]i$$

$$= 2 + 2\sqrt{3}i.$$

Figure 3

Example 2   Plot the point $z = \sqrt{2} - \sqrt{2}i$ in the complex plane and rewrite z in polar form.

Solution   The point z is plotted in Figure 4. Because the real and imaginary parts of z are $x = \sqrt{2}$ and $y = -\sqrt{2}$, it follows that

$$r = \sqrt{x^2 + y^2}$$
$$= \sqrt{(\sqrt{2})^2 + (-\sqrt{2})^2}$$
$$= \sqrt{2 + 2} = 2.$$

Figure 4

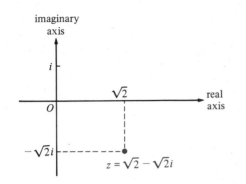

imaginary axis

Also,

$$\cos\theta = \frac{x}{r} = \frac{\sqrt{2}}{2} \quad \text{and} \quad \sin\theta = \frac{y}{r} = \frac{-\sqrt{2}}{2},$$

so $\theta = \dfrac{7\pi}{4}$. Thus,

$$z = 2\left(\cos\frac{7\pi}{4} + i\sin\frac{7\pi}{4}\right).$$

Although complex numbers are most easily added or subtracted when they are written in cartesian form, multiplication and division are more easily carried out when the numbers are expressed in polar form. The following theorem shows why.

Theorem 1   **Multiplication and Division in Polar Form**

Let $z_1 = r_1(\cos\theta_1 + i\sin\theta_1)$ and $z_2 = r_2(\cos\theta_2 + i\sin\theta_2)$. Then

(i) $z_1 z_2 = r_1 r_2[\cos(\theta_1 + \theta_2) + i\sin(\theta_1 + \theta_2)]$

and, provided that $z_2 \neq 0$,

(ii) $\dfrac{z_1}{z_2} = \dfrac{r_1}{r_2}[\cos(\theta_1 - \theta_2) + i\sin(\theta_1 - \theta_2)].$

Proof   We prove (i) and leave the proof of (ii) as an exercise (Problem 58).

$$z_1 z_2 = r_1(\cos\theta_1 + i\sin\theta_1)r_2(\cos\theta_2 + i\sin\theta_2)$$
$$= r_1 r_2(\cos\theta_1 + i\sin\theta_1)(\cos\theta_2 + i\sin\theta_2)$$
$$= r_1 r_2[(\cos\theta_1\cos\theta_2 - \sin\theta_1\sin\theta_2) + i(\sin\theta_1\cos\theta_2 + \sin\theta_2\cos\theta_1)]$$
$$= r_1 r_2[\cos(\theta_1 + \theta_2) + i\sin(\theta_1 + \theta_2)].$$

In words, part (i) of Theorem 1 says that the absolute value of the product of two complex numbers is the product of their absolute values; whereas, an argument of the product is given by the sum of their arguments. Part (ii) can be expressed in a similar way (Problem 55).

Example   If $z = 6\left(\cos\dfrac{\pi}{2} + i\sin\dfrac{\pi}{2}\right)$ and $w = 2\left(\cos\dfrac{\pi}{6} + i\sin\dfrac{\pi}{6}\right)$, find (a) zw and (b) $\dfrac{z}{w}$.

Solution   Using Theorem 1, we have

(a) $zw = (6)(2)\left[\cos\left(\dfrac{\pi}{2} + \dfrac{\pi}{6}\right) + i\sin\left(\dfrac{\pi}{2} + \dfrac{\pi}{6}\right)\right] = 12\left(\cos\dfrac{2\pi}{3} + i\sin\dfrac{2\pi}{3}\right)$

(b) $\dfrac{z}{w} = \dfrac{6}{2}\left[\cos\left(\dfrac{\pi}{2} - \dfrac{\pi}{6}\right) + i\sin\left(\dfrac{\pi}{2} - \dfrac{\pi}{6}\right)\right] = 3\left(\cos\dfrac{\pi}{3} + i\sin\dfrac{\pi}{3}\right)$

2.1 Powers and Roots of Complex Numbers

If $z = r(\cos\theta + i\sin\theta)$, then, by part (i) of Theorem 1, $z^2 = r^2(\cos 2\theta + i\sin 2\theta)$. Therefore,

$$z^3 = z^2 z$$
$$= [r^2(\cos 2\theta + i\sin 2\theta)][r(\cos\theta + i\sin\theta)]$$
$$= r^3(\cos 3\theta + i\sin 3\theta).$$

A similar computation (Problem 57) shows that

$$z^4 = r^4(\cos 4\theta + i\sin 4\theta).$$

The pattern emerging here is quite clear. It is expressed formally in the following theorem, which is attributed to Abraham De Moivre (1667–1754), a Frenchman turned Englishman.

Theorem 2   **De Moivre's Theorem**

> If $z = r(\cos\theta + i\sin\theta)$ and n is any positive integer, then
> $$z^n = r^n(\cos n\theta + i\sin n\theta).$$

In words, De Moivre's theorem can be stated as follows: To raise a complex number to a positive integer power n, you raise its absolute value to the power n and multiply its argument by n.

Examples   Use De Moivre's theorem to evaluate each expression.

1 $[2(\cos 10° + i\sin 10°)]^3$

Solution $[2(\cos 10° + i\sin 10°)]^3 = 2^3[\cos 3(10°) + i\sin 3(10°)]$
$$= 8(\cos 30° + i\sin 30°)$$
$$= 8\left[\dfrac{\sqrt{3}}{2} + i\left(\dfrac{1}{2}\right)\right] = 4\sqrt{3} + 4i.$$

2 $(-1 + \sqrt{3}i)^5$

Solution   We begin by rewriting $-1 + \sqrt{3}i$ in polar form:

$$-1 + \sqrt{3}i = 2\left(\cos\frac{2\pi}{3} + i\sin\frac{2\pi}{3}\right).$$

Therefore,

$$(-1 + \sqrt{3}i)^5 = \left[2\left(\cos\frac{2\pi}{3} + i\sin\frac{2\pi}{3}\right)\right]^5 = 2^5\left(\cos\frac{10\pi}{3} + i\sin\frac{10\pi}{3}\right)$$

$$= 32\left[\cos\left(\frac{4\pi}{3} + 2\pi\right) + i\sin\left(\frac{4\pi}{3} + 2\pi\right)\right] = 32\left(\cos\frac{4\pi}{3} + i\sin\frac{4\pi}{3}\right)$$

$$= 32\left[-\frac{1}{2} + i\left(-\frac{\sqrt{3}}{2}\right)\right] = -16 - 16\sqrt{3}i.$$

If w is a complex number and n is a positive integer, then any complex number z such that

$$z^n = w$$

is called a **complex nth root** of w. For $n = 2$ and for $n = 3$, the complex nth roots of w are called the **complex square roots** and the **complex cube roots** of w. By using the following theorem, you can find all complex nth roots of a nonzero complex number w.

Theorem 3   **Complex nth Roots**

Let $w = R(\cos\phi + i\sin\phi)$ be a nonzero complex number in polar form, and let n be a positive integer. Then there are exactly n different complex roots of w:

$$z_0, z_1, z_2, \ldots, z_{n-1},$$

where

$$z_k = \sqrt[n]{R}\left[\cos\left(\frac{\phi}{n} + \frac{2\pi k}{n}\right) + i\sin\left(\frac{\phi}{n} + \frac{2\pi k}{n}\right)\right].$$

for $k = 0, 1, 2, \ldots, n - 1$.

The fact that each of the complex numbers $z_0, z_1, \ldots, z_{n-1}$ in Theorem 3 satisfies the equation $z^n = w$ is a consequence of De Moivre's theorem (Problem 56). We omit the proof that these complex numbers are different from each other and are *all* of the nth roots of w.

Example   Find the complex fourth roots of $w = -\frac{9}{2} - \frac{9\sqrt{3}}{2}i$.

Solution   We begin by rewriting w in polar form:

$$w = 9\left(\cos\frac{4\pi}{3} + i\sin\frac{4\pi}{3}\right).$$

Then we use Theorem 3 with $R = 9$, $\phi = 4\pi/3$, and $n = 4$ to obtain the four complex fourth roots:

$$z_0 = \sqrt[4]{9}\left[\cos\left(\frac{4\pi/3}{4} + 0\right) + i\sin\left(\frac{4\pi/3}{4} + 0\right)\right] = \sqrt{3}\left(\cos\frac{\pi}{3} + i\sin\frac{\pi}{3}\right)$$

$$z_1 = \sqrt[4]{9}\left[\cos\left(\frac{4\pi/3}{4} + \frac{2\pi}{4}\right) + i\sin\left(\frac{4\pi/3}{4} + \frac{2\pi}{4}\right)\right] = \sqrt{3}\left(\cos\frac{5\pi}{6} + i\sin\frac{5\pi}{6}\right)$$

$$z_2 = \sqrt[4]{9}\left[\cos\left(\frac{4\pi/3}{4} + \frac{4\pi}{4}\right) + i\sin\left(\frac{4\pi/3}{4} + \frac{4\pi}{4}\right)\right] = \sqrt{3}\left(\cos\frac{4\pi}{3} + i\sin\frac{4\pi}{3}\right)$$

$$z_3 = \sqrt[4]{9}\left[\cos\left(\frac{4\pi/3}{4} + \frac{6\pi}{4}\right) + i\sin\left(\frac{4\pi/3}{4} + \frac{6\pi}{4}\right)\right] = \sqrt{3}\left(\cos\frac{11\pi}{6} + i\sin\frac{11\pi}{6}\right).$$

These roots can be rewritten in cartesian form as

$$z_0 = \frac{\sqrt{3}}{2} + \frac{3}{2}i, \qquad z_1 = -\frac{3}{2} + \frac{\sqrt{3}}{2}i, \qquad z_2 = -\frac{\sqrt{3}}{2} - \frac{3}{2}i, \qquad z_3 = \frac{3}{2} - \frac{\sqrt{3}}{2}i.$$

If you plot the four complex roots z_0, z_1, z_2, and z_3 of

$$w = -\frac{9}{2} - \frac{9\sqrt{3}}{2}i$$

obtained in the example above, you will find that they are equally spaced around a circle of radius $\sqrt{3}$ with center at the origin (Figure 5). More generally, Theorem 3 implies that the complex nth roots of a nonzero complex number w are equally spaced around a circle of radius $\sqrt[n]{|w|}$ with center at the origin (Problem 54).

Figure 5

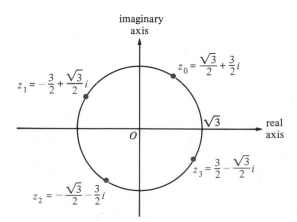

PROBLEM SET 2

In problems 1–8, plot each point z in the complex plane and rewrite z in cartesian form.

1. $z = 4(\cos 30° + i\sin 30°)$

2. $z = 6\left(\cos\dfrac{\pi}{4} + i\sin\dfrac{\pi}{4}\right)$

3. $z = 7\left(\cos\dfrac{3\pi}{4} + i\sin\dfrac{3\pi}{4}\right)$

4. $z = 8(\cos 270° + i\sin 270°)$

© **5.** $z = 6(\cos 370° + i\sin 370°)$

© **6.** $z = 6\left(\cos\dfrac{7\pi}{10} + i\sin\dfrac{7\pi}{10}\right)$

7. $z = 2\left[\cos\left(-\dfrac{11\pi}{6}\right) + i\sin\left(-\dfrac{11\pi}{6}\right)\right]$

8. $z = 12(\cos 240° + i\sin 240°)$

In problems 9–18, plot each point z in the complex plane and rewrite z in polar form.

9. $z = -1 + i$
10. $z = 3 - 3i$
11. $z = 3i$
12. $z = 4$
© **13.** $z = 5 + \sqrt{3}i$
© **14.** $z = -2 - \sqrt{3}i$
15. $z = -\sqrt{3} - i$
16. $z = \sqrt{3} - i$
17. $z = -5$
18. $z = -i$

In problems 19–26, use Theorem 1 to find (a) zw and (b) z/w. Leave the results in polar form.

19. $z = 4(\cos 70° + i\sin 70°)$, $w = 2(\cos 40° + i\sin 40°)$
20. $z = 8(\cos 80° + i\sin 80°)$, $w = 4(\cos 20° + i\sin 20°)$

21. $z = 14\left(\cos\dfrac{3\pi}{2} + i\sin\dfrac{3\pi}{2}\right)$, $w = 7\left(\cos\dfrac{5\pi}{4} + i\sin\dfrac{5\pi}{4}\right)$

22. $z = 15\left(\cos\dfrac{4\pi}{3} + i\sin\dfrac{4\pi}{3}\right)$, $w = 5\left[\cos\left(-\dfrac{\pi}{3}\right) + i\sin\left(-\dfrac{\pi}{3}\right)\right]$

23. $z = 6(\cos 90° + i\sin 90°)$, $w = 3(\cos 45° + i\sin 45°)$
24. $z = 8(\cos 85° + i\sin 85°)$, $w = 4(\cos 55° + i\sin 55°)$
25. $z = 1 + i$, $w = 1 + \sqrt{3}i$ [Change to polar form.]
26. $z = \sqrt{3} + i$, $w = 1 - i$ [Change to polar form.]

In problems 27–38, use De Moivre's theorem to evaluate each expression. Write the result in cartesian form.

27. $[2(\cos 9° + i\sin 9°)]^5$

28. $[\sqrt{2}(\cos 15° + i\sin 15°)]^{10}$

29. $\left[\sqrt{3}\left(\cos\dfrac{\pi}{10} + i\sin\dfrac{\pi}{10}\right)\right]^{10}$

30. $\left[2\left(\cos\dfrac{2\pi}{9} + i\sin\dfrac{2\pi}{9}\right)\right]^6$

31. $\left[2\left(\cos\dfrac{\pi}{12} + i\sin\dfrac{\pi}{12}\right)\right]^9$

32. $\left(\cos\dfrac{2\pi}{5} + i\sin\dfrac{2\pi}{5}\right)^{15}$

33. $(\sqrt{3} - i)^5$

34. $(-1 - \sqrt{3}i)^8$

© **35.** $(-1 + 2i)^{12}$

36. $(1 + i)^{16}$

37. $\left(\dfrac{1 + i}{\sqrt{2}}\right)^7$

© **38.** $(2 + 3i)^{11}$

In problems 39–48, use Theorem 3 to find all of the indicated complex nth roots.

39. The square roots of $\dfrac{9}{2} - \dfrac{9\sqrt{3}}{2}i$.

40. The cube roots of $-8i$.

41. The cube roots of $-1 + i$.

42. The fourth roots of $-2 + 2\sqrt{3}i$.

43. The fourth roots of $8 - 8\sqrt{3}i$.

44. The fifth roots of 1.

45. The fourth roots of -1.

46. The sixth roots of $-64i$.

47. The fifth roots of $-16\sqrt{2} - 16\sqrt{2}i$.

48. The fifth roots of $-\sqrt{3} - i$.

49. (a) Show that De Moivre's theorem (Theorem 2) holds for $n = 0$ and for $n = -1$, provided that $z \neq 0$. (b) Using the fact that $z^{-n} = 1/z^n$ for $z \neq 0$, show that De Moivre's theorem holds for all integer values of n, provided that $z \neq 0$.

50. Suppose that $z = a + bi$, where a and b are real numbers, and let $r = |z|$. (a) Show that $r + a \geq 0$. (b) Show that $r - a \geq 0$. (c) Let $x = \sqrt{\dfrac{r+a}{2}}$ and $y = \pm\sqrt{\dfrac{r-a}{2}}$, where the algebraic sign is chosen so that xy and b have the same algebraic sign. Show that $x + yi$ is a square root of z. (d) Show that $-x - yi$ is a square root of z.

51. If n is a positive integer, the complex nth roots of 1 are called the **nth roots of unity.** Use Theorem 3 to find all nth roots of unity for (a) $n = 2$, (b) $n = 3$, and (c) $n = 4$.

52. If n is a positive integer, (a) show that the product of two nth roots of unity (problem 51) is also an nth root of unity, (b) show that the reciprocal of an nth root of unity is also an nth root of unity, and (c) show that all nth roots of unity lie on the unit circle in the complex plane.

53. If z and w are points in the complex plane, give a geometric interpretation of (a) $\bar{z}$, (b) $-z$, (c) $-\bar{z}$, (d) iz, and (e) $|z - w|$.

54. Let w be a fixed nonzero complex number and let n be a positive integer. Show that the complex nth roots of w are equally spaced around a circle of radius $\sqrt[n]{|w|}$ with center at the origin.

55. Express part (ii) of Theorem 1 in words.

56. Using De Moivre's theorem, show that $z_0, z_1, \ldots, z_{n-1}$ in Theorem 3 are nth roots of w.

57. If $z = r(\cos\theta + i\sin\theta)$, show that $z^4 = r^4(\cos 4\theta + i\sin 4\theta)$ by direct use of Theorem 1.

58. Prove part (ii) of Theorem 1.

3   Complex Polynomials

If z represents a complex variable, then by a **complex polynomial** in z, we mean an expression of the form

$$f(z) = a_n z^n + a_{n-1} z^{n-1} + \cdots + a_1 z + a_0,$$

where n is a nonnegative integer and the coefficients $a_n, a_{n-1}, \ldots, a_1, a_0$ are complex numbers. If $a_n \neq 0$, we say that the polynomial has **degree** n and we call a_n the **leading coefficient.** The zero polynomial (which has no nonzero coefficient) is not assigned a degree.

Because the complex numbers have the same basic algebraic properties as the real numbers, much of our previous discussion of polynomials (Sections 3 and 4 of Chapter 1, Sections 1.1 and 3 of Chapter 2, and Sections 3 and 4 of Chapter 4) extends almost verbatim to complex polynomials. In particular, the long-division procedure (Chapter 4, page 188) works for complex polynomials, and it follows that the remainder theorem (page 194) and the factor theorem (page 195) hold as well for complex polynomials.

The following theorem, which was proved in 1799 by the great German mathematician Karl Friedrich Gauss (1777–1855), is so important that it is called the **fundamental theorem of algebra.**

Theorem 1   **The Fundamental Theorem of Algebra**

> Every complex polynomial of degree $n \geq 1$ has a complex zero.

By combining Theorem 1 with the factor theorem, we can obtain the following useful result.

Theorem 2   **The Complete Linear Factorization Theorem**

> If $f(z)$ is a complex polynomial of degree $n \geq 1$, then there is a nonzero complex number a and there are complex numbers $c_1, c_2, \ldots, c_n$ such that
> $$f(z) = a(z - c_1)(z - c_2) \cdots (z - c_n).$$

Proof   By Theorem 1, there is a complex number c_1 such that

$$f(c_1) = 0.$$

Hence, by the factor theorem, $z - c_1$ is a factor of $f(z)$; that is,

$$f(z) = (z - c_1)Q_1(z),$$

where $Q_1(z)$ is a complex polynomial of degree $n - 1$. If $n - 1 \geq 1$, we can again apply Theorem 1 and the factor theorem to $Q_1(z)$ to obtain

$$Q_1(z) = (z - c_2)Q_2(z),$$

for some complex number c_2, where $Q_2(z)$ is a complex polynomial of degree $n - 2$; therefore,

$$f(z) = (z - c_1)(z - c_2)Q_2(z).$$

If we continue this process, then, after n steps, we arrive at

$$f(z) = (z - c_1)(z - c_2) \cdots (z - c_n)Q_n,$$

where Q_n has degree $n - n = 0$; that is, Q_n is a (complex) constant. Letting $a = Q_n$, we obtain

$$f(z) = a(z - c_1)(z - c_2) \cdots (z - c_n).$$

According to Theorem 2, every complex polynomial $f(z)$ of degree $n \geq 1$ can be factored completely into a nonzero complex constant a and exactly n linear (first-degree) factors

$$(z - c_1), (z - c_2), \ldots, \text{ and } (z - c_n).$$

Evidently, the complex number a is the leading coefficient of $f(z)$ (why?) and the complex numbers $c_1, c_2, \ldots, c_n$ are the zeros of $f(z)$. There can be no other zeros (Problem 40). The zeros $c_1, c_2, \ldots, c_n$ need not be distinct from one another, but *a complex polynomial $f(z)$ of degree n can have no more than n zeros.*

That the zeros $c_1, c_2, \ldots, c_n$ of $f(z)$ need not be distinct is illustrated by

$$f(z) = z^3 - 3z^2 + 4 = (z + 1)(z - 2)(z - 2),$$

which has the number 2 as a **double zero.** More generally, a complex zero c of a complex polynomial $f(z)$ is called a **zero of multiplicity** s if the linear factor $(z - c)$ appears exactly s times in the complete factorization of $f(z)$. Thus, if a zero of multiplicity s is counted s times, we can conclude from Theorem 2 that *a complex polynomial of degree $n \geq 1$ has exactly n zeros.* For instance,

$$f(z) = 7(z - 3)^4(z - i)^3(z + i)$$

is a complex polynomial of degree 8 with the following zeros:

$$\underbrace{3, 3, 3, 3,}_{4} \underbrace{i, i, i,}_{3} \underbrace{-i.}_{1} \leftarrow \text{multiplicity}$$

The quadratic formula (Chapter 2, pages 79–80) can be easily extended to complex quadratic polynomials, since the proof by completing the square still works. With the understanding that $\pm\sqrt{b^2 - 4ac}$ stands for the two complex square roots of $b^2 - 4ac$, we can still write the quadratic formula

$$z = \frac{-b \pm \sqrt{b^2 - 4ac}}{2a}$$

for the roots of the complex quadratic equation

$$az^2 + bz + c = 0.$$

Examples   Use the quadratic formula to solve each quadratic equation.

1 $z^2 - 2z + 5 = 0$

Solution   We use the quadratic formula with $a = 1$, $b = -2$, and $c = 5$:

$$z = \frac{-(-2) \pm \sqrt{(-2)^2 - 4(1)(5)}}{2(1)} = \frac{2 \pm \sqrt{-16}}{2} = \frac{2 \pm \sqrt{16(-1)}}{2}$$

$$= \frac{2 \pm 4\sqrt{-1}}{2} = \frac{2 \pm 4i}{2} = 1 \pm 2i.$$

Thus, the two roots are the complex numbers $z = 1 + 2i$ and $z = 1 - 2i$.

2 $z^2 + (4 + 2i)z + (3 + 3i) = 0$

Solution   Here $a = 1$, $b = 4 + 2i$, and $c = 3 + 3i$. Thus,

$$b^2 - 4ac = (4 + 2i)^2 - 4(1)(3 + 3i) = 16 + 16i + 4i^2 - 12 - 12i$$

$$= 16 + 16i - 4 - 12 - 12i = 4i.$$

We are going to use Theorem 3 (page 491) to find $\pm\sqrt{b^2 - 4ac} = \pm\sqrt{4i}$; that is, the two complex square roots of $4i$. Writing $4i$ in polar form, we have

$$4i = 4(\cos 90° + i \sin 90°).$$

Therefore, the square roots of $4i$ are

$$\pm\sqrt{4i} = \pm 2(\cos 45° + i \sin 45°) = \pm 2\left(\frac{\sqrt{2}}{2} + i\frac{\sqrt{2}}{2}\right) = \pm(\sqrt{2} + \sqrt{2}i).$$

Hence,

$$z = \frac{-b \pm \sqrt{b^2 - 4ac}}{2a} = \frac{-(4 + 2i) \pm \sqrt{4i}}{2(1)}$$

$$= \frac{-(4 + 2i) \pm (\sqrt{2} + \sqrt{2}i)}{2}.$$

In other words, the two roots are the complex numbers

$$z = \frac{-(4 + 2i) + (\sqrt{2} + \sqrt{2}i)}{2} = \frac{-4 + \sqrt{2}}{2} + \frac{-2 + \sqrt{2}}{2}i$$

and

$$z = \frac{-(4 + 2i) - (\sqrt{2} + \sqrt{2}i)}{2} = \frac{-4 - \sqrt{2}}{2} + \frac{-2 - \sqrt{2}}{2}i.$$

In Example 1 above, the two roots are complex conjugates of each other, but in Example 2, they are not. The roots of a quadratic equation with *real* coefficients and a negative discriminant are complex conjugates of each other. More generally, we have the following theorem.

Theorem 3   **The Conjugate Zeros Theorem**

Let $f(z)$ be a polynomial with real coefficients. Then, if c is a zero of $f(z)$, so is $\bar{c}$, the complex conjugate of c.

Proof   Suppose that

$$f(z) = a_n z^n + a_{n-1} z^{n-1} + \cdots + a_1 z + a_0,$$

where the coefficients $a_n, a_{n-1}, \ldots, a_1$, and a_0 are real numbers. If c is a zero of $f(z)$, then $f(c) = 0$, so

$$a_n c^n + a_{n-1} c^{n-1} + \cdots + a_1 c + a_0 = 0.$$

Using the fact that complex conjugation "preserves" addition and multiplication (see page 485), we take the complex conjugate of both sides of the last equation to obtain

$$\bar{a}_n(\bar{c})^n + \bar{a}_{n-1}(\bar{c})^{n-1} + \cdots + \bar{a}_1(\bar{c}) + \bar{a}_0 = \bar{0}.$$

Because all of the coefficients are real numbers, they are equal to their own conjugates, and likewise $\bar{0} = 0$, so we have

$$a_n(\bar{c})^n + a_{n-1}(\bar{c})^{n-1} + \cdots + a_1(\bar{c}) + a_0 = 0.$$

In other words, $\bar{c}$ is also a complex zero of $f(z)$, and our proof is complete.

By Theorem 3, the complex zeros of a polynomial $f(z)$ *with real coefficients* must occur in conjugate pairs; hence, such a polynomial must have an even number of nonreal zeros. (Note: It may have 0 such nonreal zeros; but 0 is an even integer.) Consequently, a polynomial $f(z)$ of odd degree with real coefficients must have at least one real zero (Problem 41).

Example 1   Find all complex zeros of $f(z) = z^3 - 2z^2 - 3z + 10$ and factor $f(z)$ completely into linear factors.

Solution   Since $f(z)$ is a polynomial of degree 3 with real coefficients, we know that it has 3 zeros, at least one of which is a real number. By the rational zeros theorem (Chapter 4, page 196), the only possible rational zeros are ± 1, ± 2, ± 5, and ± 10. Testing these possibilities one at a time (using synthetic division for efficiency), we find that -2 is, in fact, a zero:

$$
\begin{array}{r|rrrr}
-2 & 1 & -2 & -3 & 10 \\
   &   & -2 & 8 & -10 \\
\hline
   & 1 & -4 & 5 & 0
\end{array}
$$

From this, we also see that the quotient polynomial is

$$z^2 - 4z + 5;$$

that is,

$$z^3 - 2z^2 - 3z + 10 = (z + 2)(z^2 - 4z + 5).$$

By the quadratic formula, the zeros of $z^2 - 4z + 5$ are given by

$$z = \frac{4 \pm \sqrt{16 - 20}}{2} = \frac{4 \pm \sqrt{-4}}{2}$$

$$= \frac{4 \pm 2i}{2} = 2 \pm i.$$

Therefore, the zeros of $z^3 - 2z^2 - 3z + 10$ are

$$-2, \quad 2 + i, \quad \text{and} \quad 2 - i,$$

and the polynomial factors completely into linear factors as follows:

$$z^3 - 2z^2 - 3z + 10 = [z - (-2)][z - (2 + i)][z - (2 - i)]$$

$$= (z + 2)(z - 2 - i)(z - 2 + i).$$

Example 2   Form a polynomial in z that has real coefficients, that has the smallest possible degree, and that has 2, -1, and $1 - i$ as zeros.

Solution   Since we want a polynomial of the smallest possible degree, we assume that each zero has multiplicity 1. Because we want the coefficients to be real numbers, we must include $1 + i$, the complex conjugate of $1 - i$, as a zero. The polynomial is

$$f(z) = (z - 2)[z - (-1)][z - (1 - i)][z - (1 + i)]$$
$$= (z - 2)(z + 1)[z^2 - (1 + i)z - (1 - i)z + (1 - i)(1 + i)]$$
$$= (z^2 - z - 2)(z^2 - 2z + 2)$$
$$= z^4 - 3z^3 + 2z^2 + 2z - 4.$$

PROBLEM SET 3

In problems 1–8, use the quadratic formula to find the complex roots of each quadratic equation.

1. $z^2 - 2z + 2 = 0$
2. $z^2 + iz + 1 = 0$
3. $z^2 - 2z + 10 = 0$
4. $iz^2 + 2z + i = 0$
5. $z^2 + iz - 1 = 0$
6. $4z^2 - 4z + 3 = 0$
7. $z^2 - (3 + i)z + (2 + 2i) = 0$
8. $\frac{\sqrt{3}}{8}z^2 + \frac{\sqrt{2}}{2}z + i = 0$

In problems 9–26 (a) find all of the complex zeros of each complex polynomial $f(z)$, (b) factor $f(z)$ completely into linear factors, and (c) determine the multiplicity of each zero of $f(z)$.

9. $f(z) = z^3 - 8$ [Hint: The zeros are the complex cube roots of 8.]
10. $f(z) = z^3 + 27i$ [Hint: The zeros are the complex cube roots of $-27i$.]
11. $f(z) = z^4 - 81$
12. $f(z) = z^6 + 1$
13. $f(z) = (z - 1)^3(z^2 + 1)$
14. $f(z) = z^6 - (1 + i)$
15. $f(z) = (z + 2)^2(z^2 + 4)$
16. $f(z) = (3z + 1)^5(z^2 - 5z - 6)$
17. $f(z) = (2z + 1)^3(z^2 - 2z - 3)$
18. $f(z) = (z^2 - 4)^2(z^2 + 4z + 13)$
19. $f(z) = (z^2 + 9)^4(z^2 - 4z + 5)$
20. $f(z) = z^4 + 3z^3 + 5z^2 + 12z + 4$
21. $f(z) = z^3 - z^2 + z - 1$
22. $f(z) = z^4 + 4z^3 - 4z^2 - 40z - 33$
23. $f(z) = z^3 - 5z^2 + 8z - 6$
24. $f(z) = z^4 + 4z^3 + 5z^2 + 4z + 4$
25. $f(z) = z^4 - 3z^3 + z^2 + 4$
26. $f(z) = z^4 - 3z^3 + 3z^2 - 3z + 2$

In problems 27–38, form a polynomial in z that has real coefficients, that has the smallest possible degree, and that has the indicated zeros. (Zeros are repeated to show multiplicity.)

27. $2, -2, 3$
28. $0, 0, 3, -3$
29. $0, 1, 1, 2, 2$
30. $\frac{1}{2}, \frac{1}{3}, \frac{1}{4}, \frac{1}{4}$
31. $-2, -2, -2, 1, 1, 1$
32. $\frac{1}{2}, \frac{1}{2}, 3, 3, 3, 3$
33. $3, -2, 1 + i$
34. $2, 2, -1, 2 + i$
35. $1, -1, 1 + \sqrt{3}i, i$
36. $0, -i, -i, 2, 3i$

37. $\cos\frac{2\pi}{5} + i\sin\frac{2\pi}{5}, \quad \cos\frac{4\pi}{5} + i\sin\frac{4\pi}{5}, \quad 1$
38. $\cos\frac{\pi}{3} + i\sin\frac{\pi}{3}, \quad \cos\frac{2\pi}{3} + i\sin\frac{2\pi}{3}, \quad 1, \quad -1$

39. If z_1 and z_2 are the complex roots of the quadratic equation $az^2 + bz + c = 0$, show that (a) $z_1 + z_2 = -b/a$ and (b) $z_1 z_2 = c/a$.
40. If $f(z) = a(z - c_1)(z - c_2)\cdots(z - c_n)$, where $a \neq 0$, and if c is a complex zero of $f(z)$, show that c must be one of the complex numbers $c_1, c_2, \ldots,$ or c_n.
41. Show that a polynomial with real coefficients and with odd degree must have at least one real zero.
42. Show that a nonconstant polynomial with real coefficients can be factored into linear and quadratic polynomials with real coefficients, where the quadratic factors have negative discriminants.

4   Mathematical Induction

The principle of *mathematical induction* is often illustrated by a row of oblong wooden blocks, called dominoes, set on end as in Figure 1. If we are guaranteed two conditions,

(i)   that the first domino is toppled over, and

(ii) that if any domino topples over, it will hit the next one and topple it over,

then we can be certain that *all* the dominoes will topple over.

Figure 1

The principle of mathematical induction concerns a sequence of statements

$$S_1, S_2, S_3, S_4, S_5, S_6, \ldots$$

Let's think of this sequence as corresponding to a row of dominoes; the first statement S_1 corresponds to the first domino, the second statement S_2 corresponds to the second domino, and so on. Each statement can be either true or false. If a statement proves to be true, let's think of the corresponding domino toppling over. The two conditions given above for the row of dominoes can then be interpreted as follows:

(i)   The first statement S_1 is true.

(ii) If any statement S_k is true, then the next statement S_{k+1} is also true.

If these two conditions hold, then, by analogy with the toppling over of *all* the dominoes, we conclude that *all* statements $S_1, S_2, S_3, \ldots$ are true. That this argument is valid is the **principle of mathematical induction.**

   Although the principle of mathematical induction can be established formally, the required argument is beyond the scope of this book. We ask you to accept this principle on the intuitive basis of the domino analogy. Thus, we have the following.

Procedure for Making a Proof by Mathematical Induction

Begin by clearly identifying the sequence $S_1, S_2, S_3, \ldots$ of statements to be proved. This is usually done by specifying the meaning of S_n, where n denotes an arbitrary positive integer, $n = 1, 2, 3, \ldots$. Then carry out the following two steps:

Step 1. Show that S_1 is true.

Step 2. Let k denote an arbitrary positive integer. Assume that S_k is true and show, on the basis of this assumption, that S_{k+1} is also true.

If both steps can be carried out, conclude that S_n is true for all positive integer values of n.

The assumption in step 2 that S_k is true is called the **induction hypothesis.** When you make the induction hypothesis, you're not saying that S_k is *in fact* true— you're just *supposing* that it is true to see if the truth of S_{k+1} follows from this supposition. It's as if you are checking to make sure that if the kth domino *were* to topple over, it *would* knock over the next domino.

Examples   Use mathematical induction to prove each result.

1   For every positive integer n, $1 + 2 + 3 + \cdots + n = \dfrac{n(n + 1)}{2}$.

Solution   Let S_n be the statement

$$S_n: \quad 1 + 2 + 3 + \cdots + n = \frac{n(n + 1)}{2}.$$

In other words, S_n asserts that the sum of the first n positive integers is $n(n + 1)/2$. For instance,

$$S_2 \quad \text{says that} \quad 1 + 2 = \frac{2(2 + 1)}{2} = 3$$

and

$$S_{20} \quad \text{says that} \quad 1 + 2 + 3 + \cdots + 20 = \frac{20(20 + 1)}{2} = 210.$$

Step 1. The statement

$$S_1: \quad 1 = \frac{1(1 + 1)}{2}$$

is clearly true.

Step 2. Here we must deal with the statements

$$S_k: \quad 1 + 2 + 3 + \cdots + k = \frac{k(k + 1)}{2}$$

and

$$S_{k+1}: \quad 1 + 2 + 3 + \cdots + (k + 1) = \frac{(k + 1)[(k + 1) + 1]}{2}$$

obtained by replacing n in S_n by k and then by $k + 1$. The statement S_k is our induction hypothesis, and we assume, for the sake of argument, that it is true. Our goal is to prove that, on the basis of this assumption, S_{k+1} is true. To this end, we add $k + 1$ to both sides of the equation expressing S_k to obtain

$$1 + 2 + 3 + \cdots + k + (k + 1) = \frac{k(k + 1)}{2} + (k + 1)$$

$$= \frac{k(k + 1) + 2(k + 1)}{2}$$

$$= \frac{(k + 1)(k + 2)}{2}$$

$$= \frac{(k + 1)[(k + 1) + 1]}{2},$$

which is the equation expressing S_{k+1}. This completes the proof by mathematical induction.

2   If a principal of P dollars is invested at a compound interest rate R per conversion period, the final value F dollars of the investment at the end of n conversion periods is given by $F = P(1 + R)^n$. [For *compound interest*, the interest is periodically calculated and added to the principal. The time interval between successive conversions of interest into principal is called the *conversion period*.]

Solution   Let S_n be the statement

$$S_n: \quad F = P(1 + R)^n \text{ at the end of } n \text{ conversion periods.}$$

Step 1. The statement

$$S_1: \quad F = P(1 + R) \text{ at the end of the first conversion period}$$

is true, because the interest on P dollars for one conversion period is PR dollars, so the final value of the investment at the end of the first conversion period is

$$P + PR = P(1 + R) \text{ dollars.}$$

Step 2. Assume that the statement

$$S_k: \quad F = P(1 + R)^k \text{ at the end of } k \text{ conversion periods}$$

is true. It follows that, over the $(k + 1)$st conversion period, interest at rate R is paid on $P(1 + R)^k$ dollars. Adding this interest, $P(1 + R)^k R$ dollars, to the value of the investment at the beginning of the $(k + 1)$st conversion period, we find that the value of the investment at the end of the $(k + 1)$st conversion period is given by

$$F = P(1 + R)^k + P(1 + R)^k R$$

$$= P(1 + R)^k(1 + R)$$

$$= P(1 + R)^{k+1} \text{ dollars.}$$

Hence, S_{k+1} is true, and the proof by mathematical induction is complete.

PROBLEM SET 4

In problems 1–14, use mathematical induction to prove that the assertion is true for all positive integers n.

1. $1 + 3 + 5 + \cdots + (2n - 1) = n^2$; in other words, the sum of the first n odd positive integers is n^2.

2. $2 + 4 + 6 + \cdots + 2n = n(n + 1)$; in other words, the sum of the first n even positive integers is $n(n + 1)$.

3. $1^2 + 2^2 + 3^2 + \cdots + n^2 = \frac{1}{6}n(n + 1)(2n + 1)$; that is, the sum of the first n perfect squares is $\frac{1}{6}n(n + 1)(2n + 1)$.

4. $1 + 5 + 9 + \cdots + (4n - 3) = n(2n - 1)$

5. $1^2 + 3^2 + 5^2 + \cdots + (2n - 1)^2 = \frac{1}{3}n(2n - 1)(2n + 1)$

6. $1^3 + 2^3 + 3^3 + \cdots + n^3 = \left[\dfrac{n(n + 1)}{2}\right]^2$

7. $1 \cdot 2 + 2 \cdot 3 + 3 \cdot 4 + \cdots + n(n + 1) = \frac{1}{3}n(n + 1)(n + 2)$

8. $\dfrac{1}{1 \cdot 2} + \dfrac{1}{2 \cdot 3} + \dfrac{1}{3 \cdot 4} + \cdots + \dfrac{1}{n(n + 1)} = \dfrac{n}{n + 1}$

9. $(ab)^n = a^n b^n$

10. If $h \geq 0$, then $1 + nh \leq (1 + h)^n$.

11. $2 + 2^2 + 2^3 + \cdots + 2^n = 2(2^n - 1)$

12. $r + r^2 + r^3 + \cdots + r^n = \dfrac{r(r^n - 1)}{r - 1}$

13. $\cos(n\pi) = (-1)^n$

14. De Moivre's theorem: $[r(\cos\theta + i\sin\theta)]^n = r^n(\cos n\theta + i\sin n\theta)$

C 15. Using a calculator, verify the identities in odd problems 1–7 for $n = 15$ by directly calculating both sides.

16. The domino theory was a tenet of U.S. foreign policy subscribed to by the administrations of Presidents Eisenhower, Kennedy, Johnson, and Nixon. Explain the connection, if any, between the domino theory and mathematical induction.

5 The Binomial Theorem

The principle of mathematical induction can be used to establish a general formula for the expansion of $(a + b)^n$, where the exponent n is an arbitrary positive integer. The expression $a + b$ is a binomial, so the formula is called the **binomial theorem.**

Of course, the expansion of $(a + b)^n$ for small values of n can be obtained by direct calculation. For instance,

$$(a + b)^1 = a + b$$

$$(a + b)^2 = a^2 + 2ab + b^2$$

$$(a + b)^3 = a^3 + 3a^2b + 3ab^2 + b^3$$

$$(a + b)^4 = a^4 + 4a^3b + 6a^2b^2 + 4ab^3 + b^4$$

$$(a + b)^5 = a^5 + 5a^4b + 10a^3b^2 + 10a^2b^3 + 5ab^4 + b^5,$$

and so forth.

Notice that we have written the expansions in descending powers of a and ascending powers of b. A certain pattern is already apparent. Indeed, in the expansion of $(a + b)^n$:

1. There are $n + 1$ terms, beginning with a^n and ending with b^n.

2. As we move from each term to the next, powers of a decrease by 1 and powers of b increase by 1. Therefore, in each term, the exponents of a and b add up to n.

3. The successive terms can be written in the form

$$(\text{numerical coefficient})a^{n-j}b^j$$

for $j = 0, 1, 2, \ldots, n$.

That statements 1–3 are true for every positive integer n is part of the binomial theorem. The remaining part concerns the values of the numerical coefficients, which are called the **binomial coefficients.** In order to obtain a useful formula for the binomial coefficients, we begin by introducing *factorial notation*.

Definition 1   **Factorial Notation**

> If n is a positive integer, we define $n!$, read $\textbf{\textit{n}}$ **factorial,** by
> $$n! = n(n - 1)(n - 2) \cdots 3 \cdot 2 \cdot 1.$$
> We also define
> $$0! = 1.$$

In words, the factorial of a positive integer n is the product of n and all smaller positive integers. For instance,

$$7! = 7 \cdot 6 \cdot 5 \cdot 4 \cdot 3 \cdot 2 \cdot 1 = 5040.$$

The special definition $0! = 1$ extends the usefulness of the factorial notation. As a direct consequence of Definition 1, we have the important **recursion formula** for factorials

$$(n + 1)! = (n + 1)n!,$$

which holds for every integer $n \geq 0$ (Problem 35). Using the recursion formula, you can make a table of values of $n!$ (Table 1). The table can be continued indefinitely; for instance, to obtain the next entry, use the recursion formula as follows:

$$8! = 8(7!) = 8(5040) = 40{,}320.$$

The factorials increase very rapidly. Indeed, $10!$ is over 36 million, and $70!$ is beyond the range of most calculators.

Using factorial notation, we now introduce special symbols for what, as we shall soon see, are the binomial coefficients.

Table 1

$0! = 1$
$1! = 1$
$2! = 2$
$3! = 6$
$4! = 24$
$5! = 120$
$6! = 720$
$7! = 5040$

Definition 2   **The Binomial Coefficients**

If n and j are integers with $n \geq j \geq 0$, the symbol $\binom{n}{j}$ is defined as follows:

$$\binom{n}{j} = \frac{n!}{(n-j)!\,j!}.$$

Example 1   Evaluate (a) $\binom{5}{2}$, (b) $\binom{7}{6}$, (c) $\binom{n}{0}$, and (d) $\binom{n}{n}$, if n is a nonnegative integer.

Solution   Using Definition 2, we have:

(a) $\displaystyle \binom{5}{2} = \frac{5!}{(5-2)!2!} = \frac{5!}{3!2!} = \frac{5 \cdot 4 \cdot 3 \cdot 2 \cdot 1}{(3 \cdot 2 \cdot 1)(2 \cdot 1)} = \frac{5 \cdot 4}{2} = 10$

(b) $\displaystyle \binom{7}{6} = \frac{7!}{(7-6)!6!} = \frac{7!}{1!6!} = \frac{7!}{6!} = \frac{7(6!)}{6!} = 7$

(c) $\displaystyle \binom{n}{0} = \frac{n!}{(n-0)!0!} = \frac{n!}{n!0!} = \frac{1}{0!} = \frac{1}{1} = 1$

(d) $\displaystyle \binom{n}{n} = \frac{n!}{(n-n)!n!} = \frac{1}{0!} = \frac{1}{1} = 1$

Example 2   If k and j are integers with $k \geq j > 0$, show that

$$\binom{k}{j-1} + \binom{k}{j} = \binom{k+1}{j}.$$

Solution   Using Definition 2 and the recursion formula for factorials, we have

$$\binom{k}{j-1} + \binom{k}{j} = \frac{k!}{[k-(j-1)]!(j-1)!} + \frac{k!}{(k-j)!j!}$$

$$= \frac{jk!}{(k-j+1)!j(j-1)!} + \frac{(k-j+1)k!}{(k-j+1)(k-j)!j!}$$

$$= \frac{jk!}{(k-j+1)!j!} + \frac{(k+1)k! - jk!}{(k-j+1)!j!}$$

$$= \frac{(k+1)k!}{(k-j+1)!j!} = \frac{(k+1)!}{[(k+1)-j]!j!}$$

$$= \binom{k+1}{j}.$$

We can now state and prove the binomial theorem.

Theorem 1   **The Binomial Theorem**

If n is a positive integer and if a and b are any two numbers, then

$$(a + b)^n = \binom{n}{0}a^n + \binom{n}{1}a^{n-1}b + \cdots + \binom{n}{j}a^{n-j}b^j + \cdots + \binom{n}{n}b^n.$$

Proof   The proof is by mathematical induction. Thus, for each positive integer n, let S_n be the statement

$$S_n: \quad (a + b)^n = \binom{n}{0}a^n + \binom{n}{1}a^{n-1}b + \cdots + \binom{n}{j}a^{n-j}b^j + \cdots + \binom{n}{n}b^n.$$

Step 1. The statement

$$S_1: \quad (a + b)^1 = \binom{1}{0}a^1 + \binom{1}{1}b^1$$

is true because $\binom{1}{0} = 1$ and $\binom{1}{1} = 1$.

Step 2. Assume the induction hypothesis

$$S_k: \quad (a + b)^k = \binom{k}{0}a^k + \binom{k}{1}a^{k-1}b + \cdots + \binom{k}{j}a^{k-j}b^j + \cdots + \binom{k}{k}b^k,$$

where k is a positive integer. Notice that the term immediately preceding $\binom{k}{j}a^{k-j}b^j$ in this equation is $\binom{k}{j-1}a^{k-(j-1)}b^{j-1}$; that is, $\binom{k}{j-1}a^{k-j+1}b^{j-1}$. We use this fact in the following computation:

$(a + b)^{k+1}$

$$= (a + b)(a + b)^k = a(a + b)^k + b(a + b)^k$$

$$= a\left[\binom{k}{0}a^k + \binom{k}{1}a^{k-1}b + \cdots + \binom{k}{j}a^{k-j}b^j + \cdots + \binom{k}{k}b^k\right]$$

$$+ b\left[\binom{k}{0}a^k + \binom{k}{1}a^{k-1}b + \cdots + \binom{k}{j-1}a^{k-j+1}b^{j-1} + \cdots + \binom{k}{k}b^k\right]$$

$$= \binom{k}{0}a^{k+1} + \binom{k}{1}a^kb + \cdots + \binom{k}{j}a^{k-j+1}b^j + \cdots + \binom{k}{k}ab^k$$

$$+ \binom{k}{0}a^kb + \cdots + \binom{k}{j-1}a^{k-j+1}b^j + \cdots + \binom{k}{k-1}ab^k + \binom{k}{k}b^{k+1}$$

$$= \binom{k}{0}a^{k+1} + \left[\binom{k}{1} + \binom{k}{0}\right]a^kb + \cdots + \left[\binom{k}{j} + \binom{k}{j-1}\right]a^{k+1-j}b^j$$

$$+ \cdots + \left[\binom{k}{k} + \binom{k}{k-1}\right]ab^k + \binom{k}{k}b^{k+1}.$$

This expansion of $(a + b)^{k+1}$ will fulfill the requirements of statement S_{k+1}, provided that the following three conditions hold:

(i) $\dbinom{k}{0} = \dbinom{k+1}{0}$,   (ii) $\dbinom{k}{j} + \dbinom{k}{j-1} = \dbinom{k+1}{j}$,   (iii) $\dbinom{k}{k} = \dbinom{k+1}{k+1}$.

That (i) and (iii) hold is left as an exercise (Problem 36). We just proved (ii) in Example 2 above. Hence S_{k+1} is established, and the proof by induction is complete.

By arranging the binomial coefficients in a triangle as shown in Figure 1a, we obtain the **Pascal triangle,** named in honor of its discoverer, the French mathematician Blaise Pascal (1623–1662). The numerical values of the entries in the first six horizontal rows of the Pascal triangle are shown in Figure 1b. Notice that each number in the Pascal triangle (other than those on the border, which are all equal to 1) can be obtained by adding the two numbers diagonally above it. This is called the *additive property* of the Pascal triangle, and it is a consequence of the identity established in Example 2 above.

Figure 1

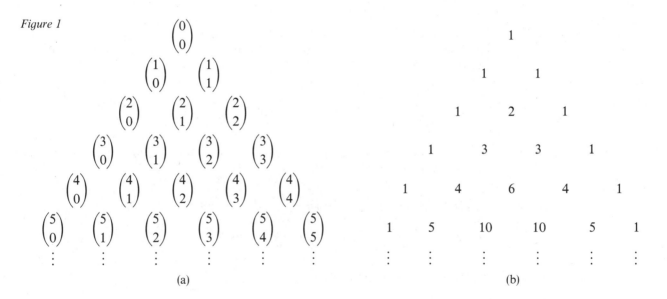

(a)   (b)

Example   Find the numerical entries in the seventh horizontal row of the Pascal triangle.

Solution   The seventh horizontal row is obtained from the sixth horizontal row in Figure 1b by using the additive property.

According to the binomial theorem, *the binomial coefficients in the expansion of $(a + b)^n$ appear in row number $n + 1$ of the Pascal triangle.* For instance, using the result of the preceding example, we have

$$(a + b)^6 = 1a^6 + 6a^5b + 15a^4b^2 + 20a^3b^3 + 15a^2b^4 + 6ab^5 + 1b^6.$$

Examples   Use the binomial theorem and the Pascal triangle to expand each of the given expressions.

1 $(2x + y)^4$

Solution   In the binomial theorem (Theorem 1), let $a = 2x$, $b = y$, and $n = 4$. The appropriate binomial coefficients (1, 4, 6, 4, 1) are found in the *fifth* horizontal row of the Pascal triangle (Figure 1b), since $n + 1 = 4 + 1 = 5$. Therefore,

$$(2x + y)^4 = 1(2x)^4 + 4(2x)^3y + 6(2x)^2y^2 + 4(2x)y^3 + 1y^4$$
$$= 16x^4 + 32x^3y + 24x^2y^2 + 8xy^3 + y^4.$$

2 $(x - y)^5$

Solution   In the binomial theorem, let $a = x$, $b = -y$, and $n = 5$. The appropriate binomial coefficients (1, 5, 10, 10, 5, 1) are found in the *sixth* horizontal row of the Pascal triangle. Thus,

$$(x - y)^5 = 1x^5 + 5x^4(-y) + 10x^3(-y)^2 + 10x^2(-y)^3 + 5x(-y)^4 + 1(-y)^5$$
$$= x^5 - 5x^4y + 10x^3y^2 - 10x^2y^3 + 5xy^4 - y^5.$$

According to the binomial theorem (Theorem 1), the $(j + 1)$st term in the expansion of $(a + b)^n$ is given by the formula

$$(j + 1)\text{st term} = \binom{n}{j}a^{n-j}b^j \quad \text{for} \quad 0 \le j \le n.$$

You can use this formula to find a specified term in a binomial expansion.

Example 1   Find the tenth term in the expansion of $(x - \frac{1}{2}y)^{12}$.

Solution   The tenth term is obtained by putting $j = 9$, $a = x$, $b = -\frac{1}{2}y$, and $n = 12$ in the formula above. Thus,

$$10\text{th term} = \binom{12}{9}a^{12-9}b^9$$
$$= \binom{12}{9}x^3(-\tfrac{1}{2}y)^9$$
$$= -\binom{12}{9}\frac{x^3y^9}{2^9}.$$

Now,

$$\binom{12}{9} = \frac{12!}{(12-9)!9!} = \frac{12!}{3!9!} = \frac{12 \cdot 11 \cdot 10(9!)}{3!9!} = \frac{12 \cdot 11 \cdot 10}{6} = 220,$$

and

$$2^9 = 512,$$

so we have

$$10\text{th term} = -220\frac{x^3y^9}{512} = -\frac{55}{128}x^3y^9.$$

Example 2   Find the term involving y^4 in the expansion of $(x + y)^8$.

Solution   With $a = x$, $b = y$, and $n = 8$, we have

$$(j + 1)\text{st term} = \binom{n}{j}a^{n-j}b^j = \binom{8}{j}x^{8-j}y^j.$$

Hence, the term involving y^4, obtained by putting $j = 4$, is the

$$5\text{th term} = \binom{8}{4}x^{8-4}y^4 = \binom{8}{4}x^4y^4.$$

Now,

$$\binom{8}{4} = \frac{8!}{(8-4)!4!} = \frac{8!}{4!4!} = \frac{8 \cdot 7 \cdot 6 \cdot 5(4!)}{4!4!} = \frac{8 \cdot 7 \cdot 6 \cdot 5}{4 \cdot 3 \cdot 2 \cdot 1} = 2 \cdot 7 \cdot 5 = 70.$$

Therefore, the term involving y^4 is the

$$5\text{th term} = 70x^4y^4.$$

PROBLEM SET 5

In problems 1–10, find the numerical value of each expression.

1. $9!$

2. $10!$

3. $\dfrac{7!}{5!3!}$

4. $\dfrac{8!6!}{5!3!}$

5. $\dbinom{7}{3}$

6. $\dbinom{12}{10}$

7. $\dbinom{5}{5}$

8. $\dbinom{12}{2}$

9. $\dbinom{6}{3}$

10. $\dbinom{50}{49}$

11. Using the additive property of the Pascal triangle and the fact that the numerical entries in the seventh horizontal row are $(1, 6, 15, 20, 15, 6, 1)$, find the numerical entries in the eighth row.

12. Construct a Pascal triangle showing the numerical entries in the first ten horizontal rows.

In problems 13–24, use the binomial theorem and Pascal triangle to expand each expression.

13. $(a + 3x)^4$

14. $(3a - 2b)^5$

15. $(x - y)^5$

16. $(x + y^2)^5$

17. $(c + 2)^6$ **18.** $(r - 3s)^5$ **19.** $(1 - c^3)^7$

20. $(b^2 + 10)^4$ **21.** $(\sqrt{x} - \sqrt{y})^6$ **22.** $\left(\dfrac{1}{x} + \dfrac{1}{y}\right)^6$

23. $\left(\dfrac{x}{2} - 2y\right)^6$ **24.** $0 = (1 - 1)^6$

In problems 25–32, find and simplify the specified term in the binomial expansion of the indicated expression.

25. The third term of $(s - t)^5$.   **26.** The fourth term of $(a^2 - b)^6$.

27. The fifth term of $\left(2x^2 + \dfrac{y^3}{4}\right)^{10}$.   **28.** The sixth term of $(\pi - \sqrt{2})^9$.

29. The term involving y^4 in $(x + y)^5$.   **30.** The term involving b^6 in $(a - 3b^3)^4$.

31. The term involving c^6 in $(c + 2d)^{10}$.   **32.** The term involving x^3 in $(\sqrt{x} + \sqrt{y})^8$.

33. Use mathematical induction to show that $2^{n+3} < (n + 3)!$ for all positive integers n.

Ⓒ **34.** Using the recursion formula and a calculator, extend Table 1 to show values of $n!$ for $n = 0, 1, 2, \ldots, 15$.

35. Verify the recursion formula for factorials: $(n + 1)! = (n + 1)n!$ for every nonnegative integer n.

36. If k is a nonnegative integer, show that $\dbinom{k}{0} = \dbinom{k + 1}{0}$ and that $\dbinom{k}{k} = \dbinom{k + 1}{k + 1}$.

37. Show that the Pascal triangle is symmetric about a vertical line through its apex by showing that for integers n and j with $n \geq j \geq 0$, $\dbinom{n}{j} = \dbinom{n}{n - j}$.

38. If n and j are integers with $n \geq j > 0$, show that

$$\binom{n}{j} = \frac{\overbrace{n(n - 1)\cdots(n - j + 1)}^{j\text{ factors}}}{\underbrace{j(j - 1)\cdots 1}_{j\text{ factors}}}.$$

39. If n is a positive integer, show that

$$(1 + x)^n = 1 + nx + \binom{n}{2}x^2 + \cdots + \binom{n}{j}x^j + \cdots + x^n.$$

40. Show that the sum of all the entries in row number $n + 1$ of the Pascal triangle is 2^n; that is,

$$\binom{n}{0} + \binom{n}{1} + \binom{n}{2} + \cdots + \binom{n}{n} = 2^n.$$

[Hint: Use the binomial theorem to expand $(1 + 1)^n$.]

41 If, in a term of the binomial expansion of $(a + b)^n$, we multiply the coefficient by the exponent of a and divide by the number of the term, show that we obtain the coefficient of the next term.

Ⓒ **42.** **Stirling's approximation,** $n! \approx \sqrt{2n\pi}\left(\dfrac{n}{e}\right)^n\left(1 + \dfrac{1}{12n - 1}\right)$, is often used to approximate $n!$

for large values of n. (a) Use Stirling's approximation to estimate 15!, and compare the result with the true value. (b) Use Stirling's approximation to estimate 50!.

6 Sequences

A **sequence** is a function

$$n \longmapsto a_n$$

whose domain is the set of positive integers. Following tradition, we write the function values as a_n rather than as $a(n)$, and call them the **terms** of the sequence. Thus, a_1 is the **first term,** a_2 is the **second term,** and, in general, a_n is the **nth term** of the sequence $n \longmapsto a_n$. The symbolism

$$a_1, a_2, a_3, \ldots, a_n, \ldots$$

is often used to denote the sequence $n \longmapsto a_n$; it suggests a never-ending list in which the terms appear in order, with the nth term in the nth position for each positive integer n. In this notation, a set of three dots is read "and so on." The more compact notation $\{a_n\}$, in which the general term is enclosed in braces, is also used to denote the sequence.

To specify a particular sequence, we give a rule by which the nth term a_n is determined. This is often done by means of a formula.

Example Find the first five terms of each sequence:

(a) $a_n = 2n - 1$

(b) $b_n = \dfrac{(-1)^n}{n(n + 1)}$

Solution (a) The first five terms of the sequence are found by substituting the positive integers 1, 2, 3, 4, and 5 for n in the formula for the general term. Thus,

$$a_1 = 2(1) - 1 = 1, \quad a_2 = 2(2) - 1 = 3, \quad a_3 = 2(3) - 1 = 5,$$
$$a_4 = 2(4) - 1 = 7, \quad a_5 = 2(5) - 1 = 9;$$

so the first five terms of the sequence $\{a_n\}$ are

$$1, 3, 5, 7, 9, \ldots.$$

(b) We have

$$b_1 = \frac{(-1)^1}{1(1 + 1)} = -\frac{1}{2}, \quad b_2 = \frac{(-1)^2}{2(2 + 1)} = \frac{1}{6}, \quad b_3 = \frac{(-1)^3}{3(3 + 1)} = -\frac{1}{12},$$
$$b_4 = \frac{(-1)^4}{4(4 + 1)} = \frac{1}{20}, \quad b_5 = \frac{(-1)^5}{5(5 + 1)} = -\frac{1}{30};$$

so the first five terms of the sequence $\{b_n\}$ are

$$-\tfrac{1}{2}, \tfrac{1}{6}, -\tfrac{1}{12}, \tfrac{1}{20}, -\tfrac{1}{30}, \ldots.$$

A formula that relates the general term a_n of a sequence to one or more of the terms that come before it is called a **recursion formula.** A sequence that is specified by giving the first term (or the first few terms) together with a recursion formula is said to be *defined recursively.*

Example   Find the first five terms of the sequence $\{a_n\}$ defined recursively by $a_1 = 1$ and $a_n = na_{n-1}$.

Solution   We are given $a_1 = 1$. By the recursion formula $a_n = na_{n-1}$, we have

$$a_2 = 2a_{2-1} = 2a_1 = 2(1) = 2.$$

Now, using the fact that $a_2 = 2$, we find that

$$a_3 = 3a_{3-1} = 3a_2 = 3(2) = 6.$$

Continuing in this way, we have

$$a_4 = 4a_{4-1} = 4a_3 = 4(6) = 24, \text{ and}$$

$$a_5 = 5a_{5-1} = 5a_4 = 5(24) = 120.$$

Thus, the first five terms of the sequence $\{a_n\}$ are

$$1, 2, 6, 24, 120, \ldots.$$

The pattern here is clear, $a_n = n!$. This can be proved by mathematical induction (Problem 30).

Arithmetic and **geometric** sequences (also called *progressions*) satisfy the following simple recursion formulas:

arithmetic sequence: $a_n = d + a_{n-1}$,	where d is a fixed number called the **common difference**
geometric sequence: $a_n = ra_{n-1}$,	where r is a fixed number known as the **common ratio**

In an arithmetic sequence, each term (after the first) is obtained by adding the fixed number d to the term just before it. Therefore, d is the common difference between successive terms; that is, $d = a_n - a_{n-1}$ for $n = 2, 3, 4, \ldots$. For instance, the sequence

$$5, 8, 11, \ldots, 3n + 2, \ldots$$

with general term $a_n = 3n + 2$ is an arithmetic sequence, and the common difference is

$$d = a_n - a_{n-1} = (3n + 2) - [3(n - 1) + 2]$$
$$= 3n + 2 - 3n + 3 - 2 = 3.$$

If $\{a_n\}$ is an arithmetic sequence with common difference d, then by the recursion formula $a_n = d + a_{n-1}$, we have

$$a_2 = a_1 + d$$
$$a_3 = a_2 + d = (a_1 + d) + d = a_1 + 2d$$
$$a_4 = a_3 + d = (a_1 + 2d) + d = a_1 + 3d,$$

and so on.

Evidently, for every positive integer n,

$$a_n = a_1 + (n - 1)d.$$

This formula for the nth term of an arithmetic sequence can be proved by mathematical induction (Problem 43).

Example   Find the seventeenth term of the arithmetic sequence 2, 5, 8, 11, ...

Solution   Here, the common difference of successive terms is $d = 3$ and the first term is $a_1 = 2$. Using the formula $a_n = a_1 + (n - 1)d$ with $n = 17$, we obtain

$$a_{17} = 2 + (17 - 1)(3) = 50.$$

In a geometric sequence, each term (after the first) is obtained by multiplying the term just before it by r. Therefore, if the terms are nonzero, r is *the common ratio of successive terms*; that is, $r = a_n/a_{n-1}$ for $n = 2, 3, 4, ...$ For instance, the sequence

$$6, 12, 24, . . . , 3(2^n), . . .$$

with general term $a_n = 3(2^n)$ is a geometric sequence and the common ratio is

$$r = \frac{3(2^n)}{3(2^{n-1})} = 2^{n-(n-1)} = 2.$$

If $\{a_n\}$ is a geometric sequence with common ratio r, then by the recursion formula $a_n = ra_{n-1}$, we have

$$a_2 = ra_1$$
$$a_3 = ra_2 = r(ra_1) = r^2a_1$$
$$a_4 = ra_3 = r(r^2a_1) = r^3a_1,$$

and so on. Evidently, for every positive integer n,

$$a_n = r^{n-1}a_1.$$

This formula for the nth term of a geometric sequence can be proved by mathematical induction (Problem 44).

Example   Find the sixth term of the geometric sequence 7, 21, 63, ...

Solution   Here, the common ratio of successive terms is $r = 3$ and the first term is $a_1 = 7$. Using the formula $a_n = r^{n-1}a_1$ with $n = 6$, we obtain

$$a_6 = 3^{6-1}(7)$$
$$= 3^5(7)$$
$$= 1701.$$

PROBLEM SET 6

In problems 1–12, find the first five terms of the sequence with the specified general term.

1. $a_n = \dfrac{1}{n+1}$

2. $b_n = \dfrac{(-1)^n}{n^2}$

3. $c_n = (n+1)^2$

4. $a_n = \dfrac{1}{n(n+1)}$

5. $b_n = (-1)^n n$

6. $c_n = \dfrac{1+(-1)^n}{1+3n}$

7. $a_n = \dfrac{2n-1}{2n+1}$

8. $b_n = \left(-\dfrac{3}{2}\right)^{n-1}$

9. $c_n = \left(\dfrac{1}{3}\right)^n$

10. $a_n = n\cos n\pi$

11. $b_n = \dfrac{n^2-1}{n^2+1}$

12. $a_n = \dfrac{1}{n^2}\sin\dfrac{n\pi}{2}$

In problems 13–18, find the first five terms of the recursively defined sequence $\{a_n\}$.

13. $a_1 = 2$ and $a_n = -3a_{n-1}$

14. $a_1 = 5$ and $a_n = 7 - 2a_{n-1}$

C 15. $a_1 = 2$ and $a_n = (a_{n-1})^{n-1}$

16. $a_1 = 1$, $a_2 = 2$, and $a_n = a_{n-1}a_{n-2}$ for $n \geq 3$

17. $a_1 = 1$ and $a_n = \dfrac{1}{n}a_{n-1}$

18. $a_1 = 0$, $a_2 = 1$, and $a_n = a_{n-1} - a_{n-2}$ for $n \geq 3$

In problems 19–29, use the first boxed formula on page 513 to find the indicated term in each arithmetic sequence.

19. The sixth term of $3, 9, 15, \ldots$

20. The eighth term of $12, 24, 36, \ldots$

21. The fifteenth term of $2, 4, 6, \ldots$

22. The tenth term of $40, 50, 60, \ldots$

23. The twentieth term of $5, 10, 15, \ldots$

24. The twenty-fifth term of $-20, -16, -12, \ldots$

25. The tenth term of $-0.8, 0, 0.8, \ldots$

26. The fifteenth term of $5\sqrt{2}, 7\sqrt{2}, 9\sqrt{2}, \ldots$

27. The eighth term of $m + r, m + 2r, m + 3r, \ldots$

28. The tenth term of $2 + b, 2 + 4b, 2 + 7b, \ldots$

29. The thirteenth term of $\sqrt{5}, 0, -\sqrt{5}, \ldots$

30. If the sequence $\{a_n\}$ has the first term $a_1 = 1$ and satisfies the recursion formula $a_n = na_{n-1}$, use mathematical induction to prove that $a_n = n!$ for all positive integers n.

In problems 31–42, use the second boxed formula on page 513 to find the indicated term in each geometric sequence.

31. The eighth term of $2, 1, \frac{1}{2}, \ldots$

32. The fourth term of $8, 4, 2, \ldots$

33. The sixth term of $\frac{2}{3}, 2, 6, \ldots$

34. The tenth term of $\frac{1}{4}, \frac{1}{8}, \frac{1}{16}, \ldots$

35. The sixteenth term of $2, 4, 8, \ldots$

36. The ninth term of $3, 9, 27, \ldots$

37. The fifth term of $-90, -9, -0.9 \ldots$

38. The seventh term of $\sqrt{2}/2, -1, \sqrt{2}, \ldots$

39. The fifteenth term of $2x, 2x^2, 2x^3, \ldots$

40. The tenth term of $c^{-4}, -c^{-2}, 1, \ldots$

41. The sixth term of $0.3, 0.03, 0.003, \ldots$

42. The fifth term of $0.12, 0.012, 0.0012, \ldots$

43. Use mathematical induction to prove that for every positive integer n, the nth term of an arithmetic sequence $\{a_n\}$ with first term a_1 and common difference d is given by $a_n = a_1 + (n-1)d$.

44. Use mathematical induction to prove that for every positive integer n, the nth term of a geometric sequence $\{a_n\}$ with first term a_1 and common ratio r is given by $a_n = r^{n-1}a_1$.

45. Suppose that the population of a certain city increases at the rate of 5% per year and that the present population is 300,000. (a) If a_n denotes the population of the city n years from now, show that $\{a_n\}$ is a geometric sequence. C (b) Find the population of the city 5 years from now.

46. The **Fibonacci sequence** $\{a_n\}$, which was introduced by Leonardo Fibonacci in 1202, is defined recursively as follows: $a_1 = 1$, $a_2 = 1$, and $a_n = a_{n-1} + a_{n-2}$ for $n \geq 3$. Find the first ten terms of the Fibonacci sequence.

7 Series

In many applications of mathematics, we have to find the sum of the terms of a sequence. Such a sum is called a **series.** Although the sum of the first n terms of a sequence $\{a_n\}$ can be written as

$$a_1 + a_2 + a_3 + \cdots + a_n,$$

a more compact notation is sometimes required. The capital Greek letter Σ (sigma), which corresponds to the letter S, is used for this purpose, and we write the sum in **sigma notation** as

$$\sum_{k=1}^{n} a_k = a_1 + a_2 + a_3 + \cdots + a_n.$$

Here, Σ indicates a sum and k is called the **index of summation.** That the summation begins with $k = 1$ and ends with $k = n$ is indicated by writing $k = 1$ below Σ and n above it. For some purposes, it is useful to allow sequences that begin with a "zeroth term" a_0, and to write

$$\sum_{k=0}^{n} a_k = a_0 + a_1 + a_2 + \cdots + a_n.$$

In any case, it is understood that the summation index runs through all *integer* values starting with the value shown below Σ and ending with the value shown above it.

Examples   Evaluate each sum.

1 $\displaystyle\sum_{k=1}^{4} (2k + 1)$

Solution

$$\sum_{k=1}^{4} (2k + 1) = [2(1) + 1] + [2(2) + 1] + [2(3) + 1] + [2(4) + 1]$$

$$= 3 + 5 + 7 + 9 = 24$$

2 $\displaystyle\sum_{k=0}^{3} \frac{2^k}{4k - 5}$

Solution

$$\sum_{k=0}^{3} \frac{2^k}{4k - 5} = \frac{2^0}{4(0) - 5} + \frac{2^1}{4(1) - 5} + \frac{2^2}{4(2) - 5} + \frac{2^3}{4(3) - 5}$$

$$= \frac{1}{-5} + \frac{2}{-1} + \frac{4}{3} + \frac{8}{7}$$

$$= -\frac{21}{105} - \frac{210}{105} + \frac{140}{105} + \frac{120}{105}$$

$$= \frac{29}{105}$$

If C is a constant, the notation $\sum\limits_{k=1}^{n} C$ is understood to mean the sum of the first n terms of the sequence $C, C, C, \ldots, C, \ldots$. Thus,

$$\sum_{k=1}^{n} C = \overbrace{C + C + C + \cdots + C}^{n \text{ terms}} = nC.$$

Example   Evaluate $\sum\limits_{k=1}^{7} 5$

Solution
$$\sum_{k=1}^{7} 5 = 7(5) = 35.$$

In calculations involving sums in sigma notation, the following properties may be used.

Properties of Summation

If $\{a_n\}$ and $\{b_n\}$ are sequences and C is a constant, then:

(i) $\sum\limits_{k=1}^{n} (a_k + b_k) = \sum\limits_{k=1}^{n} a_k + \sum\limits_{k=1}^{n} b_k$

(ii) $\sum\limits_{k=1}^{n} (a_k - b_k) = \sum\limits_{k=1}^{n} a_k - \sum\limits_{k=1}^{n} b_k$

(iii) $\sum\limits_{k=1}^{n} Ca_k = C \sum\limits_{k=1}^{n} a_k$

(iv) $\sum\limits_{k=1}^{n} C = nC$

Properties (i), (ii), and (iii) can be verified by expanding both sides of the expression or by using mathematical induction. We have already discussed Property (iv). Of course, Properties (i), (ii), and (iii) continue to hold if all summations begin with $k = 0$. In Property (iv), if we begin with $k = 0$, we obtain

$$\sum_{k=0}^{n} C = (n + 1)C$$

(Problem 42).

In addition to Properties (i) through (iv), we have

(v) $\sum\limits_{k=1}^{n} k = \dfrac{n(n + 1)}{2}$

(vi) $\sum\limits_{k=1}^{n} r^{k-1} = \dfrac{1 - r^n}{1 - r} \qquad \text{if } r \neq 0, \quad r \neq 1.$

Property (v), which is a formula for the sum of the first n positive integers, was verified by mathematical induction in Example 1, Section 4, page 501. Property (vi) can also be confirmed by mathematical induction, but the following argument is more interesting: Let $r \neq 0, r \neq 1$, and let $x = \sum_{k=1}^{n} r^{k-1}$. Then

$$x = 1 + r + r^2 + \cdots + r^{n-2} + r^{n-1},$$

so

$$rx = r + r^2 + r^3 + \cdots + r^{n-1} + r^n.$$

If we subtract the second equation from the first, we find that

$$x - rx = 1 - r^n \quad \text{or} \quad (1 - r)x = 1 - r^n.$$

Since $r \neq 1$, the last equation can be rewritten as $x = \dfrac{1 - r^n}{1 - r}$, which is Property (vi).

Using Properties (i)–(vi), we can derive formulas for the sum of the first n terms of an arithmetic or geometric sequence. We begin with the arithmetic case.

Theorem 1   **Sum of the First n Terms of an Arithmetic Sequence**

Let $\{a_n\}$ be an arithmetic sequence with first term a_1 and common difference d. Then, for every positive integer n,

$$\sum_{k=1}^{n} a_k = na_1 + \frac{n}{2}(n - 1)d = \frac{n}{2}(a_1 + a_n).$$

Proof   We have already obtained the formula $a_k = a_1 + (k - 1)d$ for the kth term of $\{a_n\}$ (page 513). Thus,

$$\sum_{k=1}^{n} a_k = \sum_{k=1}^{n} [a_1 + (k - 1)d] = \sum_{k=1}^{n} a_1 + \sum_{k=1}^{n} (k - 1)d \qquad [\text{by Property (i)}]$$

$$= na_1 + d \sum_{k=1}^{n} (k - 1) \qquad [\text{by Properties (iv) and (iii)}]$$

$$= na_1 + d[0 + 1 + 2 + \cdots + (n - 1)]$$

$$= na_1 + d\left[\frac{(n - 1)n}{2}\right] \qquad [\text{by Property (v)}]$$

$$= na_1 + \frac{n}{2}(n - 1)d,$$

which proves the first part of the formula. To obtain the remaining part, we use the fact that $a_n = a_1 + (n - 1)d$. Thus,

$$na_1 + \frac{n}{2}(n - 1)d = \left(\frac{n}{2}a_1 + \frac{n}{2}a_1\right) + \frac{n}{2}(n - 1)d$$

$$= \frac{n}{2}[a_1 + a_1 + (n - 1)d] = \frac{n}{2}(a_1 + a_n).$$

Example 1   Find the sum of the first six terms of the arithmetic sequence 4, 8, 12, 16, 20, 24, ...

Solution   We use the formula

$$\sum_{k=1}^{n} a_k = \frac{n}{2}(a_1 + a_n)$$

of Theorem 1. Here, $n = 6$, $a_1 = 4$, and $a_6 = 24$, so

$$\sum_{k=1}^{6} a_k = \frac{6}{2}(4 + 24) = 84.$$

Example 2   Find the sum of the first twelve terms of the arithmetic sequence 5, 1, -3, -7, ...

Solution   Here $d = -4$, $a_1 = 5$, and $n = 12$. Using the formula

$$\sum_{k=1}^{n} a_k = na_1 + \frac{n}{2}(n - 1)d$$

of Theorem 1, we have

$$\sum_{k=1}^{12} a_k = 12(5) + \frac{12}{2}(12 - 1)(-4) = 60 - 264 = -204.$$

Example 3   A paperboy receives $30 from a newspaper route for the first month. During each succeeding month he earns $2 more than he did the month before. If this pattern continues, how much will he earn over a period of 2 years?

Solution   The paperboy's monthly income forms an arithmetic sequence with $a_1 = 30$ dollars and $d = 2$ dollars. His total income over a 2-year period is given by the formula

$$\sum_{k=1}^{n} a_k = na_1 + \frac{n}{2}(n - 1)d$$

with $n = 24$. We have

$$\sum_{k=1}^{24} a_k = 24(30) + \frac{24}{2}(24 - 1)(2) = \$1,272.$$

Now, we turn our attention to the geometric case.

Theorem 2   **Sum of the First n Terms of a Geometric Sequence**

Let $\{a_n\}$ be a geometric sequence with first term a_1 and common ratio $r \neq 0$, $r \neq 1$. Then, for every positive integer n,

$$\sum_{k=1}^{n} a_k = \sum_{k=1}^{n} a_1 r^{k-1} = a_1 \frac{1 - r^n}{1 - r}.$$

Proof   We have already obtained the formula $a_k = r^{k-1}a_1 = a_1 r^{k-1}$ for the kth term of a geometric sequence with common ratio r (page 513). Therefore,

$$\sum_{k=1}^{n} a_k = \sum_{k=1}^{n} a_1 r^{k-1} = a_1 \sum_{k=1}^{n} r^{k-1} \qquad \text{[by Property (iii)]}$$

$$= a_1 \frac{1-r^n}{1-r} \qquad \text{[by Property (vi)]},$$

and the proof is complete.

Example 1   Find the sum of the first eight terms of the geometric sequence $3, -6, 12, -24, \ldots$.

Solution   We use the formula of Theorem 2, with $a_1 = 3$, $r = -2$, and $n = 8$. Thus,

$$\sum_{k=1}^{n} a_k = a_1 \frac{1-r^n}{1-r} = (3)\frac{1-(-2)^8}{1-(-2)} = (3)\frac{1-256}{3} = -255.$$

Example 2   © For doing a certain job, you are offered 1¢ the first day, 3¢ the second day, 9¢ the third day, and so forth, so that your daily earnings form a geometric sequence with first term $a_1 = 1$ cent and common ratio $r = 3$. How much will you earn in 14 days of work?

Solution   By Theorem 2,

$$\sum_{k=1}^{n} a_1 r^{k-1} = \sum_{k=1}^{14} 3^{k-1} = \frac{1-3^{14}}{1-3}$$

$$= \frac{1-3^{14}}{-2} = \frac{3^{14}-1}{2}.$$

Using a calculator, we find that the total earnings amount to 2,391,484 cents or $23,914.84. (Not bad for two weeks' work!)

7.1 Infinite Geometric Series

An indicated sum of all the terms of a sequence $\{a_n\}$, such as

$$a_1 + a_2 + a_3 + \cdots + a_n + \cdots$$

is called an **infinite series** or simply a **series.** Using the sigma notation, we can write such a series more compactly as

$$\sum_{k=1}^{\infty} a_k.$$

Although we cannot literally add an infinite number of terms, it is sometimes useful to assign a numerical value as the "sum" of an infinite series. This is accomplished by using the idea of a "limit," which is studied in calculus.

Although the general idea of the "sum" of an infinite series is beyond the scope of this textbook, you can easily get some feeling for this concept by considering an **infinite geometric series** of the form

$$\sum_{k=1}^{\infty} a_1 r^{k-1} = a_1 + a_1 r + a_1 r^2 + \cdots + a_1 r^{n-1} + \cdots$$

for the case in which $0 < |r| < 1$. By Theorem 2, the sum of the *first n terms* of this series is

$$\sum_{k=1}^{n} a_1 r^{k-1} = a_1 + a_1 r + \cdots + a_1 r^{n-1} = a_1 \frac{1 - r^n}{1 - r}.$$

Because $0 < |r| < 1$, it follows that r^n gets smaller and smaller as n gets larger and larger. (Try it on a calculator, say for $r = 0.75$ with $n = 10$, $n = 50$, $n = 100$, and so forth.) Therefore, as n gets larger and larger, r^n gets closer and closer to 0 and

$$a_1 \frac{1 - r^n}{1 - r} \quad \text{comes closer and closer to} \quad a_1 \frac{1 - 0}{1 - r} = \frac{a_1}{1 - r}.$$

In other words, as you add more and more terms of the geometric series, the sum comes closer and closer to $a_1/(1 - r)$. This suggests the following definition.

Definition 1   **Sum of an Infinite Geometric Series**

If $|r| < 1$, we define

$$\sum_{k=1}^{\infty} a_1 r^{k-1} = a_1 + a_1 r + a_1 r^2 + \cdots + a_1 r^{n-1} + \cdots = \frac{a_1}{1 - r}.$$

Examples   Find the sum of each infinite geometric series.

1 $\dfrac{1}{2} + \dfrac{1}{4} + \dfrac{1}{8} + \cdots$

Solution   Here $a_1 = \frac{1}{2}$ and $r = \frac{1}{2}$, so by Definition 1,

$$\frac{1}{2} + \frac{1}{4} + \frac{1}{8} + \cdots = \frac{a_1}{1 - r} = \frac{\frac{1}{2}}{1 - \frac{1}{2}} = \frac{\frac{1}{2}}{\frac{1}{2}} = 1.$$

2 $\displaystyle\sum_{k=1}^{\infty} 4\left(-\frac{1}{3}\right)^{k-1}$

Solution   Here $a_1 = 4$ and $r = -\frac{1}{3}$, so by Definition 1,

$$\sum_{k=1}^{\infty} 4\left(-\frac{1}{3}\right)^{k-1} = \frac{a_1}{1 - r}$$

$$= \frac{4}{1 - (-\frac{1}{3})} = \frac{4}{1 + \frac{1}{3}} = \frac{4}{\frac{4}{3}} = 3.$$

When we write an infinite decimal

$$x = 0.d_1 d_2 d_3 d_4 \cdots d_n \cdots$$

we are actually forming the sum of an infinite series

$$x = \frac{d_1}{10} + \frac{d_2}{100} + \frac{d_3}{1000} + \frac{d_4}{10,000} + \cdots + \frac{d_n}{10^n} + \cdots.$$

Thus, you can use the formula in Definition 1 to convert a repeating decimal to a quotient of integers.

Example   Rewrite the repeating decimal $0.\overline{31}$ as a quotient of integers.

Solution $0.\overline{31} = 0.31313131 \cdots = 0.31 + 0.0031 + 0.000031 + 0.00000031 + \cdots,$

which is an infinite geometric series with first term $a_1 = 0.31$ and common ratio $r = 0.01$. Hence, by Definition 1,

$$0.\overline{31} = \frac{a_1}{1 - r} = \frac{0.31}{1 - 0.01} = \frac{0.31}{0.99} = \frac{31}{99}.$$

(Note that the same answer is obtained by using the method in Chapter 2, page 65.)

PROBLEM SET 7

In problems 1–6, write out and evaluate each sum.

1. $\displaystyle\sum_{k=1}^{4} (3k - 2)$

2. $\displaystyle\sum_{k=1}^{5} [2 + (-1)^k]$

3. $\displaystyle\sum_{k=1}^{5} \left(\frac{1}{4}k + 3\right)$

4. $\displaystyle\sum_{k=2}^{6} (-1)^k 3^k$

5. $\displaystyle\sum_{k=0}^{5} \frac{2^k}{3k + 1}$

6. $\displaystyle\sum_{j=0}^{4} \frac{5^j}{2j - 1}$

In problems 7–12, use Properties (i)–(vi) of summation to evaluate each expression.

7. $\displaystyle\sum_{k=1}^{40} 3k$

8. $\displaystyle\sum_{k=1}^{30} (3 - 4k)$

9. $\displaystyle\sum_{k=1}^{24} (2^k + 1)$

10. $\displaystyle\sum_{k=0}^{19} 2^{-k}$

11. $\displaystyle\sum_{k=1}^{100} \frac{1}{10^k}$

12. $\displaystyle\sum_{j=0}^{100} (5^{j+1} - 5^j)$

In problems 13–24, find the sum of the first n terms of each arithmetic sequence for the given value of n.

13. $1, 2, 3, \ldots$ for $n = 10$

14. $5, 9, 13, \ldots$ for $n = 13$

15. $7, 10, 13, \ldots$ for $n = 5$

16. $14, 19, 24, \ldots$ for $n = 21$

17. $36, 48, 60, \ldots$ for $n = 8$

18. $27, 33, 39, \ldots$ for $n = 30$

19. $2, -2, -6, \ldots$ for $n = 15$

20. $0.6, 0.4, 0.2, \ldots$ for $n = 16$

21. $-3t, -5t, -7t, \ldots$ for $n = 7$

22. $-2y, 0, 2y, \ldots$ for $n = 10$

23. $x, 0, -x, \ldots$ for $n = 10$

24. $5k - 4, 3k - 3, \ldots$ for $n = 50$

25. A company had sales of \$200,000 during its first year of operation, and sales increased by \$30,000 per year during each successive year. Find the total sales of the company during its first 8 years.

26. A freely falling object dropped from rest falls $\frac{1}{2}g(2k - 1)$ units of distance during the kth second of its fall, where g is the acceleration of gravity. Find a formula for the total distance through which the body falls in n seconds.

27. A display of canned soup in a supermarket is to have the form of a pyramid with 20 cans in the bottom row, 19 cans in the next row, 18 cans in the next row, and so on, with a single can at the top. How many cans of soup will be required for the display?

28. If $\{a_n\}$ is an arithmetic sequence and if p and q are positive integers, show that

$$\sum_{k=p}^{p+q} a_k = \frac{q+1}{2}(a_p + a_{p+q}).$$

In problems 29–38, find the sum of the first n terms of each geometric sequence for the given value of n.

29. $4, 40, 400, \ldots$ for $n = 5$

30. $-3, 15, -75, \ldots$ for $n = 6$

31. $-\frac{1}{3}, -\frac{1}{9}, -\frac{1}{27}, \ldots$ for $n = 10$

32. $-\frac{1}{16}, -\frac{1}{8}, -\frac{1}{4}, \ldots$ for $n = 8$

33. $5^{10}, 5^8, 5^6, \ldots$ for $n = 9$

34. $1, -1, 1, \ldots$ for $n = 10$

35. $100, 100, 100, \ldots$ for $n = 25$

36. $0.3, 0.03, 0.003, \ldots$ for $n = 6$

37. $c^4, c^6, c^8, \ldots$ for $n = 7$

38. $1, \frac{1}{10}, \frac{1}{100}, \ldots$ for $n = 15$

C 39. Suppose that the distance traveled by a point on a pendulum in any swing is 5% less than in the previous swing. If the point travels 6 centimeters on the first swing, find the total distance traveled by the point in 7 swings.

40. If $\{a_n\}$ is a geometric sequence with common ratio r, show that $\sum_{k=1}^{n} a_k = \frac{a_1 - a_n r}{1 - r}$.

41. A ball is dropped from a height of 256 centimeters, and on each rebound it rises to $\frac{1}{2}$ the height from which it last fell. Find a formula for the total distance traveled by the ball when it hits the floor for the nth time.

42. Justify the formula $\sum_{k=0}^{n} C = (n + 1)C$. [Hint: Use $\sum_{k=0}^{n} a_k$ where $a_k = C$ for all values of k.]

In problems 43–50, find the sum of each infinite geometric series.

43. $\sum_{k=1}^{\infty} \left(\frac{2}{3}\right)^{k-1}$

44. $\sum_{k=1}^{\infty} \left(-\frac{1}{3}\right)^{k-1}$

45. $\sum_{k=1}^{\infty} \left(-\frac{2}{5}\right)^{k-1}$

46. $\sum_{k=1}^{\infty} \left(\frac{3}{7}\right)^{k-1}$

47. $\sum_{k=1}^{\infty} \left(-\frac{1}{6}\right)^{k-1}$

48. $\sum_{k=0}^{\infty} \left(-\frac{2}{9}\right)^{k}$

49. $\sum_{k=1}^{\infty} 5\left(-\frac{3}{5}\right)^{k-1}$

50. $\sum_{k=0}^{\infty} ar^k$ for $|r| < 1$

In problems 51–56, use the formula for the sum of an infinite geometric series to rewrite each infinite repeating decimal as a quotient of integers.

51. $0.\overline{4}$

52. $4.\overline{53}$

53. $7.\overline{27}$

54. $0.2\overline{79}$

55. $0.02\overline{34}$

56. $-1.9\overline{81}$

57. True or false: $0.\overline{9} < 1$. Explain your answer.

58. In the discussion leading up to Definition 1, it was assumed that $0 < |r| < 1$; however, in Definition 1, we only require that $|r| < 1$. Explain.

---

8 Permutations and Combinations

It is often necessary to calculate the number of different ways in which something can be done or can happen. For instance, in order to find the odds of being dealt a full house in a game of five-card poker, it is necessary to calculate the total number of different possible poker hands. In this section we discuss some of the methods used in making such calculations.

We begin by stating the additive and multiplicative principles of counting, which are the basic tools for the calculations made in this section.

The Additive Principle of Counting

> Let A and B be two events that cannot occur simultaneously. Then, if A can occur in a ways and B can occur in b ways, it follows that the number of ways in which either A or B can occur is $a + b$.

The additive principle of counting can be extended to three or more events in an obvious way.

Example

Suppose you have six signal flags—two red, two green, and two blue flags. By displaying a single flag, you can send three different signals. By displaying two flags, one above the other, you can send nine different signals. How many different signals can you send using one or two flags?

Solution

Let A be the event that a one-flag signal is sent, so that A can occur in $a = 3$ ways. Let B be the event that a two-flag signal is sent, so that B can occur in $b = 9$ ways. Since a one-flag signal and a two-flag signal can't be sent at the same time, the additive principle of counting applies, and $a + b = 3 + 9 = 12$ different signals can be sent using either one or two flags.

The Multiplicative Principle of Counting

> If a first event A can occur in a ways and, independently, a second event B can occur in b ways, then the number of ways in which both A and B can occur is ab.

The multiplicative principle of counting can be extended to three or more events in an obvious way.

Example

A restaurant offers a meal consisting of one of three entrees and one of four desserts at a special price. How many different meals can be ordered at this price?

Solution

By the multiplicative principle of counting, $3 \cdot 4 = 12$ different meals can be ordered.

The idea of a *permutation*, introduced in the following definition, is useful in many counting problems.

Definition 1

Permutations

> A **permutation** is an ordered arrangement of distinct objects in a row. Let S be a set of n distinct elements. An ordered arrangement of r of these elements in a row is called a *permutation of size r chosen from the n elements of S*. The symbol $_nP_r$ denotes the number of different permutations of size r chosen from a set of n elements.

Example   Let $S = \{a, b, c, d\}$. List all permutations of size 2 chosen from the 4 elements of S, and count these permutations to find $_4P_2$.

Solution   We list the permutations according to which element is in the first position:

a first	b first	c first	d first
ab	ba	ca	da
ac	bc	cb	db
ad	bd	cd	dc

Thus, $_4P_2 = 12$.

Theorem 1   **A Formula for $_nP_r$**

> If n and r are positive integers with $n \geq r$, then
> $$_nP_r = \overbrace{n(n-1)(n-2)\cdots(n-r+1)}^{r \text{ factors}}$$
> $$= \frac{n!}{(n-r)!}.$$

Proof   In forming a permutation of r elements chosen from a set of n elements, we have n choices for the element to be placed in the first position in the row. Once this choice is made, there remain $n-1$ elements to be placed in the remaining $r-1$ positions. Therefore, we have $n-1$ choices for the element to be placed in the second position. By the multiplicative principle of counting, there are $n(n-1)$ ways to choose distinct elements to place in the first two positions. Similarly, there are $n(n-1)(n-2)$ ways to choose distinct elements to place in the first three positions. Continuing in this way, we see that all r positions in the row can be filled with distinct elements in $n(n-1)(n-2)\cdots[n-(r-1)]$ ways. Therefore,

$$_nP_r = \overbrace{n(n-1)(n-2)\cdots(n-r+1)}^{r \text{ factors}}$$
$$= \frac{n(n-1)(n-2)\cdots(n-r+1)(n-r)!}{(n-r)!}$$
$$= \frac{n!}{(n-r)!}.$$

Example 1   Evaluate

(a) $_7P_1$ (b) $_5P_3$ (c) $_4P_4$ (d) $_9P_4$.

Solution   Using the formula $_nP_r = \overbrace{n(n-1)(n-2)\cdots(n-r+1)}^{r \text{ factors}}$, we have:

(a) $_7P_1 = 7$

(b) $_5P_3 = 5 \cdot 4 \cdot 3 = 60$

(c) $_4P_4 = 4 \cdot 3 \cdot 2 \cdot 1 = 24$

(d) $_9P_4 = 9 \cdot 8 \cdot 7 \cdot 6 = 3024$

Example 2 A portfolio manager knows 13 stocks that meet her investment criteria. She will choose three of these and rank them first, second, and third in order of preference. In how many ways can she do this?

Solution $_{13}P_3 = 13 \cdot 12 \cdot 11 = 1716$ ways.

Suppose that we have n distinct objects that are to be arranged in a definite order. By Theorem 1, this can be done in

$$_nP_n = n(n-1)(n-2)\cdots 1 = n!$$

different ways.

Example © In how many ways can 12 different books be arranged on a shelf?

Solution Using a calculator, we find that there are

$$12! = 479{,}001{,}600$$

different arrangements.

The value of a poker hand doesn't depend on the order in which the cards are arranged in the hand, but only on which cards are present. Thus, the idea of a permutation, which involves *order*, isn't the right tool for counting poker hands. Such situations, in which order isn't important, suggest the following definition.

Definition 2 **Combinations**

Let S be a set consisting of n distinct elements. A subset consisting of r of these elements is called a **combination of size r chosen from the n elements in S.** No importance is attached to the order of the r chosen elements. The symbol $_nC_r$ denotes the number of different combinations of size r chosen from a set of n elements.

Example Let $S = \{a, b, c, d\}$. List all combinations of size 2 chosen from the 4 elements of S, and count these combinations to find $_4C_2$.

Solution The subsets of S that consist of two elements are

$$\{a,b\}, \{a,c\}, \{a,d\}, \{b,c\}, \{b,d\}, \text{ and } \{c,d\}.$$

[Because order isn't important, the combination $\{a,b\}$ is the same as the combination $\{b,a\}$, the combination $\{a,c\}$ is the same as the combination $\{c,a\}$, and so on.] Thus, $_4C_2 = 6$.

There is a simple relationship between permutations and combinations. To obtain this relationship, we consider a set S consisting of n distinct elements. To specify a permutation of size r chosen from these n elements, we can first select the r elements that will appear in the permutation, and then we can give the order in which the selected elements are to be arranged. The first step, the selection of a combination of r elements from S, can be done in $_nC_r$ ways; the second step, the ordering of these r elements, can be accomplished in $r!$ ways. Therefore, by the multiplicative principle of counting,

$$_nP_r = (_nC_r)r!.$$

Using this relationship, we can prove the following theorem.

Theorem 2 **A Formula for $_nC_r$**

> If n and r are integers with $n \geq r \geq 0$, then
>
> $$_nC_r = \frac{\overbrace{n(n-1)(n-2)\cdots(n-r+1)}^{r \text{ factors}}}{\underbrace{r(r-1)(r-2)\cdots 1}_{r \text{ factors}}} = \frac{n!}{(n-r)!\,r!}.$$

Proof By the relationship established above and Theorem 1, we have

$$_nC_r = \frac{_nP_r}{r!} = \frac{n(n-1)(n-2)\cdots(n-r+1)}{r(r-1)(r-2)\cdots 1}.$$

Since $_nP_r = \dfrac{n!}{(n-r)!}$, we can also write

$$_nC_r = \frac{_nP_r}{r!} = \frac{n!}{(n-r)!\,r!}.$$

Notice that $_nC_r$, the number of combinations of size r chosen from a set of n elements, is the same as the binomial coefficient $\binom{n}{r}$ (Definition 2, page 505).

Example 1 Evaluate

(a) $_7C_3$ (b) $_7C_7$ (c) $_7C_1$.

Solution Using the formula

$$_nC_r = \frac{n(n-1)(n-2)\cdots(n-r+1)}{r(r-1)(r-2)\cdots 1}$$

in Theorem 2, we have

(a) $_7C_3 = \dfrac{7\cdot 6\cdot 5}{3\cdot 2\cdot 1} = 35$ (b) $_7C_7 = \dfrac{7!}{7!} = 1$ (c) $_7C_1 = \dfrac{7}{1} = 7.$

Example 2   c Find the number of different committees of four students each that can be formed from a group of 36 students.

Solution   Here the set S consists of the 36 students, so $n = 36$. A committee of 4 of these students is a subset of S consisting of $r = 4$ elements. The order of students within a committee is not important, so each committee is a combination of size $r = 4$ chosen from the $n = 36$ elements of S. The number of such committees is

$$_{36}C_4 = \frac{36 \cdot 35 \cdot 34 \cdot 33}{4 \cdot 3 \cdot 2 \cdot 1} = \frac{\overset{3}{\cancel{36}} \cdot 35 \cdot 34 \cdot 33}{\cancel{4 \cdot 3} \cdot 2} = \frac{3 \cdot 35 \cdot \overset{17}{\cancel{34}} \cdot 33}{\cancel{2}}$$

$$= 3 \cdot 35 \cdot 17 \cdot 33 = 58{,}905.$$

Example 3   c Find the number of different hands in five-card poker.

Solution   There are $n = 52$ cards in the deck, and the order of the cards in a hand isn't important. Thus, the number of different poker hands is given by

$$_{52}C_5 = \frac{52 \cdot 51 \cdot \overset{10}{\cancel{50}} \cdot 49 \cdot 48}{\cancel{5} \cdot 4 \cdot 3 \cdot 2 \cdot 1} = \frac{\overset{13}{\cancel{52}} \cdot 51 \cdot 10 \cdot 49 \cdot 48}{4 \cdot 3 \cdot 2} = \frac{13 \cdot 51 \cdot 10 \cdot 49 \cdot \overset{8}{\cancel{48}}}{\cancel{3 \cdot 2}}$$

$$= 13 \cdot 51 \cdot 10 \cdot 49 \cdot 8 = 2{,}598{,}960.$$

Suppose that we have r red balls, w white balls, and b black balls, and that balls within a color group are indistinguishable from one another. Altogether, we have $n = r + w + b$ balls. In how many **distinguishable ways** can we arrange these n balls in an ordered row; that is, how many **distinguishable permutations** are there of the n colored balls?

To specify a distinguishable arrangement of the balls in an ordered row, we first decide which of the n positions in the row will be occupied by the red balls. This amounts to specifying a subset of r positions among the n available positions, and it can be done in $_nC_r$ ways. Having made this decision, we place the red balls in the specified positions. There remain $n - r$ positions to be filled with the remaining white and black balls. We choose w of these positions for the white balls. This choice can be made in $_{n-r}C_w$ ways. The remaining $n - r - w$ positions will have to be filled with black balls—here we have no choice. Therefore, by the multiplicative principle of counting, we can specify $(_nC_r)(_{n-r}C_w)$ distinguishable ordered arrangements or permutations of the colored balls.

Using the formula in Theorem 2 and the fact that $n = r + w + b$, so that $n - r - w = b$, we have

$$(_nC_r)(_{n-r}C_w) = \frac{n!}{(n-r)!r!} \cdot \frac{(n-r)!}{[(n-r)-w]!w!}$$

$$= \frac{n!\cancel{(n-r)!}}{\cancel{(n-r)!}r!b!w!} = \frac{n!}{r!w!b!}.$$

This formula for the number of distinguishable permutations of the colored balls can be generalized as follows.

Theorem 3   **Distinguishable Permutations**

Consider a set of n objects of which n_1 are alike of one kind, n_2 are alike of another kind, ..., and n_k are alike of a last kind. Then, the number of distinguishable permutations of the objects is

$$\frac{n!}{n_1!n_2!\cdots n_k!}.$$

Example   How many distinguishable permutations are there of the letters in the word CINCINNATI?

Solution   There are $n = 10$ letters altogether. Of these, there are $n_1 = 2$ C's, $n_2 = 3$ I's, $n_3 = 3$ N's, $n_4 = 1$ A, and $n_5 = 1$ T. By Theorem 3, the number of distinguishable permutations of these letters is given by

$$\frac{n!}{n_1!n_2!n_3!n_4!n_5!} = \frac{10!}{2!3!3!1!1!} = 50,400.$$

PROBLEM SET 8

1. There are six ways of rolling a "7" with two dice, and there are two ways of rolling an "11" with two dice. In how many ways can you roll "7 or 11"?

2. There are three different roads connecting Newberry with New Mattoon, and four different roads connecting New Mattoon with Gainesburg. In how many different ways can a person drive from Newberry to Gainesburg, passing through New Mattoon on the way?

3. How many different two-letter "words" can be formed using the 26 letters of the alphabet if repeated letters are allowed? [The "words" need not be in a dictionary.]

4. How many different three-letter "words" can be formed using the 26 letters of the alphabet if repeated letters are allowed, but if the middle letter must be a, e, i, o, or u?

5. How many positive three-digit integers less than 500 can be formed using only the digits 1, 3, 5, and 7 if repetitions are allowed?

6. A combination lock for a bicycle has three levers, each of which can be set in nine positions. A thief attempting to steal the bicycle begins trying all possible arrangements of the levers. If the thief tries two arrangements every second, what is the maximum time required to open the lock?

7. List and count all permutations of size 2 chosen from the elements of $S = \{a,b,c,d,e\}$, and thus determine $_5P_2$.

8. List and count all permutations of size 3 chosen from the elements of the set $S = \{a,b,c,d\}$, and thus determine $_4P_3$.

In problems 9–16, find the value of the expression by using Theorem 1.

9. $_4P_3$	10. $_8P_6$	ⓒ 11. $_9P_9$	12. $_{12}P_1$
13. $_5P_2$	ⓒ 14. $_{13}P_{13}$	15. $_{11}P_3$	16. $_8P_2$

ⓒ 17. In how many ways can five of ten books be chosen and arranged next to each other on a shelf?

18. In how many ways can a left end, a right end, and a center for a football team be picked from among Dean, Dolores, Carlos, Carmine, Gus, and Olga?

$\boxed{c}$ **19.** In how many ways can seven people line up at a ticket window?

20. How many signals can be sent by using four distinguishable flags, one above the other, on a flag pole, if the flags can be used one, two, three, or four at a time?

$\boxed{c}$ **21.** In how many ways can a baseball team of nine players be arranged in batting order if tradition is followed and the pitcher must bat last, but if there are no other restrictions?

22. A poll consists of five questions. In how many different orders can these questions be asked? If ten runners are entered in a race for which first, second, and third prizes will be awarded, in how many different ways can the prizes be distributed? (Assume that there are no ties.)

24. How many integers that do not contain repeated digits are there between 1 and 1000, inclusive?

25. List and count all combinations of size 3 chosen from the elements of the set $S = \{a, b, c, d\}$, and thus determine $_4C_3$.

26. List and count all combinations of size 2 chosen from the elements of $S = \{a, b, c, d, e\}$, and thus determine $_5C_2$.

In problems 27–35, find the value of the expression by using Theorem 2.

27. $_4C_4$ **28.** $_{48}C_3$ **29.** $_5C_2$ $\boxed{c}$ **30.** $_{52}C_{13}$ **31.** $_6C_3$

32. $_{10}C_{10}$ **33.** $_{10}C_9$ **34.** $_{10}C_1$ **35.** $_{10}C_0$

36. Show that $_nC_0 = 1$ holds for every integer $n \geq 0$ and give an interpretation of this fact.

37. A total of how many games will be played by eight teams in a league if each team plays the other teams just once?

38. A dealer has 15 different models of television sets and wishes to display three models in the store window. Disregarding the arrangement of the three sets in the window, in how many ways can this be done?

39. An election is to be held to choose four delegates from among ten nominees to represent a district at a political convention. In how many ways can the delegation be formed?

40. A department store plans to fill ten positions with four men and six women. In how many ways can these positions be filled if there are nine men and eleven women applicants?

$\boxed{c}$ In problems 41–46, determine the number of distinguishable permutations of the letters in each word.

41. OHIO   **42.** MASSACHUSETTS   **43.** REARRANGEMENT
44. MISSISSIPPI   **45.** TENNESSEE   **46.** COMMITTEE

47. In how many distinguishable ways can eight people be arranged in a police lineup if two of them are identical twins and three are identical triplets?

48. There are three copies of a chemistry book, two copies of a biology book, and four copies of a sociology book to be placed on a shelf. In how many distinguishable ways can this be done?

9   Probability

The mathematical theory of probability has applications in nearly every area of human activity. These applications range from medical statistics to actuarial science, from statistical physics to law, and from decision making in the business world to gambling in Las Vegas. Here we can only touch the subject in the most superficial way and we can give only the simplest examples, many of which involve playing cards or dice.

Consider an experiment that has a finite number of **equally likely outcomes.** For instance, if you roll a balanced die (a small cube with different numbers of spots on its faces), the possible outcomes are 1, 2, 3, 4, 5, or 6, and they are equally likely. An **event** is something that may or may not occur, depending on which outcome is obtained. For instance, if you roll a die, the event "rolling an odd number" occurs if the outcome is 1, 3, or 5, and fails to occur if the outcome is 2, 4, or 6. The outcomes for which an event occurs are said to be *favorable* to the event. For instance, the outcomes 1, 3, and 5 are favorable to the event "rolling an odd number."

Definition 1   **The Probability of an Event**

> Let A be an event associated with an experiment with N equally likely outcomes. If $n(A)$ is the number of outcomes favorable to A, the **probability** of A, in symbols $P(A)$, is defined by
>
> $$P(A) = \frac{n(A)}{N}.$$

The number $P(A)$ in Definition 1 is the fraction of all possible outcomes that are favorable to A. Hence, $P(A)$ can be regarded as a numerical measure of the likelihood, on a scale from 0 to 1, that A will occur if the experiment is performed.

Example 1   Find the probability of rolling an odd number with a balanced die.

Solution   In Definition 1, let A denote the event "rolling an odd number." When a balanced die is rolled, there are $N = 6$ equally likely outcomes; $n(A) = 3$ of these outcomes are favorable to the event A, so

$$P(A) = \frac{n(A)}{N} = \frac{3}{6} = \frac{1}{2}.$$

We can also express $P(A)$ in decimal form as $P(A) = 0.5$, or as $P(A) = 50\%$.

Example 2   Find the probability of drawing a face card (jack, queen, or king) from a well-shuffled deck of 52 cards.

Solution   Here there are $N = 52$ possible outcomes of the experiment of drawing one card from the deck. Since the deck is well-shuffled, these outcomes are all equally likely. Of the 52 cards, there are 4 jacks, 4 queens, and 4 kings; hence, the number of face cards is $4 + 4 + 4 = 12$. Thus,

$$P(\text{"drawing a face card"}) = \frac{12}{52} = \frac{3}{13}.$$

Example 3   From a group of nine people, five male and four female, a committee of three is to be selected by chance. What is the probability that no females are on the committee?

Solution   Here the "experiment" is the selection of a committee and the outcome is the selected committee. There are

$$N = {}_9C_3 = 84$$

ways of selecting a committee of 3 from 9 people. Since the committee is to be selected by chance, all 84 of these outcomes are equally likely. The event in question, "no females on the committee," occurs if and only if the committee consists only of males. There are

$$n = {}_5C_3 = 10$$

ways of selecting a committee consisting entirely of males from among the 5 available. It follows that

$$P(\text{"no females on committee"}) = \frac{n}{N} = \frac{10}{84} \approx 11.9\%.$$

Example 4   What is the probability of being dealt a full house (three cards of one denomination—say, three 6's or three kings—and two cards of another denomination—say, two aces or two 9's) from a well-shuffled deck in 5-card poker?

Solution   Here an outcome is a 5-card hand. There are $N = {}_{52}C_5$ such hands, all of which are equally likely. Of these, we must calculate the number n of full houses. This is best done by asking ourselves in how many ways we could form a 5-card hand with 3 cards of one denomination and 2 cards of another. The common denomination of the 3 cards can be chosen in 13 ways, and the 3 cards themselves can be chosen from among the 4 cards of this denomination in ${}_4C_3$ ways. The common denomination of the other 2 cards can then be chosen in any of the remaining 12 ways, and the 2 cards themselves can be chosen from among the 4 cards of this denomination in ${}_4C_2$ ways. By the multiplicative principle of counting, we can therefore form full houses in

$$n = \frac{(13)({}_4C_3)(12)({}_4C_2)}{{}_{52}C_5}$$

different ways. It follows that

$$P(\text{"full house"}) = \frac{n}{N}$$

$$= \frac{(13)({}_4C_3)(12)({}_4C_2)}{{}_{52}C_5}$$

$$= \frac{(13)\left(\frac{4 \cdot 3 \cdot 2}{3 \cdot 2 \cdot 1}\right)(12)\left(\frac{4 \cdot 3}{2 \cdot 1}\right)}{\frac{52 \cdot 51 \cdot 50 \cdot 49 \cdot 48}{5 \cdot 4 \cdot 3 \cdot 2 \cdot 1}}$$

$$= \frac{13 \cdot 4 \cdot 12 \cdot 2 \cdot 3}{52 \cdot 51 \cdot 10 \cdot 49 \cdot 2}$$

$$= \frac{6}{4165}.$$

Example 5   What is the probability of rolling a "7" with two dice?

Solution   Call one of the dice the *first die* and the other the *second die*. An outcome of a toss of the two dice can be denoted by an ordered pair (x, y), where x is the number on the first die and y the number on the second. Here x can have any of 6 possible values and y can have any of 6 possible values, so, by the multiplicative principle of counting, there are $N = 6 \cdot 6 = 36$ possible outcomes. Of these, the outcomes favorable to the event "rolling a 7" are

$$(1,6), \quad (2,5), \quad (3,4), \quad (4,3), \quad (5,2), \quad \text{and} \quad (6,1).$$

Thus, there are $n = 6$ outcomes favorable to this event. It follows that

$$P(\text{"rolling a 7"}) = \frac{n}{N} = \frac{6}{36} = \frac{1}{6}.$$

PROBLEM SET 9

In problems 1–14, find the probability of the event.

1. Obtaining "heads" when tossing a coin.
2. Drawing a king from a deck of 52 well-shuffled cards.
3. Drawing a green ball, blindfolded, from a hat containing six red balls and eight green balls, all of the same size.
4. Failing to draw one of the kings from a deck of 52 well-shuffled cards.
5. Rolling a "5" with a single die.
6. Rolling less than a "3" with a single die.
7. [c] Being dealt four of a kind (four cards of the same denomination) in a game of five-card poker.
8. [c] Being dealt a flush (all cards of the same suit) in a game of five-card poker.
9. [c] Being dealt a royal flush (ten, jack, queen, king, and ace, all of the same suit) in a game of five-card poker.
10. [c] Being dealt a straight (five cards whose denominations form a sequence such as $7, 8, 9, 10$, jack) in a game of five-card poker. (The ace can count either as a 1 or as the denomination just above the king.)
11. Rolling a "6" with a pair of dice.
12. Rolling snake eyes ("2") with a pair of dice.
13. Rolling at least one "6" in two rolls of a single die.
14. Two people, chosen at random, having birthdays on the same day of the year. (Disregard leap year complications.)

15. A committee of five people is to be chosen at random from among ten men and three women. What is the probability that there will be three men and two women on the committee?
16. A committee of seven people is to be chosen at random from among five skilled workers, three unskilled workers, and four supervisory personnel. What is the probability that all four supervisory personnel and no unskilled workers are on the committee?
17. Two coins are tossed and you are told that at least one coin fell "heads." What is the probability that the other coin fell "heads"? [Be careful—consider the possible outcomes in the face of the given information and how many of these are favorable to the event in question.]

18. Four Eastern states have daily lotteries in which numbers from 0 to 999 are drawn. What is the probability that two (or more) of these states draw the same lucky number on the same day? [Hint: To count the number of outcomes favorable to the event, begin by counting the number of outcomes *unfavorable* to the event.]

19. Cards are drawn one at a time from a well-shuffled deck. They are not replaced after being drawn. What is the probability that the last ace in the deck is drawn before the last king in the deck?

20. Two cards are drawn from a well-shuffled deck. The first card is *not* replaced before the second card is drawn. If the second card is a face card, what is the probability that the first card was a face card?

10 The Conic Sections

If a cone with two nappes (Figure 1) is cut by a plane that does not pass through its vertex, the resulting curve of intersection will be a **circle** (Figure 1a), an **ellipse** (Figure 1b), a **parabola** (Figure 1c) or a **hyperbola** (Figure 1d). Accordingly, these curves are called **conic sections.**

Figure 1

circle	ellipse	parabola	hyperbola
(a)	(b)	(c)	(d)

10.1 Circles and Ellipses

As we saw in Section 1.2 of Chapter 3, if $r > 0$, the graph of

$$x^2 + y^2 = r^2$$

is a circle of radius r with center at the origin. This equation can be rewritten in the form

$$\frac{x^2}{r^2} + \frac{y^2}{r^2} = 1.$$

A slight modification of the last equation, namely,

$$\frac{x^2}{a^2} + \frac{y^2}{b^2} = 1,$$

where $a > 0$, $b > 0$, and $a \neq b$, produces the equation of an **ellipse** (Figure 2). The ellipse intersects the x axis at $(-a, 0)$ and $(a, 0)$; it intersects the y axis at $(0, -b)$ and $(0, b)$. (Why?) These four points are called the **vertices** of the ellipse. The two line segments joining opposite pairs of vertices are called the **axes** of the ellipse. The longer axis is called the **major axis,** the shorter is called the **minor axis,** and the two axes intersect at the **center** of the ellipse. If $a > b > 0$, the major axis is horizontal (Figure 2a); if $0 < a < b$, the major axis is vertical (Figure 2b).

Figure 2

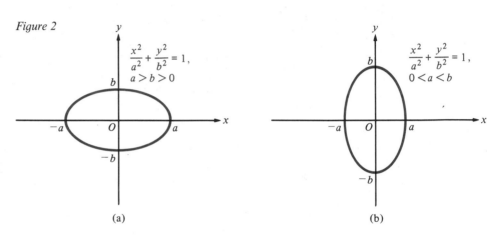

(a)   (b)

Example   Find the vertices of the ellipse $4x^2 + 9y^2 = 36$ and sketch its graph.

Solution   We divide both sides of the equation by 36 to obtain

$$\frac{x^2}{9} + \frac{y^2}{4} = 1.$$

This equation has the form

$$\frac{x^2}{a^2} + \frac{y^2}{b^2} = 1$$

with $a = 3$ and $b = 2$, so its graph is an ellipse with center at the origin and vertices $(-3, 0)$, $(3, 0)$, $(0, -2)$, and $(0, 2)$ (Figure 3).

Figure 3

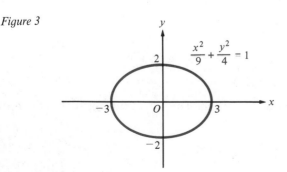

We saw in Section 1.2 of Chapter 3 that, for $r > 0$,

$$(x - h)^2 + (y - k)^2 = r^2$$

is an equation of a circle with radius r and center at (h, k). This equation can be rewritten as

$$\frac{(x - h)^2}{r^2} + \frac{(y - k)^2}{r^2} = 1.$$

Again, the slightly modified equation

$$\frac{(x - h)^2}{a^2} + \frac{(y - k)^2}{b^2} = 1,$$

where $a > 0$, $b > 0$, and $a \neq b$, is the equation of the ellipse obtained by shifting the ellipse

$$\frac{x^2}{a^2} + \frac{y^2}{b^2} = 1$$

(Figure 2) so that its center moves from the origin to the point (h, k) (Figure 4). The vertices of the shifted ellipse are $(h - a, k)$, $(h + a, k)$, $(h, k - b)$, and $(h, k + b)$; and its major and minor axes are parallel to the coordinate axes. We refer to

$$\frac{(x - h)^2}{a^2} + \frac{(y - k)^2}{b^2} = 1, \qquad a > 0, \quad b > 0, \quad a \neq b,$$

as the equation in **standard form** for an ellipse.

Figure 4

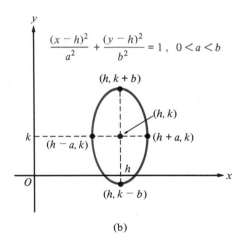

(a)   (b)

Example   Find the equation in standard form for the ellipse with vertices $(3, -2)$, $(9, -2)$, $(6, -6)$, and $(6, 2)$, and sketch the graph.

Solution   We begin by plotting the four vertices. Evidently, $(3, -2)$ and $(9, -2)$ are the endpoints of the minor axis; $(6, -6)$ and $(6, 2)$ are the endpoints of the major axis.

Figure 5

$$\frac{(x-6)^2}{9} + \frac{(y+2)^2}{16} = 1$$

Therefore, the center is the point $(h,k) = (6,-2)$. In the standard form of an equation for an ellipse, a is the distance from the center to either vertex on the horizontal axis and b is the distance from the center to either vertex on the vertical axis. In this example, $a = 3$ and $b = 4$, so the equation in standard form is

$$\frac{(x-6)^2}{3^2} + \frac{[y-(-2)]^2}{4^2} = 1$$

or

$$\frac{(x-6)^2}{9} + \frac{(y+2)^2}{16} = 1.$$

The graph is sketched in Figure 5.

10.2 Parabolas

By Theorem 1 in Section 1 of Chapter 4 (page 177), the graph of

$$y = a(x-h)^2 + k \quad \text{or} \quad y - k = a(x-h)^2,$$

where $a \neq 0$, is a parabola that has a vertical axis, that has its vertex at the point (h,k), and that opens upward if $a > 0$ and downward if $a < 0$ (Figure 6a). By interchanging the roles of x and y, we find that $x - h = a(y-k)^2$ is an equation of a parabola that has a horizontal axis, that has its vertex at the point (h,k), and that opens to the right if $a > 0$ and to the left if $a < 0$ (Figure 6b). We refer to

$$y - k = a(x-h)^2 \quad \text{and} \quad x - h = a(y-k)^2, \qquad a \neq 0,$$

as the equations in **standard form** for a parabola.

Figure 6

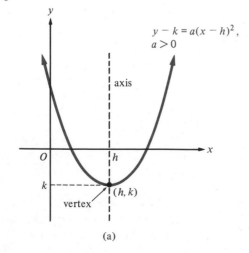

$y - k = a(x-h)^2,$
$a > 0$

(a)

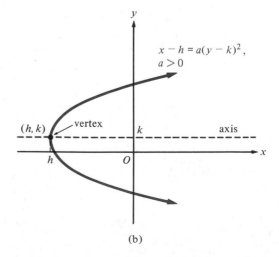

$x - h = a(y-k)^2,$
$a > 0$

(b)

Example 1 A parabola with a horizontal axis has its vertex at the origin, and contains the point $(4, 3)$. Find its equation in standard form and sketch the graph.

Solution Since the parabola has a horizontal axis, the standard form of its equation is

$$x - h = a(y - k)^2$$

Figure 7

with $a \neq 0$. Because the vertex is at the origin, $h = 0$, $k = 0$, and the equation has the form $x = ay^2$. To determine a, we substitute $x = 4$ and $y = 3$ in the equation $x = ay^2$ to obtain $4 = a(9)$. Therefore, $a = \frac{4}{9}$, and the parabola has the equation

$$x = \tfrac{4}{9}y^2 \quad \text{or} \quad 9x = 4y^2.$$

The graph is sketched in Figure 7.

Example 2 Sketch the graph of $(y + 1)^2 = -12(x - 2)$.

Solution The equation can be rewritten in the form

$$x - 2 = -\tfrac{1}{12}[y - (-1)]^2;$$

that is,

$$x - h = a(y - k)^2,$$

Figure 8

with $h = 2$, $k = -1$, and $a = -\tfrac{1}{12}$. Therefore, its graph is a parabola that has a horizontal axis, that has its vertex at $(h, k) = (2, -1)$, and that opens to the left (Figure 8).

10.3 Hyperbolas

If, in the equation

$$\frac{x^2}{a^2} + \frac{y^2}{b^2} = 1, \quad a > 0, \quad b > 0,$$

of an ellipse with center at the origin we change the plus sign to a minus sign, we obtain the equation

$$\frac{x^2}{a^2} - \frac{y^2}{b^2} = 1, \quad a > 0, \quad b > 0,$$

whose graph is a **hyperbola** (Figure 9). The hyperbola has two *branches*, one opening to the right and one opening to the left. The two points $(-a, 0)$ and $(a, 0)$ where these branches intersect the x axis are called the **vertices** of the hyperbola, and the line segment between the two vertices is called the **transverse axis.** The midpoint of the transverse axis (the origin in Figure 9) is called the **center** of the hyperbola.

Figure 9

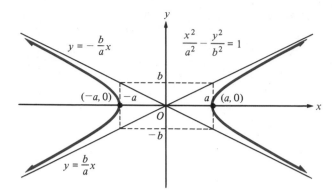

The two straight lines

$$y = \frac{b}{a}x \quad \text{and} \quad y = -\frac{b}{a}x$$

are **asymptotes** of the hyperbola

$$\frac{x^2}{a^2} - \frac{y^2}{b^2} = 1.$$

As a point (x, y) on the hyperbola moves farther and farther from the center, it comes closer and closer to one of these asymptotes. Although the asymptotes are not part of the hyperbola itself, they are useful in sketching it. You can draw them easily by sketching the so-called *fundamental rectangle* with height $2b$, horizontal base $2a$, and center at the center of the hyperbola; the straight lines through the diagonals of the rectangle are the asymptotes (Figure 9). If you use the fact that the vertices of the hyperbola are the midpoints of the left and right sides of the fundamental rectangle, and the fact that the hyperbola approaches the asymptotes as it moves away from the vertices, you can easily sketch the graph.

Example   Find the asymptotes and the vertices of the hyperbola

$$\frac{x^2}{4} - \frac{y^2}{1} = 1$$

and sketch the graph.

Solution   The equation has the form

$$\frac{x^2}{a^2} - \frac{y^2}{b^2} = 1$$

with $a = 2$ and $b = 1$. We begin by drawing the fundamental rectangle with height $2b = 2$, horizontal base $2a = 4$, and center at the origin. The diagonals of this rectangle determine the asymptotes $y = \frac{1}{2}x$ and $y = -\frac{1}{2}x$. The vertices are the midpoints $(-2, 0)$ and $(2, 0)$ of the left and right sides of the rectangle. The graph is sketched in Figure 10.

Figure 10

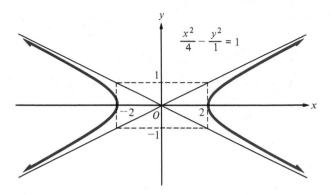

If we interchange the roles of x and y, we find that

$$\frac{y^2}{b^2} - \frac{x^2}{a^2} = 1, \quad a > 0, \quad b > 0$$

is an equation of a hyperbola with a vertical transverse axis and center at the origin (Figure 11). The vertices are the points $(0, -b)$ and $(0, b)$; the asymptotes are the lines

$$y = \frac{b}{a} x \quad \text{and} \quad y = -\frac{b}{a} x,$$

just as before.

Figure 11

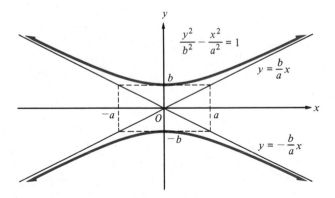

If we shift the hyperbola in Figure 9 or in Figure 11 so that the center is at the point (h, k), the resulting equation becomes either

$$\frac{(x - h)^2}{a^2} - \frac{(y - k)^2}{b^2} = 1 \quad \text{or} \quad \frac{(y - k)^2}{b^2} - \frac{(x - h)^2}{a^2} = 1,$$

depending on whether the transverse axis is horizontal or vertical. These are called equations in **standard form** for a hyperbola. For either equation, the asymptotes of the hyperbola are the lines

$$y - k = \frac{b}{a}(x - h) \quad \text{and} \quad y - k = -\frac{b}{a}(x - h)$$

(Figure 12).

Figure 12

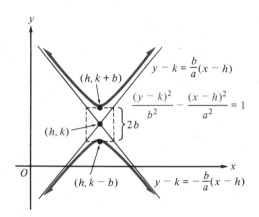

Example Find the center, the asymptotes, and the vertices, and sketch the hyperbola

$$\frac{(y - 2)^2}{9} - \frac{(x + 1)^2}{4} = 1$$

Solution The given equation

$$\frac{(y - 2)^2}{9} - \frac{(x + 1)^2}{4} = 1$$

Figure 13

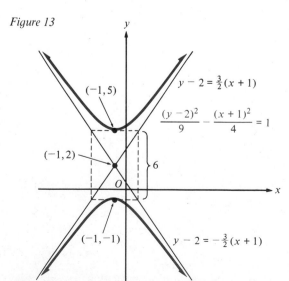

has the form

$$\frac{(y - k)^2}{b^2} - \frac{(x - h)^2}{a^2} = 1$$

with $h = -1$, $k = 2$, $a = 2$, and $b = 3$; so its graph is a hyperbola with a vertical transverse axis, center at $(-1, 2)$, asymptotes

$$y - 2 = \tfrac{3}{2}(x + 1)$$

and

$$y - 2 = -\tfrac{3}{2}(x + 1),$$

and vertices $(-1, -1)$ and $(-1, 5)$ (Figure 13).

10.4 Restoring the Standard Form by Completing the Square

If the squares of the binomials in the standard form of an equation of a circle, ellipse, parabola, or hyperbola are expanded and the resulting constant terms are combined, the equation will no longer be in standard form. By completing squares as in the following examples, the standard form can be restored.

Examples   By completing squares, rewrite each equation in standard form and sketch the graph of the conic section.

1 $25x^2 + 9y^2 - 100x - 54y - 44 = 0$

Solution   The given equation

$$25x^2 + 9y^2 - 100x - 54y - 44 = 0$$

may be rewritten as

$$25x^2 - 100x + 9y^2 - 54y = 44.$$

Thus, by completing the squares

$$25(x^2 - 4x) + 9(y^2 - 6y) = 44$$

$$25(x^2 - 4x + 4) + 9(y^2 - 6y + 9) = 44 + 25(4) + 9(9)$$

$$25(x - 2)^2 + 9(y - 3)^2 = 225$$

$$\frac{(x - 2)^2}{\left(\frac{225}{25}\right)} + \frac{(y - 3)^2}{\left(\frac{225}{9}\right)} = 1$$

$$\frac{(x - 2)^2}{9} + \frac{(y - 3)^2}{25} = 1.$$

The graph is an ellipse with center $(2, 3)$ and vertices $(-1, 3)$, $(5, 3)$, $(2, -2)$, and $(2, 8)$ (Figure 14).

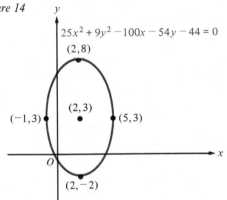

Figure 14

$25x^2 + 9y^2 - 100x - 54y - 44 = 0$

$(2, 8)$

$(2, 3)$

$(-1, 3)$ $(5, 3)$

$(2, -2)$

2 $4x^2 - 9y^2 - 24x - 90y - 225 = 0$

Solution

$$4(x^2 - 6x) - 9(y^2 + 10y) = 225$$

$$4(x^2 - 6x + 9) - 9(y^2 + 10y + 25) = 225 + 4(9) - 9(25)$$

$$4(x - 3)^2 - 9(y + 5)^2 = 36$$

$$\frac{(x - 3)^2}{\left(\frac{36}{4}\right)} - \frac{(y + 5)^2}{\left(\frac{36}{9}\right)} = 1$$

$$\frac{(x - 3)^2}{9} - \frac{(y + 5)^2}{4} = 1.$$

The graph is a hyperbola with horizontal transverse axis, center $(3, -5)$, vertices $(0, -5)$ and $(6, -5)$, and asymptotes $y + 5 = \frac{2}{3}(x - 3)$ and $y + 5 = -\frac{2}{3}(x - 3)$ (Figure 15).

Figure 15

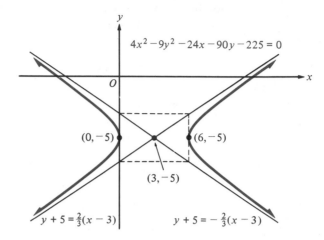

PROBLEM SET 10

In problems 1–10, determine whether the graph of the equation is a circle or an ellipse. If it is a a circle, find the radius. If it is an ellipse, find the vertices. Sketch the graph.

1. $\dfrac{x^2}{25} + \dfrac{y^2}{25} = 1$ **2.** $\dfrac{x^2}{16} + \dfrac{y^2}{4} = 1$ **3.** $\dfrac{x^2}{9} + \dfrac{y^2}{1} = 1$ **4.** $4x^2 + y^2 = 16$

5. $36x^2 + 9y^2 = 144$ **6.** $x^2 + 16y^2 = 16$ **7.** $16x^2 + 25y^2 = 400$ **8.** $x^2 + 4y^2 = 1$
9. $4x^2 + 4y^2 = 1$ **10.** $9x^2 + 36y^2 = 4$

In problems 11–16, find the equation in standard form for the ellipse that satisfies the indicated conditions.

11. Vertices at $(-8, 5), (-2, 5), (-5, 3)$, and $(-5, 7)$.
12. Vertices at $(0, -8)$ and $(0, 8)$; ellipse contains the point $(6, 0)$.
13. Vertices at $(-2, -3), (-2, 5), (-7, 1)$, and $(3, 1)$.
14. Center at the origin; ellipse contains the points $(4, 0)$ and $(3, 2)$.
15. Center at $(1, -2)$; major axis parallel to y axis and 6 units long; minor axis 4 units long.
16. Vertices at $(0, -3)$ and $(0, 3)$; ellipse contains the point $(\frac{2}{3}, 2\sqrt{2})$.

In problems 17–20, find the center and the vertices of each ellipse and sketch the graph.

17. $\dfrac{(x - 1)^2}{9} + \dfrac{(y + 2)^2}{4} = 1$ **18.** $\dfrac{(x + 2)^2}{16} + \dfrac{(y - 1)^2}{4} = 1$

19. $4(x + 3)^2 + (y - 1)^2 = 36$ **20.** $25(x + 1)^2 + 16y^2 = 400$

In problems 21–30, find the vertex of each parabola, determine whether its axis is horizontal or vertical, and sketch the graph.

21. $y^2 = 4x$ **22.** $y^2 = -9x$ **23.** $x^2 - 2y = 0$ **24.** $3x^2 - 4y = 0$
25. $x^2 + 9y = 0$ **26.** $(y - 2)^2 = 8(x + 3)$ **27.** $(x - 4)^2 = 12(y + 7)$ **28.** $(y + 1)^2 = -4(x - 1)$
29. $(x + 1)^2 = -8y$ **30.** $4x^2 + 3y + 3 = 0$

In problems 31–36, find the equation in standard form for the parabola that satisfies the indicated conditions.

31. Vertex at the origin; horizontal axis; parabola contains the point $(2, 4)$.
32. Vertex at $(3, 2)$; vertical axis; parabola contains the point $(6, 1)$.
33. Vertex at $(5, -1)$; horizontal axis; parabola contains the point $(-4, 2)$.
34. Axis coincides with the y axis; parabola contains the points $(2, 3)$ and $(-1, -2)$.
35. Vertex at $(-1, -\frac{1}{2})$; vertical axis; parabola contains the point $(2, \frac{5}{8})$.
36. Horizontal axis; parabola contains the points $(-1, 0)$, $(0, -1)$, and $(0, 1)$.

In problems 37 and 38, determine in each case whether the transverse axis of the hyperbola is horizontal or vertical.

37. (a) $\dfrac{y^2}{16} - \dfrac{x^2}{9} = 1$ **(b)** $\dfrac{y^2}{9} - \dfrac{x^2}{16} = 1$ **(c)** $\dfrac{y^2}{9} - \dfrac{x^2}{9} = 1$ **(d)** $\dfrac{x^2}{16} - \dfrac{y^2}{9} = 1$ **(e)** $\dfrac{x^2}{9} - \dfrac{y^2}{16} = 1$ **(f)** $\dfrac{x^2}{9} - \dfrac{y^2}{9} = 1$

38. (a) $\dfrac{(x-3)^2}{25} - \dfrac{(y+2)^2}{16} = 1$ **(b)** $\dfrac{(x+2)^2}{16} - \dfrac{(y-3)^2}{25} = 1$ **(c)** $\dfrac{(y-3)^2}{9} - \dfrac{(x-3)^2}{16} = 1$ **(d)** $\dfrac{(y+3)^2}{9} - \dfrac{(x+3)^2}{9} = 1$

In problems 39–48, find the center, the asymptotes, and the vertices of each hyperbola and sketch the graph.

39. $\dfrac{x^2}{9} - \dfrac{y^2}{4} = 1$ **40.** $x^2 - \dfrac{y^2}{9} = 1$ **41.** $\dfrac{y^2}{16} - \dfrac{x^2}{4} = 1$ **42.** $49x^2 - 14y^2 = 196$

43. $4x^2 - 16y^2 = 64$ **44.** $36y^2 - 10x^2 = 360$ **45.** $\dfrac{(x-1)^2}{9} - \dfrac{(y+2)^2}{4} = 1$ **46.** $(x+3)^2 - \dfrac{(y-1)^2}{9} = 1$

47. $\dfrac{(y+1)^2}{16} - \dfrac{(x+2)^2}{25} = 1$ **48.** $(y+3)^2 = 4 + 4(x-1)^2$

In problems 49–52, find the equation in standard form for the hyperbola that satisfies the indicated conditions.

49. Vertices $(-4, 0)$ and $(4, 0)$; asymptotes $y = -\frac{5}{4}x$ and $y = \frac{5}{4}x$.
50. Asymptotes $y = -2x$ and $y = 2x$; hyperbola contains the point $(1, 1)$.
51. Center at $(3, -2)$; vertices $(3, -4)$ and $(3, 0)$; asymptotes having slopes -4 and 4.
52. Asymptotes $y = x + 5$ and $y = -x + 3$; hyperbola contains the point $(2, 4)$.

In problems 53–68, rewrite the given equation in standard form and identify the conic section.

53. $x^2 + y^2 + 2x + 4y + 4 = 0$
54. $4x^2 - y^2 - 8x + 2y + 7 = 0$
55. $y^2 - 8y - 6x - 2 = 0$
56. $2y + x = y^2$
57. $x^2 + 2y^2 + 6x + 7 = 0$
58. $4x^2 + y^2 - 8x + 4y - 8 = 0$
59. $2x^2 + 5y^2 + 20x + 75 = 30y$
60. $4x^2 + 4y^2 + 8x + 1 = 4y$
61. $x^2 - 4y^2 - 4x - 8y - 4 = 0$
62. $9x^2 - 25y^2 + 72x - 100y + 269 = 0$
63. $16x^2 + 180y = 9y^2 + 612$
64. $3x^2 + 3y^2 - 6x + 9y = 27$
65. $9x^2 + 4y^2 + 18x = 16y + 11$
66. $9x^2 - 16y^2 = 223 + 90x + 256y$
67. $2x^2 + 8x - 3y + 4 = 0$
68. $x = y^2 + 10y + 21$

69. The point $P = (x, y)$ moves so that the sum of its distances from the two points $(3, 0)$ and $(-3, 0)$ is 8. Show that the point moves on an ellipse.
70. A point moves so that it is equidistant from the point $(2, 0)$ and the circle of radius 3 with center $(-2, 0)$. Find an equation of the path of the point.
71. The roadway of a suspension bridge is 400 meters long and is supported by a main cable in the shape of a parabola that is 100 meters above the roadway at the ends and 4 meters above it at the center. Vertical supporting cables run at 50-meter intervals from the roadway to the main cable. Find the lengths of each of these vertical cables.

REVIEW PROBLEM SET

In problems 1–24, express each complex number in the form $a + bi$, where a and b are real numbers.

1. $(3 + 2i) + (7 + 3i)$

2. $(2 - 3i) + (1 + 2i)$

3. $(5 - 7i) - (4 + 2i)$

4. $(3 + 5i) - (5 - 3i)$

5. $(7 - 4i) - (-6 + 4i)$

6. $(-\frac{5}{2} + 6i) + (-\frac{7}{2} - 3i)$

7. $(5 - 11i)(5 + 2i)$

8. $(5 + 2i)(7 + 3i)$

9. $(2 + 5i)(-2 + 4i)$

10. $(7 - 2i)(2 + 3i)$

11. $\dfrac{2 + 5i}{3 + 2i}$

12. $\dfrac{1 + 4i}{\sqrt{3} + 2i}$

13. $\dfrac{4 + 2i}{(2 + 3i)(4 + i)}$

14. $\dfrac{5 + 15i}{(3 - i)(1 + i)}$

15. i^{403}

16. i^{-21}

17. $\left(\dfrac{1}{3i}\right)^3$

18. $\dfrac{1}{(3 + 2i)^2}$

19. $(2 - 3i)\overline{(3 - 2i)}$

20. $\overline{(3 + 5i)}(3 + 5i)$

21. $3(\cos 45° + i \sin 45°)$

22. $6\left[\cos\left(-\dfrac{\pi}{6}\right) + i \sin\left(-\dfrac{\pi}{6}\right)\right]$

23. $16\left(\cos\dfrac{5\pi}{3} + i \sin\dfrac{5\pi}{3}\right)$

24. $\overline{\left(\cos\dfrac{71\pi}{4} + i \sin\dfrac{71\pi}{4}\right)}$

In problems 25–30, find each absolute value.

25. $|-4 - 3i|$

26. $|6 - 8i|$

27. $|8i^7|$

28. $|i^{17}|$

29. $\left|\dfrac{3 + 2i}{3 - 4i}\right|$

30. $\left|\dfrac{-3 - 2i}{-6 + 8i}\right|$

In problems 31–34, express each complex number in polar form.

31. $-4 + 4i$

32. $3 + 0i$

33. $-1 + \sqrt{3}i$

34. $7 - 7i$

In problems 35–42, find (a) zw and (b) z/w. Express the answers in polar form.

35. $z = 6(\cos 22° + i \sin 22°)$, $w = 4(\cos 8° + i \sin 8°)$

36. $z = 8(\cos 85° + i \sin 85°)$, $w = 2(\cos 95° + i \sin 95°)$

37. $z = 12(\cos 235° + i \sin 235°)$, $w = 4(\cos 125° + i \sin 125°)$

38. $z = 14(\cos 78° + i \sin 78°)$, $w = 7(\cos 18° + i \sin 18°)$

39. $z = 10\left(\cos\dfrac{13\pi}{36} + i \sin\dfrac{13\pi}{36}\right)$, $w = 5\left(\cos\dfrac{2\pi}{9} + i \sin\dfrac{2\pi}{9}\right)$

40. $z = 20\left(\cos\dfrac{143\pi}{180} + i \sin\dfrac{143\pi}{180}\right)$, $w = 4\left(\cos\dfrac{2\pi}{45} + i \sin\dfrac{2\pi}{45}\right)$

41. $z = 3 + \sqrt{3}i$, $w = \sqrt{3} + i$

42. $z = -3 - \sqrt{3}i$, $w = 2 + 2\sqrt{3}i$

In problems 43–48, use De Moivre's theorem to evaluate each expression. Write the answer in cartesian form.

43. $[2(\cos 300° + i \sin 300°)]^5$

44. $\left[2\left(\cos\dfrac{\pi}{3} + i \sin\dfrac{\pi}{3}\right)\right]^4$

45. $\left[2\left(\cos\dfrac{5\pi}{6} + i \sin\dfrac{5\pi}{6}\right)\right]^6$

46. $(1 - \sqrt{3}i)^9$

47. $(-1 + i)^6$

48. $(1 - i)^{12}$

In problems 49–52, find all of the indicated complex nth roots.

49. The cube roots of i

50. The fifth roots of $-\sqrt{3} - i$

51. The fourth roots of $\dfrac{1}{2} - \dfrac{\sqrt{3}}{2}i$

52. The seventh roots of 1

53. Find all complex roots of the equation $x^4 - 1 = 0$.

54. Find all complex roots of the equation $x^3 - i = 0$.

In problems 55–60, (a) find all complex zeros of each complex polynomial $f(z)$, (b) factor $f(z)$ completely into linear factors, and (c) determine the multiplicity of each zero of $f(z)$.

55. $z^3 - 9z - 10$ **56.** $z^4 - 8z^2 + 9z - 2$ **57.** $z^4 + 5z^3 + 8z^2 + z - 15$

58. $z^4 - 8z^2 - 4z + 3$ **59.** $z^4 - 6z^3 + 3z^2 + 24z - 28$ **60.** $z^4 - 4z^3 + 5z^2 - 4z + 4$

In problems 61–66, form a polynomial in z that has real coefficients, that has the smallest possible degree, and that has the indicated zeros. (Zeros are repeated to show multiplicity.)

61. $1, -2, 1 + i$ **62.** $0, 0, i$ **63.** $2, 2, 2, i$

64. $0, 2i, 4, -3i$ **65.** $3 + \sqrt{2}i, 1 - \sqrt{2}i$ **66.** $-1 - i, \sqrt{2}, -\sqrt{2}$

In problems 67–70, use mathematical induction to prove that the assertion is true for all positive integers n.

67. $n^2 + 3n$ is an even integer

68. 3 is an exact integral divisor of $n^3 + 6n^2 + 11n$

69. $1 \cdot 2 \cdot 3 + 2 \cdot 3 \cdot 4 + 3 \cdot 4 \cdot 5 + \cdots + n(n + 1)(n + 2) = \dfrac{1}{4} n(n + 1)(n + 2)(n + 3)$

70. $2 \cdot 4 + 4 \cdot 6 + 6 \cdot 8 + \cdots + 2n(2n + 2) = \dfrac{n}{3}(2n + 2)(2n + 4)$

In problems 71–76, use the binomial theorem and the Pascal triangle to expand each expression.

71. $(2 + x)^5$ **72.** $(3x - 4y)^4$ **C** **73.** $(3x^2 - 2y^2)^7$

74. $(2a + 3b)^6$ **C** **75.** $\left(\dfrac{3}{2}x - 1\right)^7$ **76.** $\left(2x^2 + \dfrac{3}{y}\right)^3$

In problems 77–80, find and simplify the specified term in the binomial expansion of the expression.

77. The fourth term of $(2x + y)^9$ **78.** The fifth term of $(x - 2y)^7$

79. The term containing x^5 in $(3x + y)^{10}$ **80.** The term containing x^{10} in $(2x^2 - 3)^{11}$

In problems 81 and 82, find the first five terms of the sequence with the given general term.

81. $a_n = \dfrac{(-1)^n}{n + 1}$ **82.** $b_n = \dfrac{(-1)^{n+1}}{n^2 + 4n + 1}$

In problems 83 and 84, find the first six terms of the recursively defined sequence $\{a_n\}$.

83. $a_1 = 0, a_2 = \dfrac{1}{2}$, and $a_n = \dfrac{1}{n} a_{n-2}$ **84.** $a_1 = 1, a_2 = 0$, and $(n + 1)(n + 2)a_{n+2} = -n a_n$

In problems 85–88, find the indicated term in each arithmetic sequence.

85. The twentieth term of $3, 5, 7, 9, \ldots$ **86.** The fiftieth term of $17, 14, 11, 8, \ldots$

87. The thirty-fourth term of $-5, -9, -13, -17, \ldots$ **88.** The hundredth term of $1, \frac{7}{2}, 6, \frac{17}{2}, \ldots$

In problems 89–92, find the indicated term in each geometric sequence.

89. The tenth term of $2, 4, 8, 16, \ldots$ **90.** The ninth term of $12, 6, 3, \frac{3}{2}, \ldots$

91. The eighth term of $3, -6, 12, -24, \ldots$ **92.** The seventh term of $9, -3, 1, -\frac{1}{3}, \ldots$

In problems 93–100, evaluate each sum.

93. $\displaystyle\sum_{k=1}^{3} (2k + 7)$ **94.** $\displaystyle\sum_{k=1}^{50} [3 + (-1)^k]$ **95.** $\displaystyle\sum_{k=1}^{20} \left(\dfrac{k}{3} + 2\right)$ **96.** $\displaystyle\sum_{k=1}^{20} (1 - 3k)$

97. $\displaystyle\sum_{k=0}^{3} \dfrac{7^k}{1 + 2^k}$ **98.** $\displaystyle\sum_{k=0}^{15} \dfrac{3}{2^k}$ **99.** $\displaystyle\sum_{k=0}^{2} \dfrac{(-5)^k}{1 + 3^k}$ **100.** $\displaystyle\sum_{k=0}^{25} (4^{k+1} - 4^k)$

In problems 101–104, find the sum of the first n terms of each arithmetic sequence for the given value of n.

101. $2, 6, 10, 14, \ldots$ for $n = 15$

102. $12, 13.5, 15, 16.5, \ldots$ for $n = 20$

103. $3, \frac{8}{3}, \frac{7}{3}, 2, \ldots$ for $n = 11$

104. $3x - 2, -x + 1, \ldots$ for $n = 10$

[C] In problems 105–108, find the sum of the first n terms of each geometric sequence for the given value of n.

105. $48, 96, 192, \ldots$ for $n = 8$

106. $-81, -27, -9, \ldots$ for $n = 12$

107. $\frac{3}{4}, 3, 12, \ldots$ for $n = 10$

108. $0.2, 0.002, 0.00002, \ldots$ for $n = 5$

109. A mathematics club raffles a calculator by selling 100 sealed tickets numbered in order $1, 2, 3, \ldots$. The tickets are drawn at random by purchasers who pay the number of cents indicated by the number on the ticket. How much money does the club receive if all tickets are sold?

110. A clock strikes on the hour. How many times does the clock strike between 12:00 noon on one day and 12:00 noon on the next day, inclusive?

111. Gus saved \$200 the first year he was employed. Each year thereafter, he saved \$50 more than the year before. How much did he save at the end of 8 years?

112. A car costs \$7670 and depreciates in value by 31% during the first year, by 26% during the second year, by 21% during the third year, and so on. What is the value of the car at the end of 5 years?

[C] **113.** The rungs of a ladder decrease uniformly (that is, linearly) in length from 32 inches to 18 inches. If there are 25 rungs in the ladder, find the total length of the wood in all of these rungs.

[C] **114.** In a lottery, the first ticket drawn will pay the ticket holder \$1, and each succeeding ticket will pay twice as much as the preceding one. If 15 tickets are drawn, what is the total amount paid in prize money?

115. A pyramid of cannonballs in an armaments museum stands on a square base having n cannonballs on a side. How many cannonballs are there in the pile?

116. Work problem 115 if the base has the form of an equilateral triangle.

In problems 117–120, find the sum of each infinite geometric series.

117. $\displaystyle\sum_{k=1}^{\infty} 3\left(\frac{2}{3}\right)^{k-1}$

118. $\displaystyle\sum_{k=1}^{\infty} 3\left(-\frac{2}{3}\right)^{k-1}$

119. $\displaystyle\sum_{k=1}^{\infty} \left(\sqrt{\frac{5}{7}}\right)^{k-1}$

120. $\displaystyle\sum_{k=0}^{\infty} (-1)^k \left(\frac{3}{4}\right)^k$

121. Rewrite each infinite repeating decimal as a quotient of integers:

(a) $0.\overline{83}$ (b) $0.\overline{91}$ (c) $4.65\overline{223}$ (d) $3.1\overline{9}$

[C] **122.** Suppose that on each separate swing, a pendulum describes an arc whose length is 0.98 of the length of the preceding arc. If the length of the first arc is 24 centimeters, what total distance is covered by the pendulum before it comes to rest?

In problems 123–132, find the value of the given expression.

123. $_6P_4$

124. $_7P_4$

125. $_{11}P_2$

126. $_8P_3$

127. $_{16}C_2$

[C] **128.** $_{16}C_7$

129. $_9C_3$

130. $_7C_5$

131. $_8C_8$

132. $_{19}C_0$

133. A woman has a choice of four airlines between Chicago and New York, and three airlines between New York and London. In how many ways can she fly from Chicago to London, if she stops in New York?

[C] **134.** How many different four-letter "words" can be formed using the twenty-six letters of the alphabet, if repeated letters are allowed?

135. How many different four-course meals can be ordered in a restaurant if there is a choice of three soups, four entrees, two salads, and five desserts?

136. In how many ways can the letters of the word WORTH be arranged?

C 137. In how many ways can seven of twelve books be chosen and arranged next to each other on a shelf?

138. How many different signals can be sent by arranging up to five distinguishable flags, one above the other, on a flag pole with the condition that the flags can be used one, two, three, four, or five at a time?

139. How many straight lines can be drawn through pairs of ten points, if no three points are in a straight line?

140. Twelve people meet and shake hands. If everyone shakes hands with everyone else, how many handshakes are exchanged?

C 141. A bridge hand consists of 13 cards dealt from a deck of 52. How many different bridge hands are there?

C 142. A corporation owns 20 motels. If there are 25 people eligible to be managers of these motels, in how many ways can the motel managers be appointed?

143. In how many distinguishable ways can the letters of the word MINIMUM be arranged?

144. In how many distinguishable ways can three identical racquet balls, four identical golf balls, and five identical tennis balls be placed in a row?

145. If three dimes are tossed in the air, find the probability that all three fall "heads."

146. You meet a married couple and learn that they have two children. During the course of the conversation, it becomes clear that at least one of their children is a girl. What is the probability that the other child is a girl?

147. What is the probability of rolling an "11" with two dice?

148. Four married couples draw lots to decide who will be partners in a card game. If partners must be of opposite sexes, what is the probability that no woman is paired with her own husband?

149. A bag contains four white balls, six black balls, three red balls, and eight green balls. If two balls are selected blindly, what is the probability that they are of the same color?

150. Professor Grumbles has four suits of clothes, each consisting of a vest, trousers, and a jacket. If he dresses at random, what is the probability that his clothes match?

In problems 151–154, determine whether the graph of the equation is a circle or an ellipse. If it is a circle, find the radius. If it is an ellipse, find the vertices. Sketch the graph.

151. $9x^2 + 9y^2 = 1$
152. $x^2 + \dfrac{y^2}{4} = 1$
153. $\dfrac{x^2}{16} + \dfrac{y^2}{9} = 1$
154. $3x^2 + 2y^2 = 1$

In problems 155–158, find the equation in standard form for the ellipse that satisfies the indicated conditions.

155. Center at $(0,0)$; horizontal major axis of length 16; minor axis of length 8.
156. Vertices $(0,0)$, $(6,0)$, $(3,-5)$, and $(3,5)$.
157. Vertices $(-3,1)$, $(5,1)$, $(1,-4)$, and $(1,6)$.
158. Center $(0,0)$; major axis horizontal; ellipse contains the points $(4,3)$ and $(6,2)$.

In problems 159–164, find the center and the vertices of each ellipse and sketch the graph.

159. $\dfrac{x^2}{8} + \dfrac{y^2}{12} = 1$
160. $144x^2 + 169y^2 = 24{,}336$
161. $9x^2 + 25y^2 + 18x - 50y = 191$
162. $3x^2 + 4y^2 - 28x - 16y + 48 = 0$
163. $9x^2 + 4y^2 + 72x - 48y + 144 = 0$
164. $9x^2 + 4y^2 + 36x - 24y = 252$

In problems 165–170, find the vertex of each parabola, determine whether its axis is horizontal or vertical, and sketch the graph.

165. $y^2 = 4x$
166. $-x^2 = 5y$
167. $x^2 = -4(y-1)$
168. $x^2 = 8(y+1)$
169. $x + 8 - y^2 + 2y = 0$
170. $x + 4 + 2y = y^2$

In problems 171 and 172, find the equation in standard form for the parabola that satisfies the indicated conditions.

171. Vertex at $(0, -6)$; horizontal axis; parabola contains the point $(-9, -3)$.
172. Vertical axis; parabola contains the points $(9, -1)$, $(3, -4)$, and $(-9, 8)$.

In problems 173–178, find the center, the asymptotes, and the vertices of each hyperbola, and sketch the graph.

173. $x^2 - 9y^2 = 72$ **174.** $y^2 - 9x^2 = 54$
175. $x^2 - 4y^2 + 4x + 24y - 48 = 0$ **176.** $16x^2 - 9y^2 = 96x$
177. $4y^2 - x^2 - 24y + 2x + 34 = 0$ **178.** $11y^2 - 66y - 25x^2 + 100x = 276$

In problems 179 and 180, find the equation in standard form for the hyperbola that satisfies the indicated conditions.

179. Asymptotes $y = -2x$ and $y = 2x$; hyperbola contains the point $(1, 1)$.
180. Vertices $(2, -8)$ and $(2, 2)$; asymptotes $25x - 9y = 77$ and $25x + 9y = 23$.

In problems 181–186, rewrite each equation in standard form and identify the conic section.

181. $16x^2 + y^2 - 32x + 4y - 44 = 0$ **182.** $x^2 + y^2 - 4x + 2y = 4$
183. $4x^2 + 4y^2 + 4x - 4y + 1 = 0$ **184.** $4x^2 + 9y^2 + 16x = 18y + 11$
185. $4y^2 - x^2 - 8y + 2x + 7 = 0$ **186.** $9x^2 + 25y^2 + 18x = 50y + 191$

187. A point $P = (x, y)$ moves so that the product of the slopes of the line segments $\overline{PQ}$ and $\overline{PR}$ is -6, where $Q = (3, -2)$ and $R = (-2, 1)$. Find an equation of the curve traced out by P, identify the curve, and sketch it.

188. Suppose that a point $P = (x, y)$ moves so that $|\overline{PQ}| + |\overline{PR}| = 2a$, where $Q = (-c, 0)$, $R = (c, 0)$, and $a > c > 0$. Show that P traces out the ellipse

$$\frac{x^2}{a^2} + \frac{y^2}{b^2} = 1, \quad \text{where } b = \sqrt{a^2 - c^2}.$$

189. Suppose that a point $P = (x, y)$ moves so that its distance to the point $F = (0, p)$ is the same as its distance to the line $y = -p$, where $p > 0$. Show that P traces out the parabola $x^2 = 4py$.

190. Suppose that a point $P = (x, y)$ moves so that $\left| |\overline{PQ}| - |\overline{PR}| \right| = 2a$, where $Q = (-c, 0)$, $R = (c, 0)$, and $c > a > 0$. Show that P traces out the hyperbola

$$\frac{x^2}{a^2} - \frac{y^2}{b^2} = 1, \quad \text{where } b = \sqrt{c^2 - a^2}.$$

Appendices

I

Tables of Logarithms and Exponentials

In this appendix, we present Table IA of natural logarithms, Table IB of common logarithms, Table IC of exponential functions, and examples illustrating their use.

Table IA gives values of ln x, rounded off to four decimal places, corresponding to values of x between 1 and 9.99 in steps of 0.01.

Example   Use Table IA to find the value of ln 3.47 rounded off to four decimal places.

Solution   We begin by locating the number 3.4 in the vertical column on the left side of the table. The numbers in the horizontal row to the right of 3.4 are the natural logarithms of the numbers from 3.40 to 3.49 in steps of 0.01. The entry in this horizontal row below the heading 0.07 is the natural logarithm of 3.47. Thus,

$$\ln 3.47 = 1.2442.$$

For values of x lying between two numbers whose natural logarithms are shown in Table IA, it is possible to find the approximate value of ln x (again rounded off to four decimal places) by using **linear interpolation**. The basic idea of linear interpolation is that small changes in the independent variable cause approximately proportional changes in the values of a function.

Example   Use linear interpolation and Table IA to find the approximate value of ln 2.724.

Solution   We arrange the work as follows:

$$0.01 \left[0.004 \left[\begin{array}{cc} x & \ln x \\ 2.72 & 1.0006 \\ 2.724 & \ln 2.724 \end{array} \right]d \right] 0.0037.$$

We have used Table IA to determine that ln 2.72 = 1.0006 and ln 2.73 = 1.0043. The notation

$$0.004 \left[\begin{array}{c} 2.72 \\ 2.724 \end{array} \right.$$

indicates that the difference between 2.72 and 2.724 (bottom number minus top number) is

$$2.724 - 2.72 = 0.004.$$

Other such differences are indicated similarly. For linear interpolation, we assume that corresponding differences are (approximately) proportional, so that

$$\frac{0.004}{0.01} = \frac{d}{0.0037} \quad \text{or} \quad d = 0.0037\left(\frac{0.004}{0.01}\right) = 0.00148.$$

Thus, rounded off to four decimal places, we have

$$\ln 2.724 = 1.0006 + d$$
$$= 1.0006 + 0.00148 = 1.0021.$$

Table IB gives values of $\log x$, rounded off to four decimal places, corresponding to values of x between 1 and 9.99 in steps of 0.01.

Example 1 Using Table IB, find the value of $\log 7.36$ (rounded off to four decimal places).

Solution $\log 7.36 = 0.8669$

Example 2 Use linear interpolation and Table IB to find the approximate value of $\log 5.068$.

Solution We have

$$0.01\left[0.008\begin{bmatrix}\begin{array}{cc} x & \log x \\ 5.06 & 0.7042 \\ 5.068 & \log 5.068 \\ 5.07 & 0.7050 \end{array}\end{bmatrix}d\right]0.0008.$$

Thus,

$$\frac{0.008}{0.01} = \frac{d}{0.0008} \quad \text{or} \quad d = 0.0008\left(\frac{0.008}{0.01}\right) = 0.00064;$$

hence, rounded off to four decimal places,

$$\log 5.068 = 0.7042 + d$$
$$= 0.7042 + 0.00064$$
$$= 0.7048.$$

Now, suppose that x is a positive number less than 1 or greater than 9.99. To find $\log x$, begin by writing x in scientific notation

$$x = p \times 10^n,$$

where $1 \le p < 10$ and n is an integer. (See Section 8.1 of Chapter 1.) Thus,

$$\log x = \log p + \log 10^n = \log p + n.$$

The quantity $\log p$ is called the **mantissa** of $\log x$, whereas the integer n is called the **characteristic** of $\log x$. Thus,

$$\log x = \text{mantissa} + \text{characteristic}.$$

You can find the mantissa by using Table IB.

Examples Using Table IB, find the approximate value of each logarithm.

1 $\log 371.4$

Solution In scientific notation,

$$371.4 = 3.714 \times 10^2.$$

Here, the mantissa

$$\log 3.714 = 0.5698$$

is obtained by using linear interpolation and Table IB; the characteristic is $n = 2$, and so, rounded off to four decimal places,

$$\log 371.4 = 0.5698 + 2 = 2.5698.$$

2 $\log 0.05422$

Solution In scientific notation

$$0.05422 = 5.422 \times 10^{-2}.$$

Here, the mantissa

$$\log 5.422 = 0.7342$$

is obtained by using linear interpolation and Table IB; the characteristic is $n = -2$, and so, rounded off to four decimal places,

$$\log 0.05422 = 0.7342 + (-2)$$

$$= -1.2658.$$

Although a scientific calculator gives $\log 0.05422 = -1.2658$ (rounded off to four decimal places), it is customary to leave the answer in the form

$$\log 0.05422 = 0.7342 + (-2)$$

when using a table of logarithms.

By reading Table IB "backwards," you can find 10^x (sometimes called the **antilogarithm** of x).

Examples Using Table IB, find each antilogarithm.

1 $10^{0.9159}$

Solution In the body of Table IB, we find 0.9159 and see (by reading the table "backwards") that it is the common logarithm of 8.24; hence,

$$\log 8.24 = 0.9159$$

or

$$10^{0.9159} = 8.24.$$

2 $10^{0.03}$

Solution   The number $0.03 = 0.0300$ does not appear in the body of Table IB. The numbers closest to it that do appear are 0.0294 and 0.0334. Thus, we must use linear interpolation. We have

$$0.0040\left[0.0006\left[\begin{matrix}0.0294 & 1.07 \\ 0.0300 & 10^{0.03}\end{matrix}\right]d\right]0.01.$$

$$\begin{matrix}x & 10^x\end{matrix}$$
$$0.0334 \quad 1.08$$

Thus,

$$\frac{0.0006}{0.0040} = \frac{d}{0.01} \quad \text{or} \quad d = 0.01\left(\frac{0.0006}{0.0040}\right) = 0.0015;$$

so that, rounded off to four significant digits,

$$10^{0.03} = 1.07 + d = 1.07 + 0.0015 = 1.072.$$

3 $10^{4.3312}$

Solution

$$10^{4.3312} = 10^{4+0.3312} = 10^4 \cdot 10^{0.3312}.$$

Using Table IB "backwards" with linear interpolation, we find that, rounded off to four significant digits,

$$10^{0.3312} = 2.144.$$

Therefore,

$$10^{4.3312} = 10^4 \cdot 10^{0.3312} = 10{,}000(2.144) = 21{,}440.$$

4 $10^{-5.7074}$

Solution   We begin by writing -5.7074 in the form

$$-5.7074 = \text{mantissa} + \text{characteristic},$$

where the characteristic is negative and the mantissa is between 0 and 1. Evidently, the characteristic is -6; hence,

$$\text{mantissa} = -5.7074 - (-6) = 0.2926.$$

Using Table IB "backwards" with linear interpolation, we find that, rounded off to four significant digits,

$$10^{0.2926} = 1.961.$$

Therefore, rounded off to four significant digits,

$$10^{-5.7074} = 10^{0.2926+(-6)} = 10^{0.2926} \cdot 10^{-6} = 1.961(10^{-6})$$

$$= 0.000{,}001{,}961.$$

By reading Table IA "backwards," you can find values of e^x; however, for convenience, we present a separate table of such values (Table IC).

Example   Use linear interpolation and Table IC to find the approximate value of $e^{1.96}$.

Solution   We have

$$
0.1 \left[0.06 \left[\begin{array}{cc} x & e^x \\ 1.9 & 6.6859 \\ 1.96 & e^{1.96} \end{array} \right] d \\ 2.0 \quad 7.3891 \end{array} \right] 0.7032
$$

Thus,

$$
\frac{0.06}{0.1} = \frac{d}{0.7032} \quad \text{or} \quad d = 0.7032 \left(\frac{0.06}{0.1} \right) = 0.4219,
$$

so

$$
e^{1.96} = 6.6859 + 0.4219 = 7.1078.
$$

Because Table IC is just a "short table" of values of e^x, linear interpolation may not be very accurate here. (A calculator gives $e^{1.96} = 7.0993$, rounded off to five significant digits.)

TABLE IA NATURAL LOGARITHMS

x	0.00	0.01	0.02	0.03	0.04	0.05	0.06	0.07	0.08	0.09
1.0	0.0000	0.0100	0.0198	0.0296	0.0392	0.0488	0.0583	0.0677	0.0770	0.0862
1.1	0.0953	0.1044	0.1133	0.1222	0.1310	0.1398	0.1484	0.1570	0.1655	0.1740
1.2	0.1823	0.1906	0.1989	0.2070	0.2151	0.2231	0.2311	0.2390	0.2469	0.2546
1.3	0.2624	0.2700	0.2776	0.2852	0.2927	0.3001	0.3075	0.3148	0.3221	0.3293
1.4	0.3365	0.3436	0.3507	0.3577	0.3646	0.3716	0.3784	0.3853	0.3920	0.3988
1.5	0.4055	0.4121	0.4187	0.4253	0.4318	0.4383	0.4447	0.4511	0.4574	0.4637
1.6	0.4700	0.4762	0.4824	0.4886	0.4947	0.5008	0.5068	0.5128	0.5188	0.5247
1.7	0.5306	0.5365	0.5423	0.5481	0.5539	0.5596	0.5653	0.5710	0.5766	0.5822
1.8	0.5878	0.5933	0.5988	0.6043	0.6098	0.6152	0.6206	0.6259	0.6313	0.6366
1.9	0.6419	0.6471	0.6523	0.6575	0.6627	0.6678	0.6729	0.6780	0.6831	0.6881
2.0	0.6931	0.6981	0.7031	0.7080	0.7130	0.7178	0.7227	0.7275	0.7324	0.7372
2.1	0.7419	0.7467	0.7514	0.7561	0.7608	0.7655	0.7701	0.7747	0.7793	0.7839
2.2	0.7885	0.7930	0.7975	0.8020	0.8065	0.8109	0.8154	0.8198	0.8242	0.8286
2.3	0.8329	0.8372	0.8416	0.8459	0.8502	0.8544	0.8587	0.8629	0.8671	0.8713
2.4	0.8755	0.8796	0.8838	0.8879	0.8920	0.8961	0.9002	0.9042	0.9083	0.9123
2.5	0.9163	0.9203	0.9243	0.9282	0.9322	0.9361	0.9400	0.9439	0.9478	0.9517
2.6	0.9555	0.9594	0.9632	0.9670	0.9708	0.9746	0.9783	0.9821	0.9858	0.9895
2.7	0.9933	0.9969	1.0006	1.0043	1.0080	1.0116	1.0152	1.0188	1.0225	1.0260
2.8	1.0296	1.0332	1.0367	1.0403	1.0438	1.0473	1.0508	1.0543	1.0578	1.0613
2.9	1.0647	1.0682	1.0716	1.0750	1.0784	1.0818	1.0852	1.0886	1.0919	1.0953
3.0	1.0986	1.1019	1.1053	1.1086	1.1119	1.1151	1.1184	1.1217	1.1249	1.1282
3.1	1.1314	1.1346	1.1378	1.1410	1.1442	1.1474	1.1506	1.1537	1.1569	1.1600
3.2	1.1632	1.1663	1.1694	1.1725	1.1756	1.1787	1.1817	1.1848	1.1878	1.1909
3.3	1.1939	1.1970	1.2000	1.2030	1.2060	1.2090	1.2119	1.2149	1.2179	1.2208
3.4	1.2238	1.2267	1.2296	1.2326	1.2355	1.2384	1.2413	1.2442	1.2470	1.2499
3.5	1.2528	1.2556	1.2585	1.2613	1.2641	1.2669	1.2698	1.2726	1.2754	1.2782
3.6	1.2809	1.2837	1.2865	1.2892	1.2920	1.2947	1.2975	1.3002	1.3029	1.3056
3.7	1.3083	1.3110	1.3137	1.3164	1.3191	1.3218	1.3244	1.3271	1.3297	1.3324
3.8	1.3350	1.3376	1.3403	1.3429	1.3455	1.3481	1.3507	1.3533	1.3558	1.3584
3.9	1.3610	1.3635	1.3661	1.3686	1.3712	1.3737	1.3762	1.3788	1.3813	1.3838
4.0	1.3863	1.3888	1.3913	1.3938	1.3962	1.3987	1.4012	1.4036	1.4061	1.4085
4.1	1.4110	1.4134	1.4159	1.4183	1.4207	1.4231	1.4255	1.4279	1.4303	1.4327
4.2	1.4351	1.4375	1.4398	1.4422	1.4446	1.4469	1.4493	1.4516	1.4540	1.4563
4.3	1.4586	1.4609	1.4633	1.4656	1.4679	1.4702	1.4725	1.4748	1.4770	1.4793
4.4	1.4816	1.4839	1.4861	1.4884	1.4907	1.4929	1.4952	1.4974	1.4996	1.5019
4.5	1.5041	1.5063	1.5085	1.5107	1.5129	1.5151	1.5173	1.5195	1.5217	1.5239
4.6	1.5261	1.5282	1.5304	1.5326	1.5347	1.5369	1.5390	1.5412	1.5433	1.5454
4.7	1.5476	1.5497	1.5518	1.5539	1.5560	1.5581	1.5602	1.5623	1.5644	1.5665
4.8	1.5686	1.5707	1.5728	1.5748	1.5769	1.5790	1.5810	1.5831	1.5851	1.5872
4.9	1.5892	1.5913	1.5933	1.5953	1.5974	1.5994	1.6014	1.6034	1.6054	1.6074
5.0	1.6094	1.6114	1.6134	1.6154	1.6174	1.6194	1.6214	1.6233	1.6253	1.6273
5.1	1.6292	1.6312	1.6332	1.6351	1.6371	1.6390	1.6409	1.6429	1.6448	1.6467
5.2	1.6487	1.6506	1.6525	1.6544	1.6563	1.6582	1.6601	1.6620	1.6639	1.6658
5.3	1.6677	1.6696	1.6715	1.6734	1.6752	1.6771	1.6790	1.6808	1.6827	1.6845
5.4	1.6864	1.6882	1.6901	1.6919	1.6938	1.6956	1.6974	1.6993	1.7011	1.7029
5.5	1.7047	1.7066	1.7084	1.7102	1.7120	1.7138	1.7156	1.7174	1.7192	1.7210
5.6	1.7228	1.7246	1.7263	1.7281	1.7299	1.7317	1.7334	1.7352	1.7370	1.7387
5.7	1.7405	1.7422	1.7440	1.7457	1.7475	1.7492	1.7509	1.7527	1.7544	1.7561
5.8	1.7579	1.7596	1.7613	1.7630	1.7647	1.7664	1.7682	1.7699	1.7716	1.7733
5.9	1.7750	1.7766	1.7783	1.7800	1.7817	1.7834	1.7851	1.7867	1.7884	1.7901

x	0.00	0.01	0.02	0.03	0.04	0.05	0.06	0.07	0.08	0.09
6.0	1.7918	1.7934	1.7951	1.7967	1.7984	1.8001	1.8017	1.8034	1.8050	1.8066
6.1	1.8083	1.8099	1.8116	1.8132	1.8148	1.8165	1.8181	1.8197	1.8213	1.8229
6.2	1.8245	1.8262	1.8278	1.8294	1.8310	1.8326	1.8342	1.8358	1.8374	1.8390
6.3	1.8406	1.8421	1.8437	1.8453	1.8469	1.8485	1.8500	1.8516	1.8532	1.8547
6.4	1.8563	1.8579	1.8594	1.8610	1.8625	1.8641	1.8656	1.8672	1.8687	1.8703
6.5	1.8718	1.8733	1.8749	1.8764	1.8779	1.8795	1.8810	1.8825	1.8840	1.8856
6.6	1.8871	1.8886	1.8901	1.8916	1.8931	1.8946	1.8961	1.8976	1.8991	1.9006
6.7	1.9021	1.9036	1.9051	1.9066	1.9081	1.9095	1.9110	1.9125	1.9140	1.9155
6.8	1.9169	1.9184	1.9199	1.9213	1.9228	1.9242	1.9257	1.9272	1.9286	1.9301
6.9	1.9315	1.9330	1.9344	1.9359	1.9373	1.9387	1.9402	1.9416	1.9430	1.9445
7.0	1.9459	1.9473	1.9488	1.9502	1.9516	1.9530	1.9544	1.9559	1.9573	1.9587
7.1	1.9601	1.9615	1.9629	1.9643	1.9657	1.9671	1.9685	1.9699	1.9713	1.9727
7.2	1.9741	1.9755	1.9769	1.9782	1.9796	1.9810	1.9824	1.9838	1.9851	1.9865
7.3	1.9879	1.9892	1.9906	1.9920	1.9933	1.9947	1.9961	1.9974	1.9988	2.0001
7.4	2.0015	2.0028	2.0042	2.0055	2.0069	2.0082	2.0096	2.0109	2.0122	2.0136
7.5	2.0149	2.0162	2.0176	2.0189	2.0202	2.0215	2.0229	2.0242	2.0255	2.0268
7.6	2.0282	2.0295	2.0308	2.0321	2.0334	2.0347	2.0360	2.0373	2.0386	2.0399
7.7	2.0412	2.0425	2.0438	2.0451	2.0464	2.0477	2.0490	2.0503	2.0516	2.0528
7.8	2.0541	2.0554	2.0567	2.0580	2.0592	2.0605	2.0618	2.0631	2.0643	2.0665
7.9	2.0669	2.0681	2.0694	2.0707	2.0719	2.0732	2.0744	2.0757	2.0769	2.0782
8.0	2.0794	2.0807	2.0819	2.0832	2.0844	2.0857	2.0869	2.0882	2.0894	2.0906
8.1	2.0919	2.0931	2.0943	2.0956	2.0968	2.0980	2.0992	2.1005	2.1017	2.1029
8.2	2.1041	2.1054	2.1066	2.1078	2.1090	2.1102	2.1114	2.1126	2.1138	2.1150
8.3	2.1163	2.1175	2.1187	2.1199	2.1211	2.1223	2.1235	2.1247	2.1258	2.1270
8.4	2.1282	2.1294	2.1306	2.1318	2.1330	2.1342	2.1353	2.1365	2.1377	2.1389
8.5	2.1401	2.1412	2.1424	2.1436	2.1448	2.1459	2.1471	2.1483	2.1494	2.1506
8.6	2.1518	2.1529	2.1541	2.1552	2.1564	2.1576	2.1587	2.1599	2.1610	2.1622
8.7	2.1633	2.1645	2.1656	2.1668	2.1679	2.1691	2.1702	2.1713	2.1725	2.1736
8.8	2.1748	2.1759	2.1770	2.1782	2.1793	2.1804	2.1815	2.1827	2.1838	2.1849
8.9	2.1861	2.1872	2.1883	2.1894	2.1905	2.1917	2.1928	2.1939	2.1950	2.1961
9.0	2.1972	2.1983	2.1994	2.2006	2.2017	2.2028	2.2039	2.2050	2.2061	2.2072
9.1	2.2083	2.2094	2.2105	2.2116	2.2127	2.2138	2.2148	2.2159	2.2170	2.2181
9.2	2.2192	2.2203	2.2214	2.2225	2.2235	2.2246	2.2257	2.2268	2.2279	2.2289
9.3	2.2300	2.2311	2.2322	2.2332	2.2343	2.2354	2.2364	2.2375	2.2386	2.2396
9.4	2.2407	2.2418	2.2428	2.2439	2.2450	2.2460	2.2471	2.2481	2.2492	2.2502
9.5	2.2513	2.2523	2.2534	2.2544	2.2555	2.2565	2.2576	2.2586	2.2597	2.2607
9.6	2.2618	2.2628	2.2638	2.2649	2.2659	2.2670	2.2680	2.2690	2.2701	2.2711
9.7	2.2721	2.2732	2.2742	2.2752	2.2762	2.2773	2.2783	2.2793	2.2803	2.2814
9.8	2.2824	2.2834	2.2844	2.2854	2.2865	2.2875	2.2885	2.2895	2.2905	2.2915
9.9	2.2925	2.2935	2.2946	2.2956	2.2966	2.2976	2.2986	2.2996	2.3006	2.3016

TABLE IB   COMMON LOGARITHMS

x	0.00	0.01	0.02	0.03	0.04	0.05	0.06	0.07	0.08	0.09
1.0	0.0000	0.0043	0.0086	0.0128	0.0170	0.0212	0.0253	0.0294	0.0334	0.0374
1.1	0.0414	0.0453	0.0492	0.0531	0.0569	0.0607	0.0645	0.0682	0.0719	0.0755
1.2	0.0792	0.0828	0.0864	0.0899	0.0934	0.0969	0.1004	0.1038	0.1072	0.1106
1.3	0.1139	0.1173	0.1206	0.1239	0.1271	0.1303	0.1335	0.1367	0.1399	0.1430
1.4	0.1461	0.1492	0.1523	0.1553	0.1584	0.1614	0.1644	0.1673	0.1703	0.1732
1.5	0.1761	0.1790	0.1818	0.1847	0.1875	0.1903	0.1931	0.1959	0.1987	0.2014
1.6	0.2041	0.2068	0.2095	0.2122	0.2148	0.2175	0.2201	0.2227	0.2253	0.2279
1.7	0.2304	0.2330	0.2355	0.2380	0.2405	0.2430	0.2455	0.2480	0.2504	0.2529
1.8	0.2553	0.2577	0.2601	0.2625	0.2648	0.2672	0.2695	0.2718	0.2742	0.2765
1.9	0.2788	0.2810	0.2833	0.2856	0.2878	0.2900	0.2923	0.2945	0.2967	0.2989
2.0	0.3010	0.3032	0.3054	0.3075	0.3096	0.3118	0.3139	0.3160	0.3181	0.3201
2.1	0.3222	0.3243	0.3263	0.3284	0.3304	0.3324	0.3345	0.3365	0.3385	0.3404
2.2	0.3424	0.3444	0.3464	0.3483	0.3502	0.3522	0.3541	0.3560	0.3579	0.3598
2.3	0.3617	0.3636	0.3655	0.3674	0.3692	0.3711	0.3729	0.3747	0.3766	0.3784
2.4	0.3802	0.3820	0.3838	0.3856	0.3874	0.3892	0.3909	0.3927	0.3945	0.3962
2.5	0.3979	0.3997	0.4014	0.4031	0.4048	0.4065	0.4082	0.4099	0.4116	0.4133
2.6	0.4150	0.4166	0.4183	0.4200	0.4216	0.4232	0.4249	0.4265	0.4281	0.4298
2.7	0.4314	0.4330	0.4346	0.4362	0.4378	0.4393	0.4409	0.4425	0.4440	0.4456
2.8	0.4472	0.4487	0.4502	0.4518	0.4533	0.4548	0.4564	0.4579	0.4594	0.4609
2.9	0.4624	0.4639	0.4654	0.4669	0.4683	0.4698	0.4713	0.4728	0.4742	0.4757
3.0	0.4771	0.4786	0.4800	0.4814	0.4829	0.4843	0.4857	0.4871	0.4886	0.4900
3.1	0.4914	0.4928	0.4942	0.4955	0.4969	0.4983	0.4997	0.5011	0.5024	0.5038
3.2	0.5051	0.5065	0.5079	0.5092	0.5105	0.5119	0.5132	0.5145	0.5159	0.5172
3.3	0.5185	0.5198	0.5211	0.5224	0.5237	0.5250	0.5263	0.5276	0.5289	0.5302
3.4	0.5315	0.5328	0.5340	0.5353	0.5366	0.5378	0.5391	0.5403	0.5416	0.5428
3.5	0.5441	0.5453	0.5465	0.5478	0.5490	0.5502	0.5514	0.5527	0.5539	0.5551
3.6	0.5563	0.5575	0.5587	0.5599	0.5611	0.5623	0.5635	0.5647	0.5658	0.5670
3.7	0.5682	0.5694	0.5705	0.5717	0.5729	0.5740	0.5752	0.5763	0.5775	0.5786
3.8	0.5798	0.5809	0.5821	0.5832	0.5843	0.5855	0.5866	0.5877	0.5888	0.5899
3.9	0.5911	0.5922	0.5933	0.5944	0.5955	0.5966	0.5977	0.5988	0.5999	0.6010
4.0	0.6021	0.6031	0.6042	0.6053	0.6064	0.6075	0.6085	0.6096	0.6107	0.6117
4.1	0.6128	0.6138	0.6149	0.6160	0.6170	0.6180	0.6191	0.6201	0.6212	0.6222
4.2	0.6232	0.6243	0.6253	0.6263	0.6274	0.6284	0.6294	0.6304	0.6314	0.6325
4.3	0.6335	0.6345	0.6355	0.6365	0.6375	0.6385	0.6395	0.6405	0.6415	0.6425
4.4	0.6435	0.6444	0.6454	0.6464	0.6474	0.6484	0.6493	0.6503	0.6513	0.6522
4.5	0.6532	0.6542	0.6551	0.6561	0.6571	0.6580	0.6590	0.6599	0.6609	0.6618
4.6	0.6628	0.6637	0.6646	0.6656	0.6665	0.6675	0.6684	0.6693	0.6702	0.6712
4.7	0.6721	0.6730	0.6739	0.6749	0.6758	0.6767	0.6776	0.6785	0.6794	0.6803
4.8	0.6812	0.6821	0.6830	0.6839	0.6848	0.6857	0.6866	0.6875	0.6884	0.6893
4.9	0.6902	0.6911	0.6920	0.6928	0.6937	0.6946	0.6955	0.6964	0.6972	0.6981
5.0	0.6990	0.6998	0.7007	0.7016	0.7024	0.7033	0.7042	0.7050	0.7059	0.7067
5.1	0.7076	0.7084	0.7093	0.7101	0.7110	0.7118	0.7126	0.7135	0.7143	0.7152
5.2	0.7160	0.7168	0.7177	0.7185	0.7193	0.7202	0.7210	0.7218	0.7226	0.7235
5.3	0.7243	0.7251	0.7259	0.7267	0.7275	0.7284	0.7292	0.7300	0.7308	0.7316
5.4	0.7324	0.7332	0.7340	0.7348	0.7356	0.7364	0.7372	0.7380	0.7388	0.7396
5.5	0.7404	0.7412	0.7419	0.7427	0.7435	0.7443	0.7451	0.7459	0.7466	0.7474
5.6	0.7482	0.7490	0.7497	0.7505	0.7513	0.7520	0.7528	0.7536	0.7543	0.7551
5.7	0.7559	0.7566	0.7574	0.7582	0.7589	0.7597	0.7604	0.7612	0.7619	0.7627
5.8	0.7634	0.7642	0.7649	0.7657	0.7664	0.7672	0.7679	0.7686	0.7694	0.7701
5.9	0.7709	0.7716	0.7723	0.7731	0.7738	0.7745	0.7752	0.7760	0.7767	0.7774

x	0.00	0.01	0.02	0.03	0.04	0.05	0.06	0.07	0.08	0.09
6.0	0.7782	0.7789	0.7796	0.7803	0.7810	0.7818	0.7825	0.7832	0.7839	0.7846
6.1	0.7853	0.7860	0.7868	0.7875	0.7882	0.7889	0.7896	0.7903	0.7910	0.7917
6.2	0.7924	0.7931	0.7938	0.7945	0.7952	0.7959	0.7966	0.7973	0.7980	0.7987
6.3	0.7993	0.8000	0.8007	0.8014	0.8021	0.8028	0.8035	0.8041	0.8048	0.8055
6.4	0.8062	0.8069	0.8075	0.8082	0.8089	0.8096	0.8102	0.8109	0.8116	0.8122
6.5	0.8129	0.8136	0.8142	0.8149	0.8156	0.8162	0.8169	0.8176	0.8182	0.8189
6.6	0.8195	0.8202	0.8209	0.8215	0.8222	0.8228	0.8235	0.8241	0.8248	0.8254
6.7	0.8261	0.8267	0.8274	0.8280	0.8287	0.8293	0.8299	0.8306	0.8312	0.8319
6.8	0.8325	0.8331	0.8338	0.8344	0.8351	0.8357	0.8363	0.8370	0.8376	0.8382
6.9	0.8388	0.8395	0.8401	0.8407	0.8414	0.8420	0.8426	0.8432	0.8439	0.8445
7.0	0.8451	0.8457	0.8463	0.8470	0.8476	0.8482	0.8488	0.8494	0.8500	0.8506
7.1	0.8513	0.8519	0.8525	0.8531	0.8537	0.8543	0.8549	0.8555	0.8561	0.8567
7.2	0.8573	0.8579	0.8585	0.8591	0.8597	0.8603	0.8609	0.8615	0.8621	0.8627
7.3	0.8633	0.8639	0.8645	0.8651	0.8657	0.8663	0.8669	0.8675	0.8681	0.8686
7.4	0.8692	0.8698	0.8704	0.8710	0.8716	0.8722	0.8727	0.8733	0.8739	0.8745
7.5	0.8751	0.8756	0.8762	0.8768	0.8774	0.8779	0.8785	0.8791	0.8797	0.8802
7.6	0.8808	0.8814	0.8820	0.8825	0.8831	0.8837	0.8842	0.8848	0.8854	0.8859
7.7	0.8865	0.8871	0.8876	0.8882	0.8887	0.8893	0.8899	0.8904	0.8910	0.8915
7.8	0.8921	0.8927	0.8932	0.8938	0.8943	0.8949	0.8954	0.8960	0.8965	0.8971
7.9	0.8976	0.8982	0.8987	0.8993	0.8998	0.9004	0.9009	0.9015	0.9020	0.9025
8.0	0.9031	0.9036	0.9042	0.9047	0.9053	0.9058	0.9063	0.9069	0.9074	0.9079
8.1	0.9085	0.9090	0.9096	0.9101	0.9106	0.9112	0.9117	0.9122	0.9128	0.9133
8.2	0.9138	0.9143	0.9149	0.9154	0.9159	0.9165	0.9170	0.9175	0.9180	0.9186
8.3	0.9191	0.9196	0.9201	0.9206	0.9212	0.9217	0.9222	0.9227	0.9232	0.9238
8.4	0.9243	0.9248	0.9253	0.9258	0.9263	0.9269	0.9274	0.9279	0.9284	0.9289
8.5	0.9294	0.9299	0.9304	0.9309	0.9315	0.9320	0.9325	0.9330	0.9335	0.9340
8.6	0.9345	0.9350	0.9355	0.9360	0.9365	0.9370	0.9375	0.9380	0.9385	0.9390
8.7	0.9395	0.9400	0.9405	0.9410	0.9415	0.9420	0.9425	0.9430	0.9435	0.9440
8.8	0.9445	0.9450	0.9455	0.9460	0.9465	0.9469	0.9474	0.9479	0.9484	0.9489
8.9	0.9494	0.9499	0.9504	0.9509	0.9513	0.9518	0.9523	0.9528	0.9533	0.9538
9.0	0.9542	0.9547	0.9552	0.9557	0.9562	0.9566	0.9571	0.9576	0.9581	0.9586
9.1	0.9590	0.9595	0.9600	0.9605	0.9609	0.9614	0.9619	0.9624	0.9628	0.9633
9.2	0.9638	0.9643	0.9647	0.9652	0.9657	0.9661	0.9666	0.9671	0.9675	0.9680
9.3	0.9685	0.9689	0.9694	0.9699	0.9703	0.9708	0.9713	0.9717	0.9722	0.9727
9.4	0.9731	0.9736	0.9741	0.9745	0.9750	0.9754	0.9759	0.9763	0.9768	0.9773
9.5	0.9777	0.9782	0.9786	0.9791	0.9795	0.9800	0.9805	0.9809	0.9814	0.9818
9.6	0.9823	0.9827	0.9832	0.9836	0.9841	0.9845	0.9850	0.9854	0.9859	0.9863
9.7	0.9868	0.9872	0.9877	0.9881	0.9886	0.9890	0.9894	0.9899	0.9903	0.9908
9.8	0.9912	0.9917	0.9921	0.9926	0.9930	0.9934	0.9939	0.9943	0.9948	0.9952
9.9	0.9956	0.9961	0.9965	0.9969	0.9974	0.9978	0.9983	0.9987	0.9991	0.9996

TABLE IC EXPONENTIAL FUNCTIONS

x	e^x	e^{-x}	x	e^x	e^{-x}
0.00	1.0000	1.0000	3.0	20.086	0.0498
0.05	1.0513	0.9512	3.1	22.198	0.0450
0.10	1.1052	0.9048	3.2	24.533	0.0408
0.15	1.1618	0.8607	3.3	27.113	0.0369
0.20	1.2214	0.8187	3.4	29.964	0.0334
0.25	1.2840	0.7788	3.5	33.115	0.0302
0.30	1.3499	0.7408	3.6	36.598	0.0273
0.35	1.4191	0.7047	3.7	40.447	0.0247
0.40	1.4918	0.6703	3.8	44.701	0.0224
0.45	1.5683	0.6376	3.9	49.402	0.0202
0.50	1.6487	0.6065	4.0	54.598	0.0183
0.55	1.7333	0.5769	4.1	60.340	0.0166
0.60	1.8221	0.5488	4.2	66.686	0.0150
0.65	1.9155	0.5220	4.3	73.700	0.0136
0.70	2.0138	0.4966	4.4	81.451	0.0123
0.75	2.1170	0.4724	4.5	90.017	0.0111
0.80	2.2255	0.4493	4.6	99.484	0.0101
0.85	2.3396	0.4274	4.7	109.95	0.0091
0.90	2.4596	0.4066	4.8	121.51	0.0082
0.95	2.5857	0.3867	4.9	134.29	0.0074
1.0	2.7183	0.3679	5.0	148.41	0.0067
1.1	3.0042	0.3329	5.1	164.02	0.0061
1.2	3.3201	0.3012	5.2	181.27	0.0055
1.3	3.6693	0.2725	5.3	200.34	0.0050
1.4	4.0552	0.2466	5.4	221.41	0.0045
1.5	4.4817	0.2231	5.5	244.69	0.0041
1.6	4.9530	0.2019	5.6	270.43	0.0037
1.7	5.4739	0.1827	5.7	298.87	0.0033
1.8	6.0496	0.1653	5.8	330.30	0.0030
1.9	6.6859	0.1496	5.9	365.04	0.0027
2.0	7.3891	0.1353	6.0	403.43	0.0025
2.1	8.1662	0.1225	6.5	665.14	0.0015
2.2	9.0250	0.1108	7.0	1096.6	0.0009
2.3	9.9742	0.1003	7.5	1808.0	0.0006
2.4	11.023	0.0907	8.0	2981.0	0.0003
2.5	12.182	0.0821	8.5	4914.8	0.0002
2.6	13.464	0.0743	9.0	8103.1	0.0001
2.7	14.880	0.0672	9.5	13,360	0.00007
2.8	16.445	0.0608	10.0	22,026	0.00004
2.9	18.174	0.0550			

II Tables of Trigonometric Functions

In this appendix, we present Table IIA and Table IIB of trigonometric functions of angles measured in degrees and radians, and we give examples illustrating their use.

Table IIA gives values of sine, tangent, cotangent, and cosine, rounded off to four or five significant digits, corresponding to angles between 0° and 90° in steps of 0.1°. Angles between 0° and 45° are found in the vertical columns on the left sides of the tables, and the captions at the tops of the tables apply to these angles. Angles between 45° and 90° are found in the vertical columns on the right sides of the tables, and the captions at the bottoms of the tables apply to these angles.

Examples   Use Table IIA to find the approximate value of each trigonometric function.

1 $\cos 29.3°$

Solution   Since $29.3° < 45°$, we begin by locating 29.3 in the vertical column on the *left* side of the table. As can be seen from the captions *above* the table, the numbers to the *right* of 29.3 are the sine, tangent, cotangent, and cosine of 29.3°. Looking in the vertical column with the caption "cosine," we find that

$$\cos 29.3° = 0.8721.$$

2 $\tan 128.8°$

Solution   The angle 128.8° lies in quadrant II, so its tangent is negative. The reference angle corresponding to 128.8° is

$$180° - 128.8° = 51.2°.$$

Thus,

$$\tan 128.8° = -\tan 51.2°.$$

Since $51.2° > 45°$, we locate 51.2 in the vertical column on the *right* side of the table. As can be seen from the captions *below* the table, the numbers to the *left* of 51.2 are the sine, tangent, cotangent, and cosine of 51.2°. Thus,

$$\tan 51.2° = 1.2437, \quad \text{so that} \quad \tan 128.8° = -1.2437.$$

3   csc 72.3°

Solution   Since values of the cosecant do not appear in the table, we use the fact that

$$\csc 72.3° = \frac{1}{\sin 72.3°}.$$

From the table,

$$\sin 72.3° = 0.9527, \quad \text{so} \quad \csc 72.3° = \frac{1}{0.9527} = 1.050,$$

(rounded off to four significant digits).

Linear interpolation can be used to find the values of trigonometric functions of angles lying between two angles in Table IIA. The basic idea of linear interpolation is that small changes in the angle cause approximately proportional changes in the values of the function.

Example   Use linear interpolation and Table IIA to find the approximate value of sin 19.74°.

Solution   We arrange the work as follows:

$$0.1°\left[0.04°\left[\begin{array}{cc} x & \sin x \\ 19.7° & 0.3371 \\ 19.74° & \sin 19.74° \\ 19.8° & 0.3387 \end{array}\right]d\right]0.0016.$$

Here, we have used Table IIA to determine that sin 19.7° = 0.3371 and that sin 19.8° = 0.3387. The notation

$$0.04°\left[\begin{array}{c} 19.7° \\ 19.74° \end{array}\right.$$

indicates that the difference between 19.7° and 19.74° (bottom number minus top number) is

$$19.74° - 19.7° = 0.04°.$$

Other differences are indicated similarly. For linear interpolation, we assume that corresponding differences are (approximately) proportional, so that

$$\frac{0.04°}{0.1°} = \frac{d}{0.0016} \quad \text{or} \quad d = 0.0016\left(\frac{0.04}{0.1}\right) = 0.00064.$$

Thus, rounded off to four significant digits,

$$\sin 19.74° = 0.3371 + d$$

$$= 0.3371 + 0.00064$$

$$= 0.3377.$$

By reading Table IIA "backwards," you can find values in degrees of the inverse trigonometric functions.

Examples   Using Table IIA, find the value of each inverse trigonometric function rounded off to the nearest 0.1°.

1 $\cos^{-1} 0.3714$

Solution   Looking in the body of the table, we find 0.3714 in the vertical column with the *lower* caption "cos." By reading the table "backwards," we see that

$$\cos 68.2° = 0.3714 \quad \text{or} \quad \cos^{-1} 0.3714 = 68.2°.$$

2 $\sin^{-1} 0.7320$

Solution   The number 0.7320 does not appear in the vertical sine column in Table IIA. The numbers closest to it are

$$0.7314 = \sin 47.0° \quad \text{and} \quad 0.7325 = \sin 47.1°.$$

Using linear interpolation, we have

$$0.0011 \left[0.0006 \begin{bmatrix} x & \sin^{-1} x \\ 0.7314 & 47.0° \\ 0.7320 & \sin^{-1} 0.7320 \end{bmatrix} d \right] 0.1°.$$
$$0.7325 \qquad 47.1°$$

Thus,

$$\frac{0.0006}{0.0011} = \frac{d}{0.1°} \quad \text{or} \quad d = 0.1° \left(\frac{0.0006}{0.0011} \right) = 0.05°$$

(rounded off to the nearest 0.01°), so that

$$\sin^{-1} 0.7320 = 47.0° + d$$
$$= 47.0° + 0.05°$$
$$= 47.05°.$$

Table IIB gives values of the six trigonometric functions, rounded off to four or five significant digits, corresponding to angles between 0 and $\pi/2$ radians in steps of 0.01 radian.

Examples   Use Table IIB to find the approximate value of each trigonometric function.

1 $\csc 1.02$

Solution   By Table IIB
$$\csc 1.02 = 1.174.$$

2 $\sin 3.538$

Solution   In radians, $\pi < 3.538 < (3\pi/2)$, so 3.538 is a quadrant III angle and its sine is negative. The reference angle corresponding to 3.538 is

$$3.538 - \pi = 0.396$$

(rounded off to three decimal places), so

$$\sin 3.538 = -\sin 0.396.$$

Using Table IIB and linear interpolation, we have

$$0.01\left[0.006\left[\begin{matrix} x & \sin x \\ 0.39 & 0.3802 \\ 0.396 & \sin 0.396 \\ 0.40 & 0.3894 \end{matrix}\right]d\right]0.0092.$$

Thus,

$$\frac{0.006}{0.01} = \frac{d}{0.0092} \quad \text{or} \quad d = 0.0092\left(\frac{0.006}{0.01}\right) = 0.0055;$$

so that, rounded off to four decimal places,

$$\sin 0.396 = 0.3802 + d$$
$$= 0.3802 + 0.0055$$
$$= 0.3857.$$

Therefore,

$$\sin 3.538 = -0.3857.$$

(Because of all the rounding off and the linear interpolation, this value isn't very accurate—a calculator gives

$$\sin 3.538 = -0.3861.$$

However, it's the best we can do with our table.)

By reading Table IIB "backwards," you can find values in radians of the inverse trigonometric functions.

Examples   Using Table IIB, find the value of each inverse trigonometric function rounded off to the nearest 0.01 radian.

1 $\arctan 2.427$

Solution   Looking in the body of the table in the vertical column for tangents, we find 2.427 in the position corresponding to 1.18 radians. Hence,

$$\arctan 2.427 = 1.18.$$

2 $\cos^{-1} 0.8932$

Solution   Since 0.8932 does not appear in the body of the table in the vertical column for cosines, we must use linear interpolation. Thus,

$$-0.0045\left[-0.0029\left[\begin{array}{cc} x & \cos^{-1} x \\ 0.8961 & 0.46 \\ 0.8932 & \cos^{-1} 0.8932 \end{array}\right]d \atop 0.8916 \quad 0.47 \right]0.01.$$

It follows that

$$\frac{-0.0029}{-0.0045} = \frac{d}{0.01} \quad \text{or} \quad d = 0.01\left(\frac{-0.0029}{-0.0045}\right) = 0.006$$

(rounded off to three decimal places), and so

$$\cos^{-1} 0.8932 = 0.46 + d$$
$$= 0.46 + 0.006$$
$$= 0.466.$$

TABLE IIA   TRIGONOMETRIC FUNCTIONS—DEGREE MEASURE

Deg.	sin	tan	cot	cos		Deg.	sin	tan	cot	cos	
0.0	0.00000	0.00000	∞	1.0000	90.0	5.0	0.08716	0.08749	11.430	0.9962	85.0
.1	.00175	.00175	573.0	1.0000	89.9	.1	.08889	.08925	11.205	.9960	84.9
.2	.00349	.00349	286.5	1.0000	.8	.2	.09063	.09101	10.988	.9959	.8
.3	.00524	.00524	191.0	1.0000	.7	.3	.09237	.09277	10.780	.9957	.7
.4	.00698	.00698	143.24	1.0000	.6	.4	.09411	.09453	10.579	.9956	.6
.5	.00873	.00873	114.59	1.0000	.5	.5	.09585	.09629	10.385	.9954	.5
.6	.01047	.01047	95.49	0.9999	.4	.6	.09758	.09805	10.199	.9952	.4
.7	.01222	.01222	81.85	.9999	.3	.7	.09932	.09981	10.019	.9951	.3
.8	.01396	.01396	71.62	.9999	.2	.8	.10106	.10158	9.845	.9949	.2
.9	.01571	.01571	63.66	.9999	89.1	.9	.10279	.10334	9.677	.9947	84.1
1.0	0.01745	0.01746	57.29	0.9998	89.0	6.0	0.10453	0.10510	9.514	0.9945	84.0
.1	.01920	.01920	52.08	.9998	88.9	.1	.10626	.10687	9.357	.9943	83.9
.2	.02094	.02095	47.74	.9998	.8	.2	.10800	.10863	9.205	.9942	.8
.3	.02269	.02269	44.07	.9997	.7	.3	.10973	.11040	9.058	.9940	.7
.4	.02443	.02444	40.92	.9997	.6	.4	.11147	.11217	8.915	.9938	.6
.5	.02618	.02619	38.19	.9997	.5	.5	.11320	.11394	8.777	.9936	.5
.6	.02792	.02793	35.80	.9996	.4	.6	.11494	.11570	8.643	.9934	.4
.7	.02967	.02968	33.69	.9996	.3	.7	.11667	.11747	8.513	.9932	.3
.8	.03141	.03143	31.82	.9995	.2	.8	.11840	.11924	8.386	.9930	.2
.9	.03316	.03317	30.14	.9995	88.1	.9	.12014	.12101	8.264	.9928	83.1
2.0	0.03490	0.03492	28.64	0.9994	88.0	7.0	0.12187	0.12278	8.144	0.9925	83.0
.1	.03664	.03667	27.27	.9993	87.9	.1	.12360	.12456	8.028	.9923	82.9
.2	.03839	.03842	26.03	.9993	.8	.2	.12533	.12633	7.916	.9921	.8
.3	.04013	.04016	24.90	.9992	.7	.3	.12706	.12810	7.806	.9919	.7
.4	.04188	.04191	23.86	.9991	.6	.4	.12880	.12988	7.700	.9917	.6
.5	.04362	.04366	22.90	.9990	.5	.5	.13053	.13165	7.596	.9914	.5
.6	.04536	.04541	22.02	.9990	.4	.6	.13226	.13343	7.495	.9912	.4
.7	.04711	.04716	21.20	.9989	.3	.7	.13399	.13521	7.396	.9910	.3
.8	.04885	.04891	20.45	.9988	.2	.8	.13572	.13698	7.300	.9907	.2
.9	.05059	.05066	19.74	.9987	87.1	.9	.13744	.13876	7.207	.9905	82.1
3.0	0.05234	0.05241	19.081	0.9986	87.0	8.0	0.13917	0.14054	7.115	0.9903	82.0
.1	.05408	.05416	18.464	.9985	86.9	.1	.14090	.14232	7.026	.9900	81.9
.2	.05582	.05591	17.886	.9984	.8	.2	.14263	.14410	6.940	.9898	.8
.3	.05756	.05766	17.343	.9983	.7	.3	.14436	.14588	6.855	.9895	.7
.4	.05931	.05941	16.832	.9982	.6	.4	.14608	.14767	6.772	.9893	.6
.5	.06105	.06116	16.350	.9981	.5	.5	.14781	.14945	6.691	.9890	.5
.6	.06279	.06291	15.895	.9980	.4	.6	.14954	.15124	6.612	.9888	.4
.7	.06453	.06467	15.464	.9979	.3	.7	.15126	.15302	6.535	.9885	.3
.8	.06627	.06642	15.056	.9978	.2	.8	.15299	.15481	6.460	.9882	.2
.9	.06802	.06817	14.669	.9977	86.1	.9	.15471	.15660	6.386	.9880	81.1
4.0	0.06976	0.06993	14.301	0.9976	86.0	9.0	0.15643	0.15838	6.314	0.9877	81.0
.1	.07150	.07168	13.951	.9974	85.9	.1	.15816	.16017	6.243	.9874	80.9
.2	.07324	.07344	13.617	.9973	.8	.2	.15988	.16196	6.174	.9871	.8
.3	.07498	.07519	13.300	.9972	.7	.3	.16160	.16376	6.107	.9869	.7
.4	.07672	.07695	12.996	.9971	.6	.4	.16333	.16555	6.041	.9866	.6
.5	.07846	.07870	12.706	.9969	.5	.5	.16505	.16734	5.976	.9863	.5
.6	.08020	.08046	12.429	.9968	.4	.6	.16677	.16914	5.912	.9860	.4
.7	.08194	.08221	12.163	.9966	.3	.7	.16849	.17093	5.850	.9857	.3
.8	.08368	.08397	11.909	.9965	.2	.8	.17021	.17273	5.789	.9854	.2
.9	.08542	.08573	11.664	.9963	85.1	.9	.17193	.17453	5.730	.9851	80.1
5.0	0.08716	0.08749	11.430	0.9962	85.0	10.0	0.1736	0.1763	5.671	0.9848	80.0
	cos	cot	tan	sin	Deg.		cos	cot	tan	sin	Deg.

Deg.	sin	tan	cot	cos		Deg.	sin	tan	cot	cos	
10.0	0.1736	0.1763	5.671	0.9848	80.0	15.0	0.2588	0.2679	3.732	0.9659	75.0
.1	.1754	.1781	5.614	.9845	79.9	.1	.2605	.2698	3.706	.9655	74.9
.2	.1771	.1799	5.558	.9842	.8	.2	.2622	.2717	3.681	.9650	.8
.3	.1788	.1817	5.503	.9839	.7	.3	.2639	.2736	3.655	.9646	.7
.4	.1805	.1835	5.449	.9836	.6	.4	.2656	.2754	3.630	.9641	.6
.5	.1822	.1853	5.396	.9833	.5	.5	.2672	.2773	3.606	.9636	.5
.6	.1840	.1871	5.343	.9829	.4	.6	.2689	.2792	3.582	.9632	.4
.7	.1857	.1890	5.292	.9826	.3	.7	.2706	.2811	3.558	.9627	.3
.8	.1874	.1908	5.242	.9823	.2	.8	.2723	.2830	3.534	.9622	.2
.9	.1891	.1926	5.193	.9820	79.1	.9	.2740	.2849	3.511	.9617	74.1
11.0	0.1908	0.1944	5.145	0.9816	79.0	16.0	0.2756	0.2867	3.487	0.9613	74.0
.1	.1925	.1962	5.079	.9813	78.9	.1	.2773	.2886	3.465	.9608	73.9
.2	.1942	.1980	5.050	.9810	.8	.2	.2790	.2905	3.442	.9603	.8
.3	.1959	.1998	5.005	.9806	.7	.3	.2807	.2924	3.420	.9598	.7
.4	.1977	.2016	4.959	.9803	.6	.4	.2823	.2943	3.398	.9593	.6
.5	.1994	.2035	4.915	.9799	.5	.5	.2840	.2962	3.376	.9588	.5
.6	.2011	.2053	4.872	.9796	.4	.6	.2857	.2981	3.354	.9583	.4
.7	.2028	.2071	4.829	.9792	.3	.7	.2874	.3000	3.333	.9578	.3
.8	.2045	.2089	4.787	.9789	.2	.8	.2890	.3019	3.312	.9573	.2
.9	.2062	.2107	4.745	.9785	78.1	.9	.2907	.3038	3.291	.9568	73.1
12.0	0.2079	0.2126	4.705	0.9781	78.0	17.0	0.2924	0.3057	3.271	0.9563	73.0
.1	.2096	.2144	4.665	.9778	77.9	.1	.2940	.3076	3.251	.9558	72.9
.2	.2113	.2162	4.625	.9774	.8	.2	.2957	.3096	3.230	.9553	.8
.3	.2130	.2180	4.586	.9770	.7	.3	.2974	.3115	3.211	.9548	.7
.4	.2147	.2199	4.548	.9767	.6	.4	.2990	.3134	3.191	.9542	.6
.5	.2164	.2217	4.511	.9763	.5	.5	.3007	.3153	3.172	.9537	.5
.6	.2181	.2235	4.474	.9759	.4	.6	.3024	.3172	3.152	.9532	.4
.7	.2198	.2254	4.437	.9755	.3	.7	.3040	.3191	3.133	.9527	.3
.8	.2215	.2272	4.402	.9751	.2	.8	.3057	.3211	3.115	.9521	.2
.9	.2233	.2290	4.366	.9748	77.1	.9	.3074	.3230	3.096	.9516	72.1
13.0	0.2250	0.2309	4.331	0.9744	77.0	18.0	0.3090	0.3249	3.078	0.9511	72.0
.1	.2267	.2327	4.297	.9740	76.9	.1	.3107	.3269	3.060	.9505	71.9
.2	.2284	.2345	4.264	.9736	.8	.2	.3123	.3288	3.042	.9500	.8
.3	.2300	.2364	4.230	.9732	.7	.3	.3140	.3307	3.024	.9494	.7
.4	.2317	.2382	4.198	.9728	.6	.4	.3156	.3327	3.006	.9489	.6
.5	.2334	.2401	4.165	.9724	.5	.5	.3173	.3346	2.989	.9483	.5
.6	.2351	.2419	4.134	.9720	.4	.6	.3190	.3365	2.971	.9478	.4
.7	.2368	.2438	4.102	.9715	.3	.7	.3206	.3385	2.954	.9472	.3
.8	.2385	.2456	4.071	.9711	.2	.8	.3223	.3404	2.937	.9466	.2
.9	.2402	.2475	4.041	.9707	76.1	.9	.3239	.3424	2.921	.9461	71.1
14.0	0.2419	0.2493	4.011	0.9703	76.0	19.0	0.3256	0.3443	2.904	0.9455	71.0
.1	.2436	.2512	3.981	.9699	75.9	.1	.3272	.3463	2.888	.9449	70.9
.2	.2453	.2530	3.952	.9694	.8	.2	.3289	.3482	2.872	.9444	.8
.3	.2470	.2549	3.923	.9690	.7	.3	.3305	.3502	2.856	.9438	.7
.4	.2487	.2568	3.895	.9686	.6	.4	.3322	.3522	2.840	.9432	.6
.5	.2504	.2586	3.867	.9681	.5	.5	.3338	.3541	2.824	.9426	.5
.6	.2521	.2605	3.839	.9677	.4	.6	.3355	.3561	2.808	.9421	.4
.7	.2538	.2623	3.812	.9673	.3	.7	.3371	.3581	2.793	.9415	.3
.8	.2554	.2642	3.785	.9668	.2	.8	.3387	.3600	2.778	.9409	.2
.9	.2571	.2661	3.758	.9664	75.1	.9	.3404	.3620	2.762	.9403	70.1
15.0	0.2588	0.2679	3.732	0.9659	75.0	20.0	0.3420	0.3640	2.747	0.9397	70.0
	cos	cot	tan	sin	Deg.		cos	cot	tan	sin	Deg.

TABLE IIA TRIGONOMETRIC FUNCTIONS—DEGREE MEASURE

Deg.	sin	tan	cot	cos		Deg.	sin	tan	cot	cos	
20.0	0.3420	0.3640	2.747	0.9397	70.0	25.0	0.4226	0.4663	2.145	0.9063	65.0
.1	.3437	.3659	2.733	.9391	69.9	.1	.4242	.4684	2.135	.9056	64.9
.2	.3453	.3679	2.718	.9385	.8	.2	.4258	.4706	2.125	.9048	.8
.3	.3469	.3699	2.703	.9379	.7	.3	.4274	.4727	2.116	.9041	.7
.4	.3486	.3719	2.689	.9373	.6	.4	.4289	.4748	2.106	.9033	.6
.5	.3502	.3739	2.675	.9367	.5	.5	.4305	.4770	2.097	.9026	.5
.6	.3518	.3759	2.660	.9361	.4	.6	.4321	.4791	2.087	.9018	.4
.7	.3535	.3779	2.646	.9354	.3	.7	.4337	.4813	2.078	.9011	.3
.8	.3551	.3799	2.633	.9348	.2	.8	.4352	.4834	2.069	.9003	.2
.9	.3567	.3819	2.619	.9342	69.1	.9	.4368	.4856	2.059	.8996	64.1
21.0	0.3584	0.3839	2.605	0.9336	69.0	26.0	0.4384	0.4887	2.050	0.8988	64.0
.1	.3600	.3859	2.592	.9330	68.9	.1	.4399	.4899	2.041	.8980	63.9
.2	.3616	.3879	2.578	.9323	.8	.2	.4415	.4921	2.032	.8973	.8
.3	.3633	.3899	2.565	.9317	.7	.3	.4431	.4942	2.023	.8965	.7
.4	.3649	.3919	2.552	.9311	.6	.4	.4446	.4964	2.014	.8957	.6
.5	.3665	.3939	2.539	.9304	.5	.5	.4462	.4986	2.006	.8949	.5
.6	.3681	.3959	2.526	.9298	.4	.6	.4478	.5008	1.997	.8942	.4
.7	.3697	.3979	2.513	.9291	.3	.7	.4493	.5029	1.988	.8934	.3
.8	.3714	.4000	2.500	.9285	.2	.8	.4509	.5051	1.980	.8926	.2
.9	.3730	.4020	2.488	.9278	68.1	.9	.4524	.5073	1.971	.8918	63.1
22.0	0.3746	0.4040	2.475	0.9272	68.0	27.0	0.4540	0.5095	1.963	0.8910	63.0
.1	.3762	.4061	2.463	.9265	67.9	.1	.4555	.5117	1.954	.8902	62.9
.2	.3778	.4081	2.450	.9259	.8	.2	.4571	.5139	1.946	.8894	.8
.3	.3795	.4101	2.438	.9252	.7	.3	.4586	.5161	1.937	.8886	.7
.4	.3811	.4122	2.426	.9245	.6	.4	.4602	.5184	1.929	.8878	.6
.5	.3827	.4142	2.414	.9239	.5	.5	.4617	.5206	1.921	.8870	.5
.6	.3843	.4163	2.402	.9232	.4	.6	.4633	.5228	1.913	.8862	.4
.7	.3859	.4183	2.391	.9225	.3	.7	.4648	.5250	1.905	.8854	.3
.8	.3875	.4204	2.379	.9219	.2	.8	.4664	.5272	1.897	.8846	.2
.9	.3891	.4224	2.367	.9212	67.1	.9	.4679	.5295	1.889	.8838	62.1
23.0	0.3907	0.4245	2.356	0.9205	67.0	28.0	0.4695	0.5317	1.881	0.8829	62.0
.1	.3923	.4265	2.344	.9198	66.9	.1	.4710	.5340	1.873	.8821	61.9
.2	.3939	.4286	2.333	.9191	.8	.2	.4726	.5362	1.865	.8813	.8
.3	.3955	.4307	2.322	.9184	.7	.3	.4741	.5384	1.857	.8805	.7
.4	.3971	.4327	2.311	.9178	.6	.4	.4756	.5407	1.849	.8796	.6
.5	.3987	.4348	2.300	.9171	.5	.5	.4772	.5430	1.842	.8788	.5
.6	.4003	.4369	2.289	.9164	.4	.6	.4787	.5452	1.834	.8780	.4
.7	.4019	.4390	2.278	.9157	.3	.7	.4802	.5475	1.827	.8771	.3
.8	.4035	.4411	2.267	.9150	.2	.8	.4818	.5498	1.819	.8763	.2
.9	.4051	.4431	2.257	.9143	66.1	.9	.4833	.5520	1.811	.8755	61.1
24.0	0.4067	0.4452	2.246	0.9135	66.0	29.0	0.4848	0.5543	1.804	0.8746	61.0
.1	.4083	.4473	2.236	.9128	65.9	.1	.4863	.5566	1.797	.8738	60.9
.2	.4099	.4494	2.225	.9121	.8	.2	.4879	.5589	1.789	.8729	.8
.3	.4115	.4515	2.215	.9114	.7	.3	.4894	.5612	1.782	.8721	.7
.4	.4131	.4536	2.204	.9107	.6	.4	.4909	.5635	1.775	.8712	.6
.5	.4147	.4557	2.194	.9100	.5	.5	.4924	.5658	1.767	.8704	.5
.6	.4163	.4578	2.184	.9092	.4	.6	.4939	.5681	1.760	.8695	.4
.7	.4179	.4599	2.174	.9085	.3	.7	.4955	.5704	1.753	.8686	.3
.8	.4195	.4621	2.164	.9078	.2	.8	.4970	.5727	1.746	.8678	.2
.9	.4210	.4642	2.154	.9070	65.1	.9	.4985	.5750	1.739	.8669	60.1
25.0	0.4226	0.4663	2.145	0.9063	65.0	30.0	0.5000	0.5774	1.732	0.8660	60.0
	cos	cot	tan	sin	Deg.		cos	cot	tan	sin	Deg.

Deg.	sin	tan	cot	cos		Deg.	sin	tan	cot	cos	
30.0	0.5000	0.5774	1.7321	0.8660	60.0	35.0	0.5736	0.7002	1.4281	0.8192	55.0
.1	.5015	.5797	1.7251	.8652	59.9	.1	.5750	.7028	1.4229	.8181	54.9
.2	.5030	.5820	1.7182	.8643	.8	.2	.5764	.7054	1.4176	.8171	.8
.3	.5045	.5844	1.7113	.8634	.7	.3	.5779	.7080	1.4124	.8161	.7
.4	.5060	.5867	1.7045	.8625	.6	.4	.5793	.7107	1.4071	.8151	.6
.5	.5075	.5890	1.6977	.8616	.5	.5	.5807	.7133	1.4019	.8141	.5
.6	.5090	.5914	1.6909	.8607	.4	.6	.5821	.7159	1.3968	.8131	.4
.7	.5105	.5938	1.6842	.8599	.3	.7	.5835	.7186	1.3916	.8121	.3
.8	.5120	.5961	1.6775	.8590	.2	.8	.5850	.7212	1.3865	.8111	.2
.9	.5135	.5985	1.6709	.8581	59.1	.9	.5864	.7239	1.3814	.8100	54.1
31.0	0.5150	0.6009	1.6643	0.8572	59.0	36.0	0.5878	0.7265	1.3764	0.8090	54.0
.1	.5165	.6032	1.6577	.8563	58.9	.1	.5892	.7292	1.3713	.8080	53.9
.2	.5180	.6056	1.6512	.8554	.8	.2	.5906	.7319	1.3663	.8070	.8
.3	.5195	.6080	1.6447	.8545	.7	.3	.5920	.7346	1.3613	.8059	.7
.4	.5210	.6104	1.6383	.8536	.6	.4	.5934	.7373	1.3564	.8049	.6
.5	.5225	.6128	1.6319	.8526	.5	.5	.5948	.7400	1.3514	.8039	.5
.6	.5240	.6152	1.6255	.8517	.4	.6	.5962	.7427	1.3465	.8028	.4
.7	.5255	.6176	1.6191	.8508	.3	.7	.5976	.7454	1.3416	.8018	.3
.8	.5270	.6200	1.6128	.8499	.2	.8	.5990	.7481	1.3367	.8007	.2
.9	.5284	.6224	1.6066	.8490	58.1	.9	.6004	.7508	1.3319	.7997	53.1
32.0	0.5299	0.6249	1.6003	0.8480	58.0	37.0	0.6018	0.7536	1.3270	0.7986	53.0
.1	.5314	.6273	1.5941	.8471	57.9	.1	.6032	.7563	1.3222	.7976	52.9
.2	.5329	.6297	1.5880	.8462	.8	.2	.6046	.7590	1.3175	.7965	.8
.3	.5344	.6322	1.5818	.8453	.7	.3	.6060	.7618	1.3127	.7955	.7
.4	.5358	.6346	1.5757	.8443	.6	.4	.6074	.7646	1.3079	.7944	.6
.5	.5373	.6371	1.5697	.8434	.5	.5	.6088	.7673	1.3032	.7934	.5
.6	.5388	.6395	1.5637	.8425	.4	.6	.6101	.7701	1.2985	.7923	.4
.7	.5402	.6420	1.5577	.8415	.3	.7	.6115	.7729	1.2938	.7912	.3
.8	.5417	.6445	1.5517	.8406	.2	.8	.6129	.7757	1.2892	.7902	.2
.9	.5432	.6469	1.5458	.8396	57.1	.9	.6143	.7785	1.2846	.7891	52.1
33.0	0.5446	0.6494	1.5399	0.8387	57.0	38.0	0.6157	0.7813	1.2799	0.7880	52.0
.1	.5461	.6519	1.5340	.8377	56.9	.1	.6170	.7841	1.2753	.7869	51.9
.2	.5476	.6544	1.5282	.8368	.8	.2	.6184	.7869	1.2708	.7859	.8
.3	.5490	.6569	1.5224	.8358	.7	.3	.6198	.7898	1.2662	.7848	.7
.4	.5505	.6594	1.5166	.8348	.6	.4	.6211	.7926	1.2617	.7837	.6
.5	.5519	.6619	1.5108	.8339	.5	.5	.6225	.7954	1.2572	.7826	.5
.6	.5534	.6644	1.5051	.8329	.4	.6	.6239	.7983	1.2527	.7815	.4
.7	.5548	.6669	1.4994	.8320	.3	.7	.6252	.8012	1.2482	.7804	.3
.8	.5563	.6694	1.4938	.8310	.2	.8	.6266	.8040	1.2437	.7793	.2
.9	.5577	.6720	1.4882	.8300	56.1	.9	.6280	.8069	1.2393	.7782	51.1
34.0	0.5592	0.6745	1.4826	0.8290	56.0	39.0	0.6293	0.8098	1.2349	0.7771	51.0
.1	.5606	.6771	1.4770	.8281	55.9	.1	.6307	.8127	1.2305	.7760	50.9
.2	.5621	.6796	1.4715	.8271	.8	.2	.6320	.8156	1.2261	.7749	.8
.3	.5635	.6822	1.4659	.8261	.7	.3	.6334	.8185	1.2218	.7738	.7
.4	.5650	.6847	1.4605	.8251	.6	.4	.6347	.8214	1.2174	.7727	.6
.5	.5664	.6873	1.4550	.8241	.5	.5	.6361	.8243	1.2131	.7716	.5
.6	.5678	.6899	1.4496	.8231	.4	.6	.6374	.8273	1.2088	.7705	.4
.7	.5693	.6924	1.4442	.8221	.3	.7	.6388	.8302	1.2045	.7694	.3
.8	.5707	.6950	1.4388	.8211	.2	.8	.6401	.8332	1.2002	.7683	.2
.9	.5721	.6976	1.4335	.8202	55.1	.9	.6414	.8361	1.1960	.7672	50.1
35.0	0.5736	0.7002	1.4281	0.8192	55.0	40.0	0.6428	0.8391	1.1918	0.7660	50.0
	cos	cot	tan	sin	Deg.		cos	cot	tan	sin	Deg.

TABLE IIA   TRIGONOMETRIC FUNCTIONS—DEGREE MEASURE

Deg.	sin	tan	cot	cos		Deg.	sin	tan	cot	cos	
40.0	0.6428	0.8391	1.1918	0.7660	50.0	42.5	0.6756	0.9163	1.0913	0.7373	0.5
.1	.6441	.8421	1.1875	.7649	49.9	.6	.6769	.9195	1.0875	.7361	.4
.2	.6455	.8451	1.1833	.7638	.8	.7	.6782	.9228	1.0837	.7349	.3
.3	.6468	.8481	1.1792	.7627	.7	.8	.6794	.9260	1.0799	.7337	.2
.4	.6481	.8511	1.1750	.7615	.6	.9	.6807	.9293	1.0761	.7325	47.1
.5	.6494	.8541	1.1708	.7604	.5	43.0	0.6820	0.9325	1.0724	0.7314	47.0
.6	.6508	.8571	1.1667	.7593	.4	.1	.6833	.9358	1.0686	.7302	46.9
.7	.6521	.8601	1.1626	.7581	.3	.2	.6845	.9391	1.0649	.7290	.8
.8	.6534	.8632	1.1585	.7570	.2	.3	.6858	.9424	1.0612	.7278	.7
.9	.6547	.8662	1.1544	.7559	49.1	.4	.6871	.9457	1.0575	.7266	.6
41.0	0.6561	0.8693	1.1504	0.7547	49.0	.5	.6884	.9490	1.0538	.7254	.5
.1	.6574	.8724	1.1463	.7536	48.9	.6	.6896	.9523	1.0501	.7242	.4
.2	.6587	.8754	1.1423	.7524	.8	.7	.6909	.9556	1.0464	.7230	.3
.3	.6600	.8785	1.1383	.7513	.7	.8	.6921	.9590	1.0428	.7218	.2
.4	.6613	.8816	1.1343	.7501	.6	.9	.6934	.9623	1.0392	.7206	46.1
.5	.6626	.8847	1.1303	.7490	.5	44.0	0.6947	0.9657	1.0355	0.7193	46.0
.6	.6639	.8878	1.1263	.7478	.4	.1	.6959	.9691	1.0319	.7181	45.9
.7	.6652	.8910	1.1224	.7466	.3	.2	.6972	.9725	1.0283	.7169	.8
.8	.6665	.8941	1.1184	.7455	.2	.3	.6984	.9759	1.0247	.7157	.7
.9	.6678	.8972	1.1145	.7443	48.1	.4	.6997	.9793	1.0212	.7145	.6
42.0	0.6691	0.9004	1.1106	0.7431	48.0	.5	.7009	.9827	1.0176	.7133	.5
.1	.6704	.9036	1.1067	.7420	47.9	.6	.7022	.9861	1.0141	.7120	.4
.2	.6717	.9067	1.1028	.7408	.8	.7	.7034	.9896	1.0105	.7108	.3
.3	.6730	.9099	1.0990	.7396	.7	.8	.7046	.9930	1.0070	.7096	.2
.4	.6743	.9131	1.0951	.7385	.6	.9	.7059	.9965	1.0035	.7083	45.1
42.5	0.6756	0.9163	1.0913	0.7373	0.5	45.0	0.7071	1.0000	1.0000	0.7071	45.0
	cos	cot	tan	sin	Deg.		cos	cot	tan	sin	Deg.

TABLE IIB
TRIGONOMETRIC FUNCTIONS—RADIAN MEASURE

x	$\sin x$	$\cos x$	$\tan x$	$\cot x$	$\sec x$	$\csc x$
0.00	0.0000	1.0000	0.0000	—	1.000	—
0.01	0.0100	1.0000	0.0100	99.997	1.000	100.00
0.02	0.0200	0.9998	0.0200	49.993	1.000	50.00
0.03	0.0300	0.9996	0.0300	33.323	1.000	33.34
0.04	0.0400	0.9992	0.0400	24.987	1.001	25.01
0.05	0.0500	0.9988	0.0500	19.983	1.001	20.01
0.06	0.0600	0.9982	0.0601	16.647	1.002	16.68
0.07	0.0699	0.9976	0.0701	14.262	1.002	14.30
0.08	0.0799	0.9968	0.0802	12.473	1.003	12.51
0.09	0.0899	0.9960	0.0902	11.081	1.004	11.13
0.10	0.0998	0.9950	0.1003	9.967	1.005	10.02
0.11	0.1098	0.9940	0.1104	9.054	1.006	9.109
0.12	0.1197	0.9928	0.1206	8.293	1.007	8.353
0.13	0.1296	0.9916	0.1307	7.649	1.009	7.714
0.14	0.1395	0.9902	0.1409	7.096	1.010	7.166
0.15	0.1494	0.9888	0.1511	6.617	1.011	6.692
0.16	0.1593	0.9872	0.1614	6.197	1.013	6.277
0.17	0.1692	0.9856	0.1717	5.826	1.015	5.911
0.18	0.1790	0.9838	0.1820	5.495	1.016	5.586
0.19	0.1889	0.9820	0.1923	5.200	1.018	5.295
0.20	0.1987	0.9801	0.2027	4.933	1.020	5.033
0.21	0.2085	0.9780	0.2131	4.692	1.022	4.797
0.22	0.2182	0.9759	0.2236	4.472	1.025	4.582
0.23	0.2280	0.9737	0.2341	4.271	1.027	4.386
0.24	0.2377	0.9713	0.2447	4.086	1.030	4.207
0.25	0.2474	0.9689	0.2553	3.916	1.032	4.042
0.26	0.2571	0.9664	0.2660	3.759	1.035	3.890
0.27	0.2667	0.9638	0.2768	3.613	1.038	3.749
0.28	0.2764	0.9611	0.2876	3.478	1.041	3.619
0.29	0.2860	0.9582	0.2984	3.351	1.044	3.497
0.30	0.2955	0.9553	0.3093	3.233	1.047	3.384
0.31	0.3051	0.9523	0.3203	3.122	1.050	3.278
0.32	0.3146	0.9492	0.3314	3.018	1.053	3.179
0.33	0.3240	0.9460	0.3425	2.920	1.057	3.086
0.34	0.3335	0.9428	0.3537	2.827	1.061	2.999
0.35	0.3429	0.9394	0.3650	2.740	1.065	2.916
0.36	0.3523	0.9359	0.3764	2.657	1.068	2.839
0.37	0.3616	0.9323	0.3879	2.578	1.073	2.765
0.38	0.3709	0.9287	0.3994	2.504	1.077	2.696
0.39	0.3802	0.9249	0.4111	2.433	1.081	2.630
0.40	0.3894	0.9211	0.4228	2.365	1.086	2.568
0.41	0.3986	0.9171	0.4346	2.301	1.090	2.509
0.42	0.4078	0.9131	0.4466	2.239	1.095	2.452
0.43	0.4169	0.9090	0.4586	2.180	1.100	2.399
0.44	0.4259	0.9048	0.4708	2.124	1.105	2.348

TABLE IIB
TRIGONOMETRIC FUNCTIONS—RADIAN MEASURE

x	$\sin x$	$\cos x$	$\tan x$	$\cot x$	$\sec x$	$\csc x$
0.45	0.4350	0.9004	0.4831	2.070	1.111	2.299
0.46	0.4439	0.8961	0.4954	2.018	1.116	2.253
0.47	0.4529	0.8916	0.5080	1.969	1.122	2.208
0.48	0.4618	0.8870	0.5206	1.921	1.127	2.166
0.49	0.4706	0.8823	0.5334	1.875	1.133	2.125
0.50	0.4794	0.8776	0.5463	1.830	1.139	2.086
0.51	0.4882	0.8727	0.5594	1.788	1.146	2.048
0.52	0.4969	0.8678	0.5726	1.747	1.152	2.013
$\pi/6$	0.5000	0.8660	0.5774	1.732	1.155	2.000
0.53	0.5055	0.8628	0.5859	1.707	1.159	1.978
0.54	0.5141	0.8577	0.5994	1.668	1.166	1.945
0.55	0.5227	0.8525	0.6131	1.631	1.173	1.913
0.56	0.5312	0.8473	0.6269	1.595	1.180	1.883
0.57	0.5396	0.8419	0.6410	1.560	1.188	1.853
0.58	0.5480	0.8365	0.6552	1.526	1.196	1.825
0.59	0.5564	0.8309	0.6696	1.494	1.203	1.797
0.60	0.5646	0.8253	0.6841	1.462	1.212	1.771
0.61	0.5729	0.8196	0.6989	1.431	1.220	1.746
0.62	0.5810	0.8139	0.7139	1.401	1.229	1.721
0.63	0.5891	0.8080	0.7291	1.372	1.238	1.697
0.64	0.5972	0.8021	0.7445	1.343	1.247	1.674
0.65	0.6052	0.7961	0.7602	1.315	1.256	1.652
0.66	0.6131	0.7900	0.7761	1.288	1.266	1.631
0.67	0.6210	0.7838	0.7923	1.262	1.276	1.610
0.68	0.6288	0.7776	0.8087	1.237	1.286	1.590
0.69	0.6365	0.7712	0.8253	1.212	1.297	1.571
0.70	0.6442	0.7648	0.8423	1.187	1.307	1.552
0.71	0.6518	0.7584	0.8595	1.163	1.319	1.534
0.72	0.6594	0.7518	0.8771	1.140	1.330	1.517
0.73	0.6669	0.7452	0.8949	1.117	1.342	1.500
0.74	0.6743	0.7385	0.9131	1.095	1.354	1.483
0.75	0.6816	0.7317	0.9316	1.073	1.367	1.467
0.76	0.6889	0.7248	0.9505	1.052	1.380	1.452
0.77	0.6961	0.7179	0.9697	1.031	1.393	1.437
0.78	0.7033	0.7109	0.9893	1.011	1.407	1.422
$\pi/4$	0.7071	0.7071	1.000	1.000	1.414	1.414
0.79	0.7104	0.7038	1.009	0.9908	1.421	1.408
0.80	0.7174	0.6967	1.030	0.9712	1.435	1.394
0.81	0.7243	0.6895	1.050	0.9520	1.450	1.381
0.82	0.7311	0.6822	1.072	0.9331	1.466	1.368
0.83	0.7379	0.6749	1.093	0.9146	1.482	1.355
0.84	0.7446	0.6675	1.116	0.8964	1.498	1.343
0.85	0.7513	0.6600	1.138	0.8785	1.515	1.331
0.86	0.7578	0.6524	1.162	0.8609	1.533	1.320
0.87	0.7643	0.6448	1.185	0.8437	1.551	1.308
0.88	0.7707	0.6372	1.210	0.8267	1.569	1.297
0.89	0.7771	0.6294	1.235	0.8100	1.589	1.287

x	$\sin x$	$\cos x$	$\tan x$	$\cot x$	$\sec x$	$\csc x$
0.90	0.7833	0.6216	1.260	0.7936	1.609	1.277
0.91	0.7895	0.6137	1.286	0.7774	1.629	1.267
0.92	0.7956	0.6058	1.313	0.7615	1.651	1.257
0.93	0.8016	0.5978	1.341	0.7458	1.673	1.247
0.94	0.8076	0.5898	1.369	0.7303	1.696	1.238
0.95	0.8134	0.5817	1.398	0.7151	1.719	1.229
0.96	0.8192	0.5735	1.428	0.7001	1.744	1.221
0.97	0.8249	0.5653	1.459	0.6853	1.769	1.212
0.98	0.8305	0.5570	1.491	0.6707	1.795	1.204
0.99	0.8360	0.5487	1.524	0.6563	1.823	1.196
1.00	0.8415	0.5403	1.557	0.6421	1.851	1.188
1.01	0.8468	0.5319	1.592	0.6281	1.880	1.181
1.02	0.8521	0.5234	1.628	0.6142	1.911	1.174
1.03	0.8573	0.5148	1.665	0.6005	1.942	1.166
1.04	0.8624	0.5062	1.704	0.5870	1.975	1.160
$\pi/3$	0.8660	0.5000	1.732	0.5774	2.000	1.155
1.05	0.8674	0.4976	1.743	0.5736	2.010	1.153
1.06	0.8724	0.4889	1.784	0.5604	2.046	1.146
1.07	0.8772	0.4801	1.827	0.5473	2.083	1.140
1.08	0.8820	0.4713	1.871	0.5344	2.122	1.134
1.09	0.8866	0.4625	1.917	0.5216	2.162	1.128
1.10	0.8912	0.4536	1.965	0.5090	2.205	1.122
1.11	0.8957	0.4447	2.014	0.4964	2.249	1.116
1.12	0.9001	0.4357	2.066	0.4840	2.295	1.111
1.13	0.9044	0.4267	2.120	0.4718	2.344	1.106
1.14	0.9086	0.4176	2.176	0.4596	2.395	1.101
1.15	0.9128	0.4085	2.234	0.4475	2.448	1.096
1.16	0.9168	0.3993	2.296	0.4356	2.504	1.091
1.17	0.9208	0.3902	2.360	0.4237	2.563	1.086
1.18	0.9246	0.3809	2.427	0.4120	2.625	1.082
1.19	0.9284	0.3717	2.498	0.4003	2.691	1.077
1.20	0.9320	0.3624	2.572	0.3888	2.760	1.073
1.21	0.9356	0.3530	2.650	0.3773	2.833	1.069
1.22	0.9391	0.3436	2.733	0.3659	2.910	1.065
1.23	0.9425	0.3342	2.820	0.3546	2.992	1.061
1.24	0.9458	0.3248	2.912	0.3434	3.079	1.057
1.25	0.9490	0.3153	3.010	0.3323	3.171	1.054
1.26	0.9521	0.3058	3.113	0.3212	3.270	1.050
1.27	0.9551	0.2963	3.224	0.3102	3.375	1.047
1.28	0.9580	0.2867	3.341	0.2993	3.488	1.044
1.29	0.9608	0.2771	3.467	0.2884	3.609	1.041
1.30	0.9636	0.2675	3.602	0.2776	3.738	1.038
1.31	0.9662	0.2579	3.747	0.2669	3.878	1.035
1.32	0.9687	0.2482	3.903	0.2562	4.029	1.032
1.33	0.9711	0.2385	4.072	0.2456	4.193	1.030
1.34	0.9735	0.2288	4.256	0.2350	4.372	1.027

TABLE IIB
TRIGONOMETRIC FUNCTIONS—RADIAN MEASURE

x	$\sin x$	$\cos x$	$\tan x$	$\cot x$	$\sec x$	$\csc x$
1.35	0.9757	0.2190	4.455	0.2245	4.566	1.025
1.36	0.9779	0.2092	4.673	0.2140	4.779	1.023
1.37	0.9799	0.1994	4.913	0.2035	5.014	1.021
1.38	0.9819	0.1896	5.177	0.1931	5.273	1.018
1.39	0.9837	0.1798	5.471	0.1828	5.561	1.017
1.40	0.9854	0.1700	5.798	0.1725	5.883	1.015
1.41	0.9871	0.1601	6.165	0.1622	6.246	1.013
1.42	0.9887	0.1502	6.581	0.1519	6.657	1.011
1.43	0.9901	0.1403	7.055	0.1417	7.126	1.010
1.44	0.9915	0.1304	7.602	0.1315	7.667	1.009
1.45	0.9927	0.1205	8.238	0.1214	8.299	1.007
1.46	0.9939	0.1106	8.989	0.1113	9.044	1.006
1.47	0.9949	0.1006	9.887	0.1011	9.938	1.005
1.48	0.9959	0.0907	10.983	0.0910	11.029	1.004
1.49	0.9967	0.0807	12.350	0.0810	12.390	1.003
1.50	0.9975	0.0707	14.101	0.0709	14.137	1.003
1.51	0.9982	0.0608	16.428	0.0609	16.458	1.002
1.52	0.9987	0.0508	19.670	0.0508	19.695	1.001
1.53	0.9992	0.0408	24.498	0.0408	24.519	1.001
1.54	0.9995	0.0308	32.461	0.0308	32.476	1.000
1.55	0.9998	0.0208	48.078	0.0208	48.089	1.000
1.56	0.9999	0.0108	92.620	0.0108	92.626	1.000
1.57	1.0000	0.0008	1255.8	0.0008	1255.8	1.000
$\pi/2$	1.0000	0.0000	—	0.0000	—	1.000

Answers to Selected Problems

Answers to Selected Problems

Chapter 1

Problem Set 1, page 5

1. 5^8 **3.** $3^4 \cdot 4^3$ **5.** $x^4 y^3$ **7.** $(-x)^2(-y)^3$ **9.** $(2a + 1)^3$ **11.** $z = 2(x + y)$ **13.** $A = \pi r^2$
15. $x = 0.05n$ **17.** $A = 6x^2$ **19.** $A = \frac{1}{2}bh$ **21.** $A = prt$ **23.** $6077.53   **25.** Commutative for $+$
27. Commutative for $\times$ **29.** Identity for $\times$ **31.** Associative for $+$ **33.** Identity for $+$ **35.** Distributive
37. Inverse for $+$ **39.** Associative for $\times$ **41.** Negation (i)   **43.** Negation (ii)   **45.** Cancellation for $+$
47. Zero factor (ii)   **49.** $(3 + 5)^2 = 8^2 = 64$ **51.** No
53. By the reflexive property, $ca = ca$. Since $a = b$, substitution yields $ca = cb$.

Problem Set 2, page 13

1. $\{4, 6, 8, 10\}$ **3.** $\{2, 4\}$ **5.** False   **7.** False   **9.** True   **11.** True   **13.** True   **15.** False

17.

19. (a)

(b)
(c)
(d) $\cdots$
(e)
(f)
(g)
(h)

21.   **23.**

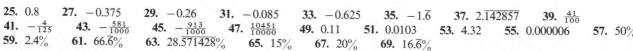

25. 0.8 **27.** -0.375 **29.** $-0.\overline{26}$ **31.** -0.085 **33.** -0.625 **35.** $-1.\overline{6}$ **37.** $2.\overline{142857}$ **39.** $\frac{41}{100}$
41. $-\frac{4}{125}$ **43.** $-\frac{581}{1000}$ **45.** $-\frac{913}{1000}$ **47.** $\frac{10451}{10000}$ **49.** 0.11 **51.** 0.0103 **53.** 4.32 **55.** 0.000006 **57.** 50%
59. 2.4% **61.** $66.\overline{6}\%$ **63.** $28.\overline{571428}\%$ **65.** 15% **67.** 20% **69.** $16.\overline{6}\%$

Problem Set 3, page 19

1. 18   **3.** 400   **5.** Monomial; polynomial in x; degree 2; coefficient -4 **7.** Trinomial; multinomial in x and y
9. Multinomial; polynomial in x and z; degree 3; coefficients 3, $-\frac{6}{11}$, -1, -1, 2
11. Multinomial; polynomial in x; degree 5; coefficients $\sqrt{2}$, π, $-\sqrt{\pi}$, $\frac{12}{13}$, -5
13. Multinomial; polynomial in x, y, z, w; degree 3; coefficients 1, 1, 1, 1   **15.** $2x + 12$ **17.** $18x^2 + x + 5$
19. $4z^2 + 10z + 3$ **21.** $3n^3 - 5n^2 - n + 7$ **23.** $2x^2y + 4xy^2 + 11xy - 3$ **25.** $-7t^2 + 14t - 9$
27. $2x^2 - 5xy + 6y^2 - 2x + y - 3$ **29.** $6x^7$ **31.** r^{13n} **33.** $(3x + y)^6$ **35.** t^{48} **37.** u^{5n^2} **39.** $324v^6$
41. $675a^7b^8$ **43.** $864(x + 3y)^{22}$ **45.** $15x^4y - 20x^4y^2 + 20x^3yz$ **47.** $-12x^2 + 7xy + 10y^2$ **49.** $x^3 + 8y^3$
51. $625c^4 - d^4$ **53.** $3x^4 + 20x^3y - 9x^2y^2 + 8xy^3 - 15x^2y - 5xy^2 + 14xy - 2x^3 + 2x^2 - y^4 + y^3 - 2y^2$
55. $4p^4q^2 - 11p^3q^3 - 7p^3q^2 + 6p^2q^4 - p^2q^3 - 15p^2q^2$ **57.** $16 + 24x + 9x^2$ **59.** $16t^4 + 8t^2 + 1 - 8st^2 - 2s + s^2$
61. $9r^2 - 4s^2$ **63.** $4a^2 - 12ab + 9b^2 - 16c^2$ **65.** $8 + t^3$ **67.** $8x^6 + 36x^4y + 54x^2y^2 + 27y^3$
69. $4x^2 + y^2 + z^2 - 4xy + 4xz - 2yz$ **71.** $t^6 - 4t^5 + 14t^4 - 16t^3 + 17t^2 + 20t + 4$
77. $a^4 + 4a^3b + 6a^2b^2 + 4ab^3 + b^4$
79. $x^2 + y^2 = (a^2 - b^2)^2 + (2ab)^2 = a^4 - 2a^2b^2 + b^4 + 4a^2b^2 = a^4 + 2a^2b^2 + b^4 = (a^2 + b^2)^2 = z^2$

Problem Set 4, page 25

1. $5x^2y(2x - y)$ **3.** $a^2b(a + 2 + b)$ **5.** $2r(x - y)(r + h)$ **7.** $(a - b)(t + r)$ **9.** $(x + 6)(x - 6)$
11. $(pq + 8)(pq - 8)$ **13.** $(n + 6m)(n - 6m)$ **15.** $(x + y + 7z)(x + y - 7z)$ **17.** $(9y^2 + z^2)(3y + z)(3y - z)$
19. $8(x + 5)(x^2 - 5x + 25)$ **21.** $(a - 3b - c)(a^2 + 3ab + ac + 9b^2 + 6bc + c^2)$
23. $(4x^2 + p - q)(16x^4 - 4x^2p + 4x^2q + p^2 - 2pq + q^2)$ **25.** $(x - 5)(x - 3)$ **27.** $(y - 5)(y + 2)$
29. $(t - 9)(t + 2)$ **31.** $(x - 5y)(x + 4y)$ **33.** $(4x + 7)(2x - 1)$ **35.** $(2v - 5)(v + 4)$ **37.** $(2r - 3s)^2$
39. $(3x^2 + 2)(2x^2 + 3)$ **41.** $(9a - 6b + 7)(3a - 2b - 2)$ **43.** Prime   **45.** $(4r - 1)(3r - 2)$ **47.** Prime
49. $(3 + y)(x - 2)$ **51.** $(x + d)(a - 1)(a + 1)$ **53.** $(2x + 3 + 3y)(2x + 3 - 3y)$ **55.** $(3v - 1)(2u + 3)(2u - 3)$
57. $(x - 3y)(x + 3y + 1)$ **59.** $(x^2 + y^2 - xy)(x^2 + y^2 + xy)$ **61.** $(t - 2)^2(t + 2)^2$
63. $(2s + 1)(a - b)(a^2 + ab + b^2)$ **65.** $(x^4 + 16y^4)(x^2 + 4y^2)(x - 2y)(x + 2y)$ **67.** $y^3z(5x + y)(2x - 3y)$
69. (a) $3(3x^2 + 7)(5 - 3x)^2(-21x^2 + 20x - 21)$
    (b) $4t(5t^2 + 1)(3t^4 + 2)^3(15t^4 + 12t^2 + 10)$

Problem Set 5, page 32

1. $\dfrac{by}{3}$ **3.** $\dfrac{c}{2}$ **5.** $\dfrac{1}{5 + 1}$ **7.** $\dfrac{1}{r + 6}$ **9.** $\dfrac{3y + 1}{y + 5}$ **11.** 1   **13.** $\dfrac{3c + 1}{2c - 1}$ **15.** $\dfrac{3r + 3t + 2}{2r + 2t - 3}$

17. $\dfrac{c + d - 3x}{c + d + x}$ **19.** $c + 2$ **21.** $\dfrac{1}{4u}$ **23.** $\dfrac{3(x^2 + 5)(x + 1)}{(x^2 + 6x + 15)(x - 1)}$ **25.** $1/(a + 1)$ **27.** $4/t$ **29.** $(x - y)^2$

31. $\dfrac{2x + 3}{x + 3}$ **33.** $\dfrac{2u - 1}{u + 2}$ **35.** $\dfrac{t + 5}{t - 6}$ **37.** $(3a - 1)/b$ **39.** $x - 1$ **41.** $\dfrac{3x - 8}{(x + 2)(x - 5)}$ **43.** $\dfrac{36u^2 - 40u - 13}{4(2u - 1)(2u + 1)}$

45. $\dfrac{2y^2 + 2y - 1}{(y - 2)(y + 1)}$ **47.** $\dfrac{x + 15}{(x - 3)(x + 3)^2}$ **49.** $\dfrac{4}{t - 2}$ **51.** $\dfrac{-u(3u^3 - u^2 + 8u + 2)}{(u^2 + 3)(u - 1)^2(u + 2)}$ **53.** $\dfrac{5x^2 - 4x + 1}{(2x - 1)^2x^3}$

55. $\dfrac{1}{cd(c + d)}$ **57.** $ab(a + b)$ **59.** $\dfrac{pq(p + q)}{p^2 + pq + q^2}$ **61.** -1 **63.** $\dfrac{x^2 - xy + 2y^2}{(x - 3y)(x + 2y)}$ **65.** $\dfrac{-2x - h}{x^2(x + h)^2}$

71. 4 isn't a factor of the numerator   **73.** c isn't a factor of the denominator
75. Middle terms $(a/b) + (b/a)$ missing

Problem Set 6, page 40

1. 11   **3.** $\frac{3}{4}$ **5.** Irrational   **7.** $\frac{2}{5}$ **9.** $\frac{1}{2}$ **11.** $\frac{3}{4}$ **13.** $3x\sqrt{3x}$ **15.** $5x^2y^4\sqrt{3y}$ **17.** $2ab\sqrt[3]{3ab^2}$
19. $(x-1)^2$ **21.** $y/2x^4$ **23.** $\sqrt{15y}/6y$ **25.** $(-3x\sqrt[3]{x})/y^7$ **27.** -66 **29.** $5t$ **31.** $x^2y^4\sqrt[4]{y^3}$ **33.** $ab^2\sqrt[3]{b}$
35. 1   **37.** $(3/2y)\sqrt[3]{xz}$ **39.** $\frac{5}{2}\sqrt[3]{x^2y}$ **41.** a^2 **43.** 98   **45.** $13\sqrt[3]{3}$ **47.** $19t^3\sqrt{5t}$ **49.** $\frac{1}{6}\sqrt[4]{6}$ **51.** $xy\sqrt[3]{xy^2}$
53. $6\sqrt{2}-6\sqrt{3}$ **55.** $-1-\sqrt{3}$ **57.** $x+3+2\sqrt{3x}$ **59.** -13 **61.** 4   **63.** $4+2\sqrt{6}$ **65.** $2a$
67. $\sqrt{10}/3$ **69.** $\dfrac{\sqrt{2(x+2y)}}{3(x+2y)}$ **71.** $4\sqrt[3]{25}$ **73.** $\dfrac{5\sqrt{3}-3}{22}$ **75.** $(7+2\sqrt{10})/3$ **77.** $\dfrac{12p+3q+13\sqrt{pq}}{16p-9q}$
79. $\dfrac{2a+5+2\sqrt{a^2+5a}}{5}$ **81.** $\dfrac{5(4+2\sqrt[3]{x}+\sqrt[3]{x^2})}{8-x}$ **83.** $\dfrac{a-2\sqrt[3]{a^2b}+2\sqrt[3]{ab^2}-b}{a+b}$ **85.** $\dfrac{1}{\sqrt{x+h}+\sqrt{x}}$
87. $\dfrac{2x+h}{\sqrt{(x+h)^2+1}+\sqrt{x^2+1}}$

Problem Set 7, page 47

1. 27   **3.** $\frac{1}{64}$ **5.** y^4/x^2z **7.** -1 **9.** $-28c^2$ **11.** m^3 **13.** $y^2/2x^5$ **15.** $9t/(9+t)$ **17.** $(a^2+b^2)/ab$
19. x^{2n-1} **21.** $1/x$ **23.** $\dfrac{p^2-3}{p^2-1}$ **25.** $\dfrac{(t-2)^2}{t+2}$ **27.** $\left(\dfrac{a}{b}\right)^{3n}$ **29.** $(a+8b)^{3+n}$ **31.** $\left(\dfrac{r+s}{c+d}\right)^4$ **33.** $-1/2q$
35. 27   **37.** -32 **39.** 8   **41.** Undefined   **43.** 4   **45.** 30   **47.** a^2 **49.** $y^{7/3}$ **51.** $1/x^{3/2}$ **53.** $x^3/8$
55. $1/x^{11}$ **57.** y^4/x^2 **59.** $2p+q$ **61.** $\dfrac{3x+2y}{4r+3t}$ **63.** $\dfrac{1}{(m^2-n^2)^{1/3}}$ **65.** $\dfrac{(x+y)^2}{\sqrt{x}}$ **67.** $\dfrac{-(p+3)}{(p-1)^3}$
69. $\dfrac{2(8t+1)}{(4t-1)^2(2t+1)^2}$ **71.** $\dfrac{3x-1}{x^{2/3}(x-1)^{1/3}}$ **81.** $2(3x+x^{-1})(6x-1)^4(63x-3+9x^{-1}+x^{-2})$
83. $2(7y+3)^{-3}(2y-1)^3(14y+19)$ **85.** Exponent not reduced to lowest terms
87. Can't apply rule when numbers are negative   **89.** Can't apply rule when number is negative

Problem Set 8, page 53

1. 1.55×10^4 **3.** 5.8761×10^7 **5.** 1.86×10^{11} **7.** 9.01×10^{-7} **9.** 33,300   **11.** 0.00004102
13. 10,010,000   **15.** 6.2×10^{-2} second   **17.** 9.29×10^7 miles, 1.92×10^{13} miles   **19.** 2.51×10^{-10} second
21. 4.8×10^8 **23.** 7×10^{31} **25.** 6.56×10^{-4} **27.** 2   **29.** 5   **31.** 5300   **33.** 0.015   **35.** 110,000
37. 2.14×10^{-13} **39.** 2.3102×10^2 **41.** 4.867×10^{-4} **43.** 1.51   **45.** 1.06×10^{27} **47.** 3.47×10^5
49. (a) $\$1.3272\times10^{12}$; (b) 9.6×10^3 dollars per person

Review Problem Set, page 55

1. 5^8x^3 **3.** $(-4)^5y^6$ **5.** $w=3xy/z$ **7.** $s=\frac{1}{2}(a+b+c)$ **9.** $P=1000(3^n)$ **11.** 315 joules
13. Commutative for $\times$ **15.** Associative for $+$ **17.** Commutative for $\times$ **19.** Identity for $\times$ **21.** Inverse for $\times$
23. Property of negation (i)   **25.** Zero factor (i)   **27.** $\frac{1}{2}+\frac{1}{3}=\frac{3}{6}+\frac{2}{6}=\frac{5}{6}$ **29.** $\{1,2\}$ **33.** $(-\frac{2}{3},5)$
35. (a) 0.22; (b) 22%   **37.** (a) -0.085; (b) -8.5% **39.** (a) 3.25; (b) 325%   **41.** (a) 0.495; (b) $\frac{99}{200}$ **43.** (a) 0.0043;
(b) $\frac{43}{10000}$ **45.** (a) 1.4; (b) $\frac{7}{5}$ **47.** 6.25%   **49.** Monomial; polynomial; degree 2; coefficient -2
51. Monomial; rational expression   **53.** Trinomial; multinomial; polynomial; degree 2; coefficients 3, -5, -1

55. Binomial; multinomial; radical expression   **57.** Binomial; multinomial; polynomial; degree 3; coefficients $\sqrt{2}$, $-\sqrt[5]{7}$
59. Binomial; multinomial; equivalent to the rational expression $(\sqrt{\pi}y + x)/y$ **61.** $3x^3 - 3x^2 + 2x + 6$
63. $3a^3 - a^2b + 6ab^2 + 3b^3$ **65.** $20y^5$ **67.** $-x^6y^4$ **69.** p^8 **71.** $-x^{14}y^7$ **73.** $a^{n+1}b^{2n}$
75. $3x^5 + 6x^3 - 9x^2$ **77.** $2t^2 - 5t - 12$ **79.** $2x^2y^4 + 7xy^2 + 3$ **81.** $2p^3 - 5p^2 - 10p + 3$
83. $2x^3 - 3x^2 - 5x + 6$ **85.** $9x^2 + 30xy + 25y^2$ **87.** $4x^4 - 20x^2yz + 25y^2z^2$
89. $4x^2 + y^2 + 9z^2 - 4xy + 12xz - 6yz$ **91.** $9t^{2n} - 121$ **93.** $27x^9 - 54x^7y + 36x^5y^2 - 8x^3y^3$
95. $3xy^2(3x - 4y^2)$ **97.** $(a + b)(a + b - c^2)c^2$ **99.** $(6c - d^2)(6c + d^2)$ **101.** $(x - y - z)(x - y + z)$
103. $(x^n - y)(x^n + y)$ **105.** $x^2(5 - 7y)(5 + 7y)$ **107.** $(2p + 3q)(4p^2 - 6pq + 9q^2)$ **109.** $4b(3a^2 + 4b^2)$
111. $(x - 4)(x + 6)$ **113.** $(a^3 + 4)^2$ **115.** $(3x - 2y)(2x + 3y)$ **117.** $(4uv + 1)(uv - 2)$ **119.** $(5 - 2x)(4 + 3x)$
121. $(p + 3q - 2)(p + 3q + 2)$ **123.** $(a^2 + 2ab + b^2 + 1 - 3a - 3b)(a^2 + 2ab + b^2 + 1 + 3a + 3b)$ **125.** $\dfrac{x + 3}{5}$
127. $\dfrac{t + 3}{t + 2}$ **129.** $\dfrac{c - 3}{c - 2}$ **131.** $\dfrac{5(x - 3)}{3x}$ **133.** $\dfrac{2(x + 1)}{x + 2}$ **135.** $2y$ **137.** $\dfrac{2(t - 5)}{t + 1}$ **139.** $2/x$ **141.** $1/t$
143. $1/(c + 3)$ **145.** $1/(a - 3)$ **147.** $-1/(a + 5)$ **149.** 13 **151.** Irrational   **153.** Not real   **155.** $2x^2$
157. $(a + b)^3/3a$ **159.** p^7 **161.** $a^2\sqrt[3]{5a}$ **163.** $12\sqrt{2a}$ **165.** $13\sqrt[3]{2p}$ **167.** $5\sqrt{6}$ **169.** $2a - 3$
171. $2a + b - 2\sqrt{a^2 + ab}$ **173.** $(3\sqrt{2x})/x$ **175.** $\dfrac{a + \sqrt{ab}}{a - b}$ **177.** $\dfrac{\sqrt{a - 1} - a + 1}{2 - a}$ **179.** $5(\sqrt[3]{4} + \sqrt[3]{2} + 1)$
181. $\dfrac{9x - y}{5(3\sqrt{x} - \sqrt{y})}$ **183.** 1 **185.** $(x + y)/x^2y^2$ **187.** $(x^2 - 1)/x^4$ **189.** $3/a^6$ **191.** x^{13}/y^6 **193.** $1/25p^{10}$
195. $\frac{4}{9}$ **197.** $\frac{1}{512}$ **199.** y **201.** a^3 **203.** $y^{3/2}$ **205.** $8^5b^{13}c^{2/3}/2$ **207.** $x^2y^{-12}(x + y)^{-4}$
209. $(y + 2)^{-2/3}(y + 1)^{-1/3}(3y + 5)$ **211.** 5.712×10^{10} **213.** 7.14×10^{-7} **215.** $17{,}320{,}000$
217. 0.0000000312 **219.** 5×10^{-6} gram   **221.** 3.584×10^8 kilometers   **223.** 14.78976 **225.** 3 **227.** 2
229. $17{,}000$ **231.** 7.23×10^5 **233.** 3.6×10^4 **235.** 2.65×10^{33} **237.** 1.8×10^{-5}
239. Yes. Consider $a + b + c$, where $a = 1$, $b = 2.4$, $c = 3.1$ and round off to the nearest unit.

Chapter 2

Problem Set 1, page 66

1. -2 **3.** $\frac{4}{3}$ **5.** $\frac{9}{2}$ **7.** -1.538 **9.** 42.57 **11.** 9 **13.** 3 **15.** 3 **17.** $-\frac{19}{2}$ **19.** 3 **21.** 4

23. 12 **25.** 7 **27.** $-\frac{5}{6}$ **29.** 1 **31.** (a) $\frac{7}{33}$; (b) $\frac{563}{165}$; (c) $\frac{13}{330}$; (d) $-\frac{9917}{9900}$; (e) $\frac{7}{900}$ **33.** 5 **35.** 4 **37.** $\frac{1}{3}$
39. $-\frac{5}{2}$ **41.** $\dfrac{5a + c}{b - 10}, b \neq 10$ **43.** $\dfrac{4(cd - b)}{4a - 1}, a \neq \frac{1}{4}$ **45.** $(a + md)/2$ **47.** $\dfrac{mn - 1}{bc + d}, d \neq -bc$ **49.** $-b, b \neq 0$
51. $V/\pi r^2$ **53.** $(A - P)/Pr$ **55.** $pq/(p + q)$ **57.** $P_2V_2T_1/P_1V_1$ **59.** Identity   **61.** Conditional
63. Conditional   **65.** No, -6 is a root of the second equation, but not the first.   **67.** Yes   **69.** No   **73.** -18.1

Problem Set 2, page 74

1. $8, 20$ **3.** $\$8500$ **5.** $\$42{,}000$ at 8.5%, $\$33{,}000$ at 9.2% **7.** $\$4760$ solar, $\$1750$ insulation   **9.** 57 hours
11. $200{,}000$ gallons of 9%, $100{,}000$ gallons of 12% **13.** 30 milliliters   **15.** $5n, 12d, 6q, 9h$ **17.** 22 km/hr
19. 2520 feet   **21.** 27 hours   **23.** $4\frac{2}{7}$ hours   **25.** 150 meters by 300 meters   **27.** $-40°C = -40°F$
29. $F = -7.\overline{27}°$ **31.** 3390 kilometers (to three significant digits)

Problem Set 3, page 84

1. $0, 7$ **3.** ± 4 **5.** $-3, 1$ **7.** $-3, 7$ **9.** $-\frac{3}{2}, 5$ **11.** $-\frac{1}{3}, \frac{5}{2}$ **13.** $\frac{1}{5}, \frac{4}{3}$ **15.** $-\frac{27}{8}, 1$ **17.** $\frac{4}{5}, \frac{4}{5}$
19. $-5, 21$ **21.** $-\frac{5}{2}, 1$ **23.** $-\frac{5}{2}, \frac{7}{2}$ **25.** $x^2 + 6x + 9$ **27.** $x^2 - 5x + \frac{25}{4}$ **29.** $x^2 + \frac{3}{4}x + \frac{9}{64}$
31. $-2 \pm \sqrt{19}$ **33.** $3 \pm \sqrt{5}$ **35.** $-1 \pm (\sqrt{21}/3)$ **37.** $-\frac{3}{5}, 2$ **39.** $(-5 \pm \sqrt{13})/6$ **41.** $(1 \pm \sqrt{17})/4$
43. $(3 \pm \sqrt{57})/8$ **45.** $\frac{1}{3}, \frac{1}{3}$ **47.** $(5 \pm \sqrt{13})/3$ **49.** $-1.85, 0.38$ **51.** $1.42, 18.25$ **53.** $\pm\sqrt{5}$ **55.** $(1 \pm \sqrt{7})/4$
57. $-3, -3$ **59.** $-2 \pm \sqrt{3}$ **61.** $\frac{1}{3}, \frac{1}{3}$ **63.** $-4, 2$ **65.** 10 **67.** 9 rows

69. $2(1 + \sqrt{161})$ feet and $2(\sqrt{161} - 2)$ feet   **71.** 1.2 seconds   **73.** $\dfrac{-R \pm \sqrt{R^2 - 4L/C}}{2L}, L \neq 0, R^2 \geq 4L/C, C \neq 0$

75. Real, unequal   **77.** Rational, double root   **79.** Complex conjugates   **81.** $(-1 \pm i\sqrt{71})/12$ **83.** $(2 \pm i\sqrt{6})/5$
85. $(-1 \pm i\sqrt{3})/2$ **87.** $(1 \pm i\sqrt{3})/4$

Problem Set 4, page 91

1. ± 3 **3.** $\pm\sqrt[6]{2}$ **5.** $\pm\frac{1}{2}$ **7.** 3 **9.** $-\sqrt[7]{4}$ **11.** -2 **13.** 8 **15.** -27 **17.** $3^{2/5}$ **19.** $\frac{33}{64}$ **21.** $\frac{8}{3}$
23. $-1, 8$ **25.** 20 **27.** $-\frac{9}{4}$ **29.** 7 **31.** -2 **33.** 3 **35.** 8 **37.** 1 **39.** 27 **41.** $-1, -\frac{3}{4}$ **43.** 2
45. 2 **47.** $\pm\sqrt{3}, \pm 2$ **49.** $-2, 1$ **51.** $1, 3^6$ **53.** 1 **55.** $\pm\frac{1}{4}, \pm 3$ **57.** $-\frac{7}{2}, 6$ **59.** $-1 \pm \sqrt{13}$

61. $-19, 61$ **63.** $-\frac{3}{2}, 3$ **65.** $-1, -1, 1$ **67.** ± 1 **69.** $-8, 1$ **71.** ± 1 **73.** $r = \sqrt{\dfrac{\sqrt{\pi^2 h^4 + 4A^2} - \pi h^2}{2\pi}}$

75. 5 cm by 12 cm   **77.** $\ell = \dfrac{d^2 + d\sqrt{d^2 + a^2}}{a}, a \neq 0$

Problem Set 5, page 97

1. (a) Addition; (b) transitive; (c) multiplication; (d) trichotomy   **3.** (a) $x + 4 > 1$; (b) $x - 4 > -7$; (c) $5x > -15$;
(d) $-5x < 15$ **5.** $(-\infty, 5)$ **7.** $(-\infty, 5]$ **9.** $(-\infty, 2)$ **11.** $[4, \infty)$ **13.** $[-\frac{22}{3}, \infty)$ **15.** $(-\infty, \frac{24}{7})$
17. $[-13, \infty)$ **19.** $\mathbb{R}$ **21.** $[2, \infty)$ **23.** $[1, 5]$ **25.** $[1, 2)$ **27.** $[-3, 2)$ **29.** $[-2, 1]$ **31.** $(-4, 4]$
33. $[\frac{10}{9}, 10]$ **35.** $[\frac{1}{3}, 1]$ **37.** $[53, 93]$ **39.** $5400 < R < 6600$ **41.** $(\frac{10}{3}, \frac{35}{6})$ **43.** $(13, 20)$
47. $-2 < 1$, but $(-2)^2 \not< 1^2$

Problem Set 6, page 104

1. 2 **3.** 2 **5.** 1 **7.** 6.043 **9.** 2.141592654 **11.** 8 **13.** 32 **15.** 8 **17.** 0 **19.** 0 **21.** ± 4
23. ± 3 **25.** 0 **27.** $-4, 5$ **29.** $2, 6$ **31.** $p \geq 0$ **33.** No solution   **35.** $\frac{4}{3}, \frac{5}{3}$ **37.** $-\frac{1}{5}, \frac{1}{3}$ **39.** $-4, -\frac{2}{3}$
41. $\frac{7}{4}, 3$ **43.** $1, 3$ **45.** $(-2, 2)$ **47.** $[-2, 2]$ **49.** $(1, 5)$ **51.** $[-5, 5]$ **53.** $(-\infty, \frac{3}{4}]$ and $[\frac{9}{4}, \infty)$
55. $(-\infty, -\frac{3}{2})$ and $(3, \infty)$ **57.** $[\frac{6}{5}, 2]$ **59.** $(-\infty, -\frac{13}{8}]$ and $[-\frac{11}{8}, \infty)$ **61.** $[\frac{4}{7}, \frac{8}{7}]$ **63.** $(-\infty, 1]$ and $[\frac{5}{2}, \infty)$
65. $(-\infty, 2)$ and $(\frac{7}{2}, \infty)$ **67.** $\mathbb{R}$ **69.** $|x| < 1$ **71.** $|x| \leq 3$

Problem Set 7, page 110

1. $x^2 - 6x + 8 < 0$ **3.** $x^2 - 10x + 25 \geq 0$ **5.** $(2, 4)$ **7.** $(-\infty, -6)$ and $(-4, \infty)$ **9.** $\{7\}$
11. $(-\infty, -5]$ and $[-\frac{7}{3}, \infty)$ **13.** $[\frac{1}{3}, \frac{1}{2}]$ **15.** $\{\frac{2}{5}\}$ **17.** $(-\infty, -\frac{2}{3})$ and $(4, \infty)$ **19.** $[-1, \frac{1}{2}]$ **21.** $(-\frac{3}{4}, \frac{1}{2})$

23. $\mathbb{R}$ **25.** $\left(-\infty, \dfrac{1-\sqrt{61}}{10}\right], \{0\}$, and $\left[\dfrac{1+\sqrt{61}}{10}, \infty\right)$ **27.** $(-1,0)$ and $(1,\infty)$ **29.** $\mathbb{R}$ **31.** $\dfrac{2x-1}{x+3} > 0$

33. $\dfrac{(x-3)(x+2)}{(x+3)(x+2)} \le 0$ **35.** All real numbers, except -2 **37.** $(-3, \frac{5}{3}]$ **39.** $(-\infty, 0)$ and $(\frac{3}{2}, \infty)$ **41.** $(-\infty, 2)$

43. $(-\infty, -2), (-1,1)$, and $(2,\infty)$ **45.** $(-\infty, -3), (-1,2)$, and $(3,\infty)$ **47.** $(-\infty, -3), [-2,-1), [1,2)$, and $[3,\infty)$

49. $(10,20)$ **51.** $\$1.25$ **53.** 5 **55.** $(-\infty, 0.203]$ and $[0.684, \infty)$ **57.** $(-\infty, -4.38)$ and $(0.14, 0.79)$

Review Problem Set, page 112

1. 5   **3.** 5   **5.** 15   **7.** -2 **9.** 32   **11.** 6   **13.** -17.5 **15.** -2 **17.** $(a+b)^2$
19. No solution unless $m=0$ **21.** 0.4650   **23.** -7.234×10^{-3} **25.** 4 liters   **27.** $\$6500$ at 7%, $\$3500$ at 8%
29. $\$750$ **31.** 30 meters   **33.** $x^2+x+\frac{1}{4}$ **35.** $x^2-9x+\frac{81}{4}$ **37.** $-1, \frac{7}{2}$ **39.** $-\frac{1}{2}, \frac{3}{5}$ **41.** $(7\pm\sqrt{89})/10$
43. $-2.867, 0.135$ **45.** $(3\pm i\sqrt{39})/8$ **47.** $(1\pm i\sqrt{15})/4$ **49.** $(-3\pm\sqrt{13})/4$ **51.** Rational, unequal
53. Complex conjugates   **55.** 450 knots   **57.** 16 inches   **59.** 39   **61.** $(15-\sqrt{33})/2$ **63.** $-\frac{10}{3}$ **65.** 3
67. -3 **69.** $\pm1, \pm2$ **71.** $\pm\sqrt{6}/3, \pm\frac{2}{3}\sqrt{3}$ **73.** 25   **75.** $-8, 1, 3\pm\sqrt{17}$ **77.** $-\frac{1}{3}, \frac{1}{2}$ **79.** -1
81. -19 **83.** $1, \pm\sqrt{2}$ **85.** $1, 1$ **87.** $-1, \pm2^{-5/2}$ **89.** 2   **91.** $\frac{4}{5}$ **93.** $L^3F/4d^3wY$
95. When the number r in the register doesn't change, $r=(r/2)+(x/2r)$ or $2r^2=r^2+x$, so $r^2=x$
97. (a) $0 \le 5-5x < 7$; (b) $-\frac{4}{5} < x-\frac{2}{5} \le \frac{3}{5}$ **99.** $[5,\infty)$ **101.** $(3,\infty)$ **103.** $(\frac{24}{13}, \infty)$ **105.** $[73,98)$ **107.** 7
109. 5.8765   **111.** $\frac{19}{6}$ **113.** True   **115.** True   **117.** True   **119.** $-\frac{3}{2}, \frac{9}{2}$ **121.** $\frac{1}{4}$ **123.** $(-\infty,1)$ and $(3,\infty)$
125. $(-\infty, -\frac{1}{2}]$ and $[2,\infty)$ **127.** $[\frac{1}{3},1]$ **129.** $(\frac{3}{2}, \frac{11}{6})$ **131.** $(-\frac{2}{3}, \frac{5}{2})$ **133.** $[-\frac{3}{2}, \frac{2}{3}]$ **135.** $(-5, \frac{1}{3})$
137. $(-\infty, -5)$ and $(-3,0)$ **139.** $(-\infty, \frac{1}{3})$ and $(\frac{5}{2}, \infty)$ **141.** $[\frac{7}{5}, \frac{13}{7}]$ **143.** $(-\infty, -3), (-2,-1)$, and $(0,2)$
145. $(-\infty, -1), (-1,1]$, and $(2,\infty)$ **147.** $[\$2, \$2.11]$ **149.** $[55,75]$ **151.** $[85,105]$

Chapter 3

Problem Set 1, page 122

1.

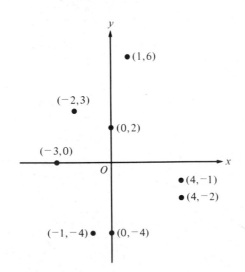

3. $Q = (3, -2); R = (-3, 2); S = (-3, -2)$ **5.** $Q = (-1, -3); R = (1, 3); S = (1, -3)$ **7.** 10   **9.** 4   **11.** $\sqrt{145}$
13. 10   **15.** 5   **17.** $\sqrt{9 + 4t^2}$ **19.** 6.224   **21.** (b) 12   **23.** (b) 6   **27.** Yes   **29.** 3, -2 **31.** Yes
33. No   **35.** (3, 4)   **37.** (a) $x^2 + y^2 = 16$ (b) $(x + 1)^2 + (y - 3)^2 = 4$
39.

41.

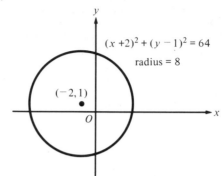

Problem Set 2, page 129

1. -1 **3.** -1 **5.** $\frac{1}{4}$ **7.** $\frac{1}{2}$ **9.** $-4.\overline{6}$
11.

15.

19. $-\frac{1}{5}$

23.

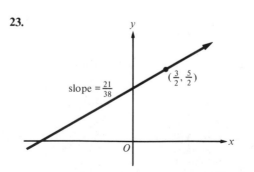

29. (a) $d = 1$ (b) $k = -\frac{10}{3}$

Problem Set 3, page 133

1. $y - 2 = \frac{3}{4}(x - 3)$ **3.** $y - 1 = -\frac{1}{7}(x - 4)$ **5.** $y - 5 = 1(x + 3)$ **7.** $y = \frac{3}{2}x - 3; m = \frac{3}{2}; b = -3$
9. $y = 3x + 1; m = 3; b = 1$ **11.** $y = -\frac{1}{3}x + 3; m = -\frac{1}{3}; b = 3$ **13.** $y = 2x + 3; m = 2; b = 3$
15. (a) $y - 2 = 4(x + 5)$ (b) $y = 4x + 22$ (c) $4x - y + 22 = 0$ **17.** (a) $y - 5 = -3(x - 0)$ (b) $y = -3x + 5$
(c) $-3x - y + 5 = 0$ **19.** (a) $y - 5 = -\frac{5}{3}(x - 0)$ (b) $y = -\frac{5}{3}x + 5$ (c) $5x + 3y - 15 = 0$ **21.** (a) $y + 4 = \frac{2}{5}(x - 4)$
(b) $y = \frac{2}{5}x - \frac{28}{5}$ (c) $2x - 5y - 28 = 0$ **23.** (a) $y - \frac{2}{3} = \frac{3}{5}(x + 3)$ (b) $y = \frac{3}{5}x + \frac{37}{15}$ (c) $9x - 15y + 37 = 0$
25. (a) $y - \frac{5}{7} = -\frac{7}{3}(x - \frac{2}{3})$ (b) $y = -\frac{7}{3}x + \frac{143}{63}$ (c) $147x + 63y - 143 = 0$ **27.** $-\frac{5}{4}$ **29.** $y = 22N + 0.2x$
31. $y = 0.25x + 3.45$; $\$5.45$ **33.** $y = -0.75x + 7$; about 1989

Problem Set 4, page 140

1. -5 **3.** 2 **5.** 3 **7.** -4 **9.** $-\frac{5}{11}$ **11.** 4.106 **13.** $-\frac{7}{18}$ **15.** 2 **17.** 3 **19.** $2a + 3$
21. $b^2 + 3b - 4$ **23.** $|8 + 5c|$ **25.** $|2 - (5/a)|$ **27.** $-2x + 1$ **29.** $\sqrt{3x^4 + 5}$ **31.** $x + 1$ **33.** x
35. $(a - b)(a + b - 3)$ **37.** $\sqrt{3x + 5} - \sqrt{5}$ **39.** 2 **41.** -3.964 **43.** $\mathbb{R}$ **45.** $x \geq 0$ **47.** $x \neq 0$
49. $(-\infty, \frac{4}{3})$ **51.** $\mathbb{R}$ **53.** $(-\infty, -2), (-2, 2)$, and $(2, \infty)$ **55.** 4 **57.** $2x + h$
59. $-1/\sqrt{x}\sqrt{x + h}(\sqrt{x} + \sqrt{x + h})$ **67.** (a) Yes (b) No (c) Yes (d) No
69. $f(0) = 32$ or $0 \overset{f}{\mapsto} 32$; $f(15) = 59$ or $15 \overset{f}{\mapsto} 59$; $f(-10) = 14$ or $-10 \overset{f}{\mapsto} 14$; $f(55) = 131$ or $55 \overset{f}{\mapsto} 131$
71. $T = -\frac{1}{250}h + 65$; $-55°$ **73.** $A = x(12 - x)$

Problem Set 5, page 148

1.   **3.**

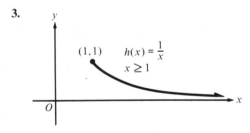

9. (a) Domain $\mathbb{R}$; range $(-\infty, 2)$; increasing on $(-\infty, -2]$ and $[0, 2]$; decreasing on $[-2, 0]$ and $[2, \infty)$; even
(b) Domain $[-5, 5]$; range $[-3, 3]$; increasing on $[-1, 1]$ and $[3, 5]$; decreasing on $[-5, -1]$ and $[1, 3]$; neither
(c) Domain $\left[-\frac{3\pi}{2}, \frac{3\pi}{2}\right]$; range $[-1, 1]$; increasing on $\left[-\frac{3\pi}{2}, -\frac{\pi}{2}\right]$ and $\left[\frac{\pi}{2}, \frac{3\pi}{2}\right]$; decreasing on $\left[-\frac{\pi}{2}, \frac{\pi}{2}\right]$; odd
(d) Domain $\mathbb{R}$; range $[-2, 1]$; constant on $(-\infty, 0]$ and $[\pi, \infty)$; increasing on $[0, \pi]$; neither
11. Even; symmetric about y axis   **13.** Odd; symmetric about origin   **15.** Even; symmetric about y axis
17. Neither   **19.** Odd; symmetric about origin   **21.** Domain $\mathbb{R}$; range $\mathbb{R}$; increasing on $\mathbb{R}$; neither even nor odd
23. Domain $\mathbb{R}$; range $\{5\}$; constant on $\mathbb{R}$; even; symmetric about y axis
25. Domain $\mathbb{R}$; range $(-\infty, 2]$; increasing on $(\infty, 0]$; decreasing on $[0, \infty)$; even; symmetric about y axis
27. Domain $\mathbb{R}$; range $\mathbb{R}$; increasing on $\mathbb{R}$; odd; symmetric about origin
29. Domain $[1, \infty)$; range $[1, \infty)$; increasing on $[1, \infty)$; neither even nor odd
31. Domain $(-\infty, 0)$ and $(0, \infty)$; range $\{-1, 1\}$; constant on $(-\infty, 0)$ and on $(0, \infty)$; odd; symmetric about origin
33. Domain $\mathbb{R}$; range $\mathbb{R}$; increasing on $\mathbb{R}$; odd; symmetric about origin
35. Domain $(-\infty, 0)$ and $(0, \infty)$; increasing on $(-\infty, 0)$; decreasing on $(0, \infty)$; even; symmetric about y axis

41.

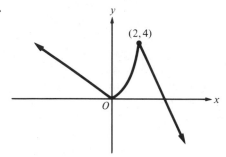

43. The graph actually falls *below* the x axis between 0 and 0.31

Problem Set 6, page 155

1.

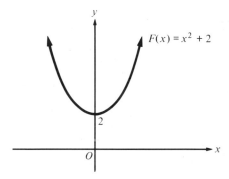

$F(x) = x^2 + 2$

5.

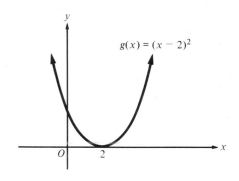

$g(x) = (x - 2)^2$

7.

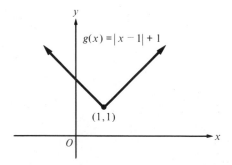

$g(x) = |x - 1| + 1$

(1,1)

15.

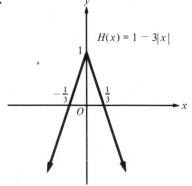

$H(x) = 1 - 3|x|$

19. Shift up 2 units.   **21.** Reflect about x axis, then shift up 1 unit.   **23.** Multiply ordinates by 2, then shift up 1 unit.
25. Shift right by 2 units, then multiply ordinates by 2, then shift up by 4 units.   **31.** Reflected across y axis.

Problem Set 7, page 161

1. (a) $7x - 3$ (b) $3x + 7$ (c) $10x^2 - 21x - 10$ (d) $\dfrac{5x + 2}{2x - 5}$ **3.** (a) $x^2 + 4$ (b) $x^2 - 4$ (c) $4x^2$ (d) $x^2/4$ **5.** (a) -1 (b) 5

(c) -6 (d) $-\frac{2}{3}$ **7.** (a) $x^2 + 2x - 4$ (b) $-x^2 + 2x - 6$ (c) $2x^3 - 5x^2 + 2x - 5$ (d) $\dfrac{2x - 5}{x^2 + 1}$ **9.** (a) $\dfrac{8x + 1}{2x - 1}$

(b) 1 if $x \neq \frac{1}{2}$ (c) $\dfrac{15x^2 + 5x}{4x^2 - 4x + 1}$ (d) $\dfrac{5x}{3x + 1}$

11.

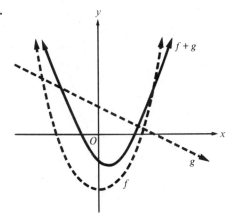

15. 17   **17.** 6.9929   **19.** -3 **21.** 2   **23.** (a) $3x + 3$ (b) $3x + 1$ (c) $9x$ **25.** (a) x (b) $|x|$ (c) x^4

27. (a) $-x^2 - 1$ (b) $x^2 - 2x + 3$ (c) x **29.** (a) $1 - 5|2x + 3|$ (b) $|5 \quad 10x|$ (c) $25x - 1$ **31.** (a) $\dfrac{1}{4x - 9}$

(b) $\dfrac{11 - 6x}{2x - 3}$ (c) $\dfrac{2x - 3}{11 - 6x}$ **33.** $h \circ g$ **35.** $h \circ f$ **37.** $f \circ h$ **39.** $h \circ h$ **41.** $f(x) = x^{-7}; g(x) = 2x^2 - 5x + 1$

43. $f(x) = x^5; g(x) = \dfrac{1 + x^2}{1 - x^2}$ **45.** $f(x) = \sqrt{x}; g(x) = \dfrac{x + 1}{x - 1}$ **47.** $f(x) = \dfrac{|x|}{x}; g(x) = x + 1$ **49.** (a) $R(x) = 10x$

(b) $P(x) = 10x - 50{,}000 - 10{,}000\sqrt[3]{x + 1}$ (c) \$56,550   **51.** $\dfrac{\sqrt{3}p^2}{36} =$ area of equilateral triangle with perimeter p

53. (a) x (b) x (c) $\frac{1}{2}x^2 - 3x + \frac{15}{2}$ (d) $\frac{1}{2}x^2 - 3x + \frac{15}{2}$ (e) $2x^2 + x + 3$ **55.** For example: $f(x) = x; g(x) = x^2 - 1$
59. $\frac{3}{2}$ **61.** -1

Problem Set 8, page 167

1. $f[g(x)] = f[(x + 3)/2] = 2[(x + 3)/2] - 3 = x$ and $g[f(x)] = g(2x - 3) = [(2x - 3) + 3]/2 = x$
7. (a) Invertible (b) Not invertible (c) Not invertible   **11.** $f^{-1}(x) = (x + 13)/7$ **13.** $f^{-1}(x) = -2x + 6$
15. $f^{-1}(x) = -\sqrt{-x}$ **17.** $f^{-1}(x) = (x - 1)^2$ **19.** $f^{-1}(x) = 1 + \sqrt{x - 1}$ **21.** (a) $f^{-1}(x) = (x + 5)/2$
23. (a) $f^{-1}(x) = \sqrt[3]{x} - 2$ **25.** (a) $f^{-1}(x) = \dfrac{\sqrt[3]{1 - x}}{2}$ **27.** (a) $f^{-1}(x) = \dfrac{-1}{1 + x}$ **29.** (a) $f^{-1}(x) = \dfrac{-x - 7}{x - 3}$

31. Not invertible   **33.** $f^{-1}(x) = \dfrac{b - dx}{cx - a}$ **39.** $C = f^{-1}(t)$

Review Problem Set, page 169

1. 5   **3.** 13   **5.** 8   **7.** 30.35   **9.** A circle of radius 5 with center $(2, -3)$ **11.** $m = -7, y - 2 = -7(x - 2)$

13. $m = \frac{3}{2}, y - 2 = \frac{3}{2}(x - 1)$ **15.** $y - 2 = -\frac{3}{5}(x - 5)$ **17.** (a) $y + 5 = \frac{3}{2}(x - 7)$ (b) $y + 5 = -\frac{2}{3}(x - 7)$

19. $y = \frac{4}{3}x + \frac{2}{3}; m = \frac{4}{3}; b = \frac{2}{3}$ **21.** (a) $y - 1 = 3(x + 7)$ (b) $y = 3x + 22$ (c) $3x - y + 22 = 0$

23. (a) $y + 2 = \frac{7}{3}(x - 1)$ (b) $y = \frac{7}{3}x - \frac{13}{3}$ (c) $7x - 3y - 13 = 0$ **25.** $y - b = -(a/b)(x - a)$

27. (a) $y - b = 2a(x - a)$ (b) $y - 8 = 4(x - 2)$ (c)

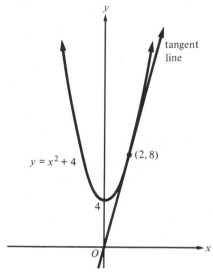

$y = x^2 + 4$ $(2, 8)$ tangent line

29. 23   **31.** 0   **33.** $3x^2 - 12$ **35.** $75x^2 - 180x + 104$ **37.** 12   **39.** $-1/[(x + k)x]$

41. $(-\infty, 1)$ and $(1, \infty)$ **43.** $[-1, \infty)$ **45.** $6x + 3h - 2$ **47.** (a) Graph of a function (b) Not the graph of a function

49. $f(74) = 99.26; f(75) = 99.63; f(76) = 100$

53. (a) Domain $\mathbb{R}$; range $[-1, 1]$; constant on $(-\infty, -2]$; decreasing on $[-2, 0]$; increasing on $[0, 2]$; constant on $[2, \infty)$; even; symmetric about y axis (b) Domain $\mathbb{R}$; range $[-3, \infty)$; decreasing on $(-\infty, -3]$; increasing on $[-3, -\frac{2}{3}]$; decreasing on $[-\frac{2}{3}, 2]$; constant on $[2, 4]$; increasing on $[4, \infty)$, neither even nor odd   **55.** Odd; symmetric about origin

57. Neither   **59.** Even; symmetric about y axis   **61.** Domain $\mathbb{R}$; range $\mathbb{R}$; increasing on $\mathbb{R}$; neither even nor odd

63. Domain $[2, \infty)$; range $[0, \infty)$; increasing on $[2, \infty)$; neither even nor odd

65. Domain $\mathbb{R}$; range $\mathbb{R}$; increasing on $\mathbb{R}$; odd; symmetric about origin   **67.** (a) Shift up 1 unit (b) Shift down 2 units

(c) Shift 1 unit to the right (d) Shift 2 units to the left (e) Reflect across x axis   **69.** 14.5 horsepower   **71.** (a) Increasing

(b) $G(p) = ps$ or $G(p) = pg(p)$ (c) At that price, producers won't supply.   **73.** (a) $4x - 2$ (b) $-2x + 6$ (c) $3x^2 + 2x - 8$

(d) $\dfrac{x + 2}{3x - 4}$ (e) $3x - 2$ **75.** (a) $\dfrac{2x}{x^2 - 1}$ (b) $\dfrac{2}{x^2 - 1}$ (c) $\dfrac{1}{x^2 - 1}$ (d) $\dfrac{x + 1}{x - 1}$ (e) $\dfrac{x + 1}{-x}$ **77.** (a) $x^4 + \sqrt{x + 1}$ (b) $x^4 - \sqrt{x + 1}$

(c) $x^4\sqrt{x + 1}$ (d) $\dfrac{x^4}{\sqrt{x + 1}}$ (e) $(x + 1)^2$ **79.** (a) $|x| - x$ (b) $|x| + x$ (c) $-x|x|$ (d) $\dfrac{|x|}{-x}$ (e) $|x|$ **81.** (a) $x^{2/3} + 1 + \sqrt{x}$

(b) $x^{2/3} + 1 - \sqrt{x}$ (c) $x^{7/6} + x^{1/2}$ (d) $\dfrac{x^{2/3} + 1}{\sqrt{x}}$ (e) $x^{1/3} + 1$ **83.** 0.8693   **85.** -11.8623 **87.** $g \circ h$ **89.** $h \circ g$

91. $f \circ h$ **93.** $f(x) = x^{-3}; g(x) = 4x^3 - 2x + 5$ **95.** $f(x) = \dfrac{2x^2 + x}{\sqrt{x}}; g(x) = x^2 + 1$ **97.** (a) $2x$ (b) 2 (c) $2x$ (d) 2

99. (a) $f(x) = \sqrt{8100 + x^2}; g(t) = 50t$ (b) $y = f[g(t)]$, so $y = (f \circ g)(t)$ (c) $(f \circ g)(t) = 10\sqrt{81 + 25t^2}$

101. (a) $P = -\frac{84}{125}p^2 + 102p - 2750$ (b) $\$875.20$ **103.** $f^{-1}(x) = \dfrac{x + 1}{3}$ **105.** $f^{-1}(x) = 5x - 25$

107. $f^{-1}(x) = \frac{1}{4}(x + 1)^2$ **111.** $f^{-1}(x) = \dfrac{1}{1 - x}$

115. $f^{-1}(x) = \dfrac{-B + \sqrt{B^2 - 4A(C - x)}}{2A}$ for $x \geq C - \dfrac{B^2}{4A}$ **117.** $N = f^{-1}(kt)$

Chapter 4

Problem Set 1, page 180

1.

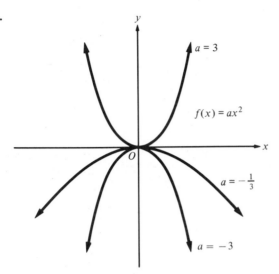

$a = 3$

$f(x) = ax^2$

$a = -\frac{1}{3}$

$a = -3$

3.

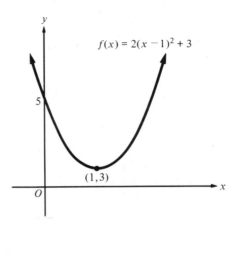

$f(x) = 2(x-1)^2 + 3$

5

$(1,3)$

7. $f(x) = (x + 1)^2 - 5$ **9.** $f(x) = 3(x - \frac{5}{3})^2 - \frac{31}{3}$ **11.** $f(x) = -2(x - \frac{3}{2})^2 + \frac{15}{2}$
13. Vertex $(1, -4)$; y intercept -3, x intercepts -1 and 3; opens upward; domain $\mathbb{R}$; range $[-4, \infty)$
15. Vertex $(-2, -4)$; y intercept 0; x intercepts 0 and -4; opens upward; domain $\mathbb{R}$; range $[-4, \infty)$
17. Vertex $(\frac{1}{4}, -\frac{119}{8})$; y intercept -15; x intercepts—none; opens downward; domain $\mathbb{R}$; range $(-\infty, \frac{-119}{8}]$
19. Vertex $(-\frac{1}{12}, -\frac{49}{24})$; y intercept -2; x intercepts $\frac{2}{3}$ and -1; opens upward; domain $\mathbb{R}$; range $[-\frac{49}{24}, \infty)$
21. Vertex $(2, 27)$; y intercept 15; x intercepts -1 and 5; opens downward; domain $\mathbb{R}$; range $(-\infty, 27]$
23. Vertex $(-1, \frac{3}{2})$; y intercept 2; x intercepts—none; opens upward; domain $\mathbb{R}$; range $[\frac{3}{2}, \infty)$
25. Vertex $(5, 7)$; y intercept 57; x intercepts—none; opens upward; domain $[0, 10]$; range $[7, 57]$
27. 50, 50   **29.** 150 meters by 150 meters   **31.** 100 thousand tons; $90   **33.** $8/month   **35.** 96 km/hr

Problem Set 2, page 186

1. Polynomial   **3.** Not   **5.** Not   **11.** y intercept 1; no x intercept   **13.** y intercept -1; x intercept $\sqrt[5]{\frac{1}{3}}$
15. y intercept 2; x intercept -1 **17.** y intercept -1; no x intercept   **19.** y intercept $-\frac{7}{8}$; x intercept $\frac{1}{2}$
21. y intercept 2593; no x intercept   **23.** 4   **25.** $\frac{3}{2}, \frac{3}{2}, 1$ **27.** $-\frac{1}{2}, \frac{1}{3}, \frac{1}{2}$ **29.** $-2, -\frac{5}{3}, \frac{5}{3}, 7$ **31.** $-1, 0, 1$
33. (a) $-1, 0, 1$ (c) $(-1, 0)$ and $(1, \infty)$ **35.** (a) $-2, 0, 1$ (c) $(-\infty, -2]$ and $[0, 1]$ **37.** (a) $-1, 1, 3$ (c) $(-\infty, -1)$ and $(1, 3)$
39. (a) $-4, -2, 1$ (c) $(-\infty, -4]$ and $[-2, 1]$ **41.**

O 4 $\frac{15}{2}$

Problem Set 3, page 193

1. $Q: x + 8; R: 30$ **3.** $Q: x^2 + x + 1; R: 0$ **5.** $Q: 3x^2 + 6x - 5; R: -1$ **7.** $Q: 2x^3 - 3x^2 + 3x - 4; R: -12$
9. $Q: x + 3; R: -x - 13$ **11.** $Q: x^2 - x + 1; R: -x^2 - 1$ **13.** $Q: (x/2) + (1/2); R: 3/2$
15. $Q: (x/3) + (1/9); R: -(5/9)x + (8/9)$ **17.** $Q: 2t + 3; R: -10$ **19.** $5x^2 + 23x + 91 + [366/(x - 4)]$
21. $5 + \dfrac{4x^2 - 28x - 16}{x^3 - 2x^2 - 8x}$ **23.** $Q: 3x^2 + 4x + 7; R: 18$ **25.** $Q: x^4 - 2x^3 - x^2 + 2x - 3; R: -10$
27. $Q: 3x^5 - 3x^4 + x^3 - x^2 + 2x - 2; R: 2$ **29.** $Q: -16x^2 - 20x - 8; R: 3$ **31.** $Q: x^2 + 2.1x + 3.31; R: 4.641$
33. $Q: 5x^2 + 4x - 2; R: -16$ **35.** $Q: -2x^2 + 2x + 10; R: -20$ **37.** 3   **39.** 15

Problem Set 4, page 199

1. 0   **3.** -456 **5.** 4   **7.** 51   **9.** $\frac{181}{27}$ **11.** Yes   **13.** No   **15.** No   **17.** No   **19.** 1, 3, 5   **21.** 3
23. $-\frac{1}{2}$ **25.** 2, 2   **27.** $-\frac{1}{2}, -\frac{1}{2}$ **29.** 3   **31.** $-\frac{3}{4}, \frac{1}{5}, 1$ **33.** $f(x) = (x - 6)(x + 1)(x - 1)$
35. $Q(x) = (x - 1)(x - 2)(x + 2)(2x - 1)$ **37.** 5 inches by 10 inches by 7 inches   **39.** 3 km, 4 km, 5 km

Problem Set 5, page 205

1. $(-\infty, 0)$ and $(0, \infty)$ **3.** $\mathbb{R}$ **5.** $\mathbb{R}$ **7.** $\mathbb{R}$ **9.** $(-\infty, -5), (-5, -3)$, and $(-3, \infty)$
11. $(-\infty, -1), (-1, 0), (0, 1)$, and $(1, \infty)$ **13.** Asymptotes: $x = 0, y = 0$; no intercepts
15. Asymptotes: $x = 0, y = 0$; no intercepts   **17.** Asymptotes: $x = 0, y = 0$; no intercepts
19. Asymptotes: $x = 0, y = 0$; no intercepts   **21.** Asymptotes; $x = 0, y = 1$; x intercept -1
23. Asymptotes: $x = 0, y = -4$; x intercepts $-\frac{1}{2}, \frac{1}{2}$ **25.** Asymptotes: $x = 3, y = 0$; y intercept $-\frac{1}{3}$
27. Asymptotes: $x = -1, y = 0$; y intercept -2 **29.** Asymptotes: $x = 2, y = 6$; x intercept $\frac{3}{2}$, y intercept $\frac{9}{2}$
31. Asymptotes: $x = 1, y = -4$; x intercept 0, y intercept 0
33. Asymptotes: $x = -3, y = -4$; x intercepts $-3 \pm \dfrac{\sqrt{5}}{2}$, y intercept $-\dfrac{31}{9}$
35. Asymptotes: $x = 1, y = 0$; y intercept 3
37. Asymptotes: $x = 2, y = 5$; x intercepts $2 \pm \dfrac{\sqrt{10}}{5}$, y intercept $\dfrac{9}{2}$
39. Asymptotes: $x = 3, y = 3$; y intercept $\frac{28}{9}$ **41.** No asymptotes; "hole" at $(-2, -4)$ **43.** Asymptote: $y = 0$
45. Asymptote: $y = 0$ **47.** Asymptote: $y = 0$ **49.** Asymptote: $x = 2$; y intercept -2 **51.** $y = \frac{5}{2}$ **53.** $y = \frac{5}{8}$
55. $y = 0$ **57.** (b) $(0, \infty)$ (c) Decreases   **59.** (b) $(0, \infty)$ (c) Increases
61.

$$N(t) = 60\left(\frac{t - 2}{t}\right)$$
for $3 \le t \le 10$

63. It has more than one horizontal asymptote.

Problem Set 6, page 214

1. $\frac{3}{4}$ **3.** $\frac{9}{5}$ **5.** $\frac{3}{200}$ **7.** $-\frac{11}{32}$ **9.** $\frac{67}{3}$ **11.** $\frac{21}{2}$ **13.** $\frac{40}{21}$ **15.** $\frac{121}{2}$

17. From $a/b = c/d$ we have $ad = bc$. Dividing by ab, we obtain $d/b = c/a$ or $c/a = d/b$.

23. $y = kt$ **25.** $V = kr^3$ **27.** $P = kv^3$ **29.** $E = k(AT/d)$ **31.** $F = k(Q_1Q_2/d^2)$ **33.** $n = k\sqrt{F}/(\ell\sqrt{d})$

35. 6   **37.** $\frac{3}{2}$ **39.** $\frac{200}{9}$ **41.** 8   **43.** $\frac{28}{3}$ **45.** 54   **47.** $P = MS/m$ **49.** 40   **51.** $F = 32 + \frac{9}{5}C$

53. $P = \frac{2}{15}R - 4000$ **55.** $A = \frac{9}{4}N^{2/3}$ **57.** $\frac{15}{8}$ days   **59.** 2.36×10^{-3} newton   **61.** $n = (kT + K)/\ell$

63. $(\sqrt{5} - 1)/2 \approx 0.618$

Review Problem Set, page 217

1. Domain $\mathbb{R}$; range $[0, \infty)$; x intercept 0; y intercept 0; vertex $(0,0)$; opens upward

3. Domain $\mathbb{R}$; range $(-\infty, 0]$; x intercept 0; y intercept 0; vertex $(0,0)$; opens downward

5. Domain $\mathbb{R}$; range $[1, \infty)$; y intercept 13; vertex $(2, 1)$; opens upward

7. Domain $\mathbb{R}$, range $\left[-\frac{1}{4}, \infty\right)$; x intercepts 1 and 2; y intercept 2; vertex $(\frac{3}{2}, -\frac{1}{4})$; opens upward

9. Domain $\mathbb{R}$; range $\left(-\infty, \frac{529}{24}\right]$; x intercepts $-\frac{5}{2}$ and $\frac{4}{3}$; y intercept 20; vertex $\left(-\frac{7}{12}, \frac{529}{24}\right)$; opens downward

11. Domain $\mathbb{R}$; range $(-\infty, 0]$; x intercept 5, y intercept -25; vertex $(5,0)$; opens downward

13. $(-\infty, 1]$ and $[2, \infty)$ **15.** $\left(-\infty, -\frac{5}{2}\right]$ and $\left[\frac{4}{3}, \infty\right)$ **17.** $\frac{21}{2}$ and $\frac{21}{2}$ **19.** $15°F$ **21.** 30 thousand

23. Polynomial   **25.** Not a polynomial   **31.** x intercept 0; y intercept 0   **33.** x intercept 1; y intercept $-\frac{31}{4}$

35. (a) $-3, 0, 3$ (c) $(-\infty, -3]$ and $[0, 3]$ **37.** (a) $-5, -2, 0$ (c) $(-5, -2)$ and $(0, \infty)$

39. (a) $-\frac{6}{5}, -1, \frac{3}{2}$ (c) $\left[-\frac{6}{5}, -1\right]$ and $\left[\frac{3}{2}, \infty\right)$ **41.** (a) $-3, 2$ (c) $(-\infty, -3)$ **43.** $Q: x + 2$; $R: -4$

45. $Q: x^2$; $R: 8x^2 - 32$ **47.** $Q: 3x^4 - x^2 + 2$; $R: -1$ **49.** $4x^3 - 3x^2 + 3x - 5 + \dfrac{8}{x + 1}$ **51.** $x + 5 + \dfrac{3x - 10}{x^2 + x + 2}$

53. $Q: 3x^3 - 4x^2 + 9x - 17$; $R: 36$ **55.** $Q: x^2 - 4x + 11$; $R: -27$ **57.** $Q: 2x^2 + 7x + 28$; $R: 119$

59. $Q: x^4 - x^3 + x^2 - x + 6$; $R: -19$ **61.** 1637.75   **63.** -27 **65.** -137 **67.** -7 **69.** Yes   **71.** No

73. $-1, 2, 7$ **75.** $-3, 2, 2, 3$ **77.** $-1, -\frac{1}{2}, 1, \frac{3}{2}$ **79.** $(x + 2)(x + 1)(x - 1)$ **81.** $(x + 1)(x - 3)(x + 3)(x - 5)$

83. 8 cm   **87.** Domain $(-\infty, 0)$ and $(0, \infty)$; asymptotes $x = 0$ and $y = 0$; no intercepts

89. Domain $(-\infty, 0)$ and $(0, \infty)$; asymptote $y = 1$; x intercepts $\pm\sqrt{3}$

91. Domain $(-\infty, 1)$ and $(1, \infty)$; asymptotes $x = 1$ and $y = 2$; x intercept 0; y intercept 0

93. Domain $(-\infty, -2)$, $(-2, 0)$, and $(0, \infty)$; asymptotes $x = 0$ and $y = 1$; no intercepts ["hole" in graph at $(-2, 0)$]

95. Domain $(-\infty, 0)$, $(0, 3)$, and $(3, \infty)$; asymptotes $x = 0$, $x = 3$, and $y = 1$; no intercepts

97. (a)   (b) Domain $[0, \infty)$; range $[0, 1)$ (c) Increases

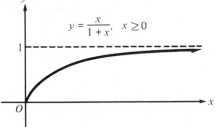

99. $\frac{240}{13}$ **101.** $\frac{1}{40}$ **103.** $\frac{25}{16}$ **105.** $-\frac{12}{13}$ **107.** $\frac{3}{2}$ **109.** 4   **111.** 5000   **113.** $P = k/V$ **115.** $V = kt + K$

117. $F = kAv^2$ **119.** $r = kN(P - N)$ **121.** $P = kAv^3$ **123.** (a) $y = \dfrac{5}{x}$ (b) $\dfrac{1}{5}$ **125.** (a) $w = \dfrac{7x}{3y}$ (b) $\dfrac{28}{3}$

127. (a) $y = \dfrac{4x^2}{z + 3}$ (b) 4   **129.** (a) $y = 2\sqrt{x} + 1$ (b) 7   **131.** 524 cubic meters   **133.** $V = (9 \times 10^{-6})T + 0.25$

135. $16,000

Chapter 5

Problem Set 1, page 228

5. 2.665144142   **7.** 8.824977830   **9.** 1.632526919   **11.** 0.292794032   **13.** $a^x a^y = 38.80960174$; $a^{x+y} = 38.80960175$
15. $(a^x)^y = 0.019378294$; $a^{xy} = 0.019378294$ **17.** $(ab)^x = 42.67011803$; $a^x b^x = 42.67011804$
19. Domain $\mathbb{R}$; range $(1, \infty)$; asymptote $y = 1$; increasing   **21.** Domain $\mathbb{R}$; range $(-1, \infty)$; asymptote $y = -1$; increasing
23. Domain $\mathbb{R}$; range $(0, \infty)$; asymptote $y = 0$; increasing
25. Domain $\mathbb{R}$; range $(-3, \infty)$; asymptote $y = -3$; decreasing

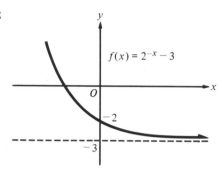

27. Domain $\mathbb{R}$; range $(0, \infty)$; asymptote $y = 0$; increasing   **29.** (a) \$2409.85 (b) 0.07   **31.** (a) \$4363.49 (b) 0.12
33. (a) \$53,471.36 (b) 0.1437   **35.** (a) \$1166.40 (b) \$1169.86 (c) \$1171.66 (d) \$1172.89 (e) \$1173.37 (f) \$1173.49 (g) \$1173.50
37. \$549.19   **39.** \$96.09

Problem Set 2, page 234

1. 0.3678794412   **3.** 20.08553692   **5.** 9.356469012   **7.** 0.0446009553   **9.** 23.14069264
13. $e^x e^y = 23.24905230$; $e^{x+y} = 23.24905230$ **15.** $(e^x)^y = 0.3166675732$; $e^{xy} = 0.3166675733$
25. (a) Discrepancy in third decimal place (b) Discrepancy in third decimal place (c) Discrepancy in second decimal place
(d) Discrepancy in first decimal place   **27.** (a) \$1645.31 (b) \$1648.72   **29.** (b) 292 bears   **31.** 2.976 grams
33. (a) 2,000,000 (b) 3,644,238   **35.** (a) 2000 (b) 8,894,134   **37.** 4.6 billion barrels

Problem Set 3, page 241

1. 3   **3.** $\frac{1}{2}$ **5.** 1   **7.** 0   **9.** $-2, 1$ **11.** 1   **13.** (a) 2 (b) 3 (c) 4 (d) 5 (e) -2 (f) -2 (g) 5   **15.** (a) $2^5 = 32$
(b) $16^{1/4} = 2$ (c) $9^{-1/2} = \frac{1}{3}$ (d) $e^1 = e$ (e) $(\sqrt{3})^4 = 9$ (f) $10^n = 10^n$ (g) $x^5 = x^5$ **17.** (a) $\log_8 1 = 0$ (b) $\log_{10} 0.0001 = -4$
(c) $\log_4 256 = 4$ (d) $\log_{27} \frac{1}{3} = -\frac{1}{3}$ (e) $\log_8 4 = \frac{2}{3}$ (f) $\log_a y = c$ **19.** 2   **21.** 5   **23.** 7   **25.** $\frac{1}{8}$ **27.** $\frac{1}{8}$ **29.** -7
31. $-\frac{2}{3}$ **33.** $\frac{1}{4}$ **35.** $\frac{9}{2}$ **37.** $\frac{85}{3}$ **39.** $-2, -1$ **41.** $-\frac{5}{3}, 1$ **43.** $\log_b x + \log_b(x + 1)$
45. $2\log_{10} x + \log_{10}(x + 1)$ **47.** $3\log_3 x + 2\log_3 y - \log_3 z$ **49.** $\frac{1}{2}\log_e x + \frac{1}{2}\log_e (x + 3)$ **51.** $\log_3 x^9$
53. $\log_5 \sqrt{\dfrac{a}{3b}}$ **55.** $\log_e(x + 1)$ **57.** $\log_3 \dfrac{x + 9}{x + 5}$
59. (a) 2.31 (b) 2.70 (c) -0.95 (d) 1.87 (e) 0.74 (f) 0.95 (g) 1.63
61. 2   **63.** $\frac{1}{9}$ **65.** 2   **67.** 67

Problem Set 4, page 248

1.

3.

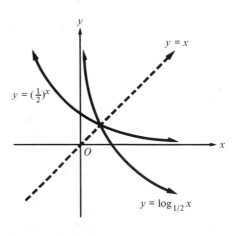

7. (a) 0.8043439185 (b) 3.090046322 (c) −1.453333975 (d) 11.48387245 (e) −8.182963774
9. (a) $\log xy = 2.214956167$; $\log x + \log y = 2.214956167$ (b) $\log xy = 0.6476648705$; $\log x + \log y = 0.6476648704$
11. (a) 8.325063694 (b) 0.9947321582 (c) −3.208826489 (d) 20.41142767 (e) −15.56881884
13. (a) $\ln e^{\pi} = \ln 23.14069264 = 3.141592654$ (b) $e^{\ln \pi} = e^{1.144729886} = 3.141592654$ **15.** (a) 4.643856190
(b) 0.6309297537 (c) 0.4808983469 (d) 1.405954306 (e) −7.551524229   **17.** (a) $(2, \infty)$ (b) $(0, \infty)$ (c) $(-\infty, 4)$
(d) $(-\infty, 0)$ and $(0, \infty)$ (e) $(0, 1)$ and $(1, \infty)$ (f) $\mathbb{R}$ (g) $\mathbb{R}$
19. Domain $(-2, \infty)$; range $\mathbb{R}$; asymptote $x = -2$; x intercept -1; y intercept $\log_3 2$; increasing on $(-2, \infty)$
21. Domain $(0, \infty)$; range $\mathbb{R}$; asymptote $x = 0$; x intercept e; increasing on $(0, \infty)$
23. Domain $(-\infty, 1)$; range $\mathbb{R}$; asymptote $x = 1$; x intercept 0; y intercept 0; decreasing on $(-\infty, 1)$
25. Domain $(1, \infty)$; range $\mathbb{R}$; asymptote $x = 1$; x intercept $1 + e^{-2}$; increasing on $(1, \infty)$
27. Domain $(-\infty, 1)$; range $\mathbb{R}$; asymptote $x = 1$; x intercept -99; y intercept 2; increasing on $(-\infty, 1)$
29. Domain $(-\infty, 0)$ and $(0, \infty)$; range $\mathbb{R}$; asymptote $x = 0$; x intercepts -1 and 1; decreasing on $(-\infty, 0)$;
increasing on $(0, \infty)$

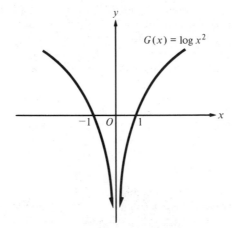

$G(x) = \log x^2$

31. Domain $(-\infty, 2)$; range $\mathbb{R}$; asymptote $x = 2$; x intercept 1; y intercept $\log_{1/3} 2$; increasing on $(-\infty, 2)$
33. $e^0 = 1$ **35.** $e^1 = e$ **37.** Property (iii)   **39.** Property (iv)   **41.** Range of ln is $\mathbb{R}$
43. All three graphs contain the point $(1, 0)$. As the base increases, the graph rises less rapidly to the right of $(1, 0)$.
47. 2   **49.** Equation of tangent line is $y = x - 1$.   **51.** $e^{x \ln y} = e^{\ln y^x} = y^x$ **53.** It's only true when $x > 0$.

Problem Set 5, page 255

1. $x = \log_4 3 = \log 3/\log 4 \approx 0.7924812504$ 3. $x = \log 2001/\log 7.07 \approx 3.886474732$
5. $x = (\log 4)/(\log(3/8)) = -1.413390105$ 7. 11.559 years $\approx$ 11 years and 29 weeks   9. (a) 7.8 (b) 4.2 (c) 6.4
11. 8863 meters   13. 5545 meters   15. 113 decibels   17. 10^{12} 19. $10^{8/5} \approx 39.8$ 21. (a) $y = y_0 e^{-0.00041857t}$
(b) 0.0418% (c) 0.9917 gram   23. 20.35 years   25. (a) $k = 1/(2 \ln 4) \approx 0.361$ (b) 0.354 or 35.4%

Problem Set 6, page 262

1. (a) $N = 10^7 e^{kt}$; $k = \ln 1.03$ (b) When $t = 20$, $N = 10^7 e^{20 \ln 1.03} = 18{,}061{,}112$ (c) $T = (\ln 2)/k = (\ln 2)/\ln 1.03 \approx 23.45$ years
3. $3/\log 2 \approx 9.97$ hours   5. (a) $N = 10 e^{(t \ln 2)/20}$ (b) 80   7. 41.42%   9. (a) $699.86 \approx 700$ (b) 421 (c) In 2.87 years
11.

Review Problem Set, page 263

1. (a) 27 (b) 1 (c) 5 (d) $\frac{1}{4}$ (e) $\sqrt{3}$ (f) $\sqrt{5}/5$ 3. Domain $\mathbb{R}$; range $(0, \infty)$; asymptote $y = 0$; increasing
5. Domain $\mathbb{R}$; range $(0, \infty)$; asymptote $y = 0$; decreasing   7. Domain $\mathbb{R}$; range $(3, \infty)$; asymptote $y = 3$; decreasing
9. Domain $\mathbb{R}$; range $(-\infty, 1)$; asymptote $y = 1$; increasing   11. \$6734.28   13. \$839.62   15. (a) 27.56714844
(b) 0.0432139183   17. Domain $\mathbb{R}$; range $(3, \infty)$; asymptote $y = 3$; increasing
19. Domain $\mathbb{R}$; range $(0, \infty)$; asymptote $y = 0$; increasing   21. Domain $\mathbb{R}$; range $(2, \infty)$; asymptote $y = 2$; increasing
23. \$11,082.57   25. (a) 66,079.39737 (b) 1505.971060 (c) $5\sqrt{35}/7$ (d) 98.17%   27. $5^x = \frac{1}{125}$; -3 29. $3^1 = 2 + x$; 1
31. $3^{5x} = 81$; $\frac{4}{5}$ 33. $x^2 = \frac{1}{49}$; $\frac{1}{7}$ 35. $3^{-2x} = \frac{1}{81}$; 2   37. $7^{|x|-1} = 49$; ± 3 39. (a) $\log_3 729 = 6$
(b) $\log_2 \frac{1}{1024} = -10$ (c) $\log_{64} 256 = \frac{4}{3}$ (d) $\log_x w = a$ (e) $\log y = x$ (f) $\ln y = x$ 41. (a) 4 (b) $\frac{1}{4}$ (c) $\frac{1}{2}$ (d) 0 (e) -3 (f) 33
(g) -1.4 43. (a) 0.4342944818 (b) 4.605170186 (c) 3.505557397 (d) $1.443225879 \times 10^{-5}$ (e) $5.923313615 \times 10^{-2}$
(f) 1.667389292 (g) 5.574941522 (h) $1.340164240 \times 10^{18}$ (i) 6.581412462   45. $(1/n) \log_b p - n \log_b R$
47. $\frac{1}{2} \log(4 - x) - \frac{1}{2} \log(4 + x)$ 49. $\log_b x^9 y^{1/c}$ 51. $\ln \dfrac{(x - 2)(2x - 1)}{(x - 4)(x - 5)}$ 53. 0   55. -2
57. (a) 3.459431618 (b) 11.07309365 (c) 0.3156023436 (d) 0.4306765581   59. $\log \sqrt{x} = 0.7606002382$; $\frac{1}{2} \log x = 0.7606002380$
61. $\ln x = 4.344855520$; $\log x/\log e = 4.344855521$ 63. $\log y^x = -28.59502498$; $x \log y = -28.59502498$
65. $\log \pi/\log e = 1.145$; $\log \pi - \log e = 0.063$ 67. $1/\log(\frac{2}{3}) = -5.679$; $\log \frac{3}{2} = 0.176$ 69. $\log_2 3 = 1.585$; $\log 8 = 0.903$
71. (a) $(\frac{3}{4}, \infty)$ (b) $(-\infty, -1)$ and $(-1, \infty)$ (c) $(-\infty, 2 - 2\sqrt{2})$ and $(2 + 2\sqrt{2}, \infty)$ (d) $\mathbb{R}$
73. Domain $(1, \infty)$; range $\mathbb{R}$; asymptote $x = 1$; x intercept $1 + (1/e)$; increasing
75. Domain $(-\infty, 4)$; range $\mathbb{R}$; asymptote $x = 4$; x intercept 3; y intercept $\log 2$; decreasing
77. Domain $(-\infty, 2)$ and $(2, \infty)$; range $\mathbb{R}$; asymptote $x = 2$; x intercepts 1 and 3; y intercept $\ln 2$; decreasing on $(-\infty, 2)$;

increasing on $(2, \infty)$ **79.** $x - 1$ **81.** $x - 5$ **83.** 1.301029996   **85.** -3.561615892 **87.** 0.653831157
89. (a) 78 (b) 35.20 (c) 25.19 months   **91.** 123 decibels   **93.** Approximately 0.9752 gram
95. (a) In 12 years, you'll have \$99,603.51. (b) In 12 years, you'll have \$167,772.16. Choose plan b.   **97.** (a) \$205.07
(b) \$1382.37   **99.** (a) \$514.31 (b) \$7062.21   **101.** (a) 4255 (b) 198 days   **103.** (b) 2209 years   **105.** (a) 75,064
(b) 57,624 (c) 8.07 years after introduction

Chapter 6

Problem Set 1, page 275

1. $60°$ **3.** $-792°$ **5.** $\frac{1}{8}$ counterclockwise   **7.** $\frac{1}{3}$ counterclockwise   **9.** $\frac{11}{12}$ clockwise   **11.** $\frac{1}{30}$ clockwise
13. $0.0\overline{3}°$ **15.** $0.000\overline{83}°$ **17.** $100.50\overline{5}°$ **19.** $62°15'$ **21.** $-0°7'30''$ **23.** $21°9'36''$ **25.** 3.3 meters
27. $\frac{4}{3}$ radians   **29.** $10\pi/3$ inches   **31.** (a) $\pi/6$ (b) $\pi/4$ (c) $\pi/2$ (d) $2\pi/3$ (e) $-5\pi/6$ (f) $26\pi/9$ (g) $\frac{2}{5}\pi$ (h) $3\pi/8$ (i) $-11\pi/6$ (j) $5\pi/2$
(k) $7\pi/60$ (l) -2π **33.** (a) $90°$ (b) $60°$ (c) $45°$ (d) $30°$ (e) $120°$ (f) $-180°$ (g) $108°$ (h) $-450°$ (i) $405°$ (j) $-67.5°$ (k) $1260°$
(l) $-\left(\frac{90}{7}\right)°$ **35.** (a) $-3\pi/4$, $-135°$ (b) $25\pi/3$, $1500°$ (c) $2\pi/3$, $120°$ **37.** $10\pi/3$ meters   **39.** $11,057°$
41. (a) $4\pi/3$ radians per hour (b) $10,240\pi \approx 32,169.91$ kilometers per hour   **43.** $760,320/13 \approx 58,486.15$ radians per hour
45. $146\pi/9 \approx 50.96$ centimeters per second   **47.** $1,550,000\pi/73 \approx 66,705$ miles per hour
49. (a) $21\pi/4$ square centimeters (b) $117\pi/2$ square inches

Problem Set 2, page 283

1. $\sin \theta = \frac{3}{5}$; $\cos \theta = \frac{4}{5}$; $\tan \theta = \frac{3}{4}$; $\cot \theta = \frac{4}{3}$; $\sec \theta = \frac{5}{4}$; $\csc \theta = \frac{5}{3}$
3. $\sin \theta = \frac{3}{4}$; $\cos \theta = \sqrt{7}/4$; $\tan \theta = 3\sqrt{7}/7$, $\cot \theta = \sqrt{7}/3$; $\sec \theta = 4\sqrt{7}/7$; $\csc \theta = \frac{4}{3}$
5. $\sin \theta = \frac{1}{2}$; $\cos \theta = \sqrt{3}/2$; $\tan \theta = \sqrt{3}/3$; $\cot \theta = \sqrt{3}$; $\sec \theta = 2\sqrt{3}/3$; $\csc \theta = 2$
7. $\sin \theta = \sqrt{2}/2$; $\cos \theta = \sqrt{2}/2$; $\tan \theta = 1$; $\cot \theta = 1$; $\sec \theta = \sqrt{2}$; $\csc \theta = \sqrt{2}$ **9.** $55°$ **11.** $3\pi/10$ **13.** $\pi/12$
15. $67.66°$ **17.** $\cos 59°$ **19.** $\cot(7\pi/18)$ **21.** $\tan(\pi/10)$ **23.** $\sec 12.97°$
25. $\sin 48° \approx 0.7431448255$ $\csc 48° \approx 1.345632730$
$\cos 48° \approx 0.6691306064$ $\sec 48° \approx 1.494476550$
$\tan 48° \approx 1.110612515$ $\cot 48° \approx 0.900404044$
27. $\sin 23°12'33'' \approx 0.3940889557$ $\csc 23°12'33'' \approx 2.537498160$
$\cos 23°12'33'' \approx 0.9190723017$ $\sec 23°12'33'' \approx 1.088053680$
$\tan 23°12'33'' \approx 0.4287899387$ $\cot 23°12'33'' \approx 2.332144273$
29. $\sin 16.19° \approx 0.2788234989$ $\csc 16.19° \approx 3.586498283$
$\cos 16.19° \approx 0.9603423642$ $\sec 16.19° \approx 1.041295310$
$\tan 16.19° \approx 0.2903376018$ $\cot 16.19° \approx 3.444266240$
31. $\sin(2\pi/7) \approx 0.7818314825$ $\csc(2\pi/7) \approx 1.279048008$
$\cos(2\pi/7) \approx 0.6234898018$ $\sec(2\pi/7) \approx 1.603875472$
$\tan(2\pi/7) \approx 1.253960338$ $\cot(2\pi/7) \approx 0.7974733887$
33. $\sin 0.7764 \approx 0.7007155788$ $\csc 0.7764 \approx 1.427112555$
$\cos 0.7764 \approx 0.7134407317$ $\sec 0.7764 \approx 1.401658127$
$\tan 0.7764 \approx 0.9821636859$ $\cot 0.7764 \approx 0.1018160226$
39. Evaluate sin 30 on the calculator; if result is 0.50, then 30 must be 30 degrees; and if result is not 0.50, then calculator must be
in radian mode.
43. $\cot \theta = \dfrac{\text{adj}}{\text{opp}} = \dfrac{1}{\text{opp/adj}} = \dfrac{1}{\tan \theta}$ **45.** $\text{adj}^2 + \text{opp}^2 = \text{hyp}^2$, so $\left(\dfrac{\text{adj}}{\text{hyp}}\right)^2 + \left(\dfrac{\text{opp}}{\text{hyp}}\right)^2 = 1$; thus, $(\cos \theta)^2 + (\sin \theta)^2 = 1$

Problem Set 3, page 292

1. Quadrant I   **3.** Quadrantal   **5.** Quadrant IV   **7.** Quadrant II   **9.** Quadrant III   **11.** $420°, -300°, 780°$
13. $7\pi/4, -9\pi/4, -17\pi/4$ **15.** $-252°, -972°, 108°$ **17.** $17\pi/6, -7\pi/6, -19\pi/6$
19. $\sin\theta = \frac{3}{5}; \cos\theta = \frac{4}{5}; \tan\theta = \frac{3}{4}; \cot\theta = \frac{4}{3}; \sec\theta = \frac{5}{4}; \csc\theta = \frac{5}{3}$
21. $\sin\theta = \frac{12}{13}; \cos\theta = -\frac{5}{13}; \tan\theta = -\frac{12}{5}; \cot\theta = -\frac{5}{12}; \sec\theta = -\frac{13}{5}; \csc\theta = \frac{13}{12}$
23. $\sin\theta = -\frac{4}{5}; \cos\theta = -\frac{3}{5}; \tan\theta = \frac{4}{3}; \cot\theta = \frac{3}{4}; \sec\theta = -\frac{5}{3}; \csc\theta = -\frac{5}{4}$
25. $\sin\theta = 3\sqrt{58}/58; \cos\theta = 7\sqrt{58}/58; \tan\theta = \frac{3}{7}; \cot\theta = \frac{7}{3}; \sec\theta = \sqrt{58}/7; \csc\theta = \sqrt{58}/3$
27. $\sin\theta = -\sqrt{10}/5; \cos\theta = -\sqrt{15}/5; \tan\theta = \sqrt{6}/3; \cot\theta = \sqrt{6}/2; \sec\theta = -\sqrt{15}/3; \csc\theta = -\sqrt{10}/2$
29. No, since $\sin^2\theta + \cos^2\theta = 1$ implies that $\sin^2\theta \le 1$; that is, $-1 \le \sin\theta \le 1$.   **31.** (a) I (b) II (c) II (d) IV (e) III (f) III
(g) IV (h) I   **33.** (a) Negative (b) Negative (c) Positive (d) Negative (e) Positive (f) Negative (g) Negative

35. $\tan\theta = \dfrac{y}{x} = \dfrac{y/r}{x/r} = \dfrac{\sin\theta}{\cos\theta}$ **37.** (a) $\sec\theta = \dfrac{13}{12}$ (b) $\csc\theta = -\dfrac{13}{5}$ (c) $\tan\theta = -\dfrac{5}{12}$ (d) $\cot\theta = -\dfrac{12}{5}$

39. $\cos\theta = \frac{3}{5}; \tan\theta = \frac{4}{3}; \cot\theta = \frac{3}{4}; \sec\theta = \frac{5}{3}; \csc\theta = \frac{5}{4}$
41. $\cos\theta = -\sqrt{7}/4; \tan\theta = 3\sqrt{7}/7; \cot\theta = \sqrt{7}/3; \sec\theta = -4\sqrt{7}/7; \csc\theta = -\frac{4}{3}$
43. $\sin\theta = -\sqrt{33}/7; \tan\theta = -\sqrt{33}/4; \cot\theta = -4\sqrt{33}/33; \sec\theta = \frac{7}{4}; \csc\theta = -7\sqrt{33}/33$
45. $\sin\theta = \frac{2}{3}; \cos\theta = \sqrt{5}/3; \tan\theta = 2\sqrt{5}/5; \cot\theta = \sqrt{5}/2; \sec\theta = 3\sqrt{5}/5$
47. $\sin\theta = \frac{4}{5}; \cos\theta = \frac{3}{5}; \cot\theta = \frac{3}{4}; \sec\theta = \frac{5}{3}; \csc\theta = \frac{5}{4}$
49. $\sin\theta = \frac{5}{13}; \cos\theta = -\frac{12}{13}; \tan\theta = -\frac{5}{12}; \sec\theta = -\frac{13}{12}; \csc\theta = \frac{13}{5}$
51. $\sin\theta = 2\sqrt{2}/3; \cos\theta = -\frac{1}{3}; \tan\theta = -2\sqrt{2}; \cot\theta = -\sqrt{2}/4; \csc\theta = 3\sqrt{2}/4$
53. $\cos\theta = \sqrt{1-\sin^2\theta} = 0.8829; \tan\theta = 0.4695/0.8829 = 0.5318; \cot\theta = 1.8805; \sec\theta = 1.1326; \csc\theta = 2.1299$
55. $\sin\theta = 1/\csc\theta = 0.3827; \cos\theta = -\sqrt{1-\sin^2\theta} = -0.9239; \tan\theta = -0.4142; \cot\theta = -2.4142; \sec\theta = -1.0824$
57. Yes, tan and cot.

Problem Set 4, page 299

3. $0°$; see Table 1   **5.** $180°$; see Table 1
7. $30°; \sin 30° = \frac{1}{2}; \cos 30° = \sqrt{3}/2; \tan 30° = \sqrt{3}/3; \cot 30° = \sqrt{3}; \sec 30° = 2\sqrt{3}/3; \csc 30° = 2$
9. $45°; \sin 45° = \sqrt{2}/2; \cos 45° = \sqrt{2}/2; \tan 45° = 1; \cot 45° = 1; \sec 45° = \sqrt{2}; \csc 45° = \sqrt{2}$ **11.** π; see Table 1
13. $\pi/3; \sin\pi/3 = \sqrt{3}/2; \cos\pi/3 = \frac{1}{2}; \tan\pi/3 = \sqrt{3}; \cot\pi/3 = \sqrt{3}/3; \sec\pi/3 = 2; \csc\pi/3 = 2\sqrt{3}/3$
17. $\theta_R = 30°; \sin\theta = \frac{1}{2}; \cos\theta = -\sqrt{3}/2; \tan\theta = -\sqrt{3}/3; \cot\theta = -\sqrt{3}; \sec\theta = -2\sqrt{3}/3; \csc\theta = 2$
19. $\theta_R = 60°; \sin\theta = -\sqrt{3}/2; \cos\theta = -\frac{1}{2}; \tan\theta = \sqrt{3}; \cot\theta = \sqrt{3}/3; \sec\theta = -2; \csc\theta = -2\sqrt{3}/3$
21. $\theta_R = 45°; \sin\theta = -\sqrt{2}/2; \cos\theta = \sqrt{2}/2; \tan\theta = -1; \cot\theta = -1; \sec\theta = \sqrt{2}; \csc\theta = -\sqrt{2}$
23. $\theta_R = 30°; \sin\theta = -\frac{1}{2}; \cos\theta = -\sqrt{3}/2; \tan\theta = \sqrt{3}/3; \cot\theta = \sqrt{3}; \sec\theta = -2\sqrt{3}/3; \csc\theta = -2$
25. $\theta_R = 60°; \sin\theta = -\sqrt{3}/2; \cos\theta = \frac{1}{2}; \tan\theta = -\sqrt{3}; \cot\theta = -\sqrt{3}/3; \sec\theta = 2; \csc\theta = -2\sqrt{3}/3$
27. $\theta_R = \pi/4; \sin\theta = -\sqrt{2}/2; \cos\theta = \sqrt{2}/2; \tan\theta = -1; \cot\theta = -1; \sec\theta = \sqrt{2}; \csc\theta = -\sqrt{2}$
29. $\theta_R = \pi/3$; see problem 19   **31.** $\theta_R = \pi/4$; see problem 21   **33.** $\theta_R = \pi/3$; see problem 25
35. $\theta_R = 60°$; see problem 25

37.

θ	$\sin\theta$	$\cos\theta$	$\tan\theta$	$\cot\theta$	$\sec\theta$	$\csc\theta$
210°	$-\frac{1}{2}$	$-\sqrt{3}/2$	$\sqrt{3}/3$	$\sqrt{3}$	$-2\sqrt{3}/3$	-2
225°	$-\sqrt{2}/2$	$-\sqrt{2}/2$	1	1	$-\sqrt{2}$	$-\sqrt{2}$
240°	$-\sqrt{3}/2$	$-\frac{1}{2}$	$\sqrt{3}$	$\sqrt{3}/3$	-2	$-2\sqrt{3}/3$
300°	$-\sqrt{3}/2$	$\frac{1}{2}$	$-\sqrt{3}$	$-\sqrt{3}/3$	2	$-2\sqrt{3}/3$
315°	$-\sqrt{2}/2$	$\sqrt{2}/2$	-1	-1	$\sqrt{2}$	$-\sqrt{2}$
330°	$-\frac{1}{2}$	$\sqrt{3}/2$	$-\sqrt{3}/3$	$-\sqrt{3}$	$2\sqrt{3}/3$	-2

39. $\sin 46° \approx 0.7193398003$ $\cos 46° \approx 0.6946583705$ $\tan 46° \approx 1.035530314$
$\cot 46° \approx 0.9656887746$ $\sec 46° \approx 1.439556540$ $\csc 46° \approx 1.390163591$

41. $\sin 143° \approx 0.6018150232$; $\cos 143° \approx -0.7986355100$; $\tan 143° \approx -0.7535540502$; $\cot 143° \approx -1.327044821$; $\sec 143° \approx -1.252135658$; $\csc 143° \approx 1.661640141$

43. $\sin(-61.37°) \approx -0.8777322126$; $\cos(-61.37°) \approx 0.4791515031$; $\tan(-61.37°) \approx -1.831846936$; $\cot(-61.37°) \approx -0.5458971382$; $\sec(-61.37°) \approx 2.087022567$; $\csc(-61.37°) \approx -1.139299647$

45. $\sin 61°35' \approx 0.8795101821$; $\cos 61°35' \approx 0.4758800684$; $\tan 61°35' \approx 1.848176128$; $\cot 61°35' \approx 0.5410739728$; $\sec 61°35' \approx 2.101369791$; $\csc 61°35' \approx 1.136996501$

47. $\sin(-97°9'8'') \approx -0.9922188692$; $\cos(-97°9'8'') \approx -0.1245058859$; $\tan(-97°9'8'') \approx 7.969252714$; $\cot(-97°9'8'') \approx 0.1254822799$; $\sec(-97°9'8'') \approx -8.031748803$; $\csc(-97°9'8'') \approx -1.007842152$

49. $\sin(-2\pi/7) \approx -0.7818314825$; $\cos(-2\pi/7) \approx 0.6234898018$; $\tan(-2\pi/7) \approx -1.253960338$; $\cot(-2\pi/7) \approx -0.7974733887$; $\sec(-2\pi/7) \approx 1.603875472$; $\csc(-2\pi/7) \approx -1.279048008$

51. $\sin(1.67) \approx 0.9950833498$; $\cos(1.67) \approx -0.0990410366$; $\tan(1.67) \approx -10.04718230$; $\cot(1.67) \approx -0.0995303927$; $\sec(1.67) \approx -10.09682485$; $\csc(1.67) \approx 1.004940943$

57. $\sin(9.673) \approx -0.2456808808$; $\cos(9.673) \approx -0.9693507646$; $\tan(9.673) \approx 0.2534488957$; $\cot(9.673) \approx 3.945568582$; $\sec(9.673) \approx -1.031618313$; $\csc(9.673) \approx -4.070320803$

61. Check that $[\sin(3\pi/5)]^2 + [\cos(3\pi/5)]^2 = 1$.

Problem Set 5, page 308

7. $-3\pi/2, \pi/2, 1$ **9.** $-3\pi/2, -\pi/2, \pi/2, 3\pi/2$ **11.** $-\pi/2, 3\pi/2, -1$
13. Consider the *number* to be a number of radians.   **15.** Amplitude 2, period 2π **17.** Amplitude $\frac{1}{2}$, period 2π
19. Amplitude $\frac{2}{3}$, period 2π; one cycle starts at $x = \pi$ and ends with $x = 3\pi$ **21.** Amplitude 1, period π
23. Amplitude 1, period $\pi/3$ **25.** Amplitude 1, period 4π
27. Amplitude π, period 2; one cycle starts at $x = 1$ and ends with $x = 3$
29. Amplitude 1, period 2π; one cycle starts at $x = \pi/6$ and ends with $x = 13\pi/6$
31. Amplitude 2, period 2π; one cycle starts at $x = \pi/3$ and ends with $x = 7\pi/3$
33. Amplitude 1, period 2π; one cycle starts at $x = -\pi/2$ and ends with $x = 3\pi/2$
35. For $0 \le bx - c \le 2\pi$, it follows that $c \le bx \le 2\pi + c$. Since $b > 0$, then $c/b \le x \le (2\pi + c)/b$.
Now $(2\pi + c)/b - c/b = (2\pi + c - c)/b = 2\pi/b$; therefore, the period is $2\pi/b$. Since $-1 \le \sin(bx - c) \le 1$ and $a > 0$, it follows that $-a \le a\sin(bx - c) \le a$, and so the amplitude is a.

Problem Set 6, page 314

1. (a)

x	$\dfrac{\pi}{12}$	$\dfrac{\pi}{6}$	$\dfrac{\pi}{4}$	$\dfrac{\pi}{3}$	$\dfrac{5\pi}{12}$	$\dfrac{\pi}{2}$	$\dfrac{7\pi}{12}$	$\dfrac{2\pi}{3}$	$\dfrac{3\pi}{4}$	$\dfrac{5\pi}{6}$	$\dfrac{11\pi}{12}$
$\cot x$	3.73	1.73	1	0.58	0.27	0	-0.27	-0.58	-1	-1.73	-3.73

11. $-2\pi, -\pi, 0, \pi, 2\pi$ **13.** $-2\pi, 0, 2\pi$; 1   **15.** π **17.** $\sec(-x) = 1/\cos(-x) = 1/\cos x = \sec x$
19. Multiply the ordinates of points on the graph of $y = \tan x$ by 2.
21. Multiply the ordinates of points on the graph of $y = \cot x$ by $\frac{1}{2}$, reflect the resulting graph about the x axis, and then shift this graph 1 unit upward.
23. Multiply the ordinates of points on the graph of $y = \sec x$ by $\frac{1}{2}$ and shift the resulting graph 1 unit upward.
25. Shift the graph of $y = \tan x$ exactly $\pi/4$ units to the right.
27. Multiply the ordinates of points on the graph of $y = \cot x$ by $\frac{2}{3}$ and shift the resulting graph $\pi/2$ units to the right.
29. Shift the graph of $y = \sec x$ exactly $\pi/6$ units to the left.
31. Multiply the ordinates of points on the graph of $y = \csc x$ by 2 and then shift the resulting graph $\pi/3$ units to the right.

Problem Set 7, page 319

1. (a) 2 (b) 1 (c) 2π (d) $\pi/3$ (e) $\pi/3$ (f) 1 (g) One cycle starts at $x = \pi/3$ and ends at $x = 7\pi/3$, with trough at $x = 4\pi/3$.

3. (a) 4 (b) 1 (c) 2π (d) $-\pi/4$ (e) $-\pi/4$ (f) 2 (g) One cycle starts at $x = -\pi/4$, ends at $x = 7\pi/4$, with trough at $x = 3\pi/4$.

5. (a) 3 (b) 3 (c) $2\pi/3$ (d) $-5\pi/2$ (e) $-5\pi/6$ (f) 0 (g) One cycle starts at $x = -5\pi/6$, ends at $x = -\pi/6$, with trough at $x = -\pi/2$.

7. (a) $\frac{3}{4}$ (b) 4 (c) $\pi/2$ (d) -12 (e) -3 (f) $-\frac{3}{4}$ (g) One cycle starts at $x = -3$, ends at $x = (\pi - 6)/2$, with trough at $x = (\pi - 12)/4$.

9. (a) 110 (b) 120π (c) $\frac{1}{60}$ (d) $-3\pi/2$ (e) $-\frac{1}{80}$ (f) 0 (g) One cycle starts at $x = -\frac{1}{80}$, ends at $x = \frac{1}{240}$, with trough at $x = -\frac{1}{240}$.

11. (l) $1/2\pi$ (3) $1/2\pi$ (5) $3/2\pi$ (7) $2/\pi$ (9) 60   13. (a) 2 (b) 2π (c) $2\pi/3$ (d) 1 (e) $2\pi/3$ (f) 3 (g) $y = 2\cos[t - (2\pi/3)] + 3$

15. (a) $\frac{3}{4}$ (b) 4 (c) $\frac{3}{4}$ (d) $\pi/2$ (e) $3\pi/8$ (f) 0 (g) $y = \frac{3}{4}\cos[(\pi/2)t - (3\pi/8)]$

17. (a) 11 (b) 4 (c) 8 (d) $\pi/2$ (e) 4π (f) 11 (g) $y = 11\cos[(\pi/2)t - 4\pi] + 11$ 19. (13) $1/2\pi$ (15) $\frac{1}{4}$ (17) $\frac{1}{4}$

21. (a) The axis is the line $y = (d + c)/2$, and the distance between the crest and the axis is $d - [(d + c)/2] = (d - c)/2$, so $a = (d - c)/2$ (b) $\omega = 2\pi/\text{period} = 2\pi/(\beta - \alpha)$ (c) $\phi = S\omega$, where $S = \alpha$; thus, $\phi = 2\pi\alpha/(\beta - \alpha)$

(d) $k = d - a = d - [(d - c)/2] = (2d - d + c)/2 = (d + c)/2$

23. (c) $\phi = S\omega = \left[\alpha + \dfrac{(\beta - \alpha)}{4}\right]\omega = \left(\dfrac{3\alpha + \beta}{4}\right)\left(\dfrac{2\pi}{\beta - \alpha}\right) = \dfrac{(3\alpha + \beta)\pi}{2(\beta - \alpha)}$

25.

27.

29.

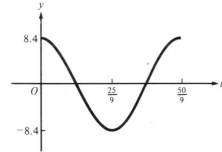

31. $T \approx \pi/4$ second, $v \approx 4/\pi$ Hz, $\omega = 8$, $y = 0.1\cos 8t$

33. Construct a pendulum of length $8/\pi^2$ feet ≈ 9.7 inches (or 0.25 meter).

35. Sketch two cycles of the graph $y = 0.04\cos 5t + k$. Choose $k = 1$, say. The first cycle starts at $x = 0$; the second cycle starts at $x = 2\pi/5$ and ends at $x = 4\pi/5$. The troughs are at $x = \pi/5$ and at $x = 3\pi/5$.

37. $E = 100\cos(10\sqrt{2}t)$, $T = (\sqrt{2}/10)\pi$ second, $v = 5\sqrt{2}/\pi$ Hz

Review Problem Set, page 322

1. $36°$ 3. $-288°$ 5. $\frac{1}{6}$ counterclockwise   7. 4 clockwise   9. $\frac{31}{12}$ counterclockwise   11. $2.0500°$

13. 55.7597 15. $87°21'$ 17. $-24°31'48''$ 19. 2.85 meters   21. 3 feet   23. $\pi/2$ radians   25. $4\pi/9$

27. $-71\pi/36$ **29.** $-31\pi/18$ **31.** $72°$ **33.** $-\frac{315°}{2}$ **35.** $2295°$ **37.** (a) 0.0873 (b) 0.4844 (c) -0.2997 (d) 0.6158
39. 6.28 feet   **41.** $45°$ **43.** $625\pi/12 \approx 163.6$ square centimeters   **45.** $0.08\pi \approx 0.25$ meter
47. $120\pi \approx 377$ feet per minute   **49.** 1207 miles above the earth   **51.** 2094 centimeters per minute
53. $\sin\theta = \sqrt{91}/10$; $\cos\theta = \frac{3}{10}$; $\tan\theta = \sqrt{91}/3$; $\cot\theta = 3\sqrt{91}/91$; $\sec\theta = \frac{10}{3}$; $\csc\theta = 10\sqrt{91}/91$ **55.** (a) $\cos 40°$
(b) $\sin(5\pi/14)$ (c) $\csc 1°$ (d) $\tan(11\pi/28)$ **57.** (a) Quadrant I (b) Quadrant II (c) Quadrant III (d) Quadrantal (e) Quadrant IV
(f) Quadrantal   **59.** (a) $345°$; $705°$; $-375°$ (b) $100°$; $820°$; $-260°$ (c) $530°$; $-190°$; $-550°$ (d) $-620°$; $-260°$; $100°$
(e) $11\pi/3$; $-(\pi/3)$; $-(7\pi/3)$ **61.** (a) $\sin\theta = 5\sqrt{34}/34$; $\cos\theta = -3\sqrt{34}/34$; $\tan\theta = -\frac{5}{3}$; $\cot\theta = -\frac{3}{5}$; $\sec\theta = -\sqrt{34}/3$;
$\csc\theta = \sqrt{34}/5$ (b) $\sin\theta = -3\sqrt{13}/13$; $\cos\theta = 2\sqrt{13}/13$; $\tan\theta = -\frac{3}{2}$; $\cot\theta = -\frac{2}{3}$; $\sec\theta = \sqrt{13}/2$; $\csc\theta = -\sqrt{13}/3$
(c) $\sin\theta = -\frac{4}{5}$; $\cos\theta = -\frac{3}{5}$; $\tan\theta = \frac{4}{3}$; $\cot\theta = \frac{3}{4}$; $\sec\theta = -\frac{5}{3}$; $\csc\theta = -\frac{5}{4}$ (d) $\sin\theta = -\frac{1}{2}$; $\cos\theta = \sqrt{3}/2$; $\tan\theta = -\sqrt{3}/3$;
$\cot\theta = -\sqrt{3}$; $\sec\theta = 2\sqrt{3}/3$; $\csc\theta = -2$ **63.** (a) II (b) III (c) IV (d) IV
65. $\tan\theta = -\frac{4}{3}$; $\cot\theta = -\frac{3}{4}$; $\sec\theta = \frac{5}{3}$; $\csc\theta = -\frac{5}{4}$
67. $\cos\theta = \frac{12}{13}$; $\tan\theta = -\frac{5}{12}$; $\cot\theta = -\frac{12}{5}$; $\sec\theta = \frac{13}{12}$; $\csc\theta = -\frac{13}{5}$
69. $\sin\theta = \frac{12}{13}$; $\cos\theta = -\frac{5}{13}$; $\tan\theta = -\frac{12}{5}$; $\cot\theta = -\frac{5}{12}$; $\sec\theta = -\frac{13}{5}$
71. $\cos\theta = -\frac{4}{5}$; $\tan\theta = -\frac{3}{4}$; $\cot\theta = -\frac{4}{3}$; $\sec\theta = -\frac{5}{4}$; $\csc\theta = \frac{5}{3}$
73. $\tan\theta = 1.6003$; $\sec\theta = -1.8870$; $\cos\theta = -0.5299$; $\sin\theta = -0.8480$; $\csc\theta = -1.1792$
75. (a) $\sin(-180°) = 0$; $\cos(-180°) = -1$; $\tan(-180°) = 0$; $\sec(-180°) = -1$
  (b) same as for part (a)
  (c) $\sin 990° = -1$; $\cos 990° = 0$; $\cot 990° = 0$; $\csc 990° = -1$
  (d) $\sin(-360°) = 0$; $\cos(-360°) = 1$; $\tan(-360°) = 0$; $\sec(-360°) = 1$
  (e) same as for part (c)
  (f) same as for part (a)
  (g) same as for part (a)
  (h) same as for part (d)
77. (a) $\sin(-150°) = -\frac{1}{2}$; $\cos(-150°) = -\sqrt{3}/2$; $\tan(-150°) = \sqrt{3}/3$; $\cot(-150°) = \sqrt{3}$; $\sec(-150°) = -2\sqrt{3}/3$;
$\csc(-150°) = -2$ (b) $\sin(-315°) = \sqrt{2}/2$; $\cos(-315°) = \sqrt{2}/2$; $\tan(-315°) = 1$; $\cot(-315°) = 1$; $\sec(-315°) = \sqrt{2}$;
$\csc(-315°) = \sqrt{2}$ (c) $\sin 780° = \sqrt{3}/2$; $\cos 780° = \frac{1}{2}$; $\tan 780° = \sqrt{3}$; $\cot 780° = \sqrt{3}/3$; $\sec 780° = 2$; $\csc 780° = 2\sqrt{3}/3$
(d) $\sin(13\pi/3) = \sqrt{3}/2$; $\cos(13\pi/3) = 1/2$; $\tan(13\pi/3) = \sqrt{3}$; $\cot(13\pi/3) = \sqrt{3}/3$; $\sec(13\pi/3) = 2$; $\csc(13\pi/3) = 2\sqrt{3}/3$
(e) $\sin(-15\pi/4) = \sqrt{2}/2$; $\cos(-15\pi/4) = \sqrt{2}/2$; $\tan(-15\pi/4) = 1$; $\cot(-15\pi/4) = 1$; $\sec(-15\pi/4) = \sqrt{2}$; $\csc(-15\pi/4) = \sqrt{2}$
79. 0.4591664533   **81.** 0.5952436037   **83.** 1.940926426   **85.** 1.042572391   **87.** 1.701301619   **89.** -26.02388181
91. 0.9346780153   **93.** -0.9545616245 **95.** 0.4440158399
97. Multiply the ordinates of points on the graph of $y = \sin x$ by $\frac{1}{3}$ and shift the resulting graph 1 unit upward.
99. Multiply the ordinates of points on the graph of $y = \cos x$ by $\frac{1}{2}$, shift the resulting graph $\pi/2$ units to the right, and then shift this resulting graph 2 units upward.
101. Shift the graph of $y = \tan x$ exactly $\pi/6$ unit to the right and then reflect the resulting graph about the x axis.
103. Shift the graph of $y = \sec x$ exactly π units to the right.
105. Multiply ordinates of points on the graph of $y = \cot x$ by $\frac{2}{3}$ and shift the resulting graph $\pi/2$ units to the right.
107. Maximum value 1 is reached at $x = -2\pi$, at $x = 0$, and at $x = 2\pi$; minimum value -1 is reached at $x = -\pi$ and at $x = \pi$.
109.

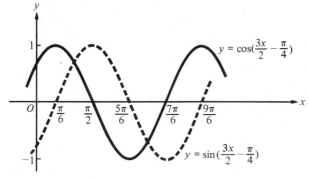

111. (a) 2 (b) π (c) 2 (d) $-\pi$ (e) -1 (f) 1 (g) $\frac{1}{2}$ (h) One cycle starts at $x = -1$ and ends at $x = 1$, with trough at $x = 0$.

113. (a) 3 (b) $\frac{1}{2}$ (c) 4π (d) $\pi/2$ (e) π (f) -3 (g) $1/4\pi$ (h) One cycle starts at $x = \pi$ and ends at $x = 5\pi$, with trough at $x = 3\pi$.

115. (a) 0.2 (b) 0.25 (c) 8π (d) π (e) 4π (f) 0 (g) $1/8\pi$ (h) One cycle starts at $x = 4\pi$ and ends at $x = 12\pi$, with trough at $x = 8\pi$.

117. $T \approx 4.6$ seconds, $v \approx 0.22$ Hz, $\omega \approx 1.4$, $y = 0.2\cos(1.4t)$; one cycle starts at $t = 0$ and ends at $t = T$, and the next cycle starts at $t = T$ and ends at $t = 2T$; the troughs are at $x = T/2$ and $x = 3T/2$.

119. $y = 0.03\cos(880\pi t - \pi/2) + 10^5$ **121.** 10,626 days $\approx$ 29.1 years

Chapter 7

Problem Set 1, page 331

1. $-\sin\theta\cos\theta$ **3.** 0   **5.** $\dfrac{1 - \csc\alpha}{1 + \cot\beta}$ **7.** $\tan\theta$ **9.** $\csc v$ **11.** $\cot\beta$ **13.** -1 **15.** $\cot^2 u$ **17.** 1

19. 1   **21.** 1   **23.** $\sin x\cos x$ **25.** $\tan t$ **27.** $2\csc t$ **29.** $\sin x$ **31.** -1 **33.** $\sin^2\alpha$ **35.** $\cos\theta + \sin\theta$

37. $\dfrac{\sin^2 u}{\cos u}$ **39.** $\pm\dfrac{\sqrt{1 - \cos^2\theta}}{\cos^3\theta}$ **41.** $\dfrac{1}{\cos^2 x}$ **43.** $-\sin\alpha$ **45.** $a\sec\theta$ **47.** $8\sin^3 t$

Problem Set 2, page 335

1. $\sin\theta\sec\theta = \sin\theta(1/\cos\theta) = \tan\theta$ **3.** $\tan x\cos x = (\sin x/\cos x)\cos x = \sin x$

5. $\csc(-t)\tan(-t) = (-1/\sin t)(-\sin t/\cos t) = 1/\cos t = \sec t$

7. $\tan\alpha\sin\alpha + \cos\alpha = \dfrac{\sin\alpha}{\cos\alpha}\sin\alpha + \cos\alpha = \dfrac{\sin^2\alpha + \cos^2\alpha}{\cos\alpha} = \dfrac{1}{\cos\alpha} = \sec\alpha$ **53.** Not true for $\theta = \pi/4$

55. Not true for $\theta = \pi/3$ **57.** Not true for $\theta = \pi/2$ **59.** Not true for $t = 2$

Problem Set 3, page 342

1. $(\sqrt{2}/4)(\sqrt{3} + 1)$ **3.** $(-\sqrt{2}/4)(\sqrt{3} - 1)$ **5.** $(-\sqrt{2}/4)(\sqrt{3} - 1)$ **7.** $(\sqrt{2}/4)(\sqrt{3} - 1)$ **9.** $2 - \sqrt{3}$

11. $-(2 + \sqrt{3})$ **13.** $\sin\theta$ **15.** $-\cos\alpha$ **17.** $-\sin s$ **19.** $-\tan t$ **21.** $-\sec\beta$ **23.** $\frac{1}{2}$ **25.** $\frac{1}{2}$ **27.** $\cos x$

29. $\cos\alpha$ **31.** $\sqrt{3}$ **33.** $\tan x$ **35.** (a) $\frac{3}{5}$ (b) $-\frac{5}{13}$ (c) $\frac{16}{65}$ (d) $-\frac{63}{65}$ (e) $-\frac{56}{65}$ (f) $-\frac{16}{63}$ (g) Quadrant II   **37.** $\frac{3}{5}$

51. $\cos(\alpha - \beta) = 0.3090169938$; $\cos\alpha\cos\beta + \sin\alpha\sin\beta = 0.3090169947$

Problem Set 4, page 350

1. $2\sin 38°\cos 38°$ **3.** $\cos^2 72° - \sin^2 72°$ **5.** $2\sin\dfrac{\pi}{9}\cos\dfrac{\pi}{9}$ **7.** $\sin 58°$ **9.** $\cos 5\theta$ **11.** $\cos\dfrac{2\pi}{17}$

13. (a) $\frac{24}{25}$ (b) $-\frac{7}{25}$ (c) $-\frac{24}{7}$ **15.** (a) $\frac{120}{169}$ (b) $-\frac{119}{169}$ (c) $-\frac{120}{119}$ **17.** (a) $-\frac{120}{169}$ (b) $-\frac{119}{169}$ (c) $\frac{120}{119}$ **19.** $\cos x$ **21.** $1 + 2\sin 2t$

23. $\dfrac{\sin 2x}{2}$ **25.** $\dfrac{1 - \cos 2\theta}{2}$ or $\sin^2\theta$ **27.** (a) $\sqrt{\dfrac{1 + \cos 30°}{2}}$ (b) $\dfrac{\sqrt{2 + \sqrt{3}}}{2}$

29. (a) $-\sqrt{\dfrac{1 + \cos(5\pi/4)}{2}}$ (b) $-\dfrac{\sqrt{2 - \sqrt{2}}}{2}$ **31.** (a) $\dfrac{1 - \cos 315°}{\sin 315°}$ (b) $1 - \sqrt{2}$ **33.** (a) $-\sqrt{\dfrac{1 - \cos 405°}{2}}$ (b) $-\dfrac{\sqrt{2 - \sqrt{2}}}{2}$

35. (a) $\dfrac{3}{5}$ (b) $\dfrac{4}{5}$ (c) $\dfrac{3}{4}$ **37.** (a) $\dfrac{3\sqrt{13}}{13}$ (b) $-\dfrac{2\sqrt{13}}{13}$ (c) $-\dfrac{3}{2}$ **39.** $\cos 125°$ **41.** $\tan 3x$ **43.** $\cos\dfrac{\pi}{5}$ **57.** $\dfrac{1 + z^2}{1 - z^2}$

59. $\dfrac{1 - z^2}{2z}$

Problem Set 5, page 358

1. $\frac{1}{2}\sin 145° + \frac{1}{2}\sin 65°$ **3.** $\cos(3\pi/4) + \cos(\pi/2)$ or $-\sqrt{2}/2$ **5.** $\frac{1}{2}\sin 8\theta - \frac{1}{2}\sin 2\theta$

7. $\frac{1}{2}\cos(7s - 5t) - \frac{1}{2}\cos(7s + 5t)$ **9.** $\frac{1}{2}\sin 4x - \frac{1}{2}\sin 2x$ **11.** $2\sin 50° \cos 30°$ **13.** $2\cos(\pi/4)\cos(\pi/8)$

15. $2\sin 3\theta \cos \theta$ **17.** $2\sin 3t \sin t$ **19.** $2\cos \dfrac{2\alpha + \beta}{2} \sin \dfrac{\beta}{2}$ **21.** (a) $-\dfrac{2\sin 45° \sin 30°}{2\sin 45° \cos 30°} = -\dfrac{\sqrt{3}}{3}$ (b) $\dfrac{2\sin 45° \cos 35°}{2\cos 45° \cos 35°} = 1$

23. $\sin 50° + \sin 10° = 2\sin 30° \cos 20° = \cos 20° = \sin 70°$

41.

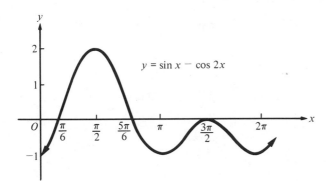

$y = \sin x - \cos 2x$

43.

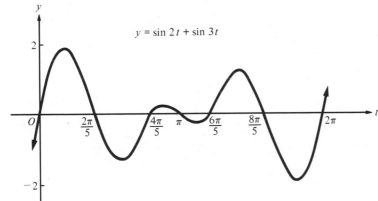

$y = \sin 2t + \sin 3t$

45. $y = \sqrt{2}\cos[2t - (\pi/4)]$ **47.** $y = 2\cos[2\pi t - (5\pi/3)]$ **49.** $y = \frac{1}{2}\sin 6x - \frac{1}{2}\sin 2x$ **51.** $y = \cos 2t - \cos 4t$

53. $y = -\frac{1}{2}\cos 3\pi x - \frac{1}{2}\cos \pi x$ **55.** 4   **57.** $2\sqrt{2}\pi$

59. The graph actually falls *below* the x axis, between 0 and about $\pi/10$.

61. No, maximum occurs a bit *to the right* of $2\pi/3$.

Problem Set 6, page 365

1. $\pi/2$ **3.** $-\pi/4$ **5.** 0   **7.** $\pi/4$ **9.** $\pi/3$ **11.** $-\pi/4$ **13.** 0.6999768749   **15.** 0.4391814802

17. 1.107148718   **19.** -1.270032196 **21.** -0.5829630403 **23.** 0.5931997761   **25.** $\frac{3}{4}$ **27.** $\pi/6$ **29.** $\frac{4}{5}$

31. $3\sqrt{10}/10$ **33.** $\frac{4}{3}$ **35.** $\frac{10}{7}$ **45.** (a) $-1 \le x \le 1$ (b) $0 \le x \le \pi$ **47.** (a) All real values of x (b) $-\pi/2 < x < \pi/2$

Problem Set 7, page 373

1. $90°, 270°$ **3.** $45°, 225°$ **5.** $\pi/6, 5\pi/6$ **7.** $2\pi/3, 5\pi/3$ **9.** $3\pi/2$ **11.** $k \cdot 180°, k = 0, \pm1, \pm2, \ldots$
13. $(\pi/4) + 2\pi k; (7\pi/4) + 2\pi k, k = 0, \pm1, \pm2, \ldots$ **15.** $(\pi/3) + k\pi, k = 0, \pm1, \pm2, \ldots$ **17.** $48.5904°, 131.4096°$
19. $0.8411, 5.4421$ **21.** $1.2900 + k\pi, k = 0, \pm1, \pm2, \ldots$ **23.** $15°, 75°, 195°, 255°$
25. $20°, 80°, 140°, 200°, 260°, 320°$ **27.** $2\pi/3, 4\pi/3$ **29.** $3\pi/2$ **31.** $210°, 330°$ **33.** $0, \pi/6, 5\pi/6, \pi$
35. $30°, 150°, 210°, 330°$ **37.** $\pi/4, 3\pi/4, 5\pi/4, 7\pi/4$ **39.** $0, \pi/3, \pi, 4\pi/3$ **41.** $0°, 45°, 180°, 225°$
43. $\pi/4, 3\pi/4, 5\pi/4, 7\pi/4$ **45.** $90°, 150°, 210°$ **47.** $0.4429, 1.5708, 2.6987$ **49.** $54.7356°, 90°, 270°, 305.2644°$
51. 1.288×10^{-3} second   **53.** $11.20°$ **55.** (a) $t = (2n + 1) \cdot 2\pi$ and $t = (4\pi/3) + (2n + 1) \cdot 2\pi, n = 0, \pm1, \pm2, \ldots$
(b) 30 or 45 wild pigs

Review Problem Set, page 374

1. $-\tan\theta$ **3.** $\sin x$ **5.** $\tan^2 t$ **7.** $1 + \tan u$ **9.** $\tan\theta$ **11.** $125\cos^3\theta$ **27.** $\cos\theta$ **29.** $-\cos\alpha$

31. $\sin t$ **33.** $\sqrt{3}/2$ **35.** $\sin(x + y)$ **37.** (a) $\dfrac{\sqrt{2}(1 + \sqrt{3})}{4}$ (b) $\dfrac{\sqrt{2}(1 - \sqrt{3})}{4}$ (c) $-(2 + \sqrt{3})$ **39.** $-\frac{36}{85}$ **41.** $-\frac{21}{221}$

43. $-\frac{304}{425}$ **45.** $\frac{297}{304}$ **47.** $-\frac{36}{325}$ **57.** (a) $\frac{336}{625}$ (b) $-\frac{527}{625}$ (c) $\frac{4}{5}$ (d) $\frac{3}{4}$ **59.** $\cos 4x$ **61.** $\sin t$ **63.** 1
65. $\frac{1}{2}\tan 2\omega t$ **67.** 1 **85.** $\frac{1}{2}\sin 2x + \frac{1}{2}\sin x$ **87.** $\frac{1}{2}\cos 30° - \frac{1}{2}\cos 45°$ or $(\sqrt{3}/4) - (\sqrt{2}/4)$
89. $2\sin 30°\cos 25°$ or $\cos 25°$ **91.** $2\cos(5\beta/2)\sin(3\beta/2)$ **105.** $y = 2\sin(7t/3) - 2\sin(t/3)$
107. $y = 2\sqrt{2}\cos[(\pi/3)t - (7\pi/4)]$ **109.** $y = 2\cos[2t - 3 + (\pi/6)]$ **111.** 30 **113.** $y = \cos 4\pi t + \cos 2\pi t; T = 1$
115. $-\pi/6$ **117.** $\pi/3$ **119.** 0.3843967745 **121.** 0.3836622700 **123.** 0.9588938924 **125.** -0.8480620790
127. 2.711892987 **129.** $\frac{3}{5}$ **131.** $-\frac{12}{13}$ **133.** $-5\pi/14$ **141.** $30°, 210°$ **143.** $\pi/6, 5\pi/6, 7\pi/6, 11\pi/6$
145. $0°, 180°, 225°, 315°$ **147.** $7\pi/6, 3\pi/2, 11\pi/6$ **149.** $15°, 60°, 105°, 150°, 195°, 240°, 285°, 330°$
151. $0°, 90°, 180°, 270°$ **153.** $26.5651°, 153.4349°, 206.5651°, 333.4349°$ **155.** $0.5236, 2.6180, 3.8713, 5.5535$ **157.** 8.135

Chapter 8

Problem Set 1, page 385

1. $c = 6.403; \alpha = 51.34°; \beta = 38.66°$ **3.** $b = 11.18; \alpha = 41.81°; \beta = 48.19°$ **5.** $a = 1031; c = 1988; \beta = 58.77°$
7. $b = 5.826; \alpha = 50.76°; \beta = 39.24°$ **9.** $c = 4331; \alpha = 0.04°; \beta = 89.96°$ **11.** $a = 9299; c = 9647; \alpha = 74.55°$
13. $a = 1.387 \times 10^{-2}$ meter; $c = 1.498 \times 10^{-2}$ meter; $\beta = 22°15'$ **15.** $a = 70.79$ miles; $\alpha = 66.53°; \beta = 23.47°$
17. $c = 6.414 \times 10^{-6}$ meter; $\alpha = 50.26°; \beta = 39.74°$ **19.** 22 feet   **21.** 34 meters   **23.** 44 meters   **25.** $35.4°$
27. 98.3 meters   **29.** 100 meters   **31.** $56°$ **33.** 162 feet   **35.** 196 feet   **37.** 24.5 miles   **39.** S20.56°W

Problem Set 2, page 393

1. $\alpha = 85°; b = 26.94; c = 19.78$ **3.** $\alpha = 121°; a = 33.41; c = 13.97$ **5.** $\beta = 73.87°; a = 36.77; c = 41.67$
7. $\gamma = 23.55°; b = 52.91; c = 51.08$ **9.** $\beta = 25.1°; \gamma = 109.9°; c = 66.49$
11. $\beta = 67.21°; \gamma = 52.79°; c = 28.51$ or $\beta = 112.79°; \gamma = 7.21°; c = 4.49$ **13.** No triangle
15. $\beta = 52.02°; \gamma = 16.48°; c = 19.97$ **17.** $\gamma = 75.65°; a = 40.28$ feet; $c = 57.38$ feet
19. $\beta = 51.44°; \gamma = 69.23°; c = 119.58$ kilometers

21. $\beta = 89.53°$; $\gamma = 68.47°$; $c = 5190$ meters or $\beta = 90.47°$; $\gamma = 67.53°$; $c = 5156$ meters   **23.** No triangle
25. $|\overline{AC}| = 316.6$ meters; $|\overline{CB}| = 345.5$ meters   **27.** 91.43 meters   **29.** 72.86 meters   **31.** 0.9906 kilometer
33. 2.11 kilometers   **35.** 88 minutes, 16 seconds   **41.** $\gamma = 38°$; $c = 5.935$; $b = 9.219$
43. $\beta = 32°$; $b = 8.048$; $c = 13.98$

Problem Set 3, page 398

1. 8.888   **3.** 44.99   **5.** 3.336   **7.** 104.48°   **9.** 44.89°   **11.** 33.72°   **13.** $\alpha = 63.95°$; $\beta = 43.60°$; $c = 347.0$
15. $\beta = 60.75°$; $\gamma = 85.50°$; $c = 559.9$ or $\beta = 119.25°$; $\gamma = 27.00°$; $c = 255.9$ **17.** $\alpha = 26.75°$; $\beta = 39.75°$; $\gamma = 113.50°$
19. 69.14 meters   **21.** 815.9 meters   **23.** 744.5 kilometers   **25.** 57.49 miles   **27.** 17.32 cm^2 **29.** 98.06 ft^2
31. 74.22 km^2 **33.** 85,733 m^2 **41.** $a + b > c$; $a + c > b$; $b + c > a$

Problem Set 4, page 407

1. $4\sqrt{2}$; 135°   **3.** 5; 36.86989765°   **5.** 2; 300°   **7.** $\frac{5}{8}$; 53.13010235°   **9.** 4; 123.9878436°   **11.** 5; 53.13010235°
13. $\langle 5\sqrt{2}/2, 5\sqrt{2}/2 \rangle$ **15.** $\langle -4, 4 \rangle$ **17.** $\langle -\sqrt{3}, -1 \rangle$ **19.** $\langle 1.789851696, 6.798332951 \rangle$ **21.** (a) $\langle 2, 8 \rangle$ (b) $\langle 4, 0 \rangle$
(c) $\langle 5, 28 \rangle$ (d) $\langle 9, -4 \rangle$ **23.** (a) $\langle -\frac{1}{2}, \frac{13}{3} \rangle$ (b) $\langle -\frac{11}{2}, -\frac{1}{3} \rangle$ (c) $\langle 1, \frac{46}{3} \rangle$ (d) $\langle -\frac{27}{2}, -3 \rangle$ **25.** (a) $\langle 11, -21 \rangle$ (b) $\langle -7, 9 \rangle$
(c) $\langle 42, -78 \rangle$ (d) $\langle -23, 33 \rangle$ **27.** (a) $-\bar{i} + 12\bar{j}$ (b) $5\bar{i} - 2\bar{j}$ (c) $-6\bar{i} + 43\bar{j}$ (d) $13\bar{i} - 11\bar{j}$ **29.** (a) $10\bar{i} - 11\bar{j}$ (b) $2\bar{i} + \bar{j}$
(c) $34\bar{i} - 39\bar{j}$ (d) $8\bar{j}$ **43.** (a) 13 (b) 18 (c) -7 (d) 0 (e) 1

Problem Set 5, page 413

21. (a) $(-5, 225°)$ (b) $(5, -315°)$ (c) $(-5, -135°)$ **23.** $(2\sqrt{2}, 2\sqrt{2})$ **25.** $(-5\sqrt{2}/2, -5\sqrt{2}/2)$ **27.** $(\sqrt{2}/2, -\sqrt{2}/2)$
29. $(2\sqrt{3}, -2)$ **31.** $(-1.495, -2.326)$ **33.** $(14.17, -10.71)$ **35.** $(2, 45°)$ **37.** $(2, 60°)$ **39.** $(3, 180°)$
41. $(13, 112.62°)$ **43.** $(20.29, -50.12°)$ **45.** $(4, -\pi/6)$ **47.** $(2, -\pi/2)$ **49.** $(116.3, -2.239)$

Problem Set 6, page 420

1. (a) 60.33 pounds (b) 27.96°   **3.** (a) 29.10 newtons (b) 79.11°   **5.** (a) 16.44 pounds (b) 12.29°   **7.** (a) 816.5 km/hr
(b) 2.97° (c) N92.03°E   **9.** (a) 3.6 m/sec (b) 56.3°   **11.** 13.73 kg; N32.87°W   **13.** A circle of radius 1, center at O
15. A straight line through O making the angle $\pi/3$ with the polar axis
17. A straight line through O perpendicular to the polar axis   **19.** $r = 5$ **21.** $3r\cos\theta - 2r\sin\theta = 6$ **23.** $r = 2\cos\theta$
25. $x^2 + y^2 = 1$ **27.** $y - 2x = 1$ **29.** $x^2 = y$
31.

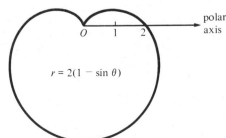

$r = 2(1 - \sin\theta)$

33.

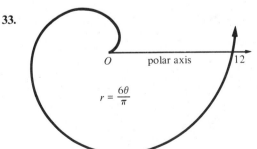

$r = \dfrac{6\theta}{\pi}$

Review Problem Set, page 422

1. $\beta = 58.75°$; $a = 103.8$ meters; $b = 171.0$ meters   **3.** $\alpha = 33.85°$; $a = 31.74$ microns; $c = 56.99$ microns
5. 25.49 meters   **7.** 658.2 meters   **9.** 264 meters/minute   **11.** 318.2 meters   **13.** 85.48 nautical miles; $S13°10'W$
15. 36.73 ft   **17.** 17.94 miles   **19.** 21.05°
21. $\alpha = 90°$; $a = 100\sqrt{3} \approx 173.2$ ft; $\gamma = 60°$ or $\alpha = 30°$; $a = 50\sqrt{3} \approx 86.6$ ft; $\gamma = 120°$
23. $\alpha = 76.28°$; $a = 66.99$ cm; $\beta = 46.47°$ **25.** No triangle   **27.** 60°   **29.** 9.006 km   **31.** 16.96 inches
33. $\alpha = 23.94°$; $c = 80.44$ km; $\beta = 115.01°$
35. $\beta = 55.58°$; $b = 0.9869$ micron; $\alpha = 69.42°$ or $\beta = 14.42°$; $b = 0.2979$ micron; $\alpha = 110.58°$
37. $\gamma = 105.40°$; $a = 1.788$ nautical miles; $c = 2.555$ nautical miles   **39.** 448.0 m²   **41.** 392,100 square miles
43. 2046 meters   **45.** 1:15 P.M.   **47.** 1275.3 meters   **49.** 83.51°   **51.** $\langle 10, -5 \rangle$ **53.** $\langle 5, 2 \rangle$ **55.** $\langle 13, 7 \rangle$
57. $\langle -13, 5 \rangle$ **59.** $4\sqrt{5}$ **61.** $2\sqrt{26}$ **63.** (a) $\langle 25\sqrt{3}, 25 \rangle$ (b) $\langle 33.94, 61.22 \rangle$ (c) $\langle 16.07, -19.15 \rangle$ (d) $\langle 125\sqrt{3}, -125 \rangle$
65. (a) 2, 60° (b) 19.21, 51.34° (c) 50, 126.87° (d) 0.3125, 53.13° (e) 4, 41.41°   **67.** (a) 30.96 pounds, 33.35°
(b) 221.2 newtons, 44.42° (c) 7.598 tonnes, 76° (d) 540.8 dynes, 303.69°   **69.** (a) 486.6 knots (b) 7.99° (c) $N37.01°W$
71. $S29°26'E$ **73.** (a) $(\sqrt{2}, \sqrt{2})$ (b) $(-\sqrt{2}, -\sqrt{2})$ (c) $(-0.1736, 0.9848)$ (d) $(1.5, 0)$ (e) $(2\sqrt{3}, -2)$ (f) $(3\sqrt{2}, -3\sqrt{2})$
75. (a) $(1, \pi)$ (b) $(7\sqrt{2}, -3\pi/4)$ (c) $(13, 1.97)$ (d) $(5, -0.93)$ (e) $(17, 2.06)$ (f) $(34, -2.65)$ **77.** $r \sin \theta = 3r \cos \theta + 1$
79. $r = 4 \sin \theta$ **81.** $r \sin^2 \theta = 6 \cos \theta$ **83.** $r^2(3 \cos^2 \theta + 1) = 4$ **85.** $(x - \frac{1}{2})^2 + (y - \frac{1}{2})^2 = \frac{1}{2}$ **87.** $x^2 - y^2 = 2$

Chapter 9

Problem Set 1, page 432

1. Consistent   **3.** Dependent   **5.** Inconsistent   **7.** Inconsistent   **9.** Consistent
11. (1,1); all points on the graph of $y = 3x - 4$; no solution   **13.** (1,3)   **15.** (4, -1)   **17.** $(\frac{7}{11}, -\frac{10}{11})$
19. (1, -1, 2)   **21.** (3,4)   **23.** (3,4)   **25.** (1,2,3)   **27.** (8,4,0)
29. (a) $x + y = 39$; $280x + 315y = 11,375$ (b) $x = 26$, $y = 13$

Problem Set 2, page 437

1. 0, 5, -3   **3.** 2   **5.** $\begin{bmatrix} \frac{3}{4} & -\frac{2}{3} & \frac{1}{7} \\ -1 & 5 & 6 \end{bmatrix}$ **7.** $\begin{bmatrix} 40 & 22 & -1 & -17 \\ 0 & 1 & 1 & 0 \\ -13 & 17 & 2 & 5 \end{bmatrix}$ **9.** $x + 3y = 0$; $2x - 4y = 1$

11. $2x + 5y + 3z = 1$; $-3x + 7y + \frac{1}{2}z = \frac{3}{4}$; $\frac{2}{3}y = -\frac{4}{5}$ **13.** $(\frac{24}{5}, -\frac{4}{5})$ **15.** (5,0)   **17.** (2,3)   **19.** (4, -2)
21. No solution   **23.** (3,0)   **25.** (4, -2, 1)   **27.** (3, -1, -2)   **29.** $(\frac{1}{4}, \frac{2}{3}, \frac{1}{6})$ **31.** $(-\frac{11}{4}, 1, -\frac{1}{4})$ **33.** (0,0,0)
35. No solution

Problem Set 3, page 447

1. (a) $\begin{bmatrix} 8 & 1 \\ 8 & -1 \end{bmatrix}$ (b) $\begin{bmatrix} -4 & -7 \\ 2 & 3 \end{bmatrix}$ (c) $\begin{bmatrix} -6 & 9 \\ -15 & -3 \end{bmatrix}$ (d) $\begin{bmatrix} 6 & 17 \\ -9 & -7 \end{bmatrix}$ **3.** (a) $\begin{bmatrix} 1 & 5 \\ 1 & 6 \\ 6 & -1 \\ -3 & 4 \end{bmatrix}$ (b) $\begin{bmatrix} 5 & -1 \\ -5 & 4 \\ -2 & 3 \\ -5 & 4 \end{bmatrix}$ (c) $\begin{bmatrix} -9 & -6 \\ 6 & -15 \\ -6 & -3 \\ 12 & -12 \end{bmatrix}$

(d) $\begin{bmatrix} -13 & 0 \\ 12 & -13 \\ 2 & -7 \\ 14 & -12 \end{bmatrix}$ 5. (a) $\begin{bmatrix} 4 & -6 & 2 & -1 \\ 0 & 4 & 0 & 6 \\ 4 & -1 & -2 & 4 \end{bmatrix}$ (b) $\begin{bmatrix} 0 & 0 & 2 & -5 \\ -6 & 0 & 2 & -4 \\ 4 & 3 & -4 & 4 \end{bmatrix}$ (c) $\begin{bmatrix} -6 & 9 & -6 & 9 \\ 9 & -6 & -3 & -3 \\ -12 & -3 & 9 & -12 \end{bmatrix}$

(d) $\begin{bmatrix} -2 & 3 & -6 & 13 \\ 15 & -2 & -5 & 7 \\ -12 & -7 & 11 & -12 \end{bmatrix}$ 7. (a) $\begin{bmatrix} 2 & -\frac{2}{3} & \sqrt{2} \\ \frac{17}{6} & \pi+1 & -2 \\ 4 & \frac{5}{2} & \frac{5}{3} \end{bmatrix}$ (b) $\begin{bmatrix} 0 & 1 & -\sqrt{2} \\ -\frac{1}{6} & \pi-1 & -2 \\ -2 & -\frac{5}{2} & \frac{5}{3} \end{bmatrix}$ (c) $\begin{bmatrix} -3 & -\frac{1}{2} & 0 \\ -4 & -3\pi & 6 \\ -3 & 0 & -5 \end{bmatrix}$

(d) $\begin{bmatrix} -1 & -\frac{13}{6} & 2\sqrt{2} \\ -1 & 2-3\pi & 6 \\ 3 & 5 & -5 \end{bmatrix}$ 21. $\begin{bmatrix} 14 & -3 \\ 6 & -2 \end{bmatrix}$ 23. $\begin{bmatrix} 2 & 6 \\ 5 & 10 \end{bmatrix}$ 25. $\begin{bmatrix} -2 & 9 \\ -3 & 1 \end{bmatrix}$ 27. $\begin{bmatrix} 4 & 0 \\ 4 & 1 \end{bmatrix}$ 29. $\begin{bmatrix} 2 & -5 & 2 \\ -2 & 0 & 6 \\ 7 & -14 & 1 \end{bmatrix}$

31. $\begin{bmatrix} -2 & 7 & 6 \\ 8 & -2 & -4 \\ 1 & -10 & -8 \end{bmatrix}$ 33. $\begin{bmatrix} 5 & -10 & 17 \\ 10 & -14 & 30 \\ 8 & -20 & 30 \end{bmatrix}$ 35. $\begin{bmatrix} 17 & 36 \\ 8 & 14 \end{bmatrix}$ 37. $\begin{bmatrix} -6 & 4 \\ -23 & 16 \\ -20 & -2 \end{bmatrix}$ 39. $\begin{bmatrix} 13 & -6 \\ 7 & -4 \end{bmatrix}$

41. Undefined 45. $\begin{bmatrix} \frac{4}{5} & \frac{1}{5} \\ \frac{1}{5} & -\frac{1}{5} \end{bmatrix}$ 47. $\begin{bmatrix} -\frac{5}{34} & \frac{1}{17} \\ \frac{1}{17} & \frac{3}{17} \end{bmatrix}$ 49. $\begin{bmatrix} \frac{1}{4} & \frac{1}{12} \\ -\frac{3}{4} & \frac{1}{12} \end{bmatrix}$ 51. No inverse 53. $\begin{bmatrix} \frac{3}{10} & -\frac{2}{5} & \frac{1}{2} \\ \frac{3}{10} & \frac{3}{5} & -\frac{1}{2} \\ -\frac{1}{10} & \frac{4}{5} & -\frac{1}{2} \end{bmatrix}$

55. $\begin{bmatrix} \frac{1}{3} & \frac{1}{3} & 0 \\ 0 & -\frac{3}{10} & \frac{1}{10} \\ -\frac{2}{3} & -\frac{4}{15} & \frac{1}{5} \end{bmatrix}$ 57. $\begin{bmatrix} -\frac{3}{2} & \frac{9}{4} & -\frac{5}{2} \\ 1 & -1 & 1 \\ -\frac{1}{2} & \frac{3}{4} & -\frac{1}{2} \end{bmatrix}$ 59. $(\frac{24}{5}, -\frac{4}{5})$ 61. $(-\frac{9}{34}, \frac{12}{17})$ 63. $(\frac{8}{3}, -\frac{10}{3})$ 65. $(-1, 2, -3)$

67. $(-\frac{11}{4}, 1, -\frac{1}{4})$

69. (a) The entry in the ith row of TX is the number of units of commodity number i used in unit time in the production of all other commodities.
(b) The entry in the ith row of $X - TX$ is the surplus number of units of commodity number i produced in unit time.

Problem Set 4, page 456

1. 5 3. 54 5. 54 7. 4 9. $(1, -1)$ 11. $(2, 1)$ 13. $(\frac{5}{4}, -\frac{1}{3})$ 15. 28 17. 38 19. 9 21. $-\frac{53}{24}$
23. $(1, -1, 2)$ 25. $(-\frac{10}{11}, \frac{18}{11}, \frac{38}{11})$ 27. No solution 29. -14 31. -80 33. 22 35. 25
37. -130.9347340 39. -12 41. 0 43. -1 45. 3 47. -3 49. -3 57. $(2 \pm \sqrt{58})/3$

Problem Set 5, page 461

1. $\dfrac{3}{x-3} - \dfrac{3}{x-2}$ 3. $\dfrac{\frac{1}{2}}{x+5} + \dfrac{\frac{1}{2}}{x-1}$ 5. $\dfrac{\frac{3}{4}}{x} - \dfrac{\frac{9}{8}}{x-2} + \dfrac{\frac{11}{8}}{x+2}$ 7. $\dfrac{-2}{x} + \dfrac{5}{x-1} - \dfrac{3}{x+1}$

9. $\dfrac{\frac{1}{6}}{x} + \dfrac{\frac{1}{3}}{x+3} + \dfrac{\frac{1}{2}}{x-2}$ 11. $\dfrac{-4}{x-3} + \dfrac{2}{x-1} + \dfrac{1}{(x-1)^2}$ 13. $-\dfrac{1}{x} - \dfrac{1}{x^2} + \dfrac{1}{x-1}$

15. $\dfrac{4}{x-1} - \dfrac{1}{x+2} + \dfrac{3}{(x+2)^2}$ 17. $\dfrac{\frac{5}{8}}{x+2} - \dfrac{\frac{13}{24}}{3x-2} + \dfrac{\frac{8}{3}}{(3x-2)^2}$ 19. $\dfrac{\frac{9}{2}}{x+1} + \dfrac{(-\frac{9}{2})x + (\frac{11}{2})}{x^2+1}$

21. $x^2 + \dfrac{\frac{1}{9}}{x} - \dfrac{(\frac{1}{9})x}{x^2+9}$ 23. $\dfrac{3}{t} + \dfrac{-3t+1}{t^2+1}$ 25. $\dfrac{\frac{1}{4}}{u-1} + \dfrac{\frac{1}{4}}{u+1} - \dfrac{(\frac{1}{2})u}{u^2+1}$ 27. $A = \frac{3}{2}; B = 0; C = -2; D = -\frac{3}{2}$

29. $A = 1; B = -1; C = -3; D = 3$ 31. 225 adults, 700 children 33. $(50°, 40°, 90°)$ 35. $(45, 20, 15)$
37. $p = 30, q = 90$ 39. $x = \$8000, y = \$9000$ 41. Coffee: \$2.80/lb; milk: \$0.60/qt; tuna: \$1.20/can

Problem Set 6, page 469

5.

7.

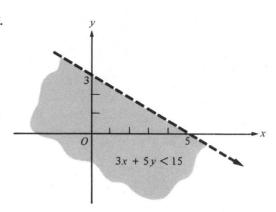

11. (b) Unbounded (c) $(2, 1)$ **13.** (b) Unbounded (c) $(\frac{8}{3}, \frac{4}{3})$ **15.** (b) Unbounded (c) $(2, 5)$
17. (b) Bounded (c) $(-6, -2)$; $(\frac{21}{2}, -2)$; $(\frac{21}{13}, \frac{40}{13})$ **19.** (b) Bounded (c) $(0, 0)$; $(0, 2)$; $(8, 0)$; $(\frac{4}{5}, \frac{18}{5})$
21. (b) Bounded (c) $(0, 0)$; $(3, 0)$; $(\frac{12}{5}, \frac{9}{5})$; $(0, 3)$ **23.** (b) Bounded (c) $(0, 0)$; $(7, 0)$; $(0, \frac{3}{8})$; $(\frac{25}{9}, \frac{19}{9})$ **25.** max 5, min 2
27. max 25, min 6   **29.** max 32, min -28 **31.** max 11, min 5   **33.** max 30, min -5
35. 50 units of A, 40 units of B **37.** 80 units of A, 0 units of B **39.** Center I open 6 days; Center II open 2 days
41. $2b = 3a$

Problem Set 7, page 475

1. $(2, 2)$; $(-5, \frac{25}{2})$ **3.** $(3, 3)$; $(\frac{3}{4}, -\frac{3}{2})$ **5.** $(0, -2)$; $(\frac{8}{5}, -\frac{6}{5})$ **7.** $\left(\frac{-1 + \sqrt{11}}{2}, \sqrt{11} - 2\right)$; $\left(\frac{-1 - \sqrt{11}}{2}, -\sqrt{11} - 2\right)$
9. No real solution   **11.** $(2, 3)$; $(2, -3)$; $(-2, 3)$; $(-2, -3)$ **13.** $(\sqrt{7}, \frac{1}{2}\sqrt{6})$; $(-\sqrt{7}, \frac{1}{2}\sqrt{6})$; $(\sqrt{7}, -\frac{1}{2}\sqrt{6})$; $(-\sqrt{7}, -\frac{1}{2}\sqrt{6})$
15. $(1, 2)$; $(1, -2)$; $(-1, 2)$; $(-1, -2)$ **17.** $(3, 7)$; $(-1, -1)$ **19.** $(2, 2)$; $(-2, -2)$; $(2\sqrt{2}, \sqrt{2})$; $(-2\sqrt{2}, -\sqrt{2})$ **21.** $(25, 4)$
23. $(16, 2)$ **25.** $(\frac{5}{2}, -\frac{3}{2})$; $(-\frac{3}{2}, \frac{5}{2})$; $(\frac{3}{2}, -\frac{5}{2})$; $(-\frac{5}{2}, \frac{3}{2})$ **27.** $(-1, 1 + \log_6 2)$ **29.** No solution   **31.** $(5, 1)$
33. 18 and 48   **35.** $b = 6, h = 10$ **37.** 4000   **39.** (a) $p = 7, q = 1$ (b) $7000

Review Problem Set, page 476

1. Inconsistent   **3.** Consistent   **5.** Dependent   **7.** $(2, 1)$ **9.** $(\frac{11}{7}, \frac{8}{7})$ **11.** Inconsistent   **13.** $(1, 1)$
15. $(3, 4, -6)$ **17.** $(1, -1)$ **19.** $(5, 6)$ **21.** $(1, -3, 2)$ **23.** $(-1, 2, -2)$ **25.** $\begin{bmatrix} -16 & 4 \\ 8 & 5 \end{bmatrix}$; $\begin{bmatrix} -18 & 2 \\ 4 & 10 \end{bmatrix}$

27. $\begin{bmatrix} 6 & 16 \\ 26 & 0 \\ -16 & 2 \end{bmatrix}$; $\begin{bmatrix} -5 & -8 \\ -7 & 8 \\ 8 & -11 \end{bmatrix}$ **29.** $\begin{bmatrix} 0 & 3 \\ -1 & 20 \end{bmatrix}$; $\begin{bmatrix} 4 & 1 & 5 \\ 16 & 6 & 6 \\ 15 & 5 & 10 \end{bmatrix}$ **31.** $\begin{bmatrix} 7 & -7 \\ -5 & -4 \\ 9 & 10 \end{bmatrix}$ **33.** $\begin{bmatrix} \frac{7}{26} & \frac{2}{13} \\ -\frac{3}{26} & \frac{1}{13} \end{bmatrix}$

35. $\begin{bmatrix} \frac{2}{3} & 0 & -\frac{1}{3} \\ \frac{1}{3} & 0 & -\frac{2}{3} \\ -\frac{2}{3} & 1 & \frac{1}{3} \end{bmatrix}$ **37.** 6   **41.** $(3, 1)$ **43.** $(-1, 3, 4)$

45. $x = \frac{1}{13}(4a - b - 6c)$; $y = \frac{1}{13}(-4a + b - 7c)$; $z = \frac{1}{13}(5a + 2b + 12c)$ **47.** $\frac{11}{3}$ **49.** 133   **51.** 0
53. $2x^2 - 3x$ **55.** -20 **57.** 37   **59.** 2   **61.** 16   **63.** -2 **65.** $(\frac{26}{19}, \frac{29}{19})$ **67.** $(3, -3, 5)$ **69.** $(1, -1, 2)$

73. $\dfrac{1}{x-1} + \dfrac{1}{x+1}$ **75.** $\dfrac{3}{x} - \dfrac{1}{x+3} + \dfrac{2}{x-1}$ **77.** $\dfrac{4}{x} - \dfrac{3x}{x^2+1}$ **79.** $(\frac{14}{5}, \frac{6}{5})$

81. \$13,000 at 8.5%; \$27,000 at 11.2%   **83.** Byron: \$6; Jason: \$4; Adrian: \$2   **85.** (b) Unbounded (c) (0, 5)
87. (b) Unbounded (c) $(-\frac{1}{4}, \frac{3}{4})$ **89.** (b) Bounded (c) (2, 0); $(-\frac{2}{3}, 0)$; $(\frac{2}{5}, \frac{16}{5})$ **91.** (b) Bounded (c) (0, 5); (10, 5); (10, −1); (2, 3)
93. max 60, min 36   **95.** max 20, min 0   **97.** 400 units A; 800 units B

99. $\left(\dfrac{\sqrt{13}-2}{3}, 2\sqrt{\sqrt{13}-2}\right); \left(\dfrac{\sqrt{13}-2}{3}, -2\sqrt{\sqrt{13}-2}\right)$ **101.** $(-1, 3); (\frac{3}{2}, -2)$ **103.** $(1, 2); (-1, -2)$

105. No real solution   **107.** $\left(\dfrac{-1+\sqrt{11}}{2}, \sqrt{11}-2\right); \left(\dfrac{-1-\sqrt{11}}{2}, -\sqrt{11}-2\right)$ **109.** 7 and 24

111. 100 meters by 75 meters

Chapter 10

Problem Set 1, page 486

1. $9+i$ **3.** $10-i$ **5.** $19-14i$ **7.** $-2-2i$ **9.** $-5+2i$ **11.** $-11-4i$ **13.** $-1-14i$
15. $-3+11i$ **17.** $-3+54i$ **19.** $-7+9i$ **21.** $15-23i$ **23.** $64-40i$ **25.** -20 **27.** $\frac{13}{36}$ **29.** i
31. i **33.** $12+16i$ **35.** $-\frac{1}{2}+\frac{1}{2}\sqrt{3}i$ **37.** $\frac{2}{13}-\frac{3}{13}i$ **39.** $\frac{21}{53}-\frac{6}{53}i$ **41.** $\frac{1}{5}-\frac{11}{10}i$ **43.** $\frac{11}{34}+\frac{41}{34}i$
45. $\frac{23}{74}+\frac{27}{74}i$ **47.** $-\frac{13}{10}-\frac{19}{10}i$ **49.** $\frac{63}{290}-\frac{201}{290}i$ **51.** $\frac{6}{145}-\frac{43}{145}i$ **53.** (a) $2-i$ (b) 4 (c) $2i$ (d) 5 (e) $\sqrt{5}$
55. (a) i (b) 0 (c) $-2i$ (d) 1 (e) 1 **57.** (a) $-i$ (b) 0 (c) $2i$ (d) 1 (e) 1 **59.** (a) $-12-5i$ (b) -24 (c) $10i$ (d) 169 (e) 13
61. (a) $-\frac{1}{5}-\frac{11}{10}i$ (b) $-\frac{2}{5}$ (c) $-\frac{11}{5}i$ (d) $\frac{5}{4}$ (e) $\frac{1}{2}\sqrt{5}$ **63.** (a) 5 (b) 10 (c) 0 (d) 25 (e) 5

Problem Set 2, page 493

1. $2\sqrt{3}+2i$ **3.** $-\frac{7}{2}\sqrt{2}+\frac{7}{2}\sqrt{2}i$ **5.** $5.909+1.042i$ **7.** $\sqrt{3}+i$ **9.** $\sqrt{2}[\cos(3\pi/4)+i\sin(3\pi/4)]$
11. $3[\cos(\pi/2)+i\sin(\pi/2)]$ or $3i$ **13.** $5.292(\cos 19.11°+i\sin 19.11°)$ **15.** $2[\cos(7\pi/6)+i\sin(7\pi/6)]$
17. $5(\cos\pi+i\sin\pi)$ or -5 **19.** (a) $8(\cos 110°+i\sin 110°)$ (b) $2(\cos 30°+i\sin 30°)$
21. (a) $98[\cos(11\pi/4)+i\sin(11\pi/4)]$ (b) $2[\cos(\pi/4)+i\sin(\pi/4)]$ **23.** (a) $18(\cos 135°+i\sin 135°)$ (b) $2(\cos 45°+i\sin 45°)$
25. (a) $2\sqrt{2}[\cos(7\pi/12)+i\sin(7\pi/12)]$ (b) $(\sqrt{2}/2)[\cos(23\pi/12)+i\sin(23\pi/12)]$ **27.** $16\sqrt{2}+16\sqrt{2}i$ **29.** -243
31. $-256\sqrt{2}+256\sqrt{2}i$ **33.** $-16\sqrt{3}-16i$ **35.** $11,753-10,296i$ **37.** $(\sqrt{2}/2)-(\sqrt{2}/2)i$
39. $(3\sqrt{3}/2)-\frac{3}{2}i; -(3\sqrt{3}/2)+\frac{3}{2}i$ **41.** $(\sqrt[3]{4}/2)(1+i); (\sqrt[3]{4}/4)[(-1-\sqrt{3})+(\sqrt{3}-1)i]; (\sqrt[3]{4}/4)[(-1+\sqrt{3})+(-\sqrt{3}-1)i]$
43. $[(\sqrt{6}+\sqrt{2})/2]-[(\sqrt{6}-\sqrt{2})/2]i; -[(\sqrt{6}+\sqrt{2})/2]+[(\sqrt{6}-\sqrt{2})/2]i; [(\sqrt{6}-\sqrt{2})/2]+[(\sqrt{6}+\sqrt{2})/2]i;$
$-[(\sqrt{6}-\sqrt{2})/2]-[(\sqrt{6}+\sqrt{2})/2]i$ **45.** $(\sqrt{2}/2)(1+i); (\sqrt{2}/2)(-1+i); (\sqrt{2}/2)(1-i); (\sqrt{2}/2)(-1-i)$
47. $\sqrt{2}(1+i)$; etc.   **51.** (a) $1, -1$ (b) $1, -\frac{1}{2}+(\sqrt{3}/2)i, -\frac{1}{2}-(\sqrt{3}/2)i$ (c) $1, -1, i, -i$
53. (a) Reflection of z about real axis (b) Reflection of z through origin (c) Reflection of z about imaginary axis
(d) Rotation of z by 90° clockwise about origin (e) Distance between z and w

Problem Set 3, page 499

1. $1\pm i$ **3.** $1\pm 3i$ **5.** $\pm(\sqrt{3}/2)-(i/2)$ **7.** $1+i, 2$ **9.** (a) $2, -1+\sqrt{3}i, -1-\sqrt{3}i$ **11.** (a) $3, -3, 3i, -3i$
13. (a) $1, 1, 1, i, -i$ **15.** (a) $-2, -2, 2i, -2i$ **17.** (a) $-\frac{1}{2}, -\frac{1}{2}, -\frac{1}{2}, 3, -1$
19. (a) $3i, 3i, 3i, 3i, -3i, -3i, -3i, -3i, 2+i, 2-i$ **21.** (a) $1, i, -i$ **23.** (a) $3, 1+i, 1-i$
25. (a) $2, 2, -\frac{1}{2}+(\sqrt{3}/2)i, -\frac{1}{2}-(\sqrt{3}/2)i$ **27.** $z^3-3z^2-4z+12$ **29.** $z^5-6z^4+13z^3-12z^2+4z$
31. $z^6+3z^5-3z^4-11z^3+6z^2+12z-8$ **33.** $z^4-3z^3-2z^2+10z-12$ **35.** $z^6-2z^5+4z^4-z^2+2z-4$
37. z^5-1

Problem Set 5, page 509

1. 362,880 **3.** 7 **5.** 35 **7.** 1 **9.** 20 **11.** 1, 7, 21, 35, 35, 21, 7, 1
13. $a^4 + 12a^3x + 54a^2x^2 + 108ax^3 + 81x^4$ **15.** $x^5 - 5x^4y + 10x^3y^2 - 10x^2y^3 + 5xy^4 - y^5$
17. $c^6 + 12c^5 + 60c^4 + 160c^3 + 240c^2 + 192c + 64$ **19.** $1 - 7c^3 + 21c^6 - 35c^9 + 35c^{12} - 21c^{15} + 7c^{18} - c^{21}$
21. $x^3 - 6x^2\sqrt{xy} + 15x^2y - 20xy\sqrt{xy} + 15xy^2 - 6y^2\sqrt{xy} + y^3$
23. $\frac{1}{64}x^6 - \frac{3}{8}x^5y + \frac{15}{4}x^4y^2 - 20x^3y^3 + 60x^2y^4 - 96xy^5 + 64y^6$ **25.** $10s^3t^2$ **27.** $\frac{105}{2}x^{12}y^{12}$ **29.** $5xy^4$
31. $3360c^6d^4$

Problem Set 6, page 514

1. $\frac{1}{2}, \frac{1}{3}, \frac{1}{4}, \frac{1}{5}, \frac{1}{6}$ **3.** 4, 9, 16, 25, 36 **5.** $-1, 2, -3, 4, -5$ **7.** $\frac{1}{3}, \frac{3}{5}, \frac{5}{7}, \frac{7}{9}, \frac{9}{11}$ **9.** $\frac{1}{3}, \frac{1}{9}, \frac{1}{27}, \frac{1}{81}, \frac{1}{243}$ **11.** $0, \frac{3}{5}, \frac{4}{5}, \frac{15}{17}, \frac{12}{13}$
13. 2, −6, 18, −54, 162 **15.** 2, 2, 4, 64, 16, 777, 216 **17.** $1, \frac{1}{2}, \frac{1}{6}, \frac{1}{24}, \frac{1}{120}$ **19.** 33 **21.** 30 **23.** 100 **25.** 6.4
27. $m + 8r$ **29.** $-11\sqrt{5}$ **31.** $\frac{1}{64}$ **33.** 162 **35.** 2^{16} or 65,536 **37.** −0.009 **39.** $2x^{15}$ **41.** 0.000003
45. (b) 382,884

Problem Set 7, page 521

1. $1 + 4 + 7 + 10 = 22$ **3.** $\frac{13}{4} + \frac{7}{2} + \frac{15}{4} + 4 + \frac{17}{4} = \frac{75}{4}$ **5.** $1 + \frac{1}{2} + \frac{4}{7} + \frac{4}{5} + \frac{16}{13} + 2 = \frac{5553}{910}$ **7.** 2460
9. $2^{25} + 22$ or 33,554, 454 **11.** $\frac{1}{9}[1 - (1/10)^{100}]$ **13.** 555 **15.** 65 **17.** 624 **19.** −390 **21.** $-63t$
23. $-35x$ **25.** 2,440,000 **27.** 210 **29.** 44,444 **31.** $-\frac{1}{2}[1 - (1/3)^{10}]$ **33.** $(5^{12}/24)[1 - (1/5)^{18}]$ **35.** 2500
37. $c^4(c^{14} - 1)/(c^2 - 1)$ **39.** 36.2 cm **41.** $768 - 2^{10-n}$ **43.** 3 **45.** $\frac{5}{7}$ **47.** $\frac{6}{7}$ **49.** $\frac{25}{8}$ **51.** $\frac{4}{9}$ **53.** $\frac{80}{11}$
55. $\frac{58}{2475}$ **57.** False, $0.\overline{9} = 1$

Problem Set 8, page 528

1. 8 **3.** 676 **5.** 32 **7.** $ab, ac, ad, ae, ba, bc, bd, be, ca, cb, cd, ce, da, db, dc, de, ea, eb, ec, ed$; 20 **9.** 24
11. 9! or 362,880 **13.** 20 **15.** 990 **17.** $_{10}P_5 = 30,240$ **19.** 7! = 5040 **21.** 40,320 **23.** $_{10}P_3 = 720$
25. $\{a,b,c\}, \{a,b,d\}, \{a,c,d\}, \{b,c,d\}$; 4 **27.** 1 **29.** 10 **31.** 20 **33.** 10 **35.** 1 **37.** $_8C_2 = 28$
39. $_{10}C_4 = 210$ **41.** 12 **43.** 43,243,200 **45.** 3780 **47.** 3360

Problem Set 9, page 532

1. $\frac{1}{2}$ **3.** $\frac{4}{7}$ **5.** $\frac{1}{6}$ **7.** $(13 \cdot 48)/_{52}C_5 = 1/4165$ **9.** $4/_{52}C_5 = 1/649,740$ **11.** $\frac{5}{36}$ **13.** $\frac{11}{36}$
15. $[(_{10}C_3)(_3C_2)]/_{13}C_5 = 40/143$ **17.** $\frac{1}{3}$ **19.** $\frac{1}{2}$

Problem Set 10, page 542

1. Circle; radius 3 **3.** Ellipse; $(\pm 3, 0), (0, \pm 1)$ **5.** Ellipse; $(\pm 2, 0), (0, \pm 4)$ **7.** Ellipse; $(\pm 5, 0), (0, \pm 4)$
9. Circle; radius $\frac{1}{2}$ **11.** $[(x + 5)^2/9] + [(y - 5)^2/4] = 1$ **13.** $[(x + 2)^2/25] + [(y - 1)^2/16] = 1$
15. $[(x - 1)^2/4] + [(y + 2)^2/9] = 1$ **17.** Center $(1, -2)$; vertices $(4, -2), (-2, -2), (1, 0), (1, -4)$
19. Center $(-3, 1)$; vertices $(0, 1), (-6, 1), (-3, -5), (-3, 7)$ **21.** Vertex $(0, 0)$; axis horizontal

23. Vertex $(0,0)$; axis vertical   **25.** Vertex $(0,0)$; axis vertical   **27.** Vertex $(4,-7)$; axis vertical
29. Vertex $(-1,0)$; axis vertical   **31.** $y^2 = 8x$ **33.** $(y+1)^2 = -(x-5)$ **35.** $y + \frac{1}{2} = \frac{1}{8}(x+1)^2$
37. (a)–(c) Vertical (d)–(f) Horizontal   **39.** Center $(0,0)$; vertices $(\pm 3, 0)$; asymptotes $y = \pm\frac{2}{3}x$
41. Center $(0,0)$; vertices $(0, \pm 4)$; asymptotes $y = \pm 2x$ **43.** Center $(0,0)$; vertices $(\pm 4, 0)$; asymptotes $y = \pm\frac{1}{2}x$
45. Center $(1,-2)$; vertices $(4,-2),(-2,-2)$; asymptotes $y + 2 = \pm\frac{2}{3}(x-1)$
47. Center $(-2,-1)$; vertices $(-2,3),(-2,-5)$; asymptotes $y+1 = \pm\frac{4}{5}(x+2)$ **49.** $(x^2/16) - (y^2/25) = 1$
51. $[(y+2)^2/4] - [(x-3)^2/\frac{4}{1}] = 1$ **53.** $(x+1)^2 + (y+2)^2 = 1$; circle   **55.** $(y-4)^2 = 6(x+3)$; parabola
57. $[(x+3)^2/2] + y^2 = 1$; ellipse   **59.** $[(x+5)^2/10] + [(y-3)^2/4] = 1$; ellipse
61. $[(x-2)^2/4] - [(y+1)^2/1] = 1$; hyperbola   **63.** $[(y-10)^2/32] - (x^2/18) = 1$; hyperbola
65. $[(x+1)^2/4] + [(y-2)^2/9] = 1$; ellipse   **67.** $(x+2)^2 = \frac{3}{2}(y + \frac{4}{3})$; parabola
71. 100, 58, 28, 10, 4, 10, 28, 58, 100 meters

Review Problem Set, page 544

1. $10 + 5i$ **3.** $1 - 9i$ **5.** $13 - 8i$ **7.** $47 - 45i$ **9.** $-24 - 2i$ **11.** $\frac{16}{13} + \frac{11}{13}i$ **13.** $\frac{48}{221} - \frac{46}{221}i$ **15.** $-i$
17. $\frac{1}{27}i$ **19.** $12 - 5i$ **21.** $\frac{3}{2}\sqrt{2} + \frac{3}{2}\sqrt{2}i$ **23.** $8 - 8\sqrt{3}i$ **25.** 5 **27.** 8 **29.** $\sqrt{13}/5$
31. $4\sqrt{2}[\cos(3\pi/4) + i\sin(3\pi/4)]$ **33.** $2[\cos(2\pi/3) + i\sin(2\pi/3)]$ **35.** (a) $24(\cos 30° + i\sin 30°)$ (b) $\frac{3}{2}(\cos 14° + i\sin 14°)$
37. (a) $48(\cos 360° + i\sin 360°)$ (b) $3(\cos 110° + i\sin 110°)$ **39.** (a) $50[\cos(7\pi/12) + i\sin(7\pi/12)]$
(b) $2[\cos(5\pi/36) + i\sin(5\pi/36)]$ **41.** (a) $4\sqrt{3}[\cos(\pi/3) + i\sin(\pi/3)]$ (b) $\sqrt{3}(\cos 0 + i\sin 0)$
43. $16 + 16\sqrt{3}i$ **45.** -64 **47.** $8i$ **49.** $-i, (\sqrt{3}/2) + \frac{1}{2}i, (-\sqrt{3}/2) + \frac{1}{2}i$
51. $[(\sqrt{6} + \sqrt{2})/4] - [(\sqrt{6} - \sqrt{2})/4]i; [(\sqrt{6} - \sqrt{2})/4] + [(\sqrt{6} + \sqrt{2})/4]i; -[(\sqrt{6} + \sqrt{2})/4] + [(\sqrt{6} - \sqrt{2})/4]i;$
$-[(\sqrt{6} - \sqrt{2})/4] - [(\sqrt{6} + \sqrt{2})/4]i$ **53.** $\pm 1, \pm i$ **55.** (a) $-2, 1 \pm \sqrt{6}$ (b) $(z + 2)(z - 1 - \sqrt{6})(z - 1 + \sqrt{6})$
(c) All multiplicity 1   **57.** (a) $-3, 1, (-3 \pm \sqrt{11}i)/2$ (b) $(z - 1)(z + 3)[z + \frac{3}{2} + (\sqrt{11}/2)i][z + \frac{3}{2} - (\sqrt{11}/2)i]$
(c) All multiplicity 1   **59.** (a) $-2, 2, 3 \pm \sqrt{2}$ (b) $(z + 2)(z - 2)(z - 3 - \sqrt{2})(z - 3 + \sqrt{2})$ (c) All multiplicity 1
61. $z^4 - z^3 - 2z^2 + 6z - 4$ **63.** $z^5 - 6z^4 + 13z^3 - 14z^2 + 12z - 8$ **65.** $z^4 - 8z^3 + 26z^2 - 40z + 33$
71. $32 + 80x + 80x^2 + 40x^3 + 10x^4 + x^5$
73. $2187x^{14} - 10{,}206x^{12}y^2 + 20{,}412x^{10}y^4 - 22{,}680x^8y^6 + 15{,}120x^6y^8 - 6048x^4y^{10} + 1344x^2y^{12} - 128y^{14}$
75. $\frac{2187}{128}x^7 - \frac{5103}{64}x^6 + \frac{5103}{32}x^5 - \frac{2835}{16}x^4 + \frac{945}{8}x^3 - \frac{189}{4}x^2 + \frac{21}{2}x - 1$ **77.** $5376x^6y^3$ **79.** $61{,}236x^5y^5$
81. $-\frac{1}{2}, \frac{1}{3}, -\frac{1}{4}, \frac{1}{5}, -\frac{1}{6}$ **83.** $0, \frac{1}{2}, 0, \frac{1}{8}, 0, \frac{1}{48}$ **85.** 41 **87.** -137 **89.** $2^{10} - 1024$ **91.** -384 **93.** 33
95. 110 **97.** $\frac{4567}{90}$ **99.** $\frac{7}{4}$ **101.** 450 **103.** $\frac{44}{3}$ **105.** $12{,}240$ **107.** $(4^{10} - 1)/4$ **109.** \$50.50
111. \$3000   **113.** 625 inches or 52 feet, 1 inch   **115.** $\frac{1}{6}n(n+1)(2n+1)$ **117.** 9 **119.** $(7 + \sqrt{35})/2$ **121.** (a) $\frac{83}{99}$
(b) $\frac{91}{99}$ (c) $\frac{232{,}379}{49{,}950}$ (d) $\frac{16}{5}$ **123.** 360 **125.** 110 **127.** 120 **129.** 84 **131.** 1 **133.** 12 **135.** 120
137. $3{,}991{,}680$ **139.** 45 **141.** $_{52}C_{13} = 6.350135596 \times 10^{11}$ **143.** 420 **145.** $\frac{1}{8}$ **147.** $\frac{1}{18}$ **149.** $\frac{26}{105}$
151. Circle; radius $\frac{1}{3}$ **153.** Ellipse; $(\pm 4, 0), (0, \pm 3)$ **155.** $(x^2/64) + (y^2/16) = 1$
157. $[(x-1)^2/16] + [(y-1)^2/25] = 1$ **159.** Center $(0,0)$; $(\pm 2\sqrt{2}, 0), (0, \pm 2\sqrt{3})$
161. Center $(-1, 1)$; $(4, 1), (-6, 1), (-1, 4), (-1, -2)$ **163.** Center $(-4, 6)$; $(0, 6), (-8, 6), (-4, 12), (-4, 0)$
165. Vertex $(0,0)$; axis horizontal   **167.** Vertex $(0, 1)$; axis vertical   **169.** Vertex $(-9, 1)$; axis horizontal
171. $x = -(y + 6)^2$ **173.** Center $(0,0)$; vertices $(\pm 6\sqrt{2}, 0)$; asymptotes $y = \pm\frac{1}{3}x$
175. Center $(-2, 3)$; vertices $(2, 3), (-6, 3)$; asymptotes $y - 3 = \pm\frac{1}{2}(x+2)$
177. Center $(1, 3)$; vertices $(1, \frac{7}{2}), (1, \frac{5}{2})$; asymptotes $y - 3 = \pm\frac{1}{2}(x - 1)$ **179.** $(x^2/\frac{3}{4}) - (y^2/3) = 1$
181. $[(x-1)^2/4] + [(y+2)^2/64] = 1$; ellipse   **183.** $(x + \frac{1}{2})^2 + (y - \frac{1}{2})^2 = \frac{1}{4}$; circle
185. $[(x-1)^2/4] - [(y-1)^2/1] = 1$; hyperbola   **187.** $[(x - \frac{1}{2})^2/(\frac{53}{8})] + [(y + \frac{1}{2})^2/(\frac{159}{4})] = 1$; ellipse

Index of Applications

Index of Applications

Index

Trigonometric Functions

Acute Angles

$$\sin \theta = \frac{\text{opp}}{\text{hyp}} \qquad \csc \theta = \frac{\text{hyp}}{\text{opp}}$$

$$\cos \theta = \frac{\text{adj}}{\text{hyp}} \qquad \sec \theta = \frac{\text{hyp}}{\text{adj}}$$

$$\tan \theta = \frac{\text{opp}}{\text{adj}} \qquad \cot \theta = \frac{\text{adj}}{\text{opp}}$$

General Angles

$$\sin \theta = \frac{y}{r} \qquad \csc \theta = \frac{r}{y}$$

$$\cos \theta = \frac{x}{r} \qquad \sec \theta = \frac{r}{x}$$

$$\tan \theta = \frac{y}{x} \qquad \cot \theta = \frac{x}{y}$$

Trigonometric Identities

Fundamental Identities

1   $\csc \theta = \dfrac{1}{\sin \theta}$

2   $\sec \theta = \dfrac{1}{\cos \theta}$

3   $\cot \theta = \dfrac{1}{\tan \theta}$

4   $\tan \theta = \dfrac{\sin \theta}{\cos \theta}$

5   $\cot \theta = \dfrac{\cos \theta}{\sin \theta}$

6   $\cos^2 \theta + \sin^2 \theta = 1$

7   $1 + \tan^2 \theta = \sec^2 \theta$

8   $1 + \cot^2 \theta = \csc^2 \theta$

Even–Odd Identities

1   $\sin(-\theta) = -\sin \theta$

2   $\cos(-\theta) = \cos \theta$

3   $\tan(-\theta) = -\tan \theta$

4   $\cot(-\theta) = -\cot \theta$

5   $\sec(-\theta) = \sec \theta$

6   $\csc(-\theta) = -\csc \theta$

Addition Formulas

1   $\sin(\alpha + \beta) = \sin \alpha \cos \beta + \sin \beta \cos \alpha$

2   $\cos(\alpha + \beta) = \cos \alpha \cos\beta - \sin \alpha \sin \beta$

3   $\tan(\alpha + \beta) = \dfrac{\tan \alpha + \tan \beta}{1 - \tan \alpha \tan \beta}$

Subtraction Formulas

1   $\sin(\alpha - \beta) = \sin \alpha \cos \beta - \sin \beta \cos \alpha$

2   $\cos(\alpha - \beta) = \cos \alpha \cos \beta + \sin \alpha \sin \beta$

3   $\tan(\alpha - \beta) = \dfrac{\tan \alpha - \tan \beta}{1 + \tan \alpha \tan \beta}$